Gijsbertus de With

Liquid-State Physical Chemistry

Related Titles

Wolynes, P., Lubchenko, V. (eds.)

Structural Glasses and Supercooled Liquids

Theory, Experiment, and Applications

2012
Print ISBN: 978-1-118-20241-8; also available in electronic formats

Schäfer, R., Schmidt, P.C. (eds.)

Methods in Physical Chemistry

2012
Print ISBN: 978-3-527-32745-4; also available in electronic formats

Lalauze, R.

Physico-Chemistry of Solid–Gas Interfaces

2008
Print ISBN: 978-1-848-21041-7; also available in electronic formats

Reichardt, C., Welton, T.

Solvents and Solvent Effects in Organic Chemistry

Fourth Edition
2011
Print ISBN: 978-3-527-32473-6; also available in electronic formats

Gijsbertus de With

Liquid-State Physical Chemistry

Fundamentals, Modeling, and Applications

WILEY-VCH
Verlag GmbH & Co. KGaA

Author

Gijsbertus de With
Eindhoven Univ. of Technology
Dept. of Chemical Engineering
and Chemistry
Den Dolech 2
5612 AZ Eindhoven
The Netherlands

Cover: Martijn de With: *Disordered order*: an artist's impression of liquids, 2013.

Library of Congress Card No.: applied for

British Library Cataloguing-in-Publication Data
A catalogue record for this book is available from the British Library.

Bibliographic information published by the Deutsche Nationalbibliothek
The Deutsche Nationalbibliothek lists this publication in the Deutsche Nationalbibliografie; detailed bibliographic data are available on the Internet at <http://dnb.d-nb.de>.

Print ISBN: 978-3-527-33322-6
ePDF ISBN: 978-3-527-67678-1
ePub ISBN: 978-3-527-67677-4
Mobi ISBN: 978-3-527-67676-7
oBook ISBN: 978-3-527-67675-0

Typesetting Toppan Best-set Premedia Limited
Printing Strauss GmbH, Mörlenbach
Cover Design Formgeber, Mannheim, Germany

Printed in the Federal Republic of Germany
Printed on acid-free paper

There are two quite different approaches to a theory of the liquid state which in fact complement each other.

Henry Eyring, page 141 in Liquids: Structure, Properties, Solid Interactions, *T.J. Hughel ed., Elsevier Publ. Comp. Amsterdam, 1965.*

Contents

Preface

For many processes and applications in science and technology a basic knowledge of liquids and solutions is a must. However, the usual curriculum in chemistry, physics, and materials science pays little or no attention to this subject. It must be said that many books have been written on liquids and solutions. However, only a few of them are suitable as an introduction (many of them are far too elaborate), and most of them have been published quite some time ago, apart from the relatively recent book by Barrat and Hansen (2005). In spite of my admiration for that book I feel that it is not suitable as an introduction for chemical engineers and chemists.

In the present book a basic but as far as possible self-contained and integrated treatment of the behavior of liquids and solutions and a few of their simplest applications is presented. After introducing the fundamentals required, we try to present an overview of models of liquids giving an approximately equal weight to pure liquids, simple solutions, be it non-electrolyte, electrolyte, or polymeric solutions. Thereafter, we deal with a few special topics: reactions in solutions, surfaces, and phase transitions. Obviously, not all topics can be treated and a certain initial acquaintance with several aspects of physical chemistry is probably an advantage for the reader.

A particular feature of this book is the attempt to provide a basic but balanced presentation of the various aspects relevant to liquids and solutions, using the regular solution concept as a guide. That does not imply that we "forget" more modern approaches, but the concept is useful as a guide, in particular for engineering applications. To clarify the authors' view on the subject a bit further, it may be useful to quote Henry Eyrings' statement, as printed on the title page, more fully:

> There are two quite different approaches to a theory of the liquid state which in fact complement each other. In the deductive approach one proceeds as far as possible strictly mathematically, and when the complications cause this logical procedure to bog down, one resorts to some more or less defensible assumption such as Kirkwood's superposition principle. In the other approach one struggles to find a physical model of the liquid state which is as faithful to reality as can be devised and yet be solvable. The solution

> of the model may then proceed with considerable rigor. There are advantages and disadvantages to both procedures. In fact, either method expertly enough executed will solve the problem.

Although rigorous approaches have advanced considerably since the time this statement was made, the essence of this remark is still to the point in our view, in spite of the rebuttal by Stuart Rice:

> The second approach mentioned by professor Eyring depends on our ability to make a very accurate guess about the structure and proper parameterization of the model or models chosen. It is the adequacy of our guesses as representations of the real liquid which I question.

There is no doubt that rigorous approaches are important, much more so than in the time Eyring made his remark but, in our opinion, understanding is still very much served by using as simple as possible models.

The whole of topics the presented is conveniently described as physical chemistry or the chemical physics[1] of liquids and solutions: it describes the physicochemical behavior of liquids and solutions with applications to engineering problems and processes. Unfortunately, this description is wide, in fact too wide, and we have to limit ourselves to those topics that are most relevant to chemical engineers and chemists. This implies that we do not deal systematically with quantum liquids, molten salts, or liquid metals. Obviously, it is impossible to reflect these considerations exactly in any title so that we have chosen for a brief one, trusting that potential users will read this preface (and the introduction) so that they know what to expect. For brevity, therefore, we refer to the field as *liquid-state physical chemistry*.

We pay quite some attention to physical models since, despite all developments in simulations, they are rather useful for providing a qualitative understanding of molecular liquids and solutions. Moreover, they form the basis for a description of the behavior of polymer solutions as presently researched, and last–but not least–they provide to a considerable extent solvable models and therefore have a substantial pedagogical value. Whilst, admittedly, this approach may be characterized by some as "old-fashioned," in my opinion it is rather useful.

This book grew out of a course on the behavior of liquids and solutions, which contained already all the essential ingredients. This course, which was conducted at the Department of Chemical Engineering and Chemistry at Eindhoven University of Technology, originated from a total revision of the curriculum some 10 years ago and the introduction of liquid-state physical chemistry (or as said, equiva-

1) I refer here to the preface of *Introduction to Chemical Physics* by J.C. Slater (1939), where he states: *"It is probably unfortunate that physics and chemistry were ever separated. Chemistry is the science of atoms and the way they combine. Physics deals with the interatomic forces and with the large-scale properties of matter resulting from those forces. . . . A wide range of study is common to both subjects. The sooner we realize this, the better".*

lently, liquid-state chemical physics) some seven years ago. The overall set-up as given here has evolved during the last few years, and hopefully both the balance in topics and their presentation is improved. I am obliged to our students and instructors who have followed and used this course and have provided many useful remarks. In particular, I wish to thank my colleagues Dr Paul van der Varst, Dr Jozua Laven, and Dr Frank Peters for their careful reading of, and commenting on, several parts of the manuscript, and their discussions on many of the topics covered. Hopefully, this has led to an improvement in the presentation. I realize that a significant part of writing a book is usually done outside office hours, and this inevitably interferes considerably with one's domestic life. This text is no exception: for my wife, this is the second experience along this line, and I hope that this second book has "removed" less attention than the first. I am, therefore, indebted to my wife Ada for her patience and forbearance. Finally, I would like to thank Dr Martin Graf-Utzmann (Wiley-VCH, publisher) and Mrs Bernadette Cabo (Toppan Best-set Premedia Limited, typesetter) for all their efforts during the production of this book.

Obviously, the border between various classical disciplines is fading out nowadays. Consequently, it is hoped that these notes are useful not only for the original target audience, chemists, and chemical engineers, but also for materials scientists, mechanical engineers, physicists, and the like. Finally, we fear that the text will not be free from errors, and these are our responsibility. Hence, any comments, corrections, or indications of omissions will be appreciated.

January 2013 *Gijsbertus de With*

Acknowledgments

Wiley-VCH and the author have attempted to trace the copyright holders of all material from various websites, journals, or books reproduced in this publication to obtain permission to reproduce this material, and apologize herewith to copyright holders if permission to publish in this form has not been obtained.

Fig. 1.4a: Courtesy of Oak Ridge National Laboratory, U.S. Dept. of Energy: http://www.ornl.gov/info/ornlreview/v34_2_01/shrinking.htm

Fig. 1.4b: Source: M. Mañas at Wikimedia Commons: http://commons.wikimedia.org/wiki/File:3D_model_hydrogen_bonds_in_water.jpg

Fig. 1.5a: This image is from the *Solubility* article on the ellesmere-chemistry Wikia and is under the Creative Commons Attibution-Share Alike License: http://ellesmere-chemistry.wikia.com/wiki/File:Salt500.jpg#file

Fig. 1.5b: This image is from the YouTube movie *Hydration Shell Dynamics of a Hydrophobic Particle* by hexawater and is under the Creative Commons Attibution-Share Alike License: http://www.youtube.com/watch?v=ETMmH2trTpM&feature=related

Fig. 1.6a: This image is used with permission from J.T. Padding and W.J. Briels, Computational Biophysics, University of Twente, the Netherlands, from their website: http://cbp.tnw.utwente.nl/index.html

Fig. 1.6b: This image is used from Science, vol. 331, issue 6023, 18 March 2011, cover page by D.R. Glowachi, School of Chemistry, University of Bristol. Reprinted with permission from AAAS.

Fig. 4.1: Reprinted with permission from *Thermal Physics*, by C.B.P. Finn, Chapman & Hall, London 1993 (fig. 9.1, page 166). Copyright 1993 Francis & Taylor.

Fig. 4.2: This image is used from the *Real Gases* article by M. Gupta from the website: http://wikis.lawrence.edu/display/CHEM/Real+Gases+-+Mohit+Gupta

Fig. 4.4: Reprinted with permission from Q.J. Su (1946), Ind. Eng. Chem. Res., 38, 803 (fig. 1). Copyright 1946 American Chemical Society.

Fig. 4.5: Reprinted from *Thermodynamics: An Advanced Treatment for Chemists and Physicists*, by E.A. Guggenheim, North Holland, copyright 1967 (fig. 3.10, page 137 and fig. 3.11, page 138), with permission from Elsevier.

Problem 6.6: Reprinted from L.V. Woodcock (1971), Chem, Phys. Letters, 10, 257 (fig. 1), with permission from Elsevier.

Fig. 6.2: Reprinted from *Mechanics of Materials*, by M.A. Meyers, R.W. Armstrong, H.O.K. Kirchner, Chapter 7: Rate processes in plastic deformation of crystalline and noncrystalline solids by A.S. Argon (fig. 7.20, page 204), J. Wiley, NY 1999. Copyright Wiley-VCH Verlag GmbH & Co. KGaA. Reproduced with permission.

Fig. 6.3: Reprinted from *Mechanics of Materials*, by M.A. Meyers, R.W. Armstrong, H.O.K. Kirchner, Chapter 7: Rate processes in plastic deformation of crystalline and noncrystalline solids by A.S. Argon (fig. 7.21, page 205), J. Wiley, NY 1999. Copyright Wiley-VCH Verlag GmbH & Co. KGaA. Reproduced with permission.

Fig. 6.6: Reprinted from *Physics of Simple Liquids*, by H.N.V. Temperley, J.S. Rowlinson, G.S. Rushbrooke, Chapter 10: Structure of simple liquids by X-ray diffraction, by C.J. Pings, North Holland, Amsterdam 1968 (fig. 2, page 406), with permission from Elsevier.

Fig. 6.7a: Reprinted from *The Liquid State*, by J.A. Pryde, Hutchinson University Library, London 1966 (fig. 3.2, page 42).

Fig. 6.7b: Reprinted with permission from J.L. Yarnell *et al.* (1973), Phys. Rev. A7 [6] 2130, APS (fig. 4). Copyright 1973 American Physical Society.

Fig. 6.8a: Reprinted with kind permission from J. Phys. Soc. Japan, 64 [8] (1995) 2886, Y. Tatuhiro (fig. 3).

Fig. 6.8b: With kind permission from Springer Science+Business Media: Il Nuovo Cimento D 12 (1990) 543, J.C. Dore (fig. 2).

Fig. 6.10: With kind permission from Springer Science+Business Media: Il Nuovo Cimento D 12 (1990) 543, J.C. Dore (fig. 4 and fig. 5).

Fig. 6.11: Reprinted from *The Liquid State*, by J.A. Pryde, Hutchinson University Library, London 1966 (fig 8.3, page 139).

Fig. 7.1: Reprinted from *Liquid State Chemical Physics*, by R.O. Watts, I.J. McGee, Wiley, NY 1976 (fig. 5.1, page 137). Copyright Wiley-VCH Verlag GmbH & Co. KGaA. Reproduced with permission.

Fig. 7.2a: Reprinted from *Physical Chemistry – an Advanced Treatise* – vol. VIIIA/ *Liquid State*, by H. Eyring, D. Henderson, W. Jost, Chapter 4: Distribution functions, by R.J. Baxter, Academic Press, New York and London 1971 (fig. 7, page 297), with permission from Elsevier.

Fig. 7.2b: Reprinted from *Liquid State Chemical Physics*, by R.O. Watts, & I.J. McGee, Wiley, NY 1976 (fig. 5.2, page 140). Copyright Wiley-VCH Verlag GmbH & Co. KGaA. Reproduced with permission.

Fig. 7.5: Reprinted with permission from J.A. Barker, D. Hendersson (1967), J. Chem. Phys., 47, 4714 (fig. 5 and fig. 6). Copyright 1967 American Institute of Physics.

Fig. 7.6: From D. Chandler, J.D. Weeks (1983), Science, 220, 787. Reprinted with permission from AAAS.

Fig. 7.7: Reprinted with permission from J.A. Barker, D. Henderson (1971), Acc. Chem. Res., 4, 3031 (fig. 4). Copyright 1971 American Chemical Society.

Fig. 8.1: Reprinted with kind permission from B.A. Dasannacharya, K.R. Rao, Phys. Rev., 137 (1965) A417 (fig. 9). Copyright 1965 by American Physical Society.

Fig. 8.2: Reprinted from *The Liquid State,* by J.A. Pryde, Hutchinson University Library, London 1966 (fig. 6.1. page 99).

Fig. 8.3: Reprinted with permission from W.G. Hoover, F.H. Ree (1968), J. Chem. Phys. 49, 3609 (fig. 2 and fig. 4). Copyright 1968 American Institute of Physics.

Fig. 8.4: Reprinted with permission from D. Henderson (1962), J. Chem. Phys., 37, 631 (fig. 1 and fig. 2). Copyright 1962 American Institute of Physics.

Fig. 8.5: Reprinted with permission from *The Dynamic Liquid State,* by A.F.M. Barton, Longman, New York 1974 (fig. 12.1, page 109). Copyright 1974 Pearson.

Fig. 8.6: Reprinted with permission from H. Eyring, R.P. H., Marchi (1963), J. Chem. Ed., 40, 562 (fig. 1). Copyright 1963 American Chemical Society.

Fig. 8.7: Reprinted with permission from T.S. Ree *et al.* (1965), J. Phys. Chem., 69, 3322 (fig. 2 and fig. 3). Copyright 1965 American Chemical Society.

Fig. 9.1a: Reprinted from *Properties of Liquids and Solutions,* by J.N. Murrell, A.D. Jenkins, Wiley, Chichester 1994 (fig. 3.3, page 54). Copyright Wiley-VCH Verlag GmbH & Co. KGaA. Reproduced with permission.

Fig. 9.1b: Reprinted from *A Course on Statistical Mechanics,* by H.L. Friedman, Prentice-Hall, Englewood Cliffs, NJ 1985 (fig. 5.2, page 95). Copyright Wiley-VCH Verlag GmbH & Co. KGaA. Reproduced with permission.

Fig. 9.2: Reprinted with permission from T. Wainwright, B.J. Alder (1958), Il Nuovo Cimento 9 [1] Supplement, 116 (fig. 6 and fig. 7). Copyright 1958 Springer.

Fig. 9.3: Reprinted from *A Course on Statistical Mechanics,* by H.L. Friedman, Prentice-Hall, Englewood Cliffs, NJ 1985 (fig. 5.3, page 100 and fig. 5.5, page 107). Copyright Wiley-VCH Verlag GmbH & Co. KGaA. Reproduced with permission.

Fig. 9.4: Reprinted with permission from S. Hannongbua (2000), J. Chem. Phys., 113, 4707 (fig. 1). Copyright 2000 American Institute of Physics.

Fig. 9.5: Reprinted with permission from S. Hannongbua (2000), J. Chem. Phys., 113, 4707 (fig. 3). Copyright 2000 American Institute of Physics.

Fig. 9.6: Reprinted with permission from S. Hannongbua (2000), J. Chem. Phys., 113, 4707 (fig. 5). Copyright 2000 American Institute of Physics.

Fig. 9.7: Reprinted with permission from S. Hannongbua (2000), J. Chem. Phys., 113, 4707 (fig. 2). Copyright 2000 American Institute of Physics.

Fig. 10.1: Reprinted from *Polar Molecules,* P. Debye, The Chemical Catalog Company Inc., New York 1929 (fig. 9, page 38 and fig. 10, page 39).

Fig. 10.4: This image is from the *Colors from Vibrations* article on the *Causes of Color* exhibition, M. Douma, curator, and is under the Creative Commons Attibution-Share Alike License. http://www.webexhibits.org/causesofcolor/5B.html

Fig. 10.5: Reprinted with permission from J.A. Odutola, T.R. Dyke (1980), J. Chem. Phys., 72, 5062 (fig. 4). Copyright 1980 American Institute of Physics.

Fig. 10.6: Reprinted with permission from *Principles of Modern Chemistry,* 5th ed. by D.W. Oxtoby, M.P. Gillis, N.H. Nachtrieb (fig. 5.19, page 145). Copyright 2003 Cengage/Nelson.

Fig. 10.7a: Reprinted from *Biophysical Chemistry,* vol. 1, by J.T. Edsall, J. Wyman, Academic Press Inc. Publishers, NY 1958 (fig. 3, page 31), with permission from Elsevier.

Fig. 10.7b: This image is used with permission from the *Ice Structure* article on The Interactive Library of EdInformatics.com: http://www.edinformatics.com/interactive_molecules/ice.htm

Fig. 10.7c: This image is used with permission from the *Water Molecule Structure* article on the Water Structure and Science webpage of M. Chaplin: http://www.lsbu.ac.uk/water/molecule.html

Fig. 10.8a: Reprinted from *Properties of Liquids and Solutions*, by J.N. Murrell, A.D. Jenkins, Wiley, Chichester 1994 (fig. 8.7, page 172). Copyright Wiley-VCH Verlag GmbH & Co. KGaA. Reproduced with permission.

Fig. 10.8b: This image is used with permission from Dr L. Ojamäe, Linköpings University, Department of Physics, Chemistry and Biology, Computational Chemistry, Linköping, Sweden from his webpage on *Molecular Dynamics Simulation of Liquid Water* on the website: http://www.ifm.liu.se/compchem/former/liquid.html

Fig. 10.9: Reprinted from *Properties of Liquids and Solutions*, by J.N. Murrell, A.D. Jenkins, Wiley, Chichester 1994 (fig. 8.9, page 175). Copyright Wiley-VCH Verlag GmbH & Co. KGaA. Reproduced with permission.

Fig. 10.10: Reprinted from Springer: F. Franks (ed.), *Water – A Comprehensive Treatise*, Vol. 1: *The Physics and Physical Chemistry of Water*, Chapter 5: Raman and infrared spectral investigation of water structure, by G.E. Walraten (fig. 11, page 177, fig. 12, page 178, fig. 15, page 182). Copyright 1972 Plenum Press, with kind permission from Springer Science+Business Media B.V.

Fig. 10.11a: Reprinted with permission from J. Morgan, B.E. Warren (1938), J. Chem. Phys., 6, 666 (fig. 4). Copyright 1938 American Institute of Physics.

Fig. 10.11b: Reprinted from Springer: F. Franks (ed.) *Water – A Comprehensive Treatise*, Vol. 1: *The Physics and Physical Chemistry of Water*, Chapter 8: Liquid water: scattering of X-rays, by H. Narten, H.A. Levy (fig. 5, page 326). Copyright 1972 Plenum Press, with kind permission from Springer Science+Business Media B.V.

Fig. 10.12: Reprinted with permission from G.C. Lie, E. Clementi (1975), J. Chem. Phys., 62, 2195 (fig. 4). Copyright 1975 American Institute of Physics.

Fig. 10.13: Reprinted from *Properties of Liquids and Solutions*, by J.N. Murrell, A.D. Jenkins, Wiley, Chichester 1994 (fig. 8.15, page 183). Copyright Wiley-VCH Verlag GmbH & Co. KGaA. Reproduced with permission.

Fig.10.14: Reprinted with permission from *The Nature of Chemical Bond*, 3rd ed., by L. Pauling, Cornell University Press, Ithaca, NY 1960 (fig. 12.3, page 456 and fig. 12.2, page 455). Copyright 1960 Cornell University Press.

Fig. 11.2: Reprinted from J. Zawidsky, Z. Phys. Chem., 35 (1900), 129 (fig. 7, fig. 8, fig. 14 and fig. 15).

Fig. 11.3: This image is used with permission from Dortmund Data Base: http://www.ddbst.com

Fig. 11.5: Reprinted from *Thermodynamics: An Advanced Treatment for Chemists and Physicists*, by E.A. Guggenheim, North Holland. Copyright 1967 (fig. 4.8 and fig. 4.9, page 199), with permission from Elsevier.

Fig. 11.6: Reprinted with permission from D. Henderson, P.J. Leonard, Proc. Nat. Acad. Sci. USA, 68 (1971) 632 (fig. 1 and fig. 2). Copyright 1971 National Academy of Sciences of the United States of America.

Fig. 12.3: Reprinted from *Ions in Solutions*, by J. Burgess, Ellis Horwood, Chicester 1988 (fig. 2.2, page 31). Copyright Wiley-VCH Verlag GmbH & Co. KGaA. Reproduced with permission.

Fig. 12.4: Reprinted from Springer: F. Franks (ed.) *Water – A Comprehensive Treatise*, Vol. 6: *Recent Advances*, Chapter 1: X-ray and neutron scattering by electrolytes, by J.E. Enderby, G.W. Neilson, G. Plenum Press, London and New York 1979 (fig. 5, page 28, fig. 8, page 31, fig. 12, page 35), with kind permission from Springer Science+Business Media B.V.

Fig. 12.5: Reprinted from *Ions in Solutions*, by J. Burgess, Ellis Horwood Chicester 1988 (fig. 3.5 and fig. 3.6, page 42). Copyright Wiley-VCH Verlag GmbH & Co. KGaA. Reproduced with permission.

Fig. 12.6: Reprinted with permission from Pitzer, K.D. (1977), Acc. Chem. Res., 10, 371 (fig. 2 and fig. 3). Copyright 1977 American Chemical Society.

Fig. 12.9: This image is used with permission from the *Grotthuss Mechanism* article on the Water Structure and Science webpage of M. Chaplin: http://www.lsbu.ac.uk/water/grotthuss.html#r160

Fig. 13.1: Reprinted from *Microstructure and Wear of Materials*, by K.-H. zum Gahr, Elsevier, Amsterdam 1987 (fig. 2.5, page 13), with permission of Elsevier.

Fig. 13.4: Reprinted with permission from R.H. Boyd, P.J. Phillips (1993), *The Science of Polymer Molecules* (fig. 6.15, page 215). Copyright 1993 Cambridge University Press.

Fig. 13.6: These two images are used with permission from the *Thermodynamic Considerations for Polymer Solubility* article on the Polymer Science Learning Center webpage of Prof. L.J. Mathias; University of Southern Mississippi, Department of Polymer Science: http://www.pslc.ws/macrog/ps4.htm

Fig. 13.7: Reprinted from E.F. Casassa (1976), J. Polymer Sci. Part C: Polymer Symposia, 54, 53. Copyright 1976 Wiley. This material is reproduced with permission of John Wiley & Sons. Inc.

Fig. 13.8a: Reprinted with permission from P.J. Flory (1970), Disc. Far. Soc., 49, 7 (fig. 1). Copyright 1970 Royal Society of Chemistry.

Fig. 13.8b: Reprinted with permission from I.C. Sanchez, R.H. Lacombe (1978), Macromolecules, 11, 1145 (fig. 6). Copyright 1978 American Chemical Society.

Fig. 13.9: Reprinted from H. Xie, E. Nies, A. Stroeks, R. Simha (1992), Pol. Eng. Sci., 32, 1654 (fig. 3 and fig. 9). Copyright 1992 John Wiley & Sons.

Fig. 13.10a: Reprinted from C.S. Wu, Y.P. Chen Y.P. (1994), Fluid Phase Equilibria, 100, 103 (fig. 4), with permission from Elsevier.

Fig. 13.10b: Reprinted with permission from I.G. Economou (2002), Ind. Eng. Chem. Res., 41, 953 (fig. 5). Copyright 2002 American Chemical Society.

Fig. 13.11a: Reprinted with permission from J. Gross, G. Sadowski (2001), Ind. Eng. Chem. Res., 40, 1244 (fig. 3). Copyright 2001 American Chemical Society.

Fig. 13.11b: Reprinted from N. von Solms *et al.* (2006), Fluid Phase Equilibria 241, 344 (fig. 1), with permission from Elsevier.

Fig. 14.1: Reprinted from *Physical Chemistry*, 3rd ed., by R.J. Silbey, R.A. Alberty, J. Wiley & Sons, NY 2001 (fig. 19.2, page 709). This material is reproduced with permission of John Wiley & Sons. Inc.

Fig. 14.4: Reprinted from *The Theory of Rate Processes*, by S. Glasstone, K.J. Laidler and H. Eyring, McGraw-Hill, NY 1941 (fig. 105 and fig. 106, page 420), with permission of McGraw-Hill Higher Education.

Fig. 14.6: Reprinted from *The Theory of Rate Processes*, by S. Glasstone, K.J. Laidler and H. Eyring, McGraw-Hill, NY 1941 (fig. 108, page 429 and fig. 109, page 431), with permission of McGraw-Hill Higher Education.

Fig. 15.4: Reprinted with permission from R.F. Crul, K. Pitzer (1958), Ind. Eng. Chem., 50, 265 (fig. 4). Copyright 1958 American Chemical Society.

Fig. 16.3: This image is from the ChemEd DL video on- Critical Point of Benzene. Video from JCE Web Software, Chemistry Comes Alive: Copyright ACS Division of Chemical Education, Inc. Used with permission. http://www.jce.divched.org/

Fig. 16.6: Reprinted with permission from P. Heller (1967), Rep. Prog. Phys., 30, 731 (fig. 10). Copyright 1967 IOP Science.

Fig. 16.7: Reprinted with permission from P. Heller (1967), Rep. Prog. Phys., 30, 731 (fig. 18). Copyright 1967 IOP Science.

Fig. 16.11: Reprinted from H.E. Stanley (1971), *Introduction to Phase Transistions and Critical Phenomena*, Oxford University Press (fig. 1.5, page 7), with permission from Oxford University Press, USA.

Fig. 16.12: This image is used with permission from M.R. Foster from the percolation article on his website: http://philosophy.wisc.edu/forster/Percolation.pdf

Fig. App. C2: Reprinted from *Statististical Mechanics and Dynamics*, 2nd ed., by H. Eyring, D. Henderson, B.J. Stover, E. M. Eyring, Wiley-Interscience, Chichester 1982 (fig. 2, page 474 and fig. 1, page 473). This material is reproduced with permission of John Wiley & Sons Inc.

List of Important Symbols and Abbreviations

(X)	dimension dependent on property, (–) dimensionless property
Φ	potential energy (J)
Ξ	grand partition function (–)
Ψ	wave function (–)
Ω	external potential energy (J)
	grand potential (J)
Λ	thermal wave length (m)
α	constant [X]
	activity coefficient [–]
	thermal expansion coefficient (K^{-1})
	polarizability ($C^2\,m^2\,J^{-1}$)
β	constant (X)
	$1/kT$ (J^{-1})
	activity coefficient (–)
γ	activity coefficient (–)
δ	solubility parameter ($MPa^{1/2}$)
	Dirac delta function (–)
ε	small scalar (–)
	energy (J)
η	(bulk) viscosity ($Pa \cdot s$)
κ	inverse Debye length (m^{-1})
	compressibility ($m^2\,N^{-1}$)
λ	parameter (X)
	wave length (m)
μ	dipole moment (C m)
ν	frequency (s^{-1})
ρ	number or mass density (m^{-3} or kg m^{-3})
	radius of curvature (m)
ρ'	mass density (kg m^{-3})
σ	(hard sphere) diameter (m)
τ	characteristic time (s)
ϕ	(pair) potential energy (J)

	electrical potential ($V\,m^{-1}$)
	volume fraction (–)
χ	Flory parameter (–)
ω	circular frequency (s^{-1})
$\mathbf{E}$	electric field ($V\,m^{-1}$)
$\mathbf{F}$	force (N)
$\mathbf{Q}$	quadrupole moment ($C\,m^2$)
$\mathbf{e}$	unit vector (m)
$\mathbf{f}$	force (N)
$\mathbf{p}$	generalized momentum (X)
$\mathbf{q}$	generalized coordinate (X)
$\mathbf{r}$	coordinate (m)
$\mathbf{u}$	displacement (m)
$\mathbf{v}$	velocity ($m\,s^{-1}$)
$\mathbf{x}$	coordinate (m)
A	area (m^2)
	generalized force (X)
	constant (X)
B	constant (X)
B_i	virial coefficients (X)
C	constant (X)
D	diffusivity ($m^2\,s^{-1}$)
E	energy (J)
	electric field ($V\,m^{-1}$)
F	Helmholtz energy (J)
	force (N)
	Faraday constant ($C\,mol^{-1}$)
G	Gibbs energy (J)
H	enthalpy (J)
I	moment of inertia ($kg\,m^2$)
K	equilibrium constant (X)
L	length (m)
N	number of molecules (particles)
N_A	Avogadro's number (mol^{-1})
P	pressure (MPa)
	probability (–)
	polarization (m^3)
Q	configuration integral (–)
R	gas constant ($J\,K^{-1}\,mol^{-1}$)
	radius (m)
	distance (m)
S	entropy ($J\,K^{-1}$)
T	kinetic energy (J)
	temperature (K)
U	(internal) energy (J)

V	potential energy (J)
	volume (m^3)
W	work (J)
Z	canonical partition function (–)
a	generalized displacement (X)
	constant (X)
b	second virial coefficient (m^3)
	constant (X)
c	constant (X)
	inverse spring constant (compliance) ($m\,N^{-1}$)
e	unit charge (C)
f	(volume) fraction (–)
	specific Helmholtz energy ($J\,kg^{-1}$)
	force (N)
	spring constant ($N\,m^{-1}$)
	rate constant (X)
g	specific Gibbs energy ($J\,kg^{-1}$)
	density of states (J^{-1})
j	flux ($s^{-1}\,m^{-2}$)
k	Boltzmann constant ($J\,K^{-1}$)
	spring constant (N^{-1})
	rate constant (X)
l	length (m)
m	Mie constant (–)
	mass (kg)
n	Mie constant (–)
	material constant (X)
n_X	number of moles of component X
p_i	probability of state i (–)
q	charge (C)
r	distance (m)
	rate (X)
s	specific entropy ($J\,K^{-1}\,kg^{-1}$)
t	time (s)
u	(pair) potential energy (J)
v	volume (m^3)
w	regular solution parameter (J)
x_i	mole fraction of component i (–)
z	single particle partition function (–)
	coordination number (–)
z_i	valency of particle i (–)

BCC	body-centered cubic
CN	coordination number
DoF	degrees of freedom

DoS	density of states
EoS	equation of state
FCC	face-centered cubic
HCP	hexagonal close packed
HNC	hypernetted chain equation
HS	hard-sphere
MC	Monte Carlo
MD	molecular dynamics
LJ	Lennard-Jones
NRD	Neutron-ray diffraction
OZ	Ornstein–Zernike
PoCS	Principle of corresponding states
PY	Percus–Yevick equation
RDF	radial distribution function
SC	simple cubic
SLS	Significant liquid structure
STP	standard temperature and pressure
SW	square well
XRD	X-ray diffraction
YBG	Yvon–Born–Green equation
lhs	left-hand side
rhs	right-hand side
vdW	van der Waals
$\cong$	approximately equal
$\equiv$	identical
$\sim$	proportional to
$\Leftrightarrow$	corresponds with
$\mathcal{O}(x)$	order of magnitude x

Superscripts

E	excess
∞	infinite dilution
$\circ$	standard
$*$	pure substance
$\ddagger$	activated complex
id	ideal

Subscripts

f	formation reaction
fus	fusion
mix	mixing

m	molar
r	reaction in general
sol	solution
sub	sublimation
trs	transition
vap	vaporization

1
Introduction

Whilst liquids and solutions are known to play an essential role in many processes in science and technology, the treatment of these topics in the literature have in the past appeared to be limited to either rather basic or rather advanced levels, with intermediate-level texts being scarce, despite the practical importance of liquids and solutions. A brief outline of the differences and similarities between solids, gases, and liquids would help to clarify the reasons for this, and this book represents an attempt to remedy the situation. In this first chapter, an outline of what will be dealt with, and the reasons for these choices, will be given.

1.1
The Importance of Liquids

A brief moment of reflection will make it clear that liquids play an important role in daily life, and in the life sciences and natural sciences, as well as in technology. Hence, these areas of interest will be considered briefly in turn.

Undoubtedly, the most important liquid is *water*. Water is essential for life itself, and its interaction with other liquids, ions and polymers is vital to many life processes. The basic component of blood, the *solvent*, is water, and blood itself is an example of a highly complex dispersion of red and white blood cells within a complex mixture of water, ions, and polymeric molecules. As is well known, blood not only transports oxygen through the body but also distributes required molecules to a variety of locations in the body, as well as removes waste material. The miscibility of water with other liquids (such as alcohol) is well known and exploited in alcoholic drinks.

Two other arbitrary examples of liquids that are highly relevant to daily life are petrol – a complex mixture of several types of aliphatic molecules and other species – for cars, and milk – a dispersion of fat globules stabilized in water by a complex of large and small molecules. Without petrol, modern society would be unthinkable, while milk provides a valuable (some say indispensable) part of the nutrition of humankind.

In technology many processes use solvents. With the current drive in industry to abolish the use of volatile organic compounds, the importance of water as a

Liquid-State Physical Chemistry, First Edition. Gijsbertus de With.

solvent – though already considerable – will be increased. Nonetheless, organic solvents are still used to a considerable extent, and properties such as rates of evaporation, miscibility, viscosity, thermal expansion, compressibility and thermal conductivity feature extensively in many technological processes. The re-use of solvents is also an important issue.

From a scientific point of view, liquids have a less extensive history of understanding than do gases and solids. The reasons for this will become clear in Section 1.2: a natural reference configuration with respect to structure and energy is missing. This implies that, for almost all aspects of liquids, it is not one largely dominant factor that determines the behavior, but rather several smaller factors must be taken into account, and this renders the overall description complex. Hence, the scientific problems posed by liquids is highly challenging, and the modeling of systems may range from simple semi-empirical models via sophisticated "physical" models to *ab initio* models and computer simulations. Yet, each of these models has its own value in the understanding of liquids with regards to their complexity from a structural, dynamic, and energetic point of view.

1.2 Solids, Gases, and Liquids

In this section, the general structural features of liquids as intermediate between solids and gases, and the associated energetics, are briefly discussed. These considerations will clarify the nature and the complexity of liquids and solutions.

Starting with solids, generally two classes of solids can be distinguished: crystalline and amorphous materials. As is well known, the basic feature of crystalline solids is order. *Crystalline* solids can further be divided into *single-crystalline* or *polycrystalline* materials, in both of which a regularly ordered structure exists at the atomic scale (Figure 1.1). This structure is maintained, at least in principle, throughout the whole material in a single-crystalline material, whereas in a polycrystalline material regions of different crystallographic orientations exist. These regions are referred to as *grains*, and the boundaries between them as *grain boundaries*. Studies using X-ray diffraction have clearly revealed the long-range atomic order of these materials. In *amorphous* solids there is no long-range order (Figure 1.1), although the local coordination of a specific molecule[1)] in the amorphous state may not be that different from the coordination of the same type of molecule in the corresponding crystalline state (if it exists).

From the observation that the structure of a solid is, in essence, maintained with increasing temperature up to the melting point[2)], it already follows that the potential energy U_{pot} is more important than the kinetic energy U_{kin} because a strong

1) Although we denote for convenience the basic entities as molecules, the term is also supposed to include atoms and ions, whenever appropriate.

2) We "forget" for convenience phase transformations.

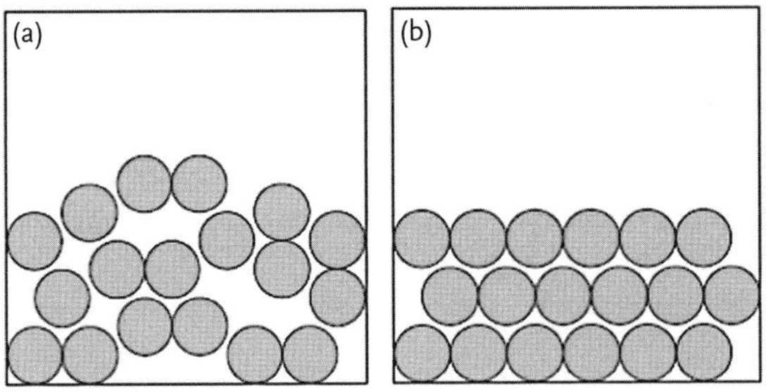

Figure 1.1 Schematics of (a) an amorphous and (b) a crystalline structure.

energetic interaction more or less immobilizes the molecules in space. Therefore, we have in general

$$U_{\text{pot}} > U_{\text{kin}} \text{ for solids} \tag{1.1}$$

This makes a regular spatial array of molecules the most suitable *reference configuration* for modeling a crystalline solid. This regularity can be described globally by the concept of lattices (long-range order) and locally by a well-defined coordination number (short-range order). Other aspects such as the kinetic energy of the molecules or defects in the regularity of the structure can be considered to be first order perturbations on this regularity. This implies that relatively simple models of particular features of the solid state – that is, models that ignore many details – can already describe the physical phenomena in solids reasonably well.

For gases, the molecules move through space almost independently of each other, as evidenced by the wide applicability of the ideal gas law $PV = nRT$, with as usual the pressure P, the volume V, the number of moles n, the gas constant R, and the temperature T. Hence, order is nearly absent and the reference configuration can be described as random. This is exactly the reverse of the situation in a solid. Thus, it can be concluded that the potential energy is small as compared with the kinetic energy and we have:

$$U_{\text{pot}} < U_{\text{kin}} \text{ for gases} \tag{1.2}$$

For gases, the reference configuration is thus a random distribution of molecules in space. In this case the influence of intermolecular interactions can be considered to first order as a perturbation, leading to some coordination of molecules with other molecules. Again, relatively simple models can provide already a good clue for the understanding of gases, as exemplified by the ideal gas model.

Liquids do have some properties akin to those of solids, and some other properties more similar to those of gases. For example, their density ρ, thermal expansion coefficient α and compressibility κ are typically not too different from the corresponding solid. As a rule of thumb, the specific volume increases from 5% to 15% upon melting (water is a well-known exception). On the other hand, liquids and gases have fluidity[3] in common, although a liquid has a meniscus while a

3) Liquids and gases together are often indicated as fluids.

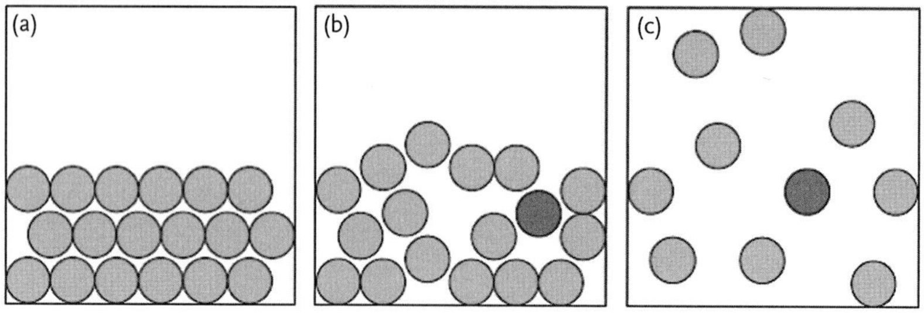

Figure 1.2 Schematic of structure and coordination of (a) a solid, (b) a liquid, and (c) a gas. While solids and liquids have comparable values of density ρ, thermal expansion coefficient α and compressibility κ, the liquid and gas have more comparable values of viscosity η, shear modulus G, and diffusion coefficient D.

gas has no such thing and the viscosity η of liquids is higher than those of gases. This indicates that the movement of molecules in fluids is relatively easy when compared to solids. This is also reflected by similar values of the shear modulus G and diffusion coefficient D for liquids and gases. Figure 1.2 shows schematically the structure of solids, liquids and gases. As might be expected, the situation for liquids with respect to energy is somewhere in the middle, and both the potential energy and kinetic energy play important roles. Therefore, we have

$$U_{\text{pot}} \sim U_{\text{kin}} \text{ for liquids} \tag{1.3}$$

This implies that neither an ordered nor a fully disordered configuration is present in a liquid. The choice of a reference configuration becomes accordingly (much) more troublesome. Although long-range order is absent, short-range order is present and the concept of coordination number is still valuable for liquids. However, because of the approximately equal importance of kinetic and potential energy, relatively simple models of liquids usually are much less reliable than for either solids or gases. This increased complexity shifts the topic to a more advanced level, implying that in education less attention is given to the subject than it deserves in view of its practical importance.

To summarize, while the dominant feature of a solid is order, and that of a gas is disorder, the liquid is somewhere in between. The static structure of solids, liquids and gases is illustrated in Figure 1.3. This is achieved by using the *pair correlation function*[4] $g(r)$, which describes the probability of finding a molecule at a distance r from a reference molecule at the origin[5]. From Figure 1.3 it is clear that for crystalline solids only at discrete distances other molecules are present, whereas for gases the probability of finding another molecule rapidly becomes constant with increasing distance. For liquids, the situation is intermediate, as evidenced by some structure in $g(r)$ for small r and the limiting behavior $g(r) = 1$

4) The pair correlation function and its properties will be discussed more extensively in Chapter 6.
5) The probability is scaled in such a way that its average value is unity.

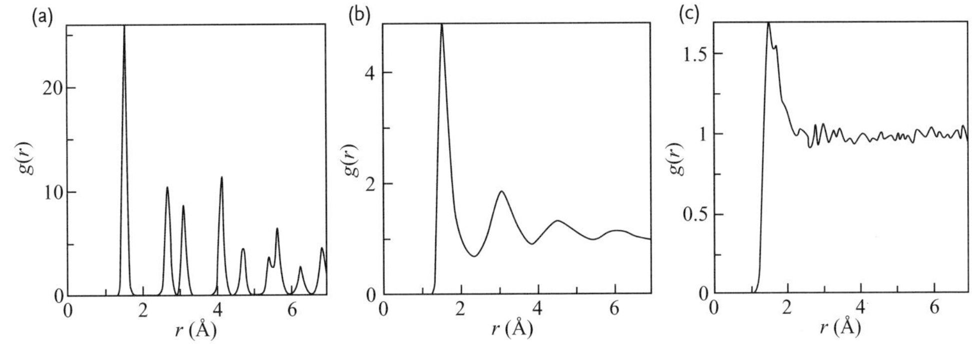

Figure 1.3 Schematic of the coordination of a solid, a liquid and a gas. (a) A solid with a regular array of molecules leading to long-range order and a well-defined coordination shell; (b) A liquid with similar density as the solid and having a random dense packing of molecules, leading only to short-range order in which the coordination number is of prime relevance; (c) A gas with also a random packing of molecules and ordering, albeit limited, due to the mutual weak attraction of the molecules.

for large r. The *coordination number*, that is, the number of nearest neighbor molecules around a reference molecule, is, however, similar to the coordination number in the corresponding solid.

1.3
Outline and Approach

From these brief descriptions of liquids and their applications, it will be clear that not all topics can be treated within reasonable limits, even at an introductory level. Hence, it has been decided to discuss those liquids and those properties that are the most relevant to chemical engineers and chemists in average daily practice. Consequently, attention will be limited to classical molecular liquids and solutions, which means that quantum fluids, molten metals, molten salts and highly concentrated ionic solutions are not dealt with. Neither are liquid crystals discussed, since that topic has been treated recently in two introductory books [1]. Here, the focus is on structure and thermodynamic aspects such as activity coefficients and phase behavior. Although some of the topics dealt with are also covered in books on physical chemistry and statistical mechanics, it was felt that a coherent presentation for the audience indicated was missing. Generally, we refrain from discussing experimental methods and focus on the underlying theoretical concepts.

Problems provide an essential part of this text, and the reader is encouraged to study these. Some of the text and some of the problems are somewhat more advanced compared to the remainder, but they are not essential to the core of this book. These sections and problems have been indicated with asterisks.

In order to be sufficiently self-contained, the book starts with the basic aspects of thermodynamics, classical mechanics and quantum mechanics (Chapter 2), after which intermolecular interactions (Chapter 3) and the phenomenology of liquids (Chapter 4) are briefly described. Thus, these chapters review the basics required for the remainder of these notes. Basic to the solution of all problems in liquids is the use of statistical thermodynamics, as discussed in Chapter 5. Subsequently, the meaning of the "structure" and "energetics" of liquids are discussed in Chapter 6. Chapters 5 and 6 largely provide the framework in which the behavior of liquids and solutions is discussed, while the various approaches to simple liquids are discussed in Chapters 7, 8, and 9. An approach that, at least in principle, is rigorous is the integral equation approach (Chapter 7) of which a sketch is provided. However, a solution to a problem in principle is not always a solution in practice, and since this is the case for the integral equation approach, other methods are called for. Physical models (Chapter 8) have been developed in the past and can still be used to advantage for basic understanding. In the present era, computers form an essential tool for scientists, and consequently modern numerical methods such as molecular dynamics and Monte Carlo simulations (Chapter 9) are outlined. Chapters 7, 8 and 9 provide the most frequently encountered approaches towards pure liquids. Polar liquids, for which dipole moment and polarizability play a dominant role, comprise the next level of complexity (Chapter 10). Water is an important example which is discussed somewhat extensively (Figure 1.4). Thereafter, the discussions relate to solutions – that is, liquid molecular mixtures. As a useful guide for mixtures, the regular solution concept is used. This concept has a long history, dating back to J.J. van Laar [2] (1860–1938), and

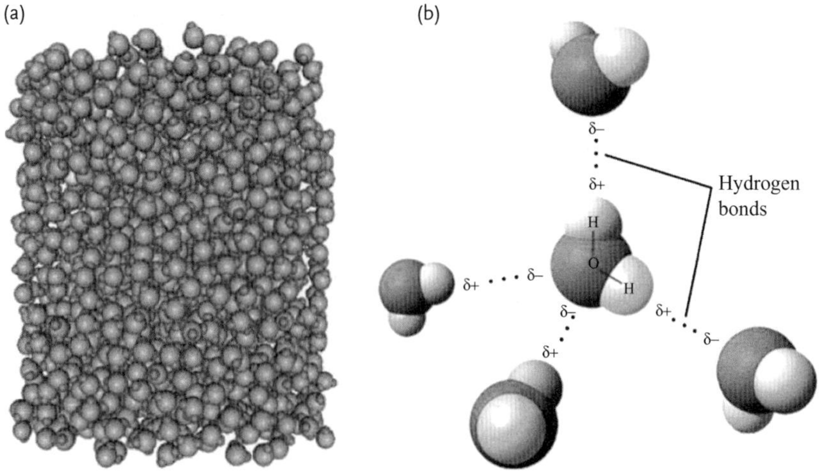

Figure 1.4 Water and its structure. (a) A snapshot of a molecular dynamics simulation of water; (b) The coordination of water molecules, indicating the hydrogen bonds.

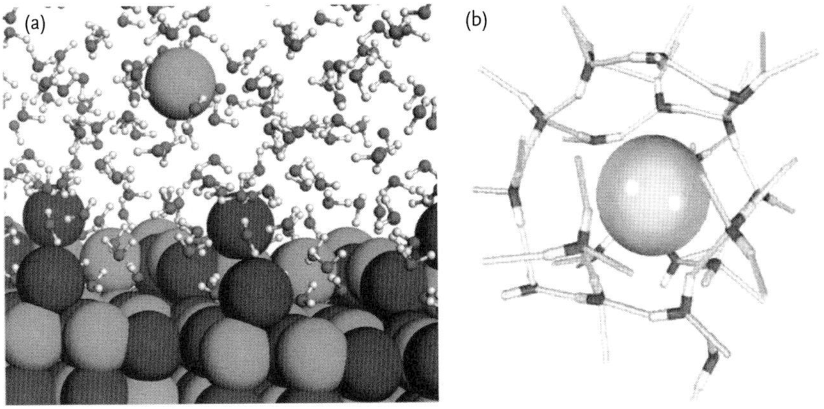

Figure 1.5 Ionic solutions. (a) Dissolution of salts by solvation; (b) The configuration of water molecules around an ion.

is particularly useful because of its conceptual simplicity, widespread use, and inherent possibilities for systematic improvement. Solutions with a comparable size of solvent and solute, and those with a rather different size of solvent and solute, can be distinguished. The first category can be subdivided in molecular (nonelectrolyte) solutions, which are treated in Chapter 11, and electrolyte solutions which are discussed in Chapter 12. In Chapter 11 the compatibility of solvents, including phase separation, forms an important part. The dissolution of salts, the structure of water around ions and the calculation of activity coefficients of ions in solution comprise a large part of Chapter 12 (Figure 1.5). The second category contains polymeric solutions and is treated in Chapter 13. The essential differences with molecular solutions and their consequences are treated (Figure 1.6).

Some specials topics are briefly introduced in the last three chapters. First, because most chemical reactions are carried out in liquids, in Chapter 14 we turn to the physical influence of solvents on chemical reactions (Figure 1.6). Second, we deal with surfaces (Chapter 15) in view of the relatively large importance of surfaces in fluids. Finally, phase transitions (Chapter 16) are treated as an example of an intimate mixture between molecular and mesoscopic arguments. The Appendices deal with physical constants, some useful mathematics, a brief review of the elements of electrostatics, the lattice gas model, some physical property data, and numerical answers to selected problems.

As with all fields of science and technology, there is an extensive literature available. In the section "Further Reading" a list of books is provided that can be selected for further study. Each chapter also contains, apart from specific references given as footnotes, a section "Further Reading" referring to small set of books relevant to that chapter. When used in the text, these are referred to by "author (year)" or "(author, year)".

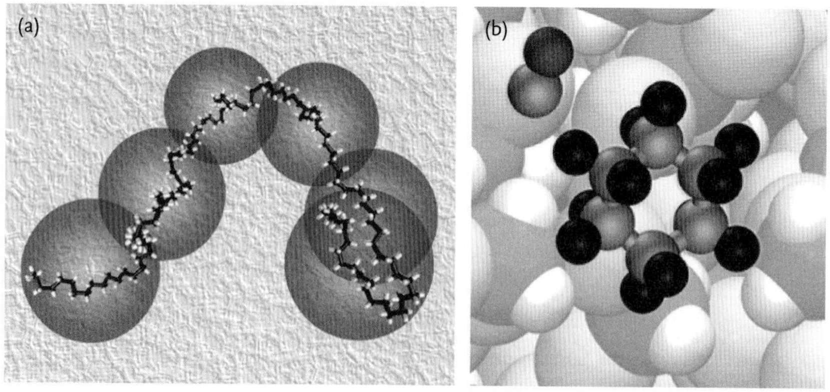

Figure 1.6 (a) Polymeric solution showing one real chain and its coarse-grained representation in a background of solvent (not shown); (b) Schematic of a reaction between cyclohexane and CN.

1.4 Notation

Within these notes we use at many occasions thermodynamics, and for that topic it is essential to agree on some conventions. For summations over particles, molecules, and so on, a lowercase Latin index, say i or j, is used, while for a summation over chemical components a lowercase Greek index, say α or β, is used. Furthermore, a superscript * is used for a pure compound, for example, the partial volume V_α^* of component α, and a superscript ° for a reference state, for example, the pressure P°, conventionally taken as 1 bar.

With respect to mathematical notation, scalars are addressed via an italic letter, say a, and vectors by an italic bold-face letter, say $\boldsymbol{a}$. Column matrices are labeled by, say a_i (index notation), or by a roman bold-face letter, say $\mathbf{a}$ (matrix notation). Similarly, square matrices are addressed by an italic letter with two subscripts, say A_{ij} or by $\mathbf{A}$. The column $\mathbf{a}$ is used as a shorthand for a collective of N quantities, that is, $\mathbf{a} = \mathbf{a}_1\mathbf{a}_2 \ldots \mathbf{a}_N$. For example, for N molecules each with coordinates $\mathbf{r}_i$ where $\mathbf{r}_i = (x_i, y_i, z_i)$, we denote the coordinates collectively by $\mathbf{r} = \mathbf{r}_1\mathbf{r}_2 \ldots \mathbf{r}_N = x_1y_1z_1x_2y_2z_2 \ldots x_Ny_Nz_N$ or in a multidimensional integral we write $\int \mathrm{d}\mathbf{r}$ where $\mathrm{d}\mathbf{r} = \mathrm{d}\mathbf{r}_1\mathrm{d}\mathbf{r}_2 \ldots \mathrm{d}\mathbf{r}_N = \mathrm{d}x_1\mathrm{d}y_1\mathrm{d}z_1\mathrm{d}x_2\mathrm{d}y_2\mathrm{d}z_2 \ldots \mathrm{d}x_N\mathrm{d}y_N\mathrm{d}z_N$. If we denote the set b_i by $\mathbf{b}$ and the set a_i by $\mathbf{a}$, we can therefore write $c = \Sigma_i b_i a_i = \mathbf{b}^\mathrm{T}\mathbf{a}$ using the transpose $\mathbf{b}^\mathrm{T}$ of $\mathbf{b}$. This allows us to write the derivatives of a function $f(a_i)$ given by $b_i = \partial f/\partial a_i$ collectively as $\mathbf{b} = \partial f/\partial \mathbf{a}$ or of a set $f_i(a_j)$ as $B_{ij} = \partial f_i/\partial a_j$ (equivalently for $\mathbf{f}$ we have $\mathbf{B} = \partial \mathbf{f}/\partial \mathbf{a}$). Note, therefore, that we distinguish between a vector $\boldsymbol{a}$ and its matrix representation $\mathbf{a}$. The inner product c of two vectors $\boldsymbol{a}$ and $\boldsymbol{b}$ is $c = \boldsymbol{a} \cdot \boldsymbol{b}$ $(= \Sigma_i b_i a_i)$ and written in matrix notation as $c = \mathbf{a}^\mathrm{T}\mathbf{b}$. For some further conventions on notation, we refer to Appendix B.

References

1 (a) Witten, T.A. (2004) *Structured Fluids*, Oxford University Press, Oxford; (b) Jones, R.A.L. (2002) *Soft Condensed Matter*, Oxford University Press, Oxford.

2 See, e.g., van Emmerik, E.P. (1991) *J.J. van Laar (1860–1938), A mathematical chemist*. Thesis, Delft University of Technology.

Further Reading

Barrat, J.-L. and Hansen, J.-P. (2005) *Basic Concepts for Simple and Complex Liquids*, Cambridge University Press, Cambridge.

Barker, J.A. (1963) *Lattice Theories of the Liquid State*, Pergamon, London.

Barton, A.F.M. (1974) *The Dynamic Liquid State*, Longman, New York.

Beck, T.L., Paulatis, M.E., and Pratt, L.R. (2006) *The Potential Distribution Theorem and Models of Molecular Solutions*, Cambridge University Press, Cambridge.

Ben-Naim, A. (1974) *Water and Aqueous Solutions: Introduction to a Molecular Theory*, Plenum, London.

Ben-Naim, A. (2006) *Molecular Theory of Solutions*, Oxford University Press, Oxford.

Croxton, C.A. (1974) *Liquid State Physics: A Statistical Mechanical Introduction*, Cambridge University Press, Cambridge.

Debenedetti, P.G. (1996) *Metastable Liquids: Concept and Principles*, Princeton University Press, Princeton.

Egelstaff, P.A. (1994) *An Introduction to the Liquid State*, 2nd edn, Clarendon, Oxford.

Eyring, H. and Jhon, M.S. (1969) *Significant Liquid Structures*, John Wiley & Sons, Ltd, London.

Fawcett, W.R. (2004) *Liquids, Solutions and Interfaces*, Oxford University Press, Oxford.

Fisher, I.Z. (1964) *Statistical Theory of Liquids*, University of Chicago Press, Chicago.

Frisch, H.L. and Salsburg, Z.W. (1968) *Simple Dense Fluids*, Academic, New York.

Frenkel, J. (1946) *Kinetic Theory of Liquids*, Oxford University Press, Oxford (see also Dover, 1953).

Guggenheim, E.A. (1952) *Mixtures*, Oxford, Clarendon.

Hansen, J.-P. and McDonald, I.R. (2006) *Theory of Simple Liquids*, 3rd edn, Academic, London (1st edn 1976, 2nd edn 1986).

Henderson, D. (ed.) (1971) *Physical Chemistry, and Advanced Treatise*, vols. VIIIa and VIIIb, Academic, New York.

Hildebrand, J.H. and Scott, R.L. (1950) *Solubility of Non-Electrolytes*, 3rd edn, Reinhold (1st edn 1924, 2nd edn 1936).

Hildebrand, J.H. and Scott, R.L. (1962) *Regular Solutions*, Prentice-Hall, Englewood Cliffs, NJ.

Hildebrand, J.H., Prausnitz, J.M., and Scott, R.L. (1970) *Regular and Related Solutions*, Van Nostrand-Reinhold, New York.

Hirschfelder, J.O., Curtiss, C.F., and Bird, R.B. (1954) *Molecular Theory of Gases and Liquids*, John Wiley & Sons, Inc., New York.

Kalikmanov, V.I. (2001) *Statistical Physics of Fluids*, Springer, Berlin.

Kohler, F. (1972) *The Liquid State*, Verlag Chemie, Weinheim.

Kruus, P. (1977) *Liquids and Solutions*, Marcel Dekker, New York.

Larson, R.G. (1999) *The Structure and Rheology of Complex Fluids*, Oxford University Press, New York.

Lucas, K. (2007) *Molecular Models of Fluids*, Cambridge University Press, Cambridge.

Lee, L.L. (1988) *Molecular Thermodynamics of Nonideal Fluids*, Butterworths, Boston.

March, N.H. and Tosi, M.P. (2002) *Introduction to the Liquid State Physics*, World Scientific, Singapore.

March, N.H. and Tosi, M.P. (1976) *Dynamics of Atoms in Liquids*, McMillan, London (see also Dover, 1991).

Marcus, Y. (1977) *Introduction to Liquid State Chemistry*, John Wiley & Sons, Ltd, London.

Murrell, J.N. and Jenkins, A.D. (1994) *Properties of Liquids and Solutions*, 2nd edn, John Wiley & Sons, Ltd, Chichester.

Prigogine, I. (1957) *The Molecular Theory of Solutions*, North-Holland, Amsterdam.

Pryde, J.A. (1966) *The Liquid State*, Hutchinson University Library, London.

Rice, S.A. and Gray, P. (1965) *The Statistical Mechanics of Simple Liquids*, Interscience, New York.

Rowlinson, J.S. and Swinton, F.L. (1982) *Liquids and Liquid Mixtures*, 3rd edn, Butterworth, London.

Temperley, H.N.V., Rowlinson, J.S., and Rushbrooke, G.S. (1968) *Physics of Simple Liquids*, North-Holland, Amsterdam.

Temperley, H.N.V. and Trevena, D.H. (1978) *Liquids and Their Properties*, Ellis Horwood, Chichester.

Ubbelohde, A.R. (1978) *The Molten State of Matter*, John Wiley & Sons, Ltd, Chichester.

Wallace, D.C. (2002) *Statistical Physics of Crystals and Liquids: A Guide to Highly Accurate Equations of State*, World Scientific, Singapore.

Watts, R.O. and McGee, I.J. (1976) *Liquid State Chemical Physics*, John Wiley & Sons, Inc., New York.

2
Basic Macroscopic and Microscopic Concepts: Thermodynamics, Classical, and Quantum Mechanics

In the description of liquids and solutions, we will need macroscopic and microscopic concepts that we do not consider as part of the subject proper but still are prerequisite to our topic. For macroscopic considerations, we will need phenomenological thermodynamics, and Section 2.1 presents a brief review of this. For microscopic considerations, we introduce in Section 2.2 some concepts from classical mechanics, and in Section 2.3 basic quantum mechanics and a few model system solutions. For further details of these matters, reference should be made to the books in the section "Further Reading" of this chapter, of which free use has been made.

2.1
Thermodynamics

Thermodynamic considerations are basic to many models and theories in science and technology, and in this section the phenomenological aspects of thermodynamics are briefly described but less briefly as in the summary given by Clausius [1]:

- Die Energie der Welt is constant.
- Die Entropie der Welt strebt einem Maximum zu[1].

2.1.1
The Four Laws

Thermodynamics is based on four well-known basic laws[2]. In order to describe these, we start with a few concepts. In thermodynamics the part of the physical world that is under consideration (the *system*) is for the sake of analysis considered to be separated from the rest (the *surroundings*). The *thermodynamic state* of the system is assumed to be determined completely by a set of macroscopic,

1) The energy of the universe is constant. The entropy of the universe strives to a maximum.
2) Frivol additions are the fourth law, which states that no experimental apparatus works the first time it is set up; the fifth law states that no experiment gives quite the expected numerical result [12].

independent *coordinates* a_i (also labeled **a**)[3] and one "extra" parameter related to the thermal condition of the system. For this parameter one has several choices, one of which is the most appropriate and to which we return later. Together, these constitute the set of *state variables. Intensive* (*extensive*) *variables* are independent of (dependent on) the size and/or quantity of the matter contained in the system. When a new set of state variable values is applied, the system adapts itself via internal processes, collectively characterized by the internal variables ξ_i (or $\boldsymbol{\xi}$). When the properties of a system do not change with time at an observable rate given certain constraints, the system is said to be in *equilibrium. Thermodynamics* is concerned with the equilibrium states available to systems, the transitions between them, and the effect of external influences upon the systems and transitions [2, 3].

Depending on the problem, one assumes a specific type of wall between the system and the surroundings, that is, a *diathermal* (contrary *adiabatic*) wall allowing influence from outside on the system in the absence of forces, a *flexible* (contrary *rigid*) wall allowing work exchange, or a *permeable* (contrary *impermeable*) wall allowing matter exchange. Systems separated by an adiabatic wall are *isolated*, while systems separated by a diathermal wall are said to be *closed*, but in *thermal contact*. Finally, systems separated by diathermal permeable walls are designated as *open*. Sometimes the wall is taken as virtual, for example, in continuum considerations.

If any two separate systems, each in equilibrium, are brought into thermal contact, the two systems will gradually adjust themselves until they do reach mutual equilibrium. The *zeroth law*[4] states that if two systems are both in thermal equilibrium with a third system, they are also in thermal equilibrium with each other–that is, they have a common value for a state variable T, called (*empirical*) *temperature*. If we consider two systems in thermal contact, one of which is much smaller than the other, the state of the larger one will only change negligibly in comparison to the state of the smaller one if heat is transferred from one system to the other. If we are primarily interested in the small system, the larger one is usually known as a *temperature bath* or *thermostat*. If, on the other hand, we are primarily interested in the large system, we regard the small system as a measuring device for registering the temperature and refer to it as a *thermometer*.

With each independent variable a_i, often indicated as *displacement*, a dependent variable is associated, generally denoted as *force*[5] A_i. A force is a quantity in the surroundings that, when multiplied with a change in displacement, yields the associated *work*. Displacement and force are often denoted as *conjugated variables*. Work done on the system is counted positive and depends on the path between the initial and the final state. For infinitesimal changes $\mathrm{d}a_i$ the work δW[6] is

3) For notation conventions, see Section 1.4.

4) The peculiar numbering of the laws of thermodynamics resulted from the fact that the importance of the zeroth law was realized only after the first law was established. The status of the zeroth law is not without dispute [13].

5) A clear presentation of the concept of force has been given by O. Redlich [13].

6) A significant objection against the use of the concept of differential δX of the variable X has been raised by Menger [14]. Essentially, his criticism boils down to the fact that $\mathrm{d}\mathbf{A}$ describes progress of the variable $\mathbf{A}$ with time as given by $(\mathrm{d}\mathbf{A}/\mathrm{d}t)\mathrm{d}t$.

$$\delta W = \sum_i A_i \mathrm{d}a_i = \mathbf{A}^\mathrm{T}\mathrm{d}\mathbf{a} \tag{2.1}$$

The most familiar example of work is the mechanical work $\delta W_{\mathrm{mec}} = -P\mathrm{d}V$, where P and V denote the pressure and volume, respectively[7]. Other examples are the chemical work $\delta W_{\mathrm{che}} = \Sigma_\alpha \mu_\alpha \mathrm{d}n_\alpha$ with the chemical potential μ_α and number of moles n_α of component α and the electrical work $\delta W_{\mathrm{ele}} = \phi \mathrm{d}q$ with the electric potential ϕ and charge q. In general, the integral over δW yields $\int \delta W = W$, where W is not a *state function*, that is, a function of state variables only, contrary to the integral over a state variable $\mathbf{a}$ yielding $\int \mathrm{d}\mathbf{a} = \Delta\mathbf{a}$. Hence, δW is not a total differential and consequently we use the notation δW instead of $\mathrm{d}W$.

When work is done on an adiabatically enclosed system it appears that the final state is independent of the type of work and its process of delivery. This leads to the state function U for which we thus have $\mathrm{d}U = \delta W$. Using other enclosures as well, the *first law* states that there exists a state function U, called *internal energy*, such that the change in the internal energy $\mathrm{d}U$ from one state to another is given by

► $$\mathrm{d}U = \delta Q + \delta W = \delta Q + \sum_i A_i \mathrm{d}a_i = \delta Q + \mathbf{A}^\mathrm{T}\mathrm{d}\mathbf{a} \tag{2.2}$$

where δQ is identified as *heat* entering the system[8] and interpreted as work done by the microscopic forces associated with the internal variables $\boldsymbol{\xi}$ in the system. Heat entering the system is counted as positive and also depends on the path between the two states. Although δQ and δW are dependent on the path between the initial and final states, $\mathrm{d}U$ is independent of the path and depends only on the initial and final states. Hence, $\mathrm{d}U$ is a total differential and the internal energy U a state variable which is the proper choice for the "extra", thermal parameter of state.

Each state is thus characterized by a set of state variables $\mathbf{a}$ and the internal energy U. If for a process $\delta Q = 0$, it is called *adiabatic*. If $\delta W = 0$ we have *isochoric* conditions and we refer to the process involved as pure *heating* or *cooling*. If both $\delta Q = 0$ and $\delta W = 0$, the system is *isolated*. The first law, Eq. (2.2), can thus be stated succinctly as follows: *for an isolated system, the energy is constant.*

The *second law* states that there exists another state function $S = S^{(\mathrm{r})} + S^{(\mathrm{i})}$, called *entropy* and composed of a reversible part $S^{(\mathrm{r})}$ and irreversible part $S^{(\mathrm{i})}$, such that for a transition between two states

► $$T\mathrm{d}S \equiv T\left(\mathrm{d}S^{(\mathrm{r})} + \mathrm{d}S^{(\mathrm{i})}\right) \geq \delta Q \tag{2.3}$$

where $T\mathrm{d}S^{(\mathrm{r})} = \delta Q$ and T is the temperature external, that is, just outside, to the volume element considered. The second law, Eq. (2.3), thus expresses that *for an isolated system the entropy can increase only or remain constant at most.* Since for any system the energy U is a characteristic of the state, the entropy S is a function of

7) It is clear that if $\mathrm{d}P = P_{\mathrm{ext}} - P_{\mathrm{int}} << P_{\mathrm{ext}}$, $\mathrm{d}W = P_{\mathrm{ext}}\mathrm{d}V \cong -P_{\mathrm{int}}\mathrm{d}V$.

8) Using the word "heat" here means, of course, that some circularity in the argument is used, since "adiabaticity" was defined as the nonability to influence the system in the absence of forces (implying no heat transfer). This circularity can be largely avoided by more elaborate considerations in the definition of adiabatic walls. See, for example, Refs [4] and [5].

U and $\mathbf{a}$ and we write $S = S(U,\mathbf{a})$. The entropy of a composite system is usually taken to be *additive*, that is, the sum of the entropies of the constituent subsystems, to be *analytical*, that is, continuously differentiable (see Appendix B), and *monotonically increasing with energy*. Additivity is true as long as the range of interactions between particles is small as compared to the size of the system. Conventionally, additivity is taken to imply *extensivity*; that is, if the extensive parameters are multiplied by λ, the entropy will obey $S(\lambda U,\lambda\mathbf{a}) = \lambda S(U,\mathbf{a})$, and the entropy is thus taken as a homogeneous function of degree 1 of the extensive parameters (see Appendix B). However, this is not necessarily true but only if the effect of the interface of the system with the walls of the enclosure can be neglected. Analyticity and continuity are also not always true. For example, for a phase transition it may be impossible to express the entropy in a Taylor series around the transition temperature and it may show a jump (discontinuous transition) or a discontinuity (continuous transition) in derivatives at that temperature. Normally, we accept analyticity and continuity.

According to the second law, for a given U and $\mathbf{a}$, the equilibrium state is obtained when S is maximized by the microscopic processes associated with the internal variables $\boldsymbol{\xi}$ of the system or $\mathrm{d}S(U,\mathbf{a}) = 0$ and $\mathrm{d}^2S(U,\mathbf{a}) < 0$ (Figure 2.1). If the equality in Eq. (2.3) holds, the process is *reversible* ($\mathrm{d}S^{(i)} = 0$), otherwise it is *irreversible* ($\mathrm{d}S^{(i)} > 0$). If for a process $\mathrm{d}S = 0$, it is called *isentropic*. If $\mathrm{d}T = 0$, we refer to it as *isothermal*. A natural process (contrary unnatural process) is a process occurring spontaneously in Nature, which proceeds towards equilibrium and in this sense a reversible process can be seen as the limiting case between natural and unnatural processes[9].

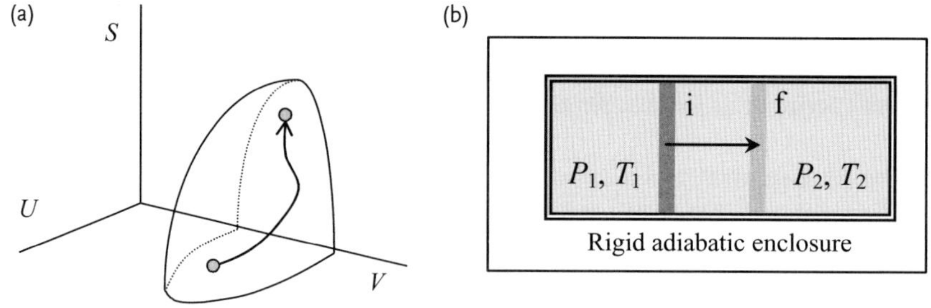

Figure 2.1 (a) Entropy S as a function of U and V and a quasi-static process from an initial to a final state. (b) The establishment of equilibrium for an isolated system consisting of a cylinder with a frictionless piston moving from initial position i to final position f.

9) A reversible process can be described by a path in the $(\mathbf{A},\mathbf{a})$ state space. The internal variables of the system, collectively indicated by $\boldsymbol{\xi}$, control the direction and rate of the processes, externally driven by $\mathbf{a}$, so that any change should actually be written as $\mathrm{d}\mathbf{A} = (\partial\mathbf{A}/\partial\boldsymbol{\xi})(\partial\boldsymbol{\xi}/\partial\mathbf{a})(\partial\mathbf{a}/\partial t)\mathrm{d}t$, keeping the criticism of Menger in mind. Limiting ourselves to quasi-static processes the system is at any moment in equilibrium, resulting in $\mathrm{d}\mathbf{A} = (\partial\mathbf{A}/\partial\mathbf{a})(\partial\mathbf{a}/\partial t)\mathrm{d}t = (\partial\mathbf{A}/\partial\mathbf{a})\,\mathrm{d}\mathbf{a}$.

The entropy expressed as $S = S(U,\mathbf{a})$ is called a *fundamental equation*. Once it is known as a function of its *natural variables* U and $\mathbf{a}$, all other thermodynamic properties can be calculated. The description $S = S(U,\mathbf{a})$ is called the *entropy representation*. Since S is a single valued continuously increasing function of U, the equation $S = S(U,\mathbf{a})$ can be inverted to $U = U(S,\mathbf{a})$ without ambiguity. For a given entropy S the equilibrium state is reached when $\mathrm{d}U(S,\mathbf{a}) = 0$. From stability considerations it follows that U is minimized by the internal processes of the system or $\mathrm{d}U(S,\mathbf{a}) = 0$ and $\mathrm{d}^2U(S,\mathbf{a}) > 0$. The description $U = U(S,\mathbf{a})$ is the *energy representation* with natural variables S and $\mathbf{a}$ and is also a fundamental equation. In equilibrium we have from[10] $\mathrm{d}U = (\partial U/\partial S)\mathrm{d}S + (\partial U/\partial \mathbf{a})^{\mathrm{T}}\mathrm{d}\mathbf{a}$ the (*thermodynamic*) *temperature* $T = \partial U/\partial S$ and the forces $\mathbf{A} = \partial U/\partial \mathbf{a}$. From $T = \partial U/\partial S$ we easily obtain the relation $T^{-1} = \partial S/\partial U$. One can show that the temperature T is independent of the properties of the thermometer used, and is consistent with the intuitive concept of (empirical) temperature. The units are *kelvins*, abbreviated as K and related to the *Centigrade* or *Celsius* scale, using °C as unit with the same size but different origin:

$$x\,^{\circ}\mathrm{C} = (273.150 + x)\,\mathrm{K} \tag{2.4}$$

The origin of the absolute scale, 0 K = –273.150 °C, is referred to as *absolute zero*.

For completeness we mention the *third law*, stating that for any isothermal process involving only phases in internal equilibrium

$$\lim_{T\to 0} \Delta S = 0 \tag{2.5}$$

This relation also holds if, instead of being in internal equilibrium, such a phase is in frozen metastable equilibrium, provided that the process does not disturb this frozen equilibrium. According to the third law, absolute zero cannot be reached. The discussion of the third law contains subtleties for which we refer to the literature.

2.1.2 Quasi-Conservative and Dissipative Forces

Only for reversible systems we can associate $T\mathrm{d}S$ with δQ and in this case $\delta W = \Sigma_i A_i \mathrm{d}a_i = \mathbf{A}^{\mathrm{T}}\mathrm{d}\mathbf{a}$ is the reversible work. For irreversible systems, $T\mathrm{d}S > \delta Q$ and the work δW contains also an irreversible or dissipative part. To show this, note that the elementary work can be written as

$$\delta W = \mathrm{d}U - \delta Q = \mathrm{d}U - T\mathrm{d}S^{(\mathrm{r})} = \mathrm{d}U - T\mathrm{d}S + T\mathrm{d}S^{(\mathrm{i})} \tag{2.6}$$

where $\mathrm{d}S = \mathrm{d}S^{(\mathrm{r})} + \mathrm{d}S^{(\mathrm{i})}$. Since $U = U(S,\mathbf{a})$ we have $\mathrm{d}U = (\partial U/\partial S)\mathrm{d}S + (\partial U/\partial \mathbf{a})^{\mathrm{T}}\mathrm{d}\mathbf{a} = T\mathrm{d}S + (\partial U/\partial \mathbf{a})^{\mathrm{T}}\mathrm{d}\mathbf{a}$. So we can also write Eq. (2.6) as

10) No subscripts or variables in the partial derivatives implies that the various state functions U, S, etc., are expressed in their natural parameters, that is, $U = U(S,\mathbf{a})$ and $\partial U/\partial S = (\partial U(S,\mathbf{a})/\partial S)_{\mathbf{a}}$, etc.

$$\delta W = \left(\frac{\partial U}{\partial \mathbf{a}}\right)^{\mathrm{T}} \mathrm{d}\mathbf{a} + T\mathrm{d}S^{(\mathrm{i})} \tag{2.7}$$

Therefore, the expression $T\mathrm{d}S^{(\mathrm{i})}$ has also the form of elementary work and we introduce forces $\mathbf{A}^{(\mathrm{d})}$ by writing

$$T\mathrm{d}S^{(\mathrm{i})} = (\mathbf{A}^{(\mathrm{d})})^{\mathrm{T}} \mathrm{d}\mathbf{a} = \sum_i A_i^{(\mathrm{d})} \mathrm{d}a_i = \delta W^{(\mathrm{d})} \quad \text{with} \quad A_i^{(\mathrm{d})} = \mathbf{A}^{(\mathrm{d})} = \mathbf{A} - \frac{\partial U}{\partial \mathbf{a}} \tag{2.8}$$

The quantity $\mathbf{A}^{(\mathrm{d})}$ is the *dissipative force* and we refer to $\delta W^{(\mathrm{d})} = \Sigma_i A_i^{(d)} \mathrm{d}a_i$ as the *dissipative work*. Writing for the total force

$$\mathbf{A} = \mathbf{A}^{(\mathrm{d})} + \mathbf{A}^{(\mathrm{q})} \tag{2.9}$$

we define the *quasi-conservative force* $\mathbf{A}^{(\mathrm{q})}$ by

$$\mathbf{A}^{(\mathrm{q})} = \partial U / \partial \mathbf{a} \tag{2.10}$$

which is the variable *conjugated* to $\mathbf{a}$. Summarizing, we have

▶
$$\begin{aligned} \mathrm{d}U &= \delta W + \delta Q = \mathbf{A}^{\mathrm{T}} \mathrm{d}\mathbf{a} + T\mathrm{d}S^{(\mathrm{r})} = \left[(\mathbf{A}^{(\mathrm{q})})^{\mathrm{T}} \mathrm{d}\mathbf{a} + (\mathbf{A}^{(\mathrm{d})})^{\mathrm{T}} \mathrm{d}\mathbf{a}\right] + T\mathrm{d}S^{(\mathrm{r})} \\ &= (\mathbf{A}^{(\mathrm{q})})^{\mathrm{T}} \mathrm{d}\mathbf{a} + (T\mathrm{d}S^{(\mathrm{i})} + T\mathrm{d}S^{(\mathrm{r})}) = (\mathbf{A}^{(\mathrm{q})})^{\mathrm{T}} \mathrm{d}\mathbf{a} + T\mathrm{d}S \end{aligned} \tag{2.11}$$

known as the *Gibbs equation*. The adjective "quasi-" stems from the fact that U is still dependent on the temperature while the adjective "conservative" indicates that U acts as a potential. Generally the adjective "quasi-conservative" is not added; using only the volume V for $\mathbf{a}$, one refers to $\mathbf{A} = -P^{(\mathrm{q})}$ as just pressure and labels it as $-P$. We will do so likewise hereafter. Hence, in the remainder we use $A_i = -P$, $a_i = V$ and write $U = U(S,V)$ so that the pressure becomes $P = -\partial U/\partial V$. The Gibbs equation then becomes

$$\mathrm{d}U = T\mathrm{d}S - P\mathrm{d}V \tag{2.12}$$

However, note that in a nonequilibrium state and for dissipative processes, $P = P^{(\mathrm{q})}$ is not the total pressure, and that in general there are more independent variables. Since for a reversible process one assumes that the system is at any moment in equilibrium, the rate $\dot{a} = \mathrm{d}a/\mathrm{d}t$ must be "infinitely" slow. If this is the case the process is *quasi-static*. It will be clear that reversible processes are quasi-static, although the reverse is not necessarily true.

2.1.3 Equation of State

Expressed in its natural variables S and V, $U(S,V)$ is a fundamental equation, just like $S(U,V)$. Again, once it is known, all other thermodynamic properties of the system can be calculated. If U is not expressed in its natural variables but, for example, as $U = U(T,V)$, we have slightly less information. In fact it follows from $T = \partial U/\partial S$ that $U = U(T,V)$ represents a differential equation for $U(S,V)$. Functional relationships such as

$$T = T(S, V) \quad \text{and} \quad P = P(S, V)$$

expressing a certain intensive state variable as a single-valued function of the remaining extensive state variables are called *equations of state* (EoS) and the intensive variable is thus described by a *state function.* If all the equations of state are known, the fundamental equation can be inferred to within an arbitrary constant. To that purpose we note that, since U is an extensive quantity, $U(S,V)$ can be taken as homogeneous of degree 1 and thus, using Euler's theorem (see Appendix B), that

$$U = \frac{\partial U}{\partial S} S + \frac{\partial U}{\partial V} V = TS - PV \tag{2.13}$$

Given $T = T(S,V)$ and $P = P(S,V)$ the energy U can be calculated. A similar argumentation for S leads to $S = (1/T)U + (P/T)V$. Only in some special cases, for example, if $\mathrm{d}S(U,V) = f(U)\mathrm{d}U + g(V)\mathrm{d}V$, the differential can be integrated directly to obtain the fundamental equation. Equation (2.13) is often referred to as an *Euler equation.*

Problem 2.1

The pressure is given by $-P = \partial U/\partial V = [\partial U(S,V)/\partial V]_S$. Show that if one uses the equation of state $U = U(T,V)$, P becomes

$$-P(T,V) = [\partial U(T,V)/\partial V]_T - T[\partial S(T,V)/\partial V]_T$$

Problem 2.2

A perfect gas is defined by $U = cnRT$ and $PV = nRT$, where c is a constant, n the number of moles and the other symbols have their usual meaning. Show that the fundamental equation for the entropy S is given by $S = S_0 + nR\ln[(U/U_0)^c(V/V_0)]$ with U_0 and V_0 a reference energy and reference volume.

Problem 2.3

Show, using a similar argumentation as for U, that the relation $S = (1/T)U + (P/T)V$ holds.

2.1.4 Equilibrium

Consider an isolated system consisting of two subsystems only capable of exchanging heat ($\mathrm{d}V_i = 0$). We now ask ourselves under what conditions thermal equilibrium would occur, and to this purpose we consider arbitrary but small variations in energy in both subsystems. From the first law we have $\mathrm{d}U = \mathrm{d}U_1 + \mathrm{d}U_2 = 0$. If the system is in *thermal equilibrium,* since S is maximal, we must also have $\mathrm{d}S = \mathrm{d}S_1 + \mathrm{d}S_2 = 0$. Because

$$\mathrm{d}S = \frac{\partial S_1}{\partial U_1}\mathrm{d}U_1 + \frac{\partial S_2}{\partial U_2}\mathrm{d}U_2 = \left(\frac{\partial S_1}{\partial U_1} - \frac{\partial S_2}{\partial U_2}\right)\mathrm{d}U_1 = \left(\frac{1}{T_1} - \frac{1}{T_2}\right)\mathrm{d}U_1 = 0 \tag{2.14}$$

and since dU_1 is arbitrary, we obtain $T_1 = T_2$ in agreement with the zeroth law. Moreover, starting from a nonequilibrium state one can show that heat flows from high to low temperature, again in agreement with intuition [3].

The procedure outlined above is in fact the general method to deal with basic equilibrium thermodynamic problems: one considers an isolated system consisting of two subsystems of which one is the object of interest and the other represents the environment. For the total, isolated system we have $dU = 0$. Moreover, in equilibrium we have $dS = 0$. Together with the *closure relations*–that is, the coupling relations between the displacements a_i of the system of interest and of the environment–a solution can be obtained.

Let us now consider hydrostatic equilibrium along these lines. Consider therefore again an isolated system consisting of two homogeneous subsystems but in this case capable of exchanging heat and work (Figure 2.1). For this system to be in equilibrium we write again $dS = dS_1 + dS_2 = 0$. However, the entropy S is now a function of U and the various values of a_i. As we are interested in hydrostatic equilibrium we take just the volume V for each subsystem. From Eq. (2.12) we obtain

$$dS = \frac{1}{T}dU + \frac{P}{T}dV \tag{2.15}$$

and thus for the equilibrium configuration

$$dS = dS_1 + dS_2 = \frac{1}{T_1}dU_1 + \frac{P_1}{T_1}dV_1 + \frac{1}{T_2}dU_2 + \frac{P_2}{T_2}dV_2 \tag{2.16}$$

Because the total system is isolated we have the closure relations $dU = dU_1 + dU_2 = 0$ and $dV = dV_1 + dV_2 = 0$. Therefore, since dU_1 and dV_1 are arbitrary, we obtain

$$T_1^{-1} = T_2^{-1} \quad \text{or} \quad T_1 = T_2 \tag{2.17}$$

regaining thermal equilibrium and

$$P_1/T_1 = P_2/T_2 \quad \text{or using } T_1 = T_2, \quad P_1 = P_2 \tag{2.18}$$

The latter condition corresponds to *hydrostatic equilibrium*. Hence, in hydrostatic equilibrium the temperature and the pressure of the subsystems are equal.

2.1.5
Auxiliary Functions

The Gibbs equation for a closed system is given by

$$dU = TdS - PdV \tag{2.19}$$

which gives the dependent variable U as a function of the independent, natural variables S and V. From consideration of the second law, the criterion for equilibrium for a closed system with fixed composition is that $dS(U,V) = 0$ or S is a maximum at constant U. Equivalently, as stated earlier, $dU(S,V) = 0$ or U is a minimum at constant S. Both descriptions are complete but not very practical, since it is difficult to keep the entropy constant and keeping the energy constant excludes interference from outside. Therefore, auxiliary functions having more practical natural variables are introduced.

If we write the energy as $U(S,V)$, the *enthalpy* H is the Legendre transform[11] of U with respect to $P = -\partial U/\partial V$, which is obtained from

$$\blacktriangleright \quad H \equiv U + PV \tag{2.20}$$

After differentiation this yields

$$dH = dU + PdV + VdP \tag{2.21}$$

Combining with $dU = TdS - PdV$ results in

$$dH = TdS + VdP \tag{2.22}$$

The natural variables for the enthalpy H are thus S and P and the equilibrium condition becomes $dH(S,P) = 0$.

Similarly, writing $H(S,P)$, the *Gibbs energy* G is the Legendre transform of H with respect to $T = \partial H/\partial S$ and given by

$$\blacktriangleright \quad G \equiv H - TS = U + PV - TS \tag{2.23}$$

On differentiation and combination with $dH = TdS + VdP$ this yields

$$dG = dH - TdS - SdT = -SdT + VdP \tag{2.24}$$

of which the natural variables are T and P. Consequently, the equilibrium condition becomes $dG(T,P) = 0$. A third transform is the *Helmholtz energy* F, defined as

$$\blacktriangleright \quad F \equiv U - TS \quad \text{resulting in} \quad dF = -SdT - PdV \tag{2.25}$$

with natural variables T and V and the corresponding equilibrium condition $dF(T,V) = 0$. It is easily shown that, for example,

$$T = \frac{\partial U}{\partial S} = \frac{\partial H}{\partial S} \quad \text{and} \quad S = -\frac{\partial F}{\partial T} = -\frac{\partial G}{\partial T} \tag{2.26}$$

$$-P = \frac{\partial U}{\partial V} = \frac{\partial F}{\partial V} \quad \text{and} \quad V = \frac{\partial H}{\partial P} = \frac{\partial G}{\partial P} \tag{2.27}$$

The functions $U(S,V)$, $H(S,P)$, $F(T,V)$ and $G(T,P)$ thus act as a potential, similar to the potential energy in mechanics, and are called *thermodynamic potentials*. The derivatives with respect to their natural, independent variables yield the conjugate, dependent variables. These potentials are also fundamental equations. A stable equilibrium state is reached when these potentials are minimized for a given set of their natural variables, that is, when $dX = 0$ and when $d^2X > 0$, where X is any of the four functions U, H, F, or G. The advantage of using F and G is obvious: While it is possible to control either the set (V,T) or (P,T) experimentally, control

11) A clear discussion of Legendre transforms is given by Callen [3]. See also Appendix B.

of either the set (V,S) or (P,S) is virtually impossible. As an aside we note that the Legendre transform can be also applied to the entropy yielding the so-called *Massieu functions*. These also act as potentials, for example, $X(V,T) = S - U/T$ or $Y(P,T) = S - U/T - PV/T$. It will be clear that X is related to F and Y to G. In practice, these functions are but limitedly used, although they were invented before the transforms of the energy.

2.1.6
Some Derivatives and Their Relationships

Apart from the potentials, we also need some of their derivatives, sometimes denotes as *response functions*, and their mutual relationships. Consider the relations

$$TdS = dU + PdV \text{ and } TdS = dH - VdP$$

The *heat capacities* are defined by

$$C_X = \partial Q/\partial T = T(\partial S/\partial T)_X \quad (X = P \text{ or } V)$$

It thus follows that

$$C_V = T(\partial S/\partial T)_V = (\partial U/\partial T)_V \quad \text{and} \quad C_P = T(\partial S/\partial T)_P = (\partial H/\partial T)_P$$

Three other derivatives that occur frequently are the *compressibilities* κ_X ($X = T$ or S) and the (*thermal*) *expansion coefficient* or (*thermal*) *expansivity* $\alpha \equiv \alpha_P$.

$$\kappa_T = -(1/V)(\partial V/\partial P)_T \quad \kappa_S = -(1/V)(\partial V/\partial P)_S \quad \alpha_P = (1/V)(\partial V/\partial T)_P$$

Moreover, if $d\phi = Xdx + Ydy$ is a *total differential*, we can use the *Maxwell relations* $(\partial X/\partial y)_x = (\partial Y/\partial x)_y$. From the expressions for dU, dH, dF and dG we obtain

$$(\partial T/\partial V)_S = -(\partial P/\partial S)_V \quad (\partial V/\partial S)_P = (\partial T/\partial P)_S$$
$$(\partial P/\partial T)_V = (\partial S/\partial V)_T \quad (\partial S/\partial P)_T = (\partial V/\partial T)_P$$

which can be used to reduce any thermodynamic quantity to a set of measurable quantities. Only three of the quantities C_V, C_P, κ_T, κ_S and α are independent because

$$\kappa_T - \kappa_S = TV\alpha^2/C_P \quad \text{and} \quad C_P - C_V = TV\alpha^2/\kappa_T \tag{2.28}$$

Using the third law, it can be shown that at $T = 0$, $C_P = C_V = 0$ and $\alpha = 0$. Using the relations $S = -\partial F/\partial T$ and $S = -\partial G/\partial T$, we have the *Gibbs–Helmholtz equations*

$$\blacktriangleright \quad U = F + TS = F - T\frac{\partial F}{\partial T} = \frac{\partial(F/T)}{\partial(1/T)} \quad \text{and}$$
$$H = G + TS = G - T\frac{\partial G}{\partial T} = \frac{\partial(G/T)}{\partial(1/T)} \tag{2.29}$$

Problem 2.4

Show for a perfect gas (see Problem 2.2) that $C_V = cNR$, $C_P = (c+1)NR$, $\kappa_T = 1/P$ and $\alpha = 1/T$.

Problem 2.5

Prove Eq. (2.28) and show that $C_P/C_V = \kappa_T/\kappa_S$.

Problem 2.6

Show that, if $F = C/T$ where C is a constant, $F = ½U$.

Problem 2.7: Blackbody radiation

For blackbody radiation in a volume V the pressure P is given $P = u/3$ with energy density $u = U/V$ and $U(T)$ the total energy being a function of temperature T only. Using $dU = TdS - PdV$, show that $u = aT^4$ where the constant a is called the Stefan–Boltzmann constant. Using $(\partial S/\partial V)_T = (\partial P/\partial T)_V$, show that $s = S/V = 4aT^3/3$ yielding T and P as functions of S. Finally, show that $S = CU^{3/4}/4V^{1/4}$ with $C = 4a^{1/3}/3$ a constant.

2.1.7
Chemical Content

The content of a system is defined by the amount of moles n_α of a chemical species α in the system which can be varied independently, often denoted as *components*. A *mixture* is a system with more than one component. A homogeneous part of a mixture, that is, with uniform properties throughout that part, is addressed as *phase* while a multicomponent phase is labeled *solution*. A system consisting of a single phase is called *homogeneous*, while one consisting of more than one phase is labeled *heterogeneous*. Moreover, a phase can be *isotropic* and *anisotropic* (properties independent or dependent on direction within a phase). The majority component of a solution is the *solvent* while *solute* refers to the minority component. For an arbitrary extensive quantity Z of a mixture, we have the molar quantities $Z_m \equiv Z/\sum_\alpha n_\alpha$ or $Z_\alpha^* \equiv Z/n_\alpha$ for a pure component α (sometimes also indicated by $Z_m(\alpha)$, z_α or even z when no confusion is possible). For mixtures, we also define for any extensive property Z the associated *partial (molar) property* Z_α as the partial derivative with respect to the number of moles n_α at constant T and P. For example, the partial volume V_α given by

$$V_\alpha \equiv (\partial V/\partial n_\alpha)_{T,P,n_\beta} \quad (\alpha \neq \beta) \tag{2.30}$$

and similarly for U_α, S_α and so on. At constant T and P we thus have

$$dZ = \sum\nolimits_\alpha (\partial Z/\partial n_\alpha) dn_\alpha = \sum\nolimits_\alpha Z_\alpha \, dn_\alpha \quad (P, T \text{ constant}) \tag{2.31}$$

Since Z is homogeneous of the first degree in n_α, we have by the Euler theorem

$$Z=\sum_\alpha(\partial Z/\partial n_\alpha)n_\alpha=\sum_\alpha Z_\alpha n_\alpha \tag{2.32}$$

Hence, we may regard Z as the sum of the contributions Z_α for each of the species α.

We note that, besides the number of moles n_α, various other quantities are used for amount of substance. We will also use the number of molecules $N_\alpha = N_A n_\alpha$ with N_A as Avogadro's number. Frequently, one is only interested in relative changes in composition, and to that purpose one uses the *mole fraction* defined by $x_\alpha = N_\alpha/\Sigma_\alpha N_\alpha = n_\alpha/\Sigma_\alpha n_\alpha = n_\alpha/n$, using $n \equiv \Sigma_\alpha n_\alpha$. In the sequel, the expressions for a binary system are given in parentheses with index 1 referring to the solvent, and index 2 to the solute.

Indicating the molar mass of the solvent by M_1, one defines the *molality* $m_\alpha \equiv n_\alpha/n_1M_1$ and since the mole fraction $x_\alpha = n_\alpha/\Sigma_\alpha n_\alpha = n_\alpha/n$ we have

$$m_\alpha/x_\alpha = n/n_1M_1 \quad (m_2/x_2=(n_1+n_2)/n_1M_1 \equiv n/n_1M_1) \tag{2.33}$$

For dilute solutions, $n_{\alpha\neq 1} \to 0$ and we obtain $m_\alpha/x_\alpha = 1/M_1$ $(m_2/x_2 = 1/M_1)$. If M_1 is expressed in kg mol^{-1}, the unit of molality becomes mol kg^{-1}, often labeled as m, for example, for $0.1\,\text{mol kg}^{-1}$ one writes 0.1 m.

In practice, *molarity* $c_\alpha \equiv n_\alpha/V$ with V the total volume of the solution, is also used. Since $\rho V = \Sigma_\alpha n_\alpha M_\alpha$ with ρ the mass density[12] of the solution, we have

$$c_\alpha=\frac{\rho n_\alpha}{\sum_\alpha n_\alpha M_\alpha} \quad \left(c_2=\frac{\rho n_2}{n_1M_1+n_2M_2}\right) \tag{2.34}$$

Using again the mole fraction x_α, the result

$$\frac{c_\alpha}{x_\alpha}=\frac{\rho n}{\sum_\alpha n_\alpha M_\alpha} \quad \left(\frac{c_2}{x_2}=\frac{\rho(n_1+n_2)}{n_1M_1+n_2M_2}=\frac{\rho n}{n_1M_1+n_2M_2}\right) \tag{2.35}$$

is easily obtained. For dilute solutions, $n_{\alpha\neq 1} \to 0$ and we obtain $c_\alpha/x_\alpha = \rho/M_1$ $(c_2/x_2 = \rho/M_1)$. If c_α is expressed in mol l^{-1}, labeled as M, one often speaks of *concentration*, for example, for $c_\alpha = 0.1\,\text{mol l}^{-1}$, one writes 0.1 M. Whatever measurement for composition is used, it is expedient to use as other independent variables P and T so that differentiation with respect to T implies constant P, and vice versa. From c_α we obtain the relation $\partial c_\alpha/\partial T = -c_\alpha\alpha_P$ with α_P the expansion coefficient, and therefore c_α is not independent of temperature. This renders theoretical calculations using c_α as independent variable often cumbersome.

For chemical problems, that is, where a change in chemical composition is involved, the fundamental equation becomes $U = U(S,V,n_\alpha)$ or $S = S(U,V,n_\alpha)$. This leads to

12) The designation *density* is used for quantities expressed per unit volume, for example, using the same symbol, the number density $\rho = N/V = N_A\Sigma_\alpha n_\alpha/V$. If confusion is possible, we use ρ' for mass density.

$$dU = \frac{\partial U}{\partial S}dS + \frac{\partial U}{\partial V}dV + \sum\nolimits_{\alpha} \frac{\partial U}{\partial N_{\alpha}}dn_{\alpha} = TdS - PdV + \sum\nolimits_{\alpha} \mu_{\alpha}\, dn_{\alpha} \qquad (2.36)$$

The partial derivative $\mu_{\alpha} = \partial U/\partial n_{\alpha}$ is called the *chemical potential* and is the conjugate intensive variable associated with the extensive variable n_{α} in the chemical work $dW_{che} = \Sigma_{\alpha}\mu_{\alpha}dn_{\alpha}$. Equilibrium is now obtained when $dU(S,V,n_{\alpha}) = 0$.

For obtaining a potential in terms of T and V, we use a Legendre transformation of U with respect to $T = \partial U/\partial S$ using the Helmholtz energy $F = U - TS$ and obtain

$$dF(T,V,n_{\alpha}) = -SdT - PdV + \sum\nolimits_{\alpha} \mu_{\alpha}\, dn_{\alpha} \qquad (2.37)$$

so that $F = F(T,V,n_{\alpha})$ and the equilibrium condition reads $dF(T,V,n_{\alpha}) = 0$.

Applying a Legendre transformation of U with respect to $T = \partial U/\partial S$ and $P = -\partial U/\partial V$ using the Gibbs energy $G = U - TS + PV$ leads to

$$dG = d(U - TS + PV) = -SdT + VdP + \sum\nolimits_{\alpha} \mu_{\alpha}\, dn_{\alpha} \qquad (2.38)$$

The Gibbs energy is thus given by $G = G(T,P,n_{\alpha})$ and the equilibrium condition becomes $dG(T,P,n_{\alpha}) = 0$. Because the chemical potential μ_{α} of the component α is defined as $\mu_{\alpha} = \partial G/\partial n_{\alpha}$, it is equal to the partial Gibbs energy G_{α}. Since G is homogeneous of the first degree in n_{α}, we have by Euler's theorem $G = \Sigma_{\alpha} G_{\alpha} n_{\alpha} = \Sigma_{\alpha} \mu_{\alpha} n_{\alpha}$. On the one hand, we thus have

$$dG = \sum\nolimits_{\alpha} \mu_{\alpha}\, dn_{\alpha} + \sum\nolimits_{\alpha} n_{\alpha}\, d\mu_{\alpha}$$

while on the other hand we know that $G = G(T,P,n_{\alpha})$ and thus that

$$dG = \frac{\partial G}{\partial T}dT + \frac{\partial G}{\partial P}dP + \sum\nolimits_{\alpha} \frac{\partial G}{\partial n_{\alpha}}dn_{\alpha} = -SdT + VdP + \sum\nolimits_{\alpha} \mu_{\alpha} dn_{\alpha}$$

Therefore, we obtain by subtraction the so-called *Gibbs–Duhem relation*

▶ $$SdT - VdP + \sum\nolimits_{\alpha} n_{\alpha} d\mu_{\alpha} = 0 \quad \text{or} \quad \sum\nolimits_{\alpha} n_{\alpha} d\mu_{\alpha} = 0 \quad (\text{constant } P \text{ and } T) \qquad (2.39)$$

implying a relation between the various differentials. This relation is here derived for T, P, and n_{α} as the only independent variables, but the extension to any number of variables is obvious. Moreover, although derived here for $\mu_{\alpha} = G_{\alpha}$, it will be clear that a similar relation can be derived for all partial quantities.

Finally, we can also apply a Legendre transformation to U with respect to $T = \partial U/\partial S$ and $\mu_{\alpha} = \partial U/\partial n_{\alpha}$ by using $\Omega = U - TS - \Sigma_{\alpha}\mu_{\alpha}n_{\alpha} = F - \Sigma_{\alpha}\mu_{\alpha}n_{\alpha}$ leading to

$$d\Omega = -SdT - PdV - \Sigma_{\alpha} n_{\alpha} d\mu_{\alpha}$$

This so-called *grand potential* Ω is mainly used in statistical thermodynamics (Chapter 5). Since we showed that $\Sigma_{\alpha}\mu_{\alpha}n_{\alpha} = G$, we identify the grand potential as

$$\Omega = F - G = -PV$$

Problem 2.8

Show that for the perfect gas (see Problem 2.2) the fundamental equation $S = S(U,V,n)$ is given by

$$S = S_0 + R\ln[(u/u_0)^c(v/v_0)(n_0/n)^{c+1}] \quad \text{with} \quad S_0 = (c+1)nR - n(\mu/T)_0$$

where u and v denote molar quantities and the subscript 0 refers to reference values. First, use the Gibbs–Duhem equation for $d\mu$ to show that the chemical potential μ is given by

$$\mu = \mu_0(T) - RT\ln[(u/u_0)^c(v/v_0)]$$

Thereafter, use the Euler equation $S = (1/T)(U + PV - \mu N)$.

2.1.8
Chemical Equilibrium

At constant P and constant T the Gibbs energy in a chemically reacting system varies with composition as $dG = \Sigma_\alpha\mu_\alpha dn_\alpha$. For a reaction to occur spontaneously $dG = \Sigma_\alpha\mu_\alpha dn_\alpha < 0$ while at equilibrium[13] $dG = \Sigma_\alpha\mu_\alpha dn_\alpha = 0$. Let us consider the reaction

$$a\text{A} + b\text{B} \leftrightarrow c\text{C} + d\text{D} \tag{2.40}$$

where $a, \ldots, d$ denote the stoichiometric coefficients of components A, . . . , D. It will be convenient to rearrange this expression to

$$0 = c\text{C} + d\text{D} - a\text{A} - b\text{B} \quad \text{or, even more compact, to} \quad 0 = \sum\nolimits_\text{X} \nu_\text{X}\text{X}$$

with a positive value for the coefficient ν_α when α is a product (C, D) and a negative value when α is a reactant (A, B). We define a factor of proportionality $d\xi(t)$ in such a way that $dn_\alpha = \nu_\alpha d\xi$. Starting at time zero with initially $n_\alpha(0)$ moles of each species the changes of the number of moles of each species in the time interval dt are

$$dn_\alpha = \nu_\alpha \frac{d\xi}{dt}dt \equiv \nu_\alpha\dot{\xi}dt \quad \text{or} \quad n_\alpha = n_\alpha(0) + \int dn_\alpha = n_\alpha(0) + \nu_\alpha\Delta\xi$$

where $\dot{\xi}$ is the *rate of reaction*. This leads to

$$dG = \sum\nolimits_\alpha \mu_\alpha dn_\alpha = \sum\nolimits_\alpha (\mu_\alpha\nu_\alpha)\dot{\xi}dt = -D\dot{\xi}dt \le 0$$

where we introduced the *affinity* $D = -\Sigma_\alpha(\nu_\alpha\mu_\alpha)$. From $dG \le 0$ we conclude that

▶ $$D\dot{\xi} \ge 0 \tag{2.41}$$

as the condition for a reaction to occur. So, D and $\dot{\xi}$ must have the same sign or be zero. At equilibrium $D = 0$. Since $\nu_\alpha = dn_\alpha/d\xi$, the affinity D is related to the

13) The chemical equilibrium conditions can also be derived similarly as those for thermal and mechanical equilibrium.

fundamental equations, the most important ones being, using $\mu_\alpha = \partial U/\partial n_\alpha = \partial G/\partial n_\alpha = -T\partial S/\partial n_\alpha$,

$$D = -\sum_\alpha \nu_\alpha \mu_\alpha = -\frac{\partial U(S,V,n_\alpha)}{\partial \xi} = -\frac{\partial G(T,P,n_\alpha)}{\partial \xi} = T\frac{\partial S(U,V,n_\alpha)}{\partial \xi} \tag{2.42}$$

All n_α must be positive or zero and the reaction goes to completion if one of the components is exhausted. This implies a lower and upper value for $\Delta\xi$. Therefore, the factor $\Delta\xi$ is sometimes normalized according to

$$\varepsilon = (\Delta\xi - \Delta\xi_{\min})/(\Delta\xi_{\max} - \Delta\xi_{\min})$$

where ε is the *degree of reaction*. Writing $D\dot{\xi}$ for the chemical *rate of work* or *power* mimics the expression for the mechanical power $\mathbf{A}^{\mathrm{T}}\dot{\mathbf{a}}$ with force $\mathbf{A}$ and rate of displacement $\dot{\mathbf{a}} = \mathrm{d}\mathbf{a}/\mathrm{d}t$. In both cases the power is positive semi-definite, that is, $\mathbf{A}^{\mathrm{T}}\dot{\mathbf{a}} \geq 0$.

Example 2.1

A vessel contains a ½ mole of H_2S, ¾ mole of H_2O, 2 moles of H_2, and 1 mole of SO_2 and is kept at constant T and P. The equilibrium condition is

$$-3\mu_{\mathrm{H_2}} - \mu_{\mathrm{SO_2}} + \mu_{\mathrm{H_2S}} + 2\mu_{\mathrm{H_2O}} = 0 \quad \text{and}$$

$$n_{\mathrm{H_2}} = 2 - 3\mathrm{d}\xi, \quad n_{\mathrm{SO_2}} = 1 - \mathrm{d}\xi, \quad n_{\mathrm{H_2S}} = \frac{1}{2} + \mathrm{d}\xi, \quad n_{\mathrm{H_2O}} = \frac{3}{4} + 2\mathrm{d}\xi$$

If the chemical potentials are known as a function of T, P and the n_α's, the solution for $\mathrm{d}\xi$ can be obtained. Suppose that the solution is $\mathrm{d}\xi = 1/4$. If $\mathrm{d}\xi = 2/3$, $n_{\mathrm{H_2}} = 0$ and therefore this is $\mathrm{d}\xi_{\mathrm{max.}}$ If $\mathrm{d}\xi = -3/8$, $n_{\mathrm{H_2O}} = 0$ and therefore this is $\mathrm{d}\xi_{\mathrm{min.}}$ Therefore, the degree of reaction $\varepsilon = [1/4 - (-3/8)]/[2/3 - (-3/8)] = 3/5$.

The above formulation leads immediately to the conventional chemical description. Let us introduce for component α the *absolute activity* $\lambda_\alpha = \exp(\mu_\alpha/RT)$ and define, considering again the reaction given by Eq. (2.40), the *reaction product*

$$\lambda_{\mathrm{C}}^{c}\lambda_{\mathrm{D}}^{d}/\lambda_{\mathrm{A}}^{a}\lambda_{\mathrm{B}}^{b} \tag{2.43}$$

Using the more compact notation introduced before we write more briefly $\prod_\alpha \lambda_\alpha^{\nu_\alpha}$. The equilibrium condition $D = -\Sigma_\alpha(\nu_\alpha\mu_\alpha) = 0$ can then be written as $\prod_\alpha \lambda_\alpha^{\nu_\alpha} = 1$.

For solid–gas reactions we distinguish between gases (α) and solids (β). This allows us to write $\prod_\alpha \lambda_\alpha^{\nu_\alpha} \prod_\beta \lambda_\beta^{\nu_\beta} = 1$, where the first product contains all the terms relating to gaseous species and the second to the solid species. Now note that the chemical potential μ_α of a gaseous component α is given by $\mu_\alpha = \mathring{\mu}_\alpha + RT\ln y_\alpha/P^\circ$, where $\mathring{\mu}_\alpha$ is the chemical potential in the standard state, R is the gas constant, $y_\alpha = f_{x,\alpha}x_\alpha P$ is the *activity* (for gases often denoted as *fugacity*), $f_{x,\alpha}$ is the (*mole fraction*) *activity coefficient*, P is the total pressure, and P° is the standard pressure (1 bar). Hence for a gas $\lambda_\alpha = \exp(\mu_\alpha/RT) = \mathring{\lambda}_\alpha y_\alpha/P^\circ$, where $\mathring{\lambda}_\alpha$ is the value of λ_α for

$P = P^{\circ}$. For gases at low pressures (hence activity coefficient $f_{x,\alpha} = 1$) the activity becomes the *partial pressure* P_α given by $P_\alpha = x_\alpha P$. For solids, on the other hand, we have $\lambda_\beta \cong \overset{\circ}{\lambda}_\beta$, only weakly dependent on the pressure. In total we have

$$\prod\nolimits_\alpha (\lambda_\alpha^{\circ})^{\nu_\alpha} x_\alpha^{\nu_\alpha} (P^{\circ})^{-\nu_\alpha} \prod\nolimits_\beta (\lambda_\beta^{\circ})^{\nu_\beta} = 1 \quad \text{or}$$

$$\prod\nolimits_\alpha x_\alpha^{\nu_\alpha} = K_P \quad \text{with} \quad K_P \equiv \prod\nolimits_\alpha (\lambda_\alpha^{\circ})^{-\nu_\alpha} (P^{\circ})^{\nu_\alpha} \prod\nolimits_\beta (\lambda_\beta^{\circ})^{-\nu_\beta}$$

The (*pressure*) *equilibrium constant* K_P is related to the standard Gibbs energy

$$\Delta\mu = c\mu_C^{\circ} + d\mu_D^{\circ} - a\mu_A^{\circ} - b\mu_B^{\circ} \quad \text{or, more generally,} \quad \Delta\mu = \sum\nolimits_\alpha \nu_\alpha \mu_\alpha^{\circ} \tag{2.44}$$

via $\Delta\mu = -RT\ln K_P$. Since $\Delta\mu$ is constant at constant temperature, the value of K_P is constant at constant temperature, which explains the name. If activities are used throughout, we refer to the activity equilibrium constant K_γ.

For reactions in solution we start again with $\prod_\alpha \lambda_\alpha^{\nu_\alpha} = 1$ and use $\lambda_{m,\alpha} = \overset{\circ}{\lambda}_{m,\alpha} m_\alpha f_{m,\alpha}$ with $f_{m,\alpha}$ the (*molality*) *activity coefficient* for molality m_α and $\overset{\circ}{\lambda}_{m,\alpha}$ the value for $m_\alpha = \overset{\circ}{m}_\alpha$. This leads to

$$\prod\nolimits_\alpha (m_\alpha)^{\nu_\alpha} \prod\nolimits_\alpha (f_{m,\alpha})^{\nu_\alpha} = K_m \quad \text{with} \quad K_m \equiv \prod\nolimits_\alpha (\overset{\circ}{\lambda}_{m,\alpha})^{-\nu_\alpha} \tag{2.45}$$

where K_m is the (*molality*) *equilibrium constant*. For ideal solutions or ideally dilute solutions $m_\alpha \to 0$, $f_{m,\alpha} \to 1$ and thus in that case $K_m = \prod_\alpha (m_\alpha)^{\nu_\alpha}$. Similarly using $\lambda_{c,\alpha} = \overset{\circ}{\lambda}_{c,\alpha} c_\alpha f_{c,\alpha}$ with $f_{c,\alpha}$ the (*molarity*) *activity coefficient* for component α for molarity c_α and $\overset{\circ}{\lambda}_{c,\alpha}$ the value for $c_\alpha = \overset{\circ}{c}_\alpha$ leads to

$$\prod\nolimits_\alpha (c_\alpha)^{\nu_\alpha} \prod\nolimits_\alpha (f_{c,\alpha})^{\nu_\alpha} = K_c \quad \text{with} \quad K_c \equiv \prod\nolimits_\alpha (\overset{\circ}{\lambda}_{c,\alpha})^{-\nu_\alpha} \tag{2.46}$$

where K_c is the (*molarity*) *equilibrium constant*. If $c_\alpha \to 0$, $f_{c,\alpha} \to 1$.

Problem 2.9

Derive the chemical equilibrium condition as done along the lines of hydrostatic and thermal equilibrium.

2.2
Classical Mechanics

Classical mechanics is used as a basis for both quantum and statistical mechanics. Therefore, we present a brief outline based on Hamilton's principle. This presentation is somewhat abstract but the most efficient. We will make extensive use of vectors (italic bold face, e.g., $\boldsymbol{r}$) and their matrix representations (roman bold face, e.g., $\mathbf{r}$).

In classical mechanics a particle is considered to be characterized by a mass m, a Cartesian position vector $\boldsymbol{r}$ and a rate of change of $\boldsymbol{r}$, that is, the velocity vector $\boldsymbol{v} = \mathrm{d}\boldsymbol{r}/\mathrm{d}t$. For N particles we need to specify N position vectors, that is, $3N$

coordinates. The number of independent coordinates is the number of *degrees of freedom* (DoF).

2.2.1 Generalized Coordinates

Starting with Cartesian coordinates we use a notation in which a symbol with no further variable indicated denotes the quantity for the system, while the variable (i) indicates a specific particle i. Hence, for example, f denotes the total force on the system and $m(1)$ the mass of particle 1. For a system of N particles with constraints, the number of DoFs does not equal $3N$ and changes in, for example, velocity $\mathrm{d}\boldsymbol{v}(i)$, are not independent. If the constraints can be expressed by $k = 3N - M$ equations (where $M < 3N$)

$$C_k(\boldsymbol{r}(1), \ldots, \boldsymbol{r}(i), \ldots, t) = c_k$$

where C_k is a function of the coordinates $\boldsymbol{r}$ and c_k is a constant, these so-called *holonomic constraints* can be taken into account by using generalized coordinates. If taken care of these constraints, the system itself is referred to as *holonomic*. We refrain from explicit considerations of constraints, refer to the literature for further discussion, and use generalized coordinates free of constraints. Within an abstract framework we find it convenient to describe the motion of systems of N particles with these N *generalized coordinates* $\mathbf{q}$ and *generalized velocities* $\dot{\mathbf{q}}$. Similar to Cartesian coordinates, the coordinates $\mathbf{q}$ are used to describe the instantaneous configuration of the system of interest, while the velocities $\dot{\mathbf{q}}$ specify the instantaneous motion of the system. With this information the system is completely specified, including the accelerations $\ddot{\mathbf{q}}$ and its subsequent motion can, in principle, be calculated. The relations between $\mathbf{q}$, $\dot{\mathbf{q}}$ and $\ddot{\mathbf{q}}$ are called *equations of motion*.

Example 2.2

The water molecule is nonlinear with an O–H bond length of 0.0958 nm and an included angle of ~105°. If the coordinates are taken as the bond lengths r_1, r_2, and the bond angle ϕ, we can approximate the vibration energy by a harmonic model and write the potential energy Φ as $\Phi = \frac{1}{2}k_r\Delta r_1^2 + \frac{1}{2}k_r\Delta r_2^2 + \frac{1}{2}k_\phi\Delta\phi^2$, where k_r and k_ϕ denote the force constants for the change in bond lengths Δr_1, Δr_2 and bond angle $\Delta\phi$, respectively. Of the other coordinates, three refer to translation and three to the rotation of the molecule as a whole.

In general, the Cartesian coordinates $\boldsymbol{r}$ are nonlinear functions of the generalized coordinates $\mathbf{q}$ which can be expressed in matrix notation by

$$\mathbf{r} = \mathbf{r}(\mathbf{q}) \quad \text{or} \quad \mathrm{d}\mathbf{r} = (\partial\mathbf{r}/\partial\mathbf{q})\mathrm{d}\mathbf{q} \equiv \mathbf{A}\mathrm{d}\mathbf{q} \tag{2.47}$$

where $\mathbf{r}^T = [r_1(1), r_2(1), r_3(1), r_1(2), \ldots, r_3(N)]$ and the matrix $\mathbf{A}$ is given explicitly by

$$\mathbf{A} = \frac{\partial \mathbf{r}}{\partial \mathbf{q}} = \begin{pmatrix} \frac{\partial r_1(1)}{\partial q_1} & \frac{\partial r_1(1)}{\partial q_2} & \cdots & \frac{\partial r_1(1)}{\partial q_{3N}} \\ \frac{\partial r_2(1)}{\partial q_1} & \frac{\partial r_2(1)}{\partial q_2} & \cdots & \cdots \\ \cdots & \cdots & \cdots & \cdots \\ \frac{\partial r_3(N)}{\partial q_1} & \cdots & \cdots & \frac{\partial r_3(N)}{\partial q_{3N}} \end{pmatrix}$$

Note that the matrix $\mathbf{A}$ is orthogonal, that is, $\mathbf{A}^{-1} = \mathbf{A}^T$. For the velocity we obtain

$$\mathbf{v} = [v_1(i), v_2(1), v_3(1), v_1(2), \ldots, v_3(N)] = \dot{\mathbf{r}} = \frac{\mathrm{d}\mathbf{r}}{\mathrm{d}t} = \frac{\mathrm{d}\mathbf{r}}{\mathrm{d}\mathbf{q}}\frac{\mathrm{d}\mathbf{q}}{\mathrm{d}t} = \mathbf{A}\dot{\mathbf{q}} \tag{2.48}$$

2.2.2
Hamilton's Principle and Lagrange's Equations

Almost always, the most economic description of physical laws is in the form of an extremum principle, and this is also the case with classical mechanics. *Hamilton's principle* [6, 7] states that, using a certain function called the *Lagrange function* $\mathcal{L}$ (and to which we come in some detail later), the *action* S defined by

► $$S \equiv \int_{t_1}^{t_2} \mathcal{L}\,\mathrm{d}t = \int_{t_1}^{t_2} \mathcal{L}(\mathbf{q}, \dot{\mathbf{q}}, t)\,\mathrm{d}t \tag{2.49}$$

is minimal for the actual path of $\mathbf{q}(t)$ given $\mathbf{q}(t_1)$ and $\mathbf{q}(t_2)$ at times t_1 and t_2. The principle thus states that the motion of an arbitrary system occurs in such a way that the variation of the action δS vanishes provided that the initial and final states are prescribed. Since $\mathcal{L}$ is a scalar, it is immaterial in which coordinate system it is expressed. In particular it may be expressed in generalized coordinates.

Evaluating Hamilton's principle using variation calculus (see Appendix B) results in

$$\delta S = \delta\int_{t_1}^{t_2} \mathcal{L}(\mathbf{q}, \dot{\mathbf{q}})\,\mathrm{d}t = \int_{t_1}^{t_2} \delta\mathcal{L}(\mathbf{q}, \dot{\mathbf{q}})\,\mathrm{d}t = \int_{t_1}^{t_2}\left[\left(\frac{\partial \mathcal{L}}{\partial \mathbf{q}}\right)^T \delta\mathbf{q} + \left(\frac{\partial \mathcal{L}}{\partial \dot{\mathbf{q}}}\right)^T \delta\dot{\mathbf{q}}\right]\mathrm{d}t = 0$$

Partial integration of the 2nd term of the previous equation using $\delta\dot{\mathbf{q}} = \mathrm{d}\delta q/\mathrm{d}t$ yields

$$\int_{t_1}^{t_2}\left(\frac{\partial \mathcal{L}}{\partial \dot{\mathbf{q}}}\right)^T \frac{\mathrm{d}}{\mathrm{d}t}\delta\mathbf{q}\,\mathrm{d}t = \left(\frac{\partial \mathcal{L}}{\partial \dot{\mathbf{q}}}\right)^T \delta\mathbf{q}\Bigg|_{t_1}^{t_2} - \int_{t_1}^{t_2} \frac{\mathrm{d}}{\mathrm{d}t}\left(\frac{\partial \mathcal{L}}{\partial \dot{\mathbf{q}}}\right)^T \delta\mathbf{q}\,\mathrm{d}t$$

The boundary term vanishes since $\delta\mathbf{q}$ vanishes at the boundary. Combining results in

$$\int_{t_1}^{t_2}\left(\frac{\partial \mathcal{L}}{\partial \mathbf{q}} - \frac{\mathrm{d}}{\mathrm{d}t}\frac{\partial \mathcal{L}}{\partial \dot{\mathbf{q}}}\right)^T \delta\mathbf{q}\,\mathrm{d}t = 0$$

and, since the variations $\delta\mathbf{q}$ are arbitrary, we obtain for the equations of motion

$$\frac{\mathrm{d}}{\mathrm{d}t}\frac{\partial \mathcal{L}}{\partial \dot{\mathbf{q}}} - \frac{\partial \mathcal{L}}{\partial \mathbf{q}} = 0 \tag{2.50}$$

These so-called *Lagrange equations*, given the Lagrange function $\mathcal{L}$, describe the behavior of a holonomic system. One can show that, using this description, $\mathcal{L}$ is defined to within an additive total time derivative of any function of the coordinates.

A basic assumption of nonrelativistic classical mechanics is that *space-time is isotropic* and *homogeneous*, that is, a so-called *inertial frame* can be found in which the laws of mechanics do not depend on the position and orientation of the system in space and are independent of the time origin. From the requirement that the behavior of a free particle is the same in any coordinate system moving with constant velocity, it follows that $\mathcal{L} \sim v^2 = |\boldsymbol{v}\cdot\boldsymbol{v}|$ or $\mathcal{L} = \frac{1}{2}mv^2$, representing the *kinetic energy*, and where $\boldsymbol{v}$ is the velocity of the particle with respect to the inertial frame and the proportionality factor m is the *mass*. For a set of free particles the Lagrange function represents the total kinetic energy T so that $\mathcal{L} = T = \frac{1}{2}\Sigma_i m_i v_i^2 = \frac{1}{2}\mathbf{v}^{\mathrm{T}}\mathbf{M}'\mathbf{v}$, where the mass matrix $\mathbf{M}'$ is defined by

$$\mathbf{M}' = \begin{pmatrix} m(1) & 0 & 0 & 0 & \cdots & 0 \\ 0 & m(1) & 0 & 0 & \cdots & 0 \\ 0 & 0 & m(1) & 0 & \cdots & 0 \\ 0 & 0 & 0 & m(2) & \cdots & 0 \\ \cdots & \cdots & \cdots & \cdots & \cdots & 0 \\ 0 & 0 & 0 & 0 & 0 & m(N) \end{pmatrix} \tag{2.51}$$

Consider now a system of particles that interact with each other but not with other bodies, described by a *potential energy* function which we take as $-\Phi(\mathbf{r})$ so that $\mathcal{L}$ becomes $\mathcal{L} = T - \Phi$. Using as generalized coordinates Cartesian coordinates, substitution in the Lagrange equations results in $m_i\mathrm{d}v_i/\mathrm{d}t = -\partial\Phi/\partial r_i \equiv f_i$ representing *Newton's Second Law*. From this law, alternatively written as $\mathrm{d}(m\mathbf{v})/\mathrm{d}t = \mathbf{f} = -\partial\Phi/\partial\mathbf{r}$, we infer *Newton's First Law*, $\mathbf{v}$ = constant if $\mathbf{f} = \mathbf{0}$. The *force* $\mathbf{f} = -\nabla\Phi(\mathbf{r})$ depends only on the coordinates $\mathbf{r}$. It is called *conservative* because the work done by $\mathbf{f}$, when moving a particle from one point to another, is independent of the path taken.

Transforming to generalized coordinates, the kinetic energy T can be expressed as

$$T = \frac{1}{2}\mathbf{v}^{\mathrm{T}}\mathbf{M}'\mathbf{v} = \frac{1}{2}\dot{\mathbf{q}}^{\mathrm{T}}\mathbf{A}^{\mathrm{T}}\mathbf{M}'\mathbf{A}\dot{\mathbf{q}} \equiv \frac{1}{2}\dot{\mathbf{q}}^{\mathrm{T}}\mathbf{M}(\mathbf{q})\dot{\mathbf{q}} \tag{2.52}$$

with $\mathbf{M} = \mathbf{A}^{\mathrm{T}}\mathbf{M}'\mathbf{A}$ the generalized mass matrix. In general therefore, T must be considered as a function of $\mathbf{q}$ and $\dot{\mathbf{q}}$, that is, $T = T(\mathbf{q},\dot{\mathbf{q}})$. We write for the set of forces $f(i)$, as before, $\mathbf{f}^{\mathrm{T}} = [f_1(1), f_2(1), f_3(1), f_1(2), \ldots, f_3(N)]$. If the force $\mathbf{f}$ is conservative, that is, if $\mathbf{f} = -(\partial\Phi/\partial\mathbf{r})$, the generalized force $\mathbf{Q}$ is conservative as well and given by

$$\mathbf{f} = -\frac{\partial \Phi}{\partial \mathbf{r}} = -\left(\frac{\partial \mathbf{q}}{\partial \mathbf{r}}\right)^{\mathrm{T}}\left(-\frac{\partial \Phi}{\partial \mathbf{q}}\right) \equiv (\mathbf{A}^{-1})^{\mathrm{T}}\mathbf{Q} = \mathbf{A}\mathbf{Q} \quad \text{or} \quad \mathbf{Q} = \mathbf{A}^{-1}\mathbf{f} = \mathbf{A}^{\mathrm{T}}\mathbf{f} \tag{2.53}$$

Using the *Lagrange function* $\mathcal{L} = T - \Phi$ and realizing that $\partial\Phi/\partial\dot{\mathbf{q}} = 0$, we may write

► $$\frac{\mathrm{d}}{\mathrm{d}t}\left(\frac{\partial T}{\partial \dot{\mathbf{q}}}\right) - \frac{\partial \Phi}{\partial \mathbf{q}} = 0 \quad \text{or equivalently} \quad \frac{\mathrm{d}}{\mathrm{d}t}\left(\frac{\partial T}{\partial \dot{q}_j}\right) - \frac{\partial \Phi}{\partial q_j} = 0 \tag{2.54}$$

which are the *Lagrange equations of motion* for a conservative system describing a system of N particles with $3N$ second-order equations.

Example 2.3: The harmonic oscillator

For a harmonic oscillator the kinetic energy T is given by $T = ½mv^2$, while the potential energy Φ is described by $\Phi = ½kx^2$. Here, m is the mass, $v = \mathrm{d}x/\mathrm{d}t$ is the velocity, k is the force constant, and x is the position coordinate of the particle. For the generalized coordinates we thus take the Cartesian coordinates and the Lagrange function reads $\mathcal{L} = T - \Phi = ½mv^2 - ½kx^2$. From the Lagrange equations $\mathrm{d}(\partial\mathcal{L}/\partial\dot{x})/\mathrm{d}t - (\partial\mathcal{L}/\partial x) = 0$, we have

$$\frac{\partial \mathcal{L}}{\partial \dot{x}} = m\dot{x}, \quad \frac{\mathrm{d}}{\mathrm{d}t}\left(\frac{\partial \mathcal{L}}{\partial \dot{x}}\right) = m\ddot{x} \quad \text{and} \quad \frac{\partial \mathcal{L}}{\partial x} = -kx$$

Combining yields $m\ddot{x} = -kx$. Solving this differential equation leads to $x = x_0\exp(\mathrm{i}\omega t)$ with $\omega = (k/m)^{1/2}$ and x_0 the amplitude. Equivalently $x = x_0\cos(\omega t + \varphi_0)$, where φ_0 is the phase.

Problem 2.10*

Show that for a free particle the Lagrange function $\mathcal{L} \sim v^2 = |\boldsymbol{v}\cdot\boldsymbol{v}|$ with $\boldsymbol{v}$ the velocity of the particle. Also show that in an inertial frame a free particle will move with constant $\boldsymbol{v}$. Finally, show that the mass $m > 0$.

Problem 2.11*

Show that the Lagrange function $\mathcal{L}$ is defined to within an additive total time derivative of any function of the coordinates.

2.2.3 Conservation Laws

In principle, there exist $3N$–1 functions of $\mathbf{q}$ and $\dot{\mathbf{q}}$, called *integrals* (or *constants*) *of the motion*, whose values remain constant during the motion. The most important of these are derived from the isotropy and homogeneity of space and time.

Let us consider *homogeneity of time*. Due to this homogeneity, the Lagrange function of a closed system is explicitly independent of time. Hence,

$$\frac{\mathrm{d}\mathcal{L}}{\mathrm{d}t}=\left(\frac{\partial \mathcal{L}}{\partial \mathbf{q}}\right)^{\mathrm{T}}\dot{\mathbf{q}}+\left(\frac{\partial \mathcal{L}}{\partial \dot{\mathbf{q}}}\right)^{\mathrm{T}}\ddot{\mathbf{q}}=\left(\frac{d}{dt}\frac{\partial \mathcal{L}}{\partial \dot{\mathbf{q}}}\right)^{\mathrm{T}}\dot{\mathbf{q}}+\left(\frac{\partial \mathcal{L}}{\partial \dot{\mathbf{q}}}\right)^{\mathrm{T}}\ddot{\mathbf{q}}=\frac{\mathrm{d}}{\mathrm{d}t}\left[\left(\frac{\partial \mathcal{L}}{\partial \dot{\mathbf{q}}}\right)^{\mathrm{T}}\dot{\mathbf{q}}\right]=0 \quad (2.55)$$

where in the 2nd step the Lagrange equations are used. So we obtain

$$\frac{\mathrm{d}}{\mathrm{d}t}\left[\left(\frac{\partial \mathcal{L}}{\partial \dot{\mathbf{q}}}\right)^{\mathrm{T}}\dot{\mathbf{q}}-\mathcal{L}\right]=0 \quad \text{or} \quad \left(\frac{\partial \mathcal{L}}{\partial \dot{\mathbf{q}}}\right)^{\mathrm{T}}\dot{\mathbf{q}}-\mathcal{L}\equiv E \quad (2.56)$$

We thus have as one integral of the motion the *energy* E of the system. This relation is also valid for systems in a constant external field since in that case $\mathrm{d}\mathcal{L}/\mathrm{d}t=0$ is still valid. If $\mathcal{L}=T(\mathbf{q},\dot{\mathbf{q}})-\Phi(\mathbf{q})$ with T a quadratic function of $\dot{\mathbf{q}}$, Euler's theorem for homogenous functions results in

$$\left(\frac{\partial \mathcal{L}}{\partial \dot{\mathbf{q}}}\right)^{\mathrm{T}}\dot{\mathbf{q}}=\left(\frac{\partial T}{\partial \dot{\mathbf{q}}}\right)^{\mathrm{T}}\dot{\mathbf{q}}=2T \quad \text{or} \quad E=T(\mathbf{q},\dot{\mathbf{q}})+\Phi(\mathbf{q}) \quad (2.57)$$

Consequently, the constant of the motion energy E equals the sum of T and Φ.

Let us now consider *homogeneity of space* and use an infinitesimal displacement $\mathrm{d}\boldsymbol{\varepsilon}$. For a parallel displacement, where every particle is moved by the same amount $\mathrm{d}\boldsymbol{\varepsilon}$, the variation in $\mathcal{L}$ should be zero and is given by

$$\delta\mathcal{L}=\left(\frac{\partial \mathcal{L}}{\partial \mathbf{r}}\right)^{\mathrm{T}}\delta\mathbf{r}=\left(\frac{\partial \mathcal{L}}{\partial \mathbf{r}}\right)^{\mathrm{T}}\mathrm{d}\boldsymbol{\varepsilon}=0 \quad (2.58)$$

Because $\mathrm{d}\boldsymbol{\varepsilon}$ is arbitrary we have $\partial\mathcal{L}/\partial\mathbf{r}=0$ and using the Lagrange equations results in

$$\frac{\partial \mathcal{L}}{\partial \mathbf{r}}=\frac{\mathrm{d}}{\mathrm{d}t}\frac{\partial \mathcal{L}}{\partial \dot{\mathbf{r}}}=0 \quad (2.59)$$

Hence, changing to direct notation, for a closed system the *momentum* vector $\boldsymbol{p}\equiv\Sigma_i\partial\mathcal{L}/\partial\boldsymbol{v}_i=\Sigma_i m_i\boldsymbol{v}_i\equiv\Sigma_i\boldsymbol{p}_i$ is conserved during the motion. It will be clear that $\boldsymbol{p}$ is additive over the individual particles i, independent of whether interaction is present or not. The interpretation follows from the expression $\partial\mathcal{L}/\partial\boldsymbol{r}_1=-\partial\Phi/\partial\boldsymbol{r}_1=\boldsymbol{f}_1$ which determines the force on particle 1. Since $\partial\mathcal{L}/\partial\mathbf{r}=0$, the total force $\boldsymbol{f}=\Sigma_i\boldsymbol{f}_i$ is zero. In particular, for two particles $\boldsymbol{f}_1+\boldsymbol{f}_2=0$, where $\boldsymbol{f}_1$ denotes the force exerted on particle 1 by particle 2. This represents *Newton's Third Law* implying that the force by particle 1 exerted on particle 2 is equal to, but is oppositely directed, to the force exerted by particle 2 on particle 1, both lying along the line joining the particles. From this result one can show that the total energy E is given by $E=\frac{1}{2}\mu\boldsymbol{V}^2+E_{\mathrm{int}}$ with total mass $\mu=\Sigma_i m_i$ and $\boldsymbol{V}=\boldsymbol{p}/\mu$ the velocity of the center of mass, the latter being given by $\boldsymbol{R}=\Sigma_i m_i\boldsymbol{r}_i/\mu$ and E_{int} the energy of the system being at rest as a whole (see Problem 2.13). The same result follows if $\mathcal{L}$ is given in generalized coordinates $\mathbf{q}$, in which case $\mathbf{p}_i$ and $\mathbf{f}_i$ are denoted as the generalized momenta and generalized forces, respectively.

Let us finally consider *isotropy of space* and use an infinitesimal angle of rotation $\mathrm{d}\boldsymbol{\phi}$ around a vector $\boldsymbol{\phi}$. The variation in radius vector $\boldsymbol{r}$ for any particle is $\delta\boldsymbol{r} = \mathrm{d}\boldsymbol{\phi} \times \boldsymbol{r}$, while the variation in velocity $\boldsymbol{v}$ reads $\delta\boldsymbol{v} = \mathrm{d}\boldsymbol{\phi} \times \boldsymbol{v}$. For a rotation where each particle is rotated by the same amount $\mathrm{d}\boldsymbol{\phi}$, the variation in $\mathcal{L}$ should be zero and in direct notation[14] reads

$$\delta\mathcal{L} = \sum\nolimits_i \left[\frac{\partial\mathcal{L}}{\partial\boldsymbol{r}_i}\cdot\delta\boldsymbol{r}_i + \left(\frac{\partial\mathcal{L}}{\partial\dot{\boldsymbol{r}}_i}\right)\cdot\delta\dot{\boldsymbol{r}}_i\right] = \sum\nolimits_i (\dot{\boldsymbol{p}}_i\cdot\delta\boldsymbol{r}_i + \boldsymbol{p}_i\cdot\delta\dot{\boldsymbol{r}}_i)$$
$$= \sum\nolimits_i [\dot{\boldsymbol{p}}_i\cdot\mathrm{d}(\boldsymbol{\phi}\times\boldsymbol{r}_i) + \boldsymbol{p}_i\cdot\mathrm{d}(\boldsymbol{\phi}\times\dot{\boldsymbol{r}}_i)] = 0 \tag{2.60}$$

or, taking $\mathrm{d}\boldsymbol{\phi}$ outside after permutation of factors,

$$\mathrm{d}\boldsymbol{\phi}\cdot\sum\nolimits_i (\boldsymbol{r}_i\times\dot{\boldsymbol{p}}_i + \dot{\boldsymbol{r}}_i\times\boldsymbol{p}_i) = \mathrm{d}\boldsymbol{\phi}\cdot\frac{\mathrm{d}}{\mathrm{d}t}\sum\nolimits_i (\boldsymbol{r}_i\times\boldsymbol{p}_i) = 0 \tag{2.61}$$

Because $\mathrm{d}\boldsymbol{\phi}$ is arbitrary, we have $\mathrm{d}\Sigma_i(\boldsymbol{r}_i\times\boldsymbol{p}_i)/\mathrm{d}t = 0$, and therefore for a closed system the *angular momentum* vector $\boldsymbol{l} \equiv \Sigma_i\boldsymbol{r}_i\times\boldsymbol{p}_i$ is conserved during the motion. Like the momentum vector $\boldsymbol{p}$, it is additive over the particles of the system.

For a collection of connected particles, three types of forces can be distinguished. First, forces acting alike on all particles due to long-range external influences. Examples of this type of force are the gravity force or forces due to externally imposed electromagnetic fields. We call them *volume* (or *body*) *forces* and indicate them again for a particle i by $f_{\mathrm{vol}}(i)$. Second, forces applied to a particle due to short-range external forces. Examples of this type are interactions with walls of an enclosure. We call them *surface* (or *contact*) *forces* and indicate them by $f_{\mathrm{sur}}(i)$. Volume and surface forces are collectively called *external forces*, $f_{\mathrm{ext}}(i)$, that is, $f_{\mathrm{ext}}(i) = f_{\mathrm{vol}}(i) + f_{\mathrm{sur}}(i)$. Third, forces due to the presence of the other particles, that is, internal loading. These *internal forces* are indicated by $f_{\mathrm{int}}(i)$. Let $f_{\mathrm{pp}}(ij)$ denote the force on particle i due to particle–particle interaction with particle j. Then according to Newton's Third Law we have

$$f_{\mathrm{pp}}(ij) = -f_{\mathrm{pp}}(ji) \tag{2.62}$$

The resultant internal force acting on particle i is then

$$f_{\mathrm{int}}(i) = \sum\nolimits_{i\neq j} f_{\mathrm{pp}}(ij) \tag{2.63}$$

and the total force on particle i is

$$f(i) = f_{\mathrm{vol}}(i) + f_{\mathrm{sur}}(i) + f_{\mathrm{int}}(i) = f_{\mathrm{ext}}(i) + \sum\nolimits_{i\neq j} f_{\mathrm{pp}}(ij) \tag{2.64}$$

Returning to the subscript notation, the system of particles is in equilibrium if the force f_i on each particle i in the system is equal to its rate of change of linear momentum $\mathrm{d}\boldsymbol{p}_i/\mathrm{d}t$, also known as the *inertial force*. Obviously, we have

$$f = \Sigma_i f_i = \Sigma_i\, \mathrm{d}\boldsymbol{p}_i/\mathrm{d}t = \mathrm{d}\boldsymbol{p}/\mathrm{d}t$$

14) We use direct instead of matrix notation, since otherwise a matrix designation for the outer product of vectors is needed. The penalty is the need of summation signs. Note that a subscript labels a particle.

Motions of the collection of all particles that leave the distances between particles unchanged are called *rigid body motions*. Clearly, according to Newton's Third Law, the work due to internal forces vanishes for rigid body motion.

Example 2.4: The harmonic oscillator again

In a harmonic oscillator an external force f acts on a mass m and is linearly related to the extension x, that is, $f = -kx$, where k is the spring constant. The force f can be obtained from the potential energy $\Phi = \frac{1}{2}kx^2$ via $f = -\partial\Phi/\partial x$. If the momentum and velocity are given by $p = mv$ and v, respectively, Newton's Second Law reads $f = \dot{p} = \mathrm{d}(mv)/\mathrm{d}t = m\ddot{x}$. Combining leads to $m\ddot{x} = -kx$. Defining the circular frequency $\omega = (k/m)^{1/2}$, the solution of this differential equation is $x = x_0\exp(-\mathrm{i}\omega t)$, as in Example 2.3.

Problem 2.12

An object of mass m moves in a plane with speed v at a constant distance r to the center of rotation. Let $\boldsymbol{x}$ be the position of the object in the plane with the origin as the center of rotation. Show that:

a) the angular speed $\omega = v/r$ and acceleration $\alpha = \dot{v}/r$,
b) the angular momentum $l = mvr = I\omega$ with $I = mr^2$ the moment of inertia,
c) the torque $q = \dot{l} = mr^2\alpha = I\dot{\omega}$, and
d) the kinetic energy $T = \frac{1}{2}I\omega^2 = l^2/2I$.

Problem 2.13

Show that the total energy E of a system of particles is given by $E = \frac{1}{2}\mu V^2 + E_{\text{int}}$ with total mass $\mu = \Sigma_i m_i$, $\boldsymbol{V} = \boldsymbol{p}/\mu$ the velocity of the center of mass (given by $\boldsymbol{R} = \Sigma_i m_i \boldsymbol{r}_i/\mu$) and E_{int} the energy of the system being at rest as a whole.

2.2.4 Hamilton's Equations

The above formalism may be put in more convenient form by introducing the *generalized momentum* $\mathbf{p}$ defined by

$$\mathbf{p} \equiv \partial\mathcal{L}/\partial\dot{\mathbf{q}} = \mathbf{M}\dot{\mathbf{q}} \tag{2.65}$$

Defining the *Hamilton function* $\mathcal{H}$ by the Legendre transform (see Appendix B)

$$\mathcal{H}(\mathbf{p}, \mathbf{q}, t) \equiv \mathbf{p}^{\mathrm{T}}\dot{\mathbf{q}} - \mathcal{L}(\mathbf{q}, \dot{\mathbf{q}})$$

we obtain for its differential

$$\mathrm{d}\mathcal{H} = \mathbf{p}^{\mathrm{T}}\mathrm{d}\dot{\mathbf{q}} + \dot{\mathbf{q}}^{\mathrm{T}}\mathrm{d}\mathbf{p} - \left(\frac{\partial \mathcal{L}}{\partial \mathbf{q}}\right)^{\mathrm{T}} \mathrm{d}\mathbf{q} - \left(\frac{\partial \mathcal{L}}{\partial \dot{\mathbf{q}}}\right)^{\mathrm{T}} \mathrm{d}\dot{\mathbf{q}}$$

$$= \left(\mathbf{p} - \frac{\partial \mathcal{L}}{\partial \dot{\mathbf{q}}}\right)^{\mathrm{T}} \mathrm{d}\dot{\mathbf{q}} + \dot{\mathbf{q}}^{\mathrm{T}}\mathrm{d}\mathbf{p} - \left(\frac{\partial \mathcal{L}}{\partial \mathbf{q}}\right)^{\mathrm{T}} \mathrm{d}\mathbf{q} - \frac{\partial \mathcal{L}}{\partial t}\mathrm{d}t = \dot{\mathbf{q}}^{\mathrm{T}}\mathrm{d}\mathbf{p} - \dot{\mathbf{p}}^{\mathrm{T}}\mathrm{d}\mathbf{q}$$

where in the last step use is made of Eqs (2.65) and (2.50). For the equations of motion we thus obtain

$$▶ \quad \dot{\mathbf{q}} = \frac{\partial \mathcal{H}}{\partial \mathbf{p}} \quad \text{and} \quad \dot{\mathbf{p}} = -\frac{\partial \mathcal{H}}{\partial \mathbf{q}} \tag{2.66}$$

which are known as *Hamilton's* equations of motion and, because of their symmetry and simplicity, also are often called the *canonical* equations of motion. In this way the system is described by $6N$ first-order equations instead of the $3N$ second-order equations. The motion of set of particles can thus be described by a trajectory in a $6N$-dimensional space, the so-called *phase space*, with coordinates $\mathbf{q}$ and $\mathbf{p}$. The phase space is an essential concept in statistical mechanics.

Formally, for a change of the Hamilton function with time t, we have

$$\frac{\mathrm{d}\mathcal{H}}{\mathrm{d}t} = \left(\frac{\partial \mathcal{H}}{\partial \mathbf{q}}\right)^{\mathrm{T}} \frac{\mathrm{d}\mathbf{q}}{\mathrm{d}t} + \left(\frac{\partial \mathcal{H}}{\partial \mathbf{p}}\right)^{\mathrm{T}} \frac{\mathrm{d}\mathbf{p}}{\mathrm{d}t} + \frac{\partial \mathcal{H}}{\partial t}$$

Using Hamilton's equations it is easy to show that the first two terms cancel so that

$$▶ \quad \frac{\mathrm{d}\mathcal{H}}{\mathrm{d}t} = \frac{\partial \mathcal{H}}{\partial t} \tag{2.67}$$

Thus, a Hamilton function explicitly independent on time (but only implicitly via $\mathbf{p}$ and $\mathbf{q}$) is constant, that is, $\mathcal{H}(\mathbf{q}, \mathbf{p}) = E$ where E is that constant. To interpret this constant for systems described by a Lagrange function $\mathcal{L} = T - \Phi$, we consider $\mathcal{H}$ and obtain

$$▶ \quad \mathcal{H} = \mathbf{p}\dot{\mathbf{q}} - \mathcal{L} = \dot{\mathbf{q}}^{\mathrm{T}} \frac{\partial \mathcal{L}}{\partial \dot{\mathbf{q}}} - T + \Phi = \dot{\mathbf{q}}^{\mathrm{T}} \frac{\partial T}{\partial \dot{\mathbf{q}}} - T + \Phi = 2T - T + \Phi = T + \Phi \tag{2.68}$$

This result shows that the constant E equals the already defined energy but now expressed in terms of position and momentum as independent coordinates. As noted before, for a system with a time-independent potential, the energy and thus the Hamilton function is explicitly independent of time and the system is *conservative*. Finally, since $\mathcal{H}$ is a function of $\mathbf{q}$ and $\mathbf{p}$, we must express T also in terms of $\mathbf{p}$. Inverting $\mathbf{p} = \mathbf{M}\dot{\mathbf{q}}$, we obtain $\dot{\mathbf{q}} = \mathbf{M}^{-1}\mathbf{p}$. Hence

$$T = \frac{1}{2}\dot{\mathbf{q}}^{\mathrm{T}}\mathbf{M}\dot{\mathbf{q}} = \frac{1}{2}(\mathbf{M}^{-1}\mathbf{p})^{\mathrm{T}}\mathbf{M}(\mathbf{M}^{-1}\mathbf{p}) = \frac{1}{2}\mathbf{p}^{\mathrm{T}}\mathbf{M}^{-1}\mathbf{p} \tag{2.69}$$

The Hamilton function thus reads $\mathcal{H} = T + \Phi = \frac{1}{2}\mathbf{p}^{\mathrm{T}}\mathbf{M}^{-1}\mathbf{p} + \Phi(\mathbf{q})$. Note that the generalized force $\mathbf{Q} = \mathbf{A}^{\mathrm{T}}\mathbf{f}$ is not equal to $\dot{\mathbf{p}}$. This is only true if $\mathbf{M}$ is independent of $\mathbf{q}$, which generally is not the case.

Example 2.5: The harmonic oscillator once more

We treat once more the oscillator with kinetic energy $T = \frac{1}{2}mv^2 = p^2/2m$ and potential energy $\Phi = \frac{1}{2}kx^2$. From Hamilton's equation $\dot{p} = -\partial\mathcal{H}/\partial x$, we get right away $m\dot{v} = m\ddot{x} = -kx$, resulting in the solution indicated in Example 2.3. From Hamilton's $\dot{x} = \partial\mathcal{H}/\partial p$ we have the identity $\dot{x} = p/m = m\dot{x}/m = \dot{x}$.

Problem 2.14

Verify Eqs (2.66) and (2.67).

Problem 2.15

Show that for a conservative system $\partial S/\partial t = -\mathcal{H}$ using $S = \int \mathcal{L}(\mathbf{q},\dot{\mathbf{q}},t)\mathrm{d}t$.

2.3 Quantum Concepts

Although atomic and molecular phenomena require quantum mechanics for their description, for our discussion we need only some principles thereof and a few exactly soluble single-particle problems.

2.3.1 Basics of Quantum Mechanics

Since the discovery of quantum mechanics we know that in the Schrödinger representation the state of a system, that is, for the atoms or molecules at hand, is given by the *wave function* $\tilde{\Psi}(\mathbf{r},t)$, where $\mathbf{r}$ stands for the full set of coordinates and t for time. This function and its gradient must be single-valued, finite, and continuous for all values of its arguments. A function satisfying these conditions is said to be *well-behaved*. It also must have a finite integral of its quadrate if integrated over the complete range of coordinates. Generally, we require that $\tilde{\Psi}(\mathbf{r},t)$ is *normalized*, that is,

$$\int \tilde{\Psi}^*(\mathbf{r},t)\tilde{\Psi}(\mathbf{r},t)\mathrm{d}\mathbf{r} = \int \left|\tilde{\Psi}(\mathbf{r},t)\right|^2 \mathrm{d}\mathbf{r} = 1$$

where the asterisk denotes the complex conjugate. Schrödinger proposed that the time development of $\tilde{\Psi}(\boldsymbol{r},t)$ for a free particle with position[15] $\boldsymbol{r}$ and mass m is given by

15) Note that $\boldsymbol{r}$ is a vector and $\mathbf{r}$ is a (column matrix of a) set of elements r_i or $\boldsymbol{r}_i$ (see Section 1.4).

$$-\frac{\hbar^2}{2m}\nabla^2\tilde{\Psi}(\mathbf{r},t)=\mathrm{i}\hbar\frac{\partial}{\partial t}\tilde{\Psi}(\mathbf{r},t) \tag{2.70}$$

where $\hbar = h/2\pi$ and h denotes *Planck's constant*. Using the de Broglie relation $\boldsymbol{p} = \hbar\boldsymbol{k}$ with $k = |\boldsymbol{k}| = 2\pi/\lambda$, where $\boldsymbol{p}$ is the momentum and $\boldsymbol{k}$ is the wave vector, we can write the wave function as $\exp(\mathrm{i}\boldsymbol{k}\cdot\boldsymbol{r}-\omega t)$ with the circular frequency $\omega = 2\pi\nu$, and, as usual, λ and ν representing the wavelength and frequency, respectively. Since the classical energy for a free particle is $E = \boldsymbol{p}^2/2m$, this suggests the operators

$$\boldsymbol{p}\rightarrow -\mathrm{i}\hbar\nabla \quad \text{and} \quad E\rightarrow \mathrm{i}\hbar\frac{\partial}{\partial t}$$

and these relations are accepted for all systems. In fact, every physical observable is represented by a linear Hermitian operator[16] acting on the wave function, for example, the position operator $\mathcal{r} = \mathbf{r}$. The only possible values, which a measurement of the observable with operator $\mathcal{A}$ can yield, are the *eigenvalues a* of the *eigenvalue equation*[17] $\mathcal{A}\tilde{\Psi}(\mathbf{r},t) = a\tilde{\Psi}(\mathbf{r},t)$ with $\tilde{\Psi}(\mathbf{r},t)$ as the *eigenfunction*. For the coordinate operator $\mathcal{r}(i) = \boldsymbol{r}(i)$ the eigenvalues ξ of a single particle i are the values for which the equation

$$\mathbf{r}\tilde{\Psi}(\mathbf{r},t)=\xi\tilde{\Psi}(\mathbf{r},t)$$

which is an ordinary algebraic equation, possesses solutions. Rewriting this as

$$(\mathbf{r}-\xi)\tilde{\Psi}(\mathbf{r},t)=0$$

it is evident that for this expression to be true either $\mathbf{r} = \xi$ or $\tilde{\Psi}(\mathbf{r},t) = 0$. This corresponds exactly to the definition of the Dirac δ-function (see Appendix B), which is thus the eigenfunction $\delta(\mathbf{r}-\xi)$ associated with the operator $\mathcal{r}$ for the eigenvalue ξ.

We note that the energy E for conservative systems is given by the Hamilton function $\mathcal{H}$ in terms of the coordinates $\mathbf{r}$ and conjugated momenta $\mathbf{p}$, that is, $E = \mathcal{H}(\mathbf{p},\mathbf{r}) = T(\mathbf{p}) + \Phi(\mathbf{r})$. We can therefore form the Hamilton operator $\mathcal{H}$ (using a symmetrized function of $\mathbf{r}$ and $\mathbf{p}$) leading to the *time-dependent Schrödinger equation*

$$\blacktriangleright \quad \mathcal{H}(\mathbf{p},\mathbf{r})\tilde{\Psi}(\mathbf{r},t)=\mathrm{i}\hbar\frac{\partial}{\partial t}\tilde{\Psi}(\mathbf{r},t) \tag{2.71}$$

For stationary solutions we may write

$$\tilde{\Psi}(\mathbf{r},t)=\Psi(\mathbf{r})f(t)$$

16) For a Hermitian operator $\mathcal{A}$ it holds that $\int f^*\mathcal{A}g\,\mathrm{d}\tau=\int g(\mathcal{A}f)^*\,\mathrm{d}\tau$, where the integration is over all of the domain of the functions f and g. Linearity implies $\mathcal{A}(c_1f+c_2g)=c_1\mathcal{A}f+c_2\mathcal{A}g$, with c_1 and c_2 as constants.

17) An operator, denoted by $\mathcal{A}$ (or by A if it is just a function), operating on any function $\Psi(\mathbf{r})$ yields another function, say $\Psi'(\mathbf{r})$, that is, $\mathcal{A}\Psi(\mathbf{r}) = \Psi'(\mathbf{r})$. If the operator operating on $\Psi(\mathbf{r})$ yields a multiple of $\Psi(\mathbf{r})$, say $a\Psi(\mathbf{r})$, the resulting equation $\mathcal{A}\Psi(\mathbf{r}) = a\Psi(\mathbf{r})$ is an *eigenvalue equation*. The function $\Psi(\mathbf{r})$ is the *eigenfunction*, and the scalar a is the *eigenvalue*. Because every physical property is real, the eigenvalue must be real and this requires the operator to be Hermitian.

which, upon substitution in Eq. (2.71), leads to

$$\frac{H\Psi(\mathbf{r})}{\Psi(\mathbf{r})} = \frac{\text{i}\hbar}{f(t)}\frac{\partial f(t)}{\partial t}$$

As both sides are independent of each other, they both must be equal to a (so-called separation) constant, say E. Solving the time-dependent part leads to

$$f(t) = C\exp(-\text{i}Et/\hbar) \quad C = \text{constant}$$

while the space-part yields the time-independent Schrödinger equation

▶ $$H(\mathbf{p},\mathbf{r})\Psi(\mathbf{r}) = E\Psi(\mathbf{r}) \tag{2.72}$$

The latter is an eigenvalue equation that shows that the eigenvalue E represents the allowed quantum energy of the system. The constant C may be chosen to normalize $\Psi(\mathbf{r})$, if necessary. In general a set of solutions is obtained where each solution characterized by a wave function $\Psi_k(\mathbf{r})$ and associated energy E_k. The various wave functions $\Psi_k(\mathbf{r})$ form, if normalized, a complete, orthonormal set, that is,

$$\int \Psi_k^* \Psi_l \,\text{d}\mathbf{r} \equiv \langle \Psi_k \,|\, \Psi_l \rangle = \delta_{kl} \quad (\delta_{kl} = 1 \text{ if } k = l \text{ and } \delta_{kl} = 0 \text{ otherwise})$$

where the Dirac *bra-ket notation*[18] introduced. This is also true for the eigenfunctions of any operator, for example, the momentum operator $-\text{i}\hbar\nabla$. Here, we have

$$-i\hbar\nabla\Psi = \boldsymbol{p}\Psi \quad \text{with solution} \quad \Psi = C\exp(i\boldsymbol{k}\cdot\boldsymbol{r}) = C\exp(i\boldsymbol{p}\cdot\boldsymbol{r}/\hbar) \tag{2.73}$$

Normalization can be done in two ways. First, by periodic box normalization where we assume that the particle is contained in an arbitrary box of volume L^3 and that the potential at $x = y = z = 0$ equals that $x = y = z = L$. We obtain, using $\langle \Psi_\text{k} | \Psi_\text{l} \rangle = \delta_\text{kl} = \delta_{kx,lx}\delta_{ky,ly}\delta_{kz,lz}$, $C = L^{-3/2}$. Second, by δ-function normalization where we employ a particular representation of the Dirac δ-function (see Appendix B), namely

$$\delta(x) = \lim_{g\to\infty}(\sin(gx)/\pi x)$$

The function $\sin(gx)/\pi x$ has a value of g/π at $x = 0$, oscillates with decreasing amplitude and period $2\pi/g$ with increasing $|x|$ and is normalized, independent of the value of g. Hence,

$$\int_{-\infty}^{+\infty} \exp[i(\mathbf{k}-\mathbf{l})\cdot\mathbf{r}]\text{d}\mathbf{r} = \lim_{g\to\infty}\int_{-g}^{+g} \exp[i(\boldsymbol{k}-\boldsymbol{l})\cdot\boldsymbol{r}]\text{d}\mathbf{r} = \lim_{g\to\infty}\frac{2\sin \text{g}\cdot(\boldsymbol{k}-\boldsymbol{l})}{\boldsymbol{k}-\boldsymbol{l}} = 2\pi\delta(\boldsymbol{k}-\boldsymbol{l})$$

Consequently, $C = 2\pi^{-3/2}$ using $\exp(\text{i}\mathbf{k}\cdot\boldsymbol{r})$ or $C = h^{-3/2}$ using $\exp(\text{i}\boldsymbol{p}\cdot\boldsymbol{r})$.

When a system is in an arbitrary state Φ the expected mean of a sequence of measurements of the observable with operator A is given by

▶ $$\langle A \rangle \equiv \int \Phi^* A\Phi \,\text{d}\mathbf{r} = \langle \Phi | A | \Phi \rangle \tag{2.74}$$

where $\langle A \rangle$ is referred to as the *expectation value* and is to be interpreted as $\langle A \rangle = \sum_i p_i a_i$, where a_i denotes the measured value and p_i its probability of

18) The association with brackets is obvious but the meaning of the bra $<\Psi|$ and ket $|\Psi>$ is much deeper, see Ref. [8].

occurrence. If we expand Φ in terms of Ψ_k's, we have $\Phi = \sum_k c_k \Psi_k$ and the expectation value becomes

$$\langle A \rangle = \left\langle \sum_k c_k \Psi_k | A | \sum_l c_l \Psi_l \right\rangle = \sum_{k,l} c_k c_l \langle \Psi_k | A | \Psi_l \rangle = \sum_k c_k^2 a_k$$

Hence, when the system is in a state Φ, a measurement of the observable A yields the value a_k with a probability c_k^2, where c_k is the expansion coefficient (or *probability amplitude*) in the expansion of Φ in terms of Ψ_ks, or

$$\rho_k = c_k^2 \tag{2.75}$$

The probability amplitude may be expressed in terms of Φ and Ψ_k as

$$\langle \Psi_k | \Phi \rangle = \left\langle \Psi_k \middle| \sum_l c_l \Psi_l \right\rangle = c_k \tag{2.76}$$

If Φ is one of the eigenfunctions Ψ_l of A, we obviously have $\rho_k = \langle \Psi_k | \Psi_l \rangle = \delta_{kl}$. In this case, the expectation value $\langle A \rangle$ is thus equal to the eigenvalue a_l. The above considerations lead directly to the interpretation of the wave function. Consider the probability that a measurement of the position of a single particle i will give the value ξ. The eigenfunction corresponding to the coordinate operator $\mathbf{r}(i)$ for the eigenvalue ξ has shown to be $\Psi_\xi = \delta(\mathbf{r} - \xi)$. From Eqs (2.75) and (2.76), we obtain

$$\rho_\xi = |\langle \delta(\mathbf{r} - \xi) | \Phi(\mathbf{r}) \rangle|^2 = |\Phi(\xi)|^2 \tag{2.77}$$

Hence, the probability of finding the particle i at $\mathbf{r}(i) = \xi$ is given by the square of its wave function evaluated at ξ. Generalization to many particles is straightforward.

An important next item is that conjugated variables, like the position coordinate $\boldsymbol{r}$ and momentum $\boldsymbol{p}$ or energy E and time t, cannot be determined precisely at the same time or, in other words, if one of the conjugated variables is exactly known, the other is fully undetermined. For the energy this is already clear: in a stationary state at *any* time t the energy is *exactly* E. In general any process that shortens the life time, broadens the energy level. Since transitions to other states are required for a change in time, there is a limited residence time Δt and this leads to a small uncertainty in energy ΔE given by $\Delta E \Delta t = ½\hbar$. More generally, if we define the variance by $(\Delta A)^2 = \langle (A - \langle A \rangle)^2 \rangle$ we have for any pair of conjugated variables a and b

$$\Delta a \Delta b \geq \hbar/2 \tag{2.78}$$

which are known as the Heisenberg *uncertainty relations*. The implication for the pair coordinate $\boldsymbol{r}$ and momentum $\boldsymbol{p}$ is that a precisely localized particle has an indeterminate momentum, and vice versa.

Another important point is that particles also possess *spin angular momentum*, for which no classical analog is available, but which obeys the same quantization rules as orbital angular momentum. For an electron the eigenvalue for the spin operator is either $½\hbar$ or $-½\hbar$ with the eigenfunctions generally denoted by α and β, respectively.

Finally, we focus our attention on systems with a Hamilton operator with additive properties. For many-particle systems the Hamilton operator H is a function of the co-ordinates and the momenta of all the particles $i,j, \ldots$, that is,

$H = H(i, j, \ldots)$. In a number of cases, however, the Hamilton operator is the sum (or can be approximated as the sum) of the operators $h(i)$ for single particles i, that is, $H = \Sigma_i h(i)$. These single-particle operators satisfy the (single particle) Schrödinger wave equation

$$h(i)\phi_k(i) = \varepsilon_k \phi_k(i) \tag{2.79}$$

where $\phi_k(i)$ and ε_k denote single-particle wave functions and eigenvalues (particle energies), respectively. The total wave function Ψ for the N-particle system can be taken as the product of the individual particle wave functions $\Psi(i) = \prod_k \phi_k(i)$, so that the N-particle Schrödinger equation reads

$$H\Psi = \sum_i h(i) \prod_k \phi_k(i) = \sum_i \varepsilon_i \prod_k \phi_k(j) = E\Psi \tag{2.80}$$

and the total energy is given by $E = \Sigma_k \varepsilon_k$. However, we note that the individual particles in a system are indistinguishable. This implies that the wave function must be either symmetrical or anti-symmetrical with respect to exchange of particle coordinates, including spin. Electrons are particles with a half-integer spin, the so-called *fermions* (or Fermi–Dirac particles), for which the wave function must be anti-symmetric in the coordinates of all electrons. A direct consequence, usually referred to as *Pauli's principle*, is that each system state can be occupied with only one electron. For an electron with spin ½ this implies either spin up ($\frac{1}{2}\hbar$) or with spin down ($-\frac{1}{2}\hbar$). Making allowance for Pauli's principle, a many-electron wave function can be expanded in anti-symmetrized products of one-electron wave functions (or *spin orbitals*) ϕ_j, each ϕ_j consisting of a spatial part (or *orbital*) and a spin function ($\sigma_j = \alpha$ or β). For this the *permutation operator* P_r' is used, which permutes the coordinates $\mathbf{r}$ of the wave function. A permutation is either odd or even, dependent on the number of binary switches in coordinates $|P_r'|$ that is required to realize that permutation. A completely anti-symmetrized wave function $\Psi^{(A)}$ is obtained by combining all $N!$ permutations $P_r'[\Psi]$ of the wave function $\Psi = \Psi(\mathbf{r}_1, \mathbf{r}_2, \ldots, \mathbf{r}_N)$, that is, by taking

$$P\Psi \equiv (N!)^{-1} \sum_{P_r} (-1)^{|P_r'|} P_r'[\Psi(\mathbf{r}_1, \mathbf{r}_2, \ldots, \mathbf{r}_N)] \equiv (N!)^{-1} \sum_{P_r} P_r[\Psi(\mathbf{r})] \tag{2.81}$$

So we have $(-1)^{|P_r'|} = -1$ or 1 for an odd or even permutation, respectively. For future reference we include the prefactor $(-1)^{|P_r'|}$ in the operator P_r' by writing P_r. A convenient form for such an anti-symmetrized product is the *Slater determinant*, in shorthand written as

$$\Psi^{(A)} = |\phi_a(1)\, \phi_b(2) \ldots \phi_n(N)| \equiv \begin{vmatrix} \phi_a(1) & \phi_a(2) & \ldots & \phi_a(N) \\ \phi_b(1) & \phi_b(2) & \ldots & \phi_b(N) \\ \ldots & \ldots & \ldots & \ldots \\ \phi_n(1) & \phi_n(2) & \ldots & \phi_n(N) \end{vmatrix} \tag{2.82}$$

where (i) denotes the coordinates of an electron including spin and N the number of electrons. Similarly, for integer spin particles, the so-called *bosons* (or Bose–Einstein particles), the wave function must be symmetric in all the coordinates of the particles and the system states may contain any number of particles. This symmetry can be realized by taking the *permanent* of individual particle wave functions indicated by

$$\Psi^{(S)} = \|\phi_a(1)\,\phi_b(2)\ldots\phi_n(N)\| \tag{2.83}$$

The permanent is constructed similarly as the determinant but upon expanding one takes all signs as positive. Hence, $(+1)^{|\mathcal{P}_r|} = 1$ always. In order to obtain a normalized wave function Ψ, if all ϕ's are orthonormal, one has to include a normalization factor $C = [N!\Pi_i(n_i!)]^{-1/2}$

$$\Psi = C|\phi_a(1)\,\phi_b(2)\ldots\phi_n(N)| \quad \text{or} \quad \Psi = C\|\phi_a(1)\,\phi_b(2)\ldots\phi_n(N)\|$$

for fermions and bosons, respectively. Note that for Fermi–Dirac particles C becomes $C = (N!)^{-1/2}$ because all n_i are either 1 or 0. The factor C is often incorporated in the definition of the (anti-)symmetrized product function so that we write for the wave function $\Psi = |\phi_a(1)\ \phi_b(2)\ \ldots\ \phi_n(N)|$ or even, assuming a fixed order of coordinates $\Psi = |\phi_a\ \phi_b\ \ldots\ \phi_n|$. The same convention is used for the symmetrized product function. A straightforward calculation shows that for the Hamilton operator with additive properties $\mathcal{H} = \Sigma_i \mathcal{h}(i)$ the N-particle system eigenvalue is still

$$E = \sum\nolimits_k \varepsilon_k$$

When $\mathcal{H} \neq \Sigma_i \mathcal{h}(i)$, the energy E is no longer the sum of the particle energies and the product function and determinant (or permanent) no longer yield the same energy.

To conclude this section we mention that the number of exact solutions for realistic systems is very limited in quantum mechanics[19]. Therefore, as has been emphasized many times before, the Schrödinger equation needs drastic approximations in almost all cases, even to obtain approximate solutions. We indicate here only the solutions of a few, exactly solvable single-particle problems, namely the particle-in-a-box, the harmonic oscillator, and the rigid rotator. Thereafter, we briefly discuss some approximation methods. For our discussion of liquids and solutions, the basic concepts and examples discussed will more than suffice. In fact, the existence of energy levels is the most important aspect for the basis of statistical thermodynamics.

Problem 2.16

Show that the eigenvalues of a Hermitian operator are real.

Problem 2.17

Prove the orthogonality of the eigenfunctions of a Hermitian operator.

19) Virtually the only realistic one is the hydrogen atom, at least if a nonrelativistic Hamilton operator is used. Fortunately, several model systems can be solved exactly.

Problem 2.18

Show that the squared deviation of the energy from its expectation value for a stationary state is zero, that is, $(\Delta H)^2 = \langle (H - \langle H \rangle)^2 \rangle = 0$.

Problem 2.19

Write explicitly the expression for two-electron determinant wave function. Show that this expression can be separated in a spatial and a spin factor.

Problem 2.20*

Verify that the total energy for an operator $H = \Sigma_i h(i)$ is $E = \Sigma_k \varepsilon_k$ for a determinant wave function.

Problem 2.21*

Calculate the energy for a two-electron product and determinant wave function using the Hamilton operator $H = h(1) + h(2) + g(1, 2)$. Show that the total energy is no longer the sum of the single particle energies for the determinant.

2.3.2
The Particle-in-a-Box

For a particle in a one-dimensional box the potential energy is given by

$$\Phi(x) = 0 \quad \text{for} \quad 0 < x < w \quad \text{and} \quad \Phi(x) = \infty \quad \text{otherwise} \tag{2.84}$$

The kinetic energy operator is

$$T(x) = -\frac{\hbar^2}{2m}\frac{\mathrm{d}^2}{\mathrm{d}x^2} \tag{2.85}$$

where m is the mass of the particle, so that the Schrödinger equation reads

$$-\frac{\hbar^2}{2m}\frac{\mathrm{d}^2}{\mathrm{d}x^2}\Psi = E\Psi \tag{2.86}$$

The solutions are

$$\Psi_n = \sqrt{\frac{2}{w}}\sin\left(\frac{2\pi nx}{w}\right) \tag{2.87}$$

where $n = 1,2, \ldots$ is a positive integer, the so-called *quantum number*, arising since only for discrete wavelengths the solution obeys the boundary conditions. These wavelengths are given by $\lambda = 2w/1, 2w/2, \ldots, 2w/n$ with allowed energies

$$E_n = \frac{h^2}{8m}\left(\frac{n^2}{w^2}\right) = \frac{\hbar^2}{2m}\left(\frac{2\pi \mathrm{n}}{2w}\right)^2 = \frac{\hbar^2}{2m}\left(\frac{\pi \mathrm{n}}{w}\right)^2 = \frac{(\hbar \mathbf{k})^2}{2m} = \frac{\mathbf{p}^2}{2m} \tag{2.88}$$

where $\boldsymbol{k}$ is the wave vector and $\boldsymbol{p}$ the momentum. In the ground state with $n = 1$, the energy is still not zero and the wave function represents a standing wave. For a three-dimensional box with potential energy $\Phi(\boldsymbol{r}) = 0$ for $\boldsymbol{0} < \boldsymbol{r} < \boldsymbol{w}$ and $\Phi(\boldsymbol{r}) = \infty$; otherwise, the energy levels are

$$E_{\mathrm{n}} = \frac{h^2}{8m}\left(\frac{n_1^2}{w_1^2} + \frac{n_2^2}{w_2^2} + \frac{n_3^2}{w_3^2}\right) \tag{2.89}$$

If one of the dimensions of the box is equal to another, the energy levels become *degenerate*, that is, there are two (or more) states with the same energy. For example, if $w_1 = w_2 = w_3$ the energies for states $\mathbf{n} = (1,2,2)$, $\mathbf{n} = (2,1,2)$ and $\mathbf{n} = (2,2,1)$ are the same, and the system is said to be three-fold degenerate. Using periodic boundary conditions $\Phi(\boldsymbol{0}) = \Phi(\boldsymbol{w}) =$ constant, instead of the fixed box boundary conditions given by Eq. (2.84), a similar procedure will lead to the same expression for E_{n} but with the factor 8 replaced by 2. In this case, $\mathbf{n} = (n_1,n_2,n_3) \in (\pm1,\pm2,\ldots)$ and the wave functions $\exp(\mathrm{i}\boldsymbol{p}\cdot\boldsymbol{r}/\hbar) = \exp(\mathrm{i}\boldsymbol{k}\cdot\boldsymbol{r})$ represent running waves.

2.3.3 The Harmonic Oscillator

For a one-dimensional harmonic oscillator the potential energy with force constant a and the kinetic energy operator are given by

$$\Phi(x) = \frac{1}{2}ax^2 \quad \text{and} \quad \mathcal{T}(x) = -\frac{\hbar^2}{2m}\frac{\mathrm{d}^2}{\mathrm{d}x^2} \tag{2.90}$$

respectively, so that the Schrödinger equation reads

$$\left[-\frac{\hbar^2}{2m}\frac{\mathrm{d}^2}{\mathrm{d}x^2} + \frac{1}{2}ax^2\right]\Psi = E\Psi \tag{2.91}$$

The solutions for the wave functions are

$$\Psi_n(\xi) = \left(\frac{\sqrt{\alpha/\pi}}{2^n n!}\right)^{1/2} H_n(\xi)\exp\left(-\frac{1}{2}\xi^2\right) \tag{2.92}$$

$$\text{with} \quad \xi = \sqrt{\alpha}x, \quad \alpha = \frac{\sqrt{ma}}{\hbar}, \quad n = 0, 1, 2, \ldots$$

and where $H_n(x)$ are known as Hermite functions defined by

$$H_n(x) = (-1)^n \exp(x^2)\frac{\mathrm{d}^{\mathrm{n}}}{\mathrm{d}x^{\mathrm{n}}}\exp(-x^2)$$

Explicitly the first few Hermite functions are

$$H_0(x) = 1 \quad H_1(x) = 2x \quad H_2(x) = 4x^2 - 2 \quad H_3(x) = 8x^3 - 12x$$

$$H_4(x) = 16x^4 - 48x^2 + 12 \quad H_5(x) = 32x^5 - 160x^3 + 120x, \quad \ldots$$

The energy levels are given by

$$E_n = \hbar\omega\left(n + \frac{1}{2}\right) \quad \text{with circular frequency } \omega = \sqrt{a/m} \tag{2.93}$$

For a three-dimensional oscillator with potential energy (shifting the origin with x_0, y_0 and z_0)

$$\Phi(x, y, z) = \frac{1}{2}a_x(x - x_0)^2 + \frac{1}{2}a_y(y - y_0)^2 + \frac{1}{2}a_z(z - z_0)^2 \tag{2.94}$$

the energy levels are given by

► $$E_n = \hbar\omega_x\left(n_x + \frac{1}{2}\right) + \hbar\omega_y\left(n_y + \frac{1}{2}\right) + \hbar\omega_z\left(n_z + \frac{1}{2}\right) \tag{2.95}$$

Again, if one of the frequencies is equal to another, the system is degenerate. For example, if $\omega_x = \omega_y = \omega_z$, for states $\mathbf{n} = (1,1,2)$, $\mathbf{n} = (1,2,1)$ and $\mathbf{n} = (2,1,1)$ the energy levels are the same and given by

$$E_\text{n} = \hbar\omega(n_x + n_y + n_z + 3/2)$$

and the system is three-fold degenerate. Finally, we note that the presence of the zero-point energy $\frac{1}{2}\hbar\omega$ for the ground state $n = 0$ is in accord with the uncertainty relations. If the oscillator had no zero-point energy, it would have zero momentum and be located exactly at the minimum of $\Phi(\mathbf{r})$. The necessary uncertainties in position and momentum thus give rise to the zero-point energy. Finally, for completeness, we mention that the selection rule for transitions between levels is given by $\Delta n = \pm 1$.

2.3.4 The Rigid Rotator

For the rigid two-particle (atom) rotator there is no potential in the conventional sense, but there is the constraint that the distance between the two particles of the rotator should remain constant. Hence, the Schrödinger equation reads

$$-\frac{\hbar^2}{2\mu}\nabla^2\Psi = E\Psi \tag{2.96}$$

where μ indicates the reduced mass of the rotator given by $1/\mu = 1/m_1 + 1/m_2$, m_1 and m_2 are the masses of the two particles. The constraint is taken into account as follows. Rigid rotation is realized if we suppose that the two rigidly connected particles rotate freely over a sphere. For this problem it is convenient to think in terms of polar coordinates, where the radial variable or radius r is fixed but the angular coordinates θ and φ are free. The Laplace operator ∇^2, expressed in polar coordinates and acting on a scalar x, is expressed by

$$\nabla^2 x = \frac{\partial^2 x}{\partial r^2} + \frac{2}{r}\frac{\partial x}{\partial r} + \frac{1}{r^2}\frac{\partial^2 x}{\partial \theta^2} + \frac{\cot\theta}{r^2}\frac{\partial x}{\partial \theta} + \frac{1}{r^2 \sin^2\theta}\frac{\partial^2 x}{\partial \varphi^2} \quad \text{or} \tag{2.97}$$

$$\nabla^2 x = r^{-1}\frac{\partial^2}{\partial r^2} rx + r^{-2}\Lambda^2 \quad \text{with} \quad \Lambda^2 = \frac{\partial^2 x}{\partial \theta^2} + \cot\theta \frac{\partial x}{\partial \theta} + \frac{1}{\sin^2\theta}\frac{\partial^2 x}{\partial \varphi^2}$$

The Legendre operator Λ^2 is the angular part of the Laplace operator. The constraint r = *constant* implies that the radial part of the Laplace operator can be skipped. Hence, the Schrödinger equation becomes

$$-\frac{\hbar^2}{2I}\Lambda^2\Psi(\theta,\varphi) = E\Psi(\theta,\varphi) \tag{2.98}$$

where $I = \mu r^2$ is the moment of inertia. The solutions for this equation are the *spherical harmonics*, conventionally indicated by $Y_{lm}(\theta,\varphi)$, and characterized by the two quantum numbers [9] l and m. The number l = 0,1,2, . . . indicates the overall angular momentum, given by $[l(l+1)]^{1/2}\hbar$, while the number $m = -l, -l+1, \ldots, l-1, l$ indicates the angular momentum in a specified direction, often taken as x, and given by $m\hbar$. The energy expression is independent of the quantum number m and becomes

$$\blacktriangleright \quad E_l = \frac{\hbar^2}{2I} l(l+1) \quad \text{with} \quad l = 0, 1, 2 \ldots \tag{2.99}$$

The above notation is the conventional one in quantum mechanics for the solution of the angular part of the Laplace operator. However, in spectroscopy the symbol for the quantum number l is usually replaced by J, and we will also do so. Each energy level is thus characterized by the quantum number J corresponding to $2J + 1$ quantum states, that is, the degeneracy is $2J + 1$. Again for completeness, we mention that the selection rule for transitions between levels is $\Delta J = \pm 1$.

2.4 Approximate Solutions

In principle, molecules and their interactions should be described by the Schrödinger equation but, since this formidable problem cannot be solved exactly, we will need approximations. Some of these are described in the following section, mainly for reference.

2.4.1 The Born–Oppenheimer Approximation

First, we must separate the electronic motion in order to arrive at the concept of potential energy between molecules. In a molecule, both the electrons and the nuclei are considered as particles. For the moment, we denote the coordinates of the electrons by $\mathbf{r}$ and the coordinates of the nuclei by $\mathbf{R}$. The Hamilton operator $\mathcal{H}(\mathbf{r},\mathbf{R})$ consists of three terms[20] representing the kinetic energy of the nuclei $\mathcal{T}^{(n)}(\mathbf{R})$ and electrons $\mathcal{T}^{(e)}(\mathbf{r})$ and the potential energy of interaction $\Phi(\mathbf{r},\mathbf{R})$:

20) Neglecting relativistic terms, spin interactions and the possibility of explicit time dependence.

$$\mathcal{H}(\mathbf{r},\mathbf{R}) = \mathcal{T}^{(n)}(\mathbf{R}) + \left(\mathcal{T}^{(e)}(\mathbf{r}) + \Phi(\mathbf{r},\mathbf{R})\right) = \mathcal{T}^{(n)}(\mathbf{R}) + \mathcal{H}^{(e)}(\mathbf{r},\mathbf{R}) \tag{2.100}$$

where the labels (n) and (e) indicate the nuclei and electrons, respectively. A further step can be made by adopting the *adiabatic* (or *Born–Oppenheimer*) *approximation*

$$\Psi(\mathbf{r},\mathbf{R}) = \Psi^{(e)}(\mathbf{r};\mathbf{R})\Psi^{(n)}(\mathbf{R}) \tag{2.101}$$

where the separate factors are obtained from

$$\mathcal{H}^{(e)}(\mathbf{r};\mathbf{R})\Psi^{(e)}(\mathbf{r};\mathbf{R}) = E(\mathbf{R})\Psi^{(e)}(\mathbf{r};\mathbf{R}) \text{ and} \tag{2.102}$$

$$\left(\mathcal{T}^{(n)}(\mathbf{R}) + E(\mathbf{R})\right)\Psi^{(n)}(\mathbf{R}) = E_{\text{tot}}\Psi^{(n)}(\mathbf{R}) \tag{2.103}$$

Here, E and E_{tot} denote the electronic and total energy, respectively. Equation (2.102) is usually solved for a fixed configuration of nuclei so that the coordinates enter the electronic wave function parametrically, hence $\Psi^{(e)}(\mathbf{r};\mathbf{R})$. From Eq. (2.102) it can be seen that in this approximation $E(\mathbf{R})$ may be considered as the *effective potential energy* for the motion of the nuclei [10]. The approximation thus allows us to discuss the electronic structure and nuclear motion separately. In fact, $E(\mathbf{R})$ is the potential energy employed in all of the discussions on intermolecular interactions (see Chapter 3).

2.4.2 The Variation Principle

Let us now focus on the electronic Schrödinger equation [11]. Dropping the label (e) and the nuclear coordinates **R**, Eq. (2.102) reads

$$\mathcal{H}(\mathbf{r})\Psi_j(\mathbf{r}) = E_j\Psi_j(\mathbf{r}) \tag{2.104}$$

where the index j refers to a specific electronic wave function Ψ_j with associated energy E_j. Indicating from now on the coordinates of electron i with $\boldsymbol{r}_i$, in general a many-electron wave function $\Psi(\boldsymbol{r}_1, \ldots, \boldsymbol{r}_N)$ can be expanded as

$$\Psi(\boldsymbol{r}_1,\ldots,\boldsymbol{r}_N) = \sum\nolimits_k c_k\Phi_k(\boldsymbol{r}_1,\ldots,\boldsymbol{r}_N) \tag{2.105}$$

where the c_ks are the coefficients of the expansion and the Φ_ks are the members of a proper complete set. For example, for many-electron systems Φ_k may be a proper anti-symmetrized product function of single-electron functions or Slater determinant

$$\Phi_k(\boldsymbol{r}_1,\ldots,\boldsymbol{r}_N) = |\phi_i(\boldsymbol{r}_1)\phi_j(\boldsymbol{r}_2)\ldots\phi_p(\boldsymbol{r}_N)|$$

where k is a collective index representing $i, \ldots, p$. The functions Φ_k form the *basis*. If we collect the basic functions Φ_k and the coefficients c_k in a column matrix, we may write $\Psi = \mathbf{\Phi}^{\text{T}}\mathbf{c}$. Now, the Hamilton operator acting on Ψ yields a new function Ψ', that is, $\mathcal{H}\Psi = \Psi'$. Expanding Ψ' similarly as Ψ we have

$$\mathcal{H}\Psi = \Psi' \quad \text{or} \quad \mathcal{H}\mathbf{\Phi}^{\text{T}}\mathbf{c} = \mathbf{\Phi}^{\text{T}}\mathbf{c}'$$

Multiplying the last equation by $\mathbf{\Phi}^*$ and integrating over all coordinates results in

$$\left(\int \mathbf{\Phi}^*\mathcal{H}\mathbf{\Phi}^{\text{T}}\,\text{d}\mathbf{r}\right)\mathbf{c} = \left(\int \mathbf{\Phi}^*\mathbf{\Phi}^{\text{T}}\,\text{d}\mathbf{r}\right)\mathbf{c}' \quad \text{or} \quad \mathcal{H}_{ij}c_j = S_{ij}c_j' \quad \text{or} \quad \mathbb{H}\mathbf{c} = \mathbf{S}\mathbf{c}'$$

where $\mathbb{H}_{ij} = \mathbb{H} = \langle \Phi_i | H | \Phi_j \rangle$ and $S_{ij} = \mathbf{S} = \langle \Phi_i | \Phi_j \rangle$ and are referred to as *Hamilton* and *overlap matrix elements*, respectively. For an orthonormal basis, $\mathbf{S} = \mathbf{1}$ and $\mathbb{H}\mathbf{c} = \mathbf{c}'$.

Consider now the Schrödinger Eq. (2.104). Applying the same procedure as before – that is, multiplying by $\mathbf{\Phi}^*$ and integrating over all coordinates – we obtain

$$\mathbb{H}\mathbf{c} = E\mathbf{S}\mathbf{c} \quad (\text{or, if } \mathbf{S} = \mathbf{1}, \mathbb{H}\mathbf{c} = E\mathbf{c}).$$

In practice we use a truncated set of n basis functions. Let us write for this case $\mathbb{H}^{(n)}\mathbf{c}^{(n)} = E^{(n)}\mathbf{S}^{(n)}\mathbf{c}^{(n)}$. This equation will have eigenvalues $E_1^{(n)}$, $E_2^{(n)}$ and so on. It can be shown that if we use a truncated set with $n + 1$ basis functions $E_1^{(n)} > E_1^{(n+1)}$ or in general

$$E_i^{(n)} > E_i^{(n+1)} > \cdots > E_i^{(\infty)} = E_0$$

where $i = 1, \ldots, n$ and E_0 the exact solution. This result implies that, by admitting one more function to an n-truncated complete set, an improved estimate is made for the eigenvalues. This is true for the ground state as well as all excited states. The conventional *variation theorem* arises by taking $n = 1$ so that c drops out and we have $H_{11} - E_1^{(1)}S_{11} = 0$ or (omitting the superscript and subscripts)

▶ $$E = \frac{\langle \Phi | H | \Phi \rangle}{\langle \Phi | \Phi \rangle} \geq E_0 \tag{2.106}$$

where E_0 is the energy of the first state of the corresponding symmetry, usually the ground state. According to this theorem, an upper bound E can be obtained from an approximate wave function. Only in cases where Φ is an exact solution will the equality sign hold; in all other cases the quotient in the middle results in a larger value than the exact value E_0. This implies that one can introduce one or more parameters $\boldsymbol{\alpha}$ in the electronic wave function so that it reads $\Phi(\mathbf{r};\boldsymbol{\alpha})$ and minimize the left-hand side of Eq. (2.106) with respect to the parameters $\boldsymbol{\alpha}$. In this way, the best approximate wave function, given a certain functional form, is obtained.

Example 2.6: The helium atom

As an example of the variation theorem we discuss briefly the helium atom having a nuclear charge $Z = 2$ and two electrons. The Hamilton operator for this atom in atomic units[21] is

21) In the *atomic unit system* the length unit is the *Bohr radius*, 1 $a_0 = (4\pi\varepsilon_0)\hbar^2/me^2 = 0.529 \times 10^{-10}$ m, and the energy unit is the *Rydberg*, 1 Ry $= me^4/2(4\pi\varepsilon_0\hbar)^2 = 2.18 \times 10^{-18}$ J $= 13.61$ eV. Here, $\hbar = h/2\pi$ with h Planck's constant, m and e the mass and the charge of an electron, respectively, and ε_0 the permittivity of the vacuum. In this system, $e^2/(4\pi\varepsilon_0) = 2$ and $\hbar^2/2m = 1$. Unfortunately, another convention for the energy atomic unit also exists, that is, the *Hartree*, 1 Ha = 2 Ry. In the Rydberg convention the kinetic energy reads $-\nabla^2$ and the electrostatic potential $2/r$, while in the Hartree convention they read $-½\nabla^2$ and $1/r$, respectively. Chemists seem to prefer Hartrees, and physicists Rydbergs.

$$H = -(\nabla_1^2 + \nabla_2^2) - Z(2/r_1 + 2/r_2) + 2/r_{12}$$

where the first term indicates the kinetic energy of electrons 1 and 2, the second term the nuclear attraction for electrons 1 and 2, and the third term the electron–electron repulsion. As a normalized trial wave function we use a product of two simple exponential functions, namely $\Psi = \phi(1)\phi(2) = (\eta^3/\pi)\exp[-\eta(r_1 + r_2)]$, where the parameter η is introduced. The expectation value of the Hamilton operator is $\langle\Psi|H|\Psi\rangle$. For the kinetic energy integral the result is $K = \eta^2$, while for the nuclear attraction integral we obtain $N = -2Z\eta$. Finally, the electron–electron repulsion integral is $I = 5\eta/4$ [11]. The total energy is thus $\langle\Psi|H|\Psi\rangle = 2(K + N) + I = 2\eta^2 - 4Z\eta + 5\eta/4$, at minimum when

$$\partial\langle\Psi|H|\Psi\rangle/\partial\eta = 0 = 4\eta - 4Z + 5/4 \quad \text{or when} \quad \eta = (4Z - 5/4)/4 = 1.688.$$

The corresponding energy is −5.70 Ry, to be compared with the experimental value of −5.81 Ry.

We now return to the original linear expansion

$$\Psi = \sum\nolimits_k c_k \Phi_k \quad \text{or} \quad \Psi = \mathbf{\Phi}^{\mathrm{T}}\mathbf{c}$$

Multiplying $H\Psi = E\Psi$ by Ψ^*, integrating over all electron coordinates, and using the expansion above results in

$$\mathbf{c}^\dagger\mathbb{H}\mathbf{c} = E\mathbf{c}^\dagger\mathbf{S}\mathbf{c} \quad \text{or} \quad \mathbf{c}^\dagger(\mathbb{H} - E\mathbf{S})\mathbf{c} = 0 \quad \text{or} \quad E = \mathbf{c}^\dagger\mathbb{H}\mathbf{c}/\mathbf{c}^\dagger\mathbf{S}\mathbf{c}$$

where $\dagger$ denotes the Hermitian transpose. Minimizing with respect to c_i yields

$$\frac{\partial}{\partial\mathbf{c}}\mathbf{c}^\dagger(\mathbb{H} - E\mathbf{S})\mathbf{c} = \mathbf{c}^\dagger(\mathbb{H} - E\mathbf{S}) + (\mathbb{H} - E\mathbf{S})\mathbf{c} = 0$$

Because the c's are independent, the determinant of $(\mathbb{H} - E\mathbf{S})$ must vanish or $\det(\mathbb{H} - E\mathbf{S}) = 0$, corresponding to the generalized eigenvalue equation $\mathbb{H}\mathbf{c} - E\mathbf{S}\mathbf{c} = 0$. This equation can be transformed to the conventional eigenvalue equation by applying a unitary transformation to the basic functions. Since $\det(\mathbf{S}) \neq 0$, we can take

$$\mathbf{\Phi}' = \mathbf{S}^{-1/2}\mathbf{\Phi} \quad \text{to obtain} \quad \langle\mathbf{\Phi}'|\mathbf{\Phi}'\rangle = 1 \quad \text{and} \quad \mathbb{H}' = \langle\mathbf{\Phi}'|H|\mathbf{\Phi}'\rangle = \mathbf{S}^{-1/2}\mathbb{H}\mathbf{S}^{-1/2}$$

In this way, we are led to the conventional eigenvalue problem $\mathbb{H}'\mathbf{c} - E\mathbf{c} = 0$, which can be solved in the usual way to find the eigenvalues. The lowest one approximates the energy of the ground state. Substituting the eigenvalues back into the secular equation results in the coefficients $\mathbf{c}$ from which the approximate wave function can be constructed. In case the functions Φ are already orthonormal, $\mathbf{S} = \mathbf{1}$ and thus $\mathbf{S}^{-1/2} = \mathbf{1}$, and we arrive directly at the standard eigenvalue problem.

Since the eigenfunctions are all orthonormal – or can be made to be so – the approximate wave functions are, like the exact ones, orthonormal. Moreover, they provide stationary solutions. As usual with eigenvalue problems, we can collect the various eigenvalues E_k in a diagonal matrix $\mathbf{E}$ where each eigenvalue takes a diagonal position. The eigenvectors can be collected in a matrix $\mathbf{C} = (\mathbf{c}_1|\mathbf{c}_2|\ldots)$. In this case we can summarize the problem by writing

$$\mathbb{H}\mathbf{C} = \mathbf{ESC} \quad \text{or equivalently} \quad \mathbf{E} = \mathbf{C}^{\dagger}\mathbb{H}\mathbf{C} \quad \text{and} \quad \mathbf{C}^{\dagger}\mathbf{SC} = \mathbf{1} \tag{2.107}$$

which shows that the matrix **C** brings $\mathbb{H}$ and **S** simultaneously to a diagonal form, **E** and **1**, respectively. This approach is used nowadays almost exclusively for the calculation of molecular wave functions, either in *ab initio* (rigorously evaluating all integrals) or semi-empirically (neglecting and/or approximating integrals by experimental data or empirical formulae).

2.4.3 Perturbation Theory

Another way to approximate wave functions is by perturbation theory. If we use the same matrix notation again we may partition the equation [11]

$$\mathbb{H}\mathbf{c} = E\mathbf{c} \quad \text{to} \quad \begin{pmatrix} \mathbb{H}_{AA} & \mathbb{H}_{AB} \\ \mathbb{H}_{BA} & \mathbb{H}_{BB} \end{pmatrix}\begin{pmatrix} \mathbf{a} \\ \mathbf{b} \end{pmatrix} = E\begin{pmatrix} \mathbf{a} \\ \mathbf{b} \end{pmatrix}$$

Solving for **b** from $\mathbb{H}_{BA}\mathbf{a} + \mathbb{H}_{BB}\mathbf{b} = E\mathbf{b}$, we obtain

$$\mathbf{b} = (E\mathbf{1}_{BB} - \mathbb{H}_{BB})^{-1}\mathbb{H}_{BA}\mathbf{a}$$

which upon substitution in $\mathbb{H}_{AA}\mathbf{a} + \mathbb{H}_{AB}\mathbf{b} = E\mathbf{a}$ yields

$$\mathbb{H}_{\text{eff}}\mathbf{a} = E\mathbf{a} \quad \text{with} \quad \mathbb{H}_{\text{eff}} = \mathbb{H}_{AA} + \mathbb{H}_{AB}(E\mathbf{1}_{BB} - \mathbb{H}_{BB})^{-1}\mathbb{H}_{BA}$$

Let us take the number of elements in **a** equal to 1. In this case, the coefficient drops out and we have $E = \mathbb{H}_{\text{eff}}(E)$. Since $\mathbb{H}_{\text{eff}}$ depends on E, we have to iterate to obtain a solution and we do so by inserting as a first approximation $E = \mathbb{H}_{AA} = \mathbb{H}_{11}$ in $\mathbb{H}_{\text{eff}}$. By expanding the inverse matrix to second order in off-diagonal elements we obtain

$$E = H_{11} + \sum\nolimits_{m\neq 1} \frac{H_{1m}H_{m1}}{H_{11} - H_{mm}} \tag{2.108}$$

This is a general form of perturbation analysis since the basis is entirely arbitrarily; in particular, it is not necessary to assume a complete set, and so far it has not been necessary to divide the Hamilton operator into a perturbed and unperturbed part.

The conventional *Rayleigh–Schrödinger perturbation* equations result if we do divide the Hamilton operator into an unperturbed part H_0 and perturbed part H' or

$$H = H_0 + \lambda H'$$

where λ is the order parameter to be used for classifying orders, for example, a term in λ^n being of nth order. The perturbation may be regarded as switched on by changing λ from 0 to 1, assuming that the energy levels and wave functions are continuous functions of λ. Furthermore, we assume that we have the exact solutions of

$$H_0\Phi_k = E_k^{(0)}\Phi_k$$

and we use these solutions Φ_k in Eq. (2.108). The matrix elements become

$$H_{11} = \langle\Phi_1|H_0 + \lambda H'|\Phi_1\rangle = E_1^{(0)} + \lambda\langle\Phi_1|H'|\Phi_1\rangle$$

$$H_{1m} = \langle\Phi_1|H_0 + \lambda H'|\Phi_m\rangle = \lambda\langle\Phi_1|H'|\Phi_m\rangle$$

Since the label 1 is arbitrary we replace it by k to obtain the final result

$$▶ \quad E_k = E_k^{(0)} + \langle\Phi_k|H'|\Phi_k\rangle + \sum\nolimits_{m\neq k} \frac{\langle\Phi_k|H'|\Phi_m\rangle\langle\Phi_m|H'|\Phi_k\rangle}{E_k^{(0)} - E_m^{(0)}} \tag{2.109}$$

where λ has been suppressed. This expression is used among others for the calculation of the scattering of radiation with matter (first-order term) and the van der Waals forces between molecules (first- and second-order terms).

Example 2.7: The helium atom revisited

As an example of perturbation theory we discuss briefly the helium atom again. The Hamilton operator in atomic units is $H = -(\nabla_1^2 + \nabla_2^2) - Z(2/r_1 + 2/r_2) + 2/r_{12}$, where the first term indicates the kinetic energy of electrons 1 and 2, the second term the nuclear attraction for electrons 1 and 2 and the third term the electron–electron repulsion. We consider the electron–electron repulsion as the perturbation, and as zeroth-order wave function we use a product of two simple exponential functions, namely $\Psi = \phi(1)\phi(2) = (Z^3/\pi)\exp[-Z(r_1 + r_2)]$, where the nuclear charge Z instead of the variation parameter η, is introduced. For the kinetic energy integral K we obtained in Example 2.6 $K = Z^2$, and for the nuclear attraction integral N we derived $N = -2Z^2$. The zeroth-order energy ε_0 is $\varepsilon_0 = 2(K + N) = 2Z^2 - 4Z^2 = -2Z^2$. For the electron–electron repulsion integral I, which in this case represents the first-order energy ε_1, we obtained $I = 5Z/4$. The total energy ε is thus $\varepsilon = \varepsilon_0 + \varepsilon_1 = -2Z^2 + 5Z/4$. The corresponding energy is −5.50 Ry, to be compared with the experimental value of −5.81 Ry and the variation result of −5.70 Ry.

A slightly different line of approach is to start with the time-dependent Schrödinger equation right away. This is particularly useful for time-dependent perturbations. In this case, we divide the Hamilton operator directly into $H = H_0 + \lambda H'$ and suppose that the solutions of $H_0\Phi_n = E_n\Phi_n$ are available. By using the same notation as before and allowing the coefficients c_k to be time-dependent, we obtain

$$\Psi = \sum\nolimits_n c_n(t)\Phi_n(\mathbf{r})\exp(-\mathrm{i}E_nt/\hbar) \quad \text{or} \quad \Psi = \tilde{\mathbf{\Phi}}^{\mathrm{T}}(\mathbf{r},t)\mathbf{c}(t)$$

where $\tilde{\Phi}_k(\mathbf{r},t) = \Phi_k(\mathbf{r})\exp(-\mathrm{i}E_kt/\hbar)$. Insertion in the Schrödinger equation leads to

$$H\tilde{\mathbf{\Phi}}^{\mathrm{T}}(\mathbf{r},t)\mathbf{c}(t) = \mathrm{i}\hbar\frac{\partial}{\partial t}\tilde{\mathbf{\Phi}}^{\mathrm{T}}(\mathbf{r},t)\mathbf{c}(t) \quad \text{or omitting the arguments for } \tilde{\mathbf{\Phi}} \text{ and } \mathbf{c},$$

$$H\tilde{\mathbf{\Phi}}^{\mathrm{T}}\mathbf{c} = \mathrm{i}\hbar\tilde{\mathbf{\Phi}}^{\mathrm{T}}\frac{\mathrm{d}}{\mathrm{d}t}\mathbf{c} + \mathrm{i}\hbar\left(\frac{\partial}{\partial t}\tilde{\mathbf{\Phi}}^{\mathrm{T}}\right)\mathbf{c} = \mathrm{i}\hbar\tilde{\mathbf{\Phi}}^{\mathrm{T}}\dot{\mathbf{c}} + \mathrm{i}\hbar\left(-\frac{\mathrm{i}}{\hbar}\mathbf{E}\tilde{\mathbf{\Phi}}^{\mathrm{T}}\mathbf{c}\right)$$

with $\mathbf{E}$ a diagonal matrix containing all eigenvalues E_k. Evaluating using

$$H\tilde{\mathbf{\Phi}}^{\mathrm{T}}\mathbf{c} = (H_0 + \lambda H')\tilde{\mathbf{\Phi}}^{\mathrm{T}}\mathbf{c} = \mathbf{E}\tilde{\mathbf{\Phi}}^{\mathrm{T}}\mathbf{c} + \lambda H'\tilde{\mathbf{\Phi}}^{\mathrm{T}}\mathbf{c}$$

leads to

$$\lambda H'\tilde{\mathbf{\Phi}}^{\mathrm{T}}\mathbf{c} = \mathrm{i}\hbar\tilde{\mathbf{\Phi}}^{\mathrm{T}}\dot{\mathbf{c}}$$

Multiplying by $\tilde{\mathbf{\Phi}}^*$ and integrating over spatial coordinates yields

$$\lambda\tilde{\mathrm{H}}'\mathbf{c}(t) = \mathrm{i}\hbar\frac{\mathrm{d}}{\mathrm{d}t}\mathbf{c}(t) = \mathrm{i}\hbar\dot{\mathbf{c}}(t) \quad \text{with} \quad \tilde{H}'_{ij} = \exp\left[\frac{\mathrm{i}(E_i - E_j)t}{\hbar}\right]\langle\Phi_i|H'|\Phi_j\rangle$$

Finally, integrating over time, we obtain the result

$$\mathbf{c}(t) = (\mathrm{i}\hbar)^{-1}\lambda\int_0^t \tilde{\mathrm{H}}'\mathbf{c}(t')\mathrm{d}t' \tag{2.110}$$

which is an equation that can be solved formally by iteration. Further particular solutions depend on the nature of the Hamilton operator and the boundary conditions.

We leave the perturbation H' undetermined, but assume that at $t = 0$ the time-dependent effect is switched on, that is, $\lambda = 1$, and that at that moment the system is in a definite state, say i. In that case $c_i(0) = 1$ and all others $c_j(0) = 0$ or equivalently $c_i = \delta_{ij}$. We thus write to first order

$$c_i(t) = (\mathrm{i}\hbar)^{-1}\int_0^t \sum_j \tilde{H}'_{ij}c_j(t')\mathrm{d}t' \;\Rightarrow\; (\mathrm{i}\hbar)^{-1}\int_0^t \tilde{H}'_{ij}\exp[\mathrm{i}\hbar^{-1}(E_i - E_j)t']\mathrm{d}t'$$

Using the abbreviation $(E_i - E_j)/\hbar = \omega_{ij}$ and evaluating leads to

$$c_i(t) = -H'_{ij}(\hbar\omega_{ij})^{-1}[\exp(\mathrm{i}\omega_{ij}t) - 1]$$

if we may assume that H'_{ij} varies but weakly (or not at all) with t. The probability that the system is in state i at the time t is given by $|c_i(t)|^2$ and this evaluates to

$$|c_i(t)|^2 = 4|H'_{ij}|^2(\hbar\omega_{ij})^{-2}\sin^2\frac{1}{2}\omega_{ij}t$$

Introducing the final density of states $g_j = \mathrm{d}n_j/\mathrm{d}E$ of states j with nearly the same energy as the initial state i, the transition probability per unit time w_j to state j is given by

▶ $$w = \frac{1}{t}\int|c_j(t)|^2 g_j\,\mathrm{d}E_j = \frac{1}{t\hbar}4|H'_{ij}|^2 g_j\int_{-\infty}^{+\infty}\frac{\sin^2\frac{1}{2}\omega_{ij}t}{\omega_{ij}^2}\mathrm{d}\omega_{ij} = \frac{2\pi}{\hbar}|H'_{ij}|^2 g_j \tag{2.111}$$

where the last but one step can be made if H'_{ij} and g_j varies slowly with energy and the last step since the integral evaluates to ½πt. This result is known as *Fermi's golden rule*.

In fact, the energy available is slightly uncertain due to the Heisenberg uncertainty principle. Therefore, the state i is actually one out of a group having nearly the same energy and labeled A. Similarly, j is one out of a set B with nearly the same energy. The occupation probability p_i of each of the states i may change in time. If a system in state i within a set of states A makes a transition to any state within a set of states B, the *one-to-many jump rate* w_{BA} is given by Eq. (2.111). Since we are interested in the *one-to-one jump rate* ν_{BA} from any state i in group A to every state j in group B, we make the *accessibility assumption*–that is, all states within a (small) energy range δE, the *accessibility range*, are equally accessible. It can be shown that the exact size of δE is unimportant for macroscopic systems (see Justification 5.2). Hence

$$\nu_{ji} = w_{BA} / g_B \delta E = 2\pi |H'_{ji}|^2 / \hbar \delta E \qquad (2.112)$$

Because $H'_{ij} = (H'_{ji})^*$, we obtain *jump rate symmetry* $\nu_{ji} = \nu_{ij}$. This result is used in conjunction with the master equation to show that entropy always increases.

References

1 Clausius, R. (1865) *Pogg. Ann. Phys. Chem.*, **125**, 353.
2 See Waldram (1985).
3 See Callen (1985).
4 See Woods (1975).
5 See Pippard (1957).
6 See Lanczos (1970).
7 See Landau and Lifshitz (1976).
8 Dirac, P.A.M. (1957) *The Principles of Quantum Mechanics*, 4th edn, Clarendon, Oxford.
9 See Pauling and Wilson (1935).
10 However, see Appendix VIII of Born, M. and Huang, K. (1954) *Dynamical Theory of Crystal Lattices*, Oxford University Press, Oxford.
11 Pilar, F.L. (1968) *Elementary Quantum Chemistry*, McGraw-Hill, London.
12 Kuhn, T.S. (1977) *The Essential Tension*, University of Chicago Press, Chicago.
13 Redlich, O. (1968) *Rev. Mod. Phys.*, **40**, 556.
14 (a) Menger, K. (1950) *Am. J. Phys.*, **18**, 89; (b) Menger, K. (1951) *Am. J. Phys.*, **19**, 476.

Further Reading

General Reference

Tolman, R.C. (1938) *The Principles of Statistical Mechanics*, Oxford University Press, Oxford (also Dover 1979).

Thermodynamics

Callen, H. (1985) *Thermodynamics and an Introduction to Thermostatistics*, John Wiley & Sons, Inc., New York.

Pippard, A.B. (1957) *The Elements of Classical Thermodynamics*, Cambridge University Press, Cambridge, reprinted and corrected edition (1964).

Guggenheim, E.A. (1967) *Thermodynamics*, 5th edn, North-Holland, Amsterdam.

Waldram, J.R. (1985) *The Theory of Thermodynamics*, Cambridge University Press, Cambridge.

Woods, L.C. (1975) *The Thermodynamics of Fluid Systems*, Oxford University Press, Oxford, reprinted and corrected edition (1986).

Classical Mechanics

Goldstein, H. (1981) *Classical Mechanics*, 2nd edn, Addison-Wesley, Amsterdam.
ter Haar, D. (1961) *Elements of Hamiltonian Mechanics*, North-Holland, Amsterdam.
Lanczos, C. (1970) *The Variational Principles of Mechanics*, 4th edn, University of Toronto Press, Toronto (also Dover, 1986).
Landau, L.D. and Lifshitz, E.M. (1976) *Mechanics*, 3rd edn, Butterworth-Heinemann, Oxford.

Quantum Mechanics

Atkins, P.W. and Friedman, R.S. (2010) *Molecular Quantum Mechanics*, 5th edn, Oxford University Press, Oxford.
McQuarrie, D.A. (2007) *Quantum Chemistry*, 2nd edn, University Science Books, Mill Valley, CA.
Merzbacher, E. (1970) *Quantum Mechanics*, 2nd edn, John Wiley & Sons, Inc., New York.
Pauling, L. and Wilson, E.B. (1935) *Introduction to Quantum Mechanics*, McGraw-Hill, New York.
Pilar, F.L. (1968) *Elementary Quantum Chemistry*, McGraw-Hill, London.
Schiff, L.I. (1955) *Quantum Mechanics*, 2nd edn, McGraw-Hill, New York.

3
Basic Energetics: Intermolecular Interactions

All bonding and molecular interaction characteristics are determined by quantum mechanics. In liquids, we deal mainly with intermolecular forces and these may be described to quite some extent via classical considerations. In this chapter, we describe the long-range electrostatic intermolecular interactions using the multipole expansion. To that we add the induction and the dispersion interactions. In this approach, short-range repulsion, due to orbital overlap, is usually treated in an empirical way. We end with a brief discussion of some often-used model potentials, together with a few remarks on hydrogen bonding and three-body interactions and the virial theorem. For further details, reference should be made to the books in the section "Further Reading" of this chapter, which have been used freely.

3.1
Preliminaries

In principle, molecules and their interactions should be described by the Schrödinger equation. As this formidable problem cannot be solved exactly we use approximate methods, and in this chapter we describe some of them for intermolecular interactions. From the *Born–Oppenheimer approximation* (see Section 2.4) we know that an *(effective) potential energy* $E(\mathbf{r})$ for the motion of the nuclei with coordinates $\mathbf{r}$ exists. In fact, this is the potential energy employed in all of the discussions on intermolecular interactions. To emphasize the fact that $E(\mathbf{r})$ acts as a potential energy, from now on, we label it as $\Phi(\mathbf{r})$.

Experiments together with quantum mechanics have made clear that all interaction potentials between molecules have a shape, as presented in Figure 3.1, which shows for short distances a strong repulsive part and for longer distances an attractive part of which the magnitude depends heavily on the type of bonding. We need thus to explain the origin of the attractive part and of the repulsive part.

The balance between the long-range, attractive interactions, collectively denoted as the *van der Waals interactions* Φ_{vdW}, and the short-ranged repulsion forces

Liquid-State Physical Chemistry, First Edition. Gijsbertus de With.

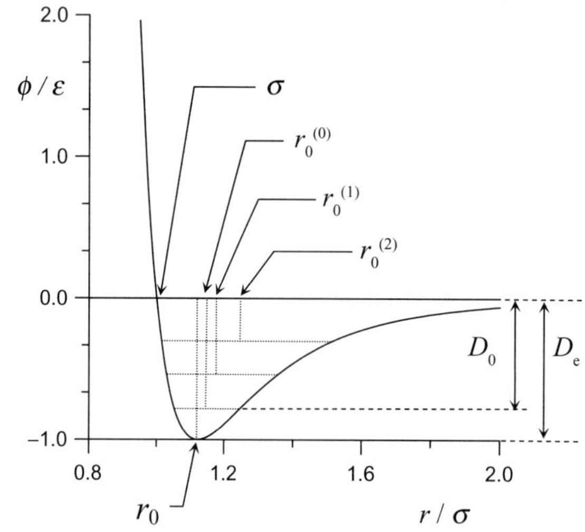

Figure 3.1 The shape of the Mie pair potential for a diatomic molecule in units of ε and σ for $n = 12$ and $m = 6$. The equilibrium distance is r_0, practically equal to $r_0^{(0)}$ for the vibrational ground state at $T = 0$ K. This distance increases with T due to the asymmetry of the potential and the increase in population of excited vibrational states, as indicated by $r_0^{(1)}$ and $r_0^{(2)}$. The *equilibrium energy* is D_e, while the *dissociation energy* is $D_0 = D_e - ½h\nu$ (vibrational state spacing exaggerated).

labeled as Φ_{rep}, determine whether a complex of molecules is stable, or not. Obviously, the total potential energy Φ is given by $\Phi = \Phi_{\text{vdW}} + \Phi_{\text{rep}}$. The equilibrium distances[1] $\mathbf{r}_0$ are given by the solution of $\partial\Phi/\partial\mathbf{r} = 0$, while the force constant matrix reads $\mathbf{K} = \partial^2\Phi/\partial\mathbf{r}\partial\mathbf{r}'|_{\mathbf{r}=\mathbf{r}_0}$ (or $K_{ij} = \partial^2\Phi/\partial r_i\partial r_j|_{\mathbf{r}=\mathbf{r}_0}$).

Before we discuss these interactions, let us spend a few words on the moments of a charge distribution. In molecular interactions the (continuum) charge distribution $\rho(\boldsymbol{r})$ of the molecule is the relevant distribution, where $\boldsymbol{r}$ is the vector of coordinates. Both, nuclei and electrons contribute to $\rho(\boldsymbol{r})$ and we assume that $\rho(\boldsymbol{r})$ is given. Alternatively, the charge distribution is described by a (discrete) set of charges q_i at positions $\boldsymbol{r}_i$. The zeroth moment of this distribution is a scalar and given by

$$q = \int_V r^0\rho(\boldsymbol{r})\mathrm{d}\boldsymbol{r} = \int_V \rho(\boldsymbol{r})\mathrm{d}\boldsymbol{r} \quad \text{or} \quad q = \sum\nolimits_i r_i^0 q_i = \sum\nolimits_i q_i \tag{3.1}$$

It is essentially the *charge*[2] q of the molecule. A molecule with $q = 0$ is called *neutral*. The first moment obviously becomes

$$\boldsymbol{\mu} = \int_V \boldsymbol{r}\rho(\boldsymbol{r})\mathrm{d}\boldsymbol{r} \quad \text{or} \quad \boldsymbol{\mu} = \sum\nolimits_i q_i\boldsymbol{r}_i \tag{3.2}$$

1) Note again that $\boldsymbol{r}$ is a vector and $\mathbf{r}$ is a (column matrix of a) set of elements r_i or $\boldsymbol{r}_i$.
2) The unit of charge is coulomb [C].

It is a vector called the *dipole moment*[3] with components μ_x, μ_y and μ_z. Essentially, the magnitude of the dipole moment is given by $\mu = |\boldsymbol{\mu}| = q\delta$, where δ is the distance between the centers of gravity of the positive and negative charges $+q$ and $-q$ in the molecule. The different location of these centers in the molecule renders the dipole moment a vector, leading to strong orientational forces. The *ideal dipole moment* is obtained by decreasing the distance δ to zero, while the charge q increases in such a way that the product $q\delta$ remains μ. A molecule with $\mu \neq 0$ ($\mu = 0$) is called *polar* (*apolar*). One also defines the *quadrupole moment* by

$$Q = \frac{1}{2}\int_V \mathbf{r}\mathbf{r}\rho(\mathbf{r})\mathrm{d}\mathbf{r} \quad \text{or} \quad Q = \frac{1}{2}\sum\nolimits_i q_i \mathbf{r}_i \mathbf{r}_i \tag{3.3}$$

It is a symmetrical tensor with components Q_{xx}, Q_{yy}, Q_{zz}, Q_{xy}, Q_{yz}, and Q_{zx}. Higher-order moments can be defined similarly, but we will need only the charge and dipole moment. The importance of the moment expansion lies in the fact that, if a full series of moments is given, it is equivalent to the original distribution. Only the first non-zero moment is independent of the location of the origin, which is usually taken as the center of gravity of the molecule.

3.2
Electrostatic Interaction

The *Coulomb energy* w of a charge q_1 in the presence of another charge q_2 is

$$w_{12}(r) = q_1\phi_2 = \mathcal{C}\frac{q_1 q_2}{r} \quad \text{with} \quad \phi_2 \equiv \mathcal{C}\frac{q_2}{r} \quad \text{and} \quad \mathcal{C} \equiv \frac{1}{4\pi\varepsilon_0} \tag{3.4}$$

Here, ϕ_2 represents the *potential* associated with the *force* $-\nabla w = \mathcal{C}q_1 q_2 \mathbf{r}/r^3$ between the charges q_1 and q_2 with $r = |\mathbf{r}|$ their distance and ε_0 the permittivity of vacuum. Molecule 1 can be represented by a charge distribution[4] $\rho_1 = \sum_i q_i^{(1)}$, and the potential ϕ_2 due to the charge distribution of molecule 2 by $\phi_2(\mathbf{r}) = \sum_j q_j^{(2)}/s_j$ where $s_j = |\mathbf{s}_j| = |\mathbf{r}-\mathbf{r}_j|$ is the distance of charge q_j to position $\mathbf{r}$ (Figure 3.2). The total interaction energy $W_{12} = \rho_1\phi_2 = \sum_j q_j^{(1)}\phi_2(\mathbf{r}_i) = \sum_j q_j^{(1)} \sum_k (q_k^{(2)}/s_{jk})$ with $s_{jk} = |\mathbf{s}_{jk}| = |\mathbf{r}_j - \mathbf{r}_k|$. After expanding the function ϕ_2, evaluating all the derivatives and collecting terms, a procedure with considerable understatement often called "a lengthy but straightforward calculation" (see Justification 3.1), one obtains the desired result. Note that, since this process takes place at constant T and V, the interaction W_{12} represents the Helmholtz energy.

3) The unit of dipole moment is coulomb × meter [C m], but frequently (for historical reasons) the Debye unit [D] is used. 1[D] = 3.336×10^{-30}[C m].

4) Because the molecule contains several charges we have to sum over them. We indicate here the charges by the subscript i and the molecule by the superscript (i). Only if we sum explicitly over charges is the molecule is indicated by the superscript; otherwise, we use a subscript.

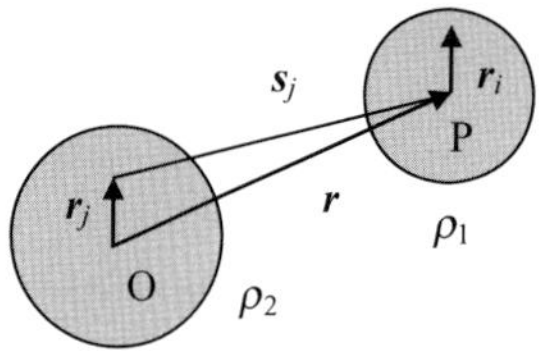

Figure 3.2 The interaction $\rho_1\phi_2$ of two charge distributions ρ_1 and ρ_2 with common origin O.

Justification 3.1: The multipole expansion*

To find the proper expression for the electrostatic interaction, let us consider an arbitrary charge distribution ρ of point charges q_j at position $\boldsymbol{r}_j$ from the origin O located at the center of mass (Figure 3.2). The potential energy ϕ at a certain point P located at $\boldsymbol{r}$ outside the sphere containing all charges is

$$\phi(\boldsymbol{r}) = \mathcal{C}\sum\nolimits_j (q_j/s_j)$$

where $s_j = |\boldsymbol{s}_j| = |\boldsymbol{r}-\boldsymbol{r}_j|$ is the distance of charge q_j to P. Developing $1/s_j$ in a Taylor series with respect to $\boldsymbol{r}_j$, we may write $1/s_j = 1/r + \boldsymbol{r}_j(\nabla 1/r)_\mathrm{O} + \cdots$ and therefore (using direct notation, Appendix B)

$$\begin{aligned}\phi(\boldsymbol{r})/\mathcal{C} &= \sum\nolimits_j \frac{q_j}{r} + \sum\nolimits_j q_j\boldsymbol{r}_j\cdot\left(\nabla\frac{1}{r}\right)_\mathrm{O} + \frac{1}{2}\sum\nolimits_j q_j\boldsymbol{r}_j\boldsymbol{r}_j : \left(\nabla\nabla\frac{1}{r}\right)_\mathrm{O} + \cdots \\ &= q\phi(0) + \boldsymbol{\mu}\cdot\phi'(0) + \boldsymbol{Q}:\phi''(0) + \cdots \\ &= \frac{q}{r} - \boldsymbol{\mu}\cdot\left(\nabla\frac{1}{r}\right)_\mathrm{P} + \boldsymbol{Q}:\left(\nabla\nabla\frac{1}{r}\right)_\mathrm{P} - \cdots \end{aligned} \quad (3.5)$$

where $r = |\boldsymbol{r}|$ is the distance of the point P to the origin O of ρ. In the second line we defined $q = \Sigma_j q_j$ as the *total charge*, $\boldsymbol{\mu} = \Sigma_j q_j\boldsymbol{r}_j$ as the *dipole moment*, and $\boldsymbol{Q} = \frac{1}{2}\Sigma_j q_j\boldsymbol{r}_j\boldsymbol{r}_j$ as the *quadrupole moment*. Moreover, we abbreviate $1/r$ by $\phi(0)$, $[\nabla(1/r)]_\mathrm{O}$ by $\varphi'(0)$, etc. The minus sign in the last step for the second, fourth, and so on terms arises because the differentiation is now made at P, and not at the origin of the potential O. This follows since for an arbitrary vector $\boldsymbol{x}$ it holds that $\nabla(1/x) = -\boldsymbol{x}/x^3$ and therefore $[\nabla(1/r)]_\mathrm{P} = -[\nabla(1/r)]_\mathrm{O}$. Equation (3.5) is usually called the *multipole expansion* of the potential. The *interaction* W of a point charge q at $\mathbf{r}$ in the potential ϕ of a charge distribution is $W = q\phi(\mathbf{r})$. Hence, to calculate the interaction energy between two charge distributions ρ_1 and ρ_2, separated by $\boldsymbol{r}$ (Figure 3.2), we express this energy as the energy of $\rho_1 = \Sigma_i q_i^{(1)}$ in the field of $\rho_2 = \Sigma_j q_j^{(2)}$. The interaction energy $W_{12} = \rho_1\phi_2$ is then

$$\begin{aligned} W_{12} &= \sum\nolimits_i q_i^{(1)}\phi_2(\boldsymbol{r}_i) = \sum\nolimits_i q_i^{(1)}\left[\phi_2(0) + \boldsymbol{r}_i^{(1)}\cdot\phi_2'(0) + \frac{1}{2}\boldsymbol{r}_i^{(1)}\boldsymbol{r}_i^{(1)}:\phi_2''(0) + \cdots\right] \\ &= q_1\phi_2(0) + \boldsymbol{\mu}_1\cdot\phi_2'(0) + \boldsymbol{Q}_1:\phi_2''(0) + \cdots \end{aligned}$$

If we take the center of mass P of ρ_1 as common origin, the potential ϕ_2 due to ρ_2 is given by the last line of Eq. (3.5), and substitution in the previous equation results in

$$\blacktriangleright \quad W_{12}/\mathcal{C} = \frac{A}{r} + B \cdot \left(\nabla \frac{1}{r} \right)_{\mathrm{P}} + \Gamma : \left(\nabla\nabla \frac{1}{r} \right)_{\mathrm{P}} + \cdots \quad \text{where} \tag{3.6}$$

$$A = q^{(1)} q^{(2)}, \quad B = \left(q^{(1)} \boldsymbol{\mu}^{(2)} - q^{(2)} \boldsymbol{\mu}^{(1)} \right), \quad \Gamma = \left(q^{(1)} \mathbf{Q}^{(2)} - \boldsymbol{\mu}^{(1)} \boldsymbol{\mu}^{(1)} + q^{(2)} \mathbf{Q}^{(1)} \right)$$

This is the general expression for the electrostatic interaction energy expressed in terms of multipole moments of the charge distributions with respect to their own center of mass [1]. For further evaluation, we need again $\nabla(1/r) = -\boldsymbol{r}/r^3$ and also $\nabla^2(1/r) = 3\boldsymbol{rr}/r^5 - \boldsymbol{I}/r^3$ with $\boldsymbol{I}$ the unit tensor, easily derived using $r^2 = x^2 + y^2 + z^2$, and leading to Eq. (3.7).

The final result reads

$$\blacktriangleright \quad W(r) = \frac{\mathcal{C} q_1 q_2}{r} + \frac{\mathcal{C}}{r^3} [(\boldsymbol{r} \cdot \boldsymbol{\mu}_1) q_2 - (\boldsymbol{r} \cdot \boldsymbol{\mu}_2) q_1] + \frac{\mathcal{C}}{r^5} [r^2 \boldsymbol{\mu}_1 \cdot \boldsymbol{\mu}_2 - 3(\boldsymbol{\mu}_1 \cdot \boldsymbol{r})(\boldsymbol{\mu}_2 \cdot \boldsymbol{r})] + \ldots \tag{3.7}$$

or explicitly (see Figure 3.3 for notation, introducing the abbreviation $\boldsymbol{\omega} = \theta,\phi$)

$$W(r, \boldsymbol{\omega}_1, \boldsymbol{\omega}_2) = \frac{\mathcal{C}}{r}(q_1 q_2) + \frac{\mathcal{C}}{r^2}(q_2 \mu_1 \cos\theta_1 - q_1 \mu_2 \cos\theta_2)$$
$$- \frac{\mathcal{C} \mu_1 \mu_2}{r^3} [2\cos\theta_1 \cos\theta_2 - \sin\theta_1 \sin\theta_2 \cos(\phi_1 - \phi_2)] + \ldots \tag{3.8}$$

In this expression $q_1 = \Sigma_i q_i^{(1)}$ is the *charge* of molecule 1, and $\boldsymbol{\mu}_1 = \Sigma_i q_i^{(1)} \boldsymbol{r}_i^{(1)}$ is the *dipole moment* with $\mu_1 = |\boldsymbol{\mu}_1|$.

The first term (proportional to q_1, q_2 and $1/r$) is just the *Coulomb interaction* (more precisely the charge–charge Coulomb interaction) between the molecules at distance r. At long distance r this interaction term will dominate, but clearly it is zero if one of the molecules is neutral.

The second term (proportional to q_1, μ_2, q_2, μ_1 and $1/r^2$; Figure 3.3) is the interaction between the charge of molecule 2 with the dipole moment of molecule 1 and the interaction between the charge of molecule 1 with the dipole moment of molecule 2. This is the *charge–dipole (Coulomb) interaction*. The charge–dipole interaction decreases as $1/r^2$, as compared to $1/r$ for the charge–charge Coulomb interaction. A simple way to obtain this result is dealt with in Problem 3.2. Because the charge–dipole interaction is orientation-dependent and molecules move generally rather rapidly, an orientational average (overbar) must be made (we will learn how to do that in Chapter 5). For the moment we just quote the result, valid for sufficiently high temperature,

$$\blacktriangleright \quad W_{\text{c-d}}(r) = \overline{W_{\text{c-d}}(r,\theta)} = -\frac{\mathcal{C}^2}{6kT}\left(\frac{q_1^2 \mu_2^2}{r^4}\right) - \frac{\mathcal{C}^2}{6kT}\left(\frac{q_2^2 \mu_1^2}{r^4}\right) \tag{3.9}$$

The third term (proportional to μ_1, μ_2 and $1/r^3$; Figure 3.3) represents the *dipole–dipole (Coulomb) interaction* between the dipole moments of molecule 1 and molecule 2, and is often addressed as the *Keesom interaction* (it also includes the charge–quadrupole interaction, but we neglect that interaction). We deal with this

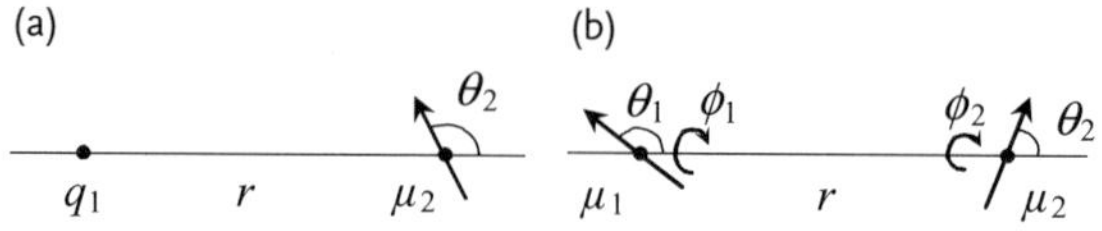

Figure 3.3 The interaction at a distance r of a point charge q_1 with a dipole μ_2 in which the charges q are separated by a distance δ (a) and the interaction of a dipole with a dipole at distance r (b).

interaction in a simplified way in Problem 3.3. This interaction is also orientation-dependent and for the orientational average one obtains, again for sufficiently high temperature and using $\boldsymbol{\omega} = (\theta,\phi)$ with θ and ϕ the orientation angles

$$\blacktriangleright \quad W_{\text{d-d}}(r) = \overline{W_{\text{d-d}}(r,\boldsymbol{\omega}_1,\boldsymbol{\omega}_2)} = -\frac{\mathcal{C}^2}{3kT}\left(\frac{\mu_1^2\mu_2^2}{r^6}\right) \tag{3.10}$$

For neutral molecules the dipole–dipole interaction is the leading term. Note (again) that the expressions given represent the Helmholtz energies and that the internal energy expressions U are given by $U = 2W$ (see Problem 2.6).

Problem 3.1

Show that for a charged molecule the dipole moment depends on the choice of the origin but is independent of this choice for a neutral molecule.

Problem 3.2: Charge–dipole interaction

A simple model to estimate the charge–dipole interaction is shown in Figure 3.3. The distances between the charge q_1 of molecule 1 and the charges of the dipole moment of molecule 2, $+q_2$ and $-q_2$, respectively, are given by

$$r(q_1,+q_2) = r + \frac{1}{2}\delta\cos\theta_2 \quad \text{and} \quad r(q_1,-q_2) = r - \frac{1}{2}\delta\cos\theta_2$$

Show, if it is assumed that the distance r is much larger than the distance δ by using Coulomb's law and the binomial expansion, that the interaction for fixed direction is $W_{\text{c-d}} = -\mathcal{C}r^{-2}q_1\mu_2\cos\theta_2$ where $\mathcal{C} = 1/4\pi\varepsilon_0$. How large should r be as compared to δ in order to have an error less than 5%?

Problem 3.3: Dipole–dipole interaction

The dipole–dipole interaction can be estimated in a similar way as for the charge–dipole problem (Figure 3.3). Show that the expression for the in-line orientation energy (orientations parallel to the connection line) is

$$W_{\text{d-d}}(r) = \pm 2\mathcal{C}r^{-3}\mu_1\mu_2 \quad \text{with, as before,} \quad \mathcal{C} = 1/4\pi\varepsilon_0$$

Here, ± indicates whether the orientation is in the same way (attraction) or in the opposite way (repulsion). Also show that the expression for the parallel orientation energy (orientations perpendicular to the connection line) is

$$W_{\text{d-d}}(r) = \pm \mathcal{C} r^{-3} \mu_1 \mu_2$$

again, dependent on whether the orientation is antiparallel (attraction) or parallel (repulsion). How large should r be as compared to δ in order to have an error less than 5%?

Problem 3.4*

Verify Equation (3.6). Alternatively, derive the expression for the electrostatic interaction up to second order using a double Taylor expansion for s_{ij}^{-1} with $s_{ij} = |\mathbf{r} + \mathbf{r}_j - \mathbf{r}_i|$ using the complete expression $W_{12} = \mathcal{C}\Sigma_i q_i^{(1)} s_{ij}^{-1} \Sigma_j q_j^{(2)}$.

3.3 Induction Interaction

Apart from the electrostatic interactions between permanent charges (c) and dipoles (d), a charge in molecule 1 will also induce a dipole moment in molecule 2 (as will a dipole in molecule 1). The induced dipole (id) in molecule 2 is proportional to the field generated by the charge (or the dipole) of molecule 1 with as proportionality factor the (isotropic[5]) *polarizability*[6] α_2 of molecule 2, that is, $\boldsymbol{\mu}_2 = \alpha_2 \mathbf{E}_1$.

The field of a charge q_1 is given by

$$\mathbf{E}_{\text{c}}(\mathbf{r}) = -\frac{\partial \phi_{\text{c}}(r)}{\partial \mathbf{r}} = -\frac{\partial}{\partial \mathbf{r}} \frac{\mathcal{C} q_1}{r} = \frac{q_1 \mathbf{r}}{4\pi\varepsilon_0 r^3}$$

The interaction between the field $\mathbf{E}_{\text{c}}$ of molecule 1 and the induced dipole moment $\boldsymbol{\mu}_{\text{id}}$ in molecule 2 can be calculated as

$$W_{\text{c-id}}(r) = -\int_0^{E_{\text{c}}} \boldsymbol{\mu}_{\text{id}} \cdot \text{d}\mathbf{E} = -\int_0^{E_{\text{c}}} \alpha_2 \mathbf{E}_1 \cdot \text{d}\mathbf{E}_1 = -\frac{1}{2} \mathcal{C}^2 \frac{\alpha_2 q_1^2}{r^4} \tag{3.11}$$

and is a contribution to the total intermolecular interaction.

The field of a dipole moment $\boldsymbol{\mu}_1$ is given by

$$\mathbf{E}_{\text{d}}(\mathbf{r}) = -\frac{\partial \phi_{\text{d}}(r)}{\partial \mathbf{r}} = -\frac{\partial}{\partial \mathbf{r}} \frac{\mathcal{C} \boldsymbol{\mu}_1 \cdot \mathbf{r}}{r^3} = \frac{1}{4\pi\varepsilon_0 r^5} \left[3(\boldsymbol{\mu}_1 \cdot \mathbf{r})\mathbf{r} - r^2 \boldsymbol{\mu}_1 \right]$$

The interaction between the field $\mathbf{E}_{\text{d}}$ of molecule 1 and the induced dipole moment $\boldsymbol{\mu}_{\text{id}}$ of molecule 2 can be calculated as

5) Isotropy implies equal response for all directions. When the polarization becomes anisotropic, these interactions become orientation-dependent and therefore temperature-dependent.
6) The unit of polarizability is $\text{C}^2\,\text{m}^2\,\text{J}^{-1}$. Incorporating $(4\pi\varepsilon_0)^{-1}$ in α via $\alpha' = \alpha/(4\pi\varepsilon_0)$ renders the dimension of α' to be m^3. Since, according to electrostatics, the polarizability of a perfectly conducting sphere of radius r in vacuum is given by $4\pi\varepsilon_0 r^3$, the molecular radius r can be estimated from $r^3 = \alpha'$. Hence, α' is known as the *polarizability volume*.

$$W_{\mathrm{d-id}}(r) = \overline{W_{\mathrm{d-id}}(r,\theta)} = -\int_0^{E_\mathrm{d}} \mu_{\mathrm{id}} \cdot \mathrm{d}E = -\int_0^{E_\mathrm{d}} \alpha_2 E_1 \cdot \mathrm{d}E_1 = -\mathbb{C}^2 \frac{\alpha_2 \mu_1^2}{r^6} \tag{3.12}$$

using $[3(\boldsymbol{\mu}\cdot\boldsymbol{r})\boldsymbol{r}-r^2\boldsymbol{\mu}]^2 = (1 + 3\cos^2\theta)\mu^2r^4$ and $\overline{\cos^2\theta} = 1/3$. This interaction is denoted as the *Debye interaction* and is always attractive. Since the Helmholtz energy W is independent of temperature (at least as derived here for an isotropic polarizability α), it also represents the internal energy U. Both interactions are due to the field of molecule 1 (originating either from its charge or dipole moment) and the polarizability of molecule 2. For the total interaction we have to add the interaction between the field of molecule 2 and polarizability of molecule 1.

Problem 3.5

Show that the molecular radius r_{pol}, as estimated from the polarization volume α' using the information of footnote 6, equals about $0.85(\sigma/2)$ as estimated from the van der Waals constant b. Derive first the relation $\sigma/2 = (3b/16\pi N_\mathrm{A})^{1/3}$, where $\sigma/2$ is the vdW radius, and use vdW and polarizability data as given in Appendix E, for example, for CH_4, N_2, and Xe.

Problem 3.6

Verify the expressions for the energy $W_{\mathrm{c\text{-}id}}$ and the energy $W_{\mathrm{d\text{-}id}}$.

3.4 Dispersion Interaction

Although the motion of electrons in a molecule is rather fast, at any moment the charge distribution of a molecule will have a (nonpermanent) dipole moment. The field associated with this dipole moment induces a (also nonpermanent) dipole moment in a neighboring molecule. Both induced dipoles interact, and this leads to the so-called *(London) dispersion interaction*, which is relatively weak but omnipresent and (in vacuum) always attractive. The exact derivation is outside the scope of these notes. However, a relatively simple model, proposed by London [19][7] and using the Drude model for atoms, yields good insight in the nature of the interaction. Moreover, it yields the correct expression apart from the exact numerical pre-factor.

In the *Drude model*, one assumes that an atom or molecule can be considered as a set of particles (electrons) with charge e_i and mass m_i. Each of these particles

7) For an overview of London's work on interactions, see Ref. [19]. The term "dispersion forces" was coined by London to indicate the analogy of the expressions derived to those which appear in the dispersion formula for the polarizability of a molecule when acted upon by an alternating field.

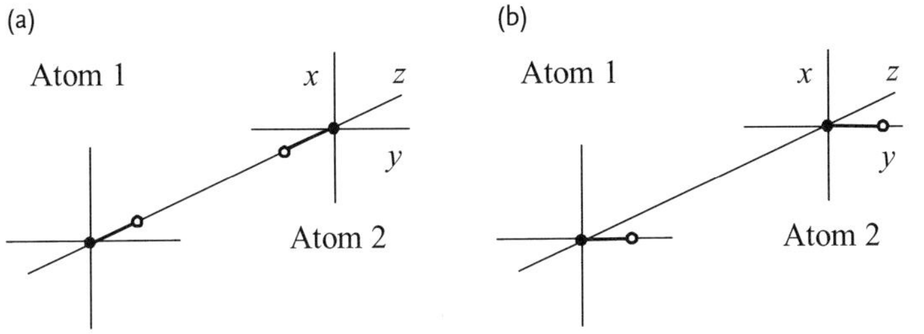

Figure 3.4 Schematic of the momentary interaction between the electrons (o) bonded to the nucleus (•) of two Drude atoms for an in-line orientation (a) and a parallel orientation (b).

is harmonically and isotropically bound to its equilibrium position. The configuration and notation as used for a single electron is sketched in Figure 3.4. By considering the in-line configuration first, the dynamics of this electron are thus given by those of a harmonic oscillator for which we have

$$U = \frac{1}{2}az^2 + \frac{1}{2}m\dot{z}^2$$

From quantum mechanics (see Section 2.2) we know that the energy expression becomes

$$U = \hbar\omega_0\left(n + \frac{1}{2}\right)$$

where n denotes the quantum number $n = 0, 1, 2, 3, \ldots$ and ω_0 the circular frequency of the oscillator. The latter is given by

$$\omega_0 = \sqrt{a/m}$$

where a is the force constant with which the electron is bound to the nucleus and m the electron mass. Without any interaction the total energy is given by

$$U = U_1(n_1) + U_2(n_2) = \hbar\omega_0\left(n_1 + \frac{1}{2}\right) + \hbar\omega_0\left(n_2 + \frac{1}{2}\right)$$

In the ground state, $n = 0$, the energy is

$$U_0 = U_1(0) + U_2(0) = \hbar\omega_0$$

At any moment electron 1 of atom 1 exerts a force on electron 2 of atom 2 and for a distance r, using Eq. (3.8), the potential is given by

$$V_{\text{int}} = -2\mathcal{C}\mu_z\mu_z/r^3 = -2\mathcal{C}z_1z_2e^2/r^3 \tag{3.13}$$

so that the resulting total energy U is

$$U = \left(\frac{1}{2}az_1^2 + \frac{1}{2}m\dot{z}_1^2\right) + \left(\frac{1}{2}az_2^2 + \frac{1}{2}m\dot{z}_2^2\right) - \left(2\mathcal{C}z_1z_2e^2/r^3\right)$$

This equation can be reduced to a sum of two independent harmonic oscillators, but with different frequencies. This is accomplished by the transformation

$$z_+ = (z_1 + z_2)/\sqrt{2} \quad \text{and} \quad z_- = (z_1 - z_2)/\sqrt{2} \quad \text{leading to}$$

$$U = \left[\frac{1}{2}\left(a - \frac{2\mathcal{C}e^2}{r^3}\right)z_+^2 + \frac{1}{2}m\dot{z}_+^2\right] + \left[\frac{1}{2}\left(a + \frac{2\mathcal{C}e^2}{r^3}\right)z_-^2 + \frac{1}{2}m\dot{z}_-^2\right]$$

Hence, the solution becomes

$$U = U_+(n_+) + U_-(n_-) = \hbar\omega_+\left(n_+ + \frac{1}{2}\right) + \hbar\omega_-\left(n_- + \frac{1}{2}\right)$$

with the new frequencies

$$\omega_+ = \sqrt{a_-/m} \quad \text{and} \quad \omega_- = \sqrt{a_+/m}$$

and force constants

$$a_- = a - (2\mathcal{C}e^2/r^3) \quad \text{and} \quad a_+ = a + (2\mathcal{C}e^2/r^3)$$

The new ground-state energy becomes

$$U_0' = U_+(0) + U_-(0) = \frac{1}{2}\hbar\omega_+ + \frac{1}{2}\hbar\omega_-$$

Expanding the square roots by the binomial theorem up to second order results in

$$U_0' = \hbar\omega_0 - (\mathcal{C}^2 e^4 \hbar\omega_0/2a^2 r^6) \tag{3.14}$$

Finally, we link the force constant a to the polarizability α using the force eE of the electric field E as compensated by the restoring force az, that is, $z = eE/a$. The induced dipole moment is given by $\mu = ez = e^2E/a$ as well as $\mu = \alpha E$, so that $\alpha = e^2/a$. The final expression for the in-line interaction energy becomes

$$W_{\text{in-line}} = U_0' - U_0 = -\mathcal{C}^2\alpha^2\hbar\omega_0/2r^6 \equiv -\mathcal{C}^2\alpha^2 I/2r^6 \tag{3.15}$$

The dispersion energy is thus proportional to the polarizability α squared, a characteristic energy $I \equiv \hbar\omega_0$ and the reciprocal sixth power of the distance r.

For the two parallel vibrations, perpendicular to the joining axis, of which the solutions are degenerated, a similar calculation leads to one-quarter of the above mentioned expression for each orientation, so that the total interaction becomes

$$W_{\text{dis}} = W_{\text{in-line}} + 2W_{\text{parallel}} = -C/r^6 \quad \text{with} \quad C = 3\mathcal{C}^2 I\alpha^2/4 \tag{3.16}$$

For dissimilar molecules, a similar but more complex calculation leads to

► $$C = \frac{3}{2}\mathcal{C}^2\frac{I_1 I_2}{I_1 + I_2}\alpha_1\alpha_2 \tag{3.17}$$

This result is thus exact for the Drude model of a molecule with only one characteristic frequency, apart from the "binomial approximation". For real molecules, the energies I should be chosen in accordance with the strongest absorption frequencies. In the absence of this information for I the ionization energies I_0 are often taken. This seems to work reasonably well for He ($I/I_0 \cong 1.20$) and H_2 ($I/I_0 \cong 1.09$). For the noble gases Ne, Ar, Kr and Xe, however, I/I_0 should be taken about 9/4 in order to match more reliable calculations [2]. A similar factor or even higher was noted for N_2, Cl_2, and CH_4. In general, although estimates for α are fairly reliable (see Chapter 10), estimates for I are less trustworthy.

Problem 3.7

Show that V_{int} for the in-line configuration and the parallel orientation using the Drude model and the expressions for the dipole interactions is given by

$$V_{\text{int}} = -2\mathcal{C}z_1z_2e^2/r^3 \quad \text{and} \quad V_{\text{int}} = \mathcal{C}x_1x_2e^2/r^3, \text{ respectively.}$$

Problem 3.8

Verify for interacting oscillators the expression for U_0' and $W_{\text{in-line}}$.

Problem 3.9*

Verify for interacting oscillators the expression for W_{parallel}.

Problem 3.10*

Derive the London dispersion interaction between dissimilar molecules.

3.5 The total Interaction

A comparison of the value of the various contributions to the *van der Waals interactions* is useful, and values as estimated for several small molecules are given in Table 3.1. Here, we limit ourselves to the Keesom, Debye and London contributions, since in many cases the interaction between neutral molecules is the most important, and three-body interactions are neglected anyway (see Section 3.7). The total Helmholtz interaction energy between two molecules is then

$$\blacktriangleright \quad \phi_{\text{vdW}} = -\mathcal{C}^2\left(\frac{\mu_1^2\mu_2^2}{3kT} + \mu_1^2\alpha_2 + \mu_2^2\alpha_1 + \frac{3}{2}\alpha_1\alpha_2\frac{I_1I_2}{I_1+I_2}\right)\frac{1}{r^6} \equiv -\frac{C_6}{r^6} \tag{3.18}$$

For this calculation we used the relatively simple model Drude model for the London interaction, with the characteristic energy estimated as the ionization potential I, implying that the estimate for the London interaction is approximate. From this comparison it becomes clear that in many cases the London dispersion

Table 3.1 The coefficient $C_6(10^{-79}\,\mathrm{J\,m^6})$ in the vdW expression for several molecules at 300 K.[a)]

Molecule	$d/2$ (pm)	μ (D)	α' ($10^{-30}\,\mathrm{m^3}$)	I (eV)	Keesom	Debye	London	Total
Ar	188	0	1.63	15.8	0	0	50.4	50.4
Kr	201	0	2.46	14.0	0	0	102	102
Xe	216	0	4.0	12.1	0	0	233	233
N_2	–	0	1.77	15.6	0	0	58.7	58.7
O_2	180	0	1.58	12.1	0	0	36.3	36.3
CO	360	0.117	1.98	13.9	0.00334	0.0570	65.5	65.5
CO_2	–	0	2.63	13.8	0	0	115	115
HI	–	0.42	5.45	10.4	0.501	1.92	371	374
HBr	190	0.80	3.61	11.7	6.59	4.62	183	194
HCl	180	1.08	2.63	12.8	21.9	6.13	106	134
HF	–	1.91	0.51	15.8	214	3.72	4.94	223
H_2O	140	1.85	1.48	12.6	189	10.1	33.2	232
NH_3	180	1.47	2.22	10.9	75.2	9.59	64.5	149
CH_4	200	0	2.60	14.4	0	0	116	116
CH_3OH	210	1.71	3.23	10.9	138	18.9	136	293

a) Most of these data are taken from Refs [20] and [21]. The parameter $d/2$ refers to the van der Waals radius as estimated from packing considerations [22].

interaction is dominating[8)]. Only for highly polar molecules does the Keesom interaction contribute significantly, for example, for H_2O and HF. As an aside, we note that incorporating higher-order terms into the analysis of the previous sections leads to $\phi_{\mathrm{vdW}} = -C_6/r^6 - C_8/r^8 + \ldots$ where the quantities C_i can all be expressed in terms of the moments of the charge distribution and polarizabilities of the molecules. We neglect these higher-order terms here.

In order to obtain the total interaction, we still need to add the repulsion acting at short distances between molecules. This interaction is due to orbital overlap, and hence is short-ranged and often called the *Born repulsion*. From quantum mechanics one can rationalize an exponential dependence, although in practice it is also often described by a power law. Hence we have

$$\phi_{\mathrm{rep}} = b\exp(-\alpha r) \quad \text{or} \quad \phi_{\mathrm{rep}} = b/r^n \tag{3.19}$$

where (b,α) or (b,n) are parameters. The parameter n is typically in the range 10 to 12.

The total interaction energy is thus given by

$$\phi = \phi_{\mathrm{rep}} + \phi_{\mathrm{vdW}} \tag{3.20}$$

8) We note that, since electromagnetic radiation has a finite velocity c, a second molecule feels the potential of the first molecule at a distance r only after a time r/c. This *retardation effect* becomes important only for distances larger than about 10 nm. The effect is important for the mutual interaction between macroscopic bodies and between a single particle and a solid surface.

Although the vdW interaction can be calculated with various degrees of sophistication, the repulsion is usually taken in to account as indicated above. In most cases the repulsion contributes significantly less than the attraction.

Problem 3.11

Calculate, using the value of C_6 between two Ar atoms and two HBr molecules (Table 3.1), the total interaction energy assuming that the molecules can be represented as attracting hard-spheres. Equating the result with kT, estimate the temperature for which dimer formation becomes significant.

3.6
Model Potentials

Although the above given description provides an insightful description of the intermolecular interactions, a more simple potential is often required. It is clear that any simplified potential should contain an attractive part and a repulsive part. However, the simplest potential that can be used is the *hard-sphere* (HS) *potential*, given by

$$\phi_{\text{HS}} = \infty \quad \text{for} \quad r < \sigma \quad \text{and} \quad \phi_{\text{HS}} = 0 \quad \text{for} \quad r \geq \sigma \tag{3.21}$$

where σ is the hard-sphere diameter. Although highly artificial, this potential is nevertheless often used in simple models and as a reference potential.

The simplest way to improve and include attractive interaction to the hard-sphere potential is to add a region with a constant attraction. This leads to the *square-well* (SW) *potential* given by

$$\begin{aligned} \phi_{\text{SW}} &= \infty \quad \text{for} \quad r < \sigma, \\ \phi_{\text{SW}} &= -\varepsilon \quad \text{for} \quad \sigma \leq r \leq R\sigma \quad \text{and} \\ \phi_{\text{SW}} &= 0 \quad \text{for} \quad r > R\sigma \end{aligned} \tag{3.22}$$

where, as before, σ is the hard-sphere diameter and ε and $R\sigma$ are the depth and range of the square-well, respectively. Typically, the value of R is ~1.5–2.0, and although simple it will appear to be versatile.

A more realistic expression is the *Mie potential* given by

$$\phi_{\text{Mie}} = br^{-n} - cr^{-m} \tag{3.23}$$

where b, c, n and m are parameters and $n > m$. An often-used choice, mainly for convenience, is the 12-6 or *Lennard-Jones* (LJ) *potential* with $n = 12$ and $m = 6$. In this case, an alternative form for Eq. (3.23) is

$$\phi_{\text{LJ}} = 4\varepsilon\left[(\sigma/r)^{12} - (\sigma/r)^{6}\right] = \varepsilon\left[(r_0/r)^{12} - 2(r_0/r)^{6}\right] \tag{3.24}$$

Here, the parameter ε describes the depth of the potential energy curve, while r_0 is the position of the minimum. For the LJ 12-6 potential we thus have $r_0 = 2^{1/6}\sigma$, $\phi_{\text{LJ}}(r_0) = -\varepsilon$, and $\phi_{\text{LJ}}(\sigma) = 0$. The parameter σ can be seen as the diameter of the molecule. The LJ potential describes the interaction energy qualitatively rather well (Figure 3.1). Typical values of parameters for the LJ potential for some molecules are given in Appendix E. It is clear that slightly different values will be obtained, depending on which properties and procedures are used for their evaluation.

For quasi-spherical molecules, such as SF_6, CF_4 and $C(CH_3)_4$, both the attractive and repulsive forces decrease more rapidly than would be expected from the Lennard-Jones potential. This implies a higher value for both the exponents 12 and 6. In fact, we can write the Mie potential, using ε and r_0, as

$$\phi_{\text{Mie}} = \frac{\varepsilon}{n-m}\left[n(r_0/r)^n - m(r_0/r)^m\right] \tag{3.25}$$

The potential for quasi-spherical molecules can be described by $(n,m) = (28,7)$, where the precise choice is largely dictated by mathematical convenience. An alternative for quasi-spherical molecules is the *Kihara potential* [3], which describes the interaction in a similar way as the Lennard-Jones potential but includes a hard core. It is given by

$$\phi_{\text{K}} = 4\varepsilon\left[\left(\frac{\sigma-\delta}{r-\delta}\right)^{12} - \left(\frac{\sigma-\delta}{r-\delta}\right)^{6}\right] = \varepsilon\left[\left(\frac{r_0-\delta}{r-\delta}\right)^{12} - 2\left(\frac{r_0-\delta}{r-\delta}\right)^{6}\right] \tag{3.26}$$

where δ is the diameter of the hard core. We thus have $\phi_{\text{K}}(r < \delta) = \infty$ and $\phi_{\text{K}}(\sigma) = 0$. Typical values for δ are $\delta/\sigma = 0.084$ for Ar, 0.194 for N_2, 0.655 for C_6H_6, and 0.900 for n-C_6H_{14}. The shape as a function of distance resembles that of the Mie 28-7 potential, taking δ equal to one-quarter of the van der Waals radius [4].

While the power-law relationship with $m = 6$ can be rationalized for the attractive part of the potential, the repulsion is on quantum-mechanical grounds expected to be better described by an exponential. This leads to the *Buckingham* or *exp-6 potential*

$$\phi_{\text{B}} = \varepsilon\left[\frac{6}{\alpha-6}\mathrm{e}^{-\alpha\left(\frac{r}{r_0}-1\right)} - \frac{\alpha}{\alpha-6}\left(\frac{r_0}{r}\right)^{-6}\right] \quad \text{and} \quad \phi_{\text{B}} = \infty \text{ for } r < r_{\text{max}} \tag{3.27}$$

where ε, r_0, and α are parameters, representing dissociation energy, equilibrium distance, and (scaled) force constant, respectively. Because this potential shows a maximum at a small value of r, the condition $\phi_{\text{B}} = \infty$ for $r < r_{\text{max}}$ is added. Another often-used expression is the *Morse potential* [5] given by

$$\begin{aligned}\phi_{\text{M}} &= \varepsilon\{\exp[-2\alpha(r-r_0)] - 2\exp[-\alpha(r-r_0)]\}\\ &= \varepsilon\{1-\exp[-\alpha(r-r_0)]\}^2 - \varepsilon\end{aligned} \tag{3.28}$$

with ε, r_0, and α parameters.

For estimates of the interaction between unlike molecules, we limit ourselves to the r^{-6} attraction. The repulsion is usually modeled as for hard-spheres, that is, we use for spheres i with diameter σ_{ii} the combination rule $\sigma_{ij} = (\sigma_{ii} + \sigma_{jj})/2$, often denoted as the *Lorentz rule*. For the attraction we learned from the London treatment that

$$\phi_{ij} = -C_{ij}/r^6 \quad \text{with} \quad C_{12} = \frac{3}{2}\frac{I_1 I_2}{I_1 + I_2}\alpha_1'\alpha_2' \tag{3.29}$$

Comparing this expression with the attractive part of the LJ potential

$$\phi_{ij} = -4\varepsilon_{ij}\sigma_{ij}^6/r^6 \equiv -\gamma_{ij}/r^6 \tag{3.30}$$

results in

$$\gamma_{ij} = \frac{3}{2}\frac{I_1 I_2}{I_1 + I_2}\alpha_1'\alpha_2' = 4\varepsilon_{12}\sigma_{12}^6 \quad \text{so that} \quad I_i = \frac{4}{3}\frac{4\varepsilon_{ii}\sigma_{ii}^6}{\alpha_i'^2} = \frac{4}{3}\frac{\gamma_{ii}}{\alpha_i'^2} \tag{3.31}$$

Substitution of I_i in Eq. (3.29) yields after some manipulation

$$\frac{\alpha_1'\alpha_2'}{\gamma_{12}} = \frac{1}{2}\left(\frac{\alpha_1'^2}{\gamma_{11}} + \frac{\alpha_2'^2}{\gamma_{22}}\right) \quad \text{or} \quad \varepsilon_{12} = \frac{2\alpha_1'\alpha_2'}{\sigma_{12}^6}\frac{\varepsilon_{11}\sigma_{11}^6\varepsilon_{22}\sigma_{22}^6}{\varepsilon_{11}\sigma_{11}^6\alpha_2'^2 + \varepsilon_{22}\sigma_{22}^6\alpha_1'^2} \tag{3.32}$$

Since reliable information on I is scarce, this is a clever way, essentially proposed by Kohler [6], to circumvent the need for an estimate of I.

If we take $\sigma_{ii} = \sigma_{jj}$ and $I_i = I_j$, we obtain the geometric mean $\varepsilon_{ij} = (\varepsilon_{ii}\varepsilon_{jj})^{1/2}$, often denoted as the *Berthelot rule*. However, while the Lorentz rule is quite acceptable, the Berthelot rule is only very approximate. From a sensitivity analysis for ε_{12} using the Lorentz–Berthelot rule it appears that the ratio σ_{11}/σ_{22} has a particularly large influence. Nevertheless, if intermolecular interactions are fitted to experimental data, the Lorentz–Berthelot rule is frequently used. It is clear that the success of Berthelot's rule is fortuitous and due to the partial canceling of errors.

Problem 3.12

Show that $r_0 = 2^{1/6}\sigma$, $\phi_{LJ}(r_0) = -\varepsilon$ and $\phi_{LJ}(\sigma) = 0$.

Problem 3.13

Show that the force constant a for the Morse and Kihara potential are given by $a = 2\alpha^2\varepsilon$ and $a = 72\varepsilon/r_0^2$, respectively.

Problem 3.14

Another well-known potential is the so-called universal bonding curve [7]

$$\phi_{uni}(r) = \varepsilon g(x), \quad g(x) = -(1 + x + 0.05x^3)\exp(-x), \quad x = (r - r_0)/l$$

Show that the force constant a for this potential $a = \varepsilon/l^2$, neglecting the x^3 term.

Problem 3.15

Show that, if $\sigma_{11} = \sigma_{22}$ and $I_1 = I_2$, Eq. (3.32) yields $\varepsilon_{12} = (\varepsilon_{11}\varepsilon_{22})^{1/2}$ and that $\varepsilon_{12} <$ $(\varepsilon_{11}\varepsilon_{22})^{1/2}$ otherwise.

Problem 3.16

Estimate ε_{12} using $\varepsilon_{12} = (\varepsilon_{11}\varepsilon_{22})^{1/2}$ as well as the more complete expression [Eq. (3.32)], taking a typical value of $\alpha_1'/\alpha_2' = 2$ for values of σ_{11}/σ_{22} and $\varepsilon_{11}/\varepsilon_{22}$ ranging from 1 to 2. How do you assess the quality of the Berthelot rule?

Problem 3.17

Show, assuming $\varepsilon_{\text{coh}} = -A/r^6$, that the average molecular cohesive energy of a fluid is given by

$$\langle \varepsilon_{\text{coh}} \rangle = -\frac{4\pi A\rho}{N-1}\left(\frac{1}{3\sigma^3} - \frac{1}{R^3}\right) \cong -\frac{4\pi A\rho}{3\sigma^3(N-1)}$$

Do so, assuming a random distribution of N molecules over a volume, that is, a uniform number density $\rho = N/V$, by integrating ε_{coh} for a reference molecule over a spherical shell for the smallest distance of approach σ to some macroscopic distance R. By taking this interaction over all pairs of molecules, show that the total molar cohesive energy U_{coh} reads

$$U_{\text{coh}} = -2\pi N_{\text{A}}^2 A/3\sigma^3 V_{\text{m}}$$

Estimate the value of U_{coh} for CCl_4 with mass density $\rho' = 1.59\,\text{g cm}^{-3}$ and molecular mass $M = 153.8\,\text{g mol}^{-1}$. Do so on the one hand by estimating σ from the polarizability volume α' and A from Eq. (3.18), and on the other hand from the LJ parameters ε and σ leading to $A = 4\varepsilon\sigma^6$, and comment on the agreement with the experimental value $U_{\text{coh}} = 32.8\,\text{kJ mol}^{-1}$.

3.7 Refinements

In this section we introduce hydrogen bonding and three-body interactions, and deal briefly with in how far empirical potentials are accurate.

3.7.1 Hydrogen Bonding

All of the above-mentioned interactions are generally valid. There is, however, a special bond, designated as *hydrogen bond*, which is an attractive interaction between two electronegative species A and B that form a link A–H···B via the covalently bonded H-atom and the lone pair of electrons (or polarizable π electrons) of B. The Lewis acid A withdraws some charge form the H-atom, leading to a partial charge δ^+ which interacts with δ^- of the Lewis base B. It is often said that this type of bonding is restricted to F, O and N atoms, but other species, such as negative ions (anions), can also participate in hydrogen bonding. For the F, O and N atoms, the interaction strength decreases from the F–H to the O–H to the N–H bond.

The hydrogen bond can also be considered as a delocalized bond in which orbitals from the A and B species as well the 1s orbital of the H atom are involved and which accommodates four electrons (two electrons from the A–H bond and two from the lone pair of atom B). These four electrons occupy the two lowest molecular orbitals, leading to an overall energy lowering. Since the hydrogen bond, of which the strength is typically about 10–40 kJ mol^{-1}, is based on orbital overlap, its strength decreases rapidly with distance. However, when this type of bonding is active it is often dominating over the other van der Waals interactions. Two types of hydrogen bond can be distinguished, namely, an *intramolecular* bond between A and B atoms within the same molecule, and an *intermolecular* bond between A and B atoms from different molecules. An example of the former is found in *ortho*-hydroxybenzoic acid, while the latter occurs, for example, in H_2O and CH_3COOH.

The hydrogen atom in the O–H···O bond has two nonequivalent minimum energy positions, each near each of the two O atoms, separated by an energy barrier. The O–H bond (typically ~0.097 nm) is shorter than the H···O bond (0.25–0.28 nm). The enthalpy of dissociation is typically 12–25 kJ mol^{-1}, but carboxylic acids in the gas-phase form an exception with ~30 kJ mol^{-1}. Hydrogen bonds provide multiple options for bonding in water, which leads to a residual entropy for ice of $R \ln(3/2)$ and anomalous behavior of water (e.g., low compressibility, high permittivity and a minimum density at 4 °C; see Chapter 10). Hydrogen bonding also leads to a number of other phenomena deviating from expected, normal behavior; for example, when hydrogen bonding is present significant deviations occur from Trouton's rule and Eötvös' rule[9]. Clearly, the infrared frequencies of the OH bond are influenced by the presence of a hydrogen bond. For example, for gaseous water the bond is characterized by a wave number of ~3756 cm^{-1}, whereas for liquid water the wave number decreases to ~3453 cm^{-1}. While the O–H···O hydrogen bond plays an important role in water, bonds of the type N–H···O=C occur in, for example, urea and proteins, playing important roles in biological systems.

In a simple model for collinear O–H···O bonds, the O–H···O bond energy $\phi_{\text{H-bond}}$ can be described by a Morse potential for the O–H bond with distance r and one for the H···O bond with distance R–r, plus the interaction between the two O atoms, $u(R)$. This expression contains several parameters: the bond dissociation energies ε_0; the force constants a_0 for these bonds unperturbed and their associated equilibrium distances r_0; the distance between the two O atoms R; and, finally, the parameters in the function $u(R)$. The values for these parameters can be obtained from the condition $(\partial\phi_{\text{H-bond}}/\partial r)_{\text{eq.}} = 0$. The IR shifts $\Delta\nu = \nu_{\text{H-bond}} - \nu_0$, with ν_0 the unperturbed bond frequency, are obtained from the force constant $a = (\partial^2\phi_{\text{H-bond}}/\partial r^2)_{\text{eq.}}$, and are to be compared with the experimental data.

9) See Section 16.2 and Section 15.2, respectively.

Hydrogen bonding affects properties to a considerable degree, but two examples will suffice at this point. Ethanol (CH_3CH_2OH) and dimethyl ether (CH_3OCH_3) are of similar size and have the same number of electrons; moreover, their dipole moments are not very different. Nevertheless, their boiling points are 78.5 °C and −24.8 °C, respectively. Pentane (C_5H_{12}) and butanol (C_4H_9OH) are also of similar size and have the same number of electrons, but in this case the dipole moment of butanol aids to increase the intermolecular interaction (C_5H_{12} is apolar) and their boiling points are 36.3 °C and 117 °C, respectively. The field of hydrogen bonding is rich and elaborate. A classical review has been provided by Pimentel and McLellan [8] while Grabowski [9] provides a more recent review.

3.7.2
Three-Body Interactions

So far, we have been discussing intermolecular interactions in terms of two-body potentials. However, there is no reason why a third body would not influence the interactions between two bodies, apart from contributing to the sum of the two-body potentials for all three bodies. The dipolar *three-body interaction* is probably [10] the most important, and has been studied by Axilrod and Teller [11] along the lines of the van der Waals potential for two bodies, resulting in

$$\phi_{123} = C_{123}(1+3\cos\theta_1 \cos\theta_2 \cos\theta_3)/r_{12}^3 r_{23}^3 r_{31}^3 \quad \text{with} \quad C_{123} \cong 3\mathcal{C}\alpha C_6/4$$

Here, θ_1 denotes the internal angle between the vector with length r_{12} between particles 1 and 2, while r_{13} is the length of the vector between particles 1 and 3. The angles θ_1 and θ_3 have a similar meaning, and the sign of this contribution depends on the geometry of the triangle formed. For acute triangles $\phi_{123} > 0$, whereas for obtuse triangles $\phi_{123} < 0$. The significance of the three-body interaction has been estimated rather differently by various authors. On the one hand, Rowlinson [12] indicated a too-negative potential energy by "10–15% of the total potential energy of a normal liquid," alleviating this to 5–10% for a typical configuration in a liquid where most close triplets of molecules form acute-angled triangles, but also indicating a substantial cancellation of multibody terms. Dymond and Alder [13], on the other hand, indicated a contribution ". . . as little as 1–2%," supported by Berendsen [14], who stated that the contribution is probably negligible. Here, attention will be paid to three-body interactions only occasionally.

3.7.3
Accurate Empirical Potentials

Although model potentials are useful, they are idealized. Figure 3.5 shows a typical set of curves for Ar [15] fitted on second virial coefficients over a temperature range of −190 °C to 600 °C. It is clear that both the depth and location of the potential vary considerably with the type of potential. Similar differences are obtained for other molecules. A rather complete evaluation on Ar dimers has been given by Parson *et al.* [16]. Only for a few cases are accurate empirical potentials available,

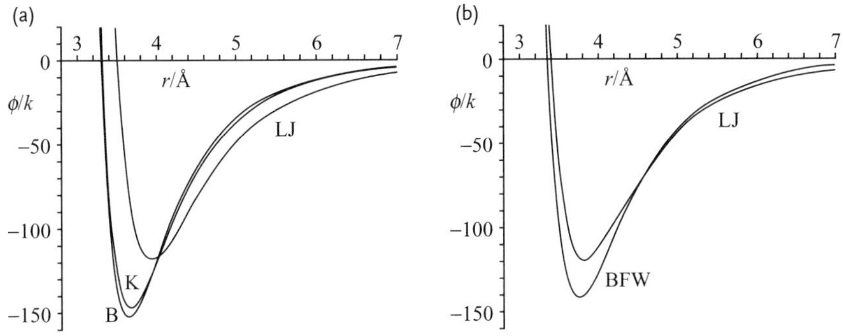

Figure 3.5 (a) Comparison of the Lennard-Jones (LJ), Kihara (K), and Buckingham (B) potential for Ar, showing the difference in minimum and depth of the various potentials; (b) Comparison of the Lennard-Jones (LJ) and Barker–Fisher–Watts (BFW) potential for Ar, showing that the LJ potential is more shallow at its minimum and longer ranged than the BFW potential.

the inert gases being such a case. Barker and coworkers [17] represented the Ar potential as a sum of pair potentials and the Axilrod–Teller three-body interaction resulting in a 15-parameter equation (!). For fitting, they used the thermodynamic and transport data of the gas, the low temperature properties of the solid, the internal energy and self-diffusion coefficient of the liquid, and the equation of state. This provided an excellent agreement with the vibration spectrum of bound Ar dimers, and with molecular beam scattering cross-section determinations [16]. It appeared also to describe the solid-state internal energy, the elastic constants, the phonon curves, and the solid-fluid coexistence curve well. Calculations of the third virial coefficient with this potential showed the contribution of the three-body interaction to be quite significant, and to bring the theoretical results into good agreement with the experimental data. Figure 3.5 shows a comparison with the optimum Lennard-Jones potential, indicating that the latter is shallower at its minimum and longer-ranged than the optimum potential. On the other hand, simple pair potential calculations for the condensed phase perform quite well in practice. So, it might be wondered why these potentials do not perform very badly. Barker *et al.* concluded that the three-body effects partially cancel out in the condensed phase, and that three-body effects are taken into account in an averaged way by using a fitted two-body potential. However, their studies showed that a single potential for Ar derived from bulk data is capable of describing all properties.

Problem 3.18

Estimate the relative contribution of the three-body interaction for a triplet of Ar atoms arranged in an equilateral triangle. Use data from Table 3.1.

3.8 The Virial Theorem*

Even without considering the details of the dynamics of a system of particles, we can connect the average kinetic energy with the average potential energy. This is achieved via a useful theorem, first introduced by Clausius [18], of which we present here a short derivation. To this purpose, it must be first noted that for a bounded function $G(t)$ with time t as variable, the time average $\langle g\rangle$ of $g = \mathrm{d}G/\mathrm{d}t$ is zero, since

$$\langle g\rangle = \lim_{t\to\infty}\frac{1}{t}\int_0^t \frac{\mathrm{d}G}{\mathrm{d}t'}\mathrm{d}t' = \lim_{t\to\infty}\frac{G(t)-G(0)}{t} = 0 \tag{3.33}$$

Attention is now turned to the kinetic energy $T = \frac{1}{2}\Sigma_i T_i = \frac{1}{2}\Sigma_i m_i(\mathrm{d}\boldsymbol{r}_i/\mathrm{d}t)^2 = \frac{1}{2}\Sigma_i m_i v_i^2 = \frac{1}{2}\Sigma_i(p_i^2/m_i)$ for a set of particles with mass m_i, position coordinates $\boldsymbol{r}_i$, velocity $\boldsymbol{v}_i$, and momentum $\boldsymbol{p}_i = m_i\boldsymbol{v}_i$. Because T_i is a homogeneous function of the second degree in $\boldsymbol{v}_i$, we have by Euler's theorem (see Appendix B)

$$(\partial T_i/\partial \boldsymbol{v}_i)\cdot \boldsymbol{v}_i = 2T_i \tag{3.34}$$

Since $\partial T_i/\partial \boldsymbol{v}_i = \boldsymbol{p}_i$, we further obtain

$$2T_i = \boldsymbol{p}_i\cdot\boldsymbol{v}_i = \mathrm{d}(\boldsymbol{p}_i\cdot\boldsymbol{r}_i)/\mathrm{d}t - (\mathrm{d}\boldsymbol{p}_i/\mathrm{d}t)\cdot\boldsymbol{r}_i \tag{3.35}$$

If the system has a finite volume, $\boldsymbol{p}_i\cdot\boldsymbol{r}_i$ is bounded and thus we have $\langle \mathrm{d}\boldsymbol{p}_i\cdot\boldsymbol{r}_i/\mathrm{d}t\rangle = 0$. The kinetic energy T is linked to the potential energy Φ via Newton's Second Law reading $\boldsymbol{f}_i = m_i\boldsymbol{a}_i$ with force $\boldsymbol{f}_i$ and acceleration $\boldsymbol{a}_i$. Using $\boldsymbol{f}_i = -\partial\Phi/\partial\boldsymbol{r}_i$, $m_i\boldsymbol{a}_i = \mathrm{d}\boldsymbol{p}_i/\mathrm{d}t$, and introducing the *virial (of force)* $\mathcal{V}_i \equiv \langle(\partial\Phi/\partial\boldsymbol{r}_i)\boldsymbol{r}_i\rangle$, we obtain the *virial theorem*

$$2\langle T_i\rangle = \mathcal{V}_i \quad \text{or} \quad 2\langle T\rangle = 2\left\langle\sum\nolimits_i T_i\right\rangle = \sum\nolimits_i \mathcal{V}_i = \mathcal{V} \tag{3.36}$$

where the second step can be made as long as $T = \Sigma_i T_i$ and $\mathcal{V} = \Sigma_i\mathcal{V}_i$. Frequently, Φ is a homogeneous function of degree n in $\boldsymbol{r}$, and from Euler's theorem we then have

$$\sum\nolimits_i(\partial\Phi/\partial\boldsymbol{r}_i)\cdot\boldsymbol{r}_i = n\Phi \quad \text{or combining with Eq. (3.36)} \quad 2\langle T\rangle = n\langle\Phi\rangle \tag{3.37}$$

Since the total energy $U = \langle T\rangle + \langle\Phi\rangle$, we have $\langle\Phi\rangle = 2U/(n+2)$ and $\langle T\rangle = nU/(n+2)$. For harmonic potentials $n = 2$ resulting in $\langle T\rangle = \langle\Phi\rangle = \frac{1}{2}U$. For the Coulomb potential $n = -1$ and since $T \geq 0$, this implies $U \leq 0$. So, for an electrostatic system to be stable the energy must be negative. Finally, we note that the virial can be used in the derivation of the equation of state (see Chapter 6).

References

1 Böttcher, C.J.F. (1973) *Theory of Dielectric Polarization*, vol. I, 2nd edn, Elsevier, Amsterdam. See also Hirschfelder *et al.* (1964).

2 Pitzer, K.S. (1959) *Adv. Chem. Phys.*, **1**, 59.

3 Kihara, T. (1953) *Rev. Mod. Phys.*, **25**, 831.

4 (a) Hammam, S.D. and Lambert, J.A. (1954) *Aust. J. Chem.*, **7**, 1; (b) Hildebrand, J. and Scott, R.L. (1962) *Regular Solutions*, Prentice-Hall, Englewood Cliffs.
5 Morse, P.M. (1929) *Phys. Rev.*, **20**, 57.
6 (a) Kohler, F. (1954) *Monatsh. Chem.*, **88**, 857; (b) Kihara, T. (1976) *Intermolecular Forces*, John Wiley & Sons, Ltd, Chichester.
7 Rose, J.H., Smith, J.R., Guinea, F., and Ferrante, J. (1984) *Phys. Rev.*, **B29**, 2963.
8 See Pimentel and McLellan (1960).
9 See Grabowski (2006).
10 Scott, R.L. (1971) *Physical Chemistry*, vol. VIIIA, Academic Press, New York, Ch. 1, p. 11.
11 (a) Axilrod, B.M. and Teller, E. (1943) *J. Chem. Phys.*, **11**, 299; (b) Axilrod, B.M. (1949) *J. Chem. Phys.* **17**, 299 and **19**, 71; (c) Midzuno, Y. and Kihara, T. (1956) *J. Phys. Soc. Jpn*, **11**, 1045.
12 (a) Rowlinson, J.S. (1965) *Discuss. Faraday Soc.*, **40**, 19; (b) Rowlinson, J.S. and Swinton, F.L. (1982) *Liquids and Liquid Mixtures*, 3rd edn, Butterworth, London.
13 Dymond, J.H. and Alder, B.J. (1968) *Chem. Phys. Lett.*, **2**, 54.
14 Berendsen, H.J.C. (2007) *Simulating the Physical World*, Cambridge University Press, p. 187.
15 Sherwood, A.E. and Prausnitz, J.M. (1964) *J. Chem. Phys.*, **41**, 429.
16 Parson, J.M., Siska, P.E., and Lee, Y.T. (1972) *J. Chem. Phys.*, **56**, 1511.
17 Barker, J.A., Fisher, R.A., and Watts, R.O. (1971) *Mol. Phys.*, **21**, 657.
18 Clausius, R.J.E. (1870) *Philos. Mag. Ser. 4*, **40**, 122.
19 London, F. (1937) *Trans. Faraday Soc.*, **33**, 8.
20 Atkins, P.W. (2002) *Physical Chemistry*, 7th edn, Oxford
21 Butt, H.-J., Graf, K., and Kappl, M. (2006) *Physics and Chemistry of Interfaces*, 2nd edn, Wiley-VCH.
22 Israelachvili, J. (1991) *Intermolecular and Surface Forces*, 2nd edn, Academic Press.

Further Reading

Grabowski, S.J. (2006) *Hydrogen Bonding–New Insights*, Springer, Dordrecht.

Hirschfelder, J.O., Curtiss, C.F., and Bird, R.B. (1954) *Molecular Theory of Gases and Liquids*, John Wiley & Sons, Inc., New York (corrected edition, 1964).

Kaplan, I.G. (2006) *Intermolecular Interactions–Physical Picture, Computational Methods and Model Potentials*, John Wiley & Sons, Ltd, Chichester.

Maitland, G.C., Rigby, M., Smith, E.B., and Wakeham, W.A. (1981) *Intermolecular Forces–Their Origin and Determination*, Clarendon, Oxford.

Margenau, H. and Kestner, N.R. (1971) *Theory of Intermolecular Forces*, Pergamon, Oxford.

Parsegian, V.A. (2006) *Van der Waals Forces: A Handbook for Biologists, Chemists, Engineers and Physicists*, Cambridge University Press, Cambridge.

Pimentel, G.C. and McLellan, A.L. (1960) *The Hydrogen Bond*, W.H. Freeman and Company, New York.

Stone, A.J. (1996) *The Theory of Intermolecular Forces*, Clarendon, Oxford.

4
Describing Liquids: Phenomenological Behavior

Before we embark on modeling the behavior of liquids, in this chapter we discuss some phenomenological properties of liquids. The content of this chapter is not essential for the remainder of the book, but has been included in order to be somewhat self-contained and for reference. A brief outline of the phase behavior and equations of state is given.

4.1
Phase Behavior

In this section we deal with the phase behavior of fluids, for which a schematic is given in Figure 4.1. For sufficiently low temperature and sufficiently high pressure a vapor (V) condenses to a liquid (L) or a solid (S), and this leads to the coexistence of vapor and liquid (V–L coexistence line), of vapor and solid (V–S coexistence line) or of liquid and solid (S–L coexistence line). Whilst, as far as is known, the fusion (S–L) curve continues to exist with increasing temperature, the vaporization (V–L) curve ends at the critical temperature (T_{cri}) or critical point (CP). Above the CP, condensation is impossible and the vapor is often addressed as a gas (G). Moreover, above the critical isotherm the difference between liquids and gases ceases to exist. Another special point here is the triple point (TP) where the gas, liquid, and solid phase are in equilibrium.

Characteristic physical properties of liquids are the (isothermal) compressibility, κ_T, and the (isobaric) volume thermal expansion coefficient, α. For organic liquids the typical range of κ_T is $(0.2–2.0) \times 10^{-9}\,\text{Pa}^{-1}$. Generally, κ_T decreases with increasing pressure, with a smaller rate at a higher pressure, and increases with increasing temperature, with a larger rate at a lower pressure. For organic liquids, the typical range of α is $(0.7–1.5) \times 10^{-3}\,\text{K}^{-1}$. The parameter α increases with increasing temperature; typical data are listed in Appendix E.

For future reference, we note that often a threefold classification of fluids is made:

- *Simple fluids* are fluids of (more or less) spherical molecules such as Ar, CH_4 and N_2, for which the angular dependence of the interaction potential is (almost) absent.

Liquid-State Physical Chemistry, First Edition. Gijsbertus de With.

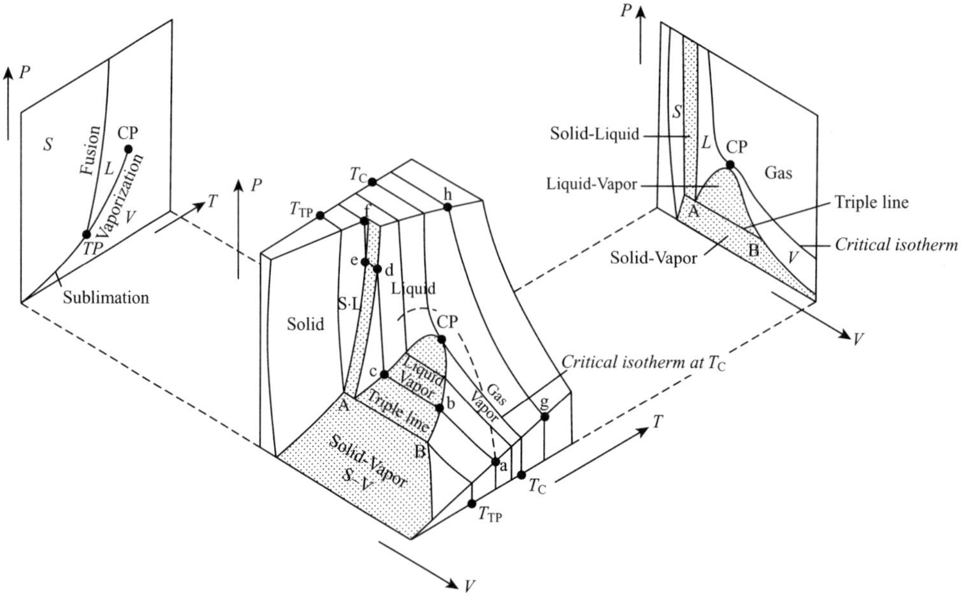

Figure 4.1 Phase relations for a simple fluid between the solid (S), liquid (L) and vapor (V) phase. The triple point (TP) and critical point (CP) are indicated. As in the figure, some authors use the word "gas" for above-critical and "vapor" for below-critical conditions. Reproduced from Ref. [16].

- *Normal fluids* are fluids without a specific interaction such as hydrogen bonding; examples are C_6H_6, CCl_4 and cyclo-C_6H_{12}.
- If these specific interactions are present, we denote these fluids as *complex fluids*, examples being H_2O, NH_3, and CH_3COOH.

4.2 Equations of State

For sufficiently low temperature and a not too-high pressure, gas is described by the well-known *perfect gas equation of state* (EoS), given by $P = nRT/V$, where the pressure P is given in terms of the gas constant R, the temperature T, and the volume V for n moles of gas. All gases obey this equation in the low pressure limit, as illustrated in Figure 4.2, where the behavior of the *compression factor* $Z = PV/RT$ is shown for H_2, N_2, and CH_4. Typically, the accuracy for P according to the perfect gas EoS is ~1% for $P < 10\,\text{bar}$. However, for somewhat higher pressures, deviations from the perfect gas law occur and the EoS becomes material-specific (Figure 4.2; see also Figure 4.4).

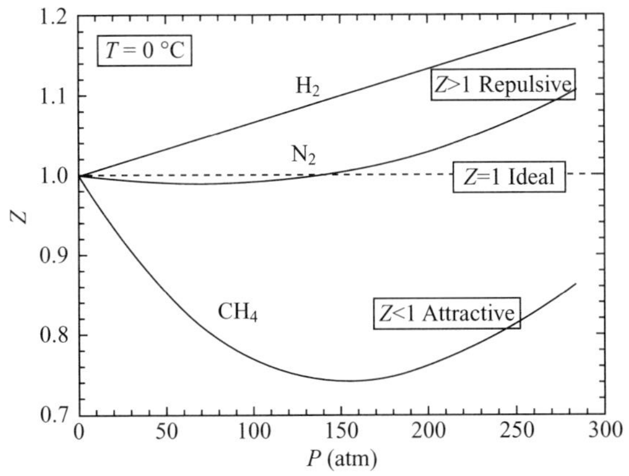

Figure 4.2 Approaching the perfect gas EoS for gases as exemplified by H_2, N_2, and CH_4. The limiting value of $PV = 2271\ \text{Pa m}^3\,\text{mol}^{-1}$ at the triple point of water, $T = 273.16\ \text{K}$. This yields $R = 8.314\ \text{Pa m}^3\,\text{mol}^{-1}\,\text{K}^{-1}$ or $8.314\ \text{J mol}^{-1}\,\text{K}^{-1}$.

An initially empirical description for a somewhat higher pressure is the "virial" equation of state. For the perfect gas the compression factor $Z = 1$. For a real gas, Z becomes a function of volume, which can be expanded as a power series in V^{-1}

$$Z(T,V) = \sum\nolimits_{j=0} B_{j+1}V^{-j} \quad \text{or} \quad Z(T,V) = 1 + \frac{B_2}{V} + \frac{B_3}{V^2} + \frac{B_4}{V^3} + \cdots \tag{4.1}$$

This equation is called the *virial equation of state*. The parameter B_j is denoted as the jth virial coefficient which is typically a function of temperature, except B_1 for which obviously holds $B_1 = 1$. Alternatively, one can expand in P

$$Z(T,P) = \sum\nolimits_{j=0} B'_{j+1}P^j \quad \text{or} \quad Z(T,P) = 1 + B'_2P + B'_3P^2 + B'_4P^3 + \cdots \tag{4.2}$$

The coefficients B_j' can be expressed in terms of the coefficients B_j by inverting the series. The first three coefficients then read

$$B'_2 = \frac{B_2}{RT}, \quad B'_3 = \frac{B_3 - B_2^2}{(RT)^2} \quad \text{and} \quad B'_4 = \frac{B_4 - 3B_2B_3 + 2B_2^3}{(RT)^3}$$

If only the second virial coefficient is available, one would be inclined to use $Z = 1 + B_2/V$. However, the expression $Z = 1 + B_2P/RT$, obtained from substituting $B_2' = B_2/RT$ in $Z = 1 + B_2'P$, is usually much more accurate. On the contrary, if the third virial coefficient is available, the expression $Z = 1 + (B_2/V) + (B_3/V^2)$ is usually much more accurate than $Z = 1 + B_2'P + B_3'P^2$.

For liquids, an expression equivalent to the perfect gas EoS does not exist. A simple EoS for liquids doing remarkably well is the *Tait equation* [1], given by

$$\frac{1}{V_0}\frac{\mathrm{d}V}{\mathrm{d}P} = -\frac{A}{B+P} \quad \text{or its integrated form} \quad V_0 - V = AV_0 \ln\left(\frac{B+P}{B+P_0}\right) \tag{4.3}$$

Table 4.1 Parameters for Tait's isotherm for H_2O and CCl_4.

Molecule	Unit	*A*	*B* (25 °C)	*B* (45 °C)	*B* (65 °C)	*B* (85 °C)
H_2O	bar	0.1368	2996	3081	3052	2939
CCl_4	atm	0.0924	869	738	622	–

Data from Ref. [18].

Table 4.2 Compressibility parameters for a few compounds.

Molecule	*T* range	κ_0 (10^{-4} atm)	β (10^{-4} K^{-1})	γ (–)
CS_2	168–353	12.55	6.73	7.5 (20 °C)
CH_3OH	175–331	18.6	6.30	8.8 (0 °C)
nC_5H_{12}	162–273	19.6	7.97	9.2 (10 °C)

Data from Ref. [18].

Here, V_0 represents the molar volume at "zero" pressure, while P_0, A and B are (positive) material-specific parameters. Although there is little theoretical justification, the Tait equation in the integrated form represents the behavior of many types of liquid quite well up to pressures of about 1000 bar. Some values for the Tait parameters are given in Table 4.1. A simple empirical expression for the compressibility κ_T reads $\kappa_T = \kappa_0(V/V_0)^\gamma$. The effect of volume on the Gibbs energy can be estimated by integrating $\mathrm{d}G = V\mathrm{d}P - S\mathrm{d}T$ at constant T, meanwhile realizing that $V\mathrm{d}P = -\kappa_T^{-1}\mathrm{d}V$. This leads to

$$\Delta G = [V/\kappa_T(\gamma-1)][1-(V/V_0)^{\gamma-1}] \tag{4.4}$$

Similarly, the temperature dependence of κ_T can be described empirically

$$\kappa_T = \kappa_0 \exp(\beta T) \tag{4.5}$$

with κ_0 and β constants. Some data for κ_0, β and γ are given in Table 4.2. Water behaves anomalously, however, and κ_T passes through a minimum at ~45 °C. For the molar volume V_{sat} of a saturated liquid–that is, a liquid in equilibrium with its vapor–several other empirical equations exist.

It will be clear that the various equations of state describe either the behavior of the gas or the liquid. As early as 1873, *van der Waals* emphasized the continuity between the liquid and the gas state and advocated the EoS named after him [2]. This EoS describes liquids and gases, and for n moles it reads

$$(P + n^2a/V^2)(V - nb) = nRT \quad \text{or} \quad (P + a\rho^2)(1 - b\rho) = \rho RT \tag{4.6}$$

with $\rho = N/V$ the number density. The parameters a and b are material-specific. The intuitive interpretation for a is that there is attraction between the molecules so that part of the pressure expected from the perfect gas law is reduced in overcoming the force of the intermolecular attraction. For b, one considers that a

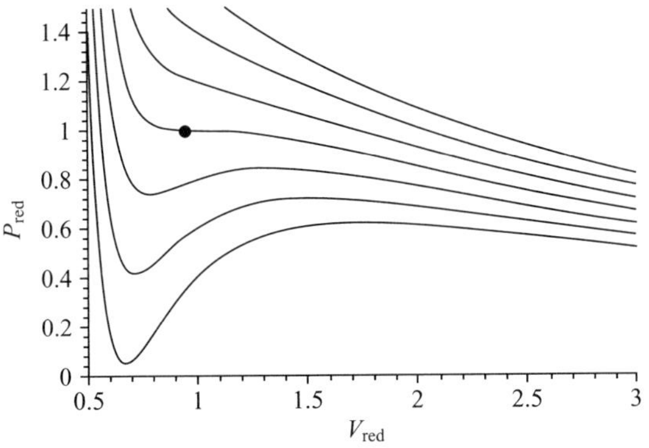

Figure 4.3 The van der Waals equation of state, as illustrated by isotherms in the *PV*-plane for T_{red} = 0.85, 0.90, 0.95, 1.00, 1.05, 1.10, and 1.15. The critical point is indicated by •.

molecule is not a point mass but rather has a certain radius, so that there is a certain excluded volume given by b. Since for a rigid particle with radius $\sigma/2$ the volume that cannot be occupied by another particle is $4\pi\sigma^3/3$, one finds $2b = 4\pi\sigma^3/3$. Figure 4.3 illustrates the van der Waals (vdW) behavior. For low densities, the vdW expression reduces to the perfect gas law, but for high densities sinuous curves are obtained that represent an unstable state of affairs. The usual interpretation is that the region of the sinuous curves corresponds to a separation of the fluid into liquid and vapor, and that the equation breaks down here. Maxwell argued that a horizontal line should be inserted so as to make the areas below and above that line equal (see Chapter 16). In this way, a general resemblance to the actual isotherms of a real fluid is obtained. For mixtures it is often assumed that $a_{ij} = (a_i a_j)^{1/2}$ and $b_{ij} = (b_i + b_j)/2$.

4.3
Corresponding States

The CP is characterized by $(\partial P/\partial V)_T = (\partial^2 P/\partial V^2)_T = 0$, and this leads for the vdW equation to (with V the molar volume V_m)

$$V_{cri} = 3b, \quad T_{cri} = 8a/27Rb \quad \text{and} \quad P_{cri} = a/27b^2 \tag{4.7}$$

Defining the reduced quantities

$$V_{red} = V/V_{cri}, \quad T_{red} = T/T_{cri} \quad \text{and} \quad P_{red} = P/P_{cri} \tag{4.8}$$

we can write the EoS in a nondimensional form, usually called the *reduced vdW equation of state*, and given by

$$(P_{red} + 3/V_{red}^2)(3V_{red} - 1) = 8T_{red} \tag{4.9}$$

As this reduced equation contains no constants characteristic of a particular fluid, the result is (theoretically) valid for all fluids, and represents an example of the

Table 4.3 Critical reduced compression factor and surface tension for several simple molecules.

Molecule	$Z_{cri,red}$	$\gamma_0 V_{cri}^{2/3}/T_{cri}$	Molecule	$Z_{cri,red}$	$\gamma_0 V_{cri}^{2/3}/T_{cri}$
Ne	0.305	4.05	N_2	0.292	4.48
Ar	0.292	4.25	O_2	0.292	4.4
Kr	0.290	–	CO	0.294	–
Xe	0.288	4.2	CH_4	0.289	4.39

Data from Ref. [19].

principle of corresponding states (PoCS). Quite apart from the applicability of the reduced vdW EoS, the PoCS is an experimental fact which asserts that, for a group of similar substances, the EoS can be written in the form

$$P/P_{cri} = \Omega(V/V_{cri}, T/T_{cri}) \tag{4.10}$$

where Ω is the same function for all the substances of the group.

In order to estimate a and b, the values P_{cri} and T_{cri} (see Appendix E) can be used. Another way of determining a and b is to fit P to the vdW equation for a range of temperatures. It will be clear that different procedures yield (slightly) different values. The (approximate) validity of the PoCS is indicated by the fact that the compression factor at critical conditions $Z_{cri} = P_{cri}V_{cri}/RT_{cri}$ is approximately constant for a set of spherical, nonpolar molecules at (0.293 ± 0.005), where $\pm$ indicates the sample standard deviation (Table 4.3). Note that for the vdW fluid $Z_{cri} = 3/8 = 0.375$. Perhaps more convincing is a plot of Z versus P_{red}, as given in Figure 4.4 for several molecules, which shows that the curves for a certain, fixed T_{red} nicely coincide.

Another example of the PoCS is shown in Figure 4.5, where the second virial coefficient B/V_{cri} versus T/T_{cri} for Ar, Kr, Xe and CH_4 is shown. The curve can be described well by the expression as derived for the square-well potential (see Chapter 3) with a hard-core diameter σ, a depth ε and a range 1.5σ, and is given by

$$B/V_{cri} = 0.440 + 1.40[1 - \exp(0.75T_{cri} / T)] \tag{4.11}$$

The parameters σ and ε are given by $\frac{2}{3}\pi N_A\sigma^3 = 0.447V_{cri}$ and $\varepsilon = 0.936kT_{cri}$. As a final example, Figure 4.5 also shows ρ/ρ_{cri} versus T/T_{cri}. The curve is well fitted by

$$(\rho_L + \rho_G)/2\rho_{cri} = 1 + 3(1 - T/T_{cri})/4 \quad \text{and} \quad (\rho_L - \rho_G)/\rho_{cri} = 7(1 - T/T_{cri})^{1/3}/2 \tag{4.12}$$

where ρ_L and ρ_G represent the densities of the liquid and gas, respectively. The first of these expressions represents the *law of rectilinear diameters*, which states that the average density of gas and liquid is a linear function of its temperature.

The PoCS implies that the intermolecular potential has a similar shape for the molecules within the class of molecules considered. For a two-parameter intermolecular potential of the form $\phi(r) = \varepsilon f(r/\sigma)$, the distances are scaled by σ, while the energy is scaled by ε. Consequently, we can scale the pressure P according to

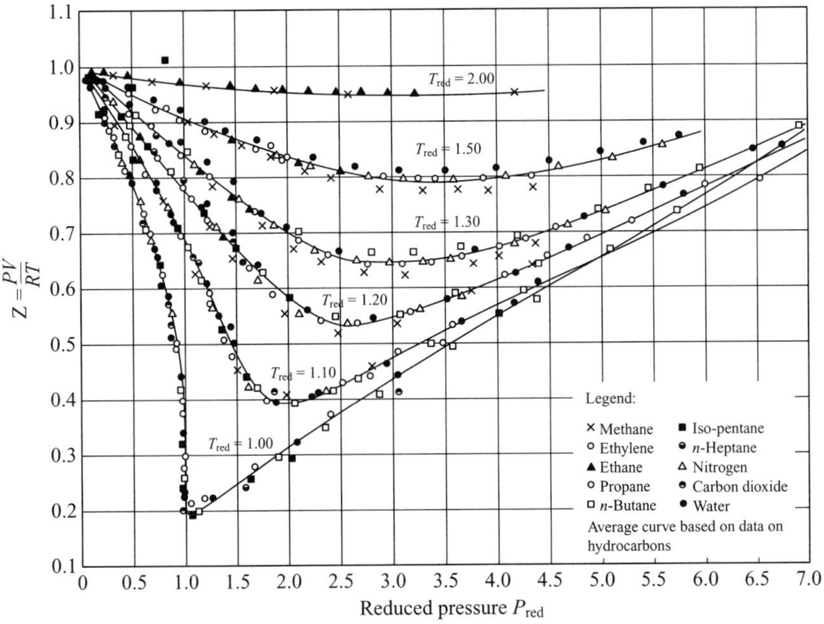

Figure 4.4 Compression factor Z as a function of the reduced pressure for various fluids, illustrating the principle of corresponding states. Data from Ref. [17].

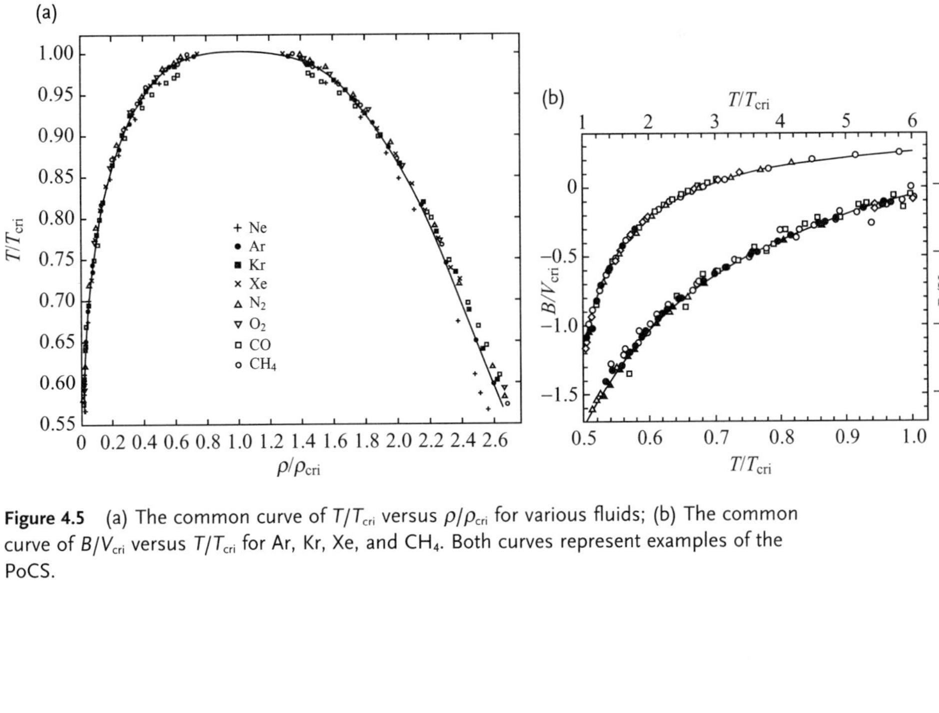

Figure 4.5 (a) The common curve of T/T_{cri} versus ρ/ρ_{cri} for various fluids; (b) The common curve of B/V_{cri} versus T/T_{cri} for Ar, Kr, Xe, and CH_4. Both curves represent examples of the PoCS.

$P^* = \sigma^3 P/\varepsilon$, the volume V according to $V^* = V/\sigma^3$, and the temperature T according to $T^* = kT/\varepsilon$. Finally, the configurational partition function (see Chapter 5) scales according to $Q^* = Q/\sigma^{3N}$. In this way, the pressure becomes

$$P = kT\left(\frac{\partial \ln Z}{\partial V}\right)_T = kT\left(\frac{\partial \ln Q}{\partial V}\right)_T \rightarrow P^*(T^*,V^*) = T^*\left(\frac{\partial \ln Q^*}{\partial V^*}\right) \tag{4.13}$$

In this dimensionless representation, the P–V curves of a set of molecules should coincide for the same T^*, as illustrated in Figure 4.4 using T_{red}.

4.3.1 Extended Principle*

The conventional corresponding state approach deals with two parameters, for example, using the compression factor $Z = Z(T_{red},P_{red})$. For simple, (more or less) spherical molecules this approach works reasonably well, as evidenced by the "universal" value of the compression factor Z [3]. In fact, any EoS that contains only two parameters can be mathematically reduced likewise. However, the reduction is essentially based on dimensional analysis with force, length and temperature as basic dimensions. In the case of categories of molecules for which extra dimensions are important, one does not expect the principle to apply. A significant improvement in describing experimental results can be obtained by extending the concept to three parameters, of which the best known approaches are due to Riedel [4] and Pitzer [5]. Both approaches use the pressure P versus temperature T. Riedel used the parameter $\alpha = (\mathrm{d}\ln P/\mathrm{d}\ln T)_{T=T_{cri}}$ taken along the saturation curve, while Pitzer *et al.* used a parameter that can be obtained if we plot the logarithm of the saturated vapor pressure in reduced units versus $1/T_{red}$. In this representation, approximately straight lines are obtained, that is, $\mathrm{d}\log P_{red}^{sat}/\mathrm{d}(1/T_{red}) = S$. The two-parameter principle predicts that the slope S is the same for all substances. If we plot the vapor pressures of simple fluids (spherical molecules such as Ar, CH_4, . . .), they indeed all fall more or less on the same line that appears to go through $\log P_{red}^{sat} = -1.0$ at $T_{red} = 0.7$. However, other, less-spherical molecules do have different slopes. These facts are used to define the so-called *acentric factor* $\omega = -1.0 - \log P_{red}^{sat}(T_{red} = 0.7)$ (Figure 4.6). This apparently arbitrary definition was

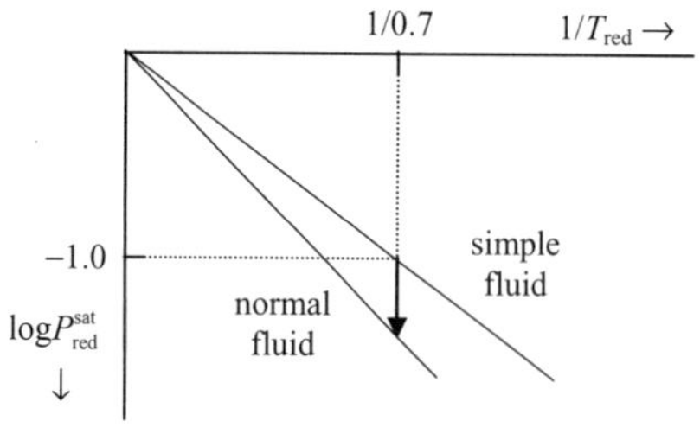

Figure 4.6 The acentric factor ω as the deviation of the slope of the vapor pressure of a normal from the fluid slope for simple fluids. Note: log = logarithm to base 10.

chosen because of the ease and precision with which this quantity can be determined (contrary to Riedel's α, which might be difficult to determine precisely). Since at T/T_{cri} = 0.7 the vapor pressure of simple fluids is close to 0.1 P_{cri}, the acentric factor is essentially zero for these compounds. The relation of the acentric factor ω to Riedel's α reads $\omega = 4.93(\alpha - 5.808)$, and therefore we limit ourselves further on to ω. The acentric factor ω is given for a variety of compounds in Appendix E [6].

The simplest correlation [7] for normal fluids is

$$Z_{cri,red} = P_{cri}V_{cri}/RT_{cri} = 0.2905 - 0.0787\omega \tag{4.14}$$

which correlates the reduced compression factor Z at critical conditions with the acentric factor ω. Pitzer *et al.* showed that, if we write

$$Z = 1 + (B_2P/RT) = 1 + (\hat{B}P_{red}/T_{red}) \tag{4.15}$$

the second virial coefficient can be represented by

$$\hat{B} = B_2P_{cri}/RT_{cri} = B^{(0)} + \omega B^{(1)} \tag{4.16}$$

where $B^{(0)}$ and $B^{(1)}$ are functions of T_{red}, reasonably well described by

$$B^{(0)} = 0.083 - \left(0.422/T_{red}^{1.6}\right) \quad \text{and} \quad B^{(1)} = 0.139 - \left(0.172/T_{red}^{4.2}\right) \tag{4.17}$$

We thus have $Z = Z(T_{red}, P_{red}, \omega)$. More generally, the compression factor can be accurately described by $Z = Z^{(0)} + \omega Z^{(1)}$, where $Z^{(0)}$ and $Z^{(1)}$ are complex, extensively tabulated [8] functions of T_{red}. Alternatively, one uses a fit for ln $P_{red} = f^{(0)} + f^{(1)}\omega$ reading

$$f^{(0)} = 5.92714 - (6.09648/T_{red}) - 1.28862\ln T_{red} + 0.169347T_{red}^6 \tag{4.18}$$

$$f^{(1)} = 15.2518 - (15.6875/T_{red}) - 13.4721\ln T_{red} + 0.43577T_{red}^6 \tag{4.19}$$

For nonpolar fluids above 1 bar, the accuracy for Z is about 2–3%, but for more polar molecules and low pressures the calculated pressure is typically too low and the deviations are larger [6], perhaps up to 10%. It should be noted that these correlations typically apply only to normal fluids, as defined in Section 4.1.

Returning to the topic of the EoS, we note that other equations of state exist, for example, the *Berthelot* and *Dieterici* equations, given by, respectively,

$$\left(P + a/TV^2\right)(V - b) = RT \quad \text{and} \quad P(V - b) = \exp(-a/RTV)RT \tag{4.20}$$

As both equations contain only two parameters, they can both be scaled to a reduced equation of state, much like the vdW equation; however, similar to the vdW equation they are not very accurate.

More complex EoS are available, for example, the Redlich–Kwong (RK) [9], Redlich–Kwong–Soave (RKS) [10], Peng–Robinson (PR) [11], and Benedict–Webb–Rubin (BWR) [12] equations. These equations use as a general form

$$P = \frac{RT\rho}{1 - b\rho} - \frac{a\alpha(T)\rho^2}{f(T)} \tag{4.21}$$

The original RK equation uses $\alpha(T) = T^{-1/2}$ and $f(\rho) = 1 + b\rho$, still obeying the two-parameter PoCS. The RKS equation differentiates between various normal liquids by using the acentric factor in the expression for $\alpha(T)$, namely $\alpha(T) = [1 + m(1 - T_{\text{red}}^{1/2})]^2$ with $m = 0.480 + 1.5574\omega - \omega^2$ in conjunction with $f(\rho)$. The PR equation also modifies $f(\rho)$ by using $f(\rho) = 1 + 2b\rho - b^2\rho^2$, but with a slightly modified $m = 0.37464 + 1.54226\omega - 0.26992\omega^2$. It appears that the RKS EoS is better suited to small values of ω, while the PR is better equipped for larger values of ω. Finally, we should mention the BWR equation and its modifications, in which the vdW repulsive term is replaced by a polynomial and exponential expression, the form being chosen also to be easily integrable to yield the Helmholtz energy. In its original form, the BWR equation uses eight parameters, but in order to maintain accuracy above $1.8\rho_{\text{cri}}$ it has been modified extensively. One particular attempt included 33 parameters [13].

A compromise between accuracy and complexity is the general cubic EoS [14] of which the vdW equation is the prototype, although cubic equations of state have limited flexibility [15]. This equation reads for one mole (omitting the subscript, m)

$$P = \frac{RT}{V-b} - \frac{\theta(V-\eta)}{(V+b)(V^2+\kappa V+\lambda)} \tag{4.22}$$

with θ, η, κ, λ, and b as parameters. The vdW equation is obtained if $\eta = b$, $\theta = a$, and $\kappa = \lambda = 0$. An often-used class is given by $\eta = b$, $\theta = a(T)$, $\kappa = (\varepsilon + \sigma)b$ and $\lambda = \varepsilon\sigma b^2$. In that case, the cubic equation reads

$$P = \frac{RT}{V-b} - \frac{a(T)}{(V+\varepsilon b)(V+\sigma b)} \tag{4.23}$$

For this equation the parameters ε and σ have the same value for all substances in the class, while $a(T)$ and $b(T)$ are substance-dependent. The vdW equation is obtained by substituting $a(T) = a$ and $\varepsilon = \sigma = 0$.

All of these equations can be developed in a power series in V in order to obtain the associated virial coefficients. For example, rewriting the van der Waals equation as

$$PV/RT = (1 - b/V)^{-1} - (a/RTV) \tag{4.24}$$

and expanding the first term on the right-hand side by the binomial series, we obtain

$$PV/RT = 1 + (b/V) + (b^2/V^2) + \cdots - (a/RTV) \quad \text{or} \tag{4.25}$$

$$B_2 = b - a/RT, \quad B_3 = b^2, \quad B_4 = b^3, \quad \text{etc.} \tag{4.26}$$

A similar procedure may be applied to other EoS. Experiments have shown that $B_2 < 0$ at low temperature, becomes positive at higher temperature, reaches a maximum, and then decreases. The vdW equation gives, correctly, $B_2 < 0$ for low temperature, changing to $B_2 > 0$ at higher temperature, but does not predict a maximum.

The virial approach, however, can be applied without reference to any analytical empirical EoS by describing the experimentally determined pressure as a function of density directly with the virial series. It appears that this, initially fully empirical approach, can be rationalized by statistical thermodynamics, as shown in Chapter 5.

Problem 4.1

Discuss why the pressure correction in the vdW equation is proportional to $(n/V)^2$. Also derive the hard sphere volume correction $b = 4N_A[4\pi(\sigma/2)^3/3]$.

Problem 4.2

Show that for the vdW fluid the Boyle temperature T_B, where the slightly nonideal fluid reacts as an ideal fluid and thus where $B_2(T_B) = 0$, is given by $T_B = a/Rb$. Also show that the Joule–Thomson inversion temperature, characterizing the temperature T_{JT} for which the Joule–Thomson coefficient $\mu_{JT} \equiv (\partial T/\partial P)_H = V(\alpha T - 1)/C_P$ changes sign, hence the cooling changes to heating and vice versa, and where $\partial(B_2/T)/\partial T = 0$, is given by $T_{JT} = 2T_B$.

Problem 4.3

Show, by integrating the pressure P over the volume V, that the vdW expression $P = RT/(V - b) - a/V^2$ yields the Helmholtz energy $F(T,V) - F°(T,V) = - RT\ln[(V - b)/V°] - a/V$, the entropy $S(T,V) - S°(T,V) = R\ln[(V - b)/V°]$, and the internal energy $U(T,V) - U°(T,V) = -a/V$, where $F°$, $S°$ and $U°$ refer to the perfect gas corresponding properties.

Problem 4.4

Show that the second and third virial coefficients of the EoS indicated are

Berthelot:	$B_2 = b - a/RT^2$	$B_3 = b^2$
Dieterici:	$B_2 = b - a/RT$	$B_3 = b^2 - ab/RT + a^2/2R^2T^2$

Problem 4.5

Show that, if the virial coefficient is given by B_n, the coefficient in the corresponding Helmholtz expression reads $B_n^F = RTVB_n/(n-1)$.

Problem 4.6*

Show that the critical points for the EoSs and the values for Z_{cri} indicated are:

Berthelot:	$P_{cri} = (aR/216b^3)^{1/3}$	$V_{cri} = 3b$	$T_{cri} = (a/27Rb)^{1/2}$	$Z_{cri} = 2/e^2$
Dieterici:	$P_{cri} = a/4e^2b^2$	$V_{cri} = 2b$	$T_{cri} = a/4Rb$	$Z_{cri} = 3/8$

Problem 4.7*

An equivalent procedure to obtain the vdW parameters is to use $(V - V_{cri})^3 = 0$, since at the critical point $V = V_{cri}$ and the vdW equation is a cubic equation in V. Compare the expansion of this equation with the standard vdW equation in its polynomial form $V^3 - \alpha V^2 + \beta V - \gamma = 0$. Show that $3V_{cri} = b + RT_{cri}/P_{cri}$, $3V_{cri}^2 = a/P_{cri}$, and $V_{cri}^3 = ab/P_{cri}$. Show that from the last two equations it follows that $a = 3P_{cri}V_{cri}^2$ and $b = V_{cri}/3$. Also show that substitution of the expression for b in the first of these three equations allows solution for V_{cri}, which then can be eliminated from a and b to obtain $V_{cri} = 3RT_{cri}/8P_{cri}$, $a = 27R^2T_{cri}^2/64P_{cri}$, and $b = RT_{cri}/8P_{cri}$.

Problem 4.8*

Show that $\dfrac{\rho_{vdW}}{\rho_{idg}} = 1 + \dfrac{3PT_{cri}}{8P_{cri}T}\left(\dfrac{9T_{cri}}{8T} - \dfrac{1}{3}\right)$, where $\rho = N/V$ is the number density, and estimate its value for Ar, N_2, and O_2 at 0 °C and 1 atm. How large is this ratio for H_2O at its equilibrium vapor pressure of 6.11 mbar at 0 °C?

References

1 (a) Tait, P.G. (1889) *Voyages of HMS Challenger*, vol. 2, part 4, HMSO, London, p. 1; (b) Dymond, J.H. and Malhotra, R. (1988) *Int. J. Thermophys.*, **9**, 941.

2 van der Waals, J.D. (1873) *Over de Continuïteit van den Gas- en Vloeistoftoestand*. Thesis.

3 Pitzer, K.S. (1929) *J. Chem. Phys.*, **7**, 583.

4 (a) Riedel, L. (1954) *Chem.-Ing. Tech. Z.*, **26**, 83, 259, 679; (b) Riedel, L. (1955) *Chem.-Ing. Tech. Z.*, **27**, 209, 475; (c) Riedel, L. (1956) *Chem.-Ing. Tech. Z.*, **28**, 557.

5 (a) Pitzer, K.S., Lippmann, D.Z., Curl, R.F., Jr, Huggins, C.M., and Petersen, D.E. (1955) *J. Am. Chem. Soc.*, **77**, 3433; (b) Pitzer, K.S. (1995) Appendix 3, in *Thermodynamics*, McGraw-Hill, New York.

6 For an extensive list, Reid, R.C., Prausnitz, J.M., and Poling, B.E. (1988) *The Properties of Gases and Liquids*, 4th edn, McGraw-Hill.

7 Schreiber, D.R. and Pitzer, K.S. (1989) *Fluid Phase Eq.*, **46**, 113.

8 Lee, B.I. and Kessler, M.G. (1975) *AIChE J.*, **21**, 510.

9 Redlich, O. and Kwong, J.N.S. (1949) *Chem. Rev.*, **44**, 233.

10 Soave, G. (1972) *Chem. Eng. Sci.*, **27**, 1197.

11 Peng, D.-Y. and Robinson, D.B. (1976) *Ind. Eng. Chem. Fundam.*, **15**, 59.

12 (a) Benedict, M., Webb, G.B., and Rubin, L.C. (1940) *J. Chem. Phys.*, **8**, 334; (b) Benedict, M., Webb, G.B., and Rubin, L.C. (1942) *J. Chem. Phys.*, **10**, 747.

13 Younglove, B.A. and Ely, J.F. (1987) *J. Phys. Chem. Ref. Data*, **16**, 577.

14 Valderrama, J.O. (2003) *Ind. Eng. Chem. Res.*, **42**, 1603.

15 (a) Abbott, M.M. (1973) *AIChE J.*, **19**, 596; (b) Chao, K.C and Robinson, R.C., Jr (eds) (1979) *Advances in Chemistry*, vol. 182, American Chemical Society, Washington, DC, p. 47.

16 Finn, C.B.P. (1993) *Thermal Physics*, Chapman & Hall, London.

17 Sen, G.-J. (1946) *Ind. Eng. Chem.*, **38**, 803.

18 Moelwyn-Hughes, E.A. (1961) *Physical Chemistry*, 2nd edn, Pergamon, Oxford.

19 Guggenheim, E.A. (1967) *Thermodynamics*, 5th edn, North-Holland, Amsterdam.

Further Reading

Murrell, J.N. and Jenkins, A.D. (1994) *Properties of Liquids and Solutions*, 2nd edn, John Wiley & Sons, Ltd, Chichester.

Pryde, J.A. (1966) *The Liquid State*, Hutchinson University Library, London.

Rowlinson, J.S. and Swinton, F.L. (1982) *Liquids and Liquid Mixtures*, 3rd edn, Butterworth, London.

5
The Transition from Microscopic to Macroscopic: Statistical Thermodynamics

As indicated in Chapter 1, in liquids a natural reference state is absent because the structure is highly irregular. Therefore, in the description of liquids and solutions, we essentially will need statistical thermodynamics, which provides the link between the microscopic considerations and concepts to the macroscopic – in particular thermodynamic – aspects. In this chapter we introduce and outline statistical thermodynamics in a concise way. First, we deal with the basic formalism and apply that to noninteracting molecules. Thereafter, the effect of intermolecular interactions is introduced, leading to the virial description. While being a reasonable approach for low- and medium-density fluids, the virial approach does not work for liquids. The reasons for this are briefly discussed.

5.1
Statistical Thermodynamics

While classical or phenomenological thermodynamics provides relations between macroscopic properties, statistical thermodynamics relates the macroscopic properties of a system (set of molecules) to the microscopic phenomena – that is, to molecular behavior.

5.1.1
Some Concepts

Let us try to describe a system by classical mechanics (CM). If the system is relatively small, say a single molecule or particle, one may proceed as follows. In classical mechanics the system is described by *Hamilton's equations* (see Section 2.2). Each system can be characterized by a set of n generalized coordinates (degrees of freedom) and the associated momenta. A state of a system thus can be depicted as a point in a $2n$-dimensional (Cartesian) space whose axes are labeled by the allowed momenta and coordinates of the particles of the system. This space is called the (molecule) *phase space* or μ-space. Example 2.2 provides a possible choice of generalized coordinates for the water molecule. If we have a system containing many (not or weakly interacting) particles, such as molecules in a

Liquid-State Physical Chemistry, First Edition. Gijsbertus de With.

volume of gas, a collective of points describes the gas in μ-space, each point describing a molecule. Similar considerations hold in quantum mechanics (QM), which yields the energy levels and associated wave functions of individual quantum systems. However, in many cases we need the time average behavior of a macroscopic system. There are four problems to this:

- The size of the macroscopic system, which contains a large number of particles. The large number of degrees of freedom present renders a general solution (both in CM and in QM) highly unlikely to be found.
- The initial conditions of such systems are unknown so that, even if a general solution was possible in principle, a particular solution cannot be obtained.
- Even if the initial conditions were known, they have a limited accuracy. Note that, for example, Avogadro's number is "only" known with an accuracy of 10^{-7}. Since it appears that the relevant equations are extremely sensitive to small changes in initial conditions, this rapidly leads to chaotic behavior.
- It is difficult to incorporate interaction between the molecules and the interaction with environment.

To overcome the aforementioned problems, in 1902 Gibbs made a major step and used the *ensemble*–a large collection of identical systems with the same Hamilton function but different initial conditions. By taking a $2nN$-dimensional space, where N is the total number of particles in the system, we can make a similar representation as before. The axes are labeled with the allowed momenta and coordinates of all the particles. This enlarged space is called the (gas) *phase space* or Γ-space. To each system in the ensemble there corresponds a *representative point* in the $2nN$-dimensional Γ-space. Since the system evolves in time, the representative point will describe a path as a function of time in Γ-space, and this path is called a *trajectory*. From general considerations of differential equations it follows that through each point in phase space (or phase point) can pass one and only one trajectory so that trajectories do not cross. The ensemble thus can be depicted as a swirl of points in Γ-space, and the macroscopic state of a system is described by the average behavior of this swirl. Since the macroscopic system considered has a large number of degrees of freedom, the density of phase points can be considered as continuous in many cases; this density is denoted by $\rho(\mathbf{p},\mathbf{q})$, where $\mathbf{p}$ and $\mathbf{q}$ denote the collective of momenta and coordinates, respectively. Obviously, $\rho(\mathbf{p},\mathbf{q}) \geq 0$ and we take it normalized, that is, $\int \rho(\mathbf{p},\mathbf{q})\mathrm{d}\,\mathbf{p}\,\mathrm{d}\,\mathbf{q} = 1$. One can prove from the equations of motion that this density is constant if we follow along with the swirl, a general theorem called *Liouville's theorem*. A function $F(\mathbf{p},\mathbf{q})$ in phase space representing a certain property is called a *phase function*, the Hamilton function $\mathcal{H}$ providing an important example. The time-average of a phase function is given by

$$\langle F \rangle_t = \lim_{t\to\infty} \frac{1}{t} \int_0^t F[\mathbf{p}(t'),\mathbf{q}(t')]\mathrm{d}t' \tag{5.1}$$

However, as stated before, this average cannot be calculated in general since neither the solution $\mathbf{p} = \mathbf{p}(t)$ and $\mathbf{q} = \mathbf{q}(t)$ nor the initial conditions $\mathbf{p} = \mathbf{p}(0)$ and

$\mathbf{q} = \mathbf{q}(0)$ are known. To obtain nevertheless estimates of properties the phase-average

$$\langle F\rangle_\Gamma = \int_\Gamma F(\mathbf{p},\mathbf{q})\rho(\mathbf{p},\mathbf{q})\mathrm{d}\mathbf{p}\,\mathrm{d}\mathbf{q} \tag{5.2}$$

is introduced. The assumption, originally introduced by Boltzmann in 1887 for μ-space and known as the *ergodic theorem*, is now that

$$\langle F\rangle_\Gamma = \langle F\rangle_t \tag{5.3}$$

and essentially implies that each trajectory visits each infinitesimal volume element of phase space. This assumption has been proven false but can be replaced by the *quasi-ergodic theorem*, proved by Birkhoff in 1931 for metrically transitive systems[1]. It states that trajectories approach all phase points as closely as desired, given sufficient time to the system. In practice, this means that we accept the equivalence of the time- and phase-average. For details (including the conditions for which the theorem is not valid) we refer to the literature.

5.1.2
Entropy and Partition Functions

We will develop statistical thermodynamics using a quantum description, since that approach is conceptually simpler, and in a later stage make the transition to classical statistical thermodynamics. However, the terminology of classical statistical mechanics is frequently used in the literature, even when dealing with quantum systems.

The first question is how to characterize a macroscopic state in terms of molecular parameters. It might be useful to recall that a macroscopic system is considered to possess a *thermodynamic* or *macro-state*, characterized by a limited set of macroscopic parameters, such as the volume V and the temperature T. However, such a system contains many particles, for example, atoms, molecules, electrons and photons. A macroscopic system is thus also a quantum system – that is, an object that contains energy and particles and is identified by a large set of distinct *quantum* or *micro-states* of the system. Identical particles – for example, the electrons of a molecule – are objects that have access to the same subset of micro-states. From these considerations we conclude that *each macro-state contains many micro-states*, and describe the macroscopic state of the system by the fraction p_i of each possible quantum state i in which the macroscopic system remains. With each quantum state i an energy E_i and a number of particles N_i is associated. It should be clear that the simple label i implicates the complete description of the micro-state while the distribution over microstates i describes the macroscopic system, and thus the whole hides a considerable complexity.

Further, it should be clear that the number of quantum states increases rapidly with increasing energy. For large systems, the states of the system will form a

1) A mechanical system is metrically transitive if the energy surface cannot be divided into two finite regions such that orbits starting from points in one region always remain in that region. See Ref. [1].

quasi-continuum that can be described by the *density of states* g, indicating the number of quantum states $\mathrm{d}n$ for a certain energy range δE, that is,

$$g(E)\delta E = (\mathrm{d}n/\mathrm{d}E)\delta E \tag{5.4}$$

where δE represents the so-called *accessibility range*, that is, an energy range that is small as compared to E but large when compared to the Heisenberg uncertainty ΔE. For a macroscopic system g is astronomically large, is related to the temperature (see Problem 5.4), and increases extremely rapidly with the size of the system (see Example 5.1).

Example 5.1: The density of states for translation*

First, we calculate the volume G of a hypersphere with radius R and of dimension D, given by $G = AR^D = A(R^2)^{D/2}$. To calculate A we consider the integral

$$I = \int \mathrm{e}^{-R^2} \frac{\mathrm{d}G}{\mathrm{d}R^2}\mathrm{d}R^2 = \frac{1}{2}AD\int \mathrm{e}^{-R^2}(R^2)^{D/2-1}\,\mathrm{d}R^2 = \frac{1}{2}AD\Gamma(D/2)$$

with $\Gamma(t)$ the gamma function. The volume can also be written as $G = \int h(R^2 - \mathbf{x}^\mathrm{T}\mathbf{x})\,\mathrm{d}\mathbf{x}$ with $h(t)$ the Heaviside function and $\mathbf{x}^\mathrm{T}\mathbf{x} = \Sigma_i x_i^2$. So, $\mathrm{d}G/\mathrm{d}R^2 = \int \delta(R^2 - \mathbf{x}^\mathrm{T}\mathbf{x})\mathrm{d}\mathbf{x}$ where the Dirac function $\delta(t) = \mathrm{d}h(t)/\mathrm{d}t$ is used. For the functions used, see Appendix B. Therefore, the integral I also becomes

$$I = \int\int \exp(-R^2)\delta(R^2 - \mathbf{x}^\mathrm{T}\mathbf{x})\mathrm{d}R^2\mathrm{d}\mathbf{x} = \int \exp(-\mathbf{x}^\mathrm{T}\mathbf{x})\,\mathrm{d}\mathbf{x} = \pi^{D/2}$$

Combining yields $A = 2\pi^{D/2}/D\Gamma(D/2) = \pi^{D/2}/\Gamma(½D + 1)$. Consider now N particles of mass m in a three-dimensional box with edges a so that the energy E can be written as $E = (h^2/8ma^2)\mathbf{n}^\mathrm{T}\mathbf{n} = (h^2/8ma^2)\Sigma_i n_i^2$ where $\mathbf{n} = n_i$ covers the set n_{1x}, n_{1y}, . . . , n_{Nz} with $n_i \in (\pm1, \pm2, \pm3, \ldots)$. The degeneracy of energy levels is given by the number of ways the integer $\mathbf{n}^\mathrm{T}\mathbf{n} = 8ma^2E/h^2$ can be written as a sum of squares of integers. To estimate the degeneracy, consider a D-dimensional space ($D = 3N$) with the axes labeled $\mathbf{n}$. The number of lattice points inside the hypersphere with radius squared $\mathbf{n}^\mathrm{T}\mathbf{n} = 8ma^2E/h^2$, gives the number of energy levels. Equating $\mathbf{n}^\mathrm{T}\mathbf{n}$ to R^2, we have

$$G = A\left(\frac{8ma^2E}{h^2}\right)^{D/2} = \frac{\pi^{D/2}}{\Gamma\left(\frac{1}{2}D+1\right)}\left(\frac{8ma^2E}{h^2}\right)^{D/2} = \frac{2^D}{\Gamma\left(\frac{1}{2}D+1\right)}\left(\frac{2\pi ma^2E}{h^2}\right)^{D/2}$$

with $D = 3N$. The final step is to realize that we need only one "octant" of the hypersphere, so we have to divide by 2^D, and that the particles are indistinguishable, so we have to divide by $N!$ (see Section 5.2). The final result for the volume G and density of states $g(E)$ thus becomes, respectively,

$$G = (2\pi ma^2E/h^2)^{D/2}\Big/N!G\left(\frac{1}{2}D+1\right) \quad \text{and} \quad g(E) = \mathrm{d}G(E)/\mathrm{d}E \sim E^{(3N/2)-1}$$

For $N = 1$, using $E = 3kT/2$, $T = 300\,\text{K}$, $m = 10^{-22}\,\text{g}$, and $a = 10^{-2}\,\text{m}$, we obtain $g \cong 10^{30}$, which is a large number. For N particles, using again $T = 300\,\text{K}$, $m = 10^{-22}\,\text{g}$ and $a = 10^{-2}\,\text{m}$, but now with $N = 6 \times 10^{23}$ and $E = 3NkT/2$, we obtain $g \cong 10^{N}$, which is an extremely large number. Finally, the expression for g shows that it increases extremely rapidly with the number of particles in the system.

Since in thermodynamics equilibrium is reached at maximum entropy $S(U,V,N)$, given energy U, volume V and number of particles N, it seems natural in statistical thermodynamics to start with defining the entropy. For the statistical representation of S various expressions can be given[2]. Since the macroscopic system is characterized by the p_is, we will have $S = S(p_i; U,V,N)$, or $S(p_i)$ for short. Probably the most direct way to obtain S for a system having n states is to require the following properties:

- $S(p_i) \le S(1/n)$. This statement implies that equi-probability for states with the same energy yield the maximum entropy and thus the equilibrium state.
- $S(p_i,0) = S(p_i)$. This statement requires that if a state cannot be occupied it will not contribute to the entropy.
- $S = S_A + S_B$ for noninteracting systems A and B. This condition delivers the usual additivity of entropy. For interacting systems the condition changes to $S = S_A + S_B^{(A)} \le S_A + S_B$, where $S_B^{(A)}$ denotes the conditional entropy of system B, that is, the entropy of system B given the entropy of system A.

These requirements have a unique solution (see Justification 5.1) apart from a multiplicative constant. This constant is chosen in such a way that the statistical entropy thus defined corresponds to the conventional thermodynamic entropy, and this constant appears to be Boltzmann's constant k. The *entropy* S, sometimes called Gibbs entropy, is then defined by

$$\blacktriangleright \quad S \equiv -k\sum\nolimits_i p_i \ln p_i \tag{5.5}$$

Justification 5.1: The Gibbs entropy*

To show that, given the three properties indicated above, Eq. (5.5) represents the unique solution for the entropy, we argue as follows [2]. First, as an abbreviation, we set

$$L(n) \equiv S(1/n, \ldots, 1/n)$$

Using properties 1 and 2 we have

$$L(n) = S(1/n, \ldots, 1/n, 0) \le S(1/(n+1), \ldots, 1/(n+1)) = L(n+1)$$

2) Formally, the statistical representation of the thermodynamic S should be denoted by another symbol, for example, $\langle S \rangle$. However, we identify right away $\langle S \rangle$ with S.

so that $L(n)$ is a nondecreasing function of n. Now, taking m and r as positive integers, consider m mutually independent systems $A_1, A_2, \ldots, A_m$, each containing r equally likely states. So, we have $S(A_k) = S(1/r, \ldots, 1/r) = L(r)$ for $1 \le k \le m$. Using property 3, since the systems are independent, we have

$$S(A_1A_2 \ldots A_m) = \Sigma_k S(A_k) = mL(r)$$

Since the combined (product) system $A_1A_2 \ldots A_m$ contains r^m equally likely events, we have for its entropy also $L(r^m)$ and therefore

$$L(r^m) = mL(r)$$

This relation holds for any other pair of integers m and r. As a next step, take three arbitrary integers r, s and n and a number m determined by

$$r^m \le s^n \le r^{m+1} \quad \text{or} \quad m \ln r \le n \ln s \le (m+1) \ln r \quad \text{or}$$

$$m/n \le \ln s / \ln r \le (m+1)/n \tag{5.6}$$

Since $L(n)$ is monotonous we have also

$$L(r^m) \le L(s^n) \le L(r^{m+1}) \quad \text{or} \quad mL(r) \le nL(s) \le (m+1)L(r) \quad \text{or}$$

$$m/n \le L(s)/L(r) \le (m+1)/n \tag{5.7}$$

By combining Eqs (5.6) and (5.7) we have

$$|L(s)/L(r) - \ln s/\ln r| \le 1/n \tag{5.8}$$

Equation (5.8) is independent of m, and since n can be arbitrarily large we obtain

$$L(s)/L(r) = \ln s/\ln r$$

which implies, since r and s are arbitrary, that $L(n) = \lambda \ln n$, with $\lambda \ge 0$ a non-negative constant as $L(n)$ is monotonous.

So far, the special case $p_i = 1/n$ $(1 \le k \le n)$ is proved and now we turn to the case where p_k can be any rational number. Suppose we have $p_k = g_k/g$ with $g = \Sigma_k g_k$ where the g_k are all positive integers. Further suppose that we have a system A consisting of n states with probabilities $p_1, p_2, \ldots, p_n$. We also have a specially devised system B, dependent on A and consisting of n groups of states such that the kth group contains g_k states. If state A_k is realized in system A, then in system B all the g_k events of the kth group have the same probability $1/g_k$, while all the events of the other groups have probability zero. In this case the system B reduces to a system of g_k equally probable states with conditional entropy $S_B^{(k)} = S(1/g_k, \ldots, 1/g_k) = L(g_k)$ $= \lambda \ln g_k$. This implies that

$$S_B^{(A)} = \sum_k S_B^{(k)} = \lambda \sum_k p_k \ln g_k = \lambda \sum_k p_k \ln p_k + \lambda \ln g$$

We turn now to the product system AB containing A_kB_l states $(1 \le k \le n, 1 \le l \le g)$. A state in this scheme is only possible if B_l belongs to the kth group, so that the number of possible states A_kB_l for a given k is g_k. The total number of states in system AB is then $\Sigma_k g_k = g$, and the probability of each possible state A_kB_l is

$p_k/g_k = 1/g$. Thus, the system AB consists of g equally likely states and thus $S = L(g) = \lambda \ln g$. We now use property 3, $S = S_A + S_B^{(A)}$, to obtain

$$\lambda \ln g = S_A + \lambda \sum\nolimits_k p_k \ln p_k + \lambda \ln g \quad \text{or} \quad S_A = -\lambda \sum\nolimits_k p_k \ln p_k \tag{5.9}$$

Since the entropy must be continuous, Eq. (5.9) is not only valid for the rational numbers $p_1, p_2, \ldots, p_n$ but for any value of its arguments. Note that, although we used a specially devised system B, the result is independent of system B. Finally, the constant λ can be identified with Boltzmann's constant k by calculating the pressure for a perfect gas, Eq. (5.33). This leads to $\lambda = k$, which we use from now on.

A short route to obtain the relevant thermodynamic expressions runs as follows. We assume an *open system*[3], that is, a system of volume V to be in contact with a thermal bath characterized by the temperature T, and a particle bath characterized by the chemical potential μ[4]. Alternatively, we may assume that the system is in contact with many other, similar systems. For such a system we may assume that we are always near equilibrium. The probabilities of the system ought to be normalized[5], that is,

$$\Sigma_i p_i = 1 \tag{5.10}$$

In addition, we require that the average energy $\langle \mathcal{H} \rangle$ of the system[6] is constant, that is,

$$\Sigma_i p_i E_i = \langle \mathcal{H} \rangle \tag{5.11}$$

and that the average number of particles $\langle N \rangle$ of the system is constant, that is,

$$\Sigma_i p_i N_i = \langle N \rangle \tag{5.12}$$

We use the expression for the entropy considered before, $S = -k\Sigma_i\, p_i \ln p_i$, and for this expression we seek, conform thermodynamics, the maximum given the above-mentioned constraints. This maximum can be obtained by using the Lagrange method of undetermined multipliers (see Appendix B). We take these multipliers as $-k\alpha$, $-k\beta$, and $-k\gamma$. The maximum is now obtained from

$$\frac{\partial}{\partial p_i}\left[S(p_i) - k\alpha\left(\sum\nolimits_i p_i - 1\right) - k\beta\left(\sum\nolimits_i p_i E_i - \langle \mathcal{H} \rangle\right) - k\gamma\left(\sum\nolimits_i p_i N_i - \langle N \rangle\right)\right] = 0 \tag{5.13}$$

3) For simplicity of notation, we assume only one type of particle.
4) This implies that the conjugate variables (pressure P, energy U, and number of particles N) fluctuate.
5) Luckily, the probability of each system being in state i can also be interpreted in classical terms as the fraction of systems in the ensemble in state i. In that case, i refers, of course, to a volume in Γ-space of size h^{nN}, labeling a particular set $(\mathbf{p},\mathbf{q})$.
6) We write $\langle \mathcal{H} \rangle$ since E represents the numerical value of the energy, while the Hamilton operator $\mathcal{H}$ represents the energy expression (see Chapter 2).

and after some calculation this expression leads to the solution

$$p_i^* = \exp[-(1+\alpha+\beta E_i+\gamma N_i)]$$

The normalization condition, Eq. (5.10), yields

$$\sum\nolimits_i p_i^* = \exp[-(1+\alpha)]\sum\nolimits_i \exp[-(\beta E_i+\gamma N_i)] = 1$$

so that we may define

$$\Xi \equiv \sum\nolimits_i \exp[-(\beta E_i+\gamma N_i)] = \exp(1+\alpha)$$

For the probabilities we thus obtain

$$p_i^* \equiv p_i^*(\mu,V,T) = \exp[-(\beta E_i+\gamma N_i)]/\Xi$$

The function Ξ is the macro-canonical or grand canonical partition function, often just called the *grand partition function*. We find for the entropy, meanwhile using the normalization condition $\Sigma_i p_i^* = 1$,

$$S = k\beta\langle \mathcal{H}\rangle + k\gamma\langle N\rangle + k\ln\Xi$$

We compare this expression with the thermodynamic expression for the entropy

$$S = (U-\mu N+PV)/T$$

as solved from $U = TS - PV + \mu N$ (see Section 2.1) with T, μ, and P as before, and U the internal energy, N the number of particles, and V the volume. If we identify the average microscopic energy $\langle \mathcal{H}\rangle$ with the internal energy U, the average number of particles $\langle N\rangle$ with the macroscopic number N, we obtain

$$\beta = 1/kT \quad \gamma = -\mu/kT \quad \text{and} \quad \ln\Xi = PV/kT \tag{5.14}$$

so that we finally have the solution

▶ $$p_i^* = \exp[(\mu N_i - E_i)/kT]/\Xi \tag{5.15}$$

Since we are only using the values of p_i corresponding to the maximum entropy, in the sequel we omit the asterisk and just write p_i. For future reference, we also write the expressions[7] for p_i and Ξ as

$$p_i = \lambda^{N_i}\exp(-\beta E_i)/\Xi \quad \text{and} \tag{5.16}$$

$$\Xi = \sum\nolimits_i \lambda^{N_i}\exp(-\beta E_i) \quad \text{with} \quad \lambda = \exp(-\gamma) = \exp(\beta\mu) \tag{5.17}$$

where λ is the *absolute activity* (see Section 2.1). The grand partition function is directly related to the grand potential Ω, since we have $-PV = \Omega$ and thus $\Omega = -kT \ln \Xi$. The thermodynamic properties can be calculated in the usual thermodynamic way by realizing that the natural variables for Ξ are T, V and μ, that is, $\Omega = \Omega(T,V,\mu)$. Hence, we have $S = -\partial\Omega/\partial T$, $P = -\partial\Omega/\partial V$, and $N = -\partial\Omega/\partial\mu$.

If the number of particles is fixed at, say N, we have a *closed system*, the constraint characterized by γ is removed, and one can take $\gamma = 0$. Therefore,

7) Even though β is identified as $1/kT$, for convenience the symbol β is still used frequently.

▶ $p_i \equiv p_i(N,V,T) = \exp(-E_i/kT)/Z_N \quad \text{with} \quad Z_N = \sum\nolimits_i \exp(-E_i/kT)$ (5.18)

Here, Z_N denotes the canonical partition function for N particles, often labeled as just Z and addressed as the *partition function*. For the entropy we have in this case

$$S = k\beta\langle \mathcal{H} \rangle + k \ln Z$$

which, upon comparison with

$$S = (U + PV - \mu N)/T = (U - F)/T$$

results again in $\beta = 1/kT$, as expected, and $F = -\,kT \ln Z$. From the Helmholtz energy F we obtain the internal energy U, the entropy S, pressure P, and chemical potential μ

$$U = F + TS = \frac{\partial(F/T)}{\partial(1/T)} = kT^2\left(\frac{\partial \ln Z}{\partial T}\right)_V \qquad S = -\frac{\partial F}{\partial T} = k\ln Z + kT\left(\frac{\partial \ln Z}{\partial T}\right)_V$$

$$P = -\left(\frac{\partial F}{\partial V}\right)_T = kT\left(\frac{\partial \ln Z}{\partial V}\right)_T \qquad \mu = \left(\frac{\partial F}{\partial N}\right)_{T,V} = -kT\left(\frac{\partial \ln Z}{\partial N}\right)_{T,V}$$

We note that using Z_N we can write $\Xi = \Sigma_N \lambda^N Z_N$, where the original index i has been replaced by the pair (N,j) with j the index used in Z_N.

Finally, if also the energy is fixed, we have an *isolated system*, the constraint characterized by β is also removed and one can take $\beta = 0$ as well. In this case the probabilities p_i reduces to the *micro-canonical distribution*

▶ $p_i \equiv p_i(N,V,E) = 1/W$

where $W = \Sigma_i(1)$ is the number of accessible states for the macroscopic system at fixed energy E[8]. In this case, the entropy $S(p_i) = -k\ \Sigma_i\ p_i\ \ln p_i$ becomes the Boltzmann relation

$$S = k \ln W \tag{5.19}$$

Writing this expression as $-TS = -kT \ln W$ renders the expressions for the relation between the potentials ($\Omega = -PV = U - TS - \mu N$, $F = U - TS$, $-TS$) and the partition functions (Ξ, Z, W) all similar. Once having calculated the p_is, the *mean* or *average value* for any property X of the system can be calculated using

8) The uncertainty relations tell us that the energy E has an uncertainty ΔE. Moreover, exactly defined energies are not realizable experimentally and cover a range δE. Typically, it holds that $\delta E >> \Delta E$. Hence, we have actually to write $W = g(E)\delta E$ with $g(E) = \mathrm{d}n/\mathrm{d}E$ the density of states and δE the *accessibility range* for the energy E in which we can take $g(E)$ as constant. However, the value of the logarithm of δE is usually completely negligible as compared to $\ln g(E)$ and is therefore omitted. By the way, W is sometimes denoted as the "thermodynamic probability". It will be clear that the use of the word "probability" is completely unjustified here.

$$\langle X\rangle = \Sigma_i p_i X_i \tag{5.20}$$

where X_i denotes the value in microstate i. This expression is valid for all three distributions discussed. Note that adopting this view we can write for the entropy $S = -k\,\Sigma_i p_i \ln p_i = -k\langle \ln p\rangle$.

To show that the entropy just defined (for all situations) can be identified as the thermodynamic entropy (defined for equilibrium situations only), we calculate $\mathrm{d}S$ for a closed system using $\ln p_i = -E_i/kT - \ln Z$ and obtain

$$\mathrm{d}S = -k\sum\nolimits_i \mathrm{d}(p_i \ln p_i) = -k\sum\nolimits_i (\ln p_i + 1)\mathrm{d}p_i = \sum\nolimits_i E_i\,\mathrm{d}p_i\,/\,T \tag{5.21}$$

Here, we have used the definition of the Boltzmann distribution and twice the fact that $\Sigma_i\,\mathrm{d}p_i = 0$. Hence, for reversible heat flow we have $\delta Q = T\,\mathrm{d}S = \Sigma_i\,E_i\,\mathrm{d}p_i$. The total energy of the system of interest $\langle \mathcal{H}\rangle = \Sigma_i E_i p_i$ and its differential $\mathrm{d}\langle \mathcal{H}\rangle = \Sigma_i(E_i\,\mathrm{d}p_i + p_i\,\mathrm{d}E_i)$. From thermodynamics we know that the internal energy U is given by the first law of thermodynamics $\mathrm{d}U = \delta W + \delta Q$, where δW and δQ denote the increment in work and heat, respectively. If we again identify U with $\langle \mathcal{H}\rangle$, we can associate $\Sigma_i\,p_i\mathrm{d}E_i$ with δW since $\Sigma_i\,E_i\mathrm{d}p_i$ corresponds to $T\mathrm{d}S = \delta Q$. Finally, from Eq. (5.21) one easily can show that $\partial S/\partial U = 1/T$ (see Problem 5.1).

One property not discussed so far is that, for the thermodynamic entropy S, we have $\mathrm{d}S \geq 0$ for an isolated system. One can show that for the statistical entropy we also have $\mathrm{d}S \geq 0$ always (see Justification 5.2). Finally, one might wonder what the partition function represents. In fact, it just provides the average number of states that is thermally accessible for the system. This is particularly clear for the micro-canonical ensemble ($S = k\ln W$).

Justification 5.2: $\mathrm{d}S \geq 0$ always*

One property of the thermodynamic entropy is that for arbitrary processes it will increase unless the process is reversible. To prove that our statistical entropy behaves similarly, we first have to make the meaning of micro-states somewhat more precise. In fact, for any macroscopic system it is impossible to obtain the exact eigenstates of the system. Moreover, even if we would know them at a certain moment, soon afterwards, because of the interaction between the particles in the system, they would be unknown. For example, for an ideal gas the eigenstates are the described by the momentum eigenfunctions of structureless point particles. During equilibration the particles collide and exchange momentum. The uncertainty in energy E thus becomes $\Delta E \geq \hbar/2\Delta t$, where Δt is the mean time between collisions. As indicated in footnote 7, experimentally we are limited to δE with $\delta E >> \Delta E$. Hence, generally we have groups of approximate eigenstates associated with a small energy range, the *accessibility range* δE, small as compared to E but large as compared to the Heisenberg uncertainty ΔE and in which the density of states is essentially constant. The probabilities p_i therefore refer to these approximate states and, when δE is small enough, can be considered as constant within the accessibility range. It then makes no difference whether we consider transitions from exact state i to exact state j as restricted by ΔE or transitions from one approximate state (group) i to another state (group) j as restricted by δE: in either case, the value of dp_i/dt will be the same. This assumption is equivalent to the ergodic assumption as it also realizes mixing of all states. Sometimes it is referred to as the *accessibility* or *equal a priori probability* assumption.

To discuss equilibration we need the jump rate v_{ij} from state j to state i. For the rate at which a state i with probability p_i is changing, we have two contributions. First, we have the contribution from state i to any state j, given by $\Sigma_j v_{ji}\, p_i$ and, second, the contribution from any state j to state i, given by $\Sigma_j v_{ij}\, p_j$. Quantum mechanics tells us that $v_{ij} = v_{ji}$, that is, the principle of *jump rate symmetry* (see Section 2.4), and therefore the total rate becomes $\mathrm{d}p_i/\mathrm{d}t = \Sigma_j v_{ij}(p_j - p_i)$, referred to as the *master equation.*

We now consider the entropy differential $\mathrm{d}S/\mathrm{d}t = -k\,\Sigma_i \ln p_i\,(\mathrm{d}p_i/\mathrm{d}t)$. For an isolated system we have for the effect of jumps between two particular states i and j a contribution to $\mathrm{d}p_i/\mathrm{d}t$ reading $v_{ij}(p_j - p_i)$ as well as a contribution $v_{ij}(p_i - p_j)$ to $\mathrm{d}p_j/\mathrm{d}t$. The total rate is thus, taking in to account all jumps, $\mathrm{d}S/\mathrm{d}t = k\Sigma_{i,j} v_{ij}(p_j - p_i)(\ln p_j - \ln p_i)$. Since the terms in brackets have the same sign, $\mathrm{d}S/\mathrm{d}t$ cannot be negative. Moreover, if $p_i = p_j$, $\mathrm{d}S/\mathrm{d}t = 0$, and so we conclude that always $\mathrm{d}S \geq 0$. We refer to the literature for a further discussion of the foundations of statistical thermodynamics [3].

5.1.3 Fluctuations*

We note that the difference between the results from the grand partition function Ξ and the partition function Z is usually small for macroscopic systems, that is, fluctuations in the number of molecules and therefore energy, are generally not terribly important. This can be seen as follows. The grand partition function is

$$\Xi = \sum\nolimits_N \exp[\beta\mu N)]Z_N = \sum\nolimits_N \lambda^N Z_N \quad \text{with} \quad \lambda = \exp(\beta\mu) \tag{5.22}$$

where Z_N is the canonical partition function for N particles, so that the probability to have N particles in the system is

$$p_N = \lambda^N Z_N / \Xi \tag{5.23}$$

Hence, it follows that

$$\langle N\rangle = \Xi^{-1}\sum\nolimits_{N=0}^{N=\infty} N\lambda^N Z_N \quad \text{and} \quad \langle N^2\rangle = \Xi^{-1}\sum\nolimits_{N=0}^{N=\infty} N^2\lambda^N Z_N \tag{5.24}$$

Noting that since $\langle N\rangle = \partial\ln\Xi/\partial\ln\lambda$, differentiation of $\langle N\rangle$ with respect to μ yields

$$\partial\langle N\rangle / \partial\mu = \beta(\langle N^2\rangle - \langle N\rangle^2) = \beta\sigma_N^2$$

where the variance $\sigma_X^2 \equiv \langle X^2\rangle - \langle X\rangle^2$ is used. We also have

$$\left(\frac{\partial\langle N\rangle}{\partial\mu}\right)_{V,T} = \left(\frac{\partial N}{\partial\mu}\right)_{V,T} = \left(\frac{\partial N}{\partial P}\right)_{V,T}\left(\frac{\partial P}{\partial\mu}\right)_{V,T} \quad \text{and} \quad \left(\frac{\partial P}{\partial\mu}\right)_{V,T} = \frac{N}{V}$$

Since the derivative $(\partial N/\partial P)_{V,T}$ is at constant V, we have $(\partial N/\partial P)_{V,T} = V[\partial(N/V)/\partial P]_{V,T} = V[\partial(N/V)/\partial P]_{N,T} = NV[\partial(1/V)/\partial P]_{N,T} = N\kappa_T$, where the second step can be made because N/V is an intensive quantity. Therefore, we have $(\partial N/\partial \mu)_{V,T} = N\kappa_T^2/V$. Hence, the relative fluctuation in number of molecules in the system is given by

$$(\sigma_N/\langle N\rangle)^2 = kT\kappa_T/V$$

and therefore for large N is negligible. The fluctuation in energy for an open system can be derived similarly, although the process is somewhat more complex, and we quote (leaving the derivation for Problem 5.6)

$$\sigma_E^2 = kT^2C_V + (\partial\langle H\rangle/\partial\langle N\rangle)\sigma_N^2$$

Problem 5.1

Show, using Eq. (5.21), that $\partial S/\partial U = 1/T$.

Problem 5.2

Verify from the thermodynamic expression for $\mathrm{d}U$ and the statistical expression for $\mathrm{d}S$ that δW in the grand canonical ensemble is given by

$$\delta W = \Sigma_i p_i \mathrm{d}E_i + \mu\Sigma_i N_i \mathrm{d}p_i \quad \text{corresponding to} \quad \delta W = -P\mathrm{d}V + \mu\mathrm{d}N.$$

Problem 5.3: The two-state model

A very simple model in statistical mechanics is the two-state model with (obviously) two states, state 1 with energy 0 and state 2 with energy ε.

a) Give the expression for the partition function Z.
b) For temperature $T = \varepsilon/k$, where k is Boltzmann's constant, calculate the occupation probabilities of state 1, p_1, and of state 2, p_2.
c) Calculate the internal energy U.
d) Calculate the heat capacity C_V and sketch the behavior of $C_V(T)$.

Problem 5.4: Temperature and the density-of-states

Show that the temperature T is related to the density of states $g(E)$ for a system with energy E_s via $1/kT = \partial \ln g(E)/\partial E|_{E=E_s}$.

Problem 5.5

Show in detail for macro-systems, although we have $S = g(E)\delta E$, that S is essentially independent of δE. Consider to that purpose the value of δE based on the quantum uncertainty relations and a typical experimental uncertainty.

Problem 5.6*

Derive $\sigma_E^2 = kT^2C_V$ for a closed system in a way similar as the relation $(\sigma_N/\langle N\rangle)^2 = kT\kappa_T/V$ is derived for the open system. For an open system, show that $\sigma_E^2 = kT^2C_V + (\partial\langle H\rangle/\partial\langle N\rangle)\sigma_N^2$.

5.2 Perfect Gases

The concept of a perfect gas is well known, and its behavior is succinctly described for one mole of gas by the equation of state

$$PV_\mathrm{m} = RT$$

where P is pressure, V_m the molar volume, R the gas constant, and T the (absolute) temperature. Originally, this equation was obtained empirically.

Let us apply the statistical mechanics ideas to the perfect gas. The simplest model that we have for a perfect gas is that of particles without internal structure and without any mutual interaction contained in a box. For evaluation of the partition function, we first need the partition function for a single particle, and in order to obtain that result we need an expression for the energy of a particle. Thereafter, we have to extend the result to many particles.

5.2.1 Single Particle

From quantum mechanics we know that for a particle of mass m in a 3D box with edge l and volume $V = l^3$, we have for the energy levels

$$\varepsilon_\mathrm{n} = \frac{h^2}{8ml^2}(n_1^2+n_2^2+n_3^2) = \frac{\hbar^2}{8ml^2}(2\pi\mathbf{n})^2 = \frac{\hbar^2}{2m}\mathbf{k}^2 = \frac{\mathbf{p}^2}{2m} \quad \text{where} \tag{5.25}$$

$\hbar = h/2\pi =$ Planck's constant$/2\pi$ and
$\mathbf{n} = (n_1, n_2, n_3)$ with $(n_1, n_2, n_3) \in (\pm1, \pm2, \pm3, \ldots)$

The momentum $\mathbf{p} = \hbar\mathbf{k}$ of such a particle is given by $\mathbf{k} = 2\pi\mathbf{n}/2l$, where $\mathbf{k}$ is the wave vector. The single particle partition function z is thus

$$z = \sum\nolimits_\mathrm{n} \exp(-\varepsilon_\mathrm{n}/kT) = \sum\nolimits_{n_1}\sum\nolimits_{n_2}\sum\nolimits_{n_3} \exp\left(-\frac{h^2}{8ml^2}(n_1^2+n_2^2+n_3^2)/kT\right)$$

The spacing of energy levels is close so that the summation may be replaced by an integration leading to the standard integrals (see Appendix B)

$$z \cong \int_{\mathrm{n}=0}^{\mathrm{n}=\infty} \exp(-\varepsilon_\mathrm{n}/kT)\,\mathrm{d}\mathbf{n} \tag{5.26}$$

$$= \int_{n_1=0}^{n_1=\infty}\int_{n_2=0}^{n_2=\infty}\int_{n_3=0}^{n_3=\infty} \exp\left[-\frac{h^2}{8ml^2}(n_1^2+n_2^2+n_3^2)/kT\right]\mathrm{d}n_1\,\mathrm{d}n_2\,\mathrm{d}n_3 \tag{5.27}$$

$$= \left[\int_{n=0}^{n=\infty} \exp(-h^2n^2/8ml^2kT)\mathrm{d}n\right]^3 = V\left(2\pi mkT/h^2\right)^{3/2} \equiv V/\Lambda^3 \tag{5.28}$$

with the *thermal wavelength* Λ defined by $\Lambda \equiv (h^2/2\pi mkT)^{1/2}$.

5.2.2
Many Particles

In many cases, such as in the case of the ideal gas, the total energy E_j in state j can be written as the sum of the energies of particles ε_i, that is, the interaction energy between particles is negligible as compared with to the total energy, and we have

$$E_j = \varepsilon_{j1} + \varepsilon_{j2} + \ldots \varepsilon_{jN} = \sum\nolimits_i \varepsilon_{ji} \tag{5.29}$$

where the indices $i = 1, 2, \ldots, N$ now indicate each of the N particles in state j. The partition function becomes

$$\begin{aligned} Z_N &= \sum\nolimits_j \exp(-E_j/kT) = \sum\nolimits_j \exp[-(\varepsilon_{j1}+\varepsilon_{j2}+\ldots\varepsilon_{jN})/kT] \\ &= \sum\nolimits_{j1}\exp(-\varepsilon_{j1}/kT)\sum\nolimits_{j2}\exp(-\varepsilon_{j2}/kT)\ldots = z_1z_2\ldots z_N = z^N \end{aligned} \tag{5.30}$$

The total partition function Z_N is thus the product of the single particle partition functions z, that is, $Z_N = z^N$.

If the gas consists of molecules instead of atoms, we must realize that molecules have an internal motion with an associated energy. To a good first approximation, the total energy can be taken as the sum of energies of the contributing mechanisms. For polyatomic molecules these mechanisms would be (overall) translation, (overall and internal) rotation, (internal) vibration and electronic transitions. So, we have

$$\varepsilon_j = \varepsilon_{j\alpha} + \varepsilon_{j\beta} + \cdots$$

where each of the indices $\alpha, \beta, \ldots$ indicates a mechanism. In this case we have

$$\begin{aligned} z &= \sum\nolimits_j \exp(-\varepsilon_j/kT) = \sum\nolimits_{j\alpha,j\beta,\ldots} \exp[-(\varepsilon_{j\alpha}+\varepsilon_{j\beta}+\ldots)/kT] \\ &= \sum\nolimits_{j\alpha}\exp(-\varepsilon_{j\alpha}/kT)\sum\nolimits_{j\beta}\exp(-\varepsilon_{j\beta}/kT)\ldots = z_\alpha z_\beta\ldots \end{aligned}$$

Thus, the total single partition function z is thus also the product of the partition functions z_α for each of the mechanisms.

There is one more aspect that we have to discuss for systems with N particles, namely that particles are indistinguishable, leading to a correction for the N-particle partition function. A simple consideration often given is that for each of the N particles we could choose another one, leading to overcounting by $N!$ and this leads to extra factor $1/N!$ in the partition function[9]. The real reason originates from the quantum nature of molecules, indeed related to indistinguishability, and, after a somewhat complex, approximate reasoning leading to the factor $1/N!$. Here, we accept that result (see Justification 5.3) and replace the expression for partition function Z_N of N independent particles $Z = z^N$ by

$$\blacktriangleright \quad Z_N = z^N/N! \tag{5.31}$$

where z is the total single particle partition function. If we apply this approximation, we usually say that we use *Boltzmann statistics*. We will see in the next paragraph that the factor $N!$ does not influence the pressure P, that is, the EoS, but does influence the entropy S (see Problem 5.8).

5.2.3
Pressure and Energy

The Helmholtz function F is related to the partition function Z via

$$F = -kT\ln Z = -kT\ln(z^N/N!) \tag{5.32}$$

So, let us try to calculate the mechanical equation of state (EoS) for an ideal gas. The mechanical EoS expresses the pressure $P = -\partial F/\partial V$ as a function of the independent variables. Here we find, using the Stirling approximation[10] for $\ln N!$,

$$\begin{aligned} F &= -kT\ln Z = -kT\ln(z^N/N!) = -kT\{N\ln z - \ln N!\} \\ &= -kT\{[N\ln(V\Lambda^{-3}) - [N\ln N - N]\} \quad \text{and with} \quad N = N_A \\ &= -RT[\ln V + \ln(\Lambda^{-3}) - \ln(N_A/e)] \quad \text{or} \end{aligned}$$

$$\blacktriangleright \quad P = -\frac{\partial F}{\partial V} = RT\frac{\partial \ln V}{\partial V} + \frac{\partial \ldots}{\partial V} = \frac{RT}{V}\frac{\partial V}{\partial V} \quad \text{or} \quad P = \frac{RT}{V} \tag{5.33}$$

In this last expression we recognize the well-known perfect gas law.

Because from the Helmholtz function F all other thermodynamic expressions can be derived, the same is true for the partition function as the two are related via the expression $F = -kT \ln Z$. For the entropy S and energy U we find, respectively,

$$S = -\partial F/\partial T = k\partial T\ln Z/\partial T \quad \text{and} \quad U = F + TS = -kT\ln Z + kT\partial T\ln Z/\partial T \tag{5.34}$$

9) This correction is overcorrecting the situation though, because in the double sum, as indicated in Eq. (5.30), the contribution ij ($i \neq j$) is counted twice because ji represents the same physical configuration while double counting is not done for the contribution ii. This shows right away that the correction is only valid for noncondensed systems.

10) The Stirling approximation for factorials reads $\ln x! = x \ln x - x + ½ \ln(2\pi x) + \ldots$. This approximation is excellent even for $x = 3$, the difference with the exact value being about 2%. Often, the term $½ \ln(2\pi x)$ is neglected. Although the latter approximation is considerably less accurate, for $x = 50$ it deviates only about 2% from the exact value. Since typically much larger numbers are used, the approximation $\ln x! = x \ln x - x$, or alternatively $x! = (x/e)^x$, is usually quite sufficient.

We can also calculate U directly from Z using to the Gibbs–Helmholtz expression $U = \partial(F/T)/\partial(1/T) = kT^2\, \partial \ln Z/\partial T$. The total energy U is entirely kinetic and reads

$$U = 3RT/2 \tag{5.35}$$

and hence the capacity C_V is given by $C_V = (\partial U/\partial T)_V = 3R/2$. Since we have the general relation $C_P - C_V = TV\alpha_P^2/\kappa_T$, we obtain $C_P = 5R/2$.

Problem 5.7

Show that the internal energy U for an ideal gas for the ideal gas consisting of N structureless particles reads $U = 3NkT/2$. Also show that $C_V = 3R/2$, and hence that $C_P = 5R/2$.

Problem 5.8

Show that the entropy S for an ideal gas is given by the so-called *Sackur–Tetrode* equation

$$S = Nk \ln(ze^{5/2}\,/\,N) \quad \text{with} \quad z = V/\Lambda^3 = (2\pi mkT/h^2)^{3/2} V$$

Similarly, show that the chemical potential $\mu = kT \ln(N/z)$.

Problem 5.9: *N*! again*

According to quantum mechanics a system wave function should be either antisymmetric (FD particles) or symmetric (BE particles) in its coordinates (see Section 2.3). For weakly interacting particles in a system we may assume the existence of single-particle states. Show that for N particles in a system with q single-particle states ($q \geq N$, generally $q >> N$), the number of system states for FD particles is $W_{FD} = q!/(q - N)!N!$, while for BE particles $W_{BE} = (q + N - 1)!/N!(q - 1)! \cong (q + N)!/N!(q)!$. Also show that for $q >> N$, both W_{FD} and W_{BE} reduces to the number of Boltzmann states $W_{Bo} = q^N/N!$.

Problem 5.10*

Show that for the perfect gas, using $U = \langle H \rangle$, the fluctuations are given by $\sigma_U^2 = (2U/3N)^{1/2}$ and thus vanish in the thermodynamic limit.

5.3 The Semi-Classical Approximation

The scheme for structureless, independent particles outlined so far comprises the calculation of:

- the quantum energy levels ε_i,
- the partition function, say Z (Ξ or W may be used as well), either exact via summation or approximate via integration,
- the corresponding thermodynamic potential, say F (or Ω or S), and
- the remaining properties (e.g., S, P . . .) using phenomenological thermodynamics.

The scheme also applies to other situations, but a few remarks are appropriate. We mentioned "structureless" particles, but molecules do contain internal structure, and the effect of internal structure is discussed in Section 5.5. We also said "independent" particles; however, interaction is important and we deal with the effect of interactions in Section 5.6. The next point relates to the energy. In many cases, the energy E_{QM} as calculated from the Schrödinger equation $H\Psi = E\Psi$ has to be approximated by the classical mechanics expression $E_{CM} = \mathcal{H}(\mathbf{p},\mathbf{q})$, where $\mathcal{H}$ is the Hamilton function. In fact, statistical thermodynamics was largely developed before the introduction of quantum mechanics. The classical development leads to the partition function

$$Z_{CM} = \int \ldots \int \exp(-\mathcal{H}/kT)\mathrm{d}\mathbf{p}\,\mathrm{d}\mathbf{q} \quad \text{with} \quad \mathcal{H} = T(\mathbf{p}) + \Phi(\mathbf{q})$$

where $\mathbf{p} = \mathbf{p}_1 \ldots \mathbf{p}_N = p_{1x}p_{1y}p_{1z}p_{2x} \ldots p_{Nz}$, $\mathbf{q} = \mathbf{q}_1 \ldots \mathbf{q}_N = q_{1x}q_{1y}q_{1z}q_{2x} \ldots q_{Nz}$ and $T(\mathbf{p})$ and $\Phi(\mathbf{q})$ represent the kinetic and potential energy, respectively. Here, $\mathbf{p}_i$ is the (generalized) momentum and $\mathbf{q}_i$ the (generalized) coordinate of particle i. The choice of coordinates is to some extent arbitrary, since for conjugated coordinates we have $\mathrm{d}\mathbf{p}\mathrm{d}\mathbf{q} = \mathrm{d}\mathbf{p}'\mathrm{d}\mathbf{q}'$. The connection between Z_{QM} and Z_{CM} has been made [4], but the process is rather complex and we illustrate here the result by example only. Since the factor between Z_{CM} and Z_{QM} should depend only on fundamental constants, this is in principle also sufficient. Consider a single particle with three degrees of freedom (DoF) for which the classical energy expression $\varepsilon = \mathcal{H}(\mathbf{p},\mathbf{q})$ $(\mathbf{p},\mathbf{q})$ is given by $\mathcal{H} = \mathbf{p}^2/2m$, so that the classical partition function becomes

$$z_{CM} = \int_0^1 \mathrm{d}\mathbf{q} \int_0^\infty \exp(-\mathcal{H}/kT)\mathrm{d}\mathbf{p} = V\int_0^\infty \exp(-\mathbf{p}^2/2mkT)\mathrm{d}\mathbf{p} = V(2\pi mkT)^{3/2} \tag{5.36}$$

We see that z_{CM} is similar to z_{QM} except for a factor h^{-3}, which is missing in the classical expression. So, for each DoF an extra factor h^{-1} is required to match the quantum expression. This result appears to be general: if we calculate the partition function using a classical expression for the energy, we need an extra factor h^{-1} for every DoF (see Justification 5.3). For a single DoF z_{1D} reads

$$z_{1D} = l(2\pi mkT/h^2)^{1/2} \quad \text{or} \quad z_{1D} = l/\Lambda \quad \text{with} \quad \Lambda = (h^2/2\pi mkT)^{1/2} \tag{5.37}$$

The partition function z_{1D} is thus dimensionless. As long as $z_{1D} >> 1$, this DoF can be described by classical mechanics. As soon as $z_{1D} \cong 1$, a full quantum description for that particular DoF is required. This applies also to many particles if l is taken to be the mean particle distance $l = (V/N)^{1/3}$. To apply classical mechanics to a gas thus requires that $\Lambda^3 << V/N$. Since $\mu = kT\ln(N/z)$ with $z = V/\Lambda^3$, this implies that $\exp(\beta\mu) >> 1$, which is exactly the condition used to arrive at classical statistics. For translation it can be easily shown that a semi-classical description is nearly always sufficient. As will be seen in the next section, this is not generally true for the internal contributions. It is thus quite possible, that classical evaluation is sufficient for one mechanism, while a quantum evaluation is necessary for another. As long as energies are additive, this presents no special problems.

Example 5.2: The harmonic oscillator

A single harmonic oscillator provides a good demonstration for the factor h. In quantum mechanics the energy for an oscillator with spring constant a and mass m is given by $E_n = \hbar\omega(n + \frac{1}{2})$ with $\omega = \sqrt{a/m}$ so that the energy difference ΔE between two successive states is $\Delta E = \hbar\omega$. In classical mechanics the total energy can be written as $\mathcal{H}(p,q) = p^2/2m + \frac{1}{2}m\omega^2q^2$, where p is the momentum and q is the coordinate. We may also write $p^2/\alpha^2 + q^2/\beta^2 = 1$ with $\alpha = (2mE)^{1/2}$ and $\beta = (2E/m\omega^2)^{1/2}$. If we plot constant energy curves in μ-space, which is in this case a two-dimensional space, we obtain ellipses. The area enclosed by such an ellipse is given by the integral $I = \oint p\,dq$, where the integration is over one period, or $I = \pi\alpha\beta = 2\pi E/\omega$. Alternatively, we have $q = q_0 \sin(\omega t)$, $p = m\dot{q} = m\omega q_0 \cos(\omega t)$ and $I = \int_0^{2\pi/\omega} p\dot{q}\,\mathrm{d}t = \pi m\omega q_0^2 = 2\pi E/\omega$. Let us draw ellipses with energies corresponding to $n - 1$, n, and $n + 1$ (Figure 5.1). The area between two successive ellipses is the area in classical phase space associated with one quantum state, and this area corresponds to $2\pi\Delta E/\omega = 2\pi\hbar\omega/\omega = h$. For large n, the classical energy is almost constant in the phase region between $n - \frac{1}{2}$ and $n + \frac{1}{2}$, so that the approximation $\iint \exp[-\mathcal{H}(p,q)/kT]\mathrm{d}p\mathrm{d}q \cong h\Sigma_n \exp(-E_n/kT)$ can be made. For $T \rightarrow \infty$ and $n \rightarrow \infty$ the argument becomes exact.

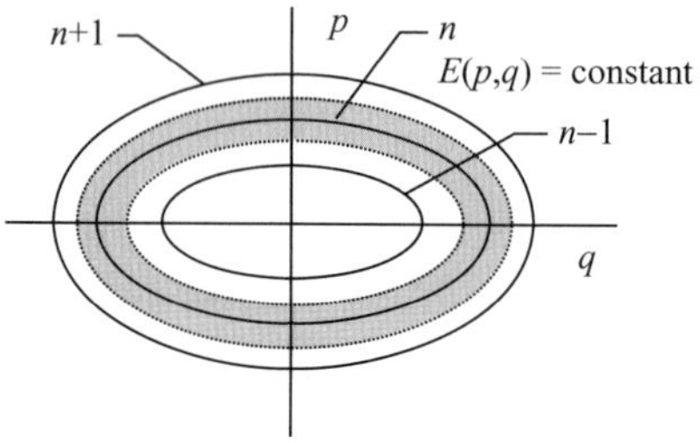

Figure 5.1 Phase space for a 1D oscillator.

Overall, the result for N particles leads to the introduction of the factor $(h^{3N}N!)^{-1}$ so that we have in total, using the Hamilton function $\mathcal{H}$ and potential energy Φ,

$$Z_N = (h^{3N}N!)^{-1}\int\ldots\int e^{-\beta\mathcal{H}}\,\mathrm{d}\mathbf{p}\,\mathrm{d}\mathbf{q} = (\Lambda^{3N}N!)^{-1}\int\ldots\int e^{-\beta\Phi}\,\mathrm{d}\mathbf{q} \equiv (\Lambda^{3N}N!)^{-1}Q_N \tag{5.38}$$

This result is the most frequently encountered expression for the partition function, also indicated as the *semi-classical partition function*. We have seen that in Eq. (5.36) for the single-particle partition function the integration over the momenta and position coordinates separate. The factor V, due to the integration over the position coordinates, is denoted as the *configurational part* while the factor $\Lambda = (h^2/2\pi mkT)^{1/2}$, due to the integration over the momenta, is denoted as the *kinetic part*. The latter part leads to $\frac{1}{2}kT$ for the kinetic energy of a classical DoF. Because in classical mechanics energy expressions the kinetic energy is generally only dependent on $\mathbf{p}$ and the potential energy is generally only dependent on $\mathbf{q}$, this separation is also generally possible and it can be concluded that in classical models the kinetic energy per DoF is (nearly) always $\frac{1}{2}kT$. For many particles this separation is shown in the last part of Eq. (5.38). For N independent particles the configurational part is deceptively simple ($Q_N = V^N$), but we will see that this integration becomes the major problem for interacting particles[11] (see Section 5.6).

Finally, we note that the expectation value for the canonical average value of a property $X(\mathbf{p},\mathbf{q})$ in the semi-classical approximation becomes

$$\blacktriangleright \quad \langle X \rangle = (Z_N)^{-1} \int \ldots \int X(\mathbf{p},\mathbf{q}) \mathrm{e}^{-\beta \mathcal{H}} \, \mathrm{d}\mathbf{p}\,\mathrm{d}\mathbf{q} = (Q_N)^{-1} \int \ldots \int X(\mathbf{q}) \mathrm{e}^{-\beta \Phi} \, \mathrm{d}\mathbf{q} \tag{5.39}$$

where the last step can be made if X depends only on the coordinates $\mathbf{q}$.

Justification 5.3: The introduction of h and $N!$*

First we note that for the Hamilton operator H operating on wave function Ψ and corresponding to the classical Hamilton function $\mathcal{H}$ we have

$$\exp(H)|\Psi\rangle = (1 + H + \ldots)|\Psi\rangle = (1 + \mathcal{H} + \ldots)|\Psi\rangle \cong \exp(\mathcal{H})|\Psi\rangle$$

Second, a state $|\Psi\rangle$ for N particles can be expanded in any complete set of the proper symmetry and we use here the antisymmetrized eigenstates of the momentum operator $\mathbf{p}$ using the δ-function normalization (see Section 2.2)

$$|\Psi\rangle = (N!h^{3N/2})^{-1} \int_{-\infty}^{+\infty} A(\mathbf{p}) \sum\nolimits_{P_\mathrm{r}} P_\mathrm{r}[|\exp(\mathrm{i}\mathbf{p}^\mathrm{T}\mathbf{r})\rangle]\mathrm{d}\mathbf{p}$$

with the permutation operator P_r for coordinates $\mathbf{r}$ (see Section 2.2) and $A(\mathbf{p})$ the corresponding coefficient. Note that $|\exp(\mathrm{i}\mathbf{p}^\mathrm{T}\mathbf{r})\rangle$ denotes the product function $\Pi_i|\exp(\mathrm{i}\mathbf{p}_i^\mathrm{T}\mathbf{r}_i)\rangle$ of the individual eigenfunctions $|\exp(\mathrm{i}\mathbf{p}_i^\mathrm{T}\mathbf{r}_i)\rangle$ for particle i. Third, we will need for an arbitrary function $F(\mathbf{p})$ of $\mathbf{p}$ the relation

$$\int_{-\infty}^{+\infty} F(\mathbf{p})\mathrm{d}\mathbf{p} = (N!)^{-1} \int_{-\infty}^{+\infty} \sum\nolimits_{P_\mathrm{p}} P_\mathrm{p}[F(\mathbf{p})]\mathrm{d}\mathbf{p}$$

Using bra-ket notation the partition function $Z_N = \Sigma_j \exp(-\beta\varepsilon_j)$ is written as

$$Z_N = \sum\nolimits_j \langle \Psi_j | \Psi_j \rangle \exp(-\beta\varepsilon_j) = \sum\nolimits_j \langle \Psi_j | \exp(-\beta H) | \Psi_j \rangle$$

11) The symbolism for Z_N and Q_N is not uniformly used. In some cases the factor $1/N!$ is included in Q_N while in other cases the reverse designation for Z_N and Q_N is used.

with Ψ_j the energy eigenfunction corresponding to state j. We see that this result can be interpreted as the sum of the diagonal elements of a matrix with as elements $\langle\Psi_i\,|\exp(-\beta H)\,|\,\Psi_j\rangle$. We expand Ψ_j in the aforementioned antisymmetrized eigenfunctions of the momentum operator $\mathbf{p}$ to obtain

$$\langle\Psi_j|\exp(-\beta H)|\Psi_j\rangle$$
$$=(N!h^{3N/2})^{-2}\int_{-\infty}^{+\infty}\int_{-\infty}^{+\infty}\left\langle A(\mathbf{p})\sum\nolimits_{P_r}P_r[\mathrm{e}^{i\mathbf{p}^T\mathbf{r}}]|\mathrm{e}^{-\beta H}|A(\mathbf{p}')\sum\nolimits_{P_r}P_r[\mathrm{e}^{i\mathbf{p}'^T\mathbf{r}}]\right\rangle\mathrm{d}\mathbf{p}\,\mathrm{d}\mathbf{p}'$$

The partition function Z_N becomes $Z_N=\sum_j\langle\Psi_j|\exp(-\beta H)|\Psi_j\rangle$

$$=\frac{1}{(N!h^{3N/2})^2}\sum\nolimits_j\int_{-\infty}^{+\infty}\int_{-\infty}^{+\infty}\left\langle A_j(\mathbf{p})\sum\nolimits_{P_r}P_r[\mathrm{e}^{i\mathbf{p}_j^T\mathbf{r}}]|\mathrm{e}^{-\beta H}|\ldots\right.$$
$$\left.\ldots|A_j(\mathbf{p}')\sum\nolimits_{P_r}P_r[\mathrm{e}^{i\mathbf{p}_j'^T\mathbf{r}}]\right\rangle\mathrm{d}\mathbf{p}_j\,\mathrm{d}\mathbf{p}_j'$$

Now, using $P_r=P_p$ for $\exp(i\mathbf{p}'^T\mathbf{r})$, $N!F(\mathbf{p})=\Sigma_P P_p[F(\mathbf{p})]$, $\Sigma_j A_j(\mathbf{p})A_j(\mathbf{p}')=\delta(\mathbf{p}'-\mathbf{p})$ and $\exp(H)|\Psi\rangle\cong\exp(\mathcal{H})|\Psi\rangle$, we obtain

$$Z_N=\frac{1}{N!^2h^{3N}}\sum\nolimits_j\int_{-\infty}^{+\infty}\int_{-\infty}^{+\infty}\left\langle A_j(\mathbf{p})\sum\nolimits_{P_r}P_r[\mathrm{e}^{i\mathbf{p}_j^T\mathbf{r}}]|\mathrm{e}^{-\beta H}|\ldots\right.$$
$$\left.\ldots|\sum\nolimits_{P_p}P_p[A_j(\mathbf{p}')\mathrm{e}^{i\mathbf{p}_j'^T\mathbf{r}}]\right\rangle\mathrm{d}\mathbf{p}_j\,\mathrm{d}\mathbf{p}_j'$$
$$=\frac{1}{N!h^{3N}}\sum\nolimits_j\int_{-\infty}^{+\infty}\int_{-\infty}^{+\infty}\left\langle A_j(\mathbf{p})\sum\nolimits_{P_r}P_r[\mathrm{e}^{i\mathbf{p}_j^T\mathbf{r}}]|\mathrm{e}^{-\beta H}|A_j(\mathbf{p}')\mathrm{e}^{i\mathbf{p}_j'^T\cdot\mathbf{r}}\right\rangle\mathrm{d}\mathbf{p}_j\,\mathrm{d}\mathbf{p}_j'$$
$$=\frac{1}{N!h^{3N}}\int_{-\infty}^{+\infty}\int_{-\infty}^{+\infty}\sum\nolimits_j A_j(\mathbf{p})A_j(\mathbf{p}')\left\langle\sum\nolimits_{P_r}P_r[\mathrm{e}^{i\mathbf{p}_j^T\mathbf{r}}]|\mathrm{e}^{-\beta H}|\mathrm{e}^{i\mathbf{p}_j'^T\mathbf{r}}\right\rangle\mathrm{d}\mathbf{p}_j\,\mathrm{d}\mathbf{p}_j'$$
$$=\frac{1}{N!h^{3N}}\int_{-\infty}^{+\infty}\langle\mathrm{e}^{i\mathbf{p}^T\mathbf{r}}|\mathrm{e}^{-\beta H}|\mathrm{e}^{i\mathbf{p}^T\mathbf{r}}\rangle\mathrm{d}\mathbf{p}\quad\text{or}$$

► $$Z_{\mathrm{CM}}\cong\frac{1}{N!h^{3N}}\int_{-\infty}^{+\infty}\ldots\int_{-\infty}^{+\infty}\exp(-\beta\mathcal{H})\mathrm{d}\mathbf{r}\,\mathrm{d}\mathbf{p}$$

The classical partition function Z_{CM} is thus $(N!h^{3N})^{-1}$ times the phase integral over $\exp(-\beta\mathcal{H})$. This analysis not only shows the origin of the factor h^{-3N} but also of the factor $(N!)^{-1}$. Extending the analysis further – in principle a little bit but in practice considerably [5] – by introducing second-order terms in $\exp(H)|\Psi\rangle\cong\exp(\mathcal{H})|\Psi\rangle$ leads to the first-order quantum correction on the classical result which reads, using $\langle\ldots\rangle$ as the canonical average,

$$Z_{\mathrm{QM}}=Z_{\mathrm{CM}}\left(1-\frac{\beta^2\hbar^2}{24}\sum_{a=1}^{3N}\frac{\Phi_{aa}}{m_a}+\ldots\right)\quad\text{with}\quad\Phi_{aa}\equiv\left\langle\frac{\partial^2\Phi(\mathbf{r})}{\partial x_a^2}\right\rangle$$

so that the Helmholtz energy becomes

$$\beta F_{\mathrm{QM}}=\beta F_{\mathrm{CM}}+\frac{\beta^2\hbar}{24}\sum_{a=1}^{3N}\frac{\Phi_{aa}}{m_a}+\ldots=\beta F_{\mathrm{CM}}+\left(\frac{298.2}{T}\right)^2\sum_{a=1}^{3N}\left(\frac{\bar{\omega}_a}{1015}\right)^2+\ldots$$

The quantity Φ_{aa} is an effective force constant so that $\omega_a = (\Phi_{aa}/m_a)^{1/2}$ is the corresponding angular frequency. At room temperature for vibrations with wave number[12] $\bar{\omega} \sim 1000\ \text{cm}^{-1}$ or more, the correction becomes significant.

Problem 5.11: The thermal wavelength

Calculate Λ for He and Ar and compare these values with the average spacing between the atoms as estimated from the densities at the triple point. Is the classical approximation valid?

Problem 5.12: Angle-averaged potentials

For arbitrary molecules the interaction is dependent on their distance r and both their orientations, indicated by $\boldsymbol{\omega} = (\theta,\phi)$ where θ and ϕ are the angles for the polar co-ordinate system ($0 \le \theta \le \pi$ and $0 \le \phi \le 2\pi$). The differential element $\text{d}\boldsymbol{\omega}$ is $\text{d}\boldsymbol{\omega} = \sin\theta\ \text{d}\theta\text{d}\phi$. Using the multipole expression for $W(r)$ for fixed distance r, expanding the exponentials and integrating term-by-term, show that,

$$W(r) = -\frac{1}{2}\beta\overline{W^2(r,\boldsymbol{\omega})} \quad \text{for} \quad \beta W \ll 1 \quad \text{with}$$

$$\overline{W^n(r,\boldsymbol{\omega})} = \int W^n(r,\boldsymbol{\omega}_1,\boldsymbol{\omega}_2)\,\text{d}\boldsymbol{\omega}_1\,\text{d}\boldsymbol{\omega}_2 \Big/ \int \text{d}\boldsymbol{\omega}_1\,\text{d}\boldsymbol{\omega}_2$$

Discuss in physical terms why the terms with n odd are identically zero. Show also that the angle-averaged internal energy

$$U(r) = \langle W(r,\boldsymbol{\omega})\rangle_{\boldsymbol{\omega}} = -\beta\overline{W^2(r,\boldsymbol{\omega})}$$

where the subscript $\boldsymbol{\omega}$ indicates averaging over $\boldsymbol{\omega}$ only, and that this result is consistent with Problem 2.6.

Problem 5.13: Charge–dipole interaction

Show that, by using the result of Problem 5.12, the charge–dipole contribution to the intermolecular interaction is given by

$$W_{\text{c-d}}(r) = \overline{W_{\text{c-d}}(r,\theta)} = -\mathcal{C}^2\left[\left(q_1^2\mu_2^2/6kT\right) + \left(q_2^2\mu_1^2/6kT\right)\right]r^{-4} \tag{3.9}$$

Problem 5.14: Dipole-induced dipole interaction

Show that the dipole-induced dipole contribution is given by

$$W_{\text{d-id}}(r) = \overline{W_{\text{d-id}}(r,\theta)} = -\mathcal{C}^2\left(\alpha_1\mu_2^2 + \alpha_2\mu_1^2\right)r^{-6} \tag{3.12}$$

12) The energy is often characterized by the frequency $\nu = \varepsilon/h$ in s^{-1} (with $B = h/8\pi^2 I$) or by the wave number $\varpi = \varepsilon/hc$ in cm^{-1} (associated with $B = h/8\pi^2 Ic$).

Problem 5.15: Dipole–dipole interaction*

Similarly, show that the dipole–dipole contribution is given by

$$W_{\text{d-d}}(r) = \overline{W_{\text{d-d}}(r, \omega_1, \omega_2)} = -\mathcal{C}^2 (\mu_1^2 \mu_2^2 / 3kT) r^{-6} \qquad (3.10)$$

5.4
A Few General Aspects

In this section, a few general aspects are dealt with that could have been made at various points in the foregoing discussion. Since the line-of-thought has been interrupted several times before, and these remarks would have interrupted this even more, we present them here together.

First, the presented line-of-thought takes the mode of the distribution as representing the average behavior, while one might think that one should calculate the average behavior directly. Apart from the fact that phenomenological thermodynamics in a way also calculates the mode (i.e., maximizes S), one can show that the distribution is extremely sharp and that mode and average are virtually the same. Essentially, one writes $\Sigma_i \exp(-\beta\varepsilon_i) \cong \int \exp(-\beta\varepsilon) \mathrm{d}n(\varepsilon) = \int \exp(-\beta\varepsilon) g(\varepsilon) \mathrm{d}\varepsilon$ with $g(\varepsilon) = \mathrm{d}n(\varepsilon)/\mathrm{d}\varepsilon$. The sharpness is due to the compensating effects of the rapidly increasing value of the density of states $g(\varepsilon)$ and the rapidly decreasing value of the Boltzmann factor $\exp(-\beta\varepsilon)$ (see Problem 5.16). In the so-called *thermodynamic limit* ($N \to \infty$ and $V \to \infty$ but N/V remains finite) mode and average are identical. The fact that mode and mean are virtually the same leads also to the (surprising) maximum-term method (see Justification 5.4) which allows one to calculate the value of a logarithmic sum by taking just the largest term. A completely rigorous calculation of average values is via the Darwin–Fowler method (see, e.g., Ref. [6]).

Second, note the difference in the behavior of conjugated variables in statistical thermodynamics as compared to phenomenological thermodynamics. Whilst in the latter case both members of a pair of conjugated variables can be fixed, in the former case they behave complementarily. For example, if the temperature is prescribed via a temperature bath (closed system), the energy of the system will fluctuate, whereas if the energy is fixed (isolated system) the temperature will fluctuate. The same applies to other pairs of conjugated variables, although for macroscopic systems these fluctuations are normally small. Fluctuations resulting from our statistical considerations must not be confused with the intrinsic uncertainty of quantum mechanics, as described by the Heisenberg uncertainty relations (Eq. 2.61), in which both members of a conjugated pair of variables can have an uncertainty. Our statistical fluctuations obviously add to the quantum uncertainty, but usually they are much larger.

Third and finally, we note that other ensembles can be defined by using other constraints to the energy expression given by the Hamilton operator $\mathcal{H}$. For example, using constant pressure P instead of constant volume V, we can calculate directly $G(P) = -kT\ln\Delta$, where $\Delta = \Sigma_i \exp[-\beta(E_i + PV)]$ using $\mathcal{H} + PV$ as energy expression [7] instead of calculating $G(P) = F(P) + PV(P)$ after solving V from $P = -\partial F(V)/\partial V$. Using an obvious indication of the independent variables, the partition function Δ refers to the N-P-T or *pressure ensemble*, while Z refers to the N-V-T or canonical ensemble. Although, strictly speaking, $\mathbb{H} + PV$ and similar expressions do not represent internal energy, in statistical thermodynamics they are still often referred to as Hamiltonians.

Justification 5.4: The maximum-term method

Surprisingly, in statistical thermodynamics the logarithm of a sum often can be approximated by the logarithm of its maximum term. Consider a typical statistical thermodynamics expression such as

$$S = \sum_{N=0}^{N=M} t_N \quad \text{where} \quad t_N = M!x^N / N!(M-N)!$$

and $x = \mathcal{O}(1)$ and $M = \mathcal{O}(10^{20})$, where $\mathcal{O}(x)$ indicates the order of magnitude of x. The sum S can be evaluated exactly as

$$S = (1+x)^M \quad \text{or} \quad \ln S = M\ln(1+x)$$

The largest term for S with index N^* is also the largest term for $\ln S$. Since N^* and $M - N^*$ are both large, we can use Stirling's approximation for the factorials and write

$$\ln t_N = M\ln M - N\ln N - (M-N)\ln(M-N) + N\ln x, \quad \text{so that} \tag{5.40}$$

$$\partial \ln t_N / \partial N = -\ln N + \ln(M-N) + \ln x = 0, \quad \text{which solves to}$$

$$N^*/(M-N^*) = x \quad \text{or} \quad N^* = xM/(1+x)$$

The maximum term evaluates to $\ln t_{N^*} = M\ln(1+x)$, identical to the sum itself! To better understand this behavior, we expand $\ln t_N$ about N^* to obtain

$$\ln t_N = \ln t_{N^*} + \frac{1}{2}(\partial^2 \ln t_N / \partial N^2)_{N=N^*}(N-N^*)^2 + \ldots$$

where the linear term is missing because $(\partial \ln t_N / \partial N)_{N=N^*} = 0$. From Eq. (5.40)

$$\frac{\partial^2 \ln t_N}{\partial N^2} = -\frac{1}{N} - \frac{1}{M-N} \quad \text{or} \quad \ln t_N = \ln t_{N^*} - \frac{(1+x)^2(N-N^*)^2}{2xM} + \ldots$$

Equivalently,

$$t_N = t_{N^*}\exp[-(N-N^*)^2 / 2\sigma^2] \quad \text{where} \quad \sigma = x^{1/2}M^{1/2}/(1+x)$$

This is a <u>very</u> sharp Gaussian distribution centered about N^*, since the relative standard deviation $\sigma/N^* = (xM)^{-1/2} = \mathcal{O}(M^{-1/2}) = \mathcal{O}(10^{-10})$. Higher-order terms would yield a completely negligible contribution, and we have $\langle N \rangle = N^*$.

Problem 5.16: The sharpness of the microstate distribution

Consider as system X a particle in a 3D box with density of states (DoS) $g_X(\varepsilon_X) \sim \varepsilon_X^{1/2}$. Take as reservoir R also such a system, so that the total energy becomes $\varepsilon = \varepsilon_X + \varepsilon_R$. Maximize the total DoS $g = g_X g_R$ using $\partial g/\partial \varepsilon_X = 0$, and show that $\varepsilon_X = \frac{1}{2}\varepsilon$. Is this the expected result, and is the associated energy distribution sharp? Next, consider a many-particle system with N_X particles with DoS $g_X \sim E_X^\alpha$ with $\alpha = (3N_X/2) - 1$. Couple this system to a many-particle reservoir with N_R particles with DoS $g_R \sim E_R^\beta$ with $\beta = (3N_X/2) - 1$. Show that in this case that g is extremely sharply peaked for large N_X and N_R with mode $E_X = \alpha E/(\alpha + \beta)$, where $E = E_X + E_R$.

5.5 Internal Contributions

So far, we have limited the discussion to molecules without internal degrees of freedom, that is, molecules without any other motion than translation. In Section 5.2 we saw that the contributions from internal mechanisms in molecules, as long as their energy is additive, can be incorporated in the overall partition function Z as a factor, for example, $Z = N!^{-1} z_{tra} z_{vib} z_{rot} z_{ele}$, where z_{tra}, z_{vib}, z_{rot} and z_{ele} denote the translation, vibration, rotation and electronic partition function, respectively. In this section we will address the vibration, the (external and internal) rotation, and (briefly) the electronic transitions that can occur in molecules. The pattern of how to calculate thermodynamic properties will be clear by now: find the energy expression, evaluate the partition function; and calculate from the partition function the thermodynamic properties.

5.5.1 Vibrations

The simplest molecule with internal structure is a diatomic molecule with m_1 and m_2 as the masses. The vibrations of a diatomic molecule can be described by the harmonic oscillator model with spring constant a and using the reduced mass $\mu = m_1 m_2/(m_1 + m_2)$ of the two atoms as the mass in the model. The energy of the oscillator using the circular frequency ω, as given in Section 2.2, is

$$\varepsilon_n = \hbar\omega\left(n + \frac{1}{2}\right) \quad \text{with} \quad n = 0, 1, 2, \ldots, \quad \text{and} \quad \omega = (a/\mu)^{1/2}$$

Hence, for the average energy $U = \langle \varepsilon \rangle$ we find

$$U = z_{vib}^{-1} \sum\nolimits_n \exp\left[-\frac{\hbar\omega}{kT}\left(n + \frac{1}{2}\right)\right] \hbar\omega\left(n + \frac{1}{2}\right) \quad \text{with}$$

$$z_{vib} = \sum\nolimits_n \exp\left[-\hbar\omega\left(n + \frac{1}{2}\right)\Big/kT\right]$$

If we use $\beta = (kT)^{-1}$, the partition function z_{vib} for a single vibrator can be written as

$$z_{\text{vib}} = \sum_n \exp\left[-\beta\hbar\omega\left(n+\frac{1}{2}\right)\right]$$

and, since this is a geometric series, we can evaluate the sum as

$$\blacktriangleright \quad z_{\text{vib}} = \exp\left(-\frac{1}{2}\hbar\omega\beta\right)[1-\exp(-\hbar\omega\beta)]^{-1} = \left[\exp\left(+\frac{1}{2}\hbar\omega\beta\right) - \exp\left(-\frac{1}{2}\hbar\omega\beta\right)\right]^{-1} \tag{5.41}$$

This expression cannot be used for a too-high temperature for two reasons. First, at higher temperature T, anharmonicity occurs and the harmonic oscillator model no longer applies. Second, in summing z_{vib} we assumed that $n \to \infty$, but in reality n is finite. Knox [8] estimated that one should have $kT < 0.1D$, where D is the dissociation energy. For a typical value of $D = 4$ eV, this results in $T < 4600$ K.

Since $z_{\text{vib}}\langle\varepsilon\rangle = \Sigma_n \exp[-\hbar\omega\beta(n+½)]\, \hbar\omega\,(n+½) = -\partial z_{\text{vib}}/\partial\beta$, we can write

$$U = -\frac{\partial z_{\text{vib}}/\partial\beta}{z} = -\frac{\partial \ln z_{\text{vib}}}{\partial\beta} = \frac{1}{2}\hbar\omega + \frac{\hbar\omega}{\exp(\hbar\omega/kT)-1} \tag{5.42}$$

The Helmholtz energy F is given by

$$F = -kT\ln z_{\text{vib}} = \frac{1}{2}\hbar\omega + kT\ln[1-\exp(-\hbar\omega/kT)] \tag{5.43}$$

so that the entropy S is given by

$$S = -\frac{\partial F}{\partial T} = -k\ln[1-\exp(-\hbar\omega/kT)] + \frac{k(\hbar\omega/kT)\exp(-\hbar\omega/kT)}{1-\exp(-\hbar\omega/kT)} \tag{5.44}$$

The energy can also be calculated from $U = F + TS$. The heat capacity C_V reads

$$C_V = \frac{\partial U}{\partial T} = \frac{k(\hbar\omega/kT)^2\exp(\hbar\omega/kT)}{[\exp(\hbar\omega/kT)-1]^2} \tag{5.45}$$

At low and high temperature the expression for C_V expands to, respectively,

$$C_V = k\left(\frac{\hbar\omega}{kT}\right)^2 \exp\left(\frac{\hbar\omega}{kT}\right) + \ldots \quad \text{and} \quad C_V = k\left[1 + \frac{1}{12}\left(\frac{\hbar\omega}{kT}\right)^2 + \ldots\right]$$

The behavior of the harmonic oscillator is characterized by ω. Equivalently, we use the *characteristic (vibration) temperature* θ_{vib}, given as $\theta_{\text{vib}} = \hbar\omega/k$. When $T \gg \theta_{\text{vib}}$, the behavior can be classified as classical and the energy becomes $U \cong kT$. When $T \ll \theta_{\text{vib}}$, expansion of the expression results in $U \cong ½\hbar\omega$. When $T \cong \theta_{\text{vib}}$, the full expression must be used. Since the characteristic temperature for vibration is typically hundreds of degrees, this implies that in most cases the quantum expression for the partition function must be applied. Values for θ_{vib} as well as the equilibrium distance r_{eq} and dissociation energy D_{eq} for some molecules are given in Table 5.1.

Table 5.1 Vibration, rotation, and dissociation data for several diatomic molecules.[a]

Molecule	θ_{vib} (K)	θ_{rot} (K)	r_{eq} (Å)	D_{eq} (kJ mol^{-1})	g_0 (–)
H_2	66 332	85.3	0.740	457.6	1
CO	3103	2.77	1.13	1085	1
NO	2719	2.39	1.15	638.1	2
N_2	3374	2.88	1.10	953.0	1
O_2	2256	2.07	1.204	503.0	3
HF	5840	30.3	0.917	565.4	1
HCl	4227	15.22	1.28	445.2	1
HBr	3787	12.02	1.41	377.7	1
HI	3266	9.25	1.60	308.6	1
F_2	1328	1.29	1.41	153.4	1
Cl_2	805	0.351	1.99	242.3	1
Br_2	463	0.116	2.28	191.9	1
I_2	308	0.0537	2.67	150.3	1

a) Data from Refs. [13a] and [16].

For a N-atomic molecule we can describe the vibrational behavior by $3N - 6$ *normal coordinates* ($3N - 5$ for a linear molecule), each of which can be modeled as a harmonic oscillator[13]. These normal coordinates are independent, and hence the behavior can be described as the sum of the behaviors of the individual normal coordinates. In fact, they are an example of generalized coordinates. For example, for H_2O the three vibrational temperatures are 2294, 5262, and 5404 K, respectively, while they read 960, 960, 132, and 3380 K for CO_2.

As an aside, we note that the harmonic oscillator results are directly applicable to the *Einstein model* for solids in which each atom vibrates independently of the others with the same (Einstein) frequency ω_E. In this case, we have for N atoms $Z_{vib} = z_{vib}^N$, that is, without the factor $N!^{-1}$ because, although the atoms are indistinguishable, the lattice sites distinguishable and the total number of configurations is $N!$.

Problem 5.17

Show that the entropy S for the harmonic oscillator is given by Eq. (5.44).

Problem 5.18

Show that the specific heat C_V for the harmonic oscillator is given by Eq. (5.45). Show also that for $T >> \theta_{vib}$, $U \cong kT$ and $C_V \cong k$.

13) Any N-atom molecule in principle has $3N - 6$ internal degrees of freedom. For linear molecules this number reduces to $3N - 5$ because rotation around the length axis, although possible in principle, will require extremely high energy in view of the small moment of inertia around that axis.

Problem 5.19

Show that at 300 K most molecules are in the vibrational ground state.

Problem 5.20: The anharmonic oscillator*

Anharmonicity changes the energy expression and selection rule for the oscillator to

$$\varepsilon_n / hc = \bar{\omega}\left(n+\frac{1}{2}\right) - x\bar{\omega}\left(n+\frac{1}{2}\right)^2 \quad \text{and} \quad \Delta n = \pm 1, \pm 2, \ldots$$

where ε_n/hc is the energy in cm^{-1}, $\bar{\omega}$ the vibrational wave number (see footnote 11; typically 100 to 4000 cm^{-1}), and x is the *anharmonicity constant* (typically ~0.01). Estimate the contribution to C_V from anharmonicity.

5.5.2
Rotations

The rotation behavior of molecules is somewhat more complex than their vibration behavior. We start with homonuclear and heteronuclear diatomic molecules, and thereafter deal briefly with polyatomic molecules and internal rotation.

The energy of the diatomic rigid rotator, as given in Section 2.2, is

$$\varepsilon_J = BJ(J+1) \text{ with } J = 0, 1, 2, \ldots \quad \text{and} \quad B = h^2/8\pi^2 I \equiv k\theta_{\text{rot}} \tag{5.46}$$

Here, the moment of inertia $I = \mu r^2$ with the reduced mass $\mu = m_1 m_2/(m_1 + m_2)$ and m_1 and m_2 the masses of the particles. The constant B is the *rotational constant*[12], alternatively expressed as the *characteristic (rotation) temperature* $\theta_{\text{rot}} = B/k$. Each of the energy levels J has a $(2J + 1)$ degeneracy. At low temperature only the first few terms are contributing, and we have

$$z_{\text{rot}} = \sum\nolimits_J (2J+1)\exp\left[-\frac{BJ(J+1)}{kT}\right] = 1 + 3\exp\left(-\frac{2\theta_{\text{rot}}}{T}\right) + 5\exp\left(-\frac{6\theta_{\text{rot}}}{T}\right) + \ldots$$

At high temperature the summation can be approximated by integration and thus

$$\blacktriangleright \quad z_{\text{rot}} = \sum\nolimits_J (2J+1)\exp\left[-\frac{B}{kT}J(J+1)\right] \cong \int_0^\infty (2J+1)\exp\left[-\frac{B}{kT}J(J+1)\right]\text{d}J$$
$$= \int_0^\infty \exp\left[-\frac{B}{kT}J(J+1)\right]\text{d}[J(J+1)] = kT/B = T/\theta_{\text{rot}} \tag{5.47}$$

Actually, we have to divide z_{rot} by the symmetry number σ, denoting the number of ways the molecule can be rotated into a configuration indistinguishable from the original configuration. Obviously, $\sigma = 1$ for AB and $\sigma = 2$ for AA molecules. The origin of this factor lies in the symmetry of the overall wave function and for the details of the derivation we refer to the literature (e.g., McQuarrie

[9]). A more complete analysis using the Euler–McLaurin summation expression results in

$$z_{\text{rot}} = \left(\frac{T}{\sigma\theta_{\text{rot}}}\right)\left[1 + \frac{1}{3}\frac{\theta_{\text{rot}}}{T} + \frac{1}{15}\left(\frac{\theta_{\text{rot}}}{T}\right)^2 + \frac{4}{315}\left(\frac{\theta_{\text{rot}}}{T}\right)^3 + \ldots\right]$$

which is good to within 1% for $\theta_{\text{rot}} < T$. Replacing the summation by integration is thus only justified at sufficiently elevated temperature, say for $T > 5\theta_{\text{rot}}$. For a lower T, one must sum z_{rot} term by term. Typical values for θ_{rot} are given in Table 5.1.

If a polyatomic molecule is linear, the same expression as for diatomic molecules applies, of course, with adjusted moments of inertia. For a nonlinear polyatomic molecule a similar, but more complex, reasoning leads to

$$z_{\text{rot}} = \left(\frac{8\pi^2 kT}{h^2}\right)^{3/2} \left(I_x I_y I_z\right)^{1/2} \frac{\sqrt{\pi}}{\sigma} = \sqrt{\frac{T}{\theta_x}}\sqrt{\frac{T}{\theta_y}}\sqrt{\frac{T}{\theta_z}}\frac{\sqrt{\pi}}{\sigma} \tag{5.48}$$

where I_x, I_y and I_z denote the principal moments of inertia or, equivalently, θ_x, θ_y and θ_z the rotational temperatures. The symmetry number has the same meaning as before. As examples we quote $\sigma(H_2O) = 2$, $\sigma(NH_3) = 3$, $\sigma(CH_4) = 12$, $\sigma(CH_3Cl) = 3$, $\sigma(C_6H_6) = 12$, and $\sigma(C_6H_5Cl) = 2$. For polyatomic molecules the semi-classical partition function is generally valid for $T > 100\,\text{K}$.

Internal rotations, such as the transition from the staggered to eclipsed conformation in CH_3–CH_3, also contribute. In the case of molecules for which the associated barrier is relatively large, say $> 10kT$, the two groups only vibrate with respect to each other and the harmonic oscillator approximation can be used. If the barrier to internal rotation is relatively low, say $< kT$, one could consider the internal rotation as a free rotation for which the partition function is

▶ $$z_{\text{int rot}} = \left(8\pi^2 I' kT/h^2\right)^{1/2} \tag{5.49}$$

where I' is the moment of inertia for the internal rotation. In the case of ethane, however, the barrier is ~12.1 kJ mol^{-1}, while $N_A kT$ at 300 K is ~2.5 kJ mol^{-1} and thus the neither the harmonic oscillator nor the free rotation approximation may be used. The internal rotation contribution is often the main source of error in calculating thermodynamic data (see, e.g., Ref. [10] for a more detailed discussion).

5.5.3
Electronic Transitions

Generally, the spacing between electronic states in molecules is large as compared with kT. This implies that transitions from the ground state to an excited state are rare, and only the electronic ground state has to be taken into account. Hence the electronic partition function is generally closely approximated by

$$z_{\text{ele}} = g_0 \exp(-\varepsilon_0/kT) + g_1 \exp(-\varepsilon_1/kT) + \ldots \cong g_0 \exp(-\varepsilon_0/kT) \tag{5.50}$$

where g_i indicates the electronic state degeneracy, for the ground state often $g_0 = 1$. Taking the zero level at the electronic ground state the approximation $z_{\text{ele}} = g_0$ is in most cases sufficiently accurate for both atoms and molecules. One exception is formed by the systems atoms where the electronic ground state is split by spin-orbit coupling. For example, for the F atom the ground state is split in a lower $^3P_{3/2}$ level with $g_0 = 4$ and an upper $^2P_{1/2}$ level with $g_1 = 2$. The energy gap between these levels is 404 cm^{-1}, while the gap with the next electronic level is much larger than $kT = 207\,\text{cm}^{-1}$ at 298 K. Another example is the NO molecule with an unpaired electron in a π^* orbital giving rise to a $^2\Pi_{1/2}$ ($g_0 = 2$) and a $^2\Pi_{3/2}$ ($g_1 = 2$) level with a gap of 121 cm^{-1}. Again, the gap to the first excited (in this case vibration) level is much larger than kT.

Problem 5.21

Estimate the temperature above which the replacement of the summation in the calculation of the rotational partition function by integration is justified. Take typical numbers from Table 5.1.

Problem 5.22

Calculate the contribution of the $^3P_{3/2}$ and $^2P_{1/2}$ levels for the F atom to C_V at room temperature and compare the value obtained with the one for translation.

Problem 5.23

Show that for a diatomic molecule the contribution of rotation to the heat capacity C_V is given by $C_V = R[1+(\theta_{\text{rot}}^2/45T^2)+\ldots]$ and to the entropy S by $R[1-\ln(\sigma\theta_{\text{rot}}/T)-(\theta_{\text{rot}}^2/90T^2)+\ldots]$.

Problem 5.24

Show that the most probable rotational state for a diatomic molecule is given by $J_{\max} \cong (T/2\sigma\theta_{\text{rot}})^{1/2} - 1/2$ by treating J as a continuous variable (high temperature approximation).

Problem 5.25: The non-rigid rotor*

Diatomic molecules are not rigid rotors, and the energy is more appropriately described by

$$\varepsilon_J/hc = BJ(J+1) + DJ^2(J+1)^2, \quad B = h/8\pi^2 Ic, \quad D = 4B^3/\bar{\omega}^2$$

and $\bar{\omega}$ the vibrational wave number. For the HCl molecule the rotational constant B and *centrifugal distortion constant* D are 10.59 cm^{-1} and 5.28×10^{-4} cm^{-1}, respectively. Calculate the C_V according the rigid rotor energy expression, Eq. 5.47. Estimate the contribution of the centrifugal distortion on the energy levels and compare the associated C_V with the total C_V.

5.6 Real Gases

Statistical thermodynamics provides a useful description of (nearly) perfect gases, that is, low density gases, proof being that the values of the standard entropy $S_m{}^\circ$ can be calculated in precise agreement with experimental data[14]. Example 10.2 illustrates this. We need, of course, also a description of nonperfect gases, and to that purpose we need to introduce the interaction potential in the partition function.

5.6.1 Single Particle

Let us recapitulate the results for the partition function z of a single particle with mass m, momentum $\mathbf{p}_1$ and (Cartesian) coordinate $\mathbf{r}_1$ enclosed in a box with edge l (and volume $V = l^3$). The energy E is only kinetic and given by the Hamilton function $\mathcal{H}$ so that

$$E = \mathcal{H}(\mathbf{p}_1, \mathbf{r}_1) = \mathbf{p}_1^2/2m$$

The partition function z is given by

$$\begin{aligned} z &= h^{-3}(1!)^{-1}\iint \exp[-\mathcal{H}(\mathbf{p}_1,\mathbf{r}_1)/kT]\mathrm{d}\mathbf{p}_1\,\mathrm{d}\mathbf{r}_1 = h^{-3}(1!)^{-1}\int \mathrm{d}\mathbf{r}_1 \int \exp(-\mathbf{p}_1^2/2mkT)\mathrm{d}\mathbf{p}_1 \\ &= h^{-3}(1!)^{-1}\int \mathrm{d}x_1 \int \mathrm{d}y_1 \int \mathrm{d}z_1 \times \iiint \exp[-(p_x^2 + p_y^2 + p_z^2)/2mkT]\mathrm{d}p_x\,\mathrm{d}p_y\,\mathrm{d}p_z \\ &= h^{-3}(1!)^{-1}\int \mathrm{d}x_1 \int \mathrm{d}y_1 \int \mathrm{d}z_1 \\ &\quad \times \int \exp(-p_x^2/2mkT)\mathrm{d}p_x \times \int \exp(-p_y^2/2mkT)\mathrm{d}p_y \times \int \exp(-p_z^2/2mkT)\mathrm{d}p_z \end{aligned}$$

Evaluating this expression results, as before, in the partition function

$$z = V \times (2\pi mkT/h^2)^{3/2} = (l/\Lambda)^3 \qquad (5.51)$$

with $\Lambda = (h^2/2\pi mkT)^{1/2}$ the *thermal wavelength*.

5.6.2 Interacting Particles

Let us now evaluate the partition function for two interacting particles, characterized by the set of variables $(\mathbf{p},\mathbf{r}) = (\mathbf{p}_1,\mathbf{r}_1,\mathbf{p}_2,\mathbf{r}_2)$ and their potential energy of interaction $\phi(\mathbf{r}_1,\mathbf{r}_2)$. In this case, the energy is given by

14) These calculations are essentially based on translation, harmonic vibration and rigid rotation as well as on the third law, that is, the residual entropy $S_{res} = 0$. For molecules such as CO with only a small dipole moment, (almost) random ordering in the lattice occurs, resulting in $W = 2^N$ possible configurations and thus to an entropy contribution $S_{res} = k \ln W = 5.8\,\mathrm{J\,K^{-1}mol^{-1}}$, in essential agreement with the experimental residual entropy for CO, determined as $4.5\,\mathrm{J\,K^{-1}mol^{-1}}$.

$$E = \mathcal{H}(\mathbf{p},\mathbf{r}) = \mathcal{H}(\mathbf{p}_1,\mathbf{r}_1,\mathbf{p}_2,\mathbf{r}_2) = \mathbf{p}_1^2/2m + \mathbf{p}_2^2/2m + \phi(\mathbf{r}_1,\mathbf{r}_2)$$

Actually, the potential energy ϕ can be written as $\phi(\mathbf{r}_1,\mathbf{r}_2) = \phi(r)$ with $r = |\mathbf{r}_2 - \mathbf{r}_1|$ if we assume that ϕ is spherically symmetrical and thus only depends on the distance r between the two particles. Thus, we obtain for the partition function

$$\begin{aligned} Z &= h^{-3N}(N!)^{-1}\iint \exp[-\mathcal{H}(\mathbf{p},\mathbf{r})/kT]\,\mathrm{d}\mathbf{p}\,\mathrm{d}\mathbf{r} \\ &= h^{-6}(2!)^{-1}\int \exp(-\mathbf{p}_1^2/2mkT)\,\mathrm{d}\mathbf{p}_1 \times \int \exp(-\mathbf{p}_2^2/2mkT)\,\mathrm{d}\mathbf{p}_2 \\ &\quad \times \iint \exp(-\phi(r)/kT)\,\mathrm{d}\mathbf{r}_1\,\mathrm{d}\mathbf{r}_2 \\ &= (2!)^{-1}\Lambda^{-3} \times \Lambda^{-3} \times V\int \exp[-\phi(r)/kT)]4\pi r^2\,\mathrm{d}r \\ &\equiv (2!)^{-1}\Lambda^{-6} \times Q_{12} \end{aligned} \tag{5.52}$$

The factor Q_{12} is usually addressed as the *(two particle) configuration integral*.

This result can be easily generalized to N particles, characterized by $(\mathbf{p},\mathbf{r}) = (\mathbf{p}_1, \mathbf{r}_1, \mathbf{p}_2, \mathbf{r}_2, \ldots, \mathbf{p}_N, \mathbf{r}_N)$ and $\Phi(\mathbf{r}) = \Phi(\mathbf{r}_1,\mathbf{r}_2, \ldots, \mathbf{r}_N)$. The total energy is

$$E = \mathcal{H}(\mathbf{p},\mathbf{r}) = \mathcal{H}(\mathbf{p}_1,\mathbf{r}_1,\mathbf{p}_2,\mathbf{r}_2,\ldots,\mathbf{p}_N,\mathbf{r}_N) = \Sigma_i \mathbf{p}_i^2/2m + \Phi(\mathbf{r}_1,\mathbf{r}_2,\ldots,\mathbf{r}_N)$$

Assuming again spherically symmetrical interactions between the particles we have

$$\Phi(\mathbf{r}) = \Phi(\mathbf{r}_1,\mathbf{r}_2,\ldots,\mathbf{r}_N) = \Phi(r_{12},\ldots,r_{1N},r_{23},\ldots,r_{2N},\ldots r_{N-1,N})$$

where r_{ij} denotes the distance between particle i and particle j. The partition function becomes

$$\begin{aligned} Z &= h^{-3N}(N!)^{-1}\iint \exp(-\mathcal{H}/kT)\,\mathrm{d}\mathbf{p}\,\mathrm{d}\mathbf{r} \\ \blacktriangleright\quad &= h^{-3N}(N!)^{-1}\Pi_i \int \exp(-\mathbf{p}_i^2/2mkT)\,\mathrm{d}\mathbf{p}_i \times \iint \exp[-\Phi(r)/kT]\,\mathrm{d}\mathbf{r}_1 \ldots \mathrm{d}\mathbf{r}_N \\ &= (N!)^{-1}(\Lambda^{-3})^N \times \iint \exp[-\Phi(r)/kT]\,\mathrm{d}\mathbf{r}_1 \ldots \mathrm{d}\mathbf{r}_N \equiv (N!)^{-1}\Lambda^{-3N} \times Q \end{aligned} \tag{5.53}$$

with Q the *configuration integral*. Our remaining task is thus to evaluate this many-dimensional integral.

5.6.3 The Virial Expansion: Canonical Method

The method we use in this paragraph to evaluate Q is denoted as the *virial expansion*. There are several ways to do this and we use the simplest possible. The first thing we assume is that we can write the potential Φ in the expression for

$$Q = \int \ldots \int \exp[-\Phi(r)/kT]\,\mathrm{d}\mathbf{r}_1 \ldots \mathrm{d}\mathbf{r}_N \tag{5.54}$$

as $\Phi(r) = \Sigma_{i<j}\phi(r_{ij}) = \frac{1}{2}\Sigma_{i,j}\phi(r_{ij})$. This allows us to rearrange the exponential of the sum as the product of exponentials since $\exp(-\Sigma_i a_i) = \Pi_i \exp(-a_i)$. Applying this

operation to the configuration integral, meanwhile using $\exp(ab) = (\exp a)^b$, we have

$$Q = \int \ldots \int \Pi_{ij} \exp\left[-\frac{1}{2}\phi(r_{ij})/kT\right] \mathrm{d}\mathbf{r}_1 \ldots \mathrm{d}\mathbf{r}_N$$
$$= \int \ldots \int \Pi_{ij} \{\exp[-\phi(r_{ij})/kT]\}^{1/2} \mathrm{d}\mathbf{r}_1 \ldots \mathrm{d}\mathbf{r}_N$$

The trick is now to introduce the *Mayer function* $f_{ij} = \exp[-\phi(r_{ij})/kT] - 1$ so that

$$Q = \int \ldots \int \Pi_{ij} (1 + f_{ij})^{1/2} \mathrm{d}\mathbf{r}_1 \ldots \mathrm{d}\mathbf{r}_N$$
$$= \int \ldots \int \left(1 + \frac{1}{2}\Sigma_{ij} f_{ij} - \frac{1}{8}\Sigma_{ij}\Sigma_{kl \neq ij} f_{ij} f_{kl} + \ldots\right) \mathrm{d}\mathbf{r}_1 \ldots \mathrm{d}\mathbf{r}_N$$
$$\cong \int \ldots \int \left(1 + \frac{1}{2}\Sigma_{ij} f_{ij}\right) \mathrm{d}\mathbf{r}_1 \ldots \mathrm{d}\mathbf{r}_N = V^N + \frac{1}{2}N(N-1)V^{N-2}Q_{12} + \ldots \tag{5.55}$$

$$Q_{12} = \iint f_{12}\, \mathrm{d}\mathbf{r}_1\, \mathrm{d}\mathbf{r}_2 = \int \mathrm{d}\mathbf{r}_1 \int f(r) 4\pi r^2\, \mathrm{d}r = 4\pi V \int f(r) r^2\, \mathrm{d}r \equiv 2Vb_2 \tag{5.56}$$

where in the last line particle 1 is taken as the origin so that the integration over particle 1 yields the volume and the remaining integration is only over the distance r. The final result, assuming $N(N-1)b_2/V << 1$, is

$$Q = V^N \left[1 + \frac{N(N-1)}{V} b_2 + \ldots\right] \quad \text{with}$$
$$b_2 \equiv 2\pi \int f(r) r^2\, \mathrm{d}r = 2\pi \int \{\exp[-\phi(r)/kT] -\} r^2\, \mathrm{d}r \tag{5.57}$$

In the canonical ensemble the pressure P can be calculated using $P = -\partial F/\partial V = kT\, \partial \ln Z/\partial V = kT\, \partial \ln Q/\partial V$ and reads $PV/nRT = 1 + B_2/V + \ldots$ with $B_2 = -b_2$ (Problem 5.27) or $\beta P = \rho + B_2\rho^2 + \ldots$ with $\rho = N/V$ and $\beta = 1/kT$. However, in general the expression for ϕ is complex so that analytical evaluation of b_2 is almost never possible and we have to resort to numerical integration.

Example 5.3: The hard-sphere second virial coefficient

A simple potential energy expression for which an analytical evaluation of b_2 is possible is the hard-sphere model,

$$\phi = \infty \quad \text{for} \quad r < \sigma, \quad \phi = 0 \text{ otherwise}$$

We split the integration for b_2 from 0 to ∞ in 0 to σ and from σ to ∞. The first integral reads $2\pi \int_0^\sigma (\mathrm{e}^{-\beta\infty} - 1) r^2\, \mathrm{d}r = -2\pi\sigma^3/3$, while the second integral results in $2\pi \int_\sigma^\infty (\mathrm{e}^{-\beta 0} - 1) r^2\, \mathrm{d}r = 0$. The total result is thus $B_2 = -b_2 = 2\pi\sigma^3/3$.

From Example 5.3 we note that $b_2 < 0$, and this appears generally to be the case (see Chapter 4). For convenience often the quantity $b \equiv B_2 = -b_2 = 2\pi\sigma^3/3$ is defined and used as a reference value in more extended calculations, and we do so likewise. Guggenheim refers to gases which can be described reasonably well by using only the second virial coefficient B_2 as *"slightly imperfect gases"*.

The excess Helmholtz energy $F^{\mathrm{E}} = F - F^{\mathrm{idg}}$ can be written as $-\beta F^{\mathrm{E}}/V = B_2^{\mathrm{F}}\rho^2 + B_3^{\mathrm{F}}\rho^3 + \ldots$ with the ideal gas Helmholtz energy $\beta F^{\mathrm{idg}} = N\ln(\rho/\rho°)$ where $\rho°$ is a reference density. The coefficients are related to the pressure virial coefficients, and can be shown by integration to be

$$B_{\mathrm{n}} = -(n-1)B_{\mathrm{n}}^{\mathrm{F}} \tag{5.58}$$

5.6.4
The Virial Expansion: Grand Canonical Method*

There is a problem with the derivation given in the previous section. It is assumed that the correction terms are small. For the first correction term with respect to the perfect gas this implies that $(N(N-1)/V)b << 1$. Assuming b to be of the order of the molecular volume, one can easily calculate that this is not the case. Another approach is thus required and the shortest is via the grand canonical partition function.

Recall that the grand canonical partition function reads

$$\Xi = \exp(\beta PV) = \sum\nolimits_N \lambda^N Z_N \quad \text{where} \quad \lambda = \exp(\beta\mu) \tag{5.59}$$

and Z_N is the canonical partition function for N particles, which thus includes all energies belonging to the state containing N particles. For consistency, Z_0 is defined as $Z_0 = 1$, while Z_1 represents the situation containing one particle and is thus $Z_1 = Vz_{\mathrm{int}}/\Lambda^3$, where z_{int} represents the internal contributions. We will neglect the latter and thus use $Z_1 = V/\Lambda^3 = Q_1/\Lambda^3$, since for a single particle the configurational partition function equals the volume. This implies that the partition function becomes

$$Z_N = \frac{1}{N!}\left(\frac{1}{\Lambda^3}\right)^N Q_N \tag{5.60}$$

where Q_N is the configurational partition function for N particles and that Ξ reads

$$\Xi = \sum\nolimits_N \frac{1}{N!}\lambda^N\left(\frac{1}{\Lambda^3}\right)^N Q_N = \sum\nolimits_N \xi^N \frac{Q_N}{N!} \quad \text{with} \quad \xi = \frac{\lambda}{\Lambda^3} \tag{5.61}$$

Let us now assume that we may expand the pressure in a power series in ξ, that is,

$$\beta P = \sum\nolimits_j b_j \xi^j \tag{5.62}$$

so that the grand partition function, using $\exp(x) = 1 + x + x^2/2 + \ldots$, becomes

$$\Xi = \exp(\beta PV) \tag{5.63}$$

$$= 1 + V\left(b_1\xi + b_2\xi^2 + b_3\xi^3 + \cdots\right) + \frac{V^2}{2}\left(b_1\xi + b_2\xi^2 + \cdots\right)^2 + \frac{V^3}{6}\left(b_1\xi + \cdots\right)^3 + \cdots \tag{5.64}$$

$$= 1 + Vb_1\xi + \left(Vb_2 + \frac{V^2b_1^2}{2}\right)\xi^2 + \left(Vb_3 + b_1b_2V^2 + \frac{V^3b_1^3}{6}\right)\xi^3 + \cdots \tag{5.65}$$

Comparing terms with $\Xi = 1 + \xi Q_1 + (\xi^2 Q_2/2!) + (\xi^3 Q_3/3!) + \cdots$, Eq. (5.61), yields

$$Vb_1 = Q_1 \quad \rightarrow \quad b_1 = Q_1/V = 1 \tag{5.66}$$

$$\left(Vb_2 + V^2b_1^2/2\right) = Q_2/2 \quad \rightarrow \quad b_2 = \left(Q_2 - Q_1^2\right)/2V \tag{5.67}$$

$$Vb_3 + b_1b_2V^2 + \frac{V^3b_1^3}{6} = \frac{Q_3}{6} \quad \rightarrow \quad b_3 = \left(Q_3 - 3Q_1Q_2 + 2Q_1^3\right)/6V \tag{5.68}$$

From the expression for the number of particles

$$N = \lambda\frac{\partial \ln \Xi}{\partial \lambda} = \xi\frac{\partial \ln \Xi}{\partial \xi} = \xi\frac{\partial (PV/kT)}{\partial \xi} = \xi V\frac{\partial}{\partial \xi}\sum\nolimits_j b_j\xi^j = V\sum\nolimits_j jb_j\xi^j \tag{5.69}$$

we obtain the number density

$$\rho = \frac{N}{V} = b_1\xi + 2b_2\xi^2 + 3b_3\xi^3 + \cdots \tag{5.70}$$

Inverting this series we write

$$\xi = a_1\rho + a_2\rho^2 + a_3\rho^3 + \cdots \tag{5.71}$$

and substitute this expression in the one for ρ, Eq. (5.70), to obtain

$$\begin{aligned} \rho &= b_1(a_1\rho + a_2\rho^2 + a_3\rho^3 + \cdots) \\ &\quad + 2b_2(a_1^2\rho^2 + 2a_1a_2\rho^3 + \cdots) + 3b_3(a_1^3\rho^3 + \cdots) + \cdots \end{aligned} \tag{5.72}$$

so that the coefficients a_i become

$$a_1b_1 = 1 \quad \rightarrow \quad a_1 = b_1 = 1 \tag{5.73}$$

$$a_2 + 2b_2a_1 = 0 \quad \rightarrow \quad a_2 = -2b_2 \tag{5.74}$$

$$a_3 + 4b_2a_1^2a_2 + 3b_3a_1^3 = 0 \quad \rightarrow \quad a_3 = 8b_2^2 - 3b_3 \tag{5.75}$$

Now these coefficients are used in the series for ξ, Eq. (5.71), and we obtain

$$\xi = b_1\rho - 2b_2\rho^2 + (8b_2^2 - 3b_3\rho^3)\rho^3 + \cdots \tag{5.76}$$

The final step is made by substituting this expression in the expression for βP, Eq. (5.62), and the result reads

$$\beta P = \sum\nolimits_j b_j\xi^j = b_1\xi + b_2\xi^2 + b_3\xi^3 + \cdots \tag{5.77}$$

$$\begin{aligned} &= b_1\left[b_1\rho - 2b_2\rho^2 + (8b_2^2 - 3b_3)\rho^3 + \cdots\right] \\ &\quad + b_2(b_1^2\rho^2 - 4b_1b_2\rho^3 + \cdots) + b_3(b_1^3\rho^3 + \cdots) + \cdots \end{aligned} \tag{5.78}$$

$$= \rho - b_2\rho^2 + (4b_2^2 - 2b_3)\rho^3 + \cdots \tag{5.79}$$

Comparing this result with the virial expansion $\beta P = \rho + B_2\rho^2 + B_3\rho^3 + \cdots$, we obtain

$$B_2 = -b_2 \quad \text{and} \quad B_3 = 4b_2^2 - 2b_3 \tag{5.80}$$

For B_2 the final result is thus

$$B_2 = -\frac{Q_2 - Q_1^2}{2V} = -\frac{Q_2 - V^2}{2V} = -\frac{1}{2V}\int [\exp(-\beta\phi) - 1]\mathrm{d}\mathbf{r}_1\,\mathrm{d}\mathbf{r}_2 \tag{5.81}$$

$$= -\frac{1}{2V}\int f(r)\mathrm{d}\mathbf{r}_1\,\mathrm{d}\mathbf{r}_2 = -\frac{1}{2}\int f(r)\mathrm{d}\mathbf{r} \tag{5.82}$$

The final result for B_2 using the grand canonical partition function thus agrees with the result from the canonical partition function. However, in the former case B_2 is the coefficient of a small term while in the latter case it is the coefficient of a large term, rendering the expansion rather doubtful. It also shows that the second virial coefficient is strictly dependent only on two-body interactions. This applies also to the higher-order terms: the nth order virial coefficient is only dependent on the n-body interactions (and the lower-order ones). The virial expansion has also been derived without making the pair potential assumption [11] as well as with an entirely different method [12].

5.6.5 Critique and Some Further Remarks

A systematic extension of the virial expansion can be realized, although this is usually performed via a graphical technique known as the cluster expansion (see Ref. [13]). This extension is required when we wish to apply the virial expansion to higher densities, such as those of liquids. As mentioned in Chapter 1 (and further discussed in Chapter 6), the coordination number of a reference molecule in a liquid is generally about 10. Recalling that the nth virial coefficient deals with n-body interactions, a description of liquids in terms of the virial equation thus requires virial coefficients at least up to order 10. The calculation of high-order virial coefficients is rather complex, even for a simple potential like the hard-sphere potential. While the third and fourth virial coefficients for the hard-sphere potential can be calculated analytically, the values for the higher-order coefficients are obtained from simulations. The virial coefficients for the hard-sphere gas up to B_{10} is shown in Table 5.2. For more realistic potentials, however, usually only the second or, maybe the third or fourth coefficient is available. Hence, for liquids we encounter problems in view of the high coordination numbers.

From the data in Table 5.2 it is be clear that the convergence of the virial expansion is not rapid. This is even made clearer if we introduce the close-packed volume, that is, the volume per molecule for a FCC lattice given by $V_0/N = \rho_0^{-1} = \sigma^3/2^{1/2}$ of spheres with diameter σ, as the reference volume. Introducing the *packing fraction* (or *reduced density*) $\eta = N(4\pi/3)(\sigma/2)^3/V = \rho\pi\sigma^3/6 = \rho b/4 =$

Table 5.2 Virial coefficients of the hard-sphere fluid in terms of $B_2 = b = 2\pi\sigma^3/3$.[a)]

Term	ρ expansion	η expansion	Term	ρ expansion	η expansion
B_3	$5/8\ b^2$	10	B_7	$0.01302354(91)\ b^6$	53.34
B_4	$0.2869495\ldots b^3$	18.36	B_8	$0.0041832(11)\ b^7$	68.54
B_5	$0.110252(1)\ b^4$	28.22	B_9	$0.0013094(13)\ b^8$	85.81
B_6	$0.03888198(91)\ b^5$	39.82	B_{10}	$0.0004035(15)\ b^9$	105.8

a) Data from Ref. [17].

$2^{1/2}\pi V_0/6V$, the virial expansion can also be expressed in η instead of ρ. Note that for the packing fraction we have $0 \le \eta \le 2^{1/2}\pi/6 \cong 0.7405$.

$$\begin{aligned}\frac{P}{\rho kT} &= 1 + b\rho + 0.625b^2\rho^2 + 0.28695b^3\rho^3 + 0.1103b^4\rho^4 + \cdots \\ &= 1 + 4\eta + 10\eta^2 + 18.36\eta^3 + 28.22\eta^4 + \cdots\end{aligned} \tag{5.83}$$

For hard-spheres, $P/\rho kT \ge 1$ and continuously increases with ρ. A comparison of the virial results including B_{10} with those of computer simulations [14] shows that they begin to deviate at $\rho\sigma^3 \cong 0.5$ or $\eta \cong 0.26$. However, a 5,4 Padé approximant[15)] to the first six hard-sphere virial coefficients reading

$$\frac{P}{\rho kT} = \frac{\begin{array}{c}1 + 0.80409b\rho + 0.46662b^2\rho^2 + 0.14198b^3\rho^3 \\ + 0.029739b^4\rho^4 + 0.0028192b^5\rho^5\end{array}}{1 - 0.19591b\rho + 0.037534b^2\rho^2 - 0.060058b^3\rho^3 + 0.012303b^4\rho^4} \tag{5.84}$$

provides an excellent agreement with molecular simulation results, that is, a deviation less than 0.3% over the complete range [14].

A few problems exist when the virial expansion is applied to liquids; already mentioned are the availability and complexity of virial coefficient calculations and the slow convergence. However, real convergence problems exist with the virial expansion at certain densities and temperatures, as it converges only in some nonzero region, although in that region convergence occurs for a wide class of potentials. Essentially, $B_n \sim V_0^{n-1}$ and if $\rho \to V_0^{-1}$ the expansion diverges. Perhaps more important, the virial expansion does not predict a transition from gas to liquid at high density. Monte Carlo computer simulations [15] have shown that for the hard-sphere system a freezing transition occurs at η = 0.494 ± 0.002 ($V/V_0 = 1.50$) from a fluid, disordered phase to a solid, ordered closed-packed phase

15) In numerical analysis it is well known that a truncated power series of a function is an unsatisfactory way of approximation. A more sophisticated method is the use of Padé approximants [18] which represents the function by a ratio of polynomials. The coefficients are found by expanding the ratio, and to require they represent the first k Taylor coefficients correctly. For example, $f(x) \cong c_0 + c_1x + \ldots c_kx^k$ is approximated by the n,m approximant $f(x) \cong (a_0 + a_1x + \ldots a_{n-1}x^{n-1})/(1 + b_1x + \ldots b_{m-1}x^{m-1})$. Obviously, $n + m - 1$ should equal $k + 1$. A matrix recipe is given by Ree and Hoover [19].

with $\eta = 0.545 \pm 0.002$ ($V/V_0 = 1.36$). The melting temperature and pressure are related by $P_{mel} = (8.27 \pm 0.13)\ \rho_0 kT_{mel}$, where ρ_0 represents the number density for the closed packed phase. The virial expansion fails to reveal this. Note that $\eta = 0.50$–0.55 is much lower than $\eta_{RCP} = 0.64$ or $\eta_{CP} = 0.74$. Hence, the regular solid phase is favored thermodynamically at a much lower density than at which the transition becomes necessary for geometric reasons. For the phase transition, several thermodynamic functions are nonanalytical which cannot be described by a finite order expansion. In the past, attempts have been made to connect the radius of convergence to condensation, but nowadays it is felt that there is no such relationship.

In summary, although the virial approach is useful for gases at intermediate density, for liquids other approaches are required. These approaches include integral equation models including thermodynamic perturbation theory, physical models and simulations. These will be discussed in the next chapters.

Problem 5.26: The square-well potential

Show that for the square-well potential as given by

$$\phi_{SW} = \infty \text{ for } r < \sigma, \quad \phi_{SW} = -\varepsilon \text{ for } \sigma \le r \le R\sigma \quad \text{and} \quad \phi_{SW} = 0 \text{ for } r > R\sigma \tag{5.85}$$

where σ is the hard-sphere diameter and ε and $R\sigma$ are the depth and range of the square-well, respectively, that the (second) virial coefficient B_2 is given by

$$B_2 = 2\pi\frac{\sigma^3}{3}\{1+[1-\exp(\varepsilon/kT)](R^3-1)\} \tag{5.86}$$

Discuss the role of the attractive part in ϕ_{SW}. Show that Eq. (5.86) increases continuously with T without showing a maximum (as experimentally observed) and discuss the reason why. Calculate the Boyle temperature T_B where $B_2(T_B) = 0$.

Problem 5.27

Show, assuming $N^2b_2/V << 1$, that

$$\frac{P}{kT} = \frac{N}{V} - \left(\frac{N(N-1)}{V^2}b_2\right)\bigg/(1+N(N-1)b_2/V) \cong \frac{N}{V} - \frac{N^2}{V^2}b_2 = \rho - b_2\rho^2$$

and thus that the second virial coefficient $B_2(T) = -b_2$.

Problem 5.28

Show that at STP the term $N^2b_2/V \approx 1$.

Problem 5.29

Consider the model potential $u(r) = \infty$ for $r < \sigma$ and $u(r) = -Ar^{-n}$ for $r \ge \sigma$. Show that B_2 is only finite if $n > 3$. Use the expansion for the exponential.

Problem 5.30

Prove Eq. (5.58).

Problem 5.31

The dissociation of dimers (d) to molecules (m) can be described by $x_d \rightarrow 2x_m$, where x denotes the mole fraction and the (mole fraction) equilibrium constant $K = x_m^2/x_d$. Assume that initially N dimers are present, each with bonding energy $-\varepsilon$ with respect to molecules, and neglect the vibration and rotation of the dimers. Show that:

a) the equilibrium constant $K = 4\exp(-\beta\varepsilon)/[2 + \exp(\beta\varepsilon)]$, and
b) $x_d = 0.576$ and $x_m = 0.424$ at $T = \varepsilon/k$.

Problem 5.32*

Evaluate using a symbolic math program, such as Maple, the second virial coefficient for the Lennard-Jones potential, and show that the result describes the experimental behavior, including the presence of a maximum, quite well.

References

1 Khinchin, A.I. (1949) *Mathematical Foundations of Statistical Mechanics*, Dover, New York.
2 Khinchin, A.I. (1957) *Mathematical Foundations of Information Theory*, Dover, New York. See also Landsberg (1978).
3 (a) Tolman, R.C. (1938) *The Principles of Statistical Mechanics*, Oxford University Press, Oxford (see also Dover Publishers reprint, 1979); (b) Waldram, J.R. (1985) *The Theory of Thermodynamics*, Cambridge University Press, Cambridge; (c) van Kampen, N.G. (1993) *Physica*, **A194**, 542.
4 (a) Kirkwood, J.G. (1933) *Phys. Rev.*, **44**, 31 and (1934) **45**, 116; (b) Wigner, E.P. (1932) *Phys. Rev.*, **40**, 749; (c) Green, H.S. (1951) *J. Chem. Phys.*, **19**, 955.
5 (a) A detailed discussion is given by Hill, T.L. (1956) *Statistical Mechanics*, McGraw-Hill, New York (see also Dover Publishers reprint, 1987); (b) Landau, L.D. and Lifschitz, E.M. (1980) *Statistical Physics*, vol. 1, 3rd edn, Addison-Wesley, Reading, MA.
6 See Schrödinger (1952).
7 Hill, T.L. (1956) *Statistical Mechanics*, McGraw-Hill, London (see also Dover, 1987).
8 Knox, J.H. (1978) *Molecular Thermodynamics*, Wiley, Interscience.
9 See McQuarrie (1973).
10 See Lucas (2007).
11 Ono, S. (1951) *J. Chem. Phys.*, **19**, 504.
12 Yang, C.N. and Lee, T.D. (1952) *Phys. Rev.*, **87**, 404 and 410.
13 (a) Friedman, H.L. (1985) *A Course in Statistical Mechanics*, Prentice-Hall, Englewood Cliffs, NJ; (b) Hansen, J.-P. and McDonald, I.R. (2006) *Theory of Simple Liquids*, 3rd edn, Academic Press, London.
14 Bannerman, M.N., Lue, L., and Woodcock, L.V. (2010) *J. Chem. Phys.*, **132**, 084507.
15 Hoover, W.G. and Ree, F.H. (1968) *J. Chem. Phys.*, **49**, 3609.
16 McQuarrie, D.A. and Simon, J.D. (1999) *Molecular Thermodynamics*, University Science Books, Sausalito, CA.

17 Clisby, N. and McCoy, B.M. (2006) *J. Stat. Phys.*, **122**, 15.
18 Ralston, A. (1965) *Introduction to Numerical Analysis*, McGraw-Hill, New York.
19 Ree, F.H. and Hoover, W.G. (1964) *J. Chem. Phys.*, **40**, 939.

Further Reading

Chandler, D. (1987) *Introduction to Modern Statistical Mechanics*, Oxford University Press, Oxford.
Fowler, R.H. and Guggenheim, E.A. (1939) *Statistical Thermodynamics*, Cambridge University Press, London.
Landsberg, P.T. (1978) *Thermodynamics and Statistical Mechanics*, Oxford University Press, Oxford, UK (see also Dover, 1990).
Lucas, K. (2007) *Molecular Models of Fluids*, Cambridge University Press, Cambridge.
McQuarrie, D.A. (1973) *Statistical Mechanics*, Harper and Row, New York.
Schrödinger, E. (1952) *Statistical Thermodynamics*, 2nd edn, Cambridge University Press, Cambridge (see also Dover Publishers reprint, 1989).
Widom, B. (2002) *Statistical Mechanics, A Concise Introduction for Chemists*, Cambridge University Press, Cambridge.

6
Describing Liquids: Structure and Energetics

In Chapter 5, an outline was provided of statistical thermodynamics and its application to noninteracting and interacting molecules in gases. Essentially, the approach did not work for liquids because the reference state (the perfect gas) is too far away from the liquid structure. Here, we take the opposite view and start from solids, and therefore we review first the structure of solids rather briefly. Thereafter, a general way to deal with the complex arrangements of atoms – that is, the use of the pair correlation function – is described and applied to explain the meaning of structure for a liquid. Together with intermolecular potentials, this function leads to expressions for the energy and pressure in liquids – that is, to the equations of state of liquids.

6.1
The Structure of Solids

Before embarking on the structure of liquids, we briefly review the structure of solids [1]. Recall that a crystal can be considered as a regular stacking of *unit cells* described by three (not necessarily orthogonal) noncoplanar basis vectors. In ideal solids, each crystallographic unit cell contains the same amount of matter in the same configuration. To create a crystal structure, it is necessary to arrange points in space in such a way that each of these points has an identical neighborhood. Stacking of the unit cells in this way can be achieved in only 14 ways in 7 crystal systems, resulting in the so-called *Bravais lattices*. If these points are actually occupied by molecules[1] and no other molecules are present in the cell, we have a *primitive lattice* whereas, in the case of more molecules occupying the unit cell, we speak of a *lattice with a basis*.

The crystal structure of many compounds is based on a limited number of simple stackings. These are the simple cubic (SC), face-centered cubic (FCC), body-centered cubic (BCC), and hexagonal close-packed (HCP) structures (Figure 6.1). While the relative theoretical density ρ_{the} of the SC and BCC lattice is 0.524 and 0.680, respectively, for the FCC and HCP lattice $\rho_{\text{the}} = 0.741$, so that a

1) Remember that we use the word "molecule" as a generic term for atoms, ions, and molecules.

Liquid-State Physical Chemistry, First Edition. Gijsbertus de With.

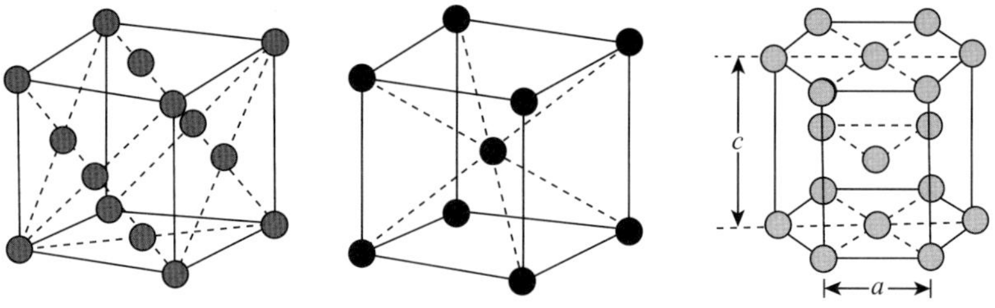

Figure 6.1 The FCC, BCC, and HCP structures (left to right).

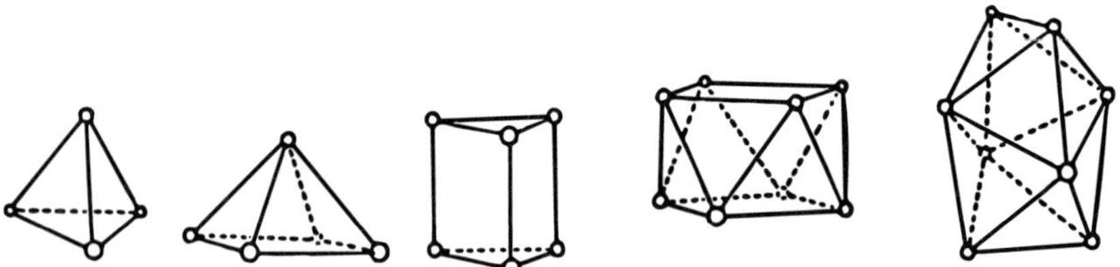

Figure 6.2 The five canonical holes of Bernal.

considerable difference in density exists. In a crystalline solid, essentially a limited number of type interstitial holes with a fixed fraction are present; for example, in the FCC lattice only tetrahedral holes are present whilst for the BCC only octahedral holes exist.

In crystalline solids defects occur and it is possible to distinguish between zero- (0-D), one- (1-D), two- (2-D), and three- (3-D) dimensional defects. Whilst for a 0-D defect (i.e., vacancies or interstitials) only a single molecule deviates from the ideal crystallographic order, for 1-D, 2-D and 3-D defects a (connected) line, plane, or volume of molecules deviates from the ideal order [1]. Although by introducing defects into solids the order is disturbed, the structure remains overwhelmingly regular. For example, near the melting point, though they are thermodynamically required, in a metal typically only about 1‰ of vacancies are present. Although the activation energy for defects in many substances is much lower than for metals, to describe a structure similar to amorphous materials requires so many defects that the use of individual defects is no longer warranted.

A useful description of amorphous materials, represented by a random dense packed structure of single-sized spheres as first presented by Bernal, is given in terms of polyhedral shapes of the interstitial holes. Five basic types of hole could be discerned (Figure 6.2), sometimes referred to as the *canonical holes*. The actual holes are somewhat distorted, but a random packing contains a certain fraction of

Table 6.1 Relative frequency of canonical holes.

Hole	Number (%)	Volume (%)	Hole	Number (%)	Volume (%)
Tetrahedron	73.0	48.4	Tetragonal dodecahedron	3.1	14.8
Half-octahedron	20.3	26.9	Archimedian antiprism	0.4	2.1
Trigonal prism	3.2	7.8			

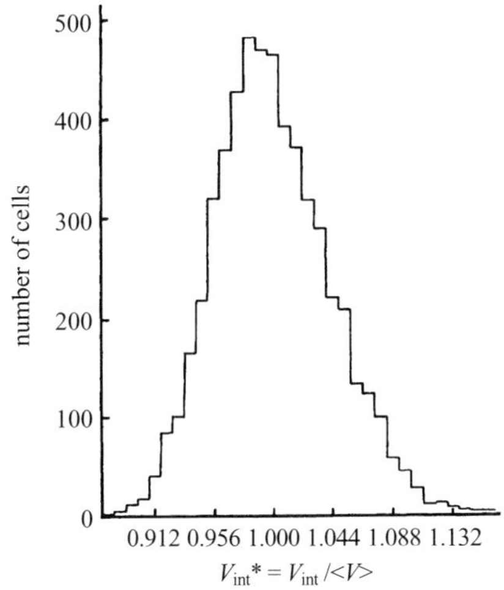

Figure 6.3 The volume per sphere distribution for a random packing of spheres with the maximum at $V_{int}^{*} = V_{int}/\langle V \rangle = 0.984$ with $\langle V \rangle = 0.637$ the random close-packed relative density.

each type [2] (Table 6.1); their frequency represents the frozen structure. The average number of nearest neighbors for each sphere is about nine, compared to 12 for the FCC lattice. The relative density for a structure of random close-packed spheres is about 0.637, compared with $\rho_{FCC} = 0.741$ and $\rho_{BCC} = 0.680$. On average, the volume per molecule is thus about 15% larger than the volume per molecule for a FCC crystal, corresponding to a lower average coordination number. The excess volume is referred to as the *free volume*. This already hints at another means of characterizing a glassy structure, namely the distribution of the volume per molecule. In an ideal crystal the volume per molecule is the same for every molecule. In an amorphous material there are molecules with a surrounding for which the volume per molecule is less than for a random close-packed structure, while others have a larger volume and there is a considerable width for this distribution [3] (Figure 6.3).

Although the picture sketched so far is static, molecules in any structure will vibrate around their equilibrium positions. Despite the molecular motions being correlated, in the simplest approach these correlations are neglected. This model was first introduced by Einstein, and accordingly denoted as the *Einstein model*. In this model it is supposed that all molecules vibrate independently around their equilibrium positions with the same circular frequency ω, denoted as the Einstein frequency. As these vibrations are treated quantum-mechanically, the energy levels of a quantum oscillator are needed to calculate its average behavior (see Chapter 2). Although this description is clearly flawed, it describes the thermal behavior of solids reasonably well to first order, and is therefore often used.

Problem 6.1

Using Figure 6.3, estimate the density of the glass relative to that of the crystal. Also estimate the local density variations that can be expected.

6.2
The Meaning of Structure for Liquids

Because the positions of molecules in liquids are neither regular nor completely random, we are forced to use a statistical description of their whereabouts. Therefore, the distribution function approach can be used to describe the structure of liquids.

6.2.1
Distributions Functions

Suppose that we want to know the distribution of specific molecules at specific positions. We denote the *specific distribution function* of N molecules by $n^{(N)}$. As an example, assume that the distribution function $n^{(3)}$ for three particles is given, and that $n^{(2)}$ is to be determined. Now, since $n^{(3)} = n^{(3)}(\mathbf{r}_1,\mathbf{r}_2,\mathbf{r}_3)$, where the independent (Cartesian) coordinates of particle i have been indicated by $\mathbf{r}_i$, we obtain

$$n^{(2)} = \int n^{(3)}(\mathbf{r}_1,\mathbf{r}_2,\mathbf{r}_3)\,\mathrm{d}\mathbf{r}_3 \quad \text{and}$$

$$n^{(1)} = \int n^{(2)}(\mathbf{r}_1,\mathbf{r}_2)\,\mathrm{d}\mathbf{r}_2 = \iint n^{(3)}(\mathbf{r}_1,\mathbf{r}_2,\mathbf{r}_3)\,\mathrm{d}\mathbf{r}_2\,\mathrm{d}\mathbf{r}_3$$

This type of relation is generally valid and leads to

$$n^{(1)} = \iiint \ldots \int n^{(N)}(\mathbf{r}_1, \mathbf{r}_2, \ldots, \mathbf{r}_N)\,\mathrm{d}\mathbf{r}_2\,\mathrm{d}\mathbf{r}_3\,\mathrm{d}\mathbf{r}_4 \ldots \mathrm{d}\mathbf{r}_N$$

$$n^{(2)} = \iint \ldots \int n^{(N)}(\mathbf{r}_1, \mathbf{r}_2, \ldots, \mathbf{r}_N)\,\mathrm{d}\mathbf{r}_3\,\mathrm{d}\mathbf{r}_4 \ldots \mathrm{d}\mathbf{r}_N$$

$$n^{(3)} = \int \ldots \int n^{(N)}(\mathbf{r}_1, \mathbf{r}_2, \ldots, \mathbf{r}_N)\,\mathrm{d}\mathbf{r}_4 \ldots \mathrm{d}\mathbf{r}_N$$

$$\ldots$$

$$n^{(M)} = \int \ldots \int n^{(N)}(\mathbf{r}_1, \mathbf{r}_2, \ldots, \mathbf{r}_N)\,\mathrm{d}\mathbf{r}_{M+1} \ldots \mathrm{d}\mathbf{r}_N \tag{6.1}$$

However, molecules are indistinguishable and we should ask for the *generic distribution functions* $\rho^{(N)}$, describing any molecule at specific positions. Let us take the same example and determine $\rho^{(2)}$, assuming that the distribution function $n^{(3)}$ for three particles is given. In this case we have $\rho^{(3)} = 3 \times 2 \times 1\; n^{(3)}(\mathbf{r}_1,\mathbf{r}_2,\mathbf{r}_3) = 3!\; n^{(3)}(\mathbf{r}_1,\mathbf{r}_2,\mathbf{r}_3)$. Further, we obtain

$$\rho^{(2)} = 3 \cdot 2 \cdot 1 \int n^{(3)}(\mathbf{r}_1, \mathbf{r}_2, \mathbf{r}_3)\,\mathrm{d}\mathbf{r}_3 \quad \text{and}$$

$$\rho^{(1)} = 2 \int n^{(2)}(\mathbf{r}_1, \mathbf{r}_2)\,\mathrm{d}\mathbf{r}_2 = 3 \iint n^{(3)}(\mathbf{r}_1, \mathbf{r}_2, \mathbf{r}_3)\,\mathrm{d}\mathbf{r}_2\,\mathrm{d}\mathbf{r}_3$$

These relations can also be generalized for N particles, and we obtain

$$\rho^{(1)} = N \iiint \ldots \int n^{(N)}(\mathbf{r}_1, \mathbf{r}_2, \ldots, \mathbf{r}_N)\,\mathrm{d}\mathbf{r}_2\,\mathrm{d}\mathbf{r}_3\,\mathrm{d}\mathbf{r}_4 \ldots \mathrm{d}\mathbf{r}_N = N n^{(1)}$$

$$\rho^{(2)} = N(N-1) \iint \ldots \int n^{(N)}(\mathbf{r}_1, \mathbf{r}_2, \ldots, \mathbf{r}_N)\,\mathrm{d}\mathbf{r}_3\,\mathrm{d}\mathbf{r}_4 \ldots \mathrm{d}\mathbf{r}_N = N(N-1) n^{(2)}$$

$$\rho^{(3)} = N(N-1)(N-2) \int \ldots \int n^{(N)}(\mathbf{r}_1, \mathbf{r}_2, \ldots, \mathbf{r}_N)\,\mathrm{d}\mathbf{r}_4 \ldots \mathrm{d}\mathbf{r}_N = N(N-1)(N-2) n^{(3)}$$

$$\ldots$$

$$\begin{aligned} \rho^{(M)} &= N(N-1)\ldots(N-M+1) \int \ldots \int n^{(N)}(\mathbf{r}_1, \mathbf{r}_2, \ldots, \mathbf{r}_N)\,\mathrm{d}\mathbf{r}_{M+1} \ldots \mathrm{d}\mathbf{r}_N \\ &= N(N-1)\ldots(N-M+1) n^{(M)} \end{aligned} \tag{6.2}$$

Equivalently, we can write $N(N-1)\ldots(N-M+1) = N!/(N-M)!$. Hence, generally,

$$▶ \quad \begin{aligned} \rho^{(M)}(\mathbf{r}_1, \ldots, \mathbf{r}_M) &= \frac{N!}{(N-M)!} \int \ldots \int n^{(N)}(\mathbf{r}_1, \ldots, \mathbf{r}_N)\,\mathrm{d}\mathbf{r}_{M+1} \ldots \mathrm{d}\mathbf{r}_N \\ &= \frac{N!}{(N-M)!} n^{(M)}(\mathbf{r}_1, \ldots, \mathbf{r}_M) \end{aligned} \tag{6.3}$$

From this expression we can also easily derive that

$$\rho^{(N)} = N! n^{(N)}, \quad (N-M)\rho^{(M)} = \int \ldots \int \rho^{(M+1)}(\mathbf{r}_1, \ldots, \mathbf{r}_{M+1})\,\mathrm{d}\mathbf{r}_{M+1} \quad \text{and} \tag{6.4}$$

$$(N-M)!\,\rho^{(M)} = \int \ldots \int \rho^{(N)}(\mathbf{r}_1, \ldots, \mathbf{r}_N)\,\mathrm{d}\mathbf{r}_{M+1} \ldots \mathrm{d}\mathbf{r}_N \tag{6.5}$$

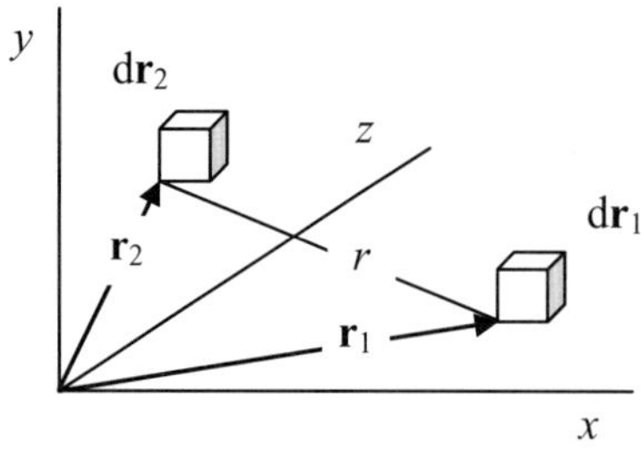

Figure 6.4 The location of infinitesimal volume element $\mathbf{dr}_1$ located at $\mathbf{r}_1$ and the volume element $\mathbf{dr}_2$ located at $\mathbf{r}_2$ with their distance r.

Extending these ideas, from position coordinates **r** to position coordinates **r** plus momenta **p**, results in

$$\rho^{(M)}(\mathbf{r}_1,\ldots,\mathbf{r}_M,\mathbf{p},\ldots,\mathbf{p}_M)=\frac{N!}{(N-M)!}\int\ldots\int n^{(N)}(\mathbf{r},\mathbf{p})\,\mathrm{d}\mathbf{r}_{M+1}\ldots\mathrm{d}\mathbf{r}_N\,\mathrm{d}\mathbf{p}_{M+1}\ldots\mathrm{d}\mathbf{p}_N \tag{6.6}$$

where, as before, $\mathbf{r} = (\mathbf{r}_1,\mathbf{r}_2,\ldots,\mathbf{r}_N)$ and $\mathbf{p} = (\mathbf{p}_1,\mathbf{p}_2,\ldots,\mathbf{p}_N)$. This function describes the probability of M molecules to be at positions **r** meanwhile having momenta **p**.

The simplest distribution function $\rho^{(h)}$ is the *singlet distribution function* with $h = 1$ that describes the probability to find any molecule in $\mathrm{d}\mathbf{r}_1$ at position $\mathbf{r}_1$. For a crystal, only specific positions $\mathbf{r}_1$ are allowed (neglecting defects and diffusion), but for a fluid all values of $\mathbf{r}_1$ are possible and hence $\rho^{(1)}$ should be independent of $\mathbf{r}_1$. So we have[2)]

$$\rho^{(1)}(\mathbf{r}_1)\mathrm{d}\mathbf{r}_1=\rho\,\mathrm{d}\mathbf{r}_1 \quad \text{and} \quad \int\rho^{(1)}(\mathbf{r}_1)\mathrm{d}\mathbf{r}_1=N \tag{6.7}$$

where $\rho = N/V$ is the number density. The *doublet distribution function* with $h = 2$ describes the probability to find one molecule in $\mathrm{d}\mathbf{r}_1$ at position $\mathbf{r}_1$ and another in $\mathrm{d}\mathbf{r}_2$ at position $\mathbf{r}_2$ (Figure 6.4). We have now

$$\rho^{(2)}(\mathbf{r}_1,\mathbf{r}_2)\mathrm{d}\mathbf{r}_1\,\mathrm{d}\mathbf{r}_2 \quad \text{and} \quad \int\rho^{(2)}\,\mathrm{d}\mathbf{r}_1\,\mathrm{d}\mathbf{r}_2=(N-1)\int\rho^{(1)}\,\mathrm{d}\mathbf{r}_1=N(N-1) \tag{6.8}$$

If we assume that the potential energy can be described as the sum of two-particle potentials which are spherically symmetrical, we have for two particles at large distance $\rho^{(2)}(\mathbf{r}_1,\mathbf{r}_2) \cong \rho^{(1)}(\mathbf{r}_1)\rho^{(1)}(\mathbf{r}_2) = \rho^2$. For more close-by distances, we need the original expression $\rho^{(2)}(\mathbf{r}_1,\mathbf{r}_2)$, which we can write as $\rho^{(2)}(\mathbf{r}_1,\mathbf{r}_2) = \rho^2 g^{(2)}(\mathbf{r}_1,\mathbf{r}_2)$ with $g^{(2)}(\mathbf{r}_1,\mathbf{r}_2)$ the *(pair) correlation function.*

2) More formally, we require *translational invariance*, i.e. $\rho^{(1)}(\mathbf{r}_1) = \rho^{(1)}(\mathbf{r}_1 + \mathbf{r}_1')$ for any $\mathbf{r}_1'$ (obviously not too close to the wall). This is only possible if $\rho^{(1)} = C$ with C a constant. Since we have on the one hand $\int\rho^{(1)}(\mathbf{r})\,\mathrm{d}\mathbf{r}_1 = N$, and on the other hand $\int\rho^{(1)}(\mathbf{r})\,\mathrm{d}\mathbf{r}_1 = C\int\mathrm{d}\mathbf{r}_1 = CV$, we obtain $CV = N$ or $\rho^{(1)} = N/V$.

For a further microscopic interpretation we note that for any microscopic property $X = X_N(\mathbf{r}) = X_N(\mathbf{r}_1,..,\mathbf{r}_N)$ depending on N (ordered) coordinates the corresponding expression for the canonical average $\langle X_N\rangle$ reads, using Eqs (5.39) and (6.4),

$$\langle X_N\rangle = Q_N^{-1}\int X_N(\mathbf{r})e^{-\beta\Phi(\mathbf{r})}\,\mathrm{d}\mathbf{r} = \int X_N n^{(N)}\,\mathrm{d}\mathbf{r} = (N!)^{-1}\int X_N \rho^{(N)}\,\mathrm{d}\mathbf{r} \qquad (6.9)$$

Suppose now that X depends on only M of the N coordinates. The number of permutations for M out of N is $N!/(N-M)!M!$. So, if we denote the microscopic property with the coordinates ordered from 1 to M by $x_M(\mathbf{r}_1,..,\mathbf{r}_M)$ and the property with any M out of N by $X_M(\mathbf{r}_1,..,\mathbf{r}_M)$, we have $X_M = [N!/(N-M)!M!]\,x_M$. From Eq. (6.5) we also know that $\int..\int \rho^{(N)}\mathrm{d}\mathbf{r}_{M+1}..\mathrm{d}\mathbf{r}_N = (N-M)!\rho^{(M)}$. So we obtain

$$\begin{aligned}\langle X_M\rangle &= Q_N^{-1}\int X_M(\mathbf{r}_1,\ldots,\mathbf{r}_M)\mathrm{e}^{-\beta\Phi}\,\mathrm{d}\mathbf{r}_1\ldots\mathrm{d}\mathbf{r}_N = \int X_M(N!)^{-1}\rho^{(N)}\,\mathrm{d}\mathbf{r}_1\ldots\mathrm{d}\mathbf{r}_N\\ &= \int \frac{N!}{M!(N-M)!}x_M(\mathbf{r}_1,\ldots,\mathbf{r}_M)\left[\frac{(N-M)!}{N!}\rho^{(M)}(\mathbf{r}_1,\ldots,\mathbf{r}_M)\right]\mathrm{d}\mathbf{r}_1\ldots\mathrm{d}\mathbf{r}_M\\ &= M!^{-1}\int x_M(\mathbf{r}_1,\ldots,\mathbf{r}_M)\rho^{(M)}(\mathbf{r}_1,\ldots,\mathbf{r}_M)\mathrm{d}\mathbf{r}_1\ldots\mathrm{d}\mathbf{r}_M\end{aligned}$$

So, if $X_1 = \Sigma_i\, x_1(\mathbf{r}_i)$ or $X_2 = \Sigma_{i<j}\, x_2(\mathbf{r}_i,\mathbf{r}_j)$, we have

$$\langle X_1\rangle = \int \rho^{(1)}(\mathbf{r})x_1(\mathbf{r}_1)\mathrm{d}\mathbf{r}_1 \quad \text{and} \quad \langle X_2\rangle = \frac{1}{2}\int \rho^{(2)}(\mathbf{r})x_2(\mathbf{r}_1,\mathbf{r}_2)\mathrm{d}\mathbf{r}_1\,\mathrm{d}\mathbf{r}_2 \qquad (6.10)$$

We now define the *density operator* $D(\mathbf{r}_i,\mathbf{r}_i') = \Sigma_i\, d_1(\mathbf{r}_i) \equiv \Sigma_i\, \delta(\mathbf{r}_i - \mathbf{r}_i')$ to describe the probability that particle i is at position $\mathbf{r}_i'$ and the *correlation operator* $C(\mathbf{r}_i,\mathbf{r}_i',\mathbf{r}_j,\mathbf{r}_j') = \frac{1}{2}\Sigma_{i,j}\, c_2(\mathbf{r}_i,\mathbf{r}_j) \equiv \frac{1}{2}\Sigma_{i,j}\, \delta(\mathbf{r}_i - \mathbf{r}_i')\delta(\mathbf{r}_j - \mathbf{r}_j')$ to describe the probability that particle i is at position $\mathbf{r}_i'$ while particle j is at position $\mathbf{r}_j'$. Here, $\delta(\mathbf{x})$ is the Dirac delta function (see Appendix 2). The canonical average $\langle D\rangle = N\langle d_1(\mathbf{r}_1)\rangle$ becomes $\rho^{(1)}(\mathbf{r}_1)$ according to

$$\langle D\rangle = \int \delta(\mathbf{r}_1 - \mathbf{r}_1')\rho^{(1)}(\mathbf{r}_1)\mathrm{d}\mathbf{r}_1 = \rho^{(1)}(\mathbf{r}_1') \qquad (6.11)$$

while the average $\langle C\rangle = \frac{1}{2}N(N-1)\langle c_2(\mathbf{r}_1,\mathbf{r}_2)\rangle$ relates to $\rho^{(2)}(\mathbf{r}_1,\mathbf{r}_2)$ according to

$$\langle C\rangle = \frac{1}{2}\int\ldots\int \delta(\mathbf{r}_1 - \mathbf{r}_1')\delta(\mathbf{r}_2 - \mathbf{r}_2')\rho^{(2)}\,\mathrm{d}\mathbf{r}_1\,\mathrm{d}\mathbf{r}_2 = \frac{1}{2}\rho^{(2)}(\mathbf{r}_1',\mathbf{r}_2') \qquad (6.12)$$

Writing $\rho^{(2)}(\mathbf{r}_1,\mathbf{r}_2) = \rho^2 g^{(2)}(\mathbf{r}_1,\mathbf{r}_2)$, we invert to obtain

$$g^{(2)}(\mathbf{r}_1,\mathbf{r}_2) = \frac{2}{\rho^2}\langle C\rangle = \frac{2}{\rho^2}\frac{N(N-1)}{2}\langle c_2(\mathbf{r}_1,\mathbf{r}_2)\rangle = (1-N^{-1})V^2\langle c_2(\mathbf{r}_1,\mathbf{r}_2)\rangle \qquad (6.13)$$

If there is no correlation $\langle c_2(\mathbf{r}_1,\mathbf{r}_2)\rangle$ reduces to V^{-2} so that $g^{(2)}(\mathbf{r}_1,\mathbf{r}_2) = (1-N^{-1})$. Hence, normalization yields $\int g^{(2)}(\mathbf{r}_1,\mathbf{r}_2)\,\mathrm{d}\mathbf{r}_1\mathrm{d}\mathbf{r}_2 = (1-N^{-1}) \cong 1$ for large N.

If we are not interested in the angular information contained in the set $(\mathbf{r}_1,\mathbf{r}_2)$, but only in the distances, we can write $\rho^{(2)}(\mathbf{r}_1,\mathbf{r}_2) = \rho^2 g^{(2)}(\mathbf{r}_1,\mathbf{r}_2) = \rho^2 g(r)$ with $g(r) \equiv g^{(2)}(r)$ and $r = |\mathbf{r}_1 - \mathbf{r}_2|$. From this expression we can calculate the total number of atoms $N(r)$ in a spherical volume of radius r by integrating over both coordinates.

Changing the coordinate system from $(\mathbf{r}_1,\mathbf{r}_2)$ to $(\mathbf{r}_1 + \mathbf{r}_2,\mathbf{r}_1 - \mathbf{r}_2) \equiv (\mathbf{R},\mathbf{r})$, integration over $\mathbf{R}$ yields the volume. The other integration over $r = |\mathbf{r}|$ (in a polar coordinate system) remains and we obtain after some manipulation

$$N(r) = \rho\int_0^r g(r)4\pi r^2\,\mathrm{d}r \tag{6.14}$$

From normalization we also conclude that $\int g(r)\,\mathrm{d}r = (1 - N^{-1}) \cong 1$. The integrand $4\pi r^2 g(r)$ is often addressed as the *radial distribution function* (RDF). The pair correlation function $g(r)$ is experimentally accessible, as elaborated briefly in Section 6.3.

In the same way as for the doublet distribution function, the triplet distribution function with $h = 3$ can be reduced, using $r_{ij} = |\mathbf{r}_i - \mathbf{r}_j|$, to

$$\rho^{(3)}(\mathbf{r}_1,\mathbf{r}_2,\mathbf{r}_3) = \rho^3 g^{(3)}(r_{12},r_{23},r_{31}) \tag{6.15}$$

Unfortunately, this function is not directly experimentally accessible. In general we have $\rho^{(h)}(\mathbf{r}_1,\ldots,\mathbf{r}_h) = \rho^h g^{(h)}(\mathbf{r}_1,\ldots,\mathbf{r}_h)$.

To illustrate these ideas, let us apply them to the perfect gas for which $\Phi = 0$ and consequently $\exp(-\beta\Phi) = 1$ and $Q_N = V^N$. Hence, for the pair distribution function

$$\begin{aligned}\rho^{(2)}\mathrm{d}\mathbf{r}_1\,\mathrm{d}\mathbf{r}_2 &= \left[\frac{N!}{(N-2)!Q_N}\int\ldots\int \mathrm{e}^{-\beta\Phi}\,\mathrm{d}\mathbf{r}_3\ldots\mathrm{d}\mathbf{r}_N\right]\mathrm{d}\mathbf{r}_1\,\mathrm{d}\mathbf{r}_2\\ &= \frac{N!}{(N-2)!}\frac{V^{N-2}}{V^N}\mathrm{d}\mathbf{r}_1\,\mathrm{d}\mathbf{r}_2 = N(N-1)\frac{\mathrm{d}\mathbf{r}_1}{V}\frac{\mathrm{d}\mathbf{r}_2}{V}\end{aligned} \tag{6.16}$$

This result matches the basic probability considerations for N particles distributed randomly over a volume V because for $N >> 1$, Eq. (6.16) reduces to

$$\rho^{(2)}(\mathbf{r}_1,\mathbf{r}_2)\mathrm{d}\mathbf{r}_1\,\mathrm{d}\mathbf{r}_2 = N^2\left(1-\frac{1}{N}\right)\frac{\mathrm{d}\mathbf{r}_1}{V}\frac{\mathrm{d}\mathbf{r}_2}{V} = \rho^2\left(1-\frac{1}{N}\right)\mathrm{d}\mathbf{r}_1\,\mathrm{d}\mathbf{r}_2 \cong \rho^2\mathrm{d}\mathbf{r}_1\,\mathrm{d}\mathbf{r}_2 \tag{6.17}$$

This shows once more that $g(r)$ is normalized to $1 - N^{-1} \cong 1$ for large N.

In summarizing, if a complete set of distribution functions could be given, the description of structure would be complete, and in that case we would have a situation comparable to that of solids. However, such a set is unavailable and we must be content with just a few of them. The pair distribution function $\rho^{(2)}(r) = \rho^2 g(r)$ with number density $\rho = N/V$, or equivalently, the pair correlation function $g(r)$, is the most important of these as it essentially describes the probability of finding two molecules at distance r. For independent molecules $g(r) = 1$ and thus $\rho^{(2)}(r) = \rho^2$, but for correlated molecules, if it is given that the reference molecule is at the origin (with "probability" ρ), the probability of finding a molecule at distance r from the origin is given by $\rho g(r)$. Alternatively, $\rho g(r)$ describes the local density around the reference molecule.

6.2.2
Two Asides

Before discussing some experimental and simulational results on the structure of liquids, two asides should first be briefly dealt with.

The first aside is that the distribution function approach provides another means of obtaining the basic result of statistical thermodynamics, namely the Boltzmann distribution. When molecules are close together, their potential energy becomes large and positive; similarly, if the momenta are large, the (always positive) kinetic energy is also large. Hence we may suspect that a relation exists between the total energy $E = \mathcal{H}(\mathbf{p},\mathbf{q})$ of N molecules and the probability of a configuration as described by $n^{(N)}$. So, we write $n^{(N)}\mathrm{d}\mathbf{p}\,\mathrm{d}\mathbf{r} = f(\mathcal{H})\mathrm{d}\mathbf{p}\,\mathrm{d}\mathbf{r}$. The form of $f(\mathcal{H})$ can be obtained by considering the distribution function for two containers, one with N molecules and one with M molecules, which are to be combined. With distribution functions being probabilities, the joint distribution function reads not only $f(\mathcal{H})\mathrm{d}\mathbf{p}\,\mathrm{d}\mathbf{r}\cdot f(\mathcal{H}')\mathrm{d}\mathbf{p}'\mathrm{d}\mathbf{r}'$ but also $f(\mathcal{H}+\mathcal{H}')\mathrm{d}\mathbf{p}\,\mathrm{d}\mathbf{r}\,\mathrm{d}\mathbf{p}'\mathrm{d}\mathbf{r}'$, so that we have $f(\mathcal{H})\cdot f(\mathcal{H}') = f(\mathcal{H}+\mathcal{H}')$. This relationship can only be satisfied by $f(\mathcal{H}) = \exp(-\beta\mathcal{H})$, where β is a positive constant and the negative sign is required because the probabilities should remain finite for arbitrarily large values for E. Normalization leads to $f(\mathcal{H}) = \exp(-\beta\mathcal{H})/Z_N$, while β can be obtained by applying these results to the simplest system possible, that is, an ideal gas (see Sections 5.2 and 5.3). This leads to $\beta = 1/kT$. We thus obtain

$$n^{(N)}(\mathbf{r},\mathbf{p}) = Z_N^{-1}\mathrm{e}^{-\beta\mathcal{H}} \quad \text{with} \quad Z_N = (N!h^{3N})^{-1}\int\ldots\int e^{-\beta\mathcal{H}}\,\mathrm{d}\mathbf{r}_1\ldots\mathrm{d}\mathbf{r}_N\,\mathrm{d}\mathbf{p}_1\ldots\mathrm{d}\mathbf{p}_N \tag{6.18}$$

If we are only interested in structure but not in dynamics, and the energy expression reads $\mathcal{H}(\mathbf{p},\mathbf{r}) = \Sigma_i\mathbf{p}_i^2/2m_i + \Phi(\mathbf{r})$, we have seen in Section 5.2 that the integrals factorize and the momenta contributions cancel from the numerator and denominator. We then have

▶ $$n^{(N)}(\mathbf{r}) = Q_N^{-1}\exp(-\beta\Phi) \quad \text{with} \quad Q_N = \int\ldots\int \mathrm{e}^{-\beta\Phi}\,\mathrm{d}\mathbf{r}_1\ldots\mathrm{d}\mathbf{r}_N \tag{6.19}$$

The second aside deals with systems in which the number of molecules is conserved as, has been the case so far. In this situation the total number of degrees of freedom (DoF) is also conserved. Bearing this in mind, we should examine the change of a distribution function with time. The conservation of the number of DoF implies conservation of the number of phase points, that is, $\mathrm{d}n^{(N)}/\mathrm{d}t = 0$, resulting in

$$\frac{\mathrm{d}}{\mathrm{d}t}n^{(N)} = \frac{\partial}{\partial t}n^{(N)} + \sum_i \frac{\partial n^{(N)}}{\partial \mathbf{r}_i}\dot{\mathbf{r}}_i + \sum_i \frac{\partial n^{(N)}}{\partial \mathbf{p}_i}\dot{\mathbf{p}}_i = 0 \quad \text{or} \tag{6.20}$$

▶ $$\frac{\partial}{\partial t}n^{(N)} = \beta\mathrm{e}^{-\beta\mathcal{H}}\left(\sum_i \frac{\partial\mathcal{H}}{\partial \mathbf{r}_i}\dot{\mathbf{r}}_i + \sum_i \frac{\partial\mathcal{H}}{\partial \mathbf{p}_i}\dot{\mathbf{p}}_i\right) = \frac{\partial}{\partial t}n^{(N)} = \frac{\partial}{\partial t}\rho^{(N)} = 0 \tag{6.21}$$

wherein the last step use has been made of $n^{(N)} = \exp(-\beta\mathcal{H})/Z_N$ and Hamilton's equations (see Section 2.2). Equation (6.21), representing *Liouville's theorem* (see Section 5.1), states that the distribution functions are only implicitly dependent on time.

6.3 The Experimental Determination of $g(r)$*

In this section we provide some background on how to obtain the correlation function from scattering experiments using X-ray diffraction (XRD) and/or neutron ray diffraction (NRD). Since these experiments are loaded with many experimental pitfalls, this topic requires significant technical and professional skills, and only the bare essentials are touched on here.

For scattering experiments using monochromatic radiation, part of the incident radiation is scattered in various directions, characterized by the angle θ between the incoming and outgoing radiation. A scattering parameter $s = (4\pi/\lambda)\sin(\theta/2)$ is defined, where λ is the wavelength of the radiation used and $s = |\mathbf{s}| = |\mathbf{s}_{\text{sca}} - \mathbf{s}_{\text{inc}}|$ is the length of the difference between the wave vector of the scattered and incident radiations. In the first Born approximation, representing single scattering, the amplitude of the outgoing radiation $A(\mathbf{s})$ is directly related to the Fourier transform of the scattering potential $\Phi(\mathbf{r})$ – that is, $A(\mathbf{s}) \sim V^{-1}\int \Phi(\mathbf{r})\exp(-i\mathbf{s}\cdot\mathbf{r})\mathrm{d}\mathbf{r}$, where V is the volume of the sample. This theory is general and can be applied to X-rays, neutrons and electrons, although the scattering matter for each is different. X-rays are scattered by electrons, neutrons are mainly scattered by nuclei, while for electron radiation both electrons and nuclei cause scattering. For X-rays, typically Kα radiation is used (Cu, $\lambda = 0.1541$ nm or Mo, $\lambda = 0.07093$ nm), while λ for thermal neutrons ($kT \cong 4\text{–}5\times 10^{-21}$ J) is ~0.1 nm. For electrons, $\lambda = h/[2m_0eV(1+eV/2m_0c^2)]^{1/2}$, where V is the acceleration voltage, m_0 and e are the rest mass and charge of the electron, respectively, and c is the speed of light; this results in 1.97 pm using 300 kV.

In general, to describe structure of a mono-atomic sample we need atomic coordinates and it is assumed that $\Phi(\mathbf{r}) = \Sigma_i\phi(\mathbf{r} - \mathbf{r}_i)$ with $\mathbf{r}_i$ the atomic coordinates. For elastic scattering $|\mathbf{s}_{\text{sca}}| = |\mathbf{s}_{\text{inc}}|$, the intensity is $I \sim |A(\mathbf{s})^2|$ and this leads to, dropping geometric factors and introducing $\rho = N/V$,

$$I \sim \left|\frac{1}{V}\int_0^\infty \Phi(\mathbf{r})\mathrm{e}^{-i\mathbf{s}\cdot\mathbf{r}}\,\mathrm{d}\mathbf{r}\right|^2 = \left|\frac{1}{V}\int_0^\infty \sum\nolimits_i \phi(\mathbf{r}-\mathbf{r}_i)\mathrm{e}^{-i\mathbf{s}\cdot\mathbf{r}}\,\mathrm{d}\mathbf{r}\right|^2$$

$$= \frac{1}{N}\left[\frac{1}{N}\sum\nolimits_{i,j}\mathrm{e}^{-i\mathbf{s}\cdot(\mathbf{r}_i-\mathbf{r}_j)}\right]\left|\rho\int_0^\infty \phi(\mathbf{r})\mathrm{e}^{-i\mathbf{s}\cdot\mathbf{r}}\,\mathrm{d}\mathbf{r}\right|^2 \equiv \frac{1}{N}S(\mathbf{s})|\phi(\mathbf{s})|^2 \tag{6.22}$$

The factor $|\phi(\mathbf{s})|$ is the Fourier transform of the atomic scattering potential, usually called the *atomic form factor*, while $S(\mathbf{s})$ is the *structure factor*. For a perfect crystal the latter represents an array of delta functions at the reciprocal lattice points. Evaluating $S(\mathbf{s})$ by taking its canonical average yields

$$S(\mathbf{s}) = 1 + \langle\exp[-i\mathbf{s}\cdot(\mathbf{r}_i - \mathbf{r}_j)]\rangle = 1 + \rho\int_0^\infty g(r)\exp(-i\mathbf{s}\cdot\mathbf{r})\mathrm{d}\mathbf{r} \tag{6.23}$$

where the term 1 is due to $i = j$ contributions. Introducing the total correlation function $h(r) = g(r) - 1$ (see Section 7.2), this expression becomes

$$S(\mathbf{s}) - 1 = \rho\int_0^\infty h(r)\exp(-i\mathbf{s}\cdot\mathbf{r})\mathrm{d}\mathbf{r} + \rho\int_0^\infty \exp(-i\mathbf{s}\cdot\mathbf{r})\mathrm{d}\mathbf{r}$$

The last integral represents a $\delta(s)$ contribution (see Appendix B) originating from the fact that for $r \to \infty$, $g(r) \to 1$ and has to be subtracted. Otherwise the Fourier inversion

$$g(r) = 1 + h(r) = 1 + (8\pi^3\rho)^{-1}\int_0^\infty [S(s)-1]\exp(-i\mathbf{s}\cdot\mathbf{r})\,\mathrm{d}\mathbf{r}$$
$$= 1 + (8\pi^3\rho)^{-1}\int_0^\infty [S(s)-1]\frac{\sin(sr)}{sr}4\pi r^2\,\mathrm{d}r \tag{6.24}$$

would not converge because $S(s) \to 1$ for $s \to \infty$. The last step can be made if the system is isotropic. The accuracy of $g(r)$ depends heavily on accurate knowledge of $S(s)$ for an as-wide range of s-values as possible, since the Fourier transform may introduce serious defects when performed with a too-limited data range.

For neutrons, an atom is characterized by $\phi(\mathbf{r}) = b\delta(\mathbf{r})$ with the *scattering length* b, typically 0.01 pm and θ-independent. As the length for isotopes is slightly different, the average value $\langle b\rangle$ is used. The scattered intensity $I(s)$ is given by $I(s) = \alpha(\theta)\langle b\rangle^2[S(s) + \Delta]$, where $\alpha(\theta)$ is a proportionality factor that is instrument-dependent but sample-independent, converting to absolute intensities. The *structure factor* $S(s)$ is due to coherent scattering, while Δ is due to incoherent scattering. The limiting values of $S(s)$ are $S(0) = \rho kT\kappa_T$, with a value of about 0.01, and $S(\infty) = 1$. Intensities at large angle $I(\infty)$ and small angle $I(0)$ can be used to eliminate $\alpha(\theta)$ and Δ, leading to

$$S(s) = [I(s) - I(s)S(0) - I(0) + I(\infty)S(0)]/[I(\infty) - I(0)]$$

For X-rays, the quantity corresponding to $\langle b\rangle$ is the *atomic scattering factor* $a(s)$, defined by the sine transform of the atomic electron density $\rho(r)$,

$$a(s) = \int_0^\infty \rho(r)4\pi r^2\frac{\sin(sr)}{sr}\,\mathrm{d}r$$

This quantity is θ-dependent. The structure factor becomes

$$S(s) = I(s)Z^2 / I_0Na^2(s)$$

where N is the number of atoms with atomic number Z. The value of $I(\infty)/I_0 \to Na^2(s)/Z^2$, so that $S(s)$ approaches 1; this can be used to normalize the data.

For molecules, one generally assumes that $\Phi(\mathbf{r}) = \Sigma_{i,\alpha}\phi_\alpha(\mathbf{r} - \mathbf{r}_{i,\alpha})$ with $\mathbf{r}_{i,\alpha}$ the atomic coordinates of the atom of type α at coordinate $\mathbf{r}_{i,\alpha}$. The Fourier transform of the scattering potential $\Phi(\mathbf{r})$ becomes

$$I \sim \left|\frac{1}{V}\int_0^\infty \Phi(\mathbf{r})\mathrm{e}^{-i\mathbf{s}\cdot\mathbf{r}}\,\mathrm{d}\mathbf{r}\right|^2 = \left|\frac{1}{V}\int_0^\infty \sum\nolimits_{i,\alpha}\phi_\alpha(\mathbf{r}-\mathbf{r}_{i,\alpha})\mathrm{e}^{-i\mathbf{s}\cdot\mathbf{r}}\,\mathrm{d}\mathbf{r}\right|^2$$
$$= \frac{1}{N}\left[\frac{1}{N}\sum\nolimits_{\alpha,\beta}\sum\nolimits_{i,\alpha,j,\beta}\mathrm{e}^{-i\mathbf{s}\cdot(\mathbf{r}_{i,\alpha}-\mathbf{r}_{j,\beta})}\right]\phi_\alpha(\mathbf{s})\phi_\beta(\mathbf{s}) \tag{6.25}$$

If we separate the terms with $i = j$ and $i,\alpha = j,\beta$, these contribute a structure-independent term, weighted with the concentrations c_α, that is, using $\bar{\phi}_\alpha(\mathbf{s}) = \sum_\alpha c_\alpha\phi_\alpha(\mathbf{s})$ leads to

$$I = \frac{1}{N}\sum\nolimits_\alpha|\phi_\alpha(\mathbf{s})|^2 = \frac{1}{N}\left[|\bar{\phi}_\alpha(\mathbf{s})|^2 + \sum\nolimits_\alpha c_\alpha|\Delta\phi_\alpha(\mathbf{s})|^2\right] \tag{6.26}$$

The second term refers to the pairs of distinct atoms (both often of the same type). Defining a partial structure factor $S_{\alpha\beta}$, again with removal of the singularity,

$$S_{\alpha\beta}(s) = 1 + \rho\int_0^\infty h_{\alpha\beta}(r)\exp(-is\cdot r)\mathrm{d}r \tag{6.27}$$

the scattered intensity becomes

$$I(s) = \sum\nolimits_{\alpha,\beta} c_\alpha c_\beta [S_{\alpha\beta}(s) - 1]\phi_\alpha(s)\phi_\beta(s) \tag{6.28}$$

Combining Eqs (6.26) and (6.28), we obtain

$$I(s) = N^{-1}\sum\nolimits_{\alpha,\beta} c_\alpha c_\beta S_{\alpha\beta}(s)\phi_\alpha(s)\phi_\beta(s) + \sum\nolimits_{\alpha} c_\alpha |\Delta\phi_\alpha(s)|^2 \tag{6.29}$$

where the first term represents the coherent scattering and the second term the incoherent scattering. In the case of NRD from an isotopic mixture, the incoherent term is just proportional to the variance of b, that is, $\langle\Delta b\rangle^2 = \langle b^2\rangle - \langle b\rangle^2$. For X-rays, $\phi_{i,\alpha}(\boldsymbol{r} - \boldsymbol{r}_{i,\alpha})$ represents the atomic electron density, while for crystals deviations from total electron density as measured, and the superposition of atomic densities due to the presence of chemical bonds, can be detected experimentally [4]. Finally, it will be clear from Eq. (6.29) that from a single measurement, whether using NRD nor XRD, the various $S_{\alpha\beta}$ cannot be determined.

6.4 The Structure of Liquids

We now discuss some structural, experimental and also simulation considerations for simple and normal liquids in terms of the correlation function. While the basics of the experimental determination of g(r) is treated in Section 6.3, we refer for simulations to Chapter 9. We first compare qualitatively the behavior of g(r) for gases, liquids, and solids (Figure 6.5). For the perfect gas the density is

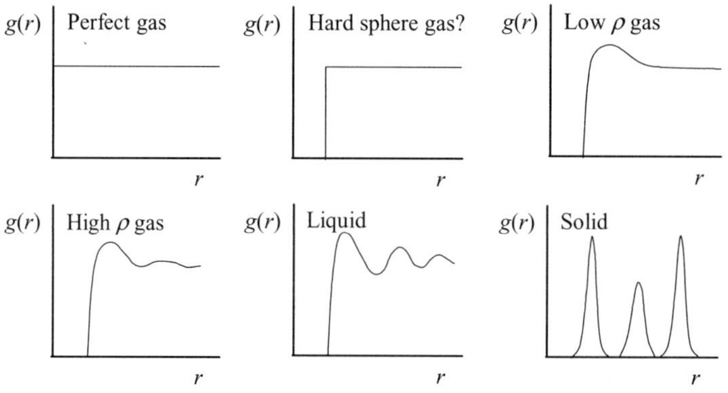

Figure 6.5 Schematic of the pair correlation function g(r) for gases, liquids, and solids.

everywhere the same, and hence no correlation is present. For a hard-sphere gas one might expect a cut-off at the hard-sphere value with no correlation for larger values of r (whether this is true or not, we will come to this point later). In a low-density real gas there is some attraction, and hence a small peak is expected at about twice the molecular radius, whereas for a high-density gas some further structure is anticipated. For a liquid, one expects even more structure as the density is comparable to that of a solid. The first peak corresponds to the first shell of atoms around the reference atom, usually indicated as the first coordination shell, while the second peak represents the second coordination shell. Finally, in the case of the solid, where atoms remain largely at their equilibrium positions, one expects clear peaks due to the largely static coordination shells.

By using Eq. (6.14), it is possible to calculate the number of molecules around a reference molecule, the so-called *coordination number* (CN). For experimentally determined distribution functions, as only the probability is known as a function of distance, determination of the coordination number is ambiguous and several methods can be applied to obtain it. Four methods, as evaluated by Pings [5], are shown in Figure 6.6. Pings concluded that determination of the CN is somewhat arbitrary, and results obtained with different methods are difficult to compare. In practice, method D in Figure 6.6 is normally used. For the first and second coordination shells, we obtain

$$N^{(1)} = \rho\int_0^{M_1} g(r)4\pi r^2\,\mathrm{d}r \quad \text{and} \quad N^{(2)} = \rho\int_{M_1}^{M_2} g(r)4\pi r^2\,\mathrm{d}r \tag{6.30}$$

where M_1 and M_2 denote the first and second minima in the correlation function. A rough estimate for $N^{(1)}$ for dense liquids can be obtained as follows. For a dense liquid, we have approximately $\rho\sigma^3 \cong 0.64/0.74$ and estimate that $M_1 \cong 2^{1/2}\sigma$. Then, by using $g(r) = 0$ for $r < \sigma$ and, say, $g(r) = 1.5$ for $\sigma < r < 2^{1/2}\sigma$, we obtain

$$N^{(1)} \cong 4\pi\rho\int_\sigma^{\sqrt{2}\sigma} g(r)r^2\,\mathrm{d}r \cong 4\pi\rho 1.5\left[\left(\sqrt{2}\sigma\right)^3 - \sigma^3\right]/3 \cong 10 \tag{6.31}$$

confirming once more the relatively large CNs for liquids.

So, close to the triple point, the density of a liquid resembles that of a random close packing of spheres with $\eta \cong 0.64$ and $N^{(1)} \cong 10$, compared with $\eta = 0.74$ and $N^{(1)} = 12$ for the FCC lattice, and implying an expansion of ~15% upon melting. It appears that, near the critical point, the intermolecular spacing for a wide range of liquids is given by $l = (V_{\text{cri}}/N)^{1/3} = (1.50 \pm 0.16)\sigma$. This implies a linear expansion from σ near the triple point to 1.5σ near the critical point, or a volume expansion by a factor of 3.4. This amount of expansion cannot occur by a rearrangement of the local coordination, but rather requires the introduction of holes of molecular size. This leads to a lower CN, although the nearest-neighbor distance is changed only slightly.

Some early experimental data, as well as more sophisticated measurements for Ar at various temperatures, are shown in Figure 6.7. This figure shows the structure anticipated with a clear first coordination shell. It is also possible to recognize a second coordination shell and, for lower temperature, also a third shell.

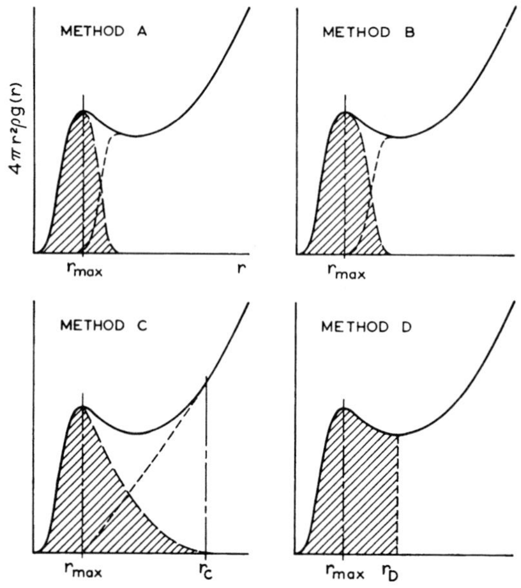

Figure 6.6 Methods of estimating the coordination number from the radial distribution function (RDF). Method A considers that the first peak in the RDF results from a symmetric $rg(r)$, while method B considers $r^2g(r)$ as symmetric. Method C uses the extension from the first maximum to the distance where the RDF is continuously increasing. Method D simply uses the first minimum in the RDF.

Thereafter, the structure becomes "fuzzy," indicating the essentially random nature of liquids. These figures also show the rapid fading out of structure with increasing temperature. Finally, when comparing Figure 6.7a with Figure 6.7b, the considerable improvement in experimental results over the years becomes very clear.

In Figure 6.8 the correlation function for a molecular liquid, namely N_2, is presented. In Figure 6.8a, the experimental correlation function, as obtained using NRD experiments, is shown, whereas in Figure 6.8b the intermolecular correlation function (this means that the intramolecular N–N distance is not shown) for this liquid is given, as obtained from a molecular dynamics (MD) simulation (see Chapter 9) of 256 rigid molecules interacting via a Lennard-Jones potential. Again, the structure fades away with increasing temperature. The CN obtained was about 12 throughout the temperature range under consideration. The structure used and the main configurations for the N_2 dimers analyzed are shown in Figure 6.9. It appeared from these calculations that the R- and T-configurations occurred each for 47–48%, while the S-configuration occurred for only about 3%, independent of the temperature (in the range of 70 to 120 K) and the assumed ratio of the moments of inertia I_z (along the molecular axis) versus I_x and I_y (perpendicular to

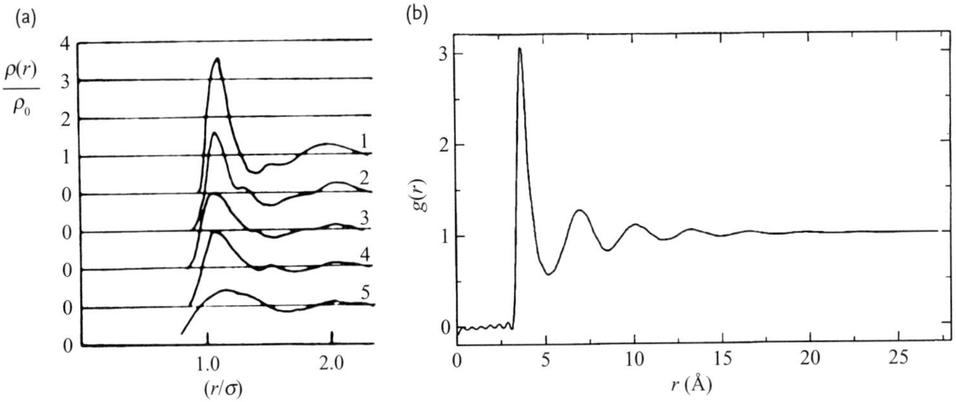

Figure 6.7 (a) Correlation of Ar as measured using X-ray diffraction [9]. Labels: 1 = 84.4 K (~ triple point); 2 = 91.8 K; 3 = 126.7 K; 4 = 144.1 K; 5 = 149.3 K (~ critical point). The size of Ar is $\sigma = 3.42$ Å; (b) Correlation function as determined using neutron ray diffraction (NRD) at 85 K [10].

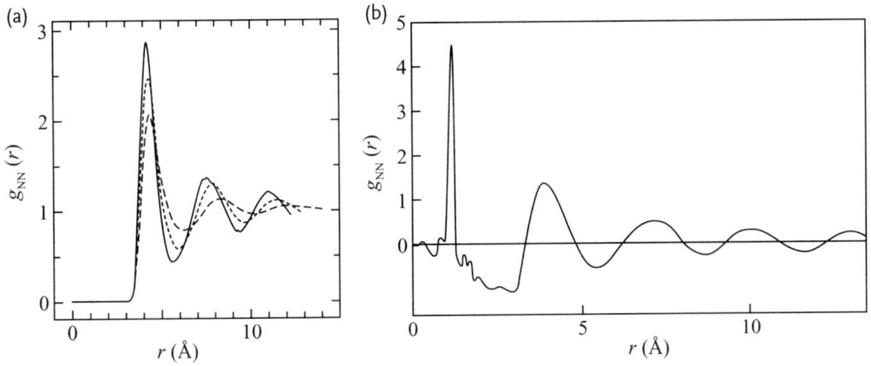

Figure 6.8 (a) The intermolecular correlation function of N_2 as obtained from rigid molecule molecular dynamics simulations at 80 K (solid line, $\rho = 0.796\,g\,cm^{-3}$), 100 K (dotted line, $\rho = 0.689\,g\,cm^{-3}$), and 120 K (dashed line, $\rho = 0.525\,g\,cm^{-3}$). The parameters used were: nuclear distance $L = 1.098$ Å, $\sigma = 3.341$ Å, and $\varepsilon = 0.6064 \times 10^{-14}$ erg [11]; (b) The (total) correlation function of N_2 as obtained from NRD experiments, showing both the intra- and intermolecular distances [12].

the molecular axis). It is clear that the interpretation of the pair correlation function of molecular liquids requires an insight into their chemical structures before conclusions can be drawn regarding the liquid structure. Results similar to those with N_2 have been obtained for several other systems; an example can be seen in Figure 6.10, which shows liquid phosphorus containing P_4 molecules.

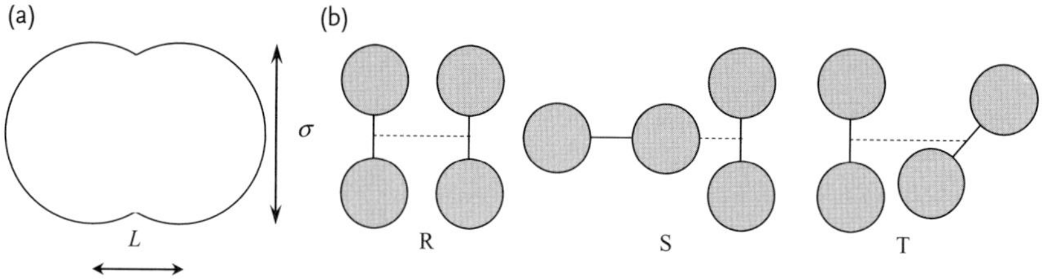

Figure 6.9 (a) The assumed structure with nuclear distance L and Lennard-Jones diameter σ; (b) Main configurations R, S, and T, as used for a molecular dynamics simulation of N_2.

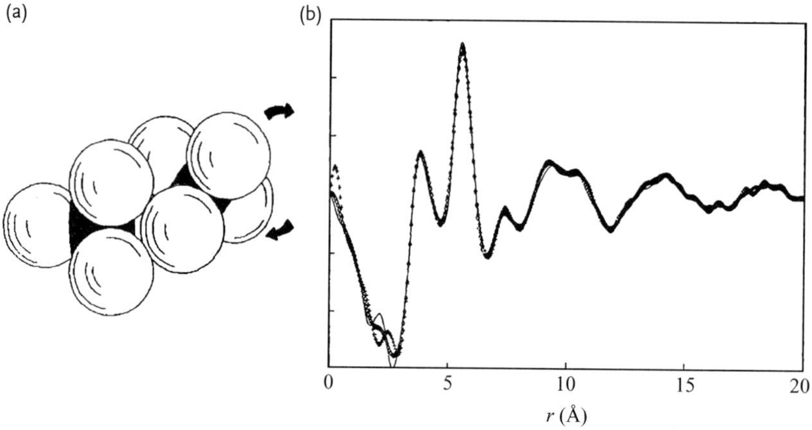

Figure 6.10 (a) The interlocking of XY_4 molecules; (b) The (total) experimental correlation function of P_4 as obtained from NRD experiments, showing as the first peak the P–P distance of the atoms in contact and as the second peak the P–P distance as arising from the interlocked configuration [12].

Before embarking on the related energetics, it should be mentioned that a number of attempts have been made in the past to simulate the (static) structure of a monoatomic liquid using *analogous models*, providing a static–though qualitatively correct–picture of the structure of liquids. In particular, packings of steel balls (as used in ball bearings) have been used. From the well-defined random closed-packed assemblies of these balls, Bernal [6] (actually his assistant) was able to determine the pair distribution function by counting their numbers with increasing distance from a reference ball, and averaging over several reference balls. The thus-obtained pair correlation function closely resembled the experimentally determined correlation functions for mono-atomic liquids, such as argon and several metals. The somewhat more elaborate results as obtained by

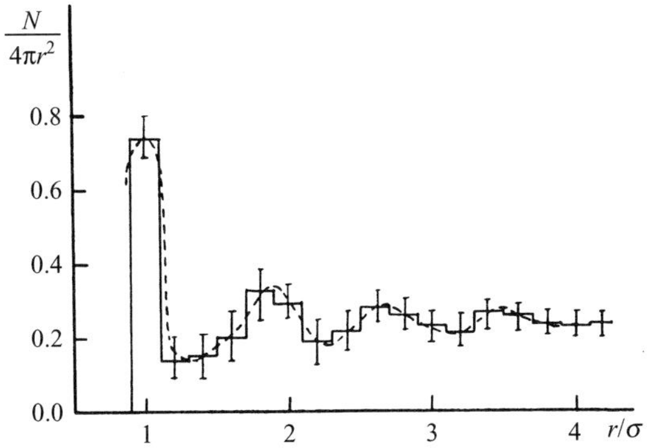

Figure 6.11 The pair correlation of the hard-sphere fluid as determined experimentally by Scott [7] from an assembly of close-packed steel balls.

Scott, are shown in Figure 6.11 [7]. The CN was also determined using an interval from 1.0σ to 1.2σ, where σ is the sphere diameter, which led to values of about 9.3 ± 0.8. In this case, determination of the CN was relatively straightforward (though tedious!), and subsequent computer simulations [3] essentially confirmed these findings. It should be borne in mind, however, that analogous models can provide a static picture of liquids, while the nature of fluids is essentially dynamic.

Problem 6.2

Verify Eqs (6.4), (6.5), and (6.14).

Problem 6.3

Calculate the pair correlation function for a solid with the simple cubic (SC) structure. Do the same for a solid with FCC structure, a solid with the HCP structure, and a solid with the BCC structure.

Problem 6.4

Calculate the CNs from the pair correlation function as given by Scott for the first and second coordination shell using method D in Figure 6.6.

Problem 6.5

Show that $f(\mathcal{H}) = \exp(-\beta\mathcal{H})$ is the only solution of $f(\mathcal{H}) \cdot f(\mathcal{H}') = f(\mathcal{H} + \mathcal{H}')$.

Problem 6.6

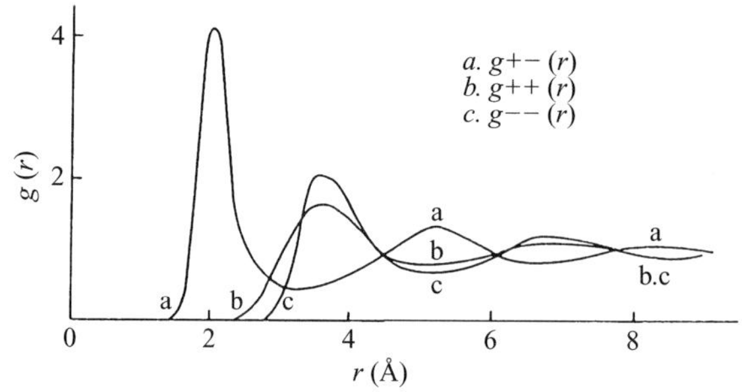

The pair distribution function [8] for molten LiCl, as calculated via MD simulations, is shown in the accompanying figure. The (Pauling) ionic radii for the Li^+ and Cl^- ions are 0.61 Å and 1.81 Å, respectively. Crystalline LiCl has the NaCl structure with lattice constant 5.14 Å.

a) What is the "nearest-neighbor" and "next-nearest-neighbor" distance' between the Li^+ and Cl^- ions (the Li–Cl pair)?

b) What are the "nearest-neighbor" distances for the pairs Li–Li and Cl–Cl?

c) At first sight, one could expect that the Li–Li distance would be about 1.2 Å, and the Cl–Cl distance about 3.6 Å. However, the pair distribution function for the molten state indicates that these "nearest-neighbor" distances in the liquid are about the same. Discuss why this is so.

6.5 Energetics

If we assume that the correlation function $g(r)$ is known and the pairwise additivity (of spherical potentials) holds for the potential energy, that is, $\Phi(\mathbf{r}) = \frac{1}{2}\Sigma_{ij}\phi_{ij}(r)$, we can write down for the internal energy $U = \langle \mathcal{H} \rangle$ almost by inspection the *energy equation*

$$\blacktriangleright \quad \frac{\langle \mathcal{H} \rangle}{N} = \frac{3kT}{2} + \frac{\rho}{2}\int \phi(r)g(r)4\pi r^2\,\mathrm{d}r \tag{6.32}$$

where the first term is the kinetic contribution. More formally, the canonical average, as given by Eq. (6.9), results in

$$\langle \mathcal{H} \rangle = N!^{-1}\int \rho^{(N)}\Phi(\mathbf{r})\mathrm{d}\mathbf{r} \quad \text{with} \quad \rho^{(N)} = N!\exp[-\beta\Phi(\mathbf{r})]/Q_N \tag{6.33}$$

and using $\Phi = \frac{1}{2}\Sigma_{ij}\phi_{ij}$ leads to $N(N-1)/2$ identical terms so that

$$\frac{\langle \mathcal{H} \rangle}{N} = \frac{N-1}{2}\int \phi(r)\rho^{(2)}4\pi r^2\,\mathrm{d}r = \frac{N-1}{2V}\int \phi(r)g(r)4\pi r^2\,\mathrm{d}r \tag{6.34}$$

to which the kinetic term has to be added. In the thermodynamic limit ($N \to \infty$, $V \to \infty$ but $N/V \to \rho$), the factor $(N-1)/V$ becomes ρ, so that the result is Eq. (6.32).

Similarly, though in a somewhat more complex fashion, an expression for the pressure P can be obtained. We start with the previously derived expression (see Chapter 5)

$$P = -\left(\frac{\partial F}{\partial V}\right)_{N,T} = kT\left(\frac{\partial \ln Z}{\partial V}\right)_{N,T} = kT\left(\frac{\partial \ln[(N!)^{-1}\Lambda^{3N}Q_N]}{\partial V}\right)_{N,T} \tag{6.35}$$

with the thermal wavelength $\Lambda \equiv (h^2/2\pi mkT)^{1/2}$. Before differentiating, we consider that the pressure is independent of the shape of the container, so that we can take a cube and employ the coordinate transformation $\mathbf{r}_i = V^{1/3}\mathbf{r}_i'$. The expression for Q_N then becomes

$$Q_N = \int_0^{V^{1/3}} \ldots \int_0^{V^{1/3}} \mathrm{e}^{-\beta\Phi}\,\mathrm{d}\mathbf{r}_1 \ldots \mathrm{d}\mathbf{r}_N = V^N \int_0^1 \ldots \int_0^1 \mathrm{e}^{-\beta\Phi}\,\mathrm{d}\mathbf{r}_1' \ldots \mathrm{d}\mathbf{r}_N' \tag{6.36}$$

while for the intermolecular distance $r_{ij} = [(x_i - x_j)^2 + (y_i - y_j)^2 + (z_i - z_j)^2]^{1/2}$ we obtain the expression $r_{ij} = V^{1/3}[(x_i' - x_j')^2 + (y_i' - y_j')^2 + (z_i' - z_j')^2]^{1/2}$. Therefore $\partial Q_N/\partial V$ becomes

$$\frac{\partial Q_N}{\partial V} = NV^{N-1}\int_0^1 \ldots \int_0^1 \mathrm{e}^{-\beta\Phi}\,\mathrm{d}\mathbf{r}_1' \ldots \mathrm{d}\mathbf{r}_N' - V^N \int_0^1 \ldots \int_0^1 \frac{\partial \Phi}{\partial V}\mathrm{e}^{-\beta\Phi}\,\mathrm{d}\mathbf{r}_1' \ldots \mathrm{d}\mathbf{r}_N' \tag{6.37}$$

$$\text{with} \quad \frac{\partial \Phi}{\partial V} = \sum_{1<i<j<N} \frac{\partial \phi}{\partial r_{ij}}\frac{\partial r_{ij}}{\partial V} = \sum_{1<i<j<N} \frac{\partial \phi}{\partial r_{ij}}\frac{r_{ij}}{3V} \tag{6.38}$$

Transforming back, noting that we have $N(N-1)/2$ identical terms and taking the thermodynamic limit, one obtains the *pressure* (or *virial*) *equation* (see Section 3.8)

$$\blacktriangleright \quad P = \rho kT - \frac{\rho^2}{6}\int r\frac{\mathrm{d}\phi(r)}{\mathrm{d}r}g(r)4\pi r^2\,\mathrm{d}r \tag{6.39}$$

In Eqs (6.32) and (6.39), the first terms are due to the momenta (i.e., the kinetics), while the second term is due to the coordinates (i.e., the configuration). Thus, both equations in principle relate a thermodynamic quantity to the structure, as represented by $g(r)$.

For future reference we note that the compressibility κ_T can be calculated using the grand canonical ensemble (see Justification 6.1). The final result

$$\blacktriangleright \quad \kappa_T = \frac{1}{kT\rho}\left\{1 + \rho\int [g(r) - 1]4\pi r^2\,\mathrm{d}r\right\} \tag{6.40}$$

is known as the *compressibility equation*, and this expression is derived without assuming any pairwise additivity for the potential energy of the system, and is thus generally valid.

Justification 6.1: The compressibility equation*

The derivation of the compressibility equation requires the grand canonical ensemble. We recall first the definition of the isothermal compressibility

$$\kappa_T = -V^{-1}(\partial V/\partial P)_T = \rho^{-1}(\partial \rho/\partial P)_{N,T} \quad \text{with} \quad \rho = N/V$$

From $PV = kT\ln\Xi$ we have, using $\Xi = \sum_{N\geq 0}\frac{\lambda^N}{N!}\int\ldots\int \exp(-\beta\Phi)\mathrm{d}\mathbf{r}_1\ldots\mathrm{d}\mathbf{r}_N$,

$$\frac{V}{kT}\left(\frac{\partial P}{\partial \rho}\right)_{N,T} = \left(\frac{\partial \ln\Xi}{\partial \rho}\right)_{N,T} = \frac{1}{\Xi}\frac{\partial}{\partial \rho}\sum_{N\geq 0}\frac{\lambda^N}{N!}\int\ldots\int \exp(-\beta\Phi)\mathrm{d}\mathbf{r}_1\ldots\mathrm{d}\mathbf{r}_N$$

We further need the absolute activity given by $\lambda = \exp(\beta\mu)$, and note that $\partial\ln\Xi/\partial\rho = (\partial\ln\Xi/\partial\lambda)/(\partial\rho/\partial\lambda)$. For the numerator $\partial\ln\Xi/\partial\lambda$ we obtain, defining $I_N \equiv \int\ldots\int\exp(-\beta\Phi)\mathrm{d}\mathbf{r}_1\ldots\mathrm{d}\mathbf{r}_N$,

$$\left(\frac{\partial \ln\Xi}{\partial \lambda}\right)_{V,T} = \frac{1}{\Xi}\frac{\partial}{\partial\lambda}\sum_{N\geq 0}\frac{\lambda^N}{N!}I_N = \frac{1}{\Xi}\sum_{N\geq 0}\frac{\lambda^{N-1}}{(N-1)!}I_N = \frac{1}{\lambda}\int\rho\,\mathrm{d}\mathbf{r}_1 = \frac{V\rho}{\lambda}$$

where the distribution as defined for the grand canonical ensemble

$$\rho^{(s)}(\mathbf{r}_1,\ldots,\mathbf{r}_s) = \Xi^{-1}\sum_{N\geq 0}\frac{\lambda^N}{(N-s)!}\int\ldots\int\exp(-\beta\Phi)\mathrm{d}\mathbf{r}_{s+1}\ldots\mathrm{d}\mathbf{r}_N$$

is used. The denominator $\partial\rho/\partial\lambda$ is given by

$$\frac{\partial\rho}{\partial\lambda} = \sum_{N\geq 1}\frac{\partial}{\partial\lambda}\frac{\lambda^{N-1}}{(N-1)!}\frac{\lambda}{\Xi}\int\ldots\int\exp(-\beta\Phi)\mathrm{d}\mathbf{r}_2\ldots\mathrm{d}\mathbf{r}_N$$

$$= \begin{pmatrix} \sum_{N\geq 2}\frac{\lambda^{N-2}}{(N-2)!}\frac{\lambda}{\Xi} + \sum_{N\geq 1}\frac{\lambda^{N-1}}{(N-1)!}\frac{1}{\Xi} \\ -\frac{1}{\Xi^2}\frac{\partial\Xi}{\partial\lambda}\sum_{N\geq 1}\frac{\lambda^N}{(N-1)!} \end{pmatrix}\int\ldots\int\exp(-\beta\Phi)\mathrm{d}\mathbf{r}_2\ldots\mathrm{d}\mathbf{r}_N$$

so that

$$\frac{\partial\rho}{\partial\lambda} = \frac{\rho^2}{\lambda}\int g(r)\mathrm{d}\mathbf{r} + \frac{\rho}{\lambda} - \frac{\rho^2}{\lambda}\int\mathrm{d}\mathbf{r} = \frac{\rho}{\lambda}\left\{1+\rho\int[g(r)-1]\mathrm{d}\mathbf{r}\right\}$$

where for the last step the definition $g(\mathbf{r}_1,\mathbf{r}_2) = \rho^{(2)}(\mathbf{r}_1,\mathbf{r}_2)/\rho^{(1)}(\mathbf{r}_1)\rho^{(2)}(\mathbf{r}_2) = \rho^{(2)}(\mathbf{r}_1,\mathbf{r}_2)/\rho^2$ is used. Combining the expressions for $\partial\ln\Xi/\partial\lambda$ and $\partial\rho/\partial\lambda$ yields

$$\frac{V}{kT}\left(\frac{\partial P}{\partial V}\right)_{V,T} = \frac{V}{1+\rho\int[g(r)-1]\mathrm{d}\mathbf{r}}$$

The isothermal compressibility is then given by

$$\kappa_T = \frac{1}{\rho}\left(\frac{\partial\rho}{\partial P}\right)_{N,T} = \frac{1}{\rho kT}\left\{1+\rho\int_0^\infty[g(r)-1]4\pi r^2\,\mathrm{d}r\right\}$$

Note that the structure factor $S(s)$ for $s = 0$ reduces to

$$S(0) = 1 + \rho\int_0^\infty [g(r) - 1]4\pi r^2\,ds$$

and therefore $S(0) = \rho kT\kappa_T$, a fact that can be used for reconstructing $g(r)$.

One might wonder whether the compressibility equation and the pressure equation yield the same result for the pressure. On integrating Eq. (6.40), one obtains

$$P_\kappa = \int \frac{\partial P}{\partial V}\mathrm{d}V = \int \frac{\partial P}{\partial \rho}\mathrm{d}\rho = \int \frac{1}{\rho\kappa_T}\mathrm{d}\rho = kT\int\left\{1 + \rho\int h(r)4\pi r^2\,\mathrm{d}r\right\}^{-1}\mathrm{d}\rho \tag{6.41}$$

to be compared with P_P as obtained from Eq. (6.39). For the ideal gas $g(r) = 1$, leading for both expressions to the ideal gas EoS $P = \rho kT$. This is no longer true for a nonideal fluid for which $\phi(r) \neq 0$, and thus $g(r)$ is unknown. Using approximations for $g(r)$ usually leads to different values for P_P and P_κ, indicating a loss of *thermodynamic consistency* (see Section 6.3).

In order to obtain a complete description of the thermodynamic state, some extra information is required. This is probably most clear from the Gibbs equation

$$T\mathrm{d}S = \mathrm{d}U + P\mathrm{d}V - \mu\mathrm{d}N \tag{6.42}$$

which shows that information is required about the energy U, the pressure P, and the chemical potential μ to obtain a complete description of the thermodynamic state of the system.

If the correlation function $g(r)$ were to be known over a wide temperature range, we could use the energy Eq. (6.32) and integrate the Gibbs–Helmholtz equation $\partial(F/T)/\partial(1/T) = U = \langle\mathcal{H}\rangle$ and so obtain the required information, since the chemical potential μ can be calculated from the Helmholtz energy F. Similarly, if $g(r)$ were to be known over a wide density range, we could use the pressure Eq. (6.39) and integrate $\partial(F/N)/\partial(1/\rho) = \rho^2 P$ (as obtained from $\mathrm{d}F = -P\mathrm{d}V - S\mathrm{d}T + \mu\mathrm{d}N$). Alas, neither the temperature nor density dependence of $g(r)$ is usually available.

As an alternative we introduce the coupling parameter approach, also known as thermodynamic integration. A *coupling parameter* ξ is defined, varying from 0 to 1, which controls the interaction between the reference molecule 1 and another, say j, by replacing ϕ_{1j} by $\xi\phi_{1j}$. The potential energy then becomes

$$\Phi(\xi) = \Phi(\mathbf{r}_1, \ldots \mathbf{r}_N, \xi) = \xi\sum_{j=2}^{N}\phi(\mathbf{r}_{1j}) + \sum_{2\le i<j\le N}^{N}\phi(\mathbf{r}_{ij}) \tag{6.43}$$

and a molecule is added to the system when ξ varies from 0 to 1. The partition function $Z_N(\rho,T)$ reads now $Z_N(\rho,T,\xi)$ while the correlation function $g(r;\rho,T)$ becomes $g(r;\rho,T,\xi)$. Because the number of particles N is very large, we can write

$$\mu = \left(\frac{\partial F}{\partial N}\right)_{V,T} \cong \frac{\Delta F}{\Delta N} = F(N,V,T) - F(N-1,V,T) \tag{6.44}$$

From $F = -kT\ln Z$, we obtain $-\beta F = \ln Q_N - \ln N! - 3N\ln \Lambda$, and hence we get

$$-\beta\mu = \ln(Q_N / Q_{N-1}) - \ln N - \ln \Lambda^3 \tag{6.45}$$

Further, we have $Q_N(\xi = 1) = Q_N$ and $Q_N(\xi = 0) = VQ_{N-1}$. Therefore,

$$\ln\left(\frac{Q_N}{Q_{N-1}}\right) = \ln\left(\frac{Q_N(\xi=1)}{Q_N(\xi=0)}\right) + \ln V = \int_0^1 \frac{\partial \ln Q_N(\xi)}{\partial \xi} d\xi + \ln V \quad \text{with} \tag{6.46}$$

$$Q_N(\xi) = \int \ldots \int \exp[-\beta\Phi(\xi)]\mathrm{d}\mathbf{r}_1 \ldots \mathrm{d}\mathbf{r}_N \tag{6.47}$$

Calculating $\partial Q_N(\xi)/\partial\xi$ using Eq. (6.43) results in

$$\frac{\partial Q_N}{\partial \xi} = -\beta\int \ldots \int \exp[-\beta\Phi(\xi)]\left[\sum\nolimits_{j=2}^{N} \phi(r_{1j})\right]\mathrm{d}\mathbf{r}_1 \ldots \mathrm{d}\mathbf{r}_N \tag{6.48}$$

Further, employing $\rho^{(N)}(\xi) = N!n^{(N)}(\xi)$ with $n^{(N)}(\xi) = Q_N(\xi)^{-1}\exp[-\beta\Phi(\xi)]$, and recognizing that we have $N - 1$ identical integrals, we obtain

$$\frac{\partial \ln Q_N}{\partial \xi} = -\frac{\beta}{N}\int \ldots \int \phi(r_{12})\rho^{(2)}(\mathbf{r}_1, \mathbf{r}_2; \xi)\mathrm{d}\mathbf{r}_1\,\mathrm{d}\mathbf{r}_2 = -\beta\rho\int_0^\infty \phi(r)g(r;\xi)4\pi r^2\,\mathrm{d}r \tag{6.49}$$

where in the last step $\rho^{(2)}(\xi) = \rho^2 g(\xi)$ is employed. Substitution in Eq. (6.45) yields the required result

$$\blacktriangleright \quad \frac{\mu}{kT} = \ln\rho\Lambda^3 + \frac{\rho}{kT}\int_0^1\int_0^\infty \phi(r)g(r;\xi)4\pi r^2\,\mathrm{d}r\,\mathrm{d}\xi \tag{6.50}$$

From $U = \langle\mathcal{H}\rangle$, P and μ, we can calculate all of the thermodynamic properties if $g(r;\xi)$ is known. The theoretical calculation of $g(r;\xi)$ is the topic of Chapter 7.

Problem 6.7*

Check the steps necessary to obtain Eqs (6.39), (6.40), and (6.50).

6.6
The Potential of Mean Force

It is useful to define a quantity $W^{(s)}(\mathbf{r}_1, \ldots, \mathbf{r}_s)$ by

$$g^{(s)}(\mathbf{r}_1, \ldots, \mathbf{r}_s) \equiv \exp[-\beta W^{(s)}(\mathbf{r}_1, \ldots, \mathbf{r}_s)] \tag{6.51}$$

where the definition of the correlation function $g^{(s)}(\mathbf{r}_1, \ldots, \mathbf{r}_s)$ is given by

$$g^{(s)}(\mathbf{r}_1, \ldots, \mathbf{r}_s) = V^s \int \ldots \int \exp(-\beta\Phi)\mathrm{d}\mathbf{r}_{s+1} \ldots \mathrm{d}\mathbf{r}_N \Big/ Q_N \tag{6.52}$$

Taking logarithms on both sides, and then taking the gradient with respect to one of the particles, say i, one obtains

$$-\nabla_i W^{(s)}(\mathbf{r}_1,\ldots,\mathbf{r}_s) = \frac{\int\ldots\int(-\nabla_i\Phi)\exp(-\beta\Phi)\mathrm{d}\mathbf{r}_{s+1}\ldots\mathrm{d}\mathbf{r}_N}{\int\ldots\int\exp(-\beta\Phi)\mathrm{d}\mathbf{r}_{s+1}\ldots\mathrm{d}\mathbf{r}_N} \tag{6.53}$$

Since $-\nabla_i\Phi$ represents the force on molecule i for a fixed configuration of $\mathbf{r}_1, \ldots, \mathbf{r}_s$, the right-hand side is the mean force $\langle F_i^{(s)}(\mathbf{r}_1,\ldots,\mathbf{r}_s)\rangle$ averaged over the configurations of all molecules $s+1, \ldots, N$ not in the fixed set $1, \ldots, s$. Therefore,

$$-\partial W^{(s)}(\mathbf{r}_1,\ldots,\mathbf{r}_s)/\partial\mathbf{r}_i = \langle F_i(\mathbf{r}_1,\ldots,\mathbf{r}_s)\rangle \tag{6.54}$$

Since a force F_j on particle j is always calculated from the corresponding potential Ψ according to $F_j = -\partial\Psi/\partial\mathbf{r}_j$, the quantity $W^{(s)}(\mathbf{r}_1, \ldots, \mathbf{r}_s)$ can be interpreted as potential of mean force (fixating s particles).

In particular, for $s = 2$ we have $W(\mathbf{r}_1,\mathbf{r}_2) \equiv W^{(2)}(\mathbf{r}_1,\mathbf{r}_2)$, representing the *potential of mean force* between one particle held at position $\mathbf{r}_1$ and another held at position $\mathbf{r}_2$. The pair correlation function $g(\mathbf{r}_1,\mathbf{r}_2)$ and the pair potential of mean force $W(\mathbf{r}_1,\mathbf{r}_2)$ are thus related by

▶ $$W(\mathbf{r}_1,\mathbf{r}_2) = -kT\ln g(\mathbf{r}_1,\mathbf{r}_2) \quad \text{or} \quad W(r) = -kT\ln g(r) \tag{6.55}$$

for isotropic systems. For r $(= |\mathbf{r}_2 - \mathbf{r}_1|) \to \infty$, $W(r) \to 0$ and $g \to 1$. From $W(r) = -kT \ln g(r)$, we see that, providing that $\beta W(r) << 1$, $g(r)$ permits the expansion

$$g(r) = 1 - \beta W(r) + \frac{1}{2}[\beta W(r)]^2 - \ldots \tag{6.56}$$

For low density ρ, $W(r) \to \phi(r)$ because the two molecules considered are no longer affected by the other molecules. Hence, at low density $g(r)$ reduces to

$$\lim_{\rho\to 0} g(r) = \exp[-\beta\phi(r)] \equiv e(r) \tag{6.57}$$

and there is a unique correspondence between $\phi(r)$ and $g(r)$. At higher density one writes, using the *background correlation function* $y(r)$,[3)]

$$g(r) = \exp[-\beta\phi(r)]y(r) = e(r)y(r) \tag{6.58}$$

Since $\mathrm{d}\exp(-\beta\phi)/\mathrm{d}r = -\beta\exp(-\beta\phi)\mathrm{d}\phi/\mathrm{d}r$, substituting Eq. (6.58) in the pressure Eq. (6.39) leads to the expression

$$\frac{\beta P}{\rho} = 1 + \frac{\rho}{6}\int r\frac{\mathrm{d}e(r)}{\mathrm{d}r}y(r)4\pi r^2\,\mathrm{d}r \tag{6.59}$$

To obtain P as a function of ρ, we need the expansion for $y(r)$ in ρ reading (and indicating but not using second-order terms)

3) Also known as cavity function, as it describes the distribution of cavities in a hard-sphere fluid. Note that taking logarithms results in $\ln g(r) = -\beta\phi(r) + \ln y(r)$ or $\ln y(r) = \ln g(r) + \beta\phi(r)$ or $\ln y(r) = -\beta[W(r) - \phi(r)] = \Delta W(r)$. Here, $\Delta W(r)$ represents of the potential of mean force in excess over the interaction potential $\phi(r)$. This permits the expansion $y(r) = \exp[\Delta W(r)] = 1 + \Delta W(r) + ½[\Delta W(r)]2 + \ldots$.

$$y(r) = 1 + A_{21}\rho + A_{22}\rho^2 + \ldots \tag{6.60}$$

so that g(r) becomes g(r) = exp[$-\beta\phi(r)$]$y(r) \equiv$ exp[$-\beta\phi(r)$](1 + $A_{21}\rho$ + $A_{22}\rho^2$ + . . .). In the same spirit, we write $g^{(3)}(r)$ = exp[$-\beta(\phi_{12} + \phi_{12} + \phi_{12})$](1 + $A_{31}\rho$ + $A_{32}\rho^2$ + . . .). Substituting these expansions in the Yvon–Born–Green (YBG) equation (Eq. [7.5]),

$$\frac{\partial g^{(2)}(\mathbf{r}_1,\mathbf{r}_2)}{\partial \mathbf{r}_1} = -\frac{g^{(2)}(\mathbf{r}_1,\mathbf{r}_2)}{kT}\frac{\partial \phi(r_{12})}{\partial \mathbf{r}_1} - \frac{\rho}{kT}\int g^{(3)}(\mathbf{r}_1,\mathbf{r}_2,\mathbf{r}_3)\frac{\partial \phi(r_{13})}{\partial \mathbf{r}_1}\mathrm{d}\mathbf{r}_3 \tag{6.61}$$

we obtain for both sides of the YBG equation a polynomial in the density ρ. The next step is to equate equal order terms in ρ, and for the first-order terms in ρ this leads to

$$\frac{\partial A_{21}(\mathbf{r}_1,\mathbf{r}_2)}{\partial \mathbf{r}_1} = -\beta\int \exp[-\beta(\phi_{13}+\phi_{23})]\frac{\partial \phi(r_{13})}{\partial \mathbf{r}_1}\mathrm{d}\mathbf{r}_3 \tag{6.62}$$

Integration yields

$$A_{21}(\mathbf{r}_1,\mathbf{r}_2) = \int \exp[-\beta(\phi_{13}+\phi_{23})]\mathrm{d}\mathbf{r}_3 + C \tag{6.63}$$

where C is the integration constant. Since for $r_{12} \to \infty$, g(r) $\to$ 1, we obtain from g(r) = exp($-\beta\phi$)(1 + $A_{21}\rho$ + . . .) as the boundary condition $A_{21}(r_{12})$ = 0 for $r_{12} \to \infty$. Using $f(r)$ = exp[$-\beta\phi(r)$] $\equiv$ $e(r)$$-$1, this boundary condition is fulfilled if we write

$$A_{21}(\mathbf{r}_1,\mathbf{r}_2) = \int f(\mathbf{r}_1,\mathbf{r}_3)f(\mathbf{r}_2,\mathbf{r}_3)\mathrm{d}\mathbf{r}_3 = \int_0^\infty f(r_{13})f(r_{23})4\pi r_3^2\,\mathrm{d}r_3 \tag{6.64}$$

Introducing Eq. (6.60) together with this result in Eq. (6.59), one obtains again the virial expression

$$P = \rho kT\left[1+\rho B_2(\phi,T)+\rho^2 B_3(\phi,T)+\ldots\right] \quad \text{where} \tag{6.65}$$

$$B_2(\phi,T) = -\frac{1}{2}\int_0^\infty f(r)4\pi r^2\,\mathrm{d}r \quad \text{and} \quad B_3(\phi,T) = \frac{1}{3}\int_0^\infty f(r)A_{21}(r)4\pi r^2\,\mathrm{d}r \tag{6.66}$$

Example 6.1: The hard-sphere fluid

To illustrate these concepts, let us apply them to a hard-sphere fluid with a potential given by Figure 6.12.

$$\phi_{\mathrm{HS}}(r) = \infty \quad \text{for} \quad r \le \sigma \quad \text{and} \quad \phi_{\mathrm{HS}}(r) = 0 \quad \text{for} \quad r > \sigma \tag{6.67}$$

Hence, we have

$$\exp(-\beta\phi_{\mathrm{HS}}) = 0 \quad \text{for} \quad r \le \sigma \quad \text{and} \quad \exp(-\beta\phi_{\mathrm{HS}}) = 1 \quad \text{for} \quad r > \sigma \tag{6.68}$$

$$f(r) = -1 \quad \text{for} \quad r \le \sigma \quad \text{and} \quad f(r) = 0 \quad \text{for} \quad r > \sigma \tag{6.69}$$

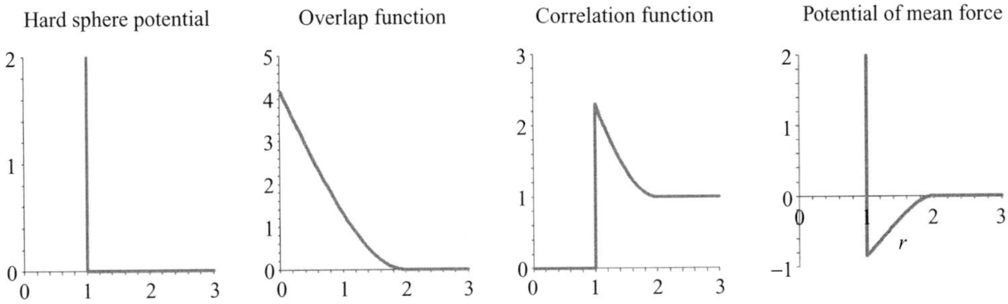

Figure 6.12 The hard-sphere potential ϕ_{HS}, the overlap function A_{HS}, the correlation function g_{HS} and the potential of mean force W_{HS} as a function of r in units σ.

The function $A_{21}(r)$ deviates from zero only if both $f(r_{12}) \neq 0$ and $f(r_{23}) \neq 0$ and for a hard-sphere fluid represents the volume of penetration of two spheres of radius $\sigma/2$ at distance r. Stereometry learns that this overlap function, labeled A_{HS}, reads (Figure 6.12)

$$A_{HS}(r) = \frac{4\pi\sigma^3}{3}\left(1 - \frac{3r}{4\sigma} + \frac{r^3}{16\sigma^3}\right) \quad \text{for} \quad r \leq 2\sigma \quad \text{and}$$
$$A_{HS}(r) = 0 \quad \text{for} \quad r > 2\sigma \tag{6.70}$$

This implies that the correlation function $g_{HS}(r)$ for hard-sphere fluid becomes

$$g_{HS}(r) = h(r-\sigma)(1 + \rho A_{HS}(r) + \ldots) \quad \text{with the Heaviside function}$$
$$h(x) = 0 \quad \text{for} \quad x < 0 \quad \text{and} \quad h(x) = 1 \quad \text{for} \quad x > 0 \tag{6.71}$$

representing $\exp(-\beta\phi_{HS})$. The second factor in Eq. (6.71) results in a peak in $g_{HS}(r)$ (Figure 6.12). The potential of mean force correlation $W_{HS}(r) = -kT \ln g_{HS}(r)$ reads

$$\beta W_{HS}(r) = \beta\phi_{HS}(r) - \ln[1 + \rho A_{HS}(r) + \ldots] \tag{6.72}$$

The net result is an effective attractive force for $\sigma < r < 2\sigma$ (Figure 6.12). This attraction is thus purely a result of geometric restrictions. Since[4] $\mathrm{d}\exp(-\beta\phi_{HS})/\mathrm{d}r = -\beta\exp(-\beta\phi_{HS})\mathrm{d}\phi_{HS}/\mathrm{d}r = \beta\exp(-\beta\phi_{HS})\delta(r-\sigma)$, the pressure P, given by Eq. (6.59), becomes

$$\frac{P}{\rho kT} = 1 + \frac{2\pi\rho}{3}\sigma^3 y(\sigma) = 1 + 4\eta y(\sigma) = 1 + 4\eta g_{HS}(\sigma^+) \tag{6.73}$$

with $\eta = \pi\sigma^3\rho/6$ the packing fraction[5]. By using Eq. (6.66), we obtain $B_2 = b \equiv 2\pi\sigma^3/3$ and $B_3 = 5b^2/8$; these results have already been derived in Chapter 5.

4) Note that because $\mathrm{d}h(r-r')/\mathrm{d}r = \delta(r-r')$, we have $\mathrm{d}\phi_{HS}(r-\sigma)/\mathrm{d}r = -\delta(r-\sigma)$. Moreover, note that direct integration of Eq. (6.39) cannot be done because $g_{HS}(r)$ is discontinuous at $r = \sigma$ but can be done using Eq. (6.59) since $y(r)$ is continuous at that point.

5) The abbreviation $g(\sigma^+)$ is defined by $g(\sigma^+) \equiv \lim_{r\to\sigma+0} g(r)$, implying that r approaches σ from the positive side. Often, $g(\sigma^+)$ is given as $g(\sigma)$.

From Example 6.1 it becomes clear that the hard-sphere correlation function is not just a step function, as one might expect naively (Figure 6.5), but rather shows an increased value near the reference molecule by the geometric restrictions imposed on the coordination shell, leading to an effective attraction. Briefly summarizing, the correlation function $g(r)$ on average should yield a value of 1. Since $g(r) = 0$ for $r < \sigma$, there should be a region where $g(r) > 1$. Because the potential of mean force $W(r) = -kT \ln g(r)$, the region where $g(r) > 1$ leads to $W(r) < 0$.

Problem 6.8*

Check the steps necessary to obtain Eqs (6.63) and (6.65).

References

1 de With, G. (2006) *Structure, Deformation and Integrity of Materials*, Wiley-VCH Verlag GmbH, Weinheim. This book provides a concise discussion on the structure, binding and defects of solids.
2 (a) Bernal, J.D. (1964) *Proc. R. Soc.* **A280**, 299; (b) Frost, H.J. (1982) *Acta Metall.*, **30**, 889.
3 Finney, J.L. (1970) *Proc. R. Soc.* **A319**, 279.
4 Coppens, P. (1997) *X-Ray Charge Densities and Chemical Bonding*, Oxford University Press.
5 Pings, C.J. (1968) *X-Ray Charge Densities and Chemical Bonding* (eds H.N.V. Temperley, J.S. Rowlinson, and G.S. Rushbrooke), North-Holland, Amsterdam, p. 389.
6 (a) Bernal, J.D. and Mason, J. (1960) *Nature*, **188**, 910; (b) Bernal, J.D., Mason, J., and Knight, K.R. (1962) *Nature*, **194**, 957; (c) For further elaboration, see Bernal, J.D. (1964) *Proc. R. Soc.*, **A280**, 299.
7 (a) Scott, G.D. (1960) *Nature*, **188**, 908; (b) Scott, G.D. (1962) *Nature*, **194**, 956.
8 (a) Woodcock, L.V. (1971) *Chem. Phys. Lett.* **10**, 257; (b) Woodcock, L.V. (1972) *Proc. R. Soc.*, **A328**, 83.
9 Eisenstein, A. and Gingrich, N.S. (1942), *Phys. Rev.*, **62**, 261.
10 Yarnell, J.L., Katz, M.J., Wenzel, R.G., and Koenig, S.H. (1973) *Phys. Rev.*, **A7**, 2130.
11 Yogi, T. (1995) *J. Phys. Soc. Jpn*, **64**, 2886.
12 Dore, J.C. (1990) *Il Nuovo Cimento*, **12D**, 543.

Further Reading

Friedman, H.L. (1985) *A Course on Statistical Mechanics*, Prentice-Hall, Englewood Cliffs, NJ.
Hansen, J.-P. and McDonald, I.R. (2006) *Theory of Simple Liquids*, 3rd edn, Academic, London.
McQuarrie, D.A. (1973) *Statistical Mechanics*, Harper and Row, New York.

7
Modeling the Structure of Liquids: The Integral Equation Approach

The statistical description using the distribution functions given in Chapter 6 led to the pair correlation function $g(r)$ for characterizing the structure of liquids and options to calculate the thermodynamic properties. In this chapter, we will see that this approach also provides options for calculating $g(r)$ using methods which are conveniently and collectively described as the "integral equation approach." Using hard-sphere results as a basis, the effect of a more realistic potential on the thermodynamic properties can be assessed via the perturbation approach. Finally, the necessary steps and associated problems to extend this approach to molecular fluids are indicated briefly.

7.1
The Vital Role of the Correlation Function

In Chapter 6 we have seen that

$$U = \langle \mathcal{H} \rangle = \frac{3NkT}{2} + \frac{N\rho}{2}\int_0^\infty \phi(r)g(r)4\pi r^2\,\mathrm{d}r, \tag{7.1}$$

$$P = \rho kT - \frac{\rho^2}{6}\int_0^\infty r\frac{\mathrm{d}\phi(r)}{\mathrm{d}r}g(r)4\pi r^2\,\mathrm{d}r \quad \text{and} \tag{7.2}$$

$$\frac{\mu}{kT} = \ln\rho\Lambda^3 + \frac{\rho}{kT}\int_0^1\int_0^\infty \phi(r)g(r,\xi)4\pi r^2\,\mathrm{d}r\,\mathrm{d}\xi \tag{7.3}$$

where the coupling parameter ξ is used in the last equation and all other symbols have their usual meaning. In these expressions, the first term is due to the momenta– that is, the kinetics– while the second term is due to the coordinates– that is, the configuration. Thus, all three equations in principle relate a thermodynamic quantity to the structure as represented by g. Now, the thought might occur that, if we could determine the correlation function $g(r)$ experimentally, we could calculate the thermodynamic properties for a liquid. However, in spite of the significant progress in experimental accuracy for determining the correlation function $g(r)$, at present these methods are insufficiently accurate to calculate

Liquid-State Physical Chemistry, First Edition. Gijsbertus de With.

reliable thermodynamic properties, and we thus must seek alternative methods. One option would be to use integral equation methods, as described in this chapter. An extensive review, dealing with several aspects of the integral equation, the associated perturbation approach and several other topics, has been provided by Barker and Henderson [1].

Problem 7.1

Indicate what are the most important reasons why Eqs (7.1), (7.2), and (7.3) yield unreliable thermodynamic results.

7.2 Integral Equations

In the previous section, as well as in Chapter 6, we have seen that the pair correlation function $g(r)$ is the essential unknown. So, let us see whether we can obtain a theoretical expression for the correlation function $g(r)$. There are several approaches to achieve this goal. We deal first with the Yvon–Born–Green equation, and continue with the Kirkwood equation. After introducing some new concepts leading to the Ornstein–Zernike equation, we introduce briefly the hypernetted chain, Percus–Yevick, and Mean Spherical Approximation approaches. Those readers not interested in the details of the various approaches should consider them simply as labels for particular approximate solutions in the paragraph comparing results, and in Section 7.3.

7.2.1 The Yvon–Born–Green Equation

The eldest– and, in principle, the most clear approach– runs as follows. We start with the pair distribution function (see Chapter 6) with, as before, $\beta = 1/kT$

$$\rho^{(2)}(\mathbf{r}_1,\mathbf{r}_2) = \frac{N!}{Q_N(N-2)!}\int \exp[-\beta\Phi(\mathbf{r})]\mathrm{d}\mathbf{r}_3 \ldots \mathrm{d}\mathbf{r}_N \tag{7.4}$$

We assume again that the potential energy is pairwise additive, $\Phi = ½\Sigma_{ij}\phi_{ij}$, and differentiate with respect to $\mathbf{r}_1$. Since the integration is over $\mathbf{r}_3$ to $\mathbf{r}_N$ and differentiation is with respect to $\mathbf{r}_1$, differentiation and integration can be exchanged to yield

$$\frac{\partial\rho^{(2)}(\mathbf{r}_1,\mathbf{r}_2)}{\partial\mathbf{r}_1} = \frac{N!}{Q_N(N-2)!}\int \frac{\partial}{\partial\mathbf{r}_1}\exp(-\beta\Phi)\mathrm{d}\mathbf{r}_3 \ldots \mathrm{d}\mathbf{r}_N$$
$$= \frac{N!}{Q_N(N-2)!}\int -\beta\exp(-\beta\Phi)\frac{\partial\Phi}{\partial\mathbf{r}_1}\mathrm{d}\mathbf{r}_3 \ldots \mathrm{d}\mathbf{r}_N$$

Because we assume that $\Phi = ½\Sigma_{ij}\phi_{ij} = \phi_{12} + \ldots + \phi_{1N} + \phi_{23} + \ldots + \phi_{N-1,N}$ we obtain $\partial\Phi/\partial\mathbf{r}_1 = \partial\phi_{12}/\partial\mathbf{r}_1 + \sum_{k=3}^{k=N}(\partial\phi_{1k}/\partial\mathbf{r}_1)$ so that

$$\frac{\partial \rho^{(2)}(\mathbf{r}_1,\mathbf{r}_2)}{\partial \mathbf{r}_1} = -\beta \frac{\partial \phi(\mathbf{r}_{12})}{\partial \mathbf{r}_1} \int \frac{N!}{Q_N(N-2)!} \exp(-\beta \Phi) d\mathbf{r}_3 \ldots d\mathbf{r}_N$$
$$-\beta \int \frac{\partial \phi(\mathbf{r}_{13})}{\partial \mathbf{r}_1} \left[\int \frac{N!(N-2)}{Q_N(N-2)!} \exp(-\beta \Phi) d\mathbf{r}_4 \ldots d\mathbf{r}_N \right] d\mathbf{r}_3$$

In the first integral we recognize $\rho^{(2)}$ and in the one in square brackets $\rho^{(3)}$ and, if we divide by ρ^2 and use $\rho^{(2)} = \rho^2 g^{(2)}$ and $\rho^{(3)} = \rho^3 g^{(3)}$, the overall result is

$$\frac{\partial g^{(2)}(\mathbf{r}_1,\mathbf{r}_2)}{\partial \mathbf{r}_1} = -\frac{g^{(2)}(\mathbf{r}_1,\mathbf{r}_2)}{kT} \frac{\partial \phi(\mathbf{r}_{12})}{\partial \mathbf{r}_1} - \frac{\rho}{kT} \int g^{(3)}(\mathbf{r}_1,\mathbf{r}_2,\mathbf{r}_3) \frac{\partial \phi(\mathbf{r}_{13})}{\partial \mathbf{r}_1} d\mathbf{r}_3 \tag{7.5}$$

This is an infinitely nested set of coupled equations that is usually denoted as the *Yvon–Born–Green* (YBG) *hierarchy* [2], and which can be solved formally by iteration. However, for a concrete solution a "closure" is required, and to that purpose we employ the potential of mean force $W^{(3)}(\mathbf{r}_1, \ldots, \mathbf{r}_3)$, as discussed in Section 6.3. Suppose that we assume– not completely unrealistically– that this potential of mean force for triplets is to a first approximation pair-wise additive, that is, $W^{(3)}(\mathbf{r}_1,\mathbf{r}_2,\mathbf{r}_3) \cong W^{(2)}(\mathbf{r}_1,\mathbf{r}_2) + W^{(2)}(\mathbf{r}_2,\mathbf{r}_3) + W^{(2)}(\mathbf{r}_3,\mathbf{r}_1)$. This results in

$$g^{(3)}(r_{12},r_{23},r_{31}) \cong g^{(2)}(r_{12})g^{(2)}(r_{23})g^{(2)}(r_{31}) \tag{7.6}$$

Using this *superposition approximation*, as originally proposed by Kirkwood and Boggs [3] as a closure relation for the YBG hierarchy, leads to an integral equation which can be solved numerically. The resulting *YBG equation* reads

$$\frac{\partial \ln g^{(2)}(\mathbf{r}_1,\mathbf{r}_2)}{\partial \mathbf{r}_1} = -\beta \frac{\partial \phi(\mathbf{r}_{12})}{\partial \mathbf{r}_1} - \beta\rho \int g^{(2)}(\mathbf{r}_3,\mathbf{r}_1) g^{(2)}(\mathbf{r}_2,\mathbf{r}_3) \frac{\partial \phi(\mathbf{r}_{13})}{\partial \mathbf{r}_1} d\mathbf{r}_3 \tag{7.7}$$

showing again that: (i) the potential of mean force $W = -kT\ln g$ is composed of the direct interaction $\phi(r_{12})$ and the weighted interaction on particle 1 exerted by all particles 3 at arbitrary positions; and (ii) $W = -kT\ln g$ reduces to ϕ for $\rho \to 0$.

Of course, this approach can be used for $\rho^{(n)}$ with $n \neq 2$, but let us here generalize slightly the meaning of Φ. So far, Φ only included the pair potential contributions, that is, $\Phi = \frac{1}{2}\Sigma_{ij}\phi_{ij}$, but in several problems there is also an external potential X acting on each molecule individually, that is, $X = \Sigma_i \chi_i$. Repeating the procedure outlined above, but now for $n = 1$ and including Φ and X, we obtain (using $\rho \equiv \rho^{(1)}$)

$$\frac{\partial \rho(\mathbf{r}_1)}{\partial \mathbf{r}_1} = -\beta\rho(\mathbf{r}) \frac{\partial \chi(\mathbf{r}_1)}{\partial \mathbf{r}_1} - \beta \int \rho^{(2)}(\mathbf{r}_1,\mathbf{r}_2) \frac{\partial \phi(\mathbf{r}_{12})}{\partial \mathbf{r}_1} d\mathbf{r}_2 \tag{7.8}$$

If Φ may be neglected, for example, as for a dilute gas, we are left with X resulting in

$$\partial \ln \rho(\mathbf{r}) / \partial \mathbf{r} = -\partial \beta \chi(\mathbf{r}) / \partial \mathbf{r} \quad \text{or} \quad \rho(\mathbf{r}) = \rho(0)\exp(-\beta\chi) \tag{7.9}$$

Taking for χ the potential due to gravity, $\chi = mgh$, we obtain the well-known *barometric formula* with $\rho(0)$ the density at $h = 0$.

7.2.2
The Kirkwood Equation*

Another relatively old approach uses thermodynamic integration in combination with the pairwise additivity assumption $\Phi = ½\Sigma_{ij}\phi_{ij}$. We start again with the distribution function (with, as before, $\beta = 1/kT$), but this function is now dependent on the coupling parameter ξ, that is,

$$\rho^{(2)}(\mathbf{r}_1,\mathbf{r}_2;\xi) = \frac{N!}{Q_N(\xi)(N-2)!}\int \exp[-\beta\Phi(\mathbf{r};\xi)]\,\mathrm{d}\mathbf{r}_3 \ldots \mathrm{d}\mathbf{r}_N \tag{7.10}$$

In this case we differentiate with respect to ξ to get

$$\begin{aligned} kT\frac{\partial\rho^{(2)}}{\partial\xi} &= \frac{\rho^{(2)}(\mathbf{r}_1,\mathbf{r}_2;\xi)}{Q_N(\xi)}\int \exp[-\beta\Phi(\mathbf{r};\xi)]\phi(\mathbf{r}_{12})\,\mathrm{d}\mathbf{r}_1 \ldots \mathrm{d}\mathbf{r}_N \\ &\quad - \frac{N!}{Q_N(\xi)(N-2)!}\sum\nolimits_{j=2}^{N}\int\ldots\int \exp[-\beta\Phi(\mathbf{r};\xi)]\phi(\mathbf{r}_{1j})\,\mathrm{d}\mathbf{r}_3 \ldots \mathrm{d}\mathbf{r}_N \end{aligned} \tag{7.11}$$

The first term can be written as

$$\frac{Q_N(\xi)}{N(N-1)}\iint \rho^{(2)}(\mathbf{r}_1,\mathbf{r}_2)\phi(\mathbf{r}_{12})\,\mathrm{d}\mathbf{r}_1\,\mathrm{d}\mathbf{r}_2 \tag{7.12}$$

while the second term involves two separate cases. For $j = 2$, the factor $\phi(\mathbf{r}_{1j})$ can be taken outside the integral resulting in

$$\phi(\mathbf{r}_{12})\frac{N!}{Q_N(\xi)(N-2)!}\int\ldots\int \exp[-\beta\Phi(\mathbf{r};\xi)]\,\mathrm{d}\mathbf{r}_3 \ldots \mathrm{d}\mathbf{r}_N = \phi(\mathbf{r}_{12})\rho^{(2)}(\mathbf{r}_1,\mathbf{r}_2) \tag{7.13}$$

For $j = 3, \ldots, N$, the result is

$$\begin{aligned} &\int \phi(\mathbf{r}_{1j})\left(\int\ldots\int \exp[-\beta\Phi(\mathbf{r};\xi)]\,\mathrm{d}\mathbf{r}_3 \ldots \mathrm{d}\mathbf{r}_{j-1}\,\mathrm{d}\mathbf{r}_{j+1} \ldots \mathrm{d}\mathbf{r}_N\right)\mathrm{d}\mathbf{r}_j \\ &\quad = \frac{Q_N(\xi)(N-3)!}{N!}\int \phi(\mathbf{r}_{1j})\rho^{(3)}(\mathbf{r}_1,\mathbf{r}_2,\mathbf{r}_j;\xi)\,\mathrm{d}\mathbf{r}_j \end{aligned} \tag{7.14}$$

Taking these results together and dividing by $\rho^{(2)}(\mathbf{r}_1,\mathbf{r}_2)$ yields

$$\begin{aligned} kT\frac{\partial \ln\rho^{(2)}}{\partial\xi} &= -\phi(\mathbf{r}_{12}) + \frac{1}{N}\iint \rho^{(2)}(\mathbf{r}_1,\mathbf{r}_2;\xi)\phi(\mathbf{r}_{12})\,\mathrm{d}\mathbf{r}_1\,\mathrm{d}\mathbf{r}_2 \\ &\quad - \int \phi(\mathbf{r}_{13})\rho^{(3)}(\mathbf{r}_1,\mathbf{r}_2,\mathbf{r}_3;\xi)[\rho^{(2)}(\mathbf{r}_1,\mathbf{r}_2;\xi)]^{-1}\,\mathrm{d}\mathbf{r}_3 \end{aligned} \tag{7.15}$$

The next step is to integrate the previous equation from 0 to ξ meanwhile using

$$\rho^{(3)}(\mathbf{r}_1,\mathbf{r}_2,0) = \frac{N!}{(N-2)!V}\frac{\int\ldots\int \exp(-\beta\Phi_{N-1})\,\mathrm{d}\mathbf{r}_3 \ldots \mathrm{d}\mathbf{r}_N}{\int\ldots\int \exp(-\beta\Phi_{N-1})\,\mathrm{d}\mathbf{r}_2 \ldots \mathrm{d}\mathbf{r}_N} = \rho\rho_{N-1}^{(1)}(\mathbf{r}_2,\ldots,\mathbf{r}_N) \tag{7.16}$$

1) We could write for $\rho_{N-1}^{(1)} = (N-1)/V$ but with keeping $\rho_{N-1}{}^{(1)}$ generalization to $n > 2$ is more obvious.

where $\rho_{N-1}^{(1)}(\mathbf{r}_2,\ldots,\mathbf{r}_N)$ denotes the distribution function for system containing $N-1$ molecules.[1)] Integration thus gives after some manipulation

$$\begin{aligned} kT\ln\rho^{(2)}(\mathbf{r}_1,\mathbf{r}_2;\xi) &= kT\ln\rho + kT\ln\rho_{N-1}^{(1)}(\mathbf{r}_2\ldots,\mathbf{r}_N) - \xi\phi(\mathbf{r}_{12}) \\ &\quad + \frac{1}{N}\int_0^{\xi}\iint \phi(\mathbf{r}_{12})\rho^{(2)}(\mathbf{r}_1,\mathbf{r}_2;\xi)\,\mathrm{d}\mathbf{r}_1\,\mathrm{d}\mathbf{r}_2\,\mathrm{d}\xi \\ &\quad - \int_0^{\xi}\int \phi(\mathbf{r}_{13})\frac{\rho^{(3)}(\mathbf{r}_1,\mathbf{r}_2,\mathbf{r}_3;\xi)}{\rho^{(2)}(\mathbf{r}_1,\mathbf{r}_2;\xi)}\mathrm{d}\mathbf{r}_3\,\mathrm{d}\xi \end{aligned} \tag{7.17}$$

Introducing the correlation functions $g^{(2)}$ and $g^{(3)}$ the final result becomes

$$\begin{aligned} -kT\ln g^{(2)}(\mathbf{r}_1,\mathbf{r}_2;\xi) &= \xi\phi(\mathbf{r}_{12}) \\ &\quad + \rho\int_0^{\xi}\int \phi(r_{13})\left[\frac{g^{(3)}(\mathbf{r}_1,\mathbf{r}_2,\mathbf{r}_3;\xi)}{g^{(2)}(\mathbf{r}_1,\mathbf{r}_2;\xi)} - g^{(2)}(\mathbf{r}_1,\mathbf{r}_3;\xi)\right]\mathrm{d}\mathbf{r}_3\,\mathrm{d}\xi \end{aligned} \tag{7.18}$$

This expression gives $g^{(2)}$ in terms of $g^{(3)}$ and provides another example of a hierarchy of coupled equations, the so-called *Kirkwood* (K) *hierarchy* [4]. Again, using the superposition approximation results in the *Kirkwood equation*

$$\begin{aligned} -kT\ln g^{(2)}(\mathbf{r}_1,\mathbf{r}_2;\xi) &= \xi\phi(r_{12}) \\ &\quad + \rho\int_0^{\xi}\int \phi(r_{13})g^{(2)}(\mathbf{r}_1,\mathbf{r}_3;\xi)\left[g^{(2)}(\mathbf{r}_2,\mathbf{r}_3)-1\right]\mathrm{d}\mathbf{r}_3\,\mathrm{d}\xi \end{aligned} \tag{7.19}$$

Also this integral equation has to be solved numerically and shows again that for $\rho \to 0$, $W = -kT\ln g \to \phi$. It can be shown that the YBG and K hierarchies are equivalent [5], but that the approximate solutions to the YBG and K equations are not.

7.2.3
The Ornstein–Zernike Equation*

For the following, it is convenient to introduce some auxiliary functions. Since we consider isotropic systems we use $r_{ij} = |r_j - r_i|$, or even r if confusion is impossible. In this notation the expressions for the pair correlation function $g(r_1,r_2)$ and the often-used Mayer function $f(r_1,r_2)$ using $e \equiv \exp[-\beta\phi(r_1,r_2)]$ become

$$g(r_{12}) = \frac{N!}{(N-2)!}\frac{1}{Q_N}\int_0^{\infty}\ldots\int_0^{\infty}\exp[-\beta\Phi(r_1,\ldots,r_N)]\,\mathrm{d}r_3\ldots\mathrm{d}r_N \quad \text{and}$$

$$f(r_{12}) \equiv \exp[-\beta\phi(r_{12})] - 1 = e(r_{12}) - 1$$

with Φ and ϕ the total potential energy and two-particle potential energy, respectively.

To introduce the auxiliary concepts, we recall that $\rho g(r_{12}) = \rho g(r_1,r_2)$ describes the probability of finding a particle at distance $r_{12} = |r_2 - r_1|$ from the origin, given that another particle is located at the origin. No correlation would intuitively imply $g \to 0$ for $r_{12} \to \infty$, but for $r_{12} \to \infty$ we actually have $g \to 1$. Therefore, we define the so-called *total correlation function* by $h(r_{12}) \equiv g(r_{12}) - 1$, for which obviously holds:

$r_{12} \to \infty$, $h \to 0$. The total correlation between particle 1 and particle 2 is supposed to be composed of two contributions:

- the direct correlation between particles 1 and 2, represented by the *direct correlation function* $c(r_{12})$; and
- the indirect correlation of particle 1 and particle 2 via the influence of particle 1 on particle 3 and the influence of particle 3 on particle 2 summed over all such third particles (with density ρ). This results in the convolution $\rho \int c(r_{13})h(r_{32})\, d\mathbf{r}_3$.

The total correlation is thus given by

$$h(r_{12}) = c(r_{12}) + \rho \int c(r_{13})h(r_{32})\,d\mathbf{r}_3 \tag{7.20}$$

This so-called *Ornstein–Zernike* (OZ) *equation*, originally used by these authors in 1914 to explain light scattering in liquids near the critical temperature, is in fact one way of defining the direct correlation function $c(r_{12})$. Although the physical interpretation of $c(r_{12})$ is not as clear as that for $g(r)$, it does have a simpler structure, and in particular it has a much shorter range. A formal solution can be obtained by iteration yielding

$$h(r_{12}) = c(r_{12}) + \rho \int c(r_{13})c(r_{32})\,d\mathbf{r}_3 + \rho^2 \int c(r_{13})c(r_{34})c(r_{42})\,d\mathbf{r}_3\,d\mathbf{r}_4 + \ldots \tag{7.21}$$

This solution can be interpreted as a sum of "chains" of direct correlation functions between $\mathbf{r}_1$ and $\mathbf{r}_2$. Since the OZ equation is a convolution, a practical solution for $h(r_{12})$ can be obtained from its Fourier transform (see Appendix B). Using

$$\tilde{h}(\mathbf{s}) = \int_0^\infty h(\mathbf{r})\exp(-i\mathbf{s}\cdot\mathbf{r})\,d\mathbf{r} \quad \text{and} \quad \tilde{c}(\mathbf{s}) = \int_0^\infty c(\mathbf{r})\exp(-i\mathbf{s}\cdot\mathbf{r})\,d\mathbf{r} \tag{7.22}$$

we obtain

$$\int h(r_{12})e^{-i\mathbf{s}\cdot\mathbf{r}_{12}}\,d\mathbf{r}_1\,d\mathbf{r}_2 = \int c(r_{12})e^{-i\mathbf{s}\cdot\mathbf{r}_{12}}\,d\mathbf{r}_1 d\mathbf{r}_2 + \rho \int\int\int c(r_{13})e^{-i\mathbf{s}\cdot(\mathbf{r}_2-\mathbf{r}_1)}h(r_{23})\,d\mathbf{r}_1\,d\mathbf{r}_2\,d\mathbf{r}_3 \quad \text{and} \tag{7.23}$$

$$\tilde{h}(s) = \tilde{c}(s) + \rho\tilde{c}(s)\tilde{h}(s) \quad \text{or} \quad 1 + \rho\tilde{h}(s) = [1 - \rho\tilde{c}(s)]^{-1} \tag{7.24}$$

The direct correlation function is connected to the compressibility equation, Eq. (6.33). This can be seen if we rewrite the compressibility equation

$$\rho\kappa_T = \left(\frac{\partial\rho}{\partial P}\right)_{N,T} = \beta\left\{1 + \rho\int_0^\infty [g(r)-1]4\pi r^2\,dr\right\} \quad \text{to} \tag{7.25}$$

$$\frac{\beta}{\rho\kappa_T} = \beta\left(\frac{\partial P}{\partial \rho}\right)_{N,T} = \frac{1}{1+\rho\int_0^\infty h(r)\,dr} = \frac{1}{1+\rho\tilde{h}(0)} = 1 - \rho\tilde{c}(0) \tag{7.26}$$

where Eq. (7.22) with $s = 0$ is used. For both $c(r)$ and $g(r) = h(r) + 1$, expansions in ρ can be formulated for which we refer to the literature (e.g., Ref. [6]).

So far, we have introduced several functions, which will be used again in the sequel, and for convenience we summarize them here. As before, ϕ denotes the pair potential, $\beta = 1/kT$, and ρ is the number density.

$$\text{Boltzmann function} \quad e = e(r) = \exp[-\beta\phi(r)] \tag{7.27}$$

$$\text{Mayer function} \quad f = f(r) = e - 1 \tag{7.28}$$

$$\text{Pair correlation function} \quad g = g(r) \tag{7.29}$$

$$\text{Potential of mean force} \quad W = W(r) = -kT \ln g \tag{7.30}$$

$$\text{Ornstein–Zernike equation} \quad h(r_{12}) = c(r_{12}) + \rho\int c(r_{13})h(r_{32})\,\mathrm{d}\mathbf{r}_3 \tag{7.31}$$

$$\text{Total correlation function} \quad h = h(r) = g - 1 \tag{7.32}$$

$$\text{Direct correlation function} \quad c = c(r) \tag{7.33}$$

$$\text{Background correlation function} \quad y = y(r) = g/e \tag{7.34}$$

In order to be concise, we omit in the following two paragraphs largely in the arguments for each of these functions, except when it is required to be precise.

7.2.4 The Percus–Yevick Equation*

The integral equations to be discussed can be derived in various ways, for example, by systematic expansion using the so-called (graphical) cluster expansion or by functional differentiation [7, 8]. Here, we use another entrance as described by McQuarrie [6], using the OZ equation. In order to solve the OZ equation for either $c(r)$ or $h(r)$, we need another relation, a "closure", between $c(r)$ and $h(r)$.

To that purpose we consider the direct correlation function $c(r)$ to be the difference between the correlation function $g(r) = \exp[-\beta W(r)]$ and an indirect correlation function $g_{\text{ind}}(r)$. Using the potential of mean force $W(r)$, we approximate $g_{\text{ind}}(r)$ by $g_{\text{ind}}(r) \cong \exp[-\beta(W-\phi)]$; that is, we assume– not unreasonably– that the potential of mean force W minus the pair potential ϕ provides a good estimate for g_{ind}. Hence, we use the approximation

$$c = g - g_{\text{ind}} \cong \exp(-\beta W) - \exp[-\beta(W-\phi)] = g - (g/e) = g - y \tag{7.35}$$

where the background correlation function $y = g/e$ is used.

The *Percus–Yevick* (PY) approximation, named after its originators, results from this approximate closure relation for the OZ equation. For c and h we then have the equivalent forms

$$c = g - y = ey - y = (e-1)y = fy \quad \text{and} \tag{7.36}$$

$$h = g - 1 = c + y - 1 = fy + y - 1$$

Inserting $h = c + y - 1$ into the OZ equation, we obtain

$$y(r_{12}) = 1 + \rho\int f(r_{13})y(r_{13})[f(r_{32})y(r_{32}) + y(r_{32}) - 1]\mathrm{d}\boldsymbol{r}_3 \tag{7.37}$$

Since f is known, this integral equation in y, known as the *PY equation*, can be solved numerically for y and, once y is obtained, this leads directly to $g = ye$.

7.2.5
The Hyper-Netted Chain Equation*

The *Hyper-Netted Chain* (HNC) equation is obtained by expanding $y = \exp[-\beta(W-\phi)]$ to first order, so that $y = 1 - \beta(W - \phi)$. Hence, from $c = g - y$ we have the various equivalent forms

$$c = g - y = g - [1 - \beta(W - \phi)] = h - \ln g - \beta\phi = h - \ln(g/e) = h - \ln y$$

Inserting $c = h - \ln y$ into the OZ equation results in one form of the HNC equation

$$\ln y(r_{12}) = \rho\int [h(r_{13}) - \ln y(r_{13})]h(r_{23})\mathrm{d}\boldsymbol{r}_3 \quad \text{or} \tag{7.38}$$

$$\ln[g(r_{12})/e(r_{12})] = \rho\int [g(r_{13}) - 1 - \ln g(r_{13}) - \beta\phi(r_{13})][g(r_{23}) - 1]\mathrm{d}\boldsymbol{r}_3 \tag{7.39}$$

This integral equation in g, since e and ϕ are known, can be solved numerically.

The HNC approximation received its name from the types of diagram that are included in the graphical cluster expansion analysis (these are not discussed here). Although from this presentation it appears that the HNC equation is an approximation to the PY equation, the more sophisticated cluster expansion analysis shows that this is not true. While the HNC approximation neglects certain terms in the expansion because of their complexity, the PY approximation neglects even more terms. However, the original, physical derivation of the PY equation is in terms of "collective coordinates," resembling the harmonic analysis of a crystalline solid [9]. While the HNC equation is thus a pure mathematical approximation, the PY equation has some physical basis.

7.2.6
The Mean Spherical Approximation*

The mean spherical approximation (MSA) consists of the core condition $g(r) = 0$ for $r < \sigma$ and the approximation $c(r) = -\beta\phi(r)$ for $r > \sigma$, together with the OZ equation. For hard-spheres with $\phi(r) = \infty$ for $r < \sigma$, where σ is the hard core diameter, this approach is identical to the PY approach and an analytical solution is possible (see Section 7.3). However, the importance of the MSA approximation is contained in the fact that it can be solved analytically for other cases. In particular, if $\phi(r)$ represents the ion–ion or dipole–dipole interaction, a solution can be found using Laplace methods. These solutions are used in the description of models for ionic solutions and polar fluids.

7.2.7 Comparison

In this paragraph we illustrate the results of some integral equation calculations. As before, we use the number density given by $\rho = N/V$ and the *packing fraction* reading $\eta = \rho b/4 = \pi\sigma^3\rho/6$ with $b = 2\pi\sigma^3/3$. The results of integral equation calculations for the hard-sphere fluid at low (η = 0.21) and high density (η = 0.47) are shown in Figure 7.1, and compared with Monte Carlo (MC) simulation results. From the figure it is clear that both the PY and HNC approximations result in a similar pair correlations function as the MC simulation. For low density, $g(r)$ falls monotonically until it reaches about 1.8σ, shows a small second maximum, and thereafter $g(r) \cong 1$. The large first peak represents the nearest-neighbor distance, and for this density it is clear that the structure becomes rapidly random. For high density there is more structure in $g(r)$ with clear second and third neighbor

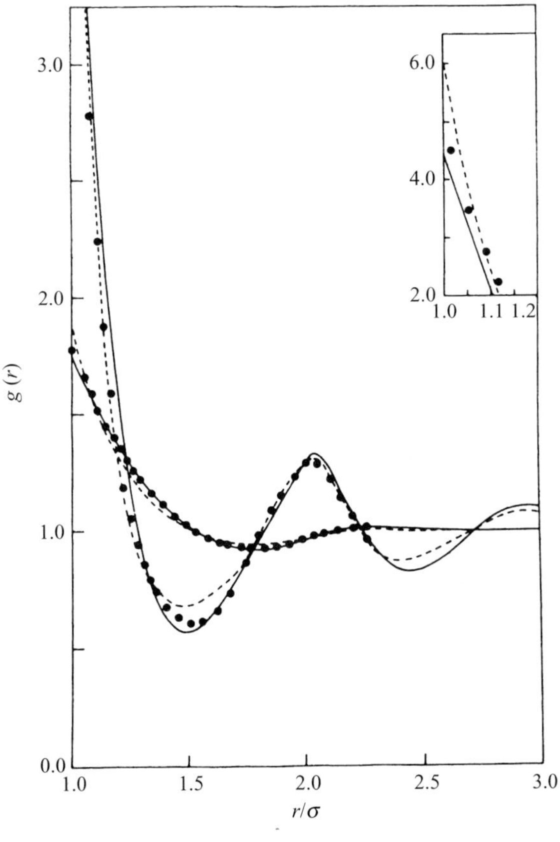

Figure 7.1 The pair correlation function for a hard-sphere fluid. Results shown are: - - -, HNC; ——, PY; ••••, Monte Carlo. The curves with the smaller intercept represent a state with $\eta = 0.21$, while the other curves represents a state with $\eta = 0.47$ (intercept shown in inset).

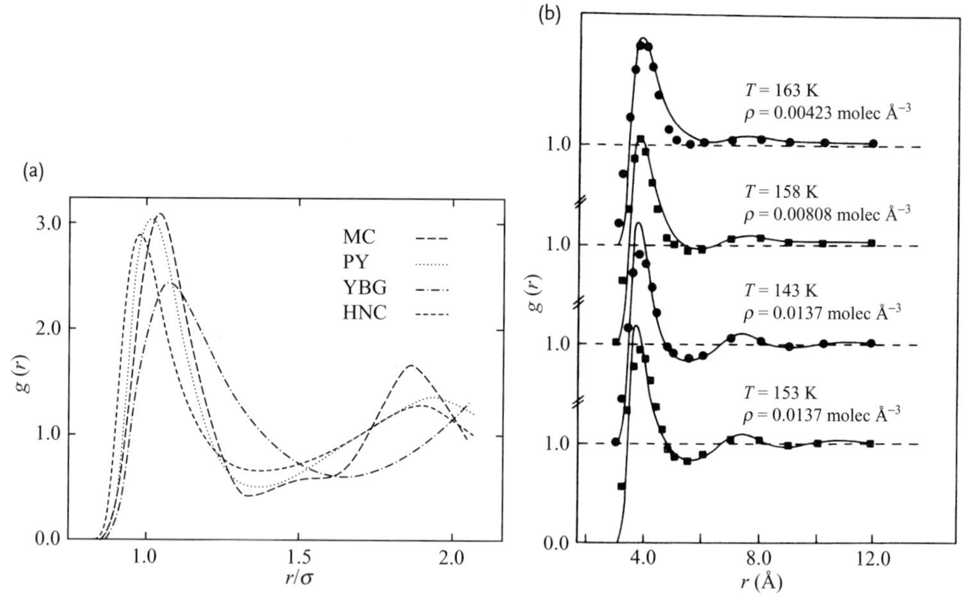

Figure 7.2 (a) The correlation function $g(r)$ for a Lennard-Jones fluid at $\eta = 0.58$ and $T_{red} = kT/\varepsilon = 2.74$, calculated using various methods; (b) The correlation function $g(r)$ for Ar using the PY approximation [29] compared to experimental data [30].

distances. Although the correlation functions using the PY and HNC approximations are qualitatively quite satisfactory, there are two defects when compared to the MC simulation. First, the value at contact is too low for the PY solution and too high for the HNC solution. Second, the oscillations are slightly out of phase for both the PY and HNC solutions.

The results for a Lennard-Jones fluid [10] at $\eta = 0.58$ and $T_{red} = kT/\varepsilon = 2.74$, using the PY, YBG, and HNC approximations, are shown in Figure 7.2, and compared with MC results. Similar to the hard-sphere fluid, the overall behavior of the calculated results mimics the MC results reasonably well, although clear differences can be noticed. The YBG result mimics the MC results poorly for the second coordination shell, and all peaks in the calculated results are displaced with respect to the MC results. Figure 7.2 also shows the comparison of PY results with experimental data for Ar. The qualitative features are well reproduced, although some differences can be noticed particularly at low densities. It is probably also clear that the influence of density is more pronounced than that of temperature, which is to be expected because the structure is largely determined by the repulsive forces.

Whilst various authors do not always agree on the merits of the PY and HNC approximations, it is clear that both are superior to the YBG approximation. For example, Watts and McGee [11] stated that the PY approximation ". . . can be derived from physical arguments and is superior to the HNC approximation for many problems". Oppositely, Fawcett [12] remarked that the "... HNC equation provides a better description of most systems than the PY equation". Nowadays, the PY approximation is often considered to be the best approximation to the solution of the integral equation describing simple liquids with strong repulsions and short-range interactions. For systems with long-range interactions, however, the HNC equation appears superior.

Problem 7.2*

Derive Eqs (7.24), (7.37), (7.38) and (7.39).

7.3
Hard-Sphere Results

Having indicated how to obtain the pair correlation function, the next step is to calculate the equation of state (EoS). This can be achieved in two ways: (i) to use $P = -\partial F/\partial V$, resulting in the so-called *pressure equation* P_P; and (ii) to use the so-called *compressibility equation* (see Justification 6.2), resulting in P_κ. We illustrate these expressions for the hard-sphere fluid, not to show for even a simple potential the resulting complexity, but rather in view of their importance as a reference for more realistic potentials.

Surprisingly, it appeared possible to solve the PY equation for the hard-sphere fluid using Laplace methods [13]. The result for the direct correlation function reads

$$c(r) = -\lambda_1 - \frac{\eta\lambda_1 r^3}{2\sigma^3} - \frac{6\eta\lambda_2 r}{\sigma} \quad (r \le \sigma) \quad \text{and} \quad c(r) = 0 \quad (r > \sigma) \quad \text{with} \tag{7.40}$$

$$\lambda_1 = \frac{(1+2\eta)^2}{(1-\eta)^4} \quad \text{and} \quad \lambda_2 = -\frac{\left(1+\frac{1}{2}\eta\right)^2}{(1-\eta)^4} \tag{7.41}$$

A piece-wise, rather complex analytical solution [14] for $g(r)$ is also available.

In Chapter 6, we saw that for the pressure equation we can write Eq. (6.39)

$$\frac{\beta P}{\rho} = 1 + \frac{\rho}{6}\int r\frac{\mathrm{d}e(r)}{\mathrm{d}r}y(r)4\pi r^2\,\mathrm{d}r \tag{7.42}$$

We also confirmed in that chapter that, for the hard-sphere potential with $e(r) = \exp[-\beta\phi_{\mathrm{HS}}(r)]$, the relationship $\mathrm{d}\,e(r)/\mathrm{d}r = \beta\,e(r)\delta(r-\sigma)$ holds. Moreover, by definition $y = g/e$ is continuous at $r = \sigma$, so that we have for the hard-sphere fluid in the PY approximation $c(r) = -y(r)$ if $r \le \sigma$ and $g(r) = y(r)$ if $r \ge \sigma$; see Eq. (7.36). Therefore $y(\sigma) = -c(\sigma) = g(\sigma)$ and we have, using, $\eta = \pi\sigma^3\rho/6$,

$$\beta P/\rho = 1 + 4\eta y(\sigma) = 1 - 4\eta c(\sigma) = 1 + 4\eta g(\sigma) \tag{7.43}$$

Inserting the expression for $c(r)$ the result for P_P can be shown to be

$$\frac{\beta P_P}{\rho} = \frac{1+2\eta+3\eta^2}{(1-\eta)^2} \tag{7.44}$$

The other option, using the compressibility equation Eq. (7.26),

$$\frac{\beta}{\rho\kappa_T} = \beta\left(\frac{\partial P}{\partial\rho}\right)_{N,T} = 1-\rho\tilde{c}(0) = 1-\rho\int_0^\infty c(r)4\pi r^2\,\mathrm{d}r \tag{7.45}$$

in combination with the solution for $c(r)$ for hard-spheres in the PY approximation, can be shown to lead to the result for P_κ which reads

$$\frac{\beta P_\kappa}{\rho} = \frac{1+\eta+\eta^2}{(1-\eta)^3} \tag{7.46}$$

It will be clear that the difference between P_P and P_κ shows a lack of thermodynamic consistency, as indicated in Section 6.3. An excellent description (which means matching the simulation results within 0.3% over the whole range $0 \le \eta \le 0.49$) of the hard-sphere EoS appears to be a weighted average of the above two results, the so-called *Carnahan–Starling* expression [15] given by

$$\frac{P_{CS}}{\rho kT} = \frac{1}{3}\frac{P_P}{\rho kT} + \frac{2}{3}\frac{P_\kappa}{\rho kT} = \frac{1+\eta+\eta^2-\eta^3}{(1-\eta)^3} \tag{7.47}$$

Since $P/\rho kT = 1 + 4\eta y$ (Eq. 6.73), we also have

$$y_{CS} = (4-2\eta)/4(1-\eta)^3 \tag{7.48}$$

The reduced pressure $P/\rho kT$ according to these three equations is shown in Figure 7.3.

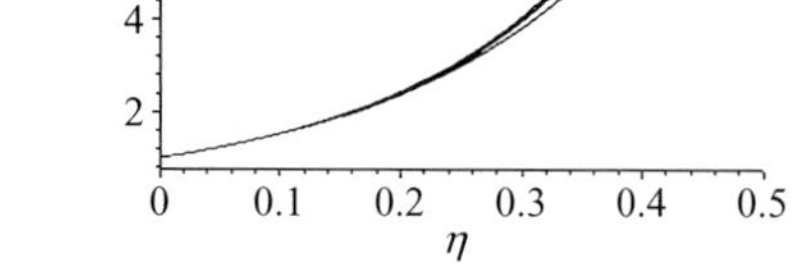

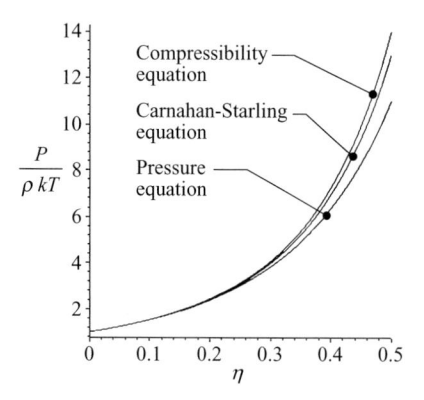

Figure 7.3 The EoS of a hard-sphere fluid according the pressure equation, the compressibility equation, and the Carnahan–Starling approximation for reduced density up to $\eta = 0.494$ ($V/V_0 = 1.50$) at which, according to simulations, the phase transition to a solid occurs.

For more realistic potentials the solution of the integral equation is only possible numerically, and consequently also the EoS can be obtained only numerically. We will not pursue this topic further, but refer the reader elsewhere (e.g., Refs [6, 7]) for a more extensive comparison.

Problem 7.3

Show that, by integrating $P = -\partial F/\partial V$ using the Carnahan–Starling EoS, the Helmholtz energy F of the hard-sphere liquid is given by

$$\frac{\beta F}{N} = \int_0^\eta \frac{\beta P}{\rho\eta'}\mathrm{d}\eta' = \ln\left(\frac{\rho\Lambda^3}{\mathrm{e}}\right) + \frac{\eta(4-3\eta)}{(1-\eta)^2}$$

Problem 7.4*

Show that, using $\mu = -\partial F/\partial N$ and the Carnahan–Starling EoS, the chemical potential μ of the hard-sphere liquid is given by

$$\beta\mu = \ln(\rho\Lambda^3) + (8\eta - 9\eta^2 + 3\eta^3)/(1-\eta)^3$$

Problem 7.5*

Show that one can obtain the Carnahan–Starling EoS by approximating the coefficients in the virial expression through B_6 as expressed in reduced units by integer numbers. Try an expression such as $an^2 + bn + c$, where n is an integer, to describe the coefficients, and sum the resulting expression analytically.

Problem 7.6: The van der Waals and Carnahan–Starling fluid*

Consider the hard-sphere fluid with a reasonably dense packing. We use the packing fraction $\eta = \pi\sigma^3\rho/6 = \rho b/4$ or using the sphere volume $v_0 = (4/3)\pi(\sigma/2)^3 = (\pi\sigma^3)/6$, $\eta = v_0\rho$. The closed-packed volume per sphere is given by $\sigma^3/2^{1/2}$. As usual, ρ represents the number density, σ the sphere diameter, k is Boltzmann's constant, and T is the temperature. The van der Waals (vdW) EoS of the hard-sphere fluid is given by

$$\beta P_{\mathrm{vdW}} = \rho/(1-b\rho)$$

with b a parameter, in the vdW approximation taken as $b = 4v_0$. The Carnahan–Starling EoS is given by Eq. (7.47).

a) The molar volume of liquid Ar at the triple point (85 K), at which equilibrium exists between gas, liquid and solid, is $V_{\mathrm{m}} = 28.39\ \mathrm{cm^3\ mol^{-1}}$. If we consider Ar as a hard-sphere molecule with a diameter of 3.4 Å, what is the volume per atom v in units of volume per atom at closed-packing?

b) Show that the vdW and CS expressions to second order in ρ agree, but differ in higher order.

c) For what value of η might these expressions become infinite? Compare your answer with the value for η for Ar at the triple point. Briefly comment on the usefulness of both approaches.

7.4
Perturbation Theory*

The EoS as obtained from the pressure and compressibility equations using the PY (and simulations) results for the hard-sphere liquid can be used as a basis for rather more accurate estimates of the EoS for more realistic potentials. This approach is usually addressed as the perturbation approach, and is based on the use of a reference potential for which the results are known and a perturbation part for which the effects are to be assessed. The starting point is the Gibbs–Bogoliubov inequality for approximate solutions, which we discuss first. Thereafter, we indicate two possible divisions of the potential in a reference and a perturbation part.

7.4.1
The Gibbs–Bogoliubov Inequality

Suppose we have a Hamilton expression consisting of a reference part $\mathcal{H}_0$ and a perturbation part $\mathcal{H}_1$ and we write, using λ as a bookkeeping label,

$$\mathcal{H} = \mathcal{H}_0 + \lambda\mathcal{H}_1 \tag{7.49}$$

We see that $\lambda = 0$ means that we have the reference potential, while $\lambda = 1$ implies that we have the full energy expression. The configurational partition function Q becomes

$$Q = \int \exp(-\beta\mathcal{H})\mathrm{d}\mathbf{r} = Q_0\left(Q_0^{-1}\int \exp[-\beta(\mathcal{H}_0 + \mathcal{H}_1)]\mathrm{d}\mathbf{r}\right) = Q_0\langle\exp(-\beta\mathcal{H}_1)\rangle_0 \tag{7.50}$$

where Q_0 is the configurational partition function of the reference system and $\langle\cdots\rangle_0$ denotes the canonical average in that reference system. The configurational Helmholtz energy is accordingly

$$\exp(-\beta F_\lambda) = V^{-N}\int \mathrm{e}^{-\beta\mathcal{H}(\mathbf{r})}\,\mathrm{d}\mathbf{r} = V^{-N}\int \mathrm{e}^{-\beta\mathcal{H}_0(\mathbf{r})}\mathrm{e}^{-\lambda\beta\mathcal{H}_1(\mathbf{r})}\,\mathrm{d}\mathbf{r} \tag{7.51}$$

The first derivative

$$\frac{\partial\exp(-\beta F_\lambda)}{\partial\lambda} = -\beta\int \mathcal{H}_1(\mathbf{r})\mathrm{e}^{-\beta\mathcal{H}(\mathbf{r})}\,\mathrm{d}\mathbf{r}\Big/\int \mathrm{e}^{-\beta\mathcal{H}(\mathbf{r})}\,\mathrm{d}\mathbf{r} = -\beta\langle\mathcal{H}_1\rangle_\lambda \tag{7.52}$$

and higher derivatives can be used to construct a Taylor series[2)] which reads

2) If we have $\langle\exp(-\beta A)\rangle = \Sigma_{n=0}(n!)^{-1}(-\beta)^n\langle A^n\rangle$ and the function A itself can be expanded as $A(-\beta) = \Sigma_{n=1}(k!)^{-1}a_k(-\beta)^k$, we generally have $a_1 = \langle A\rangle$, $a_2 = \langle A^2\rangle - \langle A\rangle^2$, $a_3 = \langle A^3\rangle - 3\langle A^2\rangle\langle A\rangle + 2\langle A\rangle^3, \ldots$. The coefficients a_n are denoted as *cumulants* for which a general expression can be derived [26]. If the cumulant, e.g. a_1, can be written as a product xy, the cumulant is zero if the factors x and y are statistically independent [27].

$$-\beta F_\lambda = -\beta F_0 + \lambda\beta\langle\mathcal{H}_1\rangle_0 + \frac{1}{2}\lambda^2\beta^2\left[\langle\mathcal{H}_1^2\rangle_0 - \langle\mathcal{H}_1\rangle_0^2\right] + \ldots \tag{7.53}$$

and is usually addressed as the *λ-expansion*. Again, $\langle f\rangle_0$ means the average over the reference ensemble using $\mathcal{H}_0$ only, but later we will use also $\langle f\rangle_1$ for the average over the complete ensemble using $\mathcal{H}$. Since $\langle\mathcal{H}_1^2\rangle_0 - \langle\mathcal{H}_1\rangle_0^2 = \langle\mathcal{H}_1 - \langle\mathcal{H}_1\rangle_0\rangle_0^2$, the λ^2 term is always positive so it appears that $F_0 - \langle\mathcal{H}_1\rangle_0$ is an upper bound for F. In fact, we can much further progress if we note that for $q \geq 0$, $q\ln q - q + 1 \geq 0$ and write $q = f/g$. For arbitrary, normalized distributions functions f and g with $\int f\mathrm{d}\mathbf{r} = \int g\mathrm{d}\mathbf{r} = 1$ we then have

$$\int g[(f/g)\ln(f/g) - (f/g) + 1]\mathrm{d}\mathbf{r} \geq 0 \tag{7.54}$$

since $g(r) \geq 0$, being a distribution function. Identifying the reference system ($\lambda = 0$) with $f = \exp[-\beta(\mathcal{H}_0 - F_0)]/V^N$ and the actual system ($\lambda > 0$) with $g = \exp[-\beta(\mathcal{H} - F_\lambda)]/V^N$ we obtain

$$-\beta(\langle\mathcal{H}_0\rangle_0 - F_0) \geq -\beta(\langle\mathcal{H}\rangle_0 - F_\lambda) \quad \text{or} \quad F_\lambda \leq F_0 + \lambda\langle\mathcal{H}_1\rangle_0 \tag{7.55}$$

Reversing the roles of f and g, the result is

$$F_\lambda \geq F_0 + \lambda\langle\mathcal{H}_1\rangle_1 \tag{7.56}$$

In total we thus have, what is called the *Gibbs–Bogoliubov inequality* [16],

▶ $$F_0 + \lambda\langle\mathcal{H}_1\rangle_1 \leq F_\lambda \leq F_0 + \lambda\langle\mathcal{H}_1\rangle_0 \tag{7.57}$$

Figure 7.4 shows the relationship between the various terms, from which it becomes clear that knowledge of $\langle\mathcal{H}_1\rangle_0$ and $\langle\mathcal{H}_1\rangle_1$ provides bounds on $F_\lambda - F_0$. Because $\langle\mathcal{H}_1\rangle_0$ is much more easily calculated than $\langle\mathcal{H}_1\rangle_1$, the upper bound is usually considered. If $\mathcal{H}_1$ contains parameters, $\langle\mathcal{H}_1\rangle_0$ can be minimized with respect to these parameters to obtain the optimum result. Some consideration makes it clear that the "flatter" $\mathcal{H}_0$, the smaller the perturbation contribution will be. In case we have a pairwise additive potential $\Phi = ½\Sigma_{ij}\phi_{ij}$, which can be split as $\phi(r) = \phi_0(r) + \phi_1(r)$ with $\phi_0(r)$ a reference part and $\phi_1(r)$ the perturbation, $\langle\mathcal{H}_1\rangle_0$ can be simplified and to first order

$$-\beta F_\lambda / V = -\beta F_0 / V + \frac{1}{2}\beta\lambda\rho^2 \int g_0(r)\phi_1(r)\mathrm{d}\mathbf{r} + \ldots \tag{7.58}$$

Higher-order terms involve three-point and higher order correlations functions, also for pair potentials. Since a great deal of information is known about the

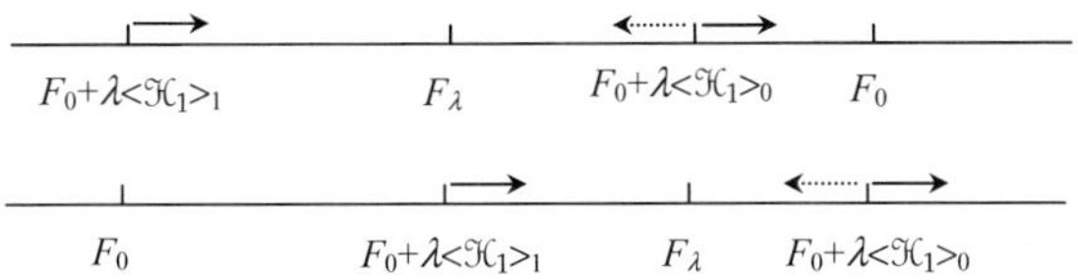

Figure 7.4 Terms in the Bogoliubov inequality for $\langle\mathcal{H}_1\rangle_0 < 0$ (upper) and $\langle\mathcal{H}_1\rangle_0 > 0$ (lower). Arrow → indicates the effect of F'', while arrow ←···· represents the effect of minimization.

hard-sphere system, this is a good reference system; however, potentials like the LJ do not contain a hard core so that an effective value for the hard-sphere diameter must be chosen. In the sequel, we discuss two ways in which to choose the reference part $\mathcal{H}_0$ and a perturbation part $\mathcal{H}_1$, so that use can be made of hard-sphere results. A more complete description of perturbation theory is given by Hansen and McDonald [7].

Problem 7.7

Verify Eq. (7.58).

Problem 7.8: The van der Waals EoS

For the vdW fluid originally an intuitive "derivation" was given leading to $[P + a/V^2](V - b) = NkT$, where a and b are parameters while P, T, and V represent the pressure, temperature, and volume, respectively, for N molecules. Using the perturbation approach, the vdW equation can be derived so that the approximations involved become clear. Use the separation $\mathcal{H} = \mathcal{H}_0 + \lambda\mathcal{H}_1$ and consider the hard-sphere model with radius σ as the reference system.

a) Show that, if $\beta\mathcal{H}_1$ is small enough, then $\langle\exp(-\beta\mathcal{H}_1)\rangle_0 \cong \exp\langle-\beta\mathcal{H}_1\rangle_0$.

b) Show, using Eq. (7.58) and approximating $g_{\text{HS}}(r)$ by $g_{\text{HS}}(r) = 0$ for $r < \sigma$ and $g_{\text{HS}}(r) = 1$ for $r > \sigma$, that $\langle\mathcal{H}_1\rangle_0 = \frac{1}{2}\rho^2 V\int_\sigma^\infty \mathcal{H}_1(r)\,\mathrm{d}\mathbf{r} \equiv -aN\rho$ where the minus sign is included to make $a > 0$.

c) Next, Z_0 has to be chosen. Approximate, with van der Waals, Z_0 by $Z_0 = (V - Nb)^N$ with excluded volume $b = ½\cdot 4\pi\sigma^3/3$, and show that this approximation leads to a pressure contribution $\beta P = \rho/(1 - b\rho)$.

d) Show that the final expression for the pressure P becomes $\beta P = \rho/(1 - b\rho) - ba\rho^2$ or $[P + a/V^2](V - b) = NkT$.

7.4.2
The Barker–Henderson Approach

For a continuous potential, like the LJ potential, a choice for the hard-sphere diameter is not unique. In order to obtain an optimum choice Barker and Henderson [17] (BH) defined a modified potential $v(r;\alpha,\gamma,d)$, containing the three parameters α, γ and d, corresponding to $\phi(r)$ by

$$
\begin{aligned}
v(r) &= \phi(d + (r-d)/\alpha) && \text{for} \quad d + (r-d)/\alpha < \sigma \\
v(r) &= 0 && \text{for} \quad \sigma < d + (r-d)/\alpha < d + (\sigma - d)/\alpha \\
v(r) &= \gamma\phi && \text{for} \quad \sigma < r
\end{aligned}
$$

Here, σ represents the point where $\phi(\sigma) = 0$. The parameters α and γ represent the "inverse steepness" and the "depth" of the potential, respectively. Note that $v(r;\alpha,\gamma,d)$ is independent of d, providing freedom to choose d optimally. Moreover $v(r;\alpha,\gamma,d)$ reduces to $\phi(r)$ if $\alpha = \gamma = 1$, and reduces to the hard-sphere potential $\phi_{\text{HS}}(r)$ with diameter d if $\alpha = \gamma = 0$. Consequently, by varying α and γ we can change the potential from $\phi_{\text{HS}}(r)$ to $\phi(r)$.

The next step taken was to expand the Helmholtz energy F in a power series in both α and γ around $\alpha = \gamma = 0$, leading to

$$F = F_0 + \alpha\left(\frac{\partial F}{\partial \alpha}\right)_{\alpha=\gamma=0} + \gamma\left(\frac{\partial F}{\partial \gamma}\right)_{\alpha=\gamma=0} + \frac{\alpha^2}{2}\left(\frac{\partial^2 F}{\partial \alpha^2}\right)_{\alpha=\gamma=0} + \cdots \tag{7.59}$$

Because the derivatives are evaluated at $\alpha = \gamma = 0$, they represent reference–that is, hard-sphere–quantities. While omitting the differentiations [17, 18], the final result is

$$\begin{aligned} F = {} & F_0 + \alpha 2\pi NkT\rho d^2 g_0(d)\left(d + \int_0^\sigma f(z)\,\mathrm{d}z\right) + \gamma 2\pi N\rho\int_\sigma^\infty g_0(r)\phi(r)r^2\,\mathrm{d}r \\ & - \gamma^2\pi N\rho\left(\frac{\partial \rho}{\partial P}\right)_0 \frac{\partial}{\partial \rho}\left[\rho\int_\sigma^\infty g_0(r)\phi^2(r)r^2\,\mathrm{d}r\right] + \mathcal{O}(\alpha^2) + \mathcal{O}(\alpha\gamma) + \cdots \end{aligned} \tag{7.60}$$

where $f(r) = \exp[-\beta\phi(r)] - 1$ denotes the Mayer function. The subscript 0 for the Helmholtz function F_0, pair correlation function $g_0(r)$ and compressibility $(\partial\rho/\partial P)_0$ indicates that they are quantities for the reference system with diameter d. The second-order term in γ has been approximated, as the three-point and four-point correlation functions are not available. Furthermore, Barker and Henderson made plausible that the second-order term in α^2 is negligible. Since d is still free, we can choose it as $d = -\int_0^\sigma f(z)\,\mathrm{d}z$. In this way, d becomes a temperature-dependent[3] diameter, but for all temperatures and densities the first-order term in α, as well as the second-order term in $\alpha\gamma$, vanishes identically. Taking $\alpha = \gamma = 1$, the result for the Helmholtz energy is

$$F = F_0 + 2\pi N\rho\int_\sigma^\infty g_0(r)\phi(r)r^2\,\mathrm{d}r - \pi N\rho\left(\frac{\partial \rho}{\partial P}\right)_0 \frac{\partial}{\partial \rho}\left[\rho\int_\sigma^\infty g_0(r)\phi^2(r)r^2\,\mathrm{d}r\right] \tag{7.61}$$

where the functions F_0, $g_0(r)$ and $(\partial\rho/\partial P)_0$ are to be evaluated for d as chosen. In spite of being approximate, the second-order term can be evaluated quite accurately [19]. The agreement with results from simulations is quite good. We only show in Figure 7.5 results for the EoS of a calculation (to second order) for the LJ potential and the density of the coexisting phases [17]. One observes an excellent agreement with the (essentially exact) results of computer simulations. For further comparison, the reader is referred elsewhere [18, 19].

3) Fitting MC results of the coexistence curve by Lu *et al.* [28] showed that $d = (a_1T + b_1)\sigma/(a_2T + b_2)$ with $a_1 = 0.56165k/\varepsilon$, $a_2 = 0.60899k/\varepsilon$, $b_1 = 0.9718$ and $b_2 = 0.92868$ describes d for a LJ fluid reasonably well.

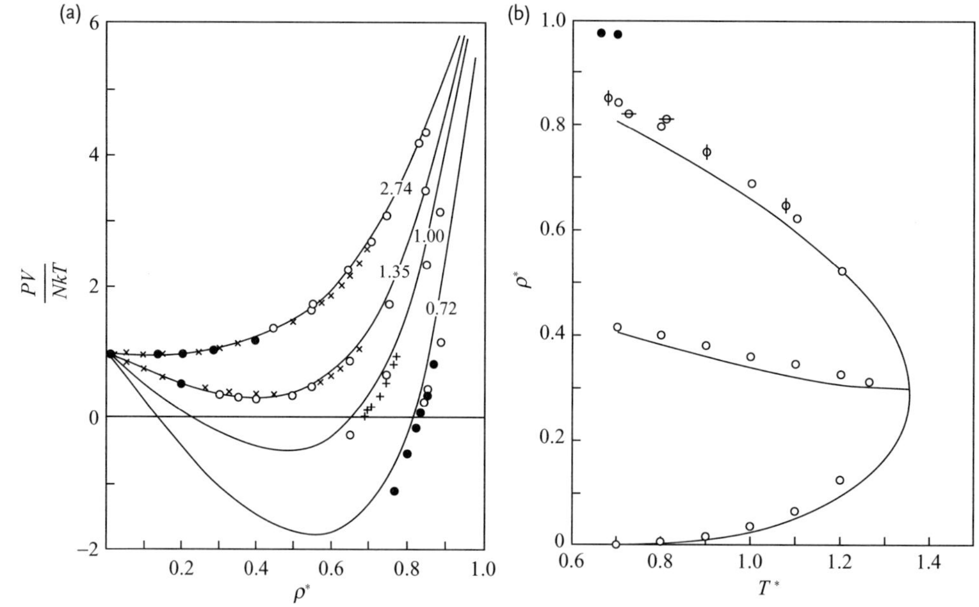

Figure 7.5 (a) The EoS and (b) density for coexisting phases for the LJ potential from second-order BH theory. The curves are labeled with T^* values, while the points indicate results from computer simulations and experimental data. The quantities ρ^* and T^* are given by $\rho\sigma^3$ and kT/ε, respectively.

7.4.3 The Weeks–Chandler–Andersen Approach

Another successful perturbation theory is due to Weeks, Chandler and Andersen [20] (WCA), who wrote the potential as $\phi(r) = \phi_0(r) + \phi_1(r)$ with

$$\phi_0(r) = \phi(r) + \varepsilon \quad \text{for} \quad r < r_0 \quad \text{and} \quad \phi_0(r) = 0 \quad \text{for} \quad r \geq r_0$$

$$\phi_1(r) = -\varepsilon \quad \text{for} \quad r < r_0 \quad \text{and} \quad \phi_1(r) = \phi(r) \quad \text{for} \quad r \geq r_0$$

where ε is the depth of the potential and r_0 is the equilibrium distance (Figure 7.6). This, in principle, is a good choice as it makes the perturbation $\phi_1(r)$ as slowly varying ("flat") as possible. WCA used the first-order perturbation expression

$$F = F_0 + 2\pi N\rho \int_\sigma^\infty g_0(r)\phi_1(r) r^2 \, dr \tag{7.62}$$

with $F_0 = F_{HS}$ and $g_0(r) = \exp(-\beta\phi_0) y_{HS}(r)$, where $y_{HS}(r)$ is the hard-sphere cavity function. Other choices for $g_0(r)$ are discussed elsewhere [21]. The effective hard-sphere diameter d, in this case a function of T and ρ, was chosen according to

$$\int_d^{r_0} y_{HS}(r) r^2 \, dr = \int_0^{r_0} y_{HS}(r) \exp[-\beta\phi_0(r)] r^2 \, dr \tag{7.63}$$

where r_0 is the equilibrium distance of the potential ϕ. An analytical expression fitted to simulation results is given by Verlet and Weis [22]. On the one hand, it is

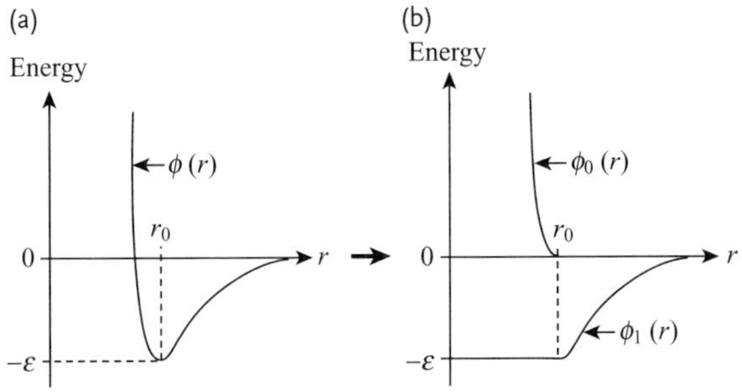

Figure 7.6 Separation of the intermolecular potential in WCA approximation. (a) The original potential; (b) Separation in a reference part $\phi_0(r)$ and a perturbation $\phi_1(r)$. Modified from Ref. [31].

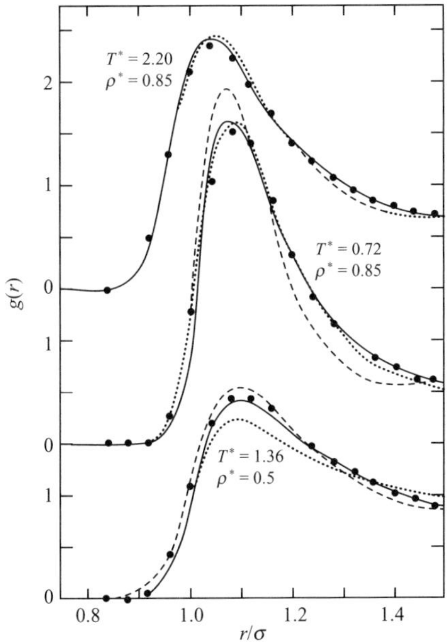

Figure 7.7 The pair correlation function $g(r)$ for the LJ fluid. The curves marked - - -, —— and . . . show results of the PY theory, the BH theory, and the WCA theory, respectively. The points refer to results of simulation studies. The quantities ρ^* and T^* are given by $\rho\sigma^3$ and kT/ε, respectively.

not immediately clear how to choose the hard-sphere reference diameter. One choice is Eq. (7.63), but on the other hand WCA approximates converges faster so that second-order terms are not required. Both, the BH and WCA theories, however, provide good approximations. Figure 7.7 shows a comparison [23] of results for the pair correlation function of the PY, BH, and WCA theories, from which a good agreement can be seen. It should be noted, however, that minor differences in $g_0(r)$ can result in relatively large differences in thermodynamic properties. The optimum choice for the reference system is therefore rather important.

7.5 Molecular Fluids*

From the above sections it will be clear that an extension of this approach to molecular fluids is complex. At least the position coordinate set must be enlarged with angular coordinates that represent the orientation of a molecule is space. Two approaches are possible for this. The first approach employs the methods as discussed before, but with much stronger intramolecular interactions as compared to the intermolecular interactions. This enlarges the problem but without introducing new concepts. Of course, the reliability depends heavily on the balance between intra- and intermolecular interactions.

The second approach recognizes upfront that intramolecular interactions are much stronger, and takes the limit to rigid molecules. The rationale for this is that the structure of sufficiently rigid molecules in the condensed phase is hardly different from that in the gas phase[4]. This approach requires extra conjugated variables, thereby introducing some additional complexity.

Apart from the difference in methods there are some further considerations to make for molecular fluids. Spectroscopy teaches us that the time for one rotation of a molecule, say N_2 at 1 bar and 25°C, is ~ 3×10^{-11} s, much shorter than the typical time between the collision of molecules, of ~ 2×10^{-10} s. In liquids, however, the molecules are much closer together and, assuming that during each vibration they actually touch, they can do so many times during one rotation because the typical vibration time of N_2 is ~1.4×10^{-14} s. In this case, the use of an uncoupled rotation partition function is less justified. Moreover, in a number of cases quantum corrections are required, in particular for the hydrogen atoms in the molecule. For example, in the case of H_2O, two internal vibrations have a wave number of about 3700 cm^{-1} which are so low that quantum corrections are required.

7.6 Final Remarks

Although the concepts as discussed in this chapter provide a direct "*ab-initio*" approach to liquid structure (and, if elaborated, to dynamics), the approach is practically troublesome. Apart from the fact that, in general, numerical solutions must be found, the extension to molecular fluids, as indicated, is not that straightforward. Starting with an EoS using the second virial coefficient, Song and Mason [25] have extended perturbation theory in a semi-empirical fashion, claiming an accuracy better than 1% for the LJ fluid. Their theory has also been applied to more realistic fluids. Nevertheless, since in most cases our interest lies more in molecular fluids than in atomic fluids, other approaches are called for.

4) See, e.g., for pyrazine, Ref. [24]. Experiments do show small but significant differences for some other, less rigid molecules.

References

1 Barker, J.A. and Henderson, D. (1976) *Rev. Mod. Phys.*, **48**, 587.
2 (a)Yvon, J. (1935) *Actualities and Scientifique et Industrielles, nr. 203*, Herman and Cie, Paris; (b)Born, M. and Green, H.S. (1946) *Proc. R. Soc. Lond.*, **A188**, 10.
3 Kirkwood, J.G. and Boggs, E.M. (1942) *J. Chem. Phys.*, **10**, 394.
4 Kirkwood, J.G. (1935) *J. Chem. Phys.*, **3**, 300.
5 Bellemans, A. (1969) *J. Chem. Phys.*, **50**, 2784.
6 See McQuarrie (1973).
7 See Hansen and McDonald (2006).
8 See Friedman (1985).
9 Percus, J.K. and Yevick, G.J. (1958) *Phys. Rev.*, **110**, 1.
10 Broyles, J. (1962) *J. Chem. Phys.*, **37**, 2462.
11 See Watts and McGee (1976).
12 Fawcett, W.R. (2004) *Liquids, Solutions and Interfaces*, Oxford University Press, Oxford.
13 (a) Thiele, E. (1963) *J. Chem. Phys.*, **39**, 474; (b) Wertheim, M.S. (1963) *Phys. Rev. Lett.*, **10**, 321; (c)Baxter, R.J. (1968) *Phys. Rev.*, **154**, 170.
14 Smith, W.R. and Henderson, D. (1970) *Mol. Phys.*, **19**, 411.
15 Carnahan, N.F. and Starling, K.E. (1965) *J. Chem. Phys.*, **51**, 635.
16 Isihara, A. (1968) *J. Phys.*, **A1**, 539; See also Friedman (1985).
17 (a) Barker, J.A. and Henderson, D. (1967) J. Chem. Phys., 47, 4714; (b) Barker, J.A. and Henderson, D. (1967) *J. Chem. Phys.*, **47**, 2856.
18 Henderson, D. and Barker, J.A. (1971) in *Physical Chemistry, an Advanced Treatise*, vol. 8A (eds H. Eyring, D. Henderson, and W. Jost), Academic Press, New York, p. 377.
19 Barker, J.A. and Henderson, D. (1972) *Annu. Rev. Phys. Chem.*, **23**, 439.
20 Weeks, J.D., Chandler, D., and Andersen, H.C. (1971) *J. Chem. Phys.*, **54**, 5237.
21 Andersen, H.C., Chandler, D., and Weeks, J.D. (1976) *Adv. Chem. Phys.*, **34**, 105.
22 Verlet, L. and Weis, J. (1972) *Phys. Rev.*, **A5**, 939.
23 Barker, J.A. and Henderson, D. (1971) *Acc. Chem. Res.*, **4**, 303.
24 Bormans, B.J.M., de With, G., and Mijlhoff, F.C. (1977) *J. Mol. Struct.*, **42**, 121.
25 Song, Y. and Mason, E.A. (1989) *J. Chem. Phys.*, **91**, 7840.
26 Zwanzig, R.W. (1954) *J. Chem. Phys.*, **22**, 1420.
27 Kubo, R. (1962) *J. Phys. Soc. Jpn*, **17**, 1100.
28 Lu, B.Q., Evans, R., and Telo da Gama, M.M. (1985) *Mol. Phys.*, **55**, 1319.
29 Watts, R.O. (1969) *J. Chem. Phys.*, **50**, 984.
30 Mikolaj, P.G. and Pings, C.J. (1967) *J. Chem. Phys.*, **46**, 1401.
31 Chandler, D.W., Weeks, J.D., and Andersen, H.C. (1983) *Science*, **220**, 787.

Further Reading

Friedman, H.L. (1985) *A course on statistical mechanics*, Prentice-Hall, Englewood Cliffs, NJ.

Hansen, J.-P. and McDonald, I.R. (2006) *Theory of Simple Liquids*, 3rd edn, Academic, London (1st ed. 1976, 2nd edition 1986).

McQuarrie, D.A. (1973) *Statistical Mechanics*, Harper and Row, New York.

Watts, R.O. and McGee, I.J. (1976) *Liquid State Chemical Physics*, John Wiley & Sons, Inc., New York.

8
Modeling the Structure of Liquids: The Physical Model Approach

In Chapters 6 and 7, an outline was provided of distribution functions and the integral equation approach. This led to a rather complicated formalism from which the pair correlation, and thus the thermodynamics of liquids, can be calculated. In many cases we are not interested in the greatest accuracy achievable, but rather to understand the overall behavior of liquids. For this purpose, physical approaches based on lattice models are quite helpful as they provide excellent examples of relatively simple models in statistical thermodynamics that can be solved analytically and thus have pedagogical value. Here, we start with some preliminaries and deal thereafter with cell and hole models, in different degrees of detail. Thereafter, we discuss briefly the theory of significant liquid structures. Although today these approaches have been largely abandoned for research purposes for simple liquids, they have nevertheless the virtue to explain liquid behavior in simple terms. However, for polymers and polymeric solutions, lattice models are still actively pursued. Finally, we address scaled-particle theory.

8.1
Preliminaries

In a lattice, the atoms are located at fixed positions, whilst in an ideal gas they can be everywhere within the enclosure of the gas. In a liquid, the atoms are much less confined than in a solid, but much more so than in a gas. Consequently, their positions change more gradually and it is tempting to use, instead of the ideal gas configuration as for the virial expansion, a lattice as reference configuration. In the so-called *lattice model*, one assumes a lattice-like structure on average for the liquid, at least for the first coordination shell. There is some (but only some) experimental justification for this. In Figure 8.1 the self-correlation function for Ar is shown [1], describing the probability of finding a molecule that at time $t = 0$ was at position $r = 0$ at a time t at position r. This function indicates how fast, on average, a molecule moves from its reference position at time $t = 0$. From these graphs it can be seen that within 3 to 10×10^{-13} s, the original structure has faded away. This time corresponds to about 3 to 10 vibrations, estimating the typical vibration frequency as 10^{13} s. So, a marginal resemblance with a lattice is present.

Liquid-State Physical Chemistry, First Edition. Gijsbertus de With.

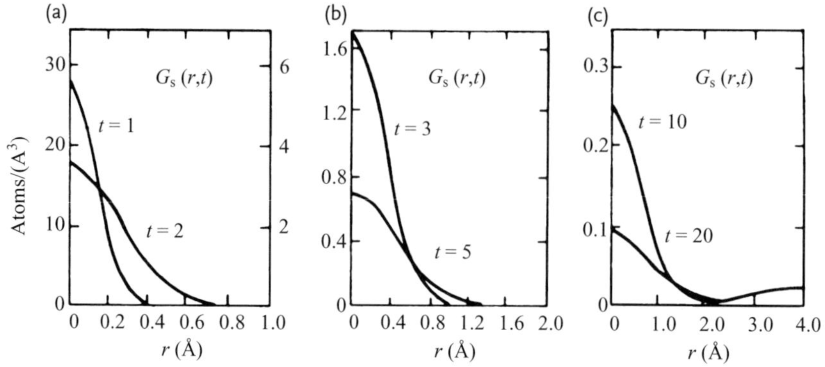

Figure 8.1 The self-correlation function for Ar, showing the probability of finding a particle at position r at time t that was at position $r = 0$ at time $t = 0$ for various times. The time is in units of 10^{-13} s.

Nevertheless, the concept is useful, both for pure liquids and solutions. The existence of this temporary local structure around a reference molecule is denoted as a *cell* for pure liquids and as a *cage* (for the solute) in solutions. The molecule we have in focus (the reference molecule) is denoted as the *wanderer*, and the coordinating molecules constitute the *wall* of the cell. In this picture, each molecule thus "rattles" in a cell of which the walls are provided by the neighboring molecules. Since any molecule can be considered as either a wanderer or as a wall molecule, this constitutes a basic inconsistency, which must be accepted in order to make progress. Lattice models in which the number of cells equals the number of molecules are addressed as *cell models*, while models in which the number of cells is larger than the number of molecules are called *hole models*. In lattice models one uses the concept of *free volume*, that is, the space available to the motion of the center of gravity of the molecules[1]. The value of this can be estimated, for example, from P_{vap} or α_P and κ_T data (see Problems 8.3 and 8.4). Pryde [2] provided a simple introduction to lattice models, while Hirschfelder *et al.* [3] and Barker [4] have reviewed many details.

8.2 Cell Models

Kirkwood contemplated how to rationalize cell models (a somewhat complex process schematically discussed in Justification 8.1), but long before that Lennard-Jones and others used the cell model to obtain both analytical and numerical results.

1) The label "free volume" can easily mean different things to different people. The present definition refers to what is also denoted as fluctuation volume [27].

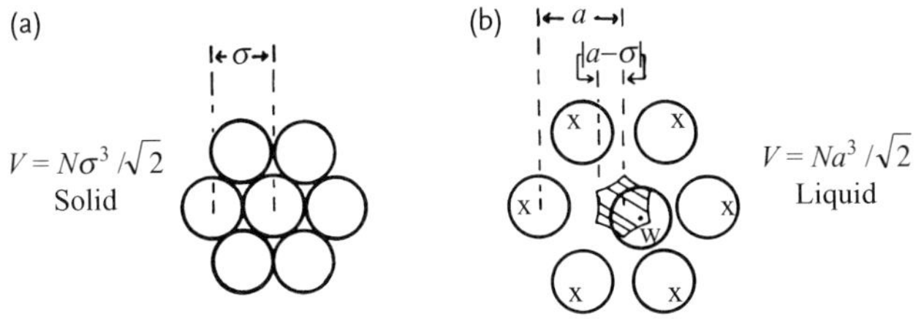

Figure 8.2 A two-dimensional representation of the FCC lattice for the solid state (a) and the liquid state (b) in which the distance between the atoms has been increased from σ to a so that the "smeared" diameter of the free volume, indicated by the hatched hexagon, becomes $a - \sigma$.

Hence, let us use a cell model with a lattice-like structure in which we introduce some free volume. As the density of a liquid deviates not too much from the solid, we use for the lattice a dense close-packed lattice such as the FCC lattice structure (Figure 8.2) with hard-sphere atoms. In a FCC lattice, the volume per molecule is $v_0 = V_0/N = \sigma^3/\gamma$, where $\gamma = 2^{1/2}$ and σ is the molecular diameter. For other lattices we will have different γ-values. Expanding this lattice slightly, we have $a^3/\gamma = v = V/N$. The configuration integral for the wanderer, labeled as molecule 1, represents the volume available to the wanderer, weighted with a Boltzmann factor, that is, the *free volume*, reads

$$Q(1) = \int \exp[-\phi(\boldsymbol{r}_1)/kT]\mathrm{d}\boldsymbol{r}_1 = \int \exp[-\phi(r)/kT]4\pi r^2 \mathrm{d}r \tag{8.1}$$

where the integration is in principle overall space, although as the wanderer cannot leave the cell, it may be taken over the cell. If the wanderer is inside the cell, the potential energy $\phi(\boldsymbol{r}_1) = 0$ because the atoms are considered as hard-spheres and therefore we have $\exp(-\phi/kT) = 1$. Since the wanderer cannot be outside the cell, for $|\boldsymbol{r}_1| > a$ the potential energy $\phi\ (|\boldsymbol{r}| > a) = \infty$, leading to $\exp(-\phi/kT) = 0$. If we assume that we may replace the cell by a sphere of radius $(a-\sigma)$, we obtain for $Q(1)$

$$Q(1) = v_\mathrm{f} \quad \text{where} \quad v_\mathrm{f} = 4\pi(a-\sigma)^3/3 = 4\pi\gamma(v^{1/3} - v_0^{1/3})^3/3$$

is the expression for the free volume per molecule for this particular model. More complex potentials will result in different expressions for the free volume. Given this result, the configurational Helmholtz energy for the wanderer becomes

$$F_\mathrm{con}(1) = -kT\ln Q(1) = -kT\ln v_\mathrm{f} \tag{8.2}$$

or since any molecule can be the wanderer we obtain the expression for the liquid by taking N times the expression for $F_\mathrm{con}(1)$

$$F_\mathrm{con}(N) = -kT\ln(v_\mathrm{f})^N = -NkT\ln\left[4\pi\gamma\left(v^{1/3} - v_0^{1/3}\right)^3/3\right] \tag{8.3}$$

Because $P = -\partial F/\partial V$ and in $F = F_\mathrm{kin} + F_\mathrm{con} \equiv -kT\ln\Lambda^{-3N} - kT\ln[Q(1)^N]$, only F_con depends on V, and we have

$$PV/NkT = \left[1-(v_0/v)^{1/3}\right]^{-1} = \left[1-(V_0/V)^{1/3}\right]^{-1} \tag{8.4}$$

The limiting behavior of this simple equation is not too bad. If $V \to V_0$, $P \to \infty$ (which means that at high pressure the behavior is approaching the solid state). If V becomes large, $P \to NkT/V$, as it should do. However, in the neighborhood of the triple point, $PV/NkT << 1$ but the expression on the right-hand side appears to be >> 1 because V_0/V is typically 0.8. The origin of this failure is simple: the model contains no attractive interaction. The simplest way to introduce attraction, based on empirical insight [5], is to use for the average potential $\phi(0)$ the expression

$$\frac{1}{2}N\phi(0) \cong -a(T)/V \text{ inside the cell and } \quad \phi = \infty \text{ outside the cell} \tag{8.5}$$

with $a(T) \geq 0$ a parameter that is dependent on temperature only, leading to

$$Q(1) = \int \exp[-\phi(\mathbf{r}_1)/kT]\mathrm{d}\mathbf{r}_1 = \exp\left[-\frac{1}{2}N\phi(0)/kT\right]v_\mathrm{f} \tag{8.6}$$

This results in

$$F_\mathrm{con}(N) = -kT\ln Q(1)^N = -NkT\ln v_\mathrm{f} + \frac{1}{2}N\phi(0) \cong -NkT\ln v_\mathrm{f} - a(T)/V \tag{8.7}$$

and via $P = -\partial F/\partial V$ to

$$P = \frac{NkT}{V}\left[1-\left(\frac{V_0}{V}\right)^{1/3}\right]^{-1} - \frac{a(T)}{V^2} \quad \text{or} \quad \left[P+\frac{a(T)}{V^2}\right]\left(V - V_0^{1/3}V^{2/3}\right) = NkT \tag{8.8}$$

This expression, often denoted as the *Eyring EoS*, is a significant improvement of the cell model, resembling the van der Waals equation, Eq. (3.5), $(P + a/V^2)(V - b) = RT$, where a and b are parameters, and gives a useful semi-empirical description of liquids.

A more sophisticated cell model was proposed by Lennard-Jones and Devonshire [6], here labeled as the LJD theory. In this model, again the FCC lattice is used but now in conjunction with the Lennard-Jones potential $\phi_\mathrm{LJ}(r)$ with equilibrium distance $r_0 = 2^{1/6}\sigma$ and depth ε. In the original LJD model the interaction of the wanderer with its $z = 12$ nearest-neighbors is calculated by supposing again that the neighbors are distributed with equal probability over a sphere with radius $a = (\gamma V/N)^{1/3} \equiv (\gamma/\rho)^{1/3}$ using a cell radius $a_1 \geq 0.5a$. For this spherical cell the potential ϕ is then a function of the distance r from the center of the cell only. At the center we have the potential

$$\phi(0) = \frac{1}{2}z\int_0^\pi \phi_\mathrm{LJ}(a)\sin\theta\,\mathrm{d}\theta \tag{8.9}$$

while for a position off-center an extra contribution $\Psi(r)$ has to be added, that is, $\phi(r) = \phi(0) + \Psi(r)$. The authors showed that, using $v_0/r_0^3 = v/a^3$,

$$\phi(0) = z\varepsilon[(r_0/a)^{12} - 2(r_0/a)^6] = z\varepsilon[(v_0/v)^4 - 2(v_0/v)^2] \quad \text{and} \tag{8.10}$$

$$\psi(r) = z\varepsilon[(r_0/a)^{12}l(r/a_1) - 2(r_0/a)^6 m(r/a_1)] \quad \text{where} \tag{8.11}$$

$$l(x) = (1+12x^2+25.2\,x^4+12x^6+x^8)(1-x^2)^{-10} - 1 \quad \text{and} \tag{8.12}$$

$$m(x) = (1+x^2)(1-x^2)^{-4} - 1 \tag{8.13}$$

The partition function $Q = Q(1)^N$ becomes (using $\beta = 1/kT$)

$$Q = \exp\left[-\frac{1}{2}N\beta\phi(0)\right]v_f^N \quad \text{with} \quad v_\text{f} = \int_0^{a_1} \exp[-\psi(r)/kT]4\pi r^2 \text{d}r \tag{8.14}$$

The choice of a_1 depends on the "smearing" assumptions. It appears that $\Psi(r)$ becomes large and positive for $r \to 0.5a$. Lennard-Jones and Devonshire therefore took $a_1 = 0.5a$. Because the introduction of the more realistic LJ potential complicates the matter[2)], the necessary integrations must be performed numerically.

The model predicts sinuous curves for the pressure below and monotonic curves above a certain temperature. Identifying this temperature as the critical temperature, the model predicts (using the LJ potential) for the critical constants

$$P_\text{cri} = 0.434\ \varepsilon/\sigma^3 \quad V_\text{cri} = 1.77\ N_\text{A}\sigma^3 \quad T_\text{cri} = 1.30\ \varepsilon/k \tag{8.15}$$

to be compared with the best experimental corresponding states estimates

$$P_\text{cri} = 0.116\ \varepsilon/\sigma^3 \quad V_\text{cri} = 3.14\ N_\text{A}\sigma^3 \quad T_\text{cri} = 1.25\ \varepsilon/k \tag{8.16}$$

We see that the value of T_cri is about right, the value for V_cri too small by a factor of about 2, and P_cri too large by a factor of about 4.

It is not very realistic to expect that the wanderer will be permanently confined to the cell at or near the critical temperature. One might expect that at low temperature, perhaps at or near the triple point, this is assumption is more realistic and agreement with experiment is better. For Ar, the triple point temperature is $0.701\varepsilon/k$, and at this temperature the pressure is so low that the molar volume may be obtained by setting $P = 0$ in the equation of state (EoS). Table 8.1 provides some data for Ar at the triple point. It is concluded that, on the whole, the values of the properties calculated with LJD theory are much closer to the experimental values for solids than for liquids, as might be expected.

Table 8.1 Data for Ar at the triple point T_tri. Experimental data taken from Ref. [4].

Origin of data	$kT_\text{tri}/\varepsilon$	$V/N\sigma^3$	$U/N\varepsilon$	S/Nk	C_V/Nk
LJD theory	0.701	1.037	−7.32	−5.51	1.11
SLS theory	0.711	1.159	−6.19	−3.89	–
Solid Ar experimental	0.701	1.035	−7.14	−5.53	1.41
Liquid Ar experimental	0.701	1.178	−5.96	−3.64	0.85

2) For more details see, e.g., Fowler and Guggenheim (1939), Hirschfelder *et al.* [3], Barker [4] or McQuarrie [7].

Later authors [8] included non-nearest-neighbor effects for $\phi(0)$ and v_f via a lattice sum[3] calculation, but this did not lead to any significant improvement. These authors also chose a_1 according to $4\pi a_1^3/3 = a^3/\gamma$, that is, the volume of the cell equals the volume per molecule, or $a_1 = 0.5527a$. The precise choice appears to be noncritical. On the whole, this conceptual improvement does not bring about any significant improvement in agreement with experiment. Also, the "smearing" approximation for v_f for hard-spheres appeared to have a limited influence [9]. Allowing the double occupancy of cells leads to somewhat improved results [10]. The thermodynamic properties as calculated by the LJD method appear to be insensitive to the precise potential used [11]. Finally, we note that cell model calculations have been made using a random close-packed structure and employing the LJ potential [12]. The results for energy compare favorably with MC and MD results, although the results for pressure appear to be rather sensitive to the correlation function used. Extending this model by using a nonzero probability for being off-center for all molecules removes the basic inconsistency mentioned in Section 8.1, and leads to somewhat improved results [13].

An important deficit of this type of models is the *communal entropy* problem, which is due to the assumption that the wanderer cannot leave the cell. To illustrate this problem, consider a gas. The correct partition function for the ideal N-atom gas is given by, using $\Lambda = (h^2/2\pi mkT)^{1/2}$,

$$Z = Z(N) = (\Lambda^{3N} N!)^{-1} V^N \tag{8.17}$$

so that the Helmholtz energy reads

$$F(N)_{\text{correct}} = -kT \ln Z(N) \tag{8.18}$$

For the cell model the partition function for the wanderer is given by

$$Z = Z(1) = (\Lambda^3 N)^{-1} V \tag{8.19}$$

so that the Helmholtz energy for N atoms in the cell model is

$$F(N)_{\text{cell model}} = -NkT \ln Z(1) \tag{8.20}$$

The difference between these two expressions constitutes the communal Helmholtz energy which we calculate from

$$\begin{aligned}\Delta_{\text{com}} F = F(N)_{\text{correct}} - F(N)_{\text{cell model}} &= -kT \ln Z(N) - kT[\ln Z(1)]^N = -kT \ln(N^N/N!) \\ &= -kT(N \ln N - N \ln N + N) = -NkT\end{aligned} \tag{8.21}$$

Hence, for one mole the cell model estimates the molar entropy too low by $\Delta_{\text{com}} S = R$. The basic reason for this is the limited volume that an atom can probe in the cell model (the cell volume) as compared to the volume probed in reality (the complete volume). Obviously, the same argument applies also to liquids, because we know that in liquids the atoms can move to any position, albeit

3) In a lattice sum the total potential energy is calculated as the sum of the interactions over all pairs of molecules, arranged according to the lattice specified.

somewhat restrictedly. For solids, where the atoms essentially remain at their positions, the approach would be correct and the cell model is thus describing a solid-like situation rather than a liquid-like situation. Eyring [14] made the rather arbitrary assumption that the communal entropy would be completely released during melting, which would lead to an (extra) communal entropy factor e^N in the partition function Z_N. In this way, the limiting behavior for gases becomes correct. The communal entropy $\Delta_{com}S$ issue was resolved by Hoover and Ree [15] by using Monte Carlo simulations (see Chapter 9) for hard-spheres. They essentially calculated $\Delta_{com}S$ by subtracting from the entropy for the unconstrained, normal hard-sphere fluid that of a constrained fluid. In the latter model each sphere can move only within its own cell, in similar fashion to the cell model and consistent with the definition of communal entropy as given by Kirkwood (see Justification 8.1). Data for varying sizes were extrapolated to those for an infinite system, and from these calculations it appeared that the hard-sphere (disordered) fluid and (ordered) solid phases are in equilibrium with densities $\rho_L = (0.667 \pm 0.003)\rho_0$ and $\rho_S = (0.736 \pm 0.003)\rho_0$, respectively, where ρ_0 is the close-packed density. The melting pressure and temperature are related by $P_M = (8.27 \pm 0.13)\rho_0 kT_M$ (Figure 8.3). Hence the hard-sphere solid expands by about 10% upon melting. The density of the solid shows a cusp at $0.637\rho_0$ where the cell walls begin to become important. At this point, the crystal becomes mechanically unstable and without the walls it would rapidly disintegrate. It also appeared that $\Delta_{com}S$ decreases approximately linearly from $\Delta_{com}S = 1.0Nk$ at $\rho = 0$, corresponding to the factor e^N in Q, to $\Delta_{com}S = 0.2Nk$ at $\rho = 0.65\ \rho_0$ (Figure 8.3). There is no simple way to either solve or circumvent the communal entropy problem for cell models, although some other approaches (e.g., hole models) do so (Section 8.3).

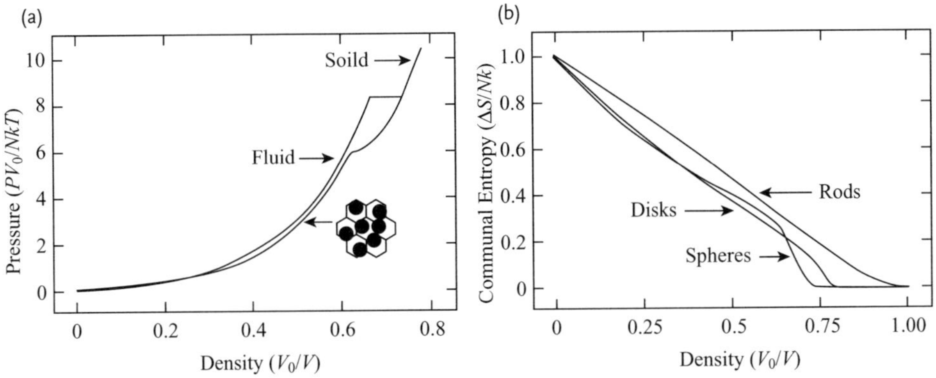

Figure 8.3 The equation of state for three-dimensional (3D) hard-spheres for the fluid and solid states, where the solid and fluid isotherms are connected by a tie line at equal chemical potential at $PV_0/NkT = 8.27$ (a) and the communal entropy for 1D, 2D, and 3D hard-spheres (b).

Justification 8.1: Kirkwood's analysis*

Consider, like Kirkwood [16], a system of N interacting identical particles with partition function $Z_N \equiv (\Lambda^{3N}N!)^{-1}Q_N$. Let us divide the coordinate space into N cells $\Delta_1,\Delta_2,\ldots,\Delta_N$, each of volume Δ, in an arbitrary manner. The configuration integral then becomes a sum of terms in which the coordinates of the molecules are confined to a particular cell. Thus, Q_N will contain N^N integrals, corresponding to the N^N ways of positioning N molecules in N cells. Hence

$$Q_N = \sum_{l_1=1}^{N}\sum_{l_2=1}^{N}\cdots\sum_{l_N=1}^{N}\int_{\Delta_{l_1}}\int_{\Delta_{l_2}}\cdots\int_{\Delta_{l_N}}\exp[-\beta\Phi(\mathbf{r})]\mathrm{d}\mathbf{r} \tag{8.22}$$

with $\Phi = \Sigma_{i<j}\phi_{ij}$. Now denote by $Q_N(m_1,m_2,\ldots,m_N)$ the various terms in Eq. (8.22) corresponding m_1 molecules in cell 1, m_2 molecules in cell 2, ... m_N molecules in cell N. Because the molecules are identical, there are $(N!/\Pi_i\, m_i!)$ ways of obtaining a particular set $\{m\} = m_1,m_2,\ldots,m_N$. Therefore, Q_N becomes

$$Q_N = \sum_{\{m\}}\frac{N!}{\prod_i m_i!}Q_N(m_1,m_2,\ldots,m_N) \quad \text{or}$$

$$Q_N/N! = \sum_{\{m\}}Q_N(m_1,m_2,\ldots,m_N)\Big/\prod_i m_i! \equiv \bar{\sigma}^N Q_N^{(1)} \quad \text{in which} \tag{8.23}$$

$$Q_N^{(1)} = Q_N(1,1,\ldots,1) \quad \text{and} \quad \bar{\sigma}^N = \sum_{\{m\}}Q_N(m_1,m_2,\ldots,m_N)\Big/Q_N^{(1)}\prod_i m_i! \tag{8.24}$$

The summation is over all sets $\{m\}$ subject to the constraint $\Sigma_i\, m_i = N$. At sufficiently high densities the repulsive interaction prevents multiple occupancy, so that all $Q_N(m_1,m_2,\ldots,m_N)$ are zero except $Q_N^{(1)}$ and $\bar{\sigma} = 1$. On the other hand, for highly dilated gases there is effectively no interaction between the molecules and all $Q_N(m_1,m_2,\ldots,m_N)$ are equal. Hence, $\bar{\sigma} = \Sigma_{\{m\}}(\Pi_i m_i!)^{-1}$ so that $\bar{\sigma}^N = N^N/N! = \mathrm{e}^N$ using Stirling's approximation, or $\bar{\sigma} = \mathrm{e}$. Hence, generally $1 \le \bar{\sigma} \le \mathrm{e}$ and its change with density represents the *communal entropy*. The derivative of $\bar{\sigma}$ with respect to V contributes to P. To make the next step we need the probability density

$$P_N^{(1)}(\mathbf{r}) = \left(Q_N^{(1)}\right)^{-1}\exp[-\beta\Phi(\mathbf{r})] \tag{8.25}$$

where $Q_N^{(1)}$ is the configurational partition function for a system of N interacting molecules, each of which is required to remain in its particular cell. We approximate $P_N^{(1)}$ by a product of functions $s(\mathbf{r}_i)$, each dependent only on the coordinates of one molecule $\mathbf{r}_i$ with respect to the origin in the cell, that is,

$$P_N^{(1)} = \prod_i s(\mathbf{r}_i) \quad \text{with normalisation condition} \quad \int_\Delta s(\mathbf{r}_i)\mathrm{d}\mathbf{r}_i = 1 \tag{8.26}$$

The optimum choice for $s(\mathbf{r}_i)$ is obtained by minimizing $F = U - TS$ at constant T and Δ subject to the normalization constraint of $s(\mathbf{r}_i)$. Here

$$\begin{aligned} F &= -kT\ln(\Lambda^{3N}N!)^{-1}Q_N = -kT\ln\Lambda^{-3N}\bar{\sigma}^N Q_N^{(1)} \\ &= -NkT\ln\Lambda^{-3}\bar{\sigma} + F^{(1)} \end{aligned} \tag{8.27}$$

where $F^{(1)} = -kT \ln Q_N^{(1)}$ is the configuration Helmholtz energy for N interacting molecules restrained to their cells. As shown in Chapter 5, the configuration energy and entropy are given by, respectively,

$$U_N^{(1)} \equiv \int_\Delta P_N^{(1)}(\mathbf{r})\Phi(\mathbf{r})\mathrm{d}\mathbf{r} \quad \text{and} \quad S_N^{(1)} \equiv -k\int_\Delta P_N^{(1)}(\mathbf{r})\ln P_N^{(1)}(\mathbf{r})\mathrm{d}\mathbf{r} \tag{8.28}$$

Therefore, $F^{(1)}$ becomes (after some manipulation)

$$F^{(1)}/NkT = \int_\Delta s(\mathbf{r})\ln s(\mathbf{r})\mathrm{d}\mathbf{r} + \frac{1}{2}\beta\int_{\Delta'}\int_\Delta Y(\mathbf{r}-\mathbf{r}')s(\mathbf{r})s(\mathbf{r}')\mathrm{d}\mathbf{r}\,\mathrm{d}\mathbf{r}' \tag{8.29}$$

with $Y(\mathbf{r}) = \Sigma_{j=2}\phi(\mathbf{R}_{1j}-\mathbf{r})$ as the contribution to the potential energy due to molecule 1 at position $\mathbf{r}$ with respect to the origin of cell 1 with all other particles located at their respective origins, and $\mathbf{R}_{1j}$ contains the origins of all j cells with respect to the origin of cell 1. To find the optimum we need to solve

$$\delta F = \int_\Delta\left[kT\ln s(\mathbf{r}) + \int_{\Delta'} Y(\mathbf{r}-\mathbf{r}')s(\mathbf{r})s(\mathbf{r}')\mathrm{d}\mathbf{r}'\right]\delta s(\mathbf{r})\mathrm{d}\mathbf{r} = 0$$

$$\text{subject to the normalisation condition} \quad \int_\Delta \delta s(\mathbf{r})\mathrm{d}\mathbf{r} = 1 \tag{8.30}$$

Using Lagrange multipliers leads (again after some manipulation) to

$$s(\mathbf{r}) = \exp[\beta(\alpha-\psi)] \quad \text{with} \quad \exp(-\beta\alpha) = \int_\Delta \exp[-\beta\psi(\mathbf{r})]\mathrm{d}\mathbf{r} \tag{8.31}$$

$$\text{where } \psi(\mathbf{r}) = \int_\Delta w(\mathbf{r}-\mathbf{r}')\exp\{\beta[\alpha-\psi(\mathbf{r}')]\}\mathrm{d}\mathbf{r}' \quad \text{and} \tag{8.32}$$

$$w(\mathbf{r}) = Y(\mathbf{r}) - Y_0 \quad \text{with} \quad Y_0 = \int_{\Delta'}\int_\Delta Y(\mathbf{r}-\mathbf{r}')s(\mathbf{r})s(\mathbf{r}')\mathrm{d}\mathbf{r}\,\mathrm{d}\mathbf{r}' \tag{8.33}$$

This solution provides the best approximation to $P_N^{(1)}(\mathbf{r})$ in terms of $s(\mathbf{r})$. The total partition function Z_N and the *free volume* v_f then become

$$Z_N = (\Lambda^{-3}\bar{\sigma}v_\mathrm{f})^N \exp\left(-\frac{1}{2}\beta NY_0\right) \quad \text{and} \quad v_\mathrm{f} \equiv \int_\Delta \exp[-\beta\psi(\mathbf{r})]\mathrm{d}\mathbf{r} \tag{8.34}$$

From these expressions the thermodynamic properties can be obtained in the usual way after a choice for $\bar{\sigma}$ and $\Psi(\mathbf{r})$ is made. As zeroth approximation, we assume that the molecules are located at the center of their cells and, using the Dirac delta function, we have $s(\mathbf{r}) = \delta(\mathbf{r})$, leading to

$$\psi(\mathbf{r}) = \sum_{l=2}^{N}[\phi(\mathbf{R}_{1l}+\mathbf{r}) - \phi(\mathbf{R}_{1l})] \quad \text{and} \quad Y_0 = \sum_{l=2}^{N}\phi(\mathbf{R}_{1l}) \tag{8.35}$$

Restricting the sums to nearest-neighbors, replacing this sum by an integral over a sphere of radius equal to the nearest-neighbor distance and taking $\bar{\sigma} = \mathrm{e}$ yields the LJD model. We have $Y_0 = z\phi(0)$ and $\Psi(r) = z[\phi(r) - \phi(0)]$ and the results become

$$Z_N = (\Lambda^{-3}\mathrm{e}v_\mathrm{f})^N \exp\left(-\frac{1}{2}\beta Nz\phi(0)\right) \quad \text{and} \quad v_\mathrm{f} = \int_\Delta \exp[-\beta\psi(\mathbf{r})]\mathrm{d}\mathbf{r} \tag{8.36}$$

Although this analysis represents a significant step in theory, it is also clear that the problem lies in evaluating $\bar{\sigma}$. An expression for $\bar{\sigma}$ may be based on the MC results of Hoover and Ree, but such an approach remains empirical.

Problem 8.1

Derive the expressions for U, S, and μ from Z for the simple cell model.

Problem 8.2: Free volume via sound velocity

The free volume in a liquid can be estimated using the speed of sound in the liquid. In a simple model we use three molecules A, B, and C in a straight line. If a sound wave travels from left to right, molecule A collides with molecule B and the signal is transferred instantaneously to the opposite side of molecule B. Thus, although molecule A travels only the distance a–d in order to strike molecule B, the sound wave effectively travels the distance a.

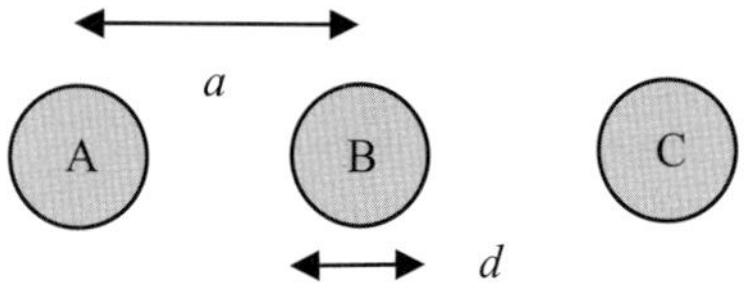

a) Estimate the ratio of the sound velocities in the gas v_G and liquid v_L.
b) Given that $v_L/v_G \cong 5$, estimate the fraction of free volume in the liquid.
c) Using this model estimate the volume expansion when the solid melts. Hint: Use the free volume expression as used in the simple cell model.

Problem 8.3: Free volume via α_P and κ_T

a) Using the Helmholtz energy F for the cell model with average attraction potential ϕ and the appropriate Gibbs–Helmholtz expression, show that the internal energy is given by $U/N_A = -\phi + 3kT/2$.

b) Using the Helmholtz energy F and $P = -\partial F/\partial V$, show that $P = \partial\phi/\partial V + kT\,\partial \ln v_f/\mathrm{d}V$.

c) Differentiating P with respect to T while keeping V (and N) constant and remembering that ϕ and v_f are supposed to be functions of V only, show that $(\partial P/\partial T)_{V,N} = k\,\partial \ln v_f/\mathrm{d}V$.

d) From the definition of the expansivity α_P and compressibility κ_T, show that $(\partial P/\partial T)_{V,N} = \alpha_P/\kappa_T$.

e) Using $v_f = 4\pi\gamma(v^{1/3} - v_0^{1/3})^3/3$, show that $\partial \ln v_f/\mathrm{d}V = (4\pi\gamma/3)^{1/3}/V^{2/3}v_f^{1/3}$.

f) Taking the average value for normal liquids $\alpha_P \cong 1 \times 10^{-3}\,\mathrm{K}^{-1}$ and $\kappa_T \cong 1 \times 10^{-4}\,\mathrm{bar}^{-1}$ together with $R = N_A k = 82\ \mathrm{bar\,cm^3\,K\,mol^{-1}}$, show that for chloroform ($V_m = 80\,\mathrm{cm^3\,mol^{-1}}$) $V_f = N_A v_f \cong 0.44\,\mathrm{cm^3\,mol^{-1}}$.

Problem 8.4: Free volume via $\Delta_{vap}H$

It appears empirically that the vapor pressure P over a wide temperature range can be given by $P = a\exp(-\Delta_{vap}H/RT)$, where $\Delta_{vap}H$ is the enthalpy of vaporization. The value of a is (nearly) constant and does not vary considerably over a range of (normal) liquids. Using the average value $a \cong 2.7 \times 10^7$ atm (correct to within a factor 2), show that $\Delta_{vap}H/kT_n = \ln a \cong 10.4$. Also show that for chloroform, using the same data as in Problem 8.3, $V_f \cong 1.0\,\text{cm}^3\,\text{mol}^{-1}$.

Problem 8.5: Vapor pressure

Consider the cell model with average attraction potential $\phi(T)$ so that the total attraction $\Phi_{lat} = N\phi$ represents the energy of evaporation of the liquid. Using the partition function given by Eq. (8.6), calculate the chemical potential μ for the cell model. Show that, by equating the chemical potential for the liquid and the (perfect) gas, the calculation for the vapor pressure P_{vap} results in $P_{vap} = (kT/v_f)\exp[-\phi/2kT]$.

Problem 8.6: Hildebrand's rule and Trouton's constant*

The result of Problem 8.5, assuming $\bar{\sigma} = \text{e}$ for the liquid phase and $\Delta U = \Delta_{vap}H - RT = a/V$, can be used to (approximately) validate Hildebrand's rule and to estimate the value of Trouton's constant (see Section 16.2).

a) Use the Eyring EoS Eq. (8.8) in combination with $v_f = 4\pi\gamma(v^{1/3} - v_0^{1/3})^3/3$ to obtain an explicit expression for v_f, neglect the external pressure P with respect to the internal pressure a/V^2 and show that the result is $\zeta PV/RT = [\Delta_{vap}H/RT - 1]^{-1}\exp(-\Delta_{vap}H/RT)$ with $\zeta = (4\pi/3)^{1/3}$.

b) This result appears to be a universal function of $\Delta_{vap}H/RT$ for many compounds. Show that this result leads to Hildebrand's rule (see Section 16.2).

c) By realizing that the factor V varies only slightly from compound to compound in comparison with the exponential factor, show that the number density ρ is a universal function of $\Delta_{vap}H/RT$.

d) Show that, by approximating $R/\zeta V$ by a constant and using the typical value $V_m = 80\,\text{cm}^3\,\text{mol}^{-1}$ at $T = 300$ K, one obtains Trouton's constant $\Delta_{vap}H = RT_n = C$ with $C \cong 10.4$, where T_n is the (normal) boiling temperature at $P = 1$ atm. Show that C is insensitive to the precise value of ζ.

8.3 Hole Models

One way to improve on the cell model is to release the requirement that each cell is always occupied by only one molecule. If we take the number of cells M somewhat larger than the number of molecules N, the volume per cell V/M and the volume per molecule V/N become different. Accepting still maximally one molecule per cell, the empty cells are addressed as holes and the whole results in *hole models* [17]. Taking instead of holes, molecules of another type, the approach is also used for mixtures.

We recall that the interparticle pair potential essentially contains a repulsive part and an attractive part. The structure of a simple liquid appears to be largely determined by the repulsive part, while the attractive part mainly provides the "internal pressure," that is, the cohesion. In order to be able to calculate the Helmholtz energy in the thermodynamic limit, the pair potential must be more repulsive than $1/r^{3+\delta}$ for $r \to 0$ and decay more rapidly than $1/r^{3+\delta'}$ for $r \to \infty$, where δ and δ' are positive constants. A simple, but nevertheless useful, model which takes both of these interactions and conditions into account is the *lattice gas model*. The basic concept is that the volume available to the fluid is divided into *cells* of molecular size. Usually, for simplicity, the cells are arranged in a regular lattice with coordination number z, for example, a simple cubic lattice with $z = 6$, a body-centered cubic lattice with $z = 8$, or a face-centered cubic lattice with $z = 12$. Although not required in principle, normally these cells are occupied by one particle at most, representing repulsion, and only nearest-neighbor cell attractive interactions are considered. The precise position of a particle in the cell is considered unimportant, so that the particle and cell positions are the same.

If we position N_1 molecules of type 1 and N_2 molecules of type 2 on a lattice, there will be N_{11} (N_{22}) nearest-neighbor pairs 1-1 (2-2) of molecules of type 1 (2). Similarly, we will have N_{12} pairs of the type in which a molecule of type 1 and type 2 combines to a pair 1-2. For any such a lattice some general relations exist. Drawing a line from each site occupied with a molecule of type 1 to the z neighboring sites, we will obtain zN lines. If the pair is of type 1-1, there will be two lines between the sites, whereas if it is of type 1-2 there is only one line. A similar consideration applies for type 2-2. Therefore,

$$zN_1 = 2N_{11} + N_{12} \quad \text{and} \quad zN_2 = 2N_{22} + N_{12} \tag{8.37}$$

Consequently,

$$N_{11} = \frac{1}{2}zN_1 - \frac{1}{2}N_{12} \quad \text{and} \quad N_{22} = \frac{1}{2}zN_2 - \frac{1}{2}N_{12} \tag{8.38}$$

If we mix N_1 molecules of type 1 with N_2 molecules of type 2, the energy is

$$E = \varepsilon_{11}N_{11} + \varepsilon_{22}N_{22} + \varepsilon_{12}N_{12} = \left(\varepsilon_{12} - \frac{1}{2}\varepsilon_{11} - \frac{1}{2}\varepsilon_{22}\right)N_{12} + \frac{1}{2}z\varepsilon_{11}N_1 + \frac{1}{2}z\varepsilon_{22}N_2 \tag{8.39}$$

The quantity $w \equiv \varepsilon_{12} - \frac{1}{2}\varepsilon_{11} - \frac{1}{2}\varepsilon_{22}$ is often denoted as the *interchange energy*. Now, we have to distinguish between a mixture of molecules of type 1 and 2 (assuming that no holes are present in the mixture) and a pure liquid with molecules (type 1) and holes (type 2). In the former case, the contribution $\frac{1}{2}\varepsilon_{11}N_1 + \frac{1}{2}\varepsilon_{22}N_2$ is a constant given by the composition, but in the latter case it is variable as the number of holes depends on the temperature. Hence, formally, we must use the canonical ensemble for the former situation describing solutions, and the grand canonical ensemble for the latter situation describing pure fluids.

8.3.1
The Basic Hole Model

We consider here the pure liquid[4] for which $\varepsilon_{11} = \varepsilon$ and $\varepsilon_{12} = \varepsilon_{22} = 0$, so that $w = -\frac{1}{2}\varepsilon_{11} = -\frac{1}{2}\varepsilon$. The energy E for lattices with M sites occupied by N molecules creating zX type 1-2 pairs becomes

$$E = -\frac{1}{2}\varepsilon_{11}N_{12} + \frac{1}{2}z\varepsilon_{11}N_1 = \varepsilon N_{11} \quad \Rightarrow \quad E = zXw + \frac{1}{2}zN\varepsilon = \frac{1}{2}\varepsilon z(N-X) \quad (8.40)$$

The last form is somewhat more convenient in calculations. The configurational partition function becomes, using the abbreviation $x = \exp(-z\beta w) = \exp(\frac{1}{2}\beta z\varepsilon)$,

$$Q_N = \sum\nolimits_X q^N g(M,N,X)\mathrm{e}^{-\beta E} = q^N \sum\nolimits_X g(M,N,X)\mathrm{e}^{-\frac{1}{2}\beta z(N-X)\varepsilon}$$
$$= q^N x^{-N} \sum\nolimits_X g(M,N,X)x^X \quad (8.41)$$

Here, q is the partition function of the molecules, assumed to be separable, and $g(M,N,X)$ denotes the number of configurations with N molecules over M sites having X pairs of type 1-2 all having the same energy (degeneracy factor). The summation Σ_X is over all possible configurations having X pairs. Recalling the activity $\lambda = \exp(\beta\mu)$ with μ the chemical potential, the grand partition function, with as usual P the pressure and V the volume, reads (see Chapter 5)

$$\Xi = \exp(\beta PV) = \sum\nolimits_N \lambda^N Q_N \quad (8.42)$$

We will use here $PV = (Pv_0)\cdot(V/v_0) \equiv \Phi\cdot M$, where v_0 is the volume of a cell and M the number of cells. Hence, Φ is the pressure in energy units.

$$\Xi = \exp(\beta\Phi M) = \sum\nolimits_N \lambda^N Q_N = \sum\nolimits_N \lambda^N q^N x^{-N} \sum\nolimits_X g(M,N,X)x^X \quad (8.43)$$

$$= \sum\nolimits_N a^N x^{-N} \sum\nolimits_X g(M,N,X)x^X \quad \text{with} \quad a \equiv q\lambda \quad (8.44)$$

This is as far as we can go without approximation. An exact solution is only known for 2D lattices, and therefore we have to approximate for the 3D case. The *zeroth approximation* (see Appendix C) considers all molecules as distributed randomly over all sites in spite of their nonzero interaction. This implies that we use the approximation

4) Normally we would indicate the number of molecules of type 1 by N_1, the number of molecules of type 2 by N_2, the total number by $N = N_1 + N_2$, and the number of pairs by N_{11} (N_{22}, N_{12}). To avoid the use of many subscripts for a pure fluid, we use for the total number of sites M and for the number of molecules N. For the number of pairs for we choose N_{12}, relabeled as zX.

$$\sum_X g(M,N,X)x^X \cong x^X \sum_X g(M,N,X) = x^X M!/(M-N)!N! \quad (8.45)$$

where the last step can be made because the summation of $g(M,N,X)$ over all possible configurations X equals the number of configurations of N molecules over M sites. In this approximation we also have $zX = N\cdot z(M-N)/M = zN - zN^2/M$, leading together to

$$\Xi = \sum_N a^N x^{-N^2/M} M!/(M-N)!N! \quad (8.46)$$

This sum is difficult to evaluate and we use the maximum-term method (see Justification 5.5). Denoting the terms in Eq. (8.46) by t_N, we thus approximate the sum by its maximum term[5] $t_{\max}$, obtained by solving $\partial t_N/\partial N = 0$ or, equivalently, $\partial \ln(t_{N-})/\partial N = 0$. After some algebra we obtain, using the density $\rho = N/M$,

$$\lambda = x^{2\rho}[\rho/(1-\rho)] \quad \text{or} \quad \beta\mu = \ln[\rho\exp(\beta\rho z\varepsilon)/(1-\rho)] \quad (8.47)$$

Substituting the solution in the expression for the pressure Φ, obtained from $\beta\Phi M = \ln\Xi = \ln t_{\max}$, the result is again after some algebra, the EoS:

$$\beta\Phi = \ln[x^{\rho^2}/(1-\rho)] \quad \text{or} \quad \beta\Phi = \frac{1}{2}\beta\rho^2 z\varepsilon - \ln(1-\rho) \quad (8.48)$$

Dependent on the temperature T, the pressure Φ shows either continuously increasing curves with increasing density ρ, or sinuous curves indicating the presence of a critical point associated with a discontinuous phase transition (see Chapter 16). One phase is the more condensed and considered as the liquid with density ρ_L, while the other phase is the less dense and considered as the gas with density ρ_G. The critical point T_{cri} can be found by realizing the symmetry between vacant and nonvacant cells. In the two-phase region $\rho_G + \rho_L = \rho$, and therefore by symmetry T_{cri} must be at $\rho_L = \rho_G = ½$. The location of T_{cri} can now be found from $\partial(\beta\Phi)/\partial\rho = 0$ evaluated at $\rho = ½$, leading to

$$z\beta\varepsilon = -4.00 \quad (8.49)$$

Although the basic physics of the liquid–gas transition described are qualitatively correct, the numerical accuracy is, as we will see shortly, not astonishing. In Chapter 11 and Appendix C an approximate solution based on the distribution of independent pairs of molecules (instead of independent single molecules) is briefly discussed. This solution is called the *first* (or *quasi-chemical*) *approximation*. In this approximation $z\varepsilon/kT_{cri} = 2z\ln[(z-2)/z]$, so that the result for a square 2D lattice with $z = 4$ is $z\varepsilon/kT_{cri} = -5.54$, while the exact result is $z\varepsilon/kT_{cri} = -7.05$.

For a pure liquid we estimate, as before using the close-packed lattice, $z = 12$. Furthermore, in order to mimic results of other models using the LJ potential, we employ this potential here also, and take the minimum LJ potential as the value for ε and the position of the minimum $r_0 = 2^{1/6}\sigma$ for size. We need also an expression for the partition function q and take it conform the cell model as $q = v_f/\Lambda^3$

5) This is equivalent to using the canonical partition function right away.

Table 8.2 Critical data comparison for several theories.

Origin of data	T^*_{cri}	V^*_{cri}	P^*_{cri}	$P^*_{cri}V^*_{cri}/T^*_{cri}$	y_{cri}
LJD theory	1.30	1.77	0.434	0.591	1.00
Cernuschi and Eyring	2.74	2.00	0.469	0.342	0.544
Henderson	1.42	3.39	0.139	0.333	0.321
SLS theory	1.31	3.36	0.141	0.362	–
Experimental	1.28	3.09	0.121	0.293	~0.5

(including kinetics). In principle, v_f should depend on the local configuration, that is, the number and arrangement of nearest-neighbors. However, here we assume that we can use the value for $z = 12$ and discuss an improved model later. For the EoS $\Phi = PV/M = Pv_0 = Pr_0^3/2^{1/2} = P\sigma^3$, we can use either the result of the zeroth or the first approximation. Employing the latter model [17], we use $z\varepsilon/kT_{cri} = 2z \ln[(z-2)/z]$ (see Appendix C) and obtain $T_{cri} = -2.74\varepsilon/k$. From the associated EoS and the critical volume $v_{cri} = 2\sigma^3$, one can also calculate that $P_{cri}\sigma^3/\varepsilon = -0.469$, and therefore $P_{cri}v_{cri}/kT_{cri} = 0.342$. Table 8.2 compares these results with those of the LJD and the SLS model (see Section 8.4).

So in conclusion, although the first approximation is a clear improvement over the zeroth approximation, probably the most important point to learn from the lattice gas model is that by taking the hard core seriously (no two particles on the same site) already provides the crudest approximation most of the realistic features of the liquid–vapor system, as indicated in the introduction to this Section.

8.3.2
An Extended Hole Model*

An obvious shortcoming of the simple hole model is that v_f does not depend on the local configuration, although it should. To improve this, we employ the LJD model, taking into account Kirkwood's considerations (see Justification 8.1), and write the partition function conforming with Eq. (8.36), using nearest-neighbor interactions only, so that we have with $\Phi_{lat} = ½zN\phi(0)$ and $\Psi(\mathbf{r}) = z[\phi(\mathbf{r}) - \phi(0)]$

$$Z_N = (\Lambda^{-3}v_f)^N \exp\left[-\frac{1}{2}\beta Nz\phi(0)\right] \quad \text{and} \quad v_f = \int_\Delta \exp[-\beta\psi(\mathbf{r})]\mathrm{d}\mathbf{r} \tag{8.50}$$

Here, the communal entropy factor e^N is omitted because the idea is to incorporate this effect via a modified free volume expression.

Choosing again the FCC lattice with coordination number $z_0 = 12$ and cell volume $a^3/\gamma = a^3/2^{1/2}$, we assume that the coordination number becomes $z = yz_0$, where y is a new parameter with a range $0 \le y \le 1$. Next, we define ω_k to be the fraction of vacant nearest-neighbor sites of molecule k. If all of the molecules are at the origin of their cells, we have for the potential energy of one molecule $z(1-\omega_k)\phi(0)$, and for the total potential energy

$$\Phi(0) = \frac{1}{2} z(N - \Omega)\phi(0) \quad \text{where} \quad \Omega = \Sigma_k \omega_k \tag{8.51}$$

When the molecules are not at their origin, we calculate, similar to the cell model, the interaction of the wanderer with the nearest-neighbor molecule located at its origin. When the wanderer is at position **r** in the cell, its potential energy is $\Psi(\mathbf{r}) = z[\phi(\mathbf{r}) - \phi(0)]$. Using the "smearing" approximation with $z(1 - \omega_j)$ nearest-neighbors, we assume that the potential energy of the wanderer depends only on the number of nearest-neighbor molecules, and not on their precise configuration. However, this is obviously an oversimplification. Accepting this we obtain for the potential energy, using an overbar for Ψ_j to indicate the smearing approximation,

$$\Phi(\mathbf{r}) = \frac{1}{2} z(N - \Omega)\phi(0) + \sum_{k=1}^{N} z(1 - \omega_k)\bar{\psi}_k(\mathbf{r}) \quad \text{with} \quad \bar{\psi}_k = \overline{\phi(\mathbf{r}_k)} - \phi(0) \tag{8.52}$$

This results for the configurational partition function Q_N in

$$Q_N = \exp\left[-\frac{1}{2}\beta z(N - \Omega)\phi(0)\right]\prod_k j(\omega_k) \tag{8.53}$$

with the generalized free volume

$$j(\omega_k) = \int_\Delta \exp[-\beta(1 - \omega_k) z\bar{\psi}_k(r)] 4\pi r^2 \, dr \tag{8.54}$$

When no nearest-neighbor molecules are missing ($\omega_k = 0$), $j(\omega_k)$ becomes the same as for the cell model using the "smearing" approximation. When all nearest-neighbor molecules are missing ($\omega_k = 1$), $j(\omega_k)$ becomes the cell size. At high and low density we obtain, respectively,

$$Q_N = [j(0)]^N \exp\left[\frac{-Nz\phi(0)}{2kT}\right] \quad \text{and} \quad Q_N = [j(1)]^N \frac{L!}{N!(L-N)!} \cong \mathrm{e}^N \left(\frac{V}{N}\right)^N \tag{8.55}$$

While the first expression is just the Einstein approximation for the partition function of a crystal, the second expression contains the factor e^N, necessary to obtain the ideal gas limit. In principle, the communal problem is thus avoided.

To calculate Q_N it is necessary to use a simple relationship for $j(\omega)$. From Eq. (8.52) we see $j(\omega)$ at T yields the same value as v_f at $T/(1 - \omega)$. We thus may use $v_\mathrm{f} = v_\mathrm{f}(T/(1 - \omega)$ to calculate $j(\omega)$. From Q_N, the thermodynamic properties can then be calculated in the usual way. Since v_f depends on the number and the arrangement of holes, the relationship between v_f and y is not simple [18]. Some calculations have been made with a linear approximation to $\ln j(\omega)$. A comparison of the results obtained teaches us that the results are not significantly better than for the original, simpler approach.

A reasonably successful approach is based on the idea that in a fluid it is plausible to have both solid-like and gas-like molecules. If the free volume of a gas-like molecule, that is, with all its neighboring sites empty, is given by $j(1)$ and the free volume of a solid-like molecule, that is, with all neighboring sites occupied, is labeled $j(0)$, we may reasonably take for the average free volume, along with Henderson [19],

$$v_f = y_i j(0) + (1 - y_i) j(1) \tag{8.56}$$

where the probability to have solid-like and gas-like properties is y_i and $1 - y_i$, respectively. If we choose for the potential ϕ the LJ potential with parameters ε and σ, the expression for $\phi(0)$ for the FCC lattice becomes

$$\phi(0) = z\varepsilon[(1/v_*^4) - (2/v_*^2)] \tag{8.57}$$

where $v_* = v/\sigma^3$ for the FCC lattice with intermolecular distance a and volume $v = a^3/2^{1/2}$. If $\omega = V/M$ and $v = V/N$, we have $y_i = N/M = \omega/v$, and the Helmholtz energy $F = -kT\ln Z_N$ evaluated using the zeroth approximation reads

$$F = -NkT \ln \Lambda^{-3} v_f + \frac{1}{2}\frac{\omega}{v} N\psi(0) + NkT \ln\left(\frac{\omega}{v}\right) + NkT\left(\frac{v}{\omega} - 1\right)\ln\left(1 - \frac{\omega}{v}\right) \tag{8.58}$$

The volume per cell is then determined by $(\partial F/\partial \omega)_{V,T} = 0$, while the EoS can be calculated from $PV = -v(\partial F/\partial v)_{\omega,T}$ which yields, after some algebra,

$$\frac{PV}{NkT} = \frac{\omega - j(0)}{v - \omega + j(0)} + \frac{\omega}{v}\frac{N\phi(0)}{2kT} - NkT\frac{v}{\omega}\ln\left(1 - \frac{\omega}{v}\right) \tag{8.59}$$

The values for $j(0)$ and its temperature derivatives needed for a numerical evaluation can be calculated from tabulated data as given Wentorff *et al.* [20]. Figure 8.4 shows the vapor pressure and the EoS for the extended hole theory as calculated

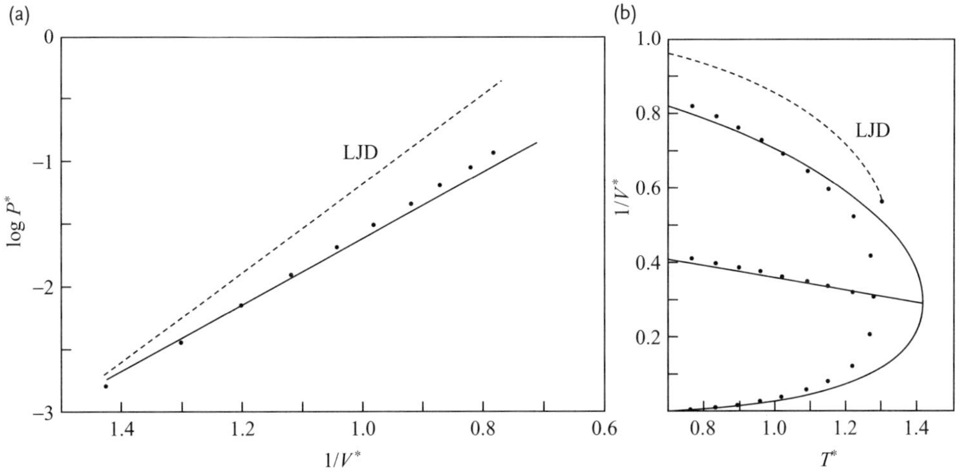

Figure 8.4 Hole theory results. (a) The logarithm of the reduced vapor pressure P^* for simple liquids versus reciprocal reduced temperature $1/T^*$; (b) The reduced density versus $1/V^*$ versus reduced temperature T^* for the hole theory ($T = (\varepsilon/k)T^*$, $V = (N/\sigma^3)V^*$ and $P = (\varepsilon/\sigma^3)P^*$). The solid line represents hole theory results, the dots represent points from the empirical Guggenheim expressions (see Chapter 4), and the dotted line represents LJD results.

by Henderson in comparison with LJD theory results, and a clear improvement can be noted. Table 8.2 presents a comparison of the critical data with the results from some other theories. From these data it is clear that, for the prediction of critical constants, the simple hole theory is inferior to the LJD theory, although the extended hole theory actually represents the experimental data quite well. Henderson [19] also calculated the internal energy, entropy and heat capacity with this model, but the results obtained were inferior to those for the EoS, the reasons for which were discussed by Henderson.

In conclusion, although hole models are capable of catching the qualitative aspects of liquids correctly, and the communal problem is avoided (in principle, at least), their ultimate success is restricted. For further details we refer to the literature.

Problem 8.7

Calculate the critical constants using the simple hole theory.

8.4
Significant Liquid Structures*

In the approach, which was advocated by Eyring and denoted by him as *significant liquid structure theory*[6] (here labeled as SLS theory), it is argued that a liquid behaves in many cases as an intermediate between a gas and a solid. From the fact that a simple liquid like Ar expands ~12% upon melting and the associated large increase in fluidity, it can be concluded that the coordination number decreases, while simultaneously introducing some free volume in the liquid. Since XRD measurement have indicated that the intermolecular distances in the liquid are essentially the same as in the solid, the resulting structure is assumed to show some heterogeneity, containing a fraction of solid-like molecules and another fraction of holes. Holes of molecular size are favored because smaller holes cannot provide easy access to the entering molecules, which limits the increase of entropy. In contrast, larger holes will require excessive energy without a compensating increase in entropy. These holes change by one or more degrees of freedom for a molecule surrounding the hole from vibration to translation, and these molecules thus act like vapor molecules. Hence, the concept does not imply that a liquid is a mixture of a solid and a gas. Rather, a molecule has solid-like properties for the time it vibrates around an equilibrium position, but transforms instantaneously to a gas-like behavior for one or more degrees of freedom if it jumps into a neighboring hole. A simple schematic of this is process shown in Figure 8.5.

6) The adjective "significant" is, of course, rather subjective. Eyring always chose particularly challenging names for his theories. Another example is the theory of 'absolute' reaction rates in which there is no element that is 'absolute' and which nowadays is often called transition state theory.

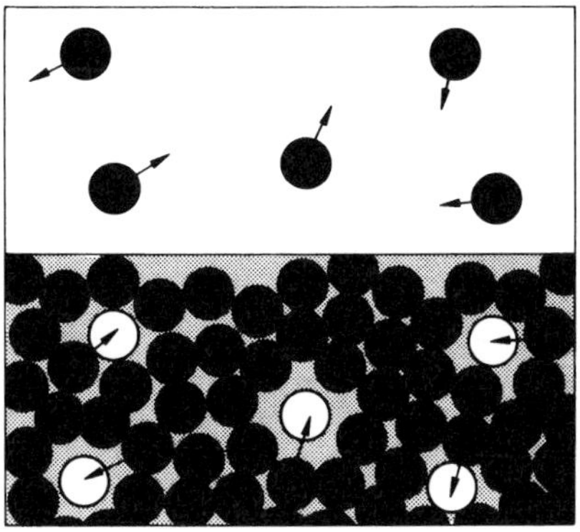

Figure 8.5 A schematic illustrating a liquid formed by removing at random molecules from a (glassy) solid. The holes so created mirror the molecules in the gas phase.

According to the *law of rectilinear diameters* (see Chapter 4), the average density of a liquid and its associated equilibrium vapor pressure is a linear function of temperature (roughly equal to $\sim\frac{1}{2}\rho_{solid}$ at T_{mp} to $\sim\frac{1}{3}\rho_{solid}$ at T_{cri}). A small decrease in temperature is expected due to thermal expansion, but this law indicates that the number density of molecules in the gas phase is about equal to the number density of vacancies in the liquid. In a simple nearest-neighbor model, the energy of vaporization per molecule is approximately $\frac{1}{2}z\phi$, where z is the coordination number and ϕ is the bond energy. In a similar way it requires energy $z\phi$ to create a hole in the liquid. The molecule removed from the interior of the liquid to the vapor can be added to the surface of the liquid, thereby regaining the energy $\frac{1}{2}z\phi$, so leaving a hole costs only the energy of vaporization. Hence, a vacancy is expected to move as freely as a molecule in the gas phase and to have about the same energy and entropy, thus explaining the law of rectilinear diameters.

So, the thermodynamic behavior in this model is described by varying the fraction of gas-like and solid-like volume elements, while the overall thermodynamic behavior is given by the sum of the contributions of these elements. In the simplest form for gas-like regions the ideal gas model is used, whereas for solid-like elements the Einstein model (see Sections 5.3 and 6.1) is used. The crux is, obviously, to determine the fractions of gas-like and solid-like regions.

Since the holes are on average of molecular size, in one mole of liquid $(V - V_s)/V_s$ moles of holes are present, where V and V_s are the molar volumes of the liquid and solid, respectively. Assuming complete randomness, the fraction of neighboring positions filled with molecules to with a hole is V_s/V. It thus follows that for N molecules there are

$$N(V_s / V)[(V - V_s)/V_s] = N(V - V_s)/V \equiv N_g \tag{8.60}$$

gas-like molecules and a remainder of $N_s \equiv NV_s/V$ solid-like molecules. According to this model, the heat capacity C_V of a mole of, say Ar, is given by sum of the contributions of the V_s/V moles of solid and $(V - V_s)/V$ moles of gas. Therefore

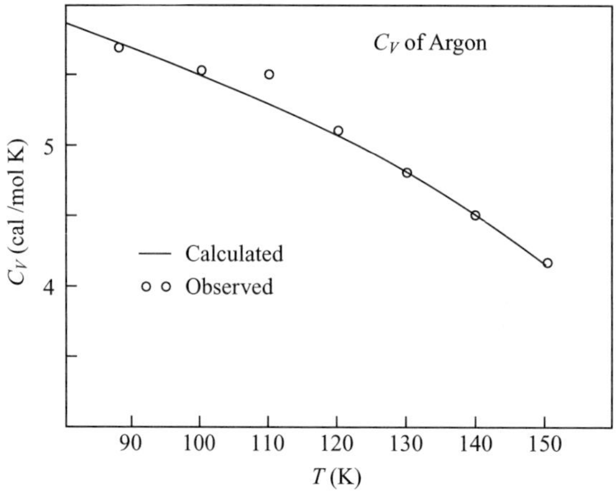

Figure 8.6 The experimental heat capacity C_V for liquid argon (circles) and the prediction according to Eq. (8.61) (solid line) using $C_{V,\text{solid}} = 3R = 25.0\,\text{J mol}^{-1}\,\text{K}^{-1}$ or $6.0\,\text{cal mol}^{-1}\,\text{K}^{-1}$ and $C_{V,\text{gas}} = 3R/2 = 12.5\,\text{J mol}^{-1}\,\text{K}^{-1}$ or $3.0\,\text{cal mol}^{-1}\,\text{K}^{-1}$.

$$C_V = C_{V,\text{solid}}\frac{V_\text{s}}{V} + C_{V,\text{gas}}\frac{V - V_\text{s}}{V} \tag{8.61}$$

for which the good agreement with experiment is shown in Figure 8.6.

Using the above concepts, the partition function Z_N can be written as

$$Z_N = (Z_\text{s})^{N_\text{s}}(Z_\text{g})^{N_\text{g}}/N_\text{g}! \tag{8.62}$$

where Z_s and Z_g represent the partition functions of the solid-like and gas-like molecules, respectively.

For calculating Z_s we consider that holes surrounding a solid-like molecule introduce positional degeneracy. The number of additional accessible sites for a solid-like molecule is the number of holes around that molecule multiplied by the probability that this molecule has the required energy ε_h to be able to jump to these holes. The number of holes n_h is proportional to the excess volume and given by

$$n_\text{h} = n(V - V_\text{s})/V_\text{s} \tag{8.63}$$

with n as proportionality factor. The energy ε_h should be inversely proportional to the excess volume and directly proportional to the energy of sublimation E_s. Therefore

$$\varepsilon_\text{h} = aE_\text{s}V_\text{s}/(V - V_\text{s}) \tag{8.64}$$

with a as proportionality factor. Since the parameter $\varepsilon_\text{h} \to \infty$ for $V \to V_\text{s}$ and $\varepsilon_\text{h} \to 0$ for $V \to \infty$, this introduces cooperative behavior in theory. The total number of positions available, that is, original plus additional, to the solid-like molecule becomes

$$1 + n_\text{h}\exp(-\varepsilon_\text{h}/kT) \tag{8.65}$$

and Z_s becomes (using Z_s' as the partition function for the pure solid-like molecule)

$$Z_s = [1 + n_h \exp(-\varepsilon_h / kT)] Z_s' \tag{8.66}$$

As noted, for Z_s' the simplest choice is the Einstein model, for which the partition function using θ as the Einstein temperature, reads

$$Z_s' = \exp(-\phi/2kT)\{\exp(-\theta/2T\,/\,[1 - \exp(-\theta/T)]\}^3 \tag{8.67}$$

The factor $\exp(-\phi/2kT)$ takes into account that in a simple nearest-neighbor model $-\phi/2$ is the potential energy of the solid-like, vibrating molecule and we interpret $E_s \equiv -½(\phi + 3k\theta)$ as the sublimation energy. For Z_g we use straightforwardly the ideal gas expression with excess volume $V - V_s$ and thermal wavelength $\Lambda = (h^2/2\pi mkT)^{½}$, that is,

$$Z_g = (V - V_s)/\Lambda^3 \tag{8.68}$$

Hence, according to Eq. (8.62) we have for the partition function Z_N of the liquid

$$Z_N = \left[e^{E_s/kT} \left(1 - e^{-\theta/T}\right)^{-3} \left(1 + n_h e^{-\varepsilon_h}\right) \right]^{N_s} (N_g!)^{-1} \left[\Lambda^{-3}(V - V_s) \right]^{N_g} \tag{8.69}$$

While the parameters E_s, θ and V_s have to be taken from solid-state data, either theoretical or experimental, the parameters a and n can be evaluated theoretically.

To calculate n, we use the molar volume V_{mp} at the melting point T_{mp} and consider that, close to T_{mp} and using the lattice coordination number z, n_h is given by

$$n_h = z\frac{(V_{mp} - V_s)}{V_{mp}} = \left(z\frac{V_s}{V_{mp}} \right) \frac{(V_{mp} - V_s)}{V_s} \tag{8.70}$$

Comparison with Eq. (8.63) yields

$$n = zV_s \,/\, V_{mp} \tag{8.71}$$

To calculate a, we consider that a solid-like molecule has a kinetic energy $3kT/2$. If a molecule is to pre-empt a neighboring position in addition to its original position, it must have additional kinetic energy equal to or in excess of that which the other $(n - 1)$ neighboring molecules would otherwise introduce into this hole. If the average molecule divides its time equally between two neighboring sites, its energy density will be halved. As the molecule will be moving for $(1/z)$th of its time in the direction of any neighbor, the average kinetic energy of the $(n - 1)$ ordinary molecules will provide a hole with kinetic energy $½(3kT/2)(n - 1)/z$. This is also the value that $\varepsilon_h = aE_sV_s/(V - V_s)$ must have at the melting temperature T_{mp}, that is,

$$\varepsilon_h = \frac{aE_sV_s}{V - V_s} = \frac{1}{2}\left(\frac{3kT_{mp}}{2} \right) \frac{n - 1}{z} \tag{8.72}$$

Since at T_{mp} it holds that $T_{mp}S_{mp} = E_{mp}$, and the entropy of melting per molecule S_{mp} for a simple liquid is about $3k/2$, we obtain $E_{mp} = 3kT_{mp}/2$ for the energy of

Table 8.3 Vapor pressure *P*, molar volume *V* and entropy of vaporization $\Delta_{vap}S$ for the noble gases Ar, Kr and Xe, and the parameters used.

Gas	*T* (K)	P_{calc} (atm)	P_{obs} (atm)	V_{calc} (cm^3)	V_{obs} (cm^3)	$\Delta_{vap}S_{calc}$ (eu)	$\Delta_{vap}S_{obs}$ (eu)
Ar (mp)	83.96	0.6874	0.6739	24.84	28.03	20.07	19.43
Ar (bp)	87.49	1.040	1.000	29.36	28.69	18.90	18.65
Kr (mp)	116.0	0.7605	0.7220	34.31	34.13	20.14	–
Kr (bp)	119.3	1.0660	1.000	34.90	–	19.15	17.99
Xe (mp)	161.3	0.8372	0.804	42.24	42.68	20.25	–
Xe (bp)	165.1	1.0623	1.000	42.84	–	19.50	18.29

	Ar	Kr	Xe
E_s (cal mol^{-1})	1888.6	2740	3897.7
V_s (cm^3 mol^{-1})	24.98	29.6	36.5
θ (K)	60.0	45.0	39.2

melting E_{mp}. Moreover, since during melting holes are introduced in the solid, essentially accompanied by a potential energy change only, we also have

$$E_{mp} = \frac{V_{mp} - V_s}{V_{mp}} E_s \quad \text{and therefore} \tag{8.73}$$

$$\varepsilon_h = \frac{1}{2} E_{mp} \frac{n-1}{z} = \frac{1}{2} \frac{V_{mp} - V_s}{V_{mp}} E_s \frac{n-1}{z} \quad \text{or} \quad a = \frac{n-1}{2z} \frac{(V_{mp} - V_s)^2}{V_{mp} V_s} \tag{8.74}$$

Equations (8.71) and (8.74) result in $a = 0.00537$ and $n = 10.7$ for Ar, using $z = 12$ and the data given in Table 8.3. The best empirical fit [21] yields $a = 0.00534$ and $n = 10.8$, so that there is good agreement. Some further data for Ar, Kr and Xe shown in Table 8.3 also demonstrate a good agreement with experimental data.

Using the same parameters as for the LJD theory, the triple point for Ar is predicted to be $0.711\varepsilon/k$, compared with $0.701\varepsilon/k$ according to LJD theory (Table 8.1). Predictions for *V*, *U* and *S* of Ar according to SLS and LJD theory are also shown in Table 8.1. The critical constants predicted are (LJD data in brackets)

$$P_{cri} = 0.141\,(0.434)\varepsilon/\sigma^3 \quad V_{cri} = 3.36\,(1.77)N\sigma^3 \quad T_{cri} = 1.31\,(1.30)\varepsilon/k$$

to be compared with the "best" experimental corresponding states estimates[7)]

$$P_{cri} = 0.116\ \varepsilon/\sigma^3 \quad V_{cri} = 3.14\ N\sigma^3 \quad T_{cri} = 1.25\ \varepsilon/k$$

Overall, we conclude that for describing liquids SLS theory is better than LJD theory, which resembles more a superheated solid than a liquid.

Some other predictions of SLS theory, which also show good agreement with experimental data, are given in Figure 8.7. For molecules that rotate freely in the

7) Different authors provide slightly different 'best' estimates. Authors even provide different estimates in the same text [22]: $P_{cri} = 0.119$ and $0.141\ \varepsilon/\sigma^3$, $V_{cri} = 3.15$ and $3.36\ N\sigma^3$, $T_{cri} = 1.28$ and $1.31\ \varepsilon/k$).

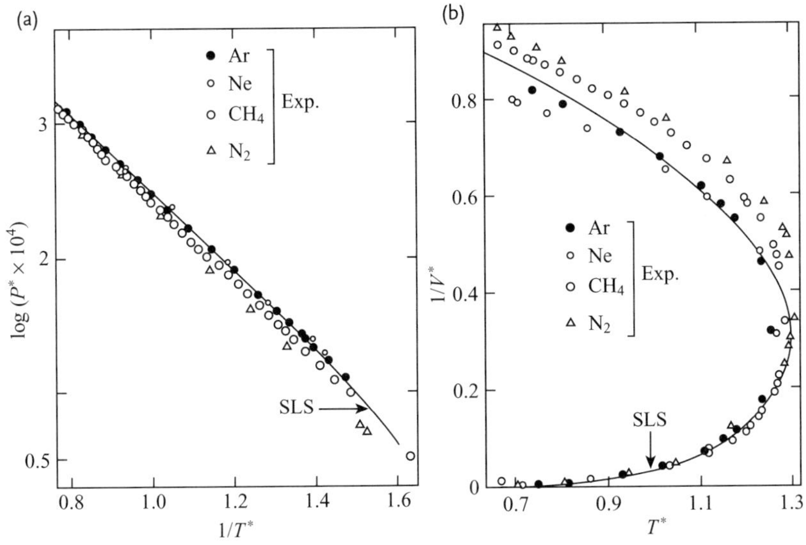

Figure 8.7 SLS theory results. (a) The logarithm of the reduced vapor pressure P^* for simple liquids versus reciprocal reduced temperature $1/T^*$; (b) The reduced density versus $1/V^*$ versus reduced temperature T^* for SLS theory ($T = (\varepsilon/k)T^*$, $V = V^*(N/\sigma^3)$ and $P = (\varepsilon/\sigma^3)P^*$).

liquid phase, such as CH_4 and N_2, Figure 8.7 shows the good agreement with experimental data. However, for those molecules that do not rotate freely in the liquid phase, such as Cl_2, some changes in Z_s must be made, for example, treating the rotation in the solid-like part as a vibration. In that case, a good agreement with experimental data was also reached. For example, for the temperature range of 180 K to 240 K the heat capacity at constant pressure C_P, as a calculated from $C_P = C_V - TV\alpha^2\beta$ with C_V the heat capacity at constant volume, $\alpha = (1/V)(\partial V/\partial T)_P$ the expansivity and $\beta = -(1/V)(\partial V/\partial P)_T$ the compressibility, all calculated from Z_N, differ by only ~2% from the experimental values. The method has been applied to molten metals, molten salts and quantum liquids such as H_2 and Ne, with fairly good results. Application of the theory to many liquids and phenomena has been reviewed by Eyring and Jhon [23].

From the previous data and figures, it is concluded that quite acceptable agreement with experiment can be obtained. Nevertheless, the basis of SLS theory is not really clear and the approach has essentially been abandoned. Although the model has been rather successful in estimating thermodynamic properties, a basic feature of liquids – the pair correlation function – is largely absent. Although at a late stage of the development of this theory attempts were made [24] to introduce the pair correlation function, the procedure used must be characterized as artificial.

Problem 8.8

Show that the partition function of the significant liquid theory reduces to the ideal gas and solid-state partition functions if $V \to \infty$ and $V \to 0$, respectively.

Problem 8.9

Calculate the vapor pressure from significant liquid structures theory for atomic fluids using the ideal gas and Einstein partition functions for the gas-like and solid-like phases, respectively.

Problem 8.10

Is it possible to calculate the RDF for the significant liquid structures theory without further approximations of additions?

8.5 Scaled-Particle Theory*

In Chapter 7 we showed that using the PY approximation for hard-spheres with diameter σ the pressure P could be described by Eq. (7.43) reading

$$\beta P/\rho = 1 + 4\eta y(\sigma) = 1 - 4\eta c(\sigma) = 1 + 4\eta g(\sigma) \tag{8.75}$$

with the background correlation function $y(r) = g(r)/e(r)$, direct correlation function $c(r)$, pair correlation function $g(r)$, Boltzmann function $e(r) = \exp[-\beta\phi(r)]$, pair potential $\phi(r)$, packing fraction $\eta = \pi\sigma^3\rho/6$, and number density ρ. We already mentioned there that $y(r)$ is also denoted as the cavity function, so in this model we need only the value of the correlation function at $r = \sigma$ and this observation led to scaled-particle theory.

Consider a spherical cavity of radius r inside a fluid and let $p_0(r)$ be the probability that no molecular center lies inside this sphere. We take the value of the number density of molecules (hard-spheres) just outside this cavity as $\rho G(r)$. Note that a cavity plays exactly the same role as a hard-sphere of diameter $2r - \sigma$. When $r = \sigma$, the cavity behaves as a typical molecule and $\rho G(\sigma)$ can be regarded as the number density at a distance σ from a specified particle, that is, we take $G(\sigma) = g(\sigma)$ leading to

$$\beta P/\rho = 1 + 2\pi\rho\sigma^3 G(\sigma)/3 \tag{8.76}$$

Consequently, instead of calculating $g(r)$, we evaluate $G(r)$ at $r = \sigma$.

Now, the probability particle of finding particle in a spherical shell between r and $r + \mathrm{d}r$ is $4\pi r^2\rho G(r)\mathrm{d}r$. Hence, the joint probability of finding an empty sphere of radius r and no particle in a shell between r and $r + \mathrm{d}r$ is given by

$$p_0(r+\mathrm{d}r) = p_0(r)[1 - 4\pi r^2\rho G(r)\mathrm{d}r] \quad \text{or} \quad \frac{\mathrm{d}\ln p_0(r)}{\mathrm{d}r} = -4\pi r^2\rho G(r) \tag{8.77}$$

where in the last step the limit $\mathrm{d}r \to 0$ is taken. For $r < \frac{1}{2}\sigma$, at most one sphere can lie in the cavity of radius r, implying that the probability $1 - p_0(r)$ is $4\pi r^3\rho/3$. The solution of Eq. (8.77) is thus

$$G(r) = [1 - 4\pi r^3\rho/3]^{-1} \quad \text{for} \quad 0 < r < \frac{1}{2}\sigma \tag{8.78}$$

One can show that $G(r)$ and its first derivative are continuous at $r = \frac{1}{2}\sigma$, but that the second derivative has a finite discontinuity due to the possible introduction of a second sphere. At $r = \sigma/\sqrt{3}$, three spheres can be included and a discontinuity occurs in the fourth derivative. With increasing r, more and more spheres can be accommodated and it appears that at the transition from accommodating L spheres to $L + 1$ spheres a discontinuity occurs in the $2L$th derivative of $G(r)$.

Considering next the behavior for large r, we recall first that

$$p_0(r) = \exp[-\beta W(r)] \tag{8.79}$$

where $W(r)$ is the work to create a cavity of radius r, that is, the potential of mean force for this process. From Eqs (8.77) and (8.79) we obtain

$$\mathrm{d}W(r) = 4\pi kTr^2\rho G(r)\mathrm{d}r \tag{8.80}$$

From thermodynamics we learn that

$$\mathrm{d}W(r) = P\mathrm{d}V + \gamma[1-(2\delta/r)]\mathrm{d}A \tag{8.81}$$

with volume increment $\mathrm{d}V = 4\pi r^2\mathrm{d}r$, surface increment $\mathrm{d}V = 8\pi r\mathrm{d}r$, pressure P, surface tension γ and Tolman length δ for a cavity of radius r (see Chapter 15). Combining Eqs (8.80) and (8.81) leads to

$$G(r) = \{P + 2\gamma[(1/r)-(2\delta/r^2)]\}/\rho kT \quad \text{for} \quad r \gg \sigma \tag{8.82}$$

Although $G(r)$ is not an analytic function, it is still continuous because the discontinuities occur in the high-order derivatives. Hence, we use $G(r)$ as an interpolation function for the complete range $\frac{1}{2}\sigma < r < \infty$ and determine the quantities P, γ and δ from Eq. (8.76) and the conditions that $G(r)$ and its first derivative should be continuous at $r = \frac{1}{2}\sigma$. This results, using $\eta = \pi\rho\sigma^3/6$, in

$$\frac{\beta P}{\rho} = \frac{(1+\eta+\eta^2)}{(1-\eta)^3}, \quad \frac{\beta\gamma}{\sigma\rho} = \frac{-3(\eta+\eta^2)}{4(1-\eta)^3} \quad \text{and} \quad \frac{\delta}{\sigma} = \frac{\eta}{4(1-\eta)} \tag{8.83}$$

Note that the expression for P is identical to that derived from the compressibility equation for hard-spheres in the PY approximation. Reiss [25] applied this model, using a temperature-dependent hard-sphere diameter, also to estimate the surface tension (actually the interface tension with a hard smooth wall). In view of the simplicity of the model the results were surprisingly in agreement with experiment (Table 8.4).

Table 8.4 Scaled-particle and experimental values for γ for various compounds.

Compound	T (K)	γ_{exp} (dyne cm^{-1})	γ_{calc} (dyne cm^{-1})
Ne	27.2	4.8	6.09
Ar	85.1	13.2	16.4
N_2	70.1	10.5	14.9
O_2	70.1	18.3	23.6
C_6H_6	273.1	29.0	34.3

The above-described information presents only the basics of this model. It appears possible to generate systematically relations [26] which improve $G(r)$, and other refinements, in particular relaxing the hard-sphere approximation, are also possible. The model has been successfully applied to estimate hydrophobic interactions of solvent molecules in hydrophobic solvents.

References

1 Dasannacharya, B.A. and Rao, K.R. (1965) *Phys. Rev.*, **137**, A417.
2 See Pryde (1966).
3 See Hirschfelder *et al.* (1954).
4 See Barker (1963).
5 Eyring H., and Hirschfelder, J.O. (1937) *J. Chem. Phys.*, **41**, 249.
6 (a)Lennard-Jones, J.E. and Devonshire, A.F. (1937) *Proc. R. Soc.*, **A163**, 53; (b)Lennard-Jones, J.E. and Devonshire, A.F. (1938) *Proc. R. Soc.*, **A165**, 1.
7 See McQuarrie (1973).
8 Wentorf, R.H., Buehler, R.J., Hirschfelder, J.O., and Curtiss, C.F. (1950) *J. Chem. Phys.*, **18**, 1484.
9 Buehler, R.J., Wentorf, R.H., Hirschfelder, J.O., and Curtiss, C.F. (1951) *J. Chem. Phys.*, **19**, 61.
10 Pople, J. (1951) *Philos. Mag.*, **12**, 459.
11 Prigogine, I. and Raulier, S. (1942) *Physica*, **9**, 396.
12 Adams, D.J. and Matheson, A.J. (1972) *J. Chem. Soc. Faraday Trans. II*, **68**.
13 Adams, D.J. and Matheson, A.J. (1973) *Chem. Phys. Lett.*, **22**, 484.
14 Hirschfelder, J.O., Stevenson, D.P., and Eyring, H. (1937) *J. Chem. Phys.*, **5**, 896.
15 Hoover, W.G. and Ree, F.H. (1968) *J. Chem. Phys.*, **49**, 3609.
16 Kirkwood, J.G. (1950) *J. Chem. Phys.*, **18**, 380.
17 Cernuschi, F. and Eyring, H. (1939) *J. Chem. Phys.*, **7**, 547.
18 Rowlinson, J.S. and Curtiss, C.F. (1951) *J. Chem. Phys.*, **19**, 1519.
19 (a) Henderson, D. (1962) *J. Chem. Phys.*, **37**, 631; (b) Henderson, D. (1963) *J. Chem. Phys.*, **39**, 54.
20 Wentorff, R.H., Buehler, R.J., Hirschfelder, J.O., and Curtiss, C.F. (1950) *J. Chem. Phys.*, **18**, 1484.
21 Fuller, E.J., Ree, T., and Eyring, H. (1962) *Proc. Natl Acad. Sci. USA*, **48**, 501.
22 Eyring H., and Jhon, M.S. (1969) *Significant Liquid Structures*, John Wiley & Sons, Ltd, London.
23 (a) Jhon, M.S. and Eyring, H. (1971) The significant structure theory of liquids, in *Physical Chemistry*, Vol. **VIIIA** (ed. D. Henderson), Academic Press, New York, Ch. 5, pp. 335–373; (b) A brief introduction is Eyring, H. and Marchi, R.P. (1963) *J. Chem. Educ.*, **40**, 562.
24 Yoon, B.J., Jhon, M.S., and Eyring, H. (1981) *Proc. Natl Acad. Sci. USA*, **78**, 6588.
25 Reiss, H. (1965) *Adv. Chem. Phys.*, **9**, 1.
26 Stillinger, F.H., Debenedetti, P.G., and Chatterjee, S. (2006) *J. Chem. Phys.*, **125**, 204504–204505.
27 Bondi, A. (1968) *Physical Properties of Molecular Crystals, Liquids and Glasses*, John Wiley & Sons, London.

Further Reading

Barker, J.A. (1963) *Lattice Theories of the Liquid State*, Pergamon, London.

Fowler, R.H. and Guggenheim, E.A. (1939) *Statistical Thermodynamics*, Cambridge University Press, London.

Hirschfelder, J.O., Curtiss, C.F., and Bird, R.B. (1954) *Molecular Theory of Gases and Liquids*, John Wiley & Sons, Inc., New York.

McQuarrie, D.A. (1973) *Statistical Thermodynamics*, Harper and Row, New York.

Pryde, J.A. (1966) *The Liquid State*, Hutchinson University Library, London.

9
Modeling the Structure of Liquids: The Simulation Approach

In Chapters 7 and 8 the integral equation and physical model approaches were introduced. Now, we outline the basics of modern simulation methods that have become feasible because nowadays sufficient computer power is available. After some preliminaries, we deal first with molecular dynamics (MD) and show the results of some early studies. Because this method essentially solves Newton's equations of motion as applied to molecules, the method is capable of calculating the structure, properties and dynamics of liquids. Next, we briefly introduce the Monte Carlo (MC) method, which essentially is a statistical method capable of calculating the static structure and properties. Finally, a typical example combining *ab-initio* quantum-mechanical results with Monte Carlo simulations is discussed.

9.1
Preliminaries

The essence of molecular simulations is to calculate from a (relatively small) number of molecules the properties of the bulk material (of course, assuming that we are not interested in the surface phenomena). The interaction of these molecules is represented by a certain potential, usually a pair-wise additive two-body potential. In reality, a macroscopic sample typically contains 10^{23} molecules, while in a simulation typically only 10^4 to 10^9 molecules can be handled. Even in the latter case several ten thousands of atoms are at the surface, and a still larger number in the immediate neighborhood of the surface, which will affect the behavior considerably. The question is thus: how can we obtain sufficiently accurate results, that is, avoiding the relatively large surface influence, without having to use excessively large numbers of molecules? Before we deal with the two main methods to provide the answers to this question–namely, MD and MC simulations–we must address two general aspects. The first point with how to describe a random structure, such as a liquid, while the second point deals with the interactions used, and their range.

Let us now focus on the first aspect. A crystalline material consists of a lattice in which molecules are positioned (see Chapter 6) and all structural information

Liquid-State Physical Chemistry, First Edition. Gijsbertus de With.

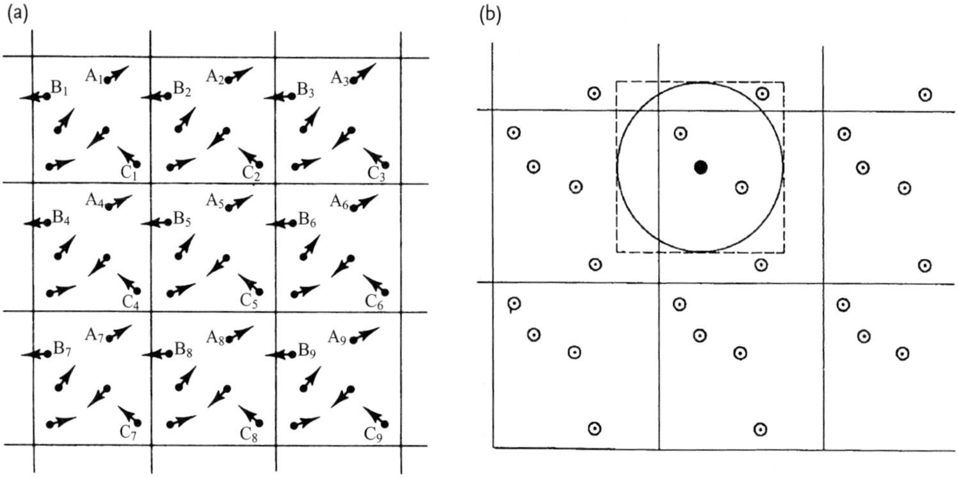

Figure 9.1 (a) Periodic boundary conditions showing identical motion of molecules in each cell (A1, A2, . . .) and leaving the cell of molecules on one side and entering on the opposite side (B1, B2, . . .); (b) Minimum image cut-off (- - -) and spherical cut-off (——), indicating the difference in counting interactions.

is known, given the content of the unit cell. In a fluid such a lattice is absent, but we will use *periodic boundary conditions* (Figure 9.1). Hereto we choose a *representative volume element* or *meso-cell* and replicate this element in the *x*-, *y*-, and *z*-directions so that in fact a "lattice" of meso-cells results. These replicas are usually addressed as *(periodic) images*. The motion of the molecules in each image is identical (e.g., molecules A1, A2, etc.) and if a molecule leaves the meso-cell on one side it enters the image again at the opposite side (e.g., molecule B1, B2, etc.). Obviously, there is no need to use a cubic cell, as in Figure 9.1, and other, more optimal choices can be made, depending on the situation at hand. Although in this way size effects are minimized (but not completely eliminated), all interactions between all molecules would have to be taken into account for the calculation of the total interaction energy, and this still represents a draconian task. Moreover, although periodic boundary conditions avoid the influence of surfaces, it adds periodicity and, therefore, also possibly artifacts.

So let us now turn to the second aspect. To avoid having to calculate all interactions between all molecules, a *cut-off value* r_c for the individual interactions is used. In order for a particular interaction to contribute, the molecules should be within the cut-off distance: contributions of molecules further away are not taken into account. Of course, the cut-off value is arbitrary and is usually taken as the value where the interaction is less than, say, 10^{-4}-fold the maximum interaction. It will be clear that the cut-off value determines the accuracy of the simulation. Typical values for the cut-off are 7 to 15 Å. Methods to correct (approximately) for the interactions outside this range exist; an example is the "tail" correction [1], which is used to calculate the potential energy contributions for $r > r_c$ by assuming that the particles are uniformly distributed. In particular for long-range interactions, special techniques are required; an example for Coulomb interactions is the Ewald summation technique [2] which essentially makes clever use of lattice sums[1)] over the periodic images. The size of the meso-cell is typically chosen as at least double the cut-off value of the intermolecular interactions, so that a specific particle only interacts with the nearest image of the particles, keeping the book-keeping simple. In this way artifacts of periodicity are avoided, although care should be exercised: sudden cut-offs cause additional noise and erroneous behavior, while smooth cut-offs modify the interaction. For the cut-off volume using a cubic cell one can take either a cube (*minimum image cut-off*) or a sphere (*spherical cut-off*), as illustrated in Figure 9.1, leading to a slightly different number of interactions to be taken in to account. In order to avoid spurious non-isotropic interactions, nowadays almost always the spherical cut-off is often taken. The shape of the cell does influence the computational time, however, and choosing an optimal space-filling *molecular-shaped cell* that minimizes the volume while the distances between the atoms of images remain larger than a specified value is an option. Overall, the best strategy for homogeneous systems seems to be to use consistent potentials by including the complete lattice sums over the periodic images, but to combine this with a study of the behavior of the system as a function of cell size [2].

Within this framework one chooses the proper potential for the molecules and uses either the MD or MC method. Initially, the hard-sphere and Lennard-Jones potential were frequently chosen, and results of simulations with Lennard-Jones potentials are generally considered to be qualitatively reliable. Nowadays, more advanced empirical potentials containing many parameters are used. Such a potential is often called a *force field*. Preferably the force field should be transferable between different molecules and applicable to a wide range of conditions and environments. However, the non-additivity of constituent terms and the omission of contributions often renders these non-transferable. Since a force field usually contains adjustable parameters, a compensation of errors may occur and the optimum for one configuration under certain conditions is then not accurate for another condition under different conditions. A particularly clear discussion on force fields is provided by Berendsen [2].

The next two sections briefly introduce the MD and MC methods. The texts by Friedman [3] and Sachs *et al.* [4] also provide a short introduction to both methods, while Allen and Tildesley [5], Frenkel and Smit [6] and Berendsen [2] provide detailed discussions.

9.2 Molecular Dynamics

The core of MD simulations is the solution of Newton's equations of motion (see Chapter 2), as applied to molecules, given the set-up of the previous paragraph.

1) See footnote 4, Chapter 8.

Denoting[2)] for particle i with mass m_i at any time t the force by f_i, the momentum by p_i and the coordinates by r_i, these equations read

$$[f_i = -\nabla_i \Phi(r) \equiv] - \frac{\partial \Phi(r)}{\partial r_i} = m \frac{\mathrm{d}^2 r_i}{\mathrm{d}t^2}\left[\equiv m_i \ddot{r}_i = \frac{\mathrm{d}p_i}{\mathrm{d}t}\right] \tag{9.1}$$

For forces that can be derived from a (time-independent) potential $\Phi(\mathbf{r})$, Newton's laws lead to a conservation of energy

$$\frac{\mathrm{d}E}{\mathrm{d}t} = \frac{\mathrm{d}}{\mathrm{d}t}\left[\sum\nolimits_i \frac{1}{2} m_i \dot{r}_i^2 + \Phi(r)\right] = \sum\nolimits_i \left[m_i \ddot{r}_i \cdot \dot{r} + \frac{\partial \Phi(r)}{\partial r_i} \cdot \dot{r}\right] = 0 \tag{9.2}$$

Correctly solving these equations leads to the trajectories of the various molecules. Since in this approach, representing the micro-canonical (or *NVE*) ensemble, the number of molecules *N*, total volume *V* and the total energy *E* are constant, the conjugate variables show fluctuations. In this representation the temperature is a fluctuating average of the kinetic energy which can be derived at any moment from the equipartition theorem, that is, via

$$3NkT/2 = \sum\nolimits_i \langle p_i \cdot p_i \rangle / 2m_i \tag{9.3}$$

where $\langle \cdots \rangle$ denotes the time average. The time average is taken to yield the same answer as the ensemble average. Although the total energy *E* is in principle conserved, in practice the unavoidable discretization of the time steps needed to solve the equations of motion will cause integrations errors and the total energy *E* will not be exactly conserved. Therefore, *T* may drift even after equilibrium has been reached.

In order to simulate the canonical (or *NVT*) ensemble, the temperature must be controlled, and to do this the system is coupled to a temperature bath or *thermostat*. This is in essence a model of the "outside world", constructed in such a way that the energy exchange is comparable to that of a system coupled to a large environment but still yielding fast answers. Introducing in the same way a *barostat* to regulate pressure, the pressure (or *NPT*) ensemble results. Finally, employing a similar strategy for particles, the grand-canonical (or μVT) ensemble can be simulated. For nonequilibrium simulations the thermostat (and other "stats") should represent the external driving forces which determine that the system responds to changes continuously. Finally, if one wants to make use of the fluctuations to calculate higher-order thermodynamic properties, the nature of the "stats" is also important. For example, the optimal choice of a thermostat will differ for different purposes, for example, to reach equilibrium or to correctly estimate fluctuations.

For practical calculations, Eq. (9.1) is replaced by an equation with finite time steps of length Δt. The algorithm for integrating the MD equations should satisfy a few basic requirements. The first of these is that on a microscopic level, time *reversal symmetry* (as required by Newton's equations) should be conserved. The

2) Recall the notation introduced in Chapter 1: if a vector $\boldsymbol{x}$ is represented by a column matrix $\mathbf{x}$, the inner product $\boldsymbol{x}\cdot\boldsymbol{y}$ is written as $\mathbf{x}^{\mathrm{T}}\mathbf{y}$ and the set $x_1x_2\ldots x_N$ ($\mathbf{x}_1\mathbf{x}_2\ldots\mathbf{x}_N$) is represented as $\boldsymbol{x}$ ($\mathbf{x}$).

second requirement, normally denoted as the *symplectic requirement*, is that upon integration the equation of motion the area and volume in phase space should be conserved. If not obeyed, the equilibrium distributions are also not conserved and may show spurious time dependences. The third requirement is that preferably for each time step only one evaluation of the force f_i is needed, because computationally this is by far the most expensive step. This is due partially to the large number of pairs, in principle $N(N-1)/2$, for which a calculation is required, and partially due to the (possibly) complex form of the potentials used.

There are various ways to satisfy these requirements, but the most commonly used algorithm is due to Verlet [7]. Using a Taylor expansion we write for $\mathbf{r}_i(t \pm \Delta t)$ around t

$$\mathbf{r}_i(t \pm \Delta t) = \mathbf{r}_i(t) \pm \Delta t \dot{\mathbf{r}}_i(t) + \frac{1}{2!}\Delta t^2 \ddot{\mathbf{r}}_i(t) \pm \frac{1}{3!}\Delta t^3 \dddot{\mathbf{r}}_i(t) + \frac{1}{4!}\Delta t^4 \ddddot{\mathbf{r}}_i(t) + \ldots \tag{9.4}$$

so that adding the forward and reverse forms leads to

$$\mathbf{r}_i(t + \Delta t) = 2\mathbf{r}_i(t) - \mathbf{r}_i(t - \Delta t) + \Delta t^2 \ddot{\mathbf{r}}_i(t) + \frac{1}{12}\Delta t^4 \ddddot{\mathbf{r}}_i(t) + \ldots \tag{9.5}$$

If Eq. (9.5) is truncated after the Δt^2 term, the resulting expression is called *Verlet's algorithm* and in essence it estimates the accelerations $\ddot{\mathbf{r}}_i$ as

$$\ddot{\mathbf{r}}_i(t) = [\mathbf{r}_i(t - \Delta t) - 2\mathbf{r}_i(t) + \mathbf{r}_i(t + \Delta t)]/\Delta t^2 \tag{9.6}$$

So, for updating the coordinates the velocities are not explicitly required. Provided that $\Delta t^2 << 12(\ddot{\mathbf{r}}_i / \ddddot{\mathbf{r}}_i)$ or, using the force $f_i = m_i \ddot{\mathbf{r}}_i$ and its second derivative, $\Delta t^2 << 12(f_i / \ddot{f}_i)$, the effect of the Δt^4 term is small as compared to the Δt^2 term. Typically, this requires Δt to be of the order of 10^{-15} s, which imposes a serious practical restraint on the total time that the system can be followed. The constancy of the energy E can be used as a check of whether Δt is small enough, but this requires the velocities to be known. The latter we easily obtain from the Verlet algorithm as

$$\dot{\mathbf{r}}_i(t) = [\mathbf{r}_i(t + \Delta t) - \mathbf{r}_i(t - \Delta t)]/2\Delta t + \mathcal{O}(\Delta t^2) \tag{9.7}$$

An estimate accurate [8] to $\mathcal{O}(\Delta t^3)$ is obtained by adding $[f_i(t - \Delta t) - f_i(t + \Delta t)]\Delta t/12m_i$ to Eq. (9.7). Using the Verlet algorithm the system propagates along a trajectory in space.

To start a simulation one needs an initial set of coordinates that can be obtained in various ways. A logical first choice would be to be minimize the potential energy, that is, a $T = 0$ K calculation. The second option is to use the final configuration of a simulation with nearby conditions. A third option is to use a lattice on which the N particles are located at $t = 0$. When using Verlet's algorithm, one clearly needs a configuration at $t = -\Delta t$. This configuration may be constructed by giving molecules a Maxwell–Boltzmann distribution for some temperature T at $t = 0$, that is,

$$f(p) = (2\pi mkT)^{-3/2} \exp[-(p \cdot p)/2mkT] \tag{9.8}$$

From this distribution we pick at random[3] values for p_i/m_i for $i = 1, 2, \ldots, N$ and use

$$r_i(-\Delta t) = r_i(0) - \Delta t(p_i / m_i) \tag{9.9}$$

which represents the distance a molecule would have traveled with a fixed velocity from $t = -\Delta t$ to $t = 0$.

After starting the calculation we compute the trajectory from one state at $t = t$ to the next at $t = t + \Delta t$. The kinetic temperature, as defined by Eq. (9.3), normally rises significantly after any sort of arbitrary initial configuration. This is due to the fact the particle has to find its way to a lower state of potential energy under the influence of the intermolecular interactions. After a sufficient number of time steps, the average temperature is reached. In case of the *NVT* ensemble, one way to control the temperature is by scaling the velocities with a factor x, typically according to

$$r_i'(t - \Delta t) = r_i(t) - x[r_i(t) - r_i(t - \Delta t)] \tag{9.10}$$

and thereafter calculate $r_i(t + \Delta t)$ from $r_i(t)$ and $r_i'(t - \Delta t)$ for $i = 1,2, \ldots, N$ leading to a change of temperature from T to $T' = x^2T$. More sophisticated thermostats are the Andersen and Nosé–Hoover thermostats. In the *Andersen thermostat*, the coupling to a heat bath is represented by stochastic impulsive forces that act by now and then on randomly selected particles. The strength of the coupling should be chosen before the simulation starts. It can be shown that this thermostat indeed generates a canonical ensemble, but it is not very useful for nonequilibrium simulations as the sudden impulse change leads to nonrealistic fluctuations. The *Nosé–Hoover thermostat* employs an extended Lagrange function that can be used in nonequilibrium simulations, but only in the case the system has only <u>one</u> constant of the motion. In nonequilibrium simulations this is the case, since in that case the total momentum is not conserved (for the system we consider) because of the externally applied force. In an equilibrium simulation this condition is fulfilled if the center of mass of the system is fixed. This restrictive condition of only one constant of the motion may be alleviated by using a sequence of Nosé–Hoover thermostats. For details, we refer to the literature[4].

Generally it will take many time steps, perhaps 10^3 to 10^4, before a situation is reached that represents the equilibrium fluctuations of the system. Logging all data provides the possibility of plotting most quantities of interest as a function of time. Often – and in particular when the time step is small enough – the configuration changes little per time step, and one uses only data at every nth step, where n may be 10 or more. Calculating the property for a number of consecutive time segments we may estimate the statistical variation. This process is comparable to an experimental repetition to obtain an estimate of the experimental errors. Advanced methods exist to eliminate time correlation[4], although systematic errors cannot be assessed so easily. For example, the use of a finite size N-particle

3) Proper random generators are nowadays standard available for any computer.
4) See, for example, Frenkel and Smit [6].

system automatically confirms that correlations for a length larger than the cell are suppressed, even when the energy expression implies such a correlation. Moreover, there is the question of how efficiently a phase space can be sampled. For sufficiently short-ranged potentials, correlation and sampling are not serious problems, but for systems where long-range correlations play an important role – for example, in crystallization or for systems with relatively deep, metastable potential wells, perhaps larger than 5 kT or 10 kT, at a density near the close-packed value – they are important.

For any property f the time average over the n steps is constructed according to

$$\langle f \rangle = n^{-1} \sum_{i=1}^{i=n} f_i \tag{9.11}$$

where f_i is the value for f at step i. Taking the potential energy as an example we have

$$\Phi_i = N^{-1} \left[\sum_{k=1}^{k-N-1} \sum_{l=k+1}^{l=N} \phi(r_{kl}) \right]_i \tag{9.12}$$

For the temperature we obtain $T_i = K_i/(3Nk/2) = [\Sigma_j(p_j^2/2m_j)]_i/(3Nk/2)$. Using the fluctuations in energy $E = \langle \mathcal{H} \rangle = \langle K + \Phi \rangle$ (see Section 5.1), the heat capacity C_V results from

$$\begin{aligned} kT^2C_V &= \langle \mathcal{H}^2 \rangle - \langle \mathcal{H} \rangle^2 = \langle K^2 \rangle - \langle K \rangle^2 + \langle \Phi^2 \rangle - \langle \Phi \rangle^2 \\ &= kT^2 C_{V,\text{kin}} + \langle \Phi^2 \rangle - \langle \Phi \rangle^2 = 3Nk^2T^2/2 + \langle \Phi^2 \rangle - \langle \Phi \rangle^2 \end{aligned} \tag{9.13}$$

Another way to calculate C_V is to use the temperature coefficient of the average internal energy E_{ave}, that is, potential plus kinetic energy, as obtained from simulations at various temperatures. This is usually a more satisfactory method. The pressure can be calculated from the virial theorem expression (see Chapter 3) reading

$$PV = NkT - \left\langle \sum_j \mathbf{r}_j \cdot \nabla_j \Phi(\mathbf{r}_1, \ldots, \mathbf{r}_N) \right\rangle \Big/ 3 \tag{9.14}$$

with ∇_j the gradient operator for molecule j. Last but not least, if we use for f_i the number of particles $n(r)$ located at a distance between r and $r + \Delta r$ from a reference particle, and taking subsequently each particle as reference, the average value $\langle n(r) \rangle$ is directly related to the pair correlation function according to

$$\langle n(r) \rangle = 4\pi\rho r^2 g(r) \Delta r \quad \text{with} \quad \rho = N/V \tag{9.15}$$

provided that Δr is small enough so that the change in $g(r)$ in this range is small.

Although the intention is to eliminate the surfaces, the sensitivity of the averages on the periodic boundary conditions depends inversely on the number of particles N, while the computational effort is of $\mathcal{O}(N)$ for simulations with a short-range potential using a cut-off. For long-range interactions, using clever summation techniques, the efforts scales as $\mathcal{O}(N \ln N)$. However, for a larger N value it takes fewer time steps n to reach a given accuracy [see Eqs (9.11) and (9.12)]. If all particles contribute directly to the average of interest, the number of data points varies

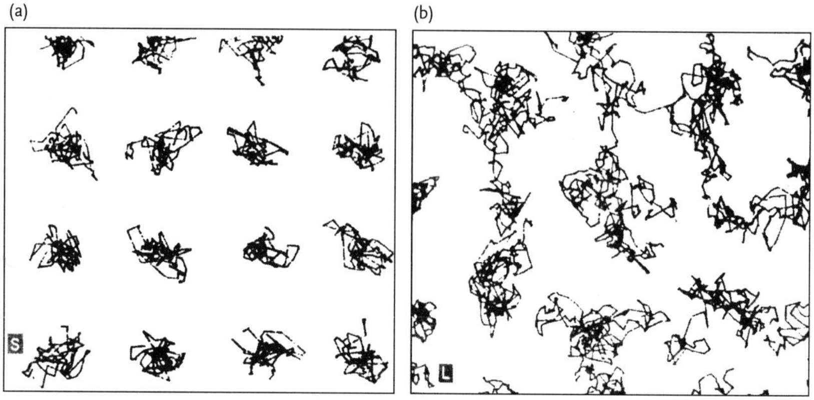

Figure 9.2 Trajectories for 3000 collisions in (a) the solid state and (b) the liquid state of a 32-particle hard-sphere system at $V/V_0 = 1.525$ projected on a plane.

as Nn, so that the cost may be more nearly proportional to N, subject to computer memory limitations. The method outlined here may be applied not only to spherical molecules such as Ar, but also to systems of more complex molecules. One can then either use the rigid molecule approximation and describe the system by "generalized coordinates" such as orientational coordinates and their associated momenta, or use flexible molecules with appropriate potentials for bond stretching, bond bending, and dihedral torsion. For details, we refer to the literature.

To illustrate this method, snapshots of the pioneering simulations by Alder and Wainwright [9] are shown in Figure 9.2. Here, Figure 9.2a shows the projected trajectories of a hard-sphere system in the solid state, and demonstrates that the atoms, on average, are largely fixed at their equilibrium position. Figure 9.2b shows the trajectories in the liquid state for those atoms which move throughout the complete volume of the cell.

As a final example, we compare in Figure 9.3 the pair-correlation function of Ar, as calculated by the MD method using the LJ potential with experimental results (note the very good agreement). Nowadays, much more elaborate simulations can be carried out, for which a number of software packages are currently available.

Problem 9.1

Verify Eq. (9.7) and its extension.

Problem 9.2

What are the advantages and disadvantages of more sophisticated updating schemes for $\boldsymbol{r}_i$?

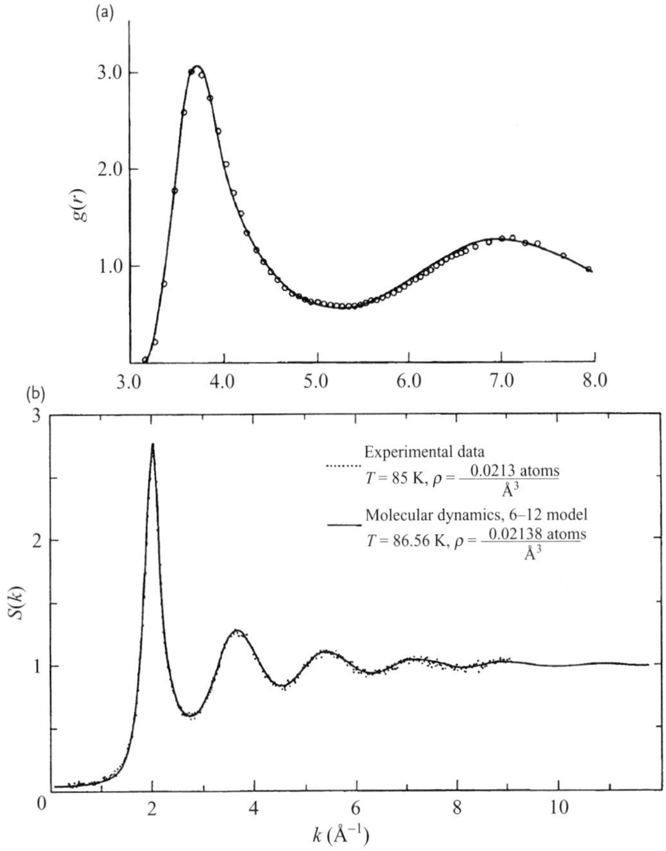

Figure 9.3 (a) The pair correlation function for liquid Ar at $T = 86.56$ K and $\rho = 0.02138$ Å^{-3}, near its triple point. The solid line represents the results of a MD simulation for a 6–12 potential ($\sigma = 3.405$ Å, $\varepsilon/k = 119.8$ K) chosen to fit the thermodynamic data for Ar. The MD result is indistinguishable from MC results, including three-body interactions. The circles represent the result of Fourier inversion of the $S(s)$ curve in panel (b); (b) Neutron diffraction data $S(s)$ for liquid Ar compared with model calculations with data as given for panel (a).

9.3 The Monte Carlo Method

Whereas, in MD the structure and dynamics are calculated, in MC calculations attention is focused on the structure alone.

The method consists essentially of distributing molecules in space, calculating the associated energy, modifying the molecular distribution, and then repeating the calculation so as to obtain eventually a sequence of, say n, molecular distributions in configuration space that samples the desired ensemble (NVE, NVT, or otherwise) properly. If f represents any observable dependent only on the configuration, then each element of the sequence will generate a sample f_j of the observable and the mean $f_m = n^{-1}\Sigma_j f_j$ is an estimator of the ensemble average $\langle f \rangle$. The essence of MC simulations is to construct the sequence in a smart fashion. The procedure is best explained by using ideas from calculating multidimensional integrals in numerical mathematics.

Suppose one has the integral

$$G = \int_a^b g(x)\mathrm{d}x \tag{9.16}$$

To evaluate this integral, the following trick is used. One writes

$$G = \int_a^b \frac{g(x)}{f(x)} f(x)\mathrm{d}x \equiv \int_a^b h(x)f(x)\mathrm{d}x \tag{9.17}$$

where $f(x) > 0$ over the interval $[a,b]$ and $f(x)$ is integrable. We treat $f(x)$ as a probability density function so that $F = \int_a^b f(x)\mathrm{d}x = 1$. Then we can write

$$\langle h(x) \rangle = G/F \tag{9.18}$$

where $<h(x)>$ is the average value over the interval $[a,b]$ assuming that x is distributed as $f(x)$. Hence, it follows that if we choose points in the interval $[a,b]$ at random[5] with probability $f(x)$ we can evaluate $<h(x)>$ and by multiplying with the normalization factor F, calculate G.

In statistical thermodynamics we may wish to evaluate integrals such as

$$\langle G \rangle = Q^{-1}N!^{-1}\int \cdots \int G(\mathbf{r}_1, \ldots, \mathbf{r}_N)\exp[-\Phi(\mathbf{r}_1, \ldots, \mathbf{r}_N)/kT]\mathrm{d}r_1 \ldots \mathrm{d}\mathbf{r}_N \tag{9.19}$$

where the configurational partition function Q acts as a normalizing factor. Consequently, when we choose position vectors according to the distribution function

$$f(\mathbf{r}_1, \ldots, \mathbf{r}_N) = \exp[-\Phi(\mathbf{r}_1, \ldots, \mathbf{r}_N)/kT] \tag{9.20}$$

we can evaluate integrals of the type of Eq. (9.19). Unfortunately, Q, when acting as normalization factor, generally cannot be evaluated, but it is still possible to determine thermodynamic properties and structure.

The most frequently used method is that of Metropolis *et al.* [10], based on Markov processes. In this method one starts with an initial configuration of molecules and changes the configuration over a long sequence. The transition probabilities T_{ij} of the transition from state[6] i to state j are taken in such a way that in the infinite limit they tend to be distributed according to $\exp[-\Phi(\mathbf{r}_1, \ldots, \mathbf{r}_N)/kT]$. Note that the total probability to end up in any state starting in state i is $\Sigma_j T_{ij}$, so

5) Whilst for a small number of dimensions, a systematic sampling of space is more efficient, for a high number of dimensions random sampling is more efficient.

6) Whereas, in the MD method, a state means specified locations $\mathbf{r}_1, \mathbf{r}_2, \ldots, \mathbf{r}_N$ and specified momenta $\mathbf{p}_1, \mathbf{p}_2, \ldots, \mathbf{p}_N$ of the N particles, in the MC method a state means specified locations $\mathbf{r}_1, \mathbf{r}_2, \ldots, \mathbf{r}_N$ only.

that we should have $\Sigma_j T_{ij} = 1$. If $T_{ij}^{(m)}$ is the probability to go from state i to state j in m steps, we have $T_{ij}^{(m)} = \Sigma_k T_{ik}^{(m-1)} T_{kj}$. It can be shown for Markov processes that, provided all states are accessible from any other state in a finite number of steps, the limit of $T_{ij}^{(m)}$ for $m \to \infty$ becomes P_j, independent of i. This implies that, after a sufficient number of steps, the state of the system is independent of the initial configuration. In the limit we have

$$P_i > 0, \quad \Sigma_i P_i = 1 \quad \text{and} \quad P_i = \Sigma_j T_{ji} P_j \tag{9.21}$$

stating that the probabilities are positive, that they are normalized, and that each state eventually is reached from any other state. We choose

$$P_j \sim \exp(-\beta\Phi_j) \quad \text{and} \quad T_{ij} / T_{ji} = \exp[-\beta(\Phi_j - \Phi_i)] \tag{9.22}$$

From Eq. (9.22) it follows that we have the detailed balance condition

$$P_i T_{ij} = P_j T_{ji} \tag{9.23}$$

If we substitute this result in Eqs (9.21)–(9.23), we obtain the $P_i = \Sigma_j T_{ji} P_j = \Sigma_j T_{ij} P_i = P_i \Sigma_j T_{ij}$ or $\Sigma_j T_{ij} = 1$, satisfying our earlier requirement. Hence, by choosing the relationship between T_{ij} and T_{ji} according to Eq. (9.22), all requirements for a Markov process are fulfilled. What remains to be done is to choose the appropriate values for T_{ij}.

Metropolis *et al.* used for T_{ij} the following recipe

$$\begin{aligned} &T_{ij} = A_{ij} \quad \text{for} \quad \Phi_i > \Phi_j \ (i \neq j), \\ &T_{ij} = A_{ij} \exp[-\beta(\Phi_j - \Phi_i)] \quad \text{for} \quad \Phi_i < \Phi_j \ (i \neq j) \quad \text{and} \quad T_{ii} = 1 - \sum\nolimits_{j \neq i} T_{ij} \end{aligned} \tag{9.24}$$

The constants A_{ij} are chosen as $A_{ij} = A_{ji}$, and determined as follows:

- A molecule is chosen at random.
- A change in position (δx, δy, δz) is selected.
- $A_{ij} = 0$ if any of $|\delta x|$, $|\delta y|$ and $|\delta z| > \Delta$, and $A_{ij} = (2\Delta)^{-3}$ otherwise.

With these rules a large number of configurations is generated with the constant Δ chosen in such a way that about half of the configurations are accepted.

Calculations are initiated by providing the positions of N molecules in the cell chosen; this starting configuration can be selected as indicated in Section 9.1. Normally the first – say 200 000 – configurations are discarded in order to limit the influence of the starting configuration, and thereafter the properties are evaluated using, say 500 000, configurations. The above-outlined procedure is for the canonical ensemble, and enables one to calculate the ensemble average by direct sampling of phase space:

$$\langle G \rangle = \sum\nolimits_j G_j(\mathbf{r}_1, \ldots, \mathbf{r}_N) \tag{9.25}$$

where $G_j(\mathbf{r}_1, \ldots, \mathbf{r}_N)$ represents the value of G in configuration j. For example, the internal energy U is given by

$$U = U_{\text{kin}} + Q^{-1}N!^{-1}\int \cdots \int \Phi(\mathbf{r}_1, \ldots, \mathbf{r}_N)\exp[-\beta\Phi(\mathbf{r}_1, \ldots, \mathbf{r}_N)]\,\mathrm{d}\mathbf{r}_1 \ldots \mathrm{d}\mathbf{r}_N \tag{9.26}$$

where U_{kin} is the kinetic energy, given for spherical molecules by $U_{\text{kin}} = 3NkT/2$ and is due to the term Λ^{-3N}. This can also be written as

$$U = U_{\text{kin}} + \langle \Phi(\mathbf{r}_1, \ldots, \mathbf{r}_N)\rangle = 3NkT/2 + \langle \Phi(\mathbf{r}_1, \ldots, \mathbf{r}_N)\rangle \tag{9.27}$$

so that the internal energy can be estimated by averaging the potential energy of all configurations in the MC sequence. The heat capacity is then

$$C_V = \frac{\partial U}{\partial T} = C_{V,\text{kin}} + \frac{1}{kT^2}\left(\langle \Phi^2\rangle - \langle \Phi\rangle^2\right) \tag{9.28}$$

with $C_{V,\text{kin}} = 3Nk/2$ the contribution of the kinetic energy to C_V. The pressure can be calculated from the virial theorem expression (see Chapter 3) reading

$$\frac{PV}{NkT} = 1 - \frac{1}{3NkT}\left\langle \sum\nolimits_i \mathbf{r}_i \cdot \nabla_i \Phi(\mathbf{r}_1, \ldots, \mathbf{r}_N)\right\rangle \tag{9.29}$$

with ∇_i the gradient operator for molecule i. Finally, using the number of particles $n(r)$ with distance between r and $r + \Delta r$ from a reference particle and taking subsequently each particle as reference, from the expression

$$\langle n(r)\rangle = 4\pi\rho r^2 g(r)\Delta r \tag{9.30}$$

one can obtain the correlation function $g(r)$.

Today, the MC method is used much like the MD method, more or less routinely, to calculate the properties of thermodynamic systems. Both, the MC and MD methods have the property of providing, in principle, exact calculations for a given interaction model, and the typical accuracy is about 1–2%. Provided that the same potentials are used and also the same conditions, the equilibrium results of the MD method agree to a large degree with those of the MC method (see Figure 9.3). Whereas, the MD method yields the full dynamics, it is somewhat more complex to eliminate time correlations when compared to the MC method.

Problem 9.3*

Is it possible to calculate Q_N in principle from a MC simulation?

9.4
An Example: Ammonia

In this section we provide, as an example of the results that can be obtained using simulation methods, a means of determining the structure of liquid NH_3. This is achieved by discussing the results of combined quantum-mechanical calculations, pair potential fitting and MC simulations, as described by

Table 9.1 Dipole moments for NH_3 calculated by various approximate Hartree–Fock (HF) calculations.

Method	Value (D)	Method	Value (D)
HF-TZP	1.87	DZP	2.35
6-31G	2.23	DZ2P	1.97
6-31G*	1.95	STO3G[a)]	1.78
6-31G**	1.89	Experimental value	1.47

a) But with wrong configuration of the dimer.
TZP = triple zeta; DZP = double zeta; STO-xG = Slater-type orbital expanded in x Gaussians.

Hannongbuo [11]. In this investigation the interaction between NH_3 molecules was obtained using HF-TZP calculations[7)] yielding a total energy for the molecule of −56.21603 Hartree and a dipole moment of 1.87 D. The latter value must be compared with other theoretical (quantum) calculations (Table 9.1), from which a dimer stabilization energy $E_{stab} = -2.53\,\text{kcal mol}^{-1}$ was obtained with an optimum N–N distance of 3.4 Å and an optimum configuration $\alpha = \beta = 12°$, where the angles are indicated in Figure 9.4. The interaction energy between two NH_3 molecules was calculated for more than 1500 configurations, leading to a detailed energy landscape. For ease of further calculation, this energy landscape for the NH_3–NH_3 interaction was fitted with pair potentials to these configurations. This resulted in

$$\Delta E_{NN}(r) = 14.85/r + 55719.0/r^{12} - 13.60/r^6$$

$$\Delta E_{HH}(r) = 1.65/r + 48.64\exp(-3.7r)$$

$$\Delta E_{NH}(r) = 4.95/r + 0.00140\{\exp[-4.6(r - 2.4)] - 2.0\exp[-2.3(r - 2.4)]\}$$

where the units are ΔE [kcal mol^{-1}] and r [Å].

MC calculations were performed using 202 rigid NH_3 molecules with the experimental structure as described by NH = 1.0124 Å and H–N–H = 106.67°. The simulations were made at an experimental density $\rho = 0.633\,\text{g cm}^{-3}$ at 277 K, which renders a box length of 20.99 Å.

These MC simulations were first made using *ab initio* HF-TZP calculations for each configuration. Although a logical choice for the cut-off value would be 10.45 Å, a value of 7.0 Å was chosen because the interaction almost disappears after 5–6 Å; the computing time was also reduced. Within a shell of 7 Å this implies a

7) These abbreviations and those used later in the comparison for the dipole moment indicate the type of basis set used in the molecular orbital calculations: HF = Hartree–Fock; TZP = triple zeta, indicating that for each occupied valence orbital in the atom three basis functions are used, plus polarization functions, indicating that nonoccupied valence orbitals in the atom are included, DZP = double zeta, similar to TZP but using two basis functions, STO-xG = Slater-type orbital expanded in x Gaussian functions. Essentially, this label indicates the quality of the basis set used.

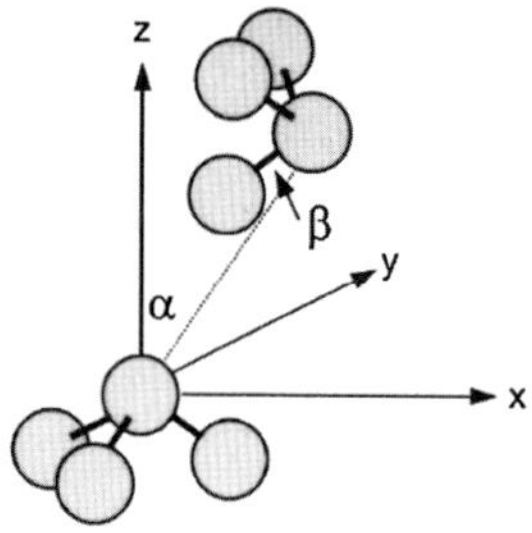

Figure 9.4 The general orientation for two NH_3 molecules with optimal configuration $\alpha = \beta = 12°$.

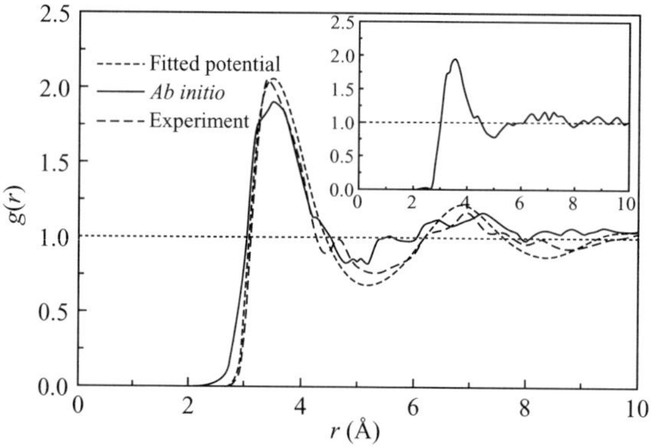

Figure 9.5 The N–N pair correlation for NH_3 as calculated from 50 000 configurations. The inset shows the results for 25 000 configurations.

calculation for about 30 molecules. The MC simulations were also made with the fitted pair potential, but in this case an expected cut-off value of 10.45 Å was used.

The N–N pair correlation function $g(r)$ was calculated using 50 000 configurations for both (*ab initio* and pair potential fitted) energy landscapes, and compared with the experimental data of Narten [12] at 277 K and 1 atm. These results are shown in Figure 9.5, where the inset shows also the results for 25 000 configurations. The shape is typical for $g(r)$. The first peak (~3.4 Å) corresponds to the nearest-neighbor N–N distance. For the *ab initio* simulations a broad shoulder between 4.0 and 4.8 Å is observed, the absence of which for the pair potential simulations is attributed to a systematic error due to using simple analytical functions to represent the complicated energy surface. In particular, it was noted that for the *ab initio* simulations multiple minima were obtained with slightly different stabilization energies. The three most stable energies were for $\alpha = 0°$ with $E_{stab} = -2.40\,\text{kcal mol}^{-1}$, $\alpha = 10°$ with $E_{stab} = -2.53\,\text{kcal mol}^{-1}$, and $\alpha = 20°$ with $E_{stab} = -2.25\,\text{kcal mol}^{-1}$ for the optimum N–N distance of 2.54 Å. For comparison,

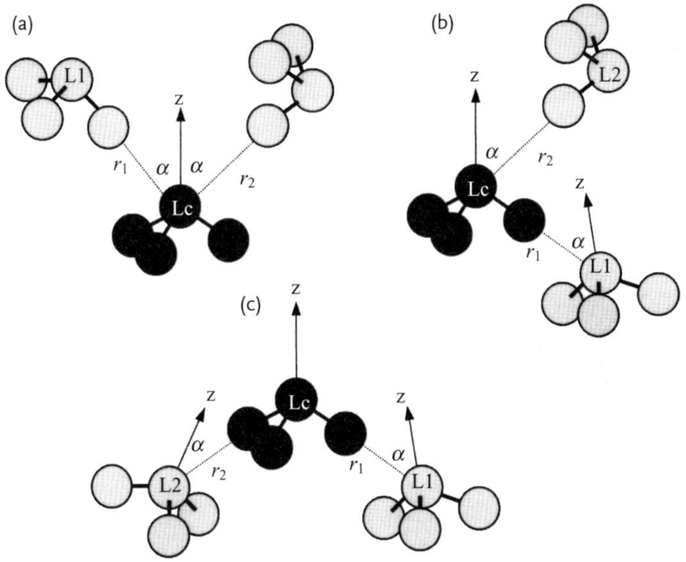

Figure 9.6 (a) Donor–donor, (b) donor–acceptor, and (c) acceptor–acceptor configurations for three NH_3 molecules.

recall that $RT \cong 0.6\,\text{kcal mol}^{-1}$. The pair potential fitting is unable to represent these minima.

The onset of the *ab initio* N–N $g(r)$ at 2.3 Å, which starts earlier than the 2.7 Å X-ray value for the experimental N–N $g(r)$, indicates less repulsion at short distances for the pair interaction derived from *ab initio* calculations during the simulation than in reality, where many-body effects are present. Another source of error which leads to discrepancy between the experimental and simulated results is the use of a fitted, nonpolarizable potential in the liquid phase simulations. It has been reported that variation of the pair potential, due to the addition of higher multipoles and of the polarization effects, influences the liquid structure by changing the dimer configurations corresponding to the local maxima of the correlation functions.

In order to investigate three-body interactions, calculations were made for the configurations as shown in Figure 9.6. The three-body contribution appeared to be important, and three configurations could be discerned within the range $r_1 \leq 2$ Å and $r_2 \geq 10$ Å. The configurations shown in Figure 9.6 can be described as donor–donor, donor–acceptor, and acceptor–acceptor, respectively. The donor–donor configuration appeared to be repulsive, while the donor–acceptor configuration was attractive, and the acceptor–acceptor configuration was strongly repulsive. At a N–N distance of 3.4 Å, the energy values for the donor–donor, acceptor–acceptor, and donor–acceptor configurations of 0.13, 0.65 and $-0.22\,\text{kcal mol}^{-1}$, respectively, led to a net effect for the E_{3bd} value of $0.36\,\text{kcal mol}^{-1}$, which equates to 14% of the global minimum of the pair interaction of $2.53\,\text{kcal mol}^{-1}$. This influence is more

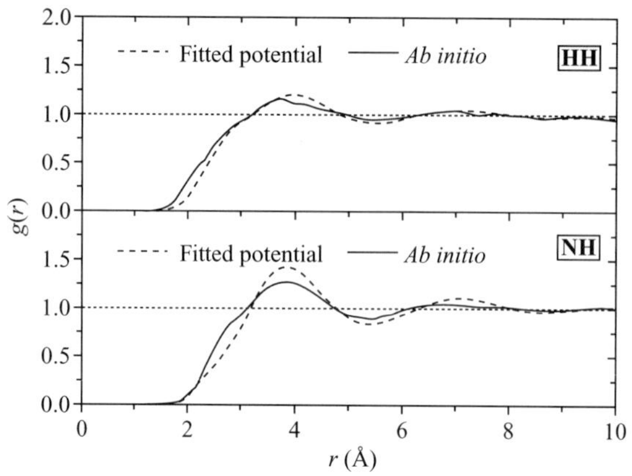

Figure 9.7 The H–H and N–H correlation functions for NH_3.

pronounced for shorter distances because the three-body energy for the acceptor–acceptor increases much faster than for the other two terms.

The partial pair correlation function for H–H was rather featureless (Figure 9.7), whereas the N–H correlation showed a main peak at 3.751 Å and a wide hump from 2.3 to 3.11 Å, indicating very weak hydrogen bonding in liquid ammonia. The calculated H coordination number was 0.97, compared to 1.2 from previous simulation results and 1.2 from experimental data. It should be noted that the hydrogen bond distance in solid ammonia is 2.35 Å, with an average of three hydrogen bonds per nitrogen atom.

Overall, the remaining discrepancies between simulations and experimental can be attributed to statistical error, due mainly to the use of simple analytical functions to represent the complicated energy surface, together with many-body and the polarization effects.

Problem 9.4

Estimate the value for the average angle β from the N–N and N–H correlations functions given in Figures 9.5 and 9.7.

References

1 Wood, W.W. and Parker, F.R. (1957) *J. Chem. Phys.*, **27**, 720.
2 See Berenden (2007).
3 See Friedman (1985).
4 See Sachs *et al.* (2006).
5 See Allen (1987).
6 See Frenkel and Smit (1987).
7 Verlet, L. (1967) *Phys. Rev.*, **A159**, 98.
8 Berendsen, H.J.C., and van Gunsteren, W.F. (1986) in *Molecular Dynamics*

Simulation of Statistical Mechanical Systems (eds G. Ciccotti and W. Hoover), North Holland, Amsterdam, p. 43.
9 Wainwright, T.E. and Alder, B.J. (1958) *Nuovo Cimento*, **9** (Suppl. 1), 116.
10 Metropolis, N., Rosenbluth, A.W., Rosenbluth, M.N., Teller, E., and Teller, A.H. (1953) *J. Chem. Phys.*, **21**, 1087.
11 Hannongbuo, S. (2000) *J. Chem. Phys.*, **113**, 4707.
12 Narten, A.H. (1977) *J. Chem. Phys.*, **66**, 3117.

Further Reading

Allen, M.P. and Tildesley, D.J. (1987) *Computer Simulation of Liquids*, Clarendon, Oxford.

Berendsen, H.J.C. (2007) *Simulating the Physical World*, Cambridge University Press, Cambridge.

Frenkel, D. and Smit, B. (2002) *Understanding Molecular Simulation: From Algorithms to Applications*, 2nd edn, Academic Press, London.

Friedman, H.L. (1985) *A Course on Statistical Mechanics*, Prentice-Hall, Englewood Cliffs, NJ.

Landau, D.P. and Binder, K. (2000) *Monte Carlo Simulations in Statistical Physics*, Cambridge University Press, Cambridge.

Leach, A.R. (2001) *Molecular Modelling, Principles and Applications*, 2nd edn, Pearson Prentice-Hall, Harlow, UK.

Sachs, I., Sen, S., and Sexton, J.C. (2006) *Statistical Mechanics*, Cambridge University Press, Cambridge.

10
Describing the Behavior of Liquids: Polar Liquids

In the previous chapters we have discussed various models to describe liquids, but we have limited ourselves so far largely to nonpolar liquids. In this chapter we extend our discussion to polar liquids. First, we discuss macroscopic (linear) dielectric behavior. Second, the dielectric behavior of polar fluids is discussed in microscopic terms. Because H_2O is the most important example of a polar liquid and behaves rather anomalously, we deal thereafter with the structure of liquid water and the relation to its properties.

10.1
Basic Aspects

Let us first briefly describe macroscopic, (linear) dielectric behavior (see also Appendix D). In macroscopic terms, if we place in vacuum on a parallel plate capacitor[1)] with electrode area A and electrode distance d an electrical voltage difference ϕ on the electrodes, a charge

$$q_0 = \int_0^\infty I(t)\,\mathrm{d}t \tag{10.1}$$

will develop, where $I(t)$ is the current during the period of charging. The capacitance

$$C_0 = \varepsilon_0 A/d \tag{10.2}$$

is the proportionality constant between the voltage applied and the charge developed, that is, $q_0 = C_0\phi$. The surface charge σ_0 is defined by

$$\sigma_0 = q_0/A \quad \text{and thus} \quad \sigma_0 = \varepsilon_0\phi/d = \varepsilon_0 E \tag{10.3}$$

where $E = \phi/d$ is the electric field strength and ε_0 is the permittivity of the vacuum, which is a universal constant. The response to the applied electric field E, the surface charge σ_0, can for isotropic media be interpreted as the dipole moment per unit volume since its dimension $C\,m^{-2}$ can be written as $(C\,m)m^{-3}$. In that case

1) We assume that edge effects can be neglected. We neglect also charge leakage through the capacitor.

Liquid-State Physical Chemistry, First Edition. Gijsbertus de With.

the response is denoted as the (di)electric displacement D, and so in vacuum we have $D = \varepsilon_0 E$.

If we put a dielectric material in the capacitor the charge changes to $q = \varepsilon_r q_0$, with ε_r the relative permittivity and thus the capacitance becomes $C = \varepsilon_r\varepsilon_0 A/d = \varepsilon A/d$ with $\varepsilon \equiv \varepsilon_r\varepsilon_0$ the permittivity. The relative permittivity ε_r is a material-specific function, dependent on temperature and pressure, and possibly also on the electric field. We can also say that the surface charge changes from $\sigma_0 = q_0/A$ to

$$\sigma = q/A = \sigma_0 + \sigma_{\text{pol}} \tag{10.4}$$

where σ_{pol} is the extra surface charge at the electrodes due to the polarization of the material put in the capacitor. In terms of the field E, the displacement D becomes

$$D = \varepsilon E = \varepsilon_r\varepsilon_0 E = \varepsilon_0 E + P_E \tag{10.5}$$

where P_E is the *polarization*.[2] Solving for P_E we obtain

$$P_E = (\varepsilon_r - 1)\varepsilon_0 E \equiv \chi\varepsilon_0 E \tag{10.6}$$

with χ the (electric) *susceptibility*, for which it always holds that $\chi \geq 0$.

The variables E and D are conjugated variables. To show this we calculate the work δW required to transfer an element of charge $\mathrm{d}q$ from the negative to the positive electrode. Because $E = \phi/d$ and $q = A\varepsilon E$, we obtain $\phi = Ed$ and $\mathrm{d}q = A\mathrm{d}(\varepsilon E)$. Therefore, the work $\delta W = \phi\,\mathrm{d}q = V_{\text{cap}}E\mathrm{d}D$ with displacement $D = \varepsilon E$ and volume of the capacitor $V_{\text{cap}} = Ad$. The first law (for a one-component system at temperature T and pressure P) thus becomes

$$\mathrm{d}U = T\mathrm{d}S - P\mathrm{d}V + \mu\mathrm{d}n + V_{\text{cap}}E\mathrm{d}D \tag{10.7}$$

Applying a Legendre transform using $G = U - TS + PV - V_{\text{cap}}ED$, we obtain

$$\mathrm{d}G = -S\mathrm{d}T + V\mathrm{d}P + \mu\mathrm{d}n - V_{\text{cap}}D\mathrm{d}E \tag{10.8}$$

For normal dielectrics at low field, the permittivity is not a function of the field E, and the integration of Eq. (10.8) results in $G = G^{(0)} - ½\varepsilon E^2 V_{\text{cap}}$, where $G^{(0)}$ represents the value of G at zero field at given temperature and pressure. From this expression, results for the entropy S, the volume V, and the energy U follow in the usual manner.

Problem 10.1

Show that the entropy S, volume V, and chemical potential energy μ for a normal dielectric, using temperature T, pressure P and number of moles n, are given by, respectively,

$$S = S^{(0)} + \frac{1}{2}\frac{\partial\varepsilon}{\partial T}E^2V_{\text{cap}}, \quad V = V^{(0)} - \frac{1}{2}\frac{\partial\varepsilon}{\partial P}E^2V_{\text{cap}}, \quad \mu = \mu^{(0)} - \frac{1}{2}\frac{\partial\varepsilon}{\partial n}E^2V_{\text{cap}}$$

2) We denote the polarization here by P_E to avoid confusion with the pressure P. In the remainder of this chapter we just use P for the polarization.

Problem 10.2

Show that the energy U for a normal dielectric, using temperature T and pressure P, is given by

$$U = U^{(0)} + \frac{1}{2}\left[\varepsilon + T\frac{\partial \varepsilon}{\partial T} + P\frac{\partial \varepsilon}{\partial P}\right]E^2 V_{\text{cap}}$$

10.2
Towards a Microscopic Interpretation

So, an applied field E introduces into a material a polarization $P = \varepsilon_0(\varepsilon_r - 1)E$., and our task is now to interpret the polarization P in microscopic terms. In general, we have several molecular contributions to the polarization. First, if the molecules are polar and have a dipole moment[3] μ, we have the alignment of these permanent dipoles due to the applied field. This process results in a polarization contribution labeled P_μ. Second, even if we have nonpolar molecules, all molecules have an electronic polarizability[4] α_{ele} and molecular polarizability α_{mol}, collectively denoted by α, that results in a dipole moment induced by the applied field. This contribution is labeled P_α. The field experienced by the molecules is not equal to the applied field E as the surrounding molecules shield the reference molecule from the applied field, and a polar reference molecule will influence the surroundings leading to a different shielding than for a nonpolar molecule. If we have N molecules in a volume V, we may write $P_\mu = (N/V)p_\mu$ and $P_\alpha = (N/V)p_\alpha$. Here, p_μ represents the (rotational) average dipole moment $\bar{\mu}$, which is dependent on the *directional field* E_{dir} experienced by the molecules. For p_α we have $p_\alpha = \alpha E_{\text{int}}$, where E_{int} is the *internal field*. Although both fields E_{dir} and E_{int} are proportional to the applied field E, they are not the same because a proportional increase in all three components of E will affect the magnitude of the induced dipole moment but not the directional effect for the permanent dipole. Hence, our task is to calculate E_{int}, E_{dir} and $\bar{\mu}$ so that P_μ and P_α can be determined.

Altogether, in microscopic terms polarization can be interpreted as the number of molecules N with (effective) dipole moment μ_{eff} in a volume V, that is

$$\varepsilon_0(\varepsilon_r - 1)E = P = \mu_{\text{eff}}(N/V) = \alpha_{\text{eff}}E_{\text{loc}}(N/V) = (N/V)p_\alpha + (N/V)p_\mu \tag{10.9}$$

Overall, the effective dipole moment is described by $\mu_{\text{eff}} = \alpha_{\text{eff}}E_{\text{loc}}$, where α_{eff} is the *effective polarizability* of the molecule and E_{loc} is the *local field*, the latter being the appropriate combination of the internal and directional field. We deal first with gases and thereafter with liquids and solutions.

3) The unit of dipole moment is C m, but often the Debye unit D is used, where $1\,\text{D} = 3.336 \times 10^{-30}\,\text{C m}$.

4) The unit of polarizability is $\text{C}^2\,\text{m}^2\,\text{J}^{-1}$. Often, the *polarizability volume* $\alpha' = \alpha/(4\pi\varepsilon_0)$ is used. Note that while α_{ele} represents the displacement of the electrons with respect to the nuclei, α_{mol} represents the displacement of the molecule as a whole.

Problem 10.3

The relative permittivity ε_r can be accurately described by $\varepsilon_r(T) = \varepsilon_{r,0}\exp(-LT)$, with as parameters $\varepsilon_{r,0}$ and L. Calculate $\varepsilon_r(298)$ for the three liquids indicated (see Table 10.1).

Table 10.1 Parameters $\varepsilon_{r,0}$ and L for H_2O, CH_3OH, and C_2H_5OH.

	H_2O	CH_3OH	C_2H_5OH
$\varepsilon_{r,0}$ (−)	311.17	157.6	146.0
L (10^{-3}K^{-1})	4.63	5.39	6.02

10.3 Dielectric Behavior of Gases

We have seen that there are two molecular contributions to the polarization. First, we have the dipoles induced by the applied field and, second, the alignment of permanent dipoles due to applied field. Let us consider each of these factors in turn.

For a dipole induced by a field E_{int} in an apolar molecule, we know that the induced dipole moment μ_{ind} is given by $\mu_{ind} = \alpha E_{int}$, where α is the polarizability. Lorentz showed that the internal field is given by $E_{int} = E(\varepsilon_r + 2)/3$ (see Justification 10.1). So, we obtain

$$P = \varepsilon_0(\varepsilon_r - 1)E = \alpha_{eff}E_{loc}(N/V) = \alpha E_{int}(N_A/V_m) = \alpha E(\varepsilon_r + 2)(N_A/3V_m) \quad \text{or}$$
$$\alpha N_A/3\varepsilon_0 V_m = (\varepsilon_r - 1)/(\varepsilon_r + 2) \tag{10.10}$$

where the molar volume $V_m = M/\rho$ is used with M the molecular mass and ρ the mass density. It is traditional (although unfortunate) to call $P_m \equiv \alpha_{eff}N_A/3\varepsilon_0$ the *molar polarization* (with dimensions of volume/mole). Using this definition, Eq. (10.10) becomes the so-called *Clausius–Mossotti equation*

$$\blacktriangleright \quad P_m/V_m \equiv (\alpha N_A/3\varepsilon_0)/V_m = (\varepsilon_r - 1)/(\varepsilon_r + 2) \tag{10.11}$$

This equation links the microscopic terms, the polarizability α and number density N_A/V_m, to a macroscopic parameter, the relative permittivity ε_r. It will be clear that this expression can be applied to nonpolar fluids only.

For polar fluids, we need the permanent dipole moment averaged over all angular orientations, and to calculate this quantity we need the potential energy Φ of dipole moment $\boldsymbol{\mu}$ in an electric field $\boldsymbol{E}$ given by

$$\Phi = -\boldsymbol{\mu}\cdot\boldsymbol{E} = -\mu E\cos\theta \tag{10.12}$$

Note that $\mu = |\boldsymbol{\mu}|$ and $E = |\boldsymbol{E}|$ and $\boldsymbol{x}\cdot\boldsymbol{y}$ denotes the inner product of the vectors $\boldsymbol{x}$ and $\boldsymbol{y}$, where θ is the angle between $\boldsymbol{x}$ and $\boldsymbol{y}$. From statistical mechanics we learned how to calculate the average dipole moment: we first determine the Hamilton function, that is, the energy in terms of coordinates and associated momenta, and thereafter we evaluate the partition function and finally calculate the expectation value of μ. Here, we restrict ourselves to diatomic molecules, but the result is more general.[5)]

Recall that for a diatomic molecule containing atom 1 (2) with mass m_1 (m_2) at distance r_1 (r_2), the center of gravity is defined by $m_1r_1 = m_2r_2$. Hence, the bond length is $a = r_1 + r_2$, while the *moment of inertia* I is given by

$$I = m_1r_1^2 + m_2r_2^2 = \frac{m_1m_2}{m_1+m_2}a^2 \tag{10.13}$$

We use polar coordinates r, θ and ϕ (see Appendix B) and calculate the kinetic energy T. For atom 1 we have the kinetic energy T_1 reading

$$T_1 = \frac{1}{2}m_1\dot{r}_1^2 = \frac{1}{2}m_1(\dot{x}_1^2+\dot{y}_1^2+\dot{z}_1^2) = \frac{1}{2}m_1r_1^2(\dot{\theta}^2+\dot{\phi}^2\sin^2\theta) \tag{10.14}$$

and similarly for atom 2 so that the total kinetic $T = T_1 + T_2$ becomes

$$T_1 = \frac{1}{2}m_1r_1^2(\dot{\theta}^2+\dot{\phi}^2\sin^2\theta) + \frac{1}{2}m_2r_2^2(\dot{\theta}^2+\dot{\phi}^2\sin^2\theta) = \frac{1}{2}I(\dot{\theta}^2+\dot{\phi}^2\sin^2\theta) \tag{10.15}$$

Denoting the momenta associated with θ and ϕ by p_θ and p_ϕ, we have

$$p_\theta = \partial T/\partial\theta = I\dot{\theta} \quad \text{and} \quad p_\phi = \partial T/\partial\phi = I\dot{\phi}\sin^2\theta \tag{10.16}$$

Expressed in momenta the (kinetic) energy becomes the (kinetic) Hamilton function $\mathcal{H}_T(\mathbf{p},\mathbf{q})$ with $\mathbf{p} = (p_\theta,p_\phi)$ and $\mathbf{q} = (\theta,\phi)$ reading

$$\mathcal{H}_T = (p_\theta^2 + p_\phi^2/\sin^2\theta)/2I \tag{10.17}$$

To obtain the associated element in phase space we rewrite Eq. (10.17) as

$$2I\mathcal{H}_T = (p_\theta^2 + p_\phi^2/\sin^2\theta) = R^2 \tag{10.18}$$

which is the equation of a circle with radius R. The element in phase space due to the momenta is thus given by

$$\mathrm{d}p_\theta\mathrm{d}(p_\phi/\sin\theta) = 2\pi R\mathrm{d}R \tag{10.19}$$

However, from Eq. (10.18) we also obtain

$$2I\mathrm{d}\mathcal{H}_T = 2R\mathrm{d}R \tag{10.20}$$

5) For a detailed general treatment, see Ref. [19]. In brief, the Hamilton function $\mathcal{H}$ can be written as $\mathcal{H} = f(\mathbf{P},\mathbf{p}) + g(\mathbf{Q},\mathbf{q},E)$ where $\mathbf{P}$ are the so-called "momentoids" (pseudo-momenta) associated with the orientation angles $\mathbf{Q}$ used as co-ordinates while $\mathbf{p}$ are the momenta associated with the other co-ordinates $\mathbf{q}$. This procedure separates the rigid body rotational motion, which is independent of E and therefore can be omitted. But since $\mathbf{P}$ are not the real momenta associated with $\mathbf{Q}$, this procedure introduces a Jacobian determinant $\sin\theta$ for $g(\mathbf{Q},\mathbf{q},E)$. This is in essence the procedure used here also.

so that

$$\mathrm{d}p_\theta \mathrm{d}p_\phi = 2\pi I \sin\theta \mathrm{d}\mathcal{H}_T \tag{10.21}$$

Since the (total) Hamilton function is the sum of the kinetic part $\mathcal{H}_T$ and potential part $\Phi = -\mu E\cos\theta$, we have for the partition function of the diatomic rotator

$$Z = 2\pi I h^{-2}\int_0^\infty \exp(-\beta\mathcal{H}_T)\mathrm{d}\mathcal{H}_T \cdot \int_0^{2\pi}\int_0^\pi \exp(\beta\mu E\cos\theta)\sin\theta \mathrm{d}\theta \mathrm{d}\phi \tag{10.22}$$

$$= 2\pi IkTh^{-2}\cdot 2\pi \cdot -\int_0^\pi \exp(\beta\mu E\cos\theta)\mathrm{d}(\cos\theta) \tag{10.23}$$

$$= \frac{4\pi^2 IkT}{h^2}\cdot\frac{(\mathrm{e}^{\beta\mu E}-\mathrm{e}^{-\beta\mu E})}{\beta\mu E} = \frac{8\pi^2 IkT}{h^2}\cdot\frac{\sinh(\beta\mu E)}{\beta\mu E} \equiv Z_T Z_\Phi \tag{10.24}$$

where p_θ and p_ϕ have the range $0 \le (p_\theta, p_\phi) \le \infty$, and θ and ϕ have the range $0 \le \theta \le \pi$ and $0 \le \phi \le 2\pi$, respectively. We see that Z separates into two factors where Z_T represents the free rotation and Z_Φ the influence of the electric field. This separation is general, and therefore the following applies to all molecules. Because, in the calculation of the average value the kinetic parts cancel, the average value of $\mu E\cos\theta$ becomes

$$\langle\mu\rangle = Z_\Phi^{-1}\int_0^\pi \mu E\cos\theta\exp(\beta\mu E\cos\theta)\sin\theta \mathrm{d}\theta$$

Using $a \equiv \beta\mu E$ the average evaluates to

$$\begin{aligned}\langle\mu\rangle &= \mu L(a) \equiv \mu\{[\exp(a)+\exp(-a)]/[\exp(a)-\exp(-a)]-1/a\}\\ &= \mu(a/3 - a^3/45+\cdots) \cong \mu(a/3) \quad \text{for } a \ll 1\end{aligned} \tag{10.25}$$

where the second line is obtained by expanding the exponentials. The function $L(a)$ is usually referred to as the *Langevin function*, for which the approximation for small a is nearly always valid. Recalling that for the local field we need the directional field E_{dir}, the total average dipole moment is thus

$$\langle\mu\rangle = (\mu^2/3kT)E_{\mathrm{dir}} \tag{10.26}$$

Debye took for the directional field E_{dir} the Lorentz internal field $E_{\mathrm{int}} = E(\varepsilon_\mathrm{r}+2)/3$. Making the same assumption and recalling that the induced dipole moment μ_{ind} is given by $\mu_{\mathrm{ind}} = \alpha E_{\mathrm{int}}$, the total dipole moment μ_{tot} becomes

$$\mu_{\mathrm{tot}} = \langle\mu\rangle + \mu_{\mathrm{ind}} = [\alpha + (\mu^2/3kT)]E_{\mathrm{int}} \equiv \alpha_{\mathrm{eff}}E_{\mathrm{int}} \tag{10.27}$$

where the effective polarizability is now $\alpha_{\mathrm{eff}} = \alpha + \mu^2/3kT$. Combining with $P = \varepsilon_0(\varepsilon_\mathrm{r} - 1)E$ leads to the *Debye equation*, similar to the Clausius–Mossotti equation,

► $$P_\mathrm{m}/V_\mathrm{m} \equiv (\alpha_{\mathrm{eff}}N_\mathrm{A}/3\varepsilon_0)/V_\mathrm{m} = (\varepsilon_\mathrm{r}-1)/(\varepsilon_\mathrm{r}+2) \tag{10.28}$$

The Debye equation links the effective polarizability α_{eff} (and number density $N_\mathrm{A}/V_\mathrm{m}$) to the relative permittivity ε_r. A plot of $P_\mathrm{m} = V_\mathrm{m}(\varepsilon_\mathrm{r}-1)/(\varepsilon_\mathrm{r}+2)$ versus $1/T$ yields a slope $\mu^2/3k$, while the intercept provides α. For gases, the values so obtained for the dipole moment compare rather well with independently

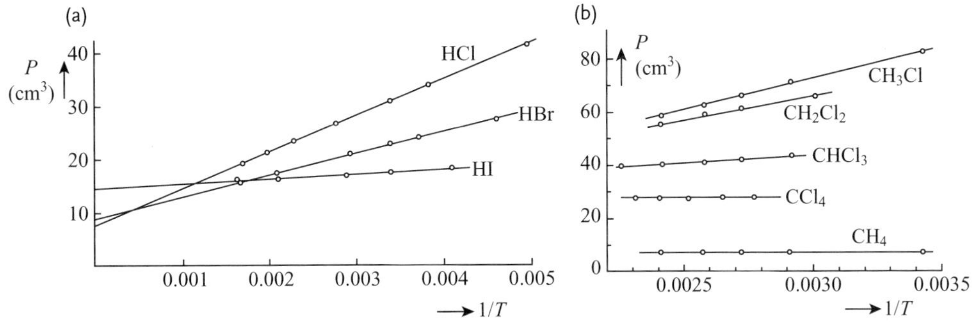

Figure 10.1 Molar polarization P of (a) HCl, HBr and HI and (b) of CH_4, CH_3Cl, CH_2Cl_2, $CHCl_3$ and CCl_4 as a function of $1/T$ Data from Ref. [20].

Table 10.2 Dipole moment for HF, HCl, HBr, and HI. Data from Ref. [20] and Appendix E.

	r (pm)	μ^* (D)	$\alpha^*/(4\pi\varepsilon_0)$ (Å^3)	μ (D)	$\alpha/(4\pi\varepsilon_0)$ (Å^3)
HF	91.7	1.91	0.80	–	–
HCl	127.4	1.03	2.63	1.052	3.01
HBr	141.4	0.79	3.61	0.807	3.54
HI	160.9	0.38	5.45	0.388	5.58

determined values, and the method can be considered as accurate. Figure 10.1 shows the graphs and Table 10.2 the numerical data for HCl, HBr and HI as determined in this way. Figure 10.1 also shows data for the series CH_4, CH_3Cl, CH_2Cl_2, $CHCl_3$ and CCl_4, revealing clearly which molecules are polar and also the magnitude of dipole moment and polarizability.

So far, we have considered only the effect of a static electric field or a low-frequency electric field. With increasing frequency the polar molecules as a whole will not be able to move fast enough to comply with the field, so that the dipole orientation remains random and we are left with only the electronic polarization (Figure 10.2). According to Maxwell's equations, the refractive index n is given by $n^2 = \varepsilon_r\mu_r$ with ε_r and μ_r the relative electric and magnetic permittivity, respectively. Since for a dielectric $\mu_r = 1$, we have at high frequency–that is, in the optical region–$\varepsilon_\infty = n^2$, where ε_∞ is the high-frequency relative permittivity. Hence, we can define the *molar refractivity*

$$R_m \equiv V_m(n^2-1)/(n^2+2) = \alpha N_A/3\varepsilon_0 \tag{10.29}$$

which, empirically, appears to be largely independent of temperature, pressure, and state of aggregation. Equation (10.29) is conventionally addressed as the *Lorentz–Lorenz equation*. Note that the refractive index must be slightly temperature-dependent because of the presence of the factor V_m in the expression for R_m. For polar molecules, we thus should have $R_m = P_m$ at high frequency. The permittivity and refractive index data for some molecules are shown in Table 10.3. It will be clear

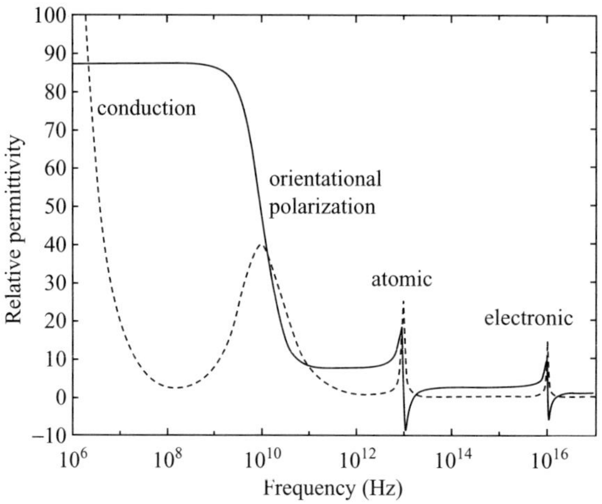

Figure 10.2 Relative permittivity as a function of frequency, indicating the orientational, atomic and electronic contribution to the polarization. The dotted line indicates the loss factor.

Table 10.3 Permittivity ε_r and refractive index n of several molecules.[a)]

Molecule	ε_r (−)	n (at 589 nm)	Molecule	ε_r (−)	n (at 589 nm)
C_6H_6	2.284 (20 °C)	1.498	H_2O (l)	80.37 (20 °C)	1.333
$C_6H_5CH_3$	2.387 (20 °C)	1.494		78.54 (25 °C)	
	2.379 (25 °C)		CH_3OH	33.62 (20 °C)	1.326
c-C_6H_{12}	2.023 (20 °C)	1.424		32.63 (25 °C)	
	2.015 (25 °C)		C_2H_5OH	25.3 (20 °C)	1.359
n-C_6H_{14}	1.874 (20 °C)	1.372		24.30 (25 °C)	
	1.890 (25 °C)		$C_6H_5NO_2$	35.74 (20 °C)	1.550
CCl_4	2.238 (20 °C)	1.468		34.82 (25 °C)	
	2.228 (25 °C)		C_6H_5Cl	5.708 (20 °C)	1.523
CS_2 (l)	2.641 (20 °C)	1.629		5.621 (25 °C)	
			NH_3	16.9 (25 °C)	–

a) Data from *Handbook of Chemistry and Physics* (ed. R.C. Weast), 60th edn, CRC Press, 1979.

that, for the left half of the table the relationship $\varepsilon_r = n^2$ is reasonably well obeyed, but for the right half this relationship is not obeyed at all. In general this is due to cooperative effects, such as dimer formation and other structural effects, contributing to the polarization only at low frequency. Whilst in this way the effect of the electronic polarizability can estimated, the contribution of the atomic polarizability is more difficult. Böttcher [1] estimates the contribution as 5%, that is, $\varepsilon_\infty = 1.05n^2$.

10.3.1 Estimating μ and α

One way to try to understand the dipole moment of a molecule is in terms of the contributions of the individual bonds, the so-called *bond moments*. As an example, consider water, the total dipole moment of which is 1.85 D, due to the vector sum of two O–H bond moments. From the total dipole moment and the geometry of the molecule, one can derive that the bond moment of an O–H bond is 1.52 D. Using this process and, on rather doubtful grounds, using a C–H bond moment of 0.4 D as a reference value, the bond moments of various bonds can be deduced. The dipole moments for several chemical bonds [2] are shown in Table 10.4, using data for saturated compounds.

These data will differ for aromatic compounds as the moment of a C(sp^2)–H bond differs from that of a C(sp^3)–H bond. Thus, taking a value of 0.7 D for the C(sp^2)–H bond, a bond moment of C(sp^2)–Cl = 0.89 D from chlorobenzene can be obtained. Different authors use different values as references; for example, using C(sp^3)–H = 0.31 D, C(sp^2)–H = 0.63 D, and C(sp)–H = 1.05 D from infrared measurements, Petro [3] calculated C(sp^3)–C(sp^2) = 0.68 D from toluene (0.37 D) and propylene (0.35 D). Moreover, he obtained C(sp^3)–C(sp) = 1.48 D from propyne (0.75D) and C(sp^2)–C(sp) = 1.15 D from phenylacetylene (0.73 D). The problem here is that the bond contributions are considered as independent, whereas in reality the molecule should be considered as one entity of which the properties should be calculated with quantum methods.

A slightly more reliable method is to use groups of atoms in molecules. For example, in the case of nitro-groups in aromatic compounds one uses the complete nitro-group, and the dipole moment of the nitro-group in nitrobenzene replaces the dipole moment of a C–H group in the benzene molecule. In general, in this method the data are taken as the values for the monosubstituted benzenes for aromatic compounds and for monosubstituted methanes for aliphatic compounds. For further details, see Minkin *et al.* [2].

Table 10.4 Dipole moments for several chemical bonds using C–H$^+$ ≡ 0.4 D as a reference.[a)]

Bond	μ (D)	Bond	μ (D)	Bond	μ (D)
C–H	0.4	C–S	0.9	H–O	1.51
C–F	1.39	C–Se	0.7	H–N	1.31
C–Cl	1.47	C=N	1.4	H–S	0.7
C–Br	1.42	C=O	2.4	Si–C	1.2
C–I	1.25	C=S	2.0	Si–H	1.0
C–N	0.45	C≡N	3.1	Si–N	1.55
C–O[b)]	0.7				

a) Data calculated from methyl derivatives (CH_3X) dissolved in benzene at 25 °C.
b) Ethers, alcohols.

Table 10.5 Group contributions to the molar R_m refractivity at 589 nm.[a)]

Group	R_m ($cm^3 mol^{-1}$)	Group	R_m ($cm^3 mol^{-1}$)
C	2.591	$C_{10}H_7$	43.00
H	1.028	–S–	7.729
=O	2.122	=S	7.921
–O–	1.643	C≡N	5.459
OH	2.553	N (primary aliphatic)	2.376
F	0.81	N (secondary aliphatic)	2.582
Cl	5.844	N (aromatic)	3.550
Br	8.741	Added ethylenic double bond contribution	1.575
I	13.954		
C_6H_5	25.463	Added acetylenic triple bond contribution	1.977

a) Data from *Handbook of Chemistry and Physics*, 56th edn, CRC Press, 1975.

Similar to the dipole moment, the molar refractivity (equivalent to the polarizability) can be estimated as the sum of the contributions due to various groups in the molecule. Table 10.5 provides data for several groups from which the molar refractivity can be estimated. In contrast to the dipole moment, the estimates for refractivity (and thus polarizability) are quite accurate. Example 10.1 provides the details for both α and μ for methanol.

Example 10.1: Methanol

The dipole moment μ of methanol can be estimated using the contributions of the CH_3 group (equivalent to a C–H bond), the C=O bond and the O–H bond. These contributions are, respectively, 0.4 D, 0.7 D, and $1.51 \cdot \cos\theta = 0.49$ D with $\theta = 109°$ the C–O–H bond angle. This results in $\mu = 1.59$ D, to be compared with the experimental value $\mu = 1.70$ D. Whilst in this case the agreement is good, this is not always the case. The polarizability α can be estimated using the contributions of the C atom ($2.591\,cm^3 mol^{-1}$), the three H atoms ($3 \times 1.029\,cm^3 mol^{-1}$) and the OH group ($2.553\,cm^3 mol^{-1}$) to the molar refractivity R_m. The result is $R_m = 8.23\,cm^3 mol^{-1}$. By using $R_m = N_A\alpha/3\varepsilon_0$, one obtains $\alpha = 3.63 \times 10^{-40}\,C^2 m^2 J^{-1}$ or $\alpha' = \alpha/4\pi\varepsilon_0 = 3.26\,Å^3$, in excellent agreement with the experimental value.

Problem 10.4

Derive Eq. (10.25). Verify that $L(a) \cong a/3$ for $a << 1$ and show that, for reasonable conditions, this approximation is nearly always satisfied. Typical values are $\mu = 1\,\text{D}$, $E = 10\,\text{kV}\,\text{cm}^{-1}$, and $T = 300\,\text{K}$.

Problem 10.5

Calculate α' and μ for C_6H_5F given the gas-phase P_m versus T data below, and assuming ideal gas behavior.

T (K)	344	371	414	453	507
P_m	69.9	66.8	62.2	59.3	55.8

Problem 10.6

Show that for NH_3 with an angle of 107.5° between the three N–H bonds, that is, with an angle of 68° between a N–H bond and the molecular axis through the N atom, that the bond moment is 1.31 D if the total dipole moment is 1.47 D.

Problem 10.7

Estimate μ and α' for acetone, and compare the estimates with the experimental values $\mu = 2.84\,\text{D}$ and $\alpha' = 6.33\,\text{Å}^3$.

Problem 10.8

Given the data below, estimate the charges on the various atoms.

$\mu(H_2O) = 1.85\,\text{D}$,	$r(\text{OH}) = 0.958\,\text{Å}$,	$\theta(\text{HOH}) = 104.5°$
$\mu(H_2S) = 0.97\,\text{D}$,	$r(\text{SH}) = 1.335\,\text{Å}$,	$\theta(\text{HSH}) = 92.25°$
$\mu(NH_3) = 1.47\,\text{D}$,	$r(\text{NH}) = 1.017\,\text{Å}$,	$\theta(\text{HNH}) = 107.5°$

Problem 10.9

Indicate why R_m is expected to be largely independent of the temperature, pressure, and state of aggregation.

10.4 Dielectric Behavior of Liquids

The question is now how well the approach as described before can be applied to liquids. Let us first consider mixtures of polar molecules in a nonpolar solvent.

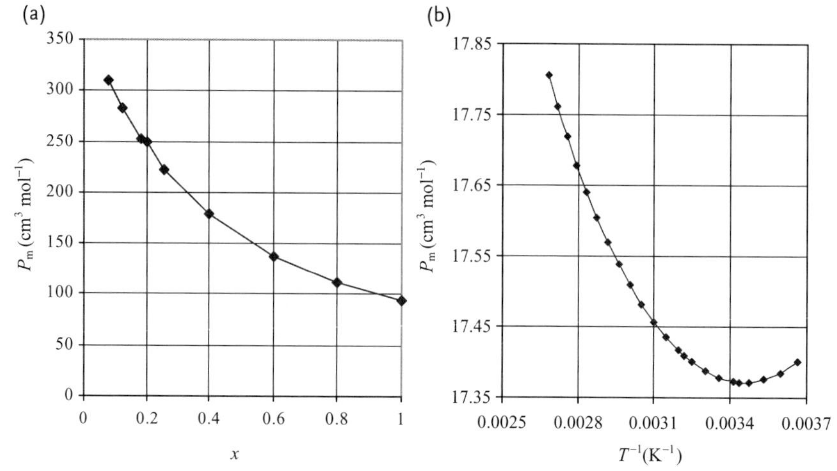

Figure 10.3 (a) Molar polarization P_m of the polar molecule $C_6H_5NO_2$ in the nonpolar solvent hexane as a function of mole fraction x (the small discontinuity at x ~0.2 is due to the merging of two data sets); (b) The anomalous behavior of the molar polarization of water as function of inverse temperature.

Figure 10.3 shows an example: $C_6H_5NO_2$ in hexane. We consider that the molar polarization of a mixture P_m is given by the rule of mixtures

$$P_m = \Sigma_i P_i x_i$$

where x_i is the mole fraction of the component i. The molar polarization P_i of each component i is, as before, given by

$$P_i = (\alpha_i + \mu_i^2/3kT)N_A/3\varepsilon_0$$

If the value for α_i for the nonpolar solvent is known, for example, using the refractive index, the value for μ_i of solute can be calculated given either the temperature-dependent data or independent data for α_i for the polar component. For the example mentioned we have $\mu = 3.9\,D$ ($\alpha \cong 0\,C\,m^2\,J^{-1}$) as compared with $\mu = 4.2\,D$ for the molecule in the gas phase. The agreement is good, which is true in general for molecules dissolved in nonpolar solvents. Because P_m is not necessarily linearly related to x_i, extrapolated values of P_m with $x_i \to 0$ are often used, although in practice the permittivity of a mixture is often reasonably well described by a rule of mixtures using volume fractions for each component (but see Figure 10.3a).

The situation is quite different for pure liquids, however. For water, it is known that the dipole moment $\mu = 1.85\,D$ and the polarizability volume $\alpha' = 1.48\,Å^3$. If we calculate from these data the molar polarization via $P_m = (\alpha + \mu^2/3kT)N_A/3\varepsilon_0$, and from that result the value of P_m/V_m, we obtain the value 4.2. Since $P_m/V_m = (\varepsilon_r - 1)/(\varepsilon_r + 2)$ should be always smaller than 1, this result is physically impossible and such behavior is, therefore, occasionally called the "polarization catastrophe." Moreover, it appears that the dielectric behavior of water (and other pure, polar liquids) is very unlike the Debye-behavior (Figure 10.3), due to shielding and correlation.

With regard to shielding, as indicated before, any molecule will experience a local field not equal to the applied field, because the surrounding molecules shield the applied field. In the Clausius–Mossotti and Debye models this effect was taken into account by using for both the internal and directional field the Lorentz internal field. However, it was argued that for the dipole contribution the directional field is different from the internal field, and taking this effect into account Onsager [4] derived an expression for the directional field. Although presenting all details of Onsager's calculations would lead us far astray, we nevertheless present an outline for a spherical molecule in a spherical cavity with radius a in Justification 10.1 (after all, this is beautiful physical chemistry!) from which we quote the result, preserving as much as possible the resemblance with the Debye equation,

$$\frac{(2\varepsilon_r + 1)(\varepsilon_r - 1)}{9\varepsilon_r} = \frac{N_A}{3\varepsilon_0 V_m}\left(\alpha^* + \frac{\mu^{*2}}{3kT}\right) \tag{10.30}$$

Here, $\mu^* = \mu/(1 - f\alpha)$ and $\alpha^* = \alpha/(1 - f\alpha)$ with $f = 2(\varepsilon_r - 1)/(2\varepsilon_r + 1)a^3$ and $4\pi N_A a^3/3V_m = 1$. This result implies, first, that the functional dependence on ε_r is changed and, second, that both the dipole moment μ^* and polarizability α^* in the condensed state are typically enhanced by 20–50% as compared to the gas state values μ and α by the near presence of other molecules. For example, for $CHCl_3$ with $\varepsilon_r = 4.8$ and $n = 1.446$ we obtain $\mu^*/\mu = 1.24$ when using for α the Lorentz–Lorenz estimate $\mathcal{C}\alpha/a^3 = (n^2 - 1)/(n^2 + 2)$. Introducing the expressions for f, a and α in Eq. (10.30), the explicit result for μ reads

$$\mu^2 = \frac{9\varepsilon_0 kTV_m}{N_A}\frac{(\varepsilon_r - n^2)(2\varepsilon_r + n^2)}{\varepsilon_r(n^2 + 2)^2} \tag{10.31}$$

which usually is denoted as the *Onsager equation*. From this equation, reasonable estimates of μ for nonassociating molecules can be obtained as compared to results from independent gas-phase experiments.

Now, on the subject of correlation, we assumed when calculating the average dipole moment that the alignment of the molecules was only due to the applied field. However, there is a correlation between neighboring molecules due to their mutual interaction, which results in a correction [5] for the value of $(\mu^2/3kT)$ by a correlation factor g, defined via $g\mu^2 \equiv \mu\cdot\mu^*$ with μ^* the dipole moment of the molecule and its immediate surroundings. More explicitly, the parameter g is determined by

$$g = 1 + \sum_{j=1}^{j=\infty} g_j \equiv 1 + \sum_{j=1}^{j=\infty} N_j\langle\cos\gamma_j\rangle \tag{10.32}$$

where N_j is the number of atoms in coordination shell j and $\langle\cos\gamma_j\rangle$ the average cosine of the angle between the dipole of the (fixed) reference molecule and the molecules in shell j. The molecular dipoles tend to line up, but are counteracted by geometric packing effects. If the dipole moment is along the long axis of the molecule, close packing forces more molecules to have $\gamma \cong 90°$ and $g < 1$. In case the dipole moment is along the short axis of the molecule, packing effects result in having more molecules with $\gamma \cong 0°$ and $g > 1$. The correlation function, however, cannot be determined experimentally in an independent way, and accurate information on structure and intermolecular interactions is required for its calculation. It should be noted from the data in Table 10.6 that there is neither correlation between g and ε, nor between g and μ. One would expect, however, that in a series of similar molecules using independent values for the dipole moments similar values for g are obtained. Table 10.7 shows that, for aliphatic ketones, this is indeed the case. Unfortunately, however, the theoretical treatment of the Kirkwood factor is involved and we have to refer to the literature.[6)]

One factor not taken into account in the previous analysis was the association of molecules, in particular via hydrogen bonding. This effect is particularly

Table 10.6 Estimates of Kirkwood's correlation factor g for several polar molecules.[a)]

Molecule	T (°C)	ρ (10^{-2} mol cm^{-3})	ε_r (–)	μ (D)	g (–)
Chlorobenzene C_2H_5Cl	25	0.9785	5.61	1.73	0.61
Nitrobenzene $C_6H_5NO_2$	25	0.9377	34.9	4.24	0.87
	200	0.8264	15.9	4.24	0.83
Monochloromethane CH_3Cl	25	1.740	9.68	1.86	0.9
Dichloromethane CH_2Cl_2	25	1.551	8.93	1.57	1.01
Chloroform $CHCl_3$	25	1.239	4.72	1.01	1.19
Trimethylamine $N(CH_3)_3$	25	1.060	2.44	0.61	1.5
Ammonia NH_3	25	3.534	16.9	1.47	1.65
Water H_2O	0	5.549	88.2	1.84	2.89
	25	5.534	78.5	1.84	2.81
	83	5.382	60.4	1.84	2.69

a) Data from Ref. [21].

Table 10.7 Values of g for aliphatic ketones [2].

Molecule	ε_r (–) at 25 °C	μ (D)	$g\mu^2$ (D^2)	g (–)
Acetone	21.45	2.85	9.82	1.21
Methyl ethyl ketone	18.51	2.77	9.97	1.30
Methyl propyl ketone	15.45	2.72	9.69	1.29
Diethyl ketone	17.00	2.71	10.40	1.42
Methyl isobutyl ketone	13.11	2.72	9.20	1.24
Methyl amyl ketone	11.95	2.70	9.10	1.25
Dipropyl ketone	12.60	2.72	9.60	1.30
Methyl hexyl ketone	10.39	2.70	8.70	1.19

important for water. If the Kirkwood model is applied to water at 0 °C using a model where every molecule is hydrogen-bonded to four neighbors and taking into account interactions up to the 4th shell, using $N_1 = 4$, $N_2 = 11$, $N_3 = 22$, the contributions g_1, g_2 and g_3 of the first, second and third shells to g are 1.20, 0.33 and 0.07, respectively, and a g-value of ~2.6 is obtained. It appears therefore that correlation with shells outside the first shell results in an appreciable contribution to the total dielectric polarization. The permittivity obtained is $\varepsilon_r = 71.9$, which is in reasonable agreement with the experimental value $\varepsilon_r = 88.0$ [6], while the dipole moment was estimated as $\mu^* = 2.17\,\text{D}$, some 17% larger than for the gas state, $\mu = 1.85\,\text{D}$. For ice [7] at 0 °C, Coulson and Eisenberg calculated $\mu^* = 2.60\,\text{D}$, which was some 40% larger than $\mu = 1.85\,\text{D}$. Whilst for water a reasonable agreement is obtained using the Kirkwood model, for aliphatic alcohols the values obtained are close to those as calculated from the Onsager model, and differ considerably from the experimental data. For further details we refer to the literature[6].

Justification 10.1: The internal, reaction, and directional field*

In this justification an outline of the Onsager calculation is given. For full details we refer to Böttcher [1]. We first quote some results from electrostatics, of which a brief discussion is given in Appendix D, and thereafter describe the internal, reaction, and directional fields leading to the Clausius–Mossotti, Debye, and Onsager equations using the abbreviation $\mathcal{C} = (4\pi\varepsilon_0)^{-1}$.

From electrostatics (see Appendix D) we borrow the expression for the field $\boldsymbol{E}_{\text{cav}}$ inside a spherical cavity with radius a, embedded in a dielectric with relative permittivity ε_r and provided with an external field $\boldsymbol{E}$, reading

$$\boldsymbol{E}_{\text{cav}} = \frac{3\varepsilon_r}{2\varepsilon_r + 1}\boldsymbol{E} \tag{10.33}$$

If a nonpolarizable spherical molecule with dipole moment $\boldsymbol{\mu}$ is placed in the center of such a cavity without any applied field $\boldsymbol{E}$, the molecule polarizes the dielectric and this polarization in turn creates a *reaction field* $\boldsymbol{R}$ on the molecule. The reaction field $\boldsymbol{R}$ appears to be uniform, is proportional to the dipole moment $\boldsymbol{\mu}$, and given by $\boldsymbol{R} = f\boldsymbol{\mu}$ with $f = 2(\varepsilon_r - 1)/(2\varepsilon_r + 1)a^3$. If the molecule is polarizable with polarizability α, the reaction field is due to the permanent dipole $\boldsymbol{\mu}$ and the induced dipole $\alpha\boldsymbol{R}$, so that we have $\boldsymbol{R} = f(\boldsymbol{\mu} + \alpha\boldsymbol{R})$. Solving for $\boldsymbol{R}$, we obtain

$$\boldsymbol{R} = \frac{f}{1 - f\alpha}\boldsymbol{\mu} \tag{10.34}$$

Now, recall that in the case where a field $\boldsymbol{E}$ is applied, the dielectric displacement $\boldsymbol{D} = \varepsilon_0\varepsilon_r\boldsymbol{E} = \varepsilon_0\boldsymbol{E} + \boldsymbol{P}$ with polarization $\boldsymbol{P} = \varepsilon_0(\varepsilon_r - 1)\boldsymbol{E}$. The total polarization is thought to be composed of the polarization $\boldsymbol{P}_\mu$ due to the dipole $\boldsymbol{\mu}$ of the molecules, and

6) A detailed discussion is given by Böttcher [1].

the polarization $\boldsymbol{P}_\alpha$ due to the polarizability α of the molecules. We assume that $\boldsymbol{P}_\alpha$ equals the product $\rho\alpha\boldsymbol{E}_{\text{int}}$ with $\boldsymbol{E}_{\text{int}} = \boldsymbol{E}_{\text{cav}} + \boldsymbol{R}$ the *internal field*, and $\rho = N/V$ the number density of the polarizable molecules. For $\boldsymbol{P}_\mu$ we have $\rho\bar{\boldsymbol{\mu}}$ with $\bar{\boldsymbol{\mu}}$ the rotational average of $\boldsymbol{\mu}$. Onsager argued that for $\boldsymbol{P}_\mu$ only the *directional field* $\boldsymbol{E}_{\text{dir}}$, defined as $\boldsymbol{E}_{\text{dir}} = \boldsymbol{E}_{\text{int}} - \bar{\boldsymbol{R}}$ with $\bar{\boldsymbol{R}}$ the rotational average of $\boldsymbol{R}$, is active. So, generally

$$\varepsilon_0(\varepsilon_{\text{r}} - 1)\boldsymbol{E} = \boldsymbol{P}_\alpha + \boldsymbol{P}_\mu = \rho\alpha\boldsymbol{E}_{\text{int}} + \rho\bar{\boldsymbol{\mu}} = \rho\alpha\boldsymbol{E}_{\text{int}} + \rho(\mu^2/3kT)\boldsymbol{E}_{\text{dir}} \tag{10.35}$$

where the Langevin estimate for $\bar{\boldsymbol{\mu}}$ is used.

Let us focus first on nonpolar but polarizable molecules. The conventional approach to estimating the internal field is due to Lorentz, who assumed that the reference molecule is surrounded by an imaginary sphere of such a size that the material outside the sphere could be treated as a continuum. If the reference molecule is removed, then the total field in the sphere would be due to the applied field $\boldsymbol{E}$, the field $\boldsymbol{E}_{\text{sphere}}$ due to free ends of the surrounding dipoles covering the sphere, and the field $\boldsymbol{E}_{\text{mol}}$ due to molecules in the near-vicinity of the reference molecule. The latter field averages to zero for a sufficiently high symmetry. The component of $\boldsymbol{E}_{\text{sphere}}$ in the direction of $\boldsymbol{E}$ is calculated by integrating the field of infinitesimal rings perpendicular to the field direction. The apparent surface charge density on these rings is $P\cos\theta$, while their surface is $2\pi r^2\sin\theta\,\text{d}\theta$, leading to a total charge on each ring of $\text{d}q = 2\pi r^2 P\sin\theta\cos\theta\,\text{d}\theta$. According to Coulomb's law, each charge element contributes to the field by $\text{d}E = \mathcal{C}r^{-2}\cos\theta\text{d}q$. Combining yields $E_{\text{sphere}} = \mathcal{C}P\int 2\pi\sin\theta\cos^2\text{d}\theta = P/3\mathcal{O} = (\varepsilon_{\text{r}} - 1)E/3$. Due to symmetry, the other contributions vanish so that $\boldsymbol{E}_{\text{int}}$ becomes $\boldsymbol{E}_{\text{int}} = \boldsymbol{E} + \boldsymbol{P}/3 = \boldsymbol{E}(\varepsilon_{\text{r}} + 2)/3$. Introducing $\boldsymbol{E}_{\text{int}}$ in $\boldsymbol{P}_\alpha = (\rho\alpha\boldsymbol{E}_{\text{int}}) = \varepsilon_0(\varepsilon_{\text{r}} - 1)\boldsymbol{E}$ yields

$$P_\alpha = \frac{3(\varepsilon_{\text{r}} - 1)}{(\varepsilon_{\text{r}} + 2)} \quad \text{or} \quad \frac{\varepsilon_{\text{r}} - 1}{\varepsilon_{\text{r}} + 2} = \frac{\rho\alpha}{3\varepsilon_0} \tag{10.36}$$

known as the *Clausius–Mossotti equation*. An alternative derivation due to Onsager uses the cavity field $\boldsymbol{E}_{\text{cav}} = 3\varepsilon_{\text{r}}\boldsymbol{E}/(2\varepsilon_{\text{r}} + 1)$, and the reaction field $\boldsymbol{R}$. The latter is in this case $\boldsymbol{R} = f(\alpha\boldsymbol{E}_{\text{cav}} + \alpha\boldsymbol{R})$. Solving for $\boldsymbol{R}$ leads to

$$\boldsymbol{R} = \frac{f\alpha}{1 - f\alpha}\boldsymbol{E}_{\text{cav}} \tag{10.37}$$

Because $\boldsymbol{E}_{\text{int}} = \boldsymbol{E}_{\text{cav}} + \boldsymbol{R}$, the result is

$$\mathbf{E}_{\text{int}} = \boldsymbol{E}_{\text{cav}} + \frac{f\alpha}{1 - f\alpha}\boldsymbol{E}_{\text{cav}} = \frac{1}{1 - f\alpha}\boldsymbol{E}_{\text{cav}} = \frac{1}{1 - f\alpha}\frac{3\varepsilon_{\text{r}}}{2\varepsilon_{\text{r}} + 1}\boldsymbol{E} \tag{10.38}$$

Onsager further assumed that the relationship $4\pi\rho a^3/3 = 1$ holds. Putting this together with the expression for f, one obtains from Eq. (10.35) with $\boldsymbol{P}_\mu = 0$ (after some manipulation) again the Clausius–Mossotti equation. This alternative derivation makes the assumption on the cavity size clear.

We now turn to polar polarizable molecules. Debye used the Lorentz internal field $\boldsymbol{E}_{\text{int}} = \boldsymbol{E}(\varepsilon_{\text{r}} + 2)/3$ for both $\boldsymbol{P}_\mu$ and $\boldsymbol{P}_\alpha$ in Eq. (10.35), and this in turn leads to the *Debye equation*

$$\frac{\varepsilon_r - 1}{\varepsilon_r + 2} = \frac{\rho}{3\varepsilon_0}\left(\alpha + \frac{\mu^2}{3kT}\right) \tag{10.39}$$

Onsager used $\boldsymbol{E}_{\text{dir}}$ and $\bar{\boldsymbol{R}}$ to calculate $\boldsymbol{E}_{\text{int}}$ and obtained

$$\bar{\boldsymbol{R}} = \frac{f}{1-f\alpha}\bar{\boldsymbol{\mu}} = \frac{f}{1-f\alpha}\frac{\mu^2}{3kT}\boldsymbol{E}_{\text{dir}} \text{ and} \tag{10.40}$$

$$\boldsymbol{E}_{\text{dir}} = \boldsymbol{E}_{\text{cav}} + f\alpha\boldsymbol{E}_{\text{dir}} \quad \text{or} \quad \boldsymbol{E}_{\text{dir}} = \frac{1}{1-f\alpha}\boldsymbol{E}_{\text{cav}} = \frac{1}{1-f\alpha}\frac{3\varepsilon_r}{2\varepsilon_r+1}\boldsymbol{E} \tag{10.41}$$

The internal field $\boldsymbol{E}_{\text{int}}$ is obtained from $\boldsymbol{E}_{\text{int}} = \boldsymbol{E}_{\text{dir}} + \bar{\boldsymbol{R}}$ resulting in

$$\begin{aligned}\boldsymbol{E}_{\text{int}} &= \boldsymbol{E}_{\text{dir}} + \bar{\boldsymbol{R}} = \boldsymbol{E}_{\text{dir}} + \frac{f}{1-f\alpha}\frac{\mu^2}{3kT}\boldsymbol{E}_{\text{dir}} = \left(1 + \frac{f}{1-f\alpha}\frac{\mu^2}{3kT}\right)\boldsymbol{E}_{\text{dir}} \\ &= \left(1 + \frac{1}{1-f\alpha}\frac{\mu^2}{3kT}\right)\frac{1}{1-f\alpha}\frac{3\varepsilon_r}{2\varepsilon_r+1}\boldsymbol{E}\end{aligned} \tag{10.42}$$

Finally, substituting $\boldsymbol{E}_{\text{dir}}$ and $\boldsymbol{E}_{\text{int}}$, together with the expression for f, in Eq. (10.35) leads, after considerable manipulation, to the *Onsager equation*

$$\frac{(\varepsilon_r - 1)(2\varepsilon_r + 1)}{3\varepsilon_r} = \frac{\rho}{\varepsilon_0}\left[\alpha^* + \frac{(\mu^*)^2}{3kT}\right] \quad \text{with} \quad \frac{\mu^*}{\mu} = \frac{\alpha^*}{\alpha} = \frac{1}{1-f\alpha} \tag{10.43}$$

Further substitution of the expression for f, the condition $4\pi\rho a^3/3 = 1$ and the Lorentz–Lorenz estimate $\mathcal{C}\alpha/a^3 = (n^2-1)/(n^2+2)$ and explicitly solving for μ leads to Eq. (10.31). Further refinements include the use of an ellipsoidal rather than a spherical cavity, the use of an ellipsoidal rather than a spherical molecule, and the use of a fixed cavity size for the Clausius–Mossotti equation [1]. These refinements all improve the agreement of the value for μ with results of independent experiments.

Problem 10.10

For methanol ($T_m = -98\,°\text{C}$, $\rho = 0.791\,\text{g}\,\text{cm}^{-3}$ at $20\,°\text{C}$), ε_r data corrected for density are given below. Determine α and μ for methanol.

T (°C)	−185	−170	−150	−140	−110	−80	−50	−20	0	20
ε_r (−)	3.2	3.6	4.0	5.1	67	57	49	43	38	34

What further information can be retrieved from these data?

Problem 10.11

Calculate μ^*/μ for propanone with $\varepsilon_r = 21.4$ and $n = 1.359$.

Problem 10.12

For C_2H_5Br ($\rho = 1.43\,\text{g}\,\text{cm}^{-3}$), $\mu = 6.67\times10^{-30}\,\text{C}\,\text{m}$, and $\alpha = 11.1\times10^{-40}\,\text{C}\,\text{m}^2\,\text{V}^{-1}$.

a. Calculate ε_r for this liquid, using the Debye equation and Onsager equation.
b. Comment on the agreement of the results of a) and b) with the experimental value $\varepsilon_r = 9.3$.

10.5 Water

Water is arguably the most important liquid; moreover, water behaves anomalously with respect to dielectric behavior and some other properties, and therefore deserves closer examination. We first discuss some aspects of the water molecule and its thermodynamics, and thereafter deal with some models for water, finally elaborating on the structure of liquid water and the relation of such structure to its properties.

Any N-atom molecule has $3N$ degrees of freedom (DoF). First, there is the external motion or "rigid body" motion, which comprises translation (DoF = 3) and rotation (DoF = 3). Second, we have the internal motion of which vibration[7] is the most important. The number of internal degrees of freedom is $3N-6$ (or $3N-5$ for linear molecules). For water, we thus have three internal vibrations (Figure 10.4) that add to the value for C_P, as elaborated in Example 10.2.

Various clusters of molecules occur in water vapor, for which theoretical estimates of bonding energy are available. For the dimer, a bonding energy of ~22 kJ mol^{-1} is calculated, with a rather broad minimum for the "tetrahedral" angle [8]. For the trimer, the result is a bonding energy of ~7 kJ mol^{-1}, whilst for a quadrimer a bonding energy of ~2 kJ mol^{-1} is obtained. Moreover, various larger clusters seem to exist. However, taking into consideration the bonding energies mentioned, dimers are clearly the most important; the associated structure is shown in Figure 10.5.

It is well known that water behaves anomalously with respect to melting: it has a negative slope for the solid–liquid transition in the P–T diagram, which implies

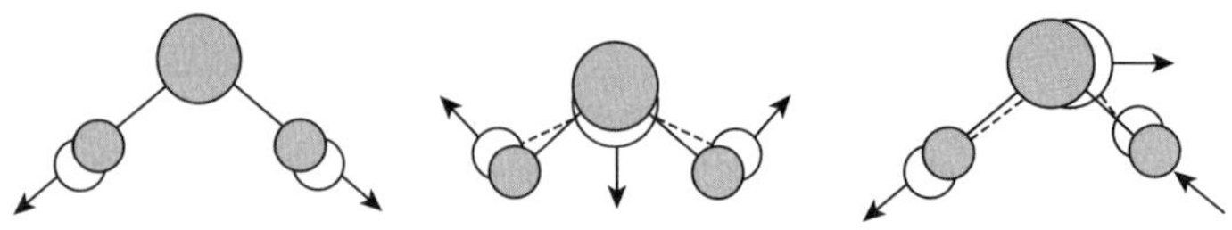

Figure 10.4 The internal modes of water and their energies 3652, 1595, and 3756 cm^{-1} (from left to right).

7) For molecules containing four or more atoms other internal mechanisms may operate; for example, inversion, as in NH_3, or internal rotation, as in H_2O_2.

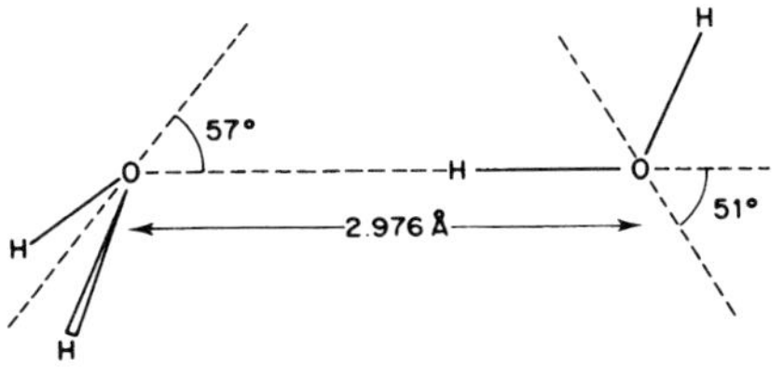

Figure 10.5 The optimum structure of the water dimer.

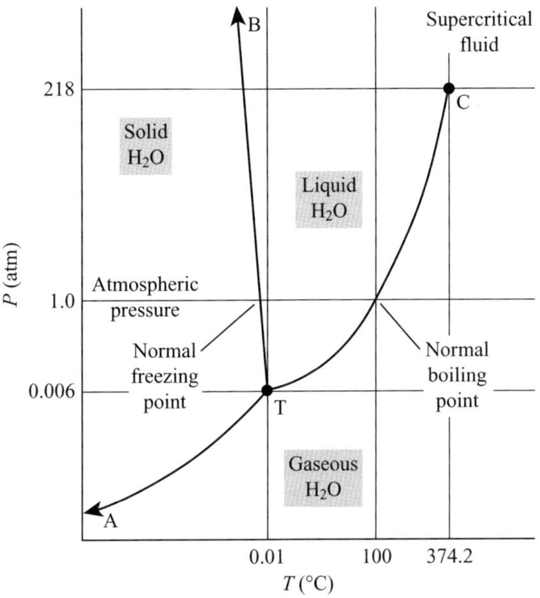

Figure 10.6 The *P–T* diagram for water.

that ice has a lower density than liquid water (Figure 10.6). On examining ice further, many polymorphs of ice exist but in our case the modification called ice I is most relevant. In the ice structure the molecules are bonded via hydrogen bonds in a tetrahedral configuration (Figure 10.7). This leads to a rather open structure that has channels, similar to diamond, and an O···O bond distance of ~2.8 Å. Because the water molecule is almost spherical, in the lattice (almost free) overall rotations are possible. Upon cooling, one might expect that these rotations freeze out so that at 0 K the molecules would be perfectly ordered. However, it appears that ice I has a residual entropy, which indicates that at 0 K the molecules are not perfectly ordered (Figure 10.8). Pauling [9] has explained this phenomenon on the basis of the structure using the *Bernal–Fowler rule*, stating that each O atom is linked to four other O atoms through hydrogen bonds, twice as a hydrogen bond donor and twice as a hydrogen bond acceptor. Since *N* molecules have 2*N* H atoms

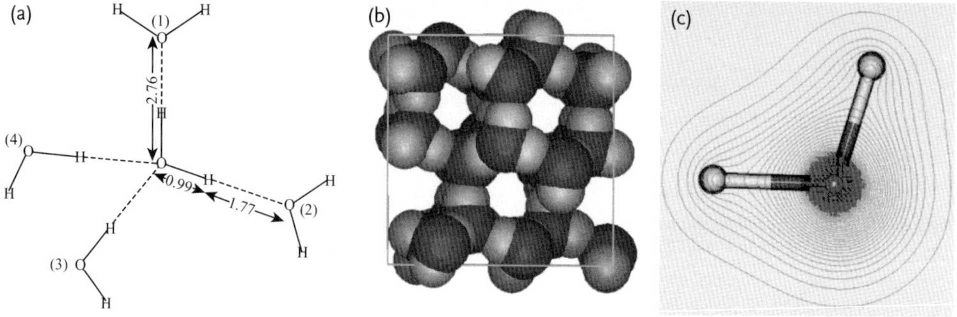

Figure 10.7 (a) The tetrahedral configuration in ice I; (b) The channels in ice I; (c) The nearly spherical shape of the water molecule. Panel (a) also shows the configuration of water as used for MC simulations. Molecules (1) and (2), as well as the central H_2O molecule, lie entirely in the plane of the paper. Molecule (3) lies above this plane, molecule (4) below it, so that the oxygen atoms (1), (2), (3), and (4) are at the corners of a tetrahedron. All distances are in Å.

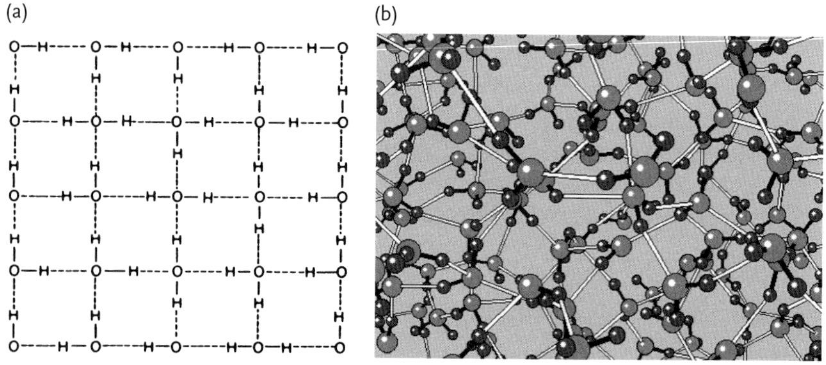

Figure 10.8 The disorder in water shown (a) schematically and (b) pictorially. The covalent O–H bonds are indicated with filled lines, while the hydrogen bonds are represented by open lines.

and each H atom can occupy two positions, we have 2^{2N} configurations. For each O atom surrounded by four H atoms, we obtain 16 distinct configurations: one H_4O^{2+}, four H_3O^+, six H_2O, four HO^-, and one O^{2-}. Of these only six are used (Bernal–Fowler rule satisfied, other configurations being charged and thus less likely). The number of configurations is

$$W = 2^{2N}(6/16)^N = (3/2)^N$$

and the residual entropy thus becomes, taking $N = N_A$,

$$S = k \ln W = R \ln(3/2) = 3.37 \text{ J K mol}^{-1} \tag{10.44}$$

This value[8], characterizing the residual disorder in ice at 0 K, agrees well with the experimental data of 3.4 J K mol^{-1} for ice and 3.2 J K mol^{-1} for heavy ice. The residual entropy for several other molecules is estimated likewise.

Example 10.2: The entropy of water

The molecular mass of H_2O is $M = 18.02\,\text{g mol}^{-1}$, while the moments of inertia are $A = 27.33\,\text{cm}^{-1}$, $B = 14.58\,\text{cm}^{-1}$ and $C = 9.50\,\text{cm}^{-1}$, and therefore $\theta_{\text{rot}} = 39.30\,\text{K}$, 20.96 K, and 13.7 K, respectively. The vibrational wave numbers are 3652 cm^{-1}, 1595 cm^{-1} and 3756 cm^{-1}, so that $\theta_{\text{vib}} = 5160\,\text{K}$, 2290 K and 5360 K, respectively. Table 10.8 indicates the contributions from translation, rotation and vibration to the statistical entropy, as well as the experimental estimate based on the various thermodynamic processes involved. Note the limited theoretical contribution from vibration (at 25 °C, $kT \cong 200\,\text{cm}^{-1}$) and the relatively large experimental contribution from vaporization, as expected because of the large difference in translational motion between liquid and gas. The results imply that $\Delta S = 0.81\,\text{cal mol}^{-1}\,\text{K}^{-1}$, while the Fowler–Pauling estimate yields $\Delta S = N_A k \ln(3/2) = 0.81\,\text{cal mol}^{-1}\,\text{K}^{-1}$, demonstrating the good agreement.

Table 10.8 The entropy for H_2O at 298.1 K and 1 atm (all data in cal mol^{-1} K^{-1}).

Translation	34.61	0–10 K Debye extrapolation	0.022
Rotation	10.48	10–273.1 K C_P(ice)	9.081
Vibration	0.00	273.1 K Melting (Λ/T)	5.257
		273.1–298.1 K C_P(water)	1.580
		298.1 K Vaporization (Λ/T)	35.220
		Compression to 1 atm ($R \ln P$)	−6.886
		Correction for gas imperfection	0.002
Total	45.09	Total	44.28 (±0.05)

10.5.1 Models of Water

For the water molecule itself various models for the structure are available for use in calculations and simulations. These models should at least represent the dipole and local charges in the molecule. Models typically use three to five sites that may atoms or dummy sites, such as the positions of the lone pair electrons. The simplest models treat the water molecule as rigid with a geometry that matches the experimental geometry and which rely only on nonbonded interactions. The

8) Note though that this model neglects the local topological constraints which slightly alter the statistical properties of the distribution in the neighbourhood of a site. The best series approximation for a 3D ice model yields $S \cong k \ln(1.5068)$ [22].

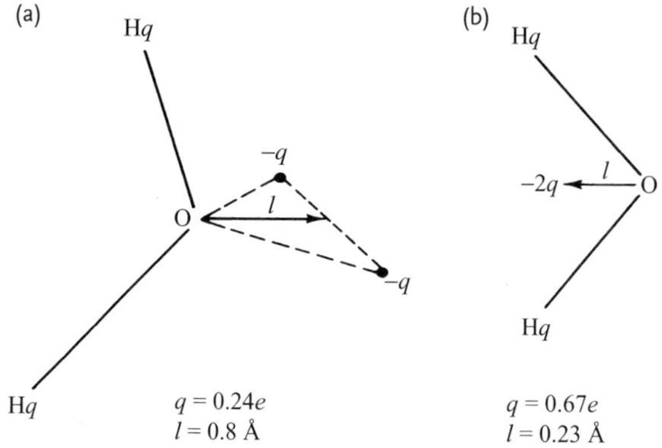

Figure 10.9 The water models of (a) Bjerrum and (b) Popkie.

electrostatic interaction is modeled using Coulomb's law, and the dispersion and repulsion forces using the LJ potential. In particular, three-site models, using the atoms as sites, are popular in view of their computational efficiency. A simple point charge (SPC) three-site model exists with both rigid and flexible geometry. The flexible SPC model is one of the most accurate three-center water models without taking into account polarization since, in MD simulations it provides the correct density and dielectric permittivity for water [10]. Nowadays, due to increased computational power, however, more complex potentials are in use.

Early models using more than three sites were suggested by Bjerrum and Popkie (Figure 10.9). In the four-site models (Figure 10.9b), a negative charge is placed on a dummy atom near the oxygen, along the bisector of the H–O–H angle, and this improves the electrostatic distribution around the water molecule. A variety of models with the generic designation TIP4, complemented with specific labels, is currently in use. In five-site models with the generic label TIP5, the negative charge is placed on dummy atoms representing the lone pairs of the oxygen atom, with a tetrahedral-like geometry (Figure 10.9a). In particular, the TIP5P model results in a reasonably good geometry for the water dimer, a more "tetrahedral" water structure that better reproduces the experimental radial distribution functions from neutron diffraction and the temperature of maximum density of water. The TIP5P-E model is a reparameterization of TIP5P for use with Ewald sums.

It should be said that a potential which describes all properties of water well does not yet exist, despite major efforts having been made. In particular, neglecting polarizability prevents an accurate description of virial coefficients, vapor pressures, critical pressure, and dielectric constant [11]. Details of many potential energy and other, more sophisticated, models are available elsewhere [12].

10.5.2 The Structure of Liquid Water

Upon melting of ice, the volume is decreased by about 9%. The heat of melting is only 13% of the sublimation energy of the solid, and this may be interpreted that a comparable percentage of hydrogen bonds rupture upon melting. Although the actual rupture percentage may be somewhat higher, since a more favorable alignment of non-nearest neighbors can occur in the liquid than in the solid, it seems safe to conclude that the majority of hydrogen bonds survive melting. In cold water, therefore, an ice-like order exists, but this disappears at higher temperature. Such disappearance can be explained by three models:

- The *mixture model* describing liquid water as a mixture of nonpermanent ice-like regions and fluid regions.
- The *interstitial model*, in which water molecules are freely moving in the empty channels of an ice-like structure.
- The *distorted network model*, in which an ice-like structure with distorted hydrogen bonds represents liquid water.

At this point, let us review the experimental data to see whether a choice can be made, or some preference can be given. The vibrational frequencies for the O–H stretching vibration are at about ~3500 cm^{-1} (D_2O ~2500 cm^{-1}), while the H–O–H bending vibration occurs at ~1600 cm^{-1}. Also librations, which represent a mixture of hindered rotation and hydrogen bond bending, occur at 300–1000 cm^{-1}. If we mix H_2O and D_2O, a rapid exchange of protons and deuterons takes place; for a mixture of 1% H_2O in D_2O, this implies that almost all of the hydrogen atoms are in HDO species. The OH stretching band for this mixture has two components, while the 1% D_2O in H_2O shows similar, though less-prominent, features for the OD stretching band. This suggests that there are two environments for the dilute isotope. The relative intensities change with temperature, and the high-frequency band is the only one present in high-density supercritical (pure) water. Only low-frequency bands are observed in ice containing both isotopes (Figure 10.10c). These features are consistent with the mixture model as well as the interstitial model. In view of the presence of two environments, the distorted model, having a wide range of configurations, seems less probable. However, in the latter model a bimodal distribution may be present, also resulting effectively in "two states."

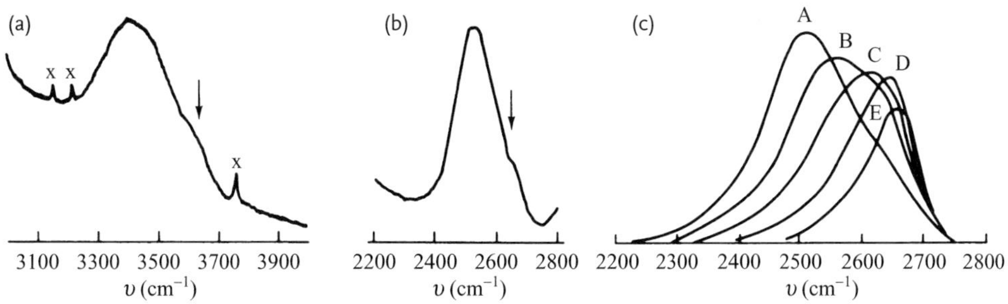

Figure 10.10 The Raman spectrum of (a) 1% H_2O in D_2O; (b) 1% D_2O in H_2O; and (c) the high-pressure Raman spectra of 11% D_2O in H_2O (A = 293 K, 1000 atm, B = 373 K, 1000 atm, C = 473 K, 2800 atm, D = 573 K, 4700 atm, E = 673 K, 3900 atm. In panels (a) and (b) the weak features are indicated by the arrows.

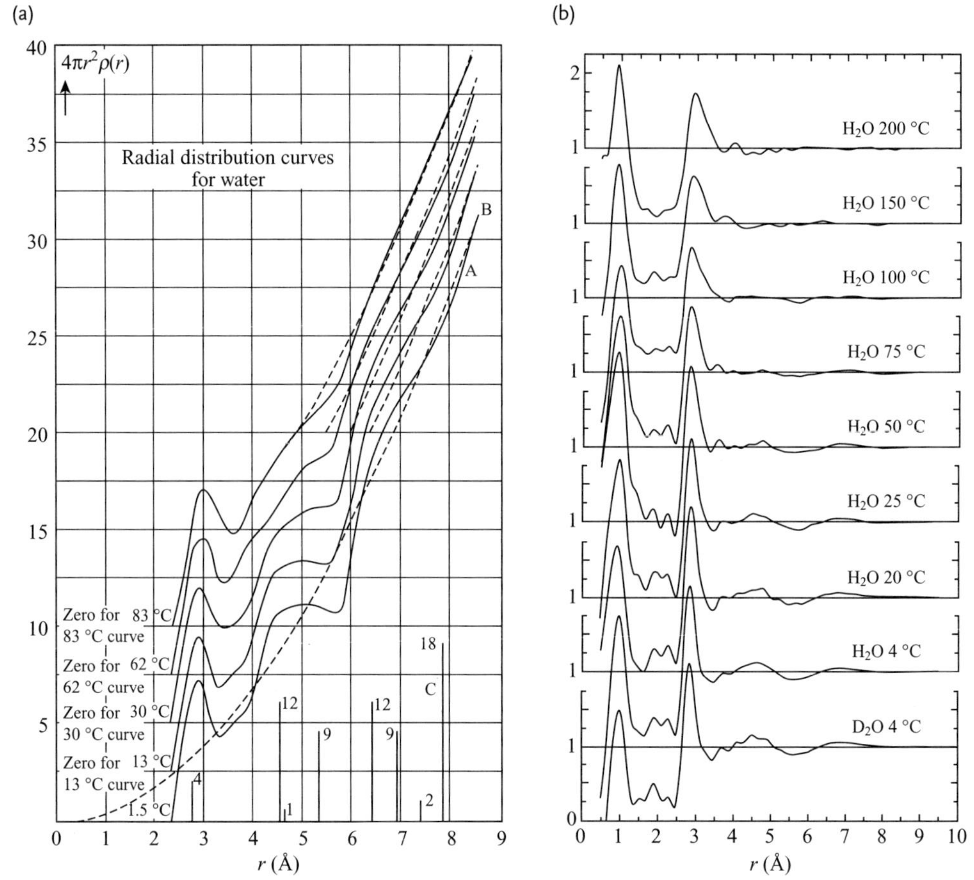

Figure 10.11 The correlation function for water at various temperatures as determined via X-ray diffraction by (a) Morgan and Warren [13] and (b) Narten and Levy [14], showing the considerable improvement in experimental data over time.

Other evidence has been provided by X-ray and neutron diffraction data of water. In Figure 10.11, the pair-correlation function for water at various temperatures, as determined by Morgan and Warren [13] and Narten and Levy [14], is shown. The scattering is mainly due to the O atoms. and only 12% results from the O···H and 2% from the H···H. The O–H bond has a length of ~1.0 Å, while the nearest neighbor O···O distance is ~ 2.9 Å, but this increases from 2.84 Å at 4 °C to 2.94 Å at 200 °C. These values must be compared with the distance in ice, which is 2.76 Å. The next nearest-neighbor O···O distances are ~4.5–7.0 Å, compared to ~4.75 and 5.2 Å in ice. Hence, the first and second neighbor distances have ice-like values, but for third neighbors the differences become appreciable. The combined diffraction evidence is strongly supportive of the persistence of tetrahedral hydrogen-bond order beyond the melting transition, but with substantial disorder present.

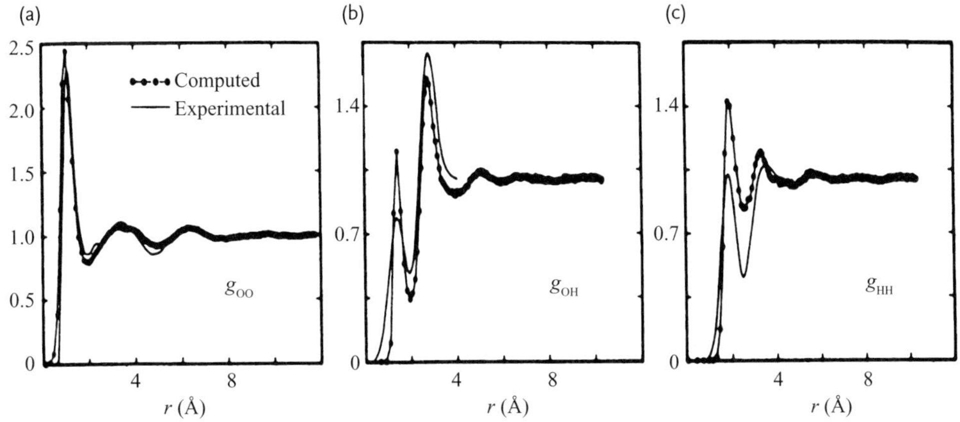

Figure 10.12 (a) O–O, (b) O–H and (c) H–H correlation functions resulting from MC calculations, as compared to experimental data.

As in the case of Ar (see Figure 6.7), the considerable improvements in experimental results over the years will be clear.

The contributions to pair correlation for water are due to the O–O distances, the O–H distances, and the H–H distances. In Figure 10.7 the basic configuration as used for a MC calculation is shown, while in Figure 10.12 a comparison is made for the partial correlation functions as obtained from this MC calculation and the experimental data. Although the latter functions are subject to interpretation, because necessarily some assumptions had to be made to distill these partial correlation functions from the total correlation function, the general agreement is good.

An important factor for water is the presence of the hydrogen bonds. It should be stated that counting hydrogen bonds is not straightforward: their number depends on the energy criterion used. In Figure 10.13 the number of hydrogen bonds as a function of O–H···O distance is plotted for various values of the bond energy chosen. Although the distribution changes with the chosen energy level, it always shows a single peak. This renders the mixture and interstitial models less probable as they would probably result in a bimodal distribution. Computer simulations have taught us that any structural model in which there is a sharp distinction between molecules that are hydrogen-bonded and those that are not, is highly unlikely. A wide range of hydrogen bond interactions exist with no clear distinction between bonded and nonbonded molecules. Therefore, nowadays, the distorted H-bond network model is preferred.

10.5.3 Properties of Water

Let us now describe a few properties of water and compare the melting point, boiling point, and enthalpy of vaporization of the isoelectronic series of molecules

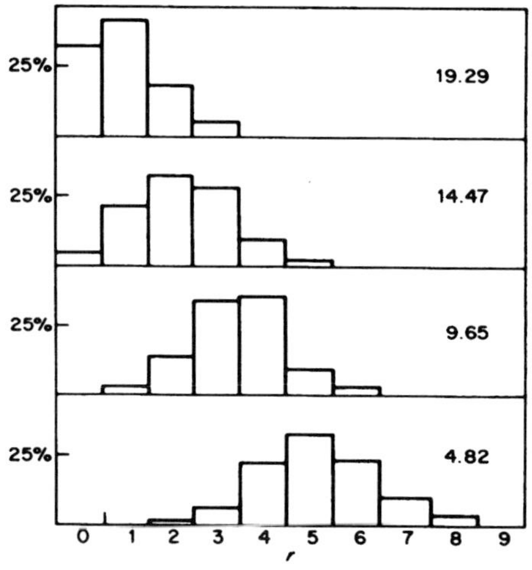

Figure 10.13 Numbers of hydrogen bonds as a function of O–H···O distance, plotted for various values of bond energy (values shown are kJ mol^{-1}). Data from Ref. [23].

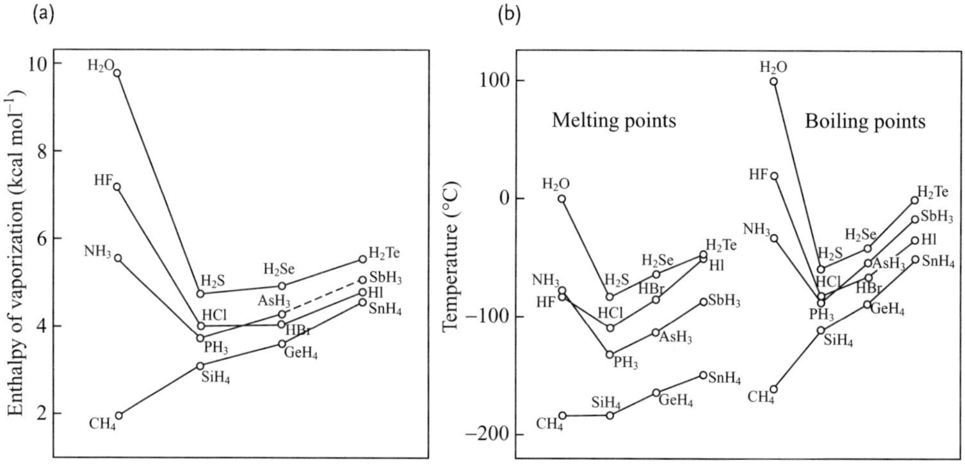

Figure 10.14 (a) Enthalpy of vaporization and (b) melting and boiling temperatures of isoelectronic series of hydrides, according to Pauling [9].

[9] CH_4, NH_3, H_2O, and HF, all of which have 10 electrons. In this series the nuclear charge of the central atom increases from 6 to 10. At first sight, one would expect a gradual increase in properties due to increasing intermolecular interactions. Figure 10.14 shows, however, that all three properties show an extremum for H_2O. A somewhat more elaborate comparison, also shown in Figure 10.14, indicates that the three properties for a series RH_n all increase with atomic weight

Table 10.9 Properties of selected liquids.

	T_n (°C)	$\Delta_{vap}H$ (cal g^{-1})	T_{mp} (°C)	$\Delta_{fus}H$ (cal g^{-1})	γ[a] (mJ m^{-2})	ε_r (–) (at T/K)	Dimer (kJ mol^{-1})
Ne	–246	415	–249	–	–	–	–
CH_4	–161	2800	–184	–	–	–	–
NH_3	–33	5550	–75	108.1	41.3	25 (195)	16
H_2O	100	9750	0	79.9	72.8	79 (298)	24
H_2S	–61	–	–86	–	–	9 (187)	8
HF	20	7220	–89	–	–	84 (273)	24
CH_3OH	65	289.2	–97	16	22.7	33.0 (298)	–
C_2H_5OH	78.3	204	–	–	22.3	24.0	–
$(CH_3)_2CO$	56	125	–95.5	23	23.7	21.4	–
C_6H_6	80	94	5.4	30	28.9	2.28	–
$C_6H_5NO_2$	151	79	5.7	22	–	–	–
CH_3COOH	118	97	16.6	45	27.6	–	–

a) γ-data at 20 °C except for NH_3 (at –75 °C, just above the melting point).

of R, provided that we ignore NH_3, H_2O, and HF. Extrapolating from the series SnH_4, GeH_4 and SiH_4 to CH_4 yields nearly the experimental value for CH_4 (this is also true for the series Xe, Kr, and Ar extrapolated to Ne). The corresponding extrapolation for molecules NH_3, H_2O and HF deviate, in particular for H_2O. Extrapolating the boiling point would yield about –90 °C instead of +100 °C, this deviation being due to the relatively strong intermolecular interactions. The bond energy of H_2O according to the reaction $H_2O \rightarrow 2H + O$ is 220 kcal mol^{-1} or 110 kcal mol^{-1} for (covalent) OH bonds.[9] Pauling has estimated the total enthalpy of sublimation of ice as 12.2 kcal mol^{-1}, and the contribution of the van der Waals interactions as one-fifth of that, leaving ~5 kcal mol^{-1} for a hydrogen bond. Although less than 5% of the value for the OH bond, it is the prime factor in the molecular interactions in hydrogen-bonded systems such as water. The small value $\Delta_{fus}H$ = 1.44 kcal mol^{-1} indicates that only about 15% of the hydrogen bonds are broken during melting.

The heat capacity C_P is high and approximately constant at 1 cal g^{-1} K^{-1}. Only C_P for NH_3 is larger, at about 1.23 cal g^{-1} K^{-1}, due mainly to the presence of hydrogen bonds. The enthalpy of vaporization $\Delta_{vap}H$ is also high (Table 10.9). Notably, it takes 1 kcal to heat 1 kg of water by 1 °C, but only 2 g of water have to evaporate to cool the rest down to the original temperature ($\Delta_{vap}H$ = 574 cal g^{-1} at 40 °C, about twice that for many other molecules). This is the main temperature-regulating mechanism for plants, animals and the environment, and is indeed a unique aspect of water. The enthalpy of fusion, 79.9 cal g^{-1}, is also larger than for many

9) The reaction $H_2O \rightarrow 2H + O$ can be described as two consecutive reactions, $H_2O \rightarrow OH + H$ with 119 kcal mol^{-1}, and $OH \rightarrow O + H$ with 101 kcal mol^{-1}. Accordingly, the average is 110 kcal mol^{-1}.

other molecules (except for NH_3 and a few fused salts such as KF, NaF and NaCl). For organic liquids a typical value is 45 cal g^{-1}.

It has already been mentioned that water expands during freezing. The density ρ of liquid water is thus higher than for ice, due mainly to the tetrahedral structure. It can be estimated that, if the H_2O molecules are treated as though they were contacting rigid spheres centered at the oxygen vertices of the lattice, the density would be only 57% of that for a close-packed arrangement of those spheres, so that a close-packed structure would result in $\rho \cong 1.75\,\text{g cm}^{-3}$. With increasing temperature, the density increases because a fraction of the hydrogen bonds break and the molecules then occupy the vacant space. However, the kinetic energy also increases, and this leads to a lower density. Overall, water shows a maximum in density at 4 °C. The surface tension γ of water is also one of the highest (except for some liquid metals and fused salts), at about 73 mJ m^{-2}, compared to typical values of about 30 mJ m^{-2} or less for organic liquids. In order to increase the surface area, molecules must be brought from the interior of the liquid to the surface, and the work required for this acts against the attractive forces. A similar consideration is valid for evaporation, so it is no wonder that compounds with a high $\Delta_{\text{vap}}H$ have also a high γ. Finally, we note that the relative permittivity ε_r, of about 80, is also higher than for many other compounds [except for liquid HCN ($\varepsilon_r = 116$) and formamide ($\varepsilon_r = 110$)], with typical values being 20 or less (compare the value for benzene, which is about 2.3). This renders the solvation capability of water for ions high, as the electrostatic forces are (very) effectively shielded. Whilst for a high value of ε_r a polar molecule is required, the cooperativity renders the permittivity of water particularly high.

Two main factors make H_2O "special": (i) H_2O is rather polar; and (ii) the number of hydrogen atoms equals the number of lone pair electrons. This, in view of the tetrahedral arrangement, determines the extensive 3D network with a high degree of cohesion. Although the N–H···N bond is somewhat weaker than the O–H···O bond, it is more important for NH_3 that it cannot form a network but only chains and rings. Similarly, although the F–H···F bond is stronger than the O–H···O bond (even in the gas phase, HF is largely polymerized), again it can form only chains and rings. Consequently the melting point, boiling point and enthalpy of vaporization are lower than for water. Although the structure of H_2S is similar to that of H_2O, the hydrogen atoms are largely buried in the large S atom sphere, rendering the S–H···S bond very weak. Hence, the structure is close-packed rather than tetrahedral.

The question to be answered finally is: Is water unique? As with many questions in science, this question deserves a nuanced answer. When comparing some of the properties of several small molecules (as listed in Table 10.9), we see that water is not exceptionable as judged by one parameter. Rather, the anomalous behavior of water is due to the combination of several properties. Unique to H_2O is the fourfold relatively strong hydrogen-bonded coordination that is the reason for the high surface energy, permittivity, and the volume expansion during freezing.[10)]

10) Other materials that expand on freezing also have fourfold coordination, including Ga, Ge, Si, Bi, and $RbNO_3$.

A compact review on the structure and some properties of water has been provided by Stillinger [15], while more extended reviews are available from Eisenberg and Kauzmann [16] and in the series edited by Franks [17]. Several other aspects have been collated in a recent report [18].

Problem 10.13

At 0 °C at zero pressure, the compressibility of H_2O is $\kappa_T = 0.51\,\text{GPa}^{-1}$, and reaches a minimum of $0.44\,\text{GPa}^{-1}$ at about 45 °C before increasing again with increasing temperature. Explain why this occurs.

Problem 10.14

Solid NO contains rectangular dimers with different side lengths. Estimate the residual entropy for NO.

References

1 See Böttcher (1973).
2 Minkin, V.I., Osipov, O.A., and Zhadanov, Y.A. (1970) *Dipole Moments in Organic Chemistry*, Plenum, New York.
3 Petro, A.J. (1958) *J. Am. Chem. Soc.*, **80**, 4230.
4 Onsager, L. (1936) *J. Am. Chem. Soc.*, **58**, 1486.
5 Kirkwood, J.G. (1939) *J. Chem. Phys.*, **7**, 911.
6 Pople, J.A. (1951) *Proc. Roy. Soc.*, **A205**, 163.
7 Coulson, C.A., and Eisenberg, D. (1966) *Proc. Roy. Soc.*, **A291**, 445.
8 Niesar, U., Corongiu, G., Clementi, E., Kneller, G.R., and Bhattacharya, D.K. (1990) *J. Phys. Chem.*, **94**, 7949.
9 Pauling, L. (1960) *The Nature of Chemical Bond*, 3rd edn, Cornell, Ithaca, NY., Ch. 12.
10 Praprotnik, M., Janezic, D., and Mavri, J. (2004) *J. Phys. Chem.*, **A108**, 11056.
11 Vega, C. and Abascal, J.L.F. (2011) *Phys. Chem. Chem. Phys.*, **13**, 19663.
12 (a) Paesani, F. and Voth, G.A. (2009) *J. Phys. Chem.*, **B113**, 5702; (b) Medders, G.R., Babin, V., and Paesani, F. (2013) *J. Chem. Theory Comp.* **9**, 1103.
13 Morgan, J. and Warren, B.E. (1938) *J. Chem. Phys.*, **6**, 666.
14 Narten, A.H. and Levy, H.A. (1972) *J. Chem. Phys.*, **55**, 2263; (1972) **56**, 5681.
15 Stillinger, F.H. (1980) *Science*, **209**, 451.
16 See Eisenberg and Kauzmann (1969).
17 See Franks (1973).
18 (a) Wernet, P., *et al.* (2004) *Science*, **304**, 995; (b) Zubavicus, Y. and Grunze, M. (2004) *Science*, **304**, 974; (c) Bukowski, R., *et al.* (2009) *Science*, **315**, 1249; (d) Malenkov, G. (2009) *J. Phys. Condens. Matter*, **21**, 83101.
19 van Vleck, J.H. (1932, reprint 1966) *The theory of electric and magnetic susceptibilities*. Oxford, p. 32.
20 Debye, P. (1929) *Polar molecules*. Chemical Catalog Company.
21 Buckingham, A.D. (1967) *Disc. Faraday Soc.*, **34**, 205.
22 Nagle, J.F. (1966) *J. Math. Phys.*, **7**, 1482, 1492.
23 Rahman, A. and Stillinger, F.H. (1971) *J. Chem. Phys.*, **55**, 3366.

Further Reading

Böttcher, C.J.F. (1973) *Theory of Electric Polarization*, vol. 1, 2nd edn, Elsevier, Amsterdam.

Eisenberg, D. and Kauzmann, W. (1969) *The Structure and Properties of Water*, Clarendon Press, Oxford.

Franks, F. (ed.) (1973) *Water, a Comprehensive Treatise*, Plenum, New York.

Fröhlich, H. (1958) *Theory of Dielectrics*, 2nd edn, Oxford University Press, Oxford.

Smyth, C.P. (1955) *Dielectric Behavior and Structure*, McGraw-Hill, New York.

11
Mixing Liquids: Molecular Solutions

So far, we have discussed pure liquids with increasing complexity. In this chapter we start with solutions and, in particular, with molecular solutions. We first deal briefly with some basic aspects, including molar and partial quantities, perfect, and ideal solutions. Thereafter, nonideal behavior is discussed, based on a treatment of the regular solution model. Finally, some possible improvements are indicated.

11.1
Basic Aspects

In this section we iterate and extend somewhat on some of the basic aspects that were introduced in Chapter 2. The content of a system is defined by the amount of moles n_α of chemical species α in the system, often denoted as components, which can be varied independently. A mixture is a system with more than one component. A homogeneous part of a mixture, that is, with uniform properties, is addressed as a phase, while a multicomponent phase is labeled a solution. The majority component of a solution is the solvent, while solute refers to the minority component. We restrict ourselves from the outset to binary mixtures and solutions and use, apart from n_α for the number of moles, N_α for the number of molecules and x_α for the mole fraction of component α. The label 1 refers always to the solvent (e.g., x_1 or $1 - x$), while the label 2 indicates the solute (e.g., x_2 or x). As usual, N_A denotes Avogadro's constant.

11.1.1
Partial and Molar Quantities

We refer for an arbitrary extensive quantity Z of a mixture to *molar quantities* Z_m defined by $Z_m = Z/\Sigma_\alpha n_\alpha$ or for single, pure component α by $Z_\alpha^* = Z/n_\alpha$. For solutions, it is also useful to discuss the situation with respect to extensive quantities Z in terms of *partial (molar) quantities* $Z_\alpha \equiv (\partial Z/\partial n_\alpha)_{P,T,n'}$, where n' indicates all components except for the component α. Generally, for $Z = Z(P,T,n_\alpha)$, we have

Liquid-State Physical Chemistry, First Edition. Gijsbertus de With.

$$dZ = (\partial Z/\partial P)dP + (\partial Z/\partial T)dT + \Sigma_\alpha Z_\alpha dn_\alpha \tag{11.1}$$

Since the extensive quantity $Z(P,T,n_\alpha)$ is a homogeneous function of degree 1 in n_α, we have by Euler's theorem[1)]

$$Z = \sum\nolimits_\alpha Z_\alpha n_\alpha \tag{11.2}$$

So, we have for dZ on the one hand Eq. (11.1), and on the other hand differentiating Eq. (11.2) $dZ = \Sigma_\alpha Z_\alpha dn_\alpha + \Sigma_\alpha n_\alpha dZ_\alpha$. Hence,

$$(\partial Z/\partial P)dP + (\partial Z/\partial T)dT = \sum\nolimits_\alpha n_\alpha dZ_\alpha \quad \text{or at constant } T \text{ and } P$$

$$\sum\nolimits_\alpha n_\alpha dZ_\alpha = 0 \tag{11.3}$$

This is the (*general*) *Gibbs–Duhem equation*, which puts a constraint on the possible changes in the properties of solutions.

For the volume, the partial volume is given by $V_\alpha = (\partial V/\partial n_\alpha)_{P,T,n'}$ and the total volume becomes at constant T and P, using Euler's theorem,

$$V_m = x_1V_1 + x_2V_2 \tag{11.4}$$

Similarly, for the Gibbs energy G we have the partial Gibbs energy $G_\alpha = (\partial G/\partial n_\alpha)_{P,T,n'}$, and since this quantity is equal to the chemical potential defined by $\mu_\alpha \equiv (\partial G/\partial n_\alpha)_{P,T,n'}$, the total Gibbs energy becomes

$$G_m = x_1G_1 + x_2G_2 = x_1\mu_1 + x_2\mu_2 \tag{11.5}$$

Note that the partial quantity G_α and chemical potential μ_α are equivalent, but that for the Helmholtz energy F one would have $\mu_\alpha = (\partial F/\partial n_\alpha)_{V,T,n'} \neq (\partial F/\partial n_\alpha)_{P,T,n'} = F_\alpha$.

For the Gibbs energy G we further obtain at constant P and T

► $$x_1 d\mu_1 + x_2 d\mu_2 = 0 \tag{11.6}$$

This is the Gibbs–Duhem equation (at constant P and T) of which another form is

$$d\mu_1 = -(x_2/x_1)d\mu_2 \quad \text{or} \quad \mu_1 - \mu_1^* = -\int_0^{\mu_2} \frac{x_2}{1-x_2} d\mu_2 = -\int_0^{x_2} \frac{x_2}{1-x_2}\frac{d\mu_2}{dx_2} dx_2 \tag{11.7}$$

where μ_1^* is the chemical potential of the pure component 1. This relation can be used for a consistency check on experimental data. If the molar Gibbs energy G_m is known, we can obtain μ_1 and μ_2 by differentiation with respect to x_2, since

$$\frac{\partial G_m}{\partial x_2} = \frac{\partial}{\partial x_2}(x_1\mu_1 + x_2\mu_2) = \mu_2 - \mu_1 \tag{11.8}$$

and solving μ_1 and μ_2 from Eqs (11.5) and (11.8) results in

1) Although we use Z_m for a *molar quantity*, often the subscript m is left out if the meaning is clear from the context. Moreover, while for an arbitrary amount n_α of component α the expression $Z = n_\alpha Z_\alpha$ holds exactly for any extensive quantity Z, it is sometimes approximated by $Z = n_\alpha Z_\alpha^*$, in particular for the volume V. Note that for a pure component, $Z_\alpha = Z_\alpha^*$.

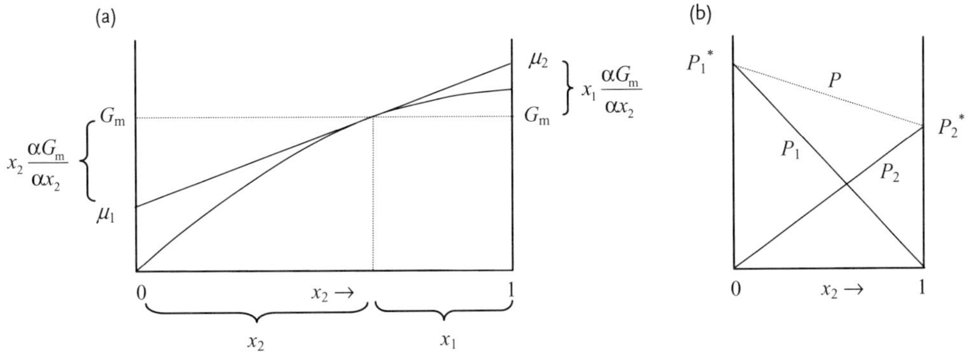

Figure 11.1 (a) Calculation of the chemical potentials μ_1 and μ_2 from the Gibbs energy G_m (although illustrated here for μ_1 and μ_2 using G_m, this calculation can be made for any partial quantity, remembering that μ_α equals G_α); (b) Pressure diagram for a perfect mixture, showing the total pressure P and the partial pressures P_1 and P_2.

$$\mu_1 = G_m + x_2 \frac{\partial G_m}{\partial x_1} = G_m - x_2 \frac{\partial G_m}{\partial x_2} \quad \text{and}$$
$$\mu_2 = G_m + x_1 \frac{\partial G_m}{\partial x_2} = G_m - x_1 \frac{\partial G_m}{\partial x_1} \tag{11.9}$$

Thus, if the dependence of G_m as function of x_2 (or x_1 for that matter) is known (Figure 11.1), the behavior of μ_1 and μ_2 can be obtained by differentiation. It will be clear that a similar reasoning applies to all partial quantities, for example, V_m. It will also be evident that G_m is less easily accessible experimentally than V_m.

11.1.2 Perfect Solutions

To describe perfect solutions we start with the perfect gas, and thereafter deal with perfect gas mixtures and generalize to perfect (fluid) mixtures. The *perfect gas* can be defined in several ways. Here, we choose a definition based on the chemical potential

$$\mu(T,P) = \mu^\circ(T) + RT\ln(P/P^\circ) \tag{11.10}$$

where P is the pressure, P° is a reference pressure (1 bar), and μ° is the value for $\mu = G_m = G/n$ if $P = P^\circ$. From this definition we obtain for an arbitrary number of moles n, $V = \partial G/\partial P = \partial(n\mu)/\partial P = nRT/P$, the well-known equation of state (EoS) for a perfect gas. Hence

$$F = G - PV = n[\mu^\circ - RT + RT\ln(nRT/VP^\circ)] \tag{11.11}$$

For *perfect gas mixtures* we assume that each component contributes independently from the others to the Helmholtz energy, so that we have

$$F_{mix} = \sum\nolimits_\alpha n_\alpha[\mu_i^\circ - RT + RT\ln(n_\alpha RT/VP^\circ)] \tag{11.12}$$

This leads to

$$G_{\text{mix}} = \sum_{\alpha} n_\alpha[\mu_\alpha^\circ + RT\ln(x_\alpha P/P^\circ)] = \sum_{\alpha} n_\alpha[\mu_\alpha^\circ + RT\ln(P_\alpha/P^\circ)] \tag{11.13}$$

where $P_\alpha \equiv x_\alpha P$ is referred to as the *partial pressure*[2]. The total pressure is accordingly given by $P = \Sigma_\alpha P_\alpha$ or, for a perfect binary mixture, by (Figure 11.2)

$$P = x_1 P + x_2 P = P_1 + P_2 \tag{11.14}$$

known as *Dalton's law of partial pressures.* For the perfect gas mixture the partial volume becomes $V_\alpha = \partial\mu_\alpha/\partial P = \partial \ln P_\alpha/\partial P = RT/P$, equal for all components α.

To calculate $\Delta_{\text{mix}}G$ for a binary gas mixture at P_1 and P_2 we subtract $n_1\mu_1 + n_2\mu_2$ at P_1' and P_2' from G_{mix} to obtain

$$\begin{aligned}\Delta_{\text{mix}}G &= n_1[\mu_1^\circ + RT\ln(P_1/P^\circ)] + n_2[\mu_2^\circ + RT\ln(P_2/P^\circ)]\\ &\quad - n_1[\mu_1^\circ + RT\ln(P_1'/P^\circ)] - n_2[\mu_2^\circ + RT\ln(P_2'/P^\circ)]\\ &= n_1RT\ln(P_1/P_1') + n_2RT\ln(P_2/P_2')\end{aligned} \tag{11.15}$$

From this expression we obtain the entropy of mixture $\Delta_{\text{mix}}S$ as

$$\begin{aligned}\Delta_{\text{mix}}S &= -\frac{\partial}{\partial T}\Delta_{\text{mix}}G = -\frac{\partial}{\partial T}[n_1RT\ln(P_1/P_1') + n_2RT\ln(P_2/P_2')]\\ &= -n_1R\ln(P_1/P_1') + n_2R\ln(P_2/P_2')\end{aligned} \tag{11.16}$$

In the special case that $P_1' = P_2' = P'$, that is, when the original pressures of each gas P_1' and P_2' are equal to the final pressure P', we have $P_\alpha/P_\alpha' = P_\alpha/P' = x_\alpha$ and

► $$\Delta_{\text{mix}}G = RT(n_1\ln x_1 + n_2\ln x_2) \quad \text{and} \quad \Delta_{\text{mix}}S = -R(n_1\ln x_1 + n_2\ln x_2) \tag{11.17}$$

A *perfect solution* is defined as having the same Gibbs energy of mixing as the perfect gas mixture for the whole composition range, that is, for one mole

$$\Delta_{\text{mix}}\mu = RT(x_1\ln x_1 + x_2\ln x_2) \quad \text{and} \quad \Delta_{\text{mix}}S_{\text{m}} = -R(x_1\ln x_1 + x_2\ln x_2) \tag{11.18}$$

Equivalently, we have

$$\mu_\alpha(T,P) = \mu_\alpha^*(T,P) + RT\ln x_\alpha \tag{11.19}$$

Equation (11.19) forms the basis for all further treatments. In practice only very few systems behave as a perfect mixture, and perfect behavior usually involves a comparable chemical nature of the solvent and solute, although this provides neither a guarantee, nor is required. Generally, strong deviations are observed. An example of both types of near-perfect solutions is shown in Figure 11.2, as are examples of a negative and positive from ideal behavior-behaving systems.

2) Note that the partial pressure is *not* a partial quantity as defined here. This is unfortunate but the designation is conventional.

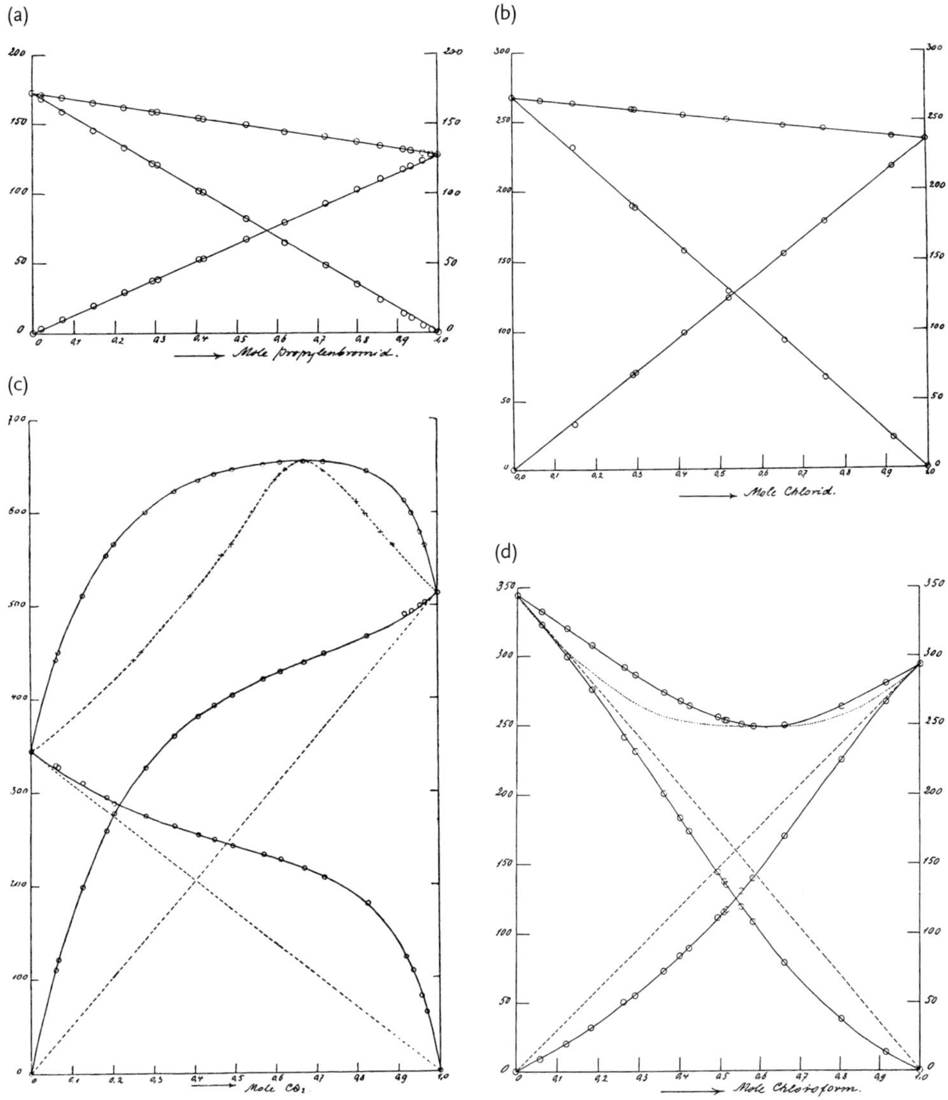

Figure 11.2 The vapor pressure of the ideally behaving system for: (a) Propylene bromide/ethylene bromide at 85 °C; (b) Ethylene chloride/benzene at 50 °C; (c) Positively-deviating-from-ideality system carbon disulfide/acetone at 35 °C; (d) Negatively-deviating-from-ideality system chloroform/acetone at 35 °C. Many books have used von Zawidski's data [20], usually without reference. It is therefore appropriate to show some of his original figures.

Problem 11.1

Recall that a real gas can be described by the virial expansion

$$PV = nRT(1 + nB_2/V + n^2B_3/V^2 + \ldots)$$

Using the inversion of the virial expansion in terms of P, show that the expression for G becomes

$$G - G^\circ = nRT\ln(P/P^\circ) + nB_2(P - P^\circ) + \frac{1}{2}n^2(B_3 - B_2^2)[P^2 - (P^\circ)^2] + \ldots$$

leading for sufficient low pressure to $\mu = \mu^\circ + RT \ln (P/P^\circ)$.

Problem 11.2

Show that for a perfect gas the energy U depends only on the temperature T.

Problem 11.3

Show that for a perfect gas mixture, $\Delta_{\text{mix}}H$ and $\Delta_{\text{mix}}V$ are both identically zero.

Problem 11.4

Verify Eq. (11.9).

Problem 11.5

Suppose that the molar volume of a binary solution of components 1 and 2 is given by $V_\text{m} = A + Bx_1 + Cx_1^2$. Calculate the partial volumes V_1 and V_2.

Problem 11.6

Suppose that the partial volume of solute 2 in solvent 1 is given by $V_2 = a + b(1 - x_2)^2$, while the molar volume V_m (1) of the pure solvent is given by c. Calculate the partial molar volume V_1 using the Gibbs–Duhem equation.

11.2 Ideal and Real Solutions

Perfect solutions behave as perfect gas mixtures over the complete concentration range. In practice, perfect solution behavior is obeyed only for a limited range of concentrations, and we label these cases as *ideal* solutions. For perfect and ideal behavior, a number of consequences follow of which we describe Raoult's and Henry's laws. Thereafter, we discuss how to deal with real solutions.

11.2.1
Raoult's and Henry's Laws

For ideal solutions we have (within the range of validity) the expression

$$\mu_\alpha(T,P) = \mu_\alpha^*(T,P) + RT\ln x_\alpha \tag{11.20}$$

while for the perfect gas the expression

$$\mu_\alpha(T,P) = \mu_\alpha^\circ(T) + RT\ln P_\alpha \tag{11.21}$$

holds. Using the equilibrium condition $\mu(\text{sol}) = \mu(\text{vap})$, we easily obtain

$$\mu_\alpha^* + RT\ln x_\alpha = \mu_\alpha^\circ + RT\ln P_\alpha \tag{11.22}$$

Solving for P_α leads to

$$P_\alpha = K_{\text{H},\alpha}x_\alpha \quad \text{with} \quad K_{\text{H},\alpha} = \exp[(\mu_\alpha^* - \mu_\alpha^\circ)/RT] \tag{11.23}$$

The parameter $K_{\text{H},\alpha}$ is independent of composition. For the solute we have

$$P_2 = K_{\text{H},2}x_2 \tag{11.24}$$

known as *Henry's law* (1803!), and stating that the partial pressure P_2 of solute 2 is proportional to the mole fraction in the liquid phase with a proportionality constant dependent on the difference in chemical potential, that is, in interaction, between solvent 1 and solute 2. For the solvent, the limiting situation is $x_1 = 1$ for which $K_1 = P_1^*$. Hence approximately

$$P_1 = P_1^* x_1 \tag{11.25}$$

known as *Raoult's law* (1886–1887), and stating that the partial pressure P_1 of solvent 1 is directly proportional to the vapor pressure P_1^* of the pure solvent. Obviously, for perfect solutions Raoult's law is valid for the complete composition range.

Some data for Henry's constant are given in Table 11.1. Example 11.1 shows how to use these concepts.

Table 11.1 Henry's constant K_H for gases dissolved in H_2O and C_6H_6 (in brackets) at 25 °C.[a)]

Gas	K_H (GPa)	Gas	K_H (GPa)	Gas	K_H (GPa)
CH_4	4.185 (0.0569)	Air	7.295	He	12.66
C_2H_2	0.135	N_2	8.7650 (0.239)	CO	5.79 (0.163)
C_2H_4	1.155	O_2	44 380	CO_2	1.670 (0.0144)
C_2H_6	3.06	H_2	7.16 (0.367)	H_2S	0.055

a) Data from Refs [21, 22].

Example 11.1

Assuming that sparkling water contains only H_2O (1) and CO_2 (2), we want to determine for a sealed can the compositions of the vapor and liquid phases and the pressure exerted at 10 °C, knowing that Henry's constant for CO_2 in H_2O at 10 °C is about 990 bar. There is only one degree of freedom according to the phase rule. We use the mole fraction x_2 of CO_2 in the liquid and assume it is 0.01. We denote the mole fraction in the gas phase by y. For the solute (2) we have Henry's law, while for the solvent (1) Raoult's law applies.

$$y_2 P = x_2 K_{H,2} \quad \text{and} \quad y_1 P = x_1 P_1^* \quad \text{or} \quad P = x_2 K_{H,2} + x_1 P_1^*$$

With $K_{H,2} = 990$ bar and $P_1^* = 0.01227$ bar (steam tables at 10 °C or Antoine's equation), $P = 9.912$ bar. By Raoult's law, $y_1 = x_1 P_2^*/P = 0.0012$. Therefore, $y_2 = 1 - y_1 = 0.9988$. As expected, the vapor is nearly pure CO_2.

11.2.2
Deviations

To describe deviations from perfect solution behavior we have a few options. The most well-known option is introducing the *activity* $a = \gamma x$, where γ is the *activity coefficient*. The activity coefficient $\gamma_\alpha = \gamma_\alpha(T,x)$ is introduced "to keep up appearances," that is, to keep the same formal expression for μ_α as for the perfect solution. Hence, $\mu_\alpha = \mu_\alpha^* + RT \ln x_\alpha$ becomes $\mu_\alpha = \mu_\alpha^* + RT \ln a_\alpha$, with activity $a_\alpha = \gamma_\alpha x_\alpha$. Obviously, $\gamma_{ideal} = 1$ always. There are two conventions for the activity coefficient.

- Convention I: $\gamma_\alpha \to 1$ for $x_\alpha \to 1$ and for $x_\alpha \to 0$. This convention is usually employed if the (liquid) solute is fully soluble in the solvent. Because, in a dilute solution of component β in α, molecules of type α are mainly surrounded by molecules of type α, component α behaves largely as if the liquid was pure. The reference state is thus the chemical potential of the pure liquid μ_α^*. This convention is also referred to as Raoult's law convention.
- Convention II: $\gamma_1 \to 1$ for $x_1 \to 1$ and $\gamma_2 \to 1$ for $x_2 \to 0$. This convention is usually employed if the (solid or liquid) solute is only partially soluble in the solvent. For the solvent this convention is identical to convention I. Because, in a dilute solution of component 2 in 1, molecules of type 2 do not interact with each other but mainly with molecules of type 1, the activity coefficient approaches a constant value, characteristic for the 1–2 interaction. Since we would like to have $\gamma_2 \to 1$ for $x_2 \to 0$, the reference state for the solute is taken as that of the pure solute in a hypothetical liquid state extrapolated to infinite dilution using Henry's law. Accordingly, this convention is referred to as Henry's law convention.

From the Gibbs–Duhem equation $x_1 d\mu_1 + x_2 d\mu_2 = 0$, we obtain (using $\mu = \mu^* + RT \ln \gamma x$) $x_1 \, d\ln\gamma_1 + x_2 \, d\ln\gamma_2 = 0$. Similarly, as for Eq. (11.7), some manipulation results in

$$\ln \gamma_2 = -\int_{\ln \gamma_{1,0}}^{\ln \gamma_1} \frac{x_1}{x_2} \mathrm{d}\ln \gamma_1 = -\int_0^{x_1} \frac{x_1}{1-x_1} \frac{\mathrm{d}\ln \gamma_1}{\mathrm{d}x_1} \mathrm{d}x_1 \tag{11.26}$$

where $\gamma_{1,0}$ is the limiting value of γ_1 for $x_1 \to 0$. Since for $x_1 \to 0$, $\gamma_1 \to 1$, the lower limit in the second integral should be taken as zero. This relation provides a check on the experimental data for γ_1 and γ_2 or allows one to calculate γ_2 if γ_1 is known.

Another way to introduce deviations from ideal behavior is by introducing the *osmotic coefficients* g and ϕ, defined for the solvent by

$$\mu_1 \equiv \mu_1^* + gRT\ln x_1 = \mu_1^* - gRT\ln(1+m_2M_1) \equiv \mu_1^* + RTm_2M_1\phi \tag{11.27}$$

where the molality $m_2 = x_2/x_1M_1$ and the mole fraction $x_1 = (1 + m_2M_1)^{-1}$ are used (why g and ϕ are addressed as osmotic coefficients will become clear in the next section). Since for high dilution we should regain ideal behavior, we require that for $x_1 \to 1$, $g \to 1$, and thus $\phi \to 1$. We have the relations $\ln\gamma_1 = (g-1)\ln x_1$ and $\phi = -(m_2M_1)^{-1}g\ln(1 + m_2M_1)^{-1} \cong g(1 - ½m_2M_1 + \ldots)$.

We also can define *excess functions* by

▶ $$Z^{\mathrm{E}} = \Delta_{\mathrm{mix}}Z - \Delta_{\mathrm{mix}}Z^{\mathrm{ideal}} \tag{11.28}$$

The partial quantities Z_α are directly related to the excess quantities Z^{E}. For example, for G we have $Z_\alpha = G_\alpha = \mu_\alpha$ and substitution of Eq. (11.7) in $G_{\mathrm{m}}^{\mathrm{E}} = x_1(\mu_1 - \mu_1^*) + x_2(\mu_2 - \mu_2^*)$, dividing by $x_1 = 1 - x_2$ and differentiating with respect to x_2 yields

$$\frac{\partial}{\partial x_2} \frac{G_{\mathrm{m}}^{\mathrm{E}}}{1-x_2} = \frac{\mu_2 - \mu_2^*}{(1-x_2)^2} \tag{11.29}$$

It is a matter of convenience whether excess or partial quantities are used. Figure 11.3 shows two examples of excess functions.

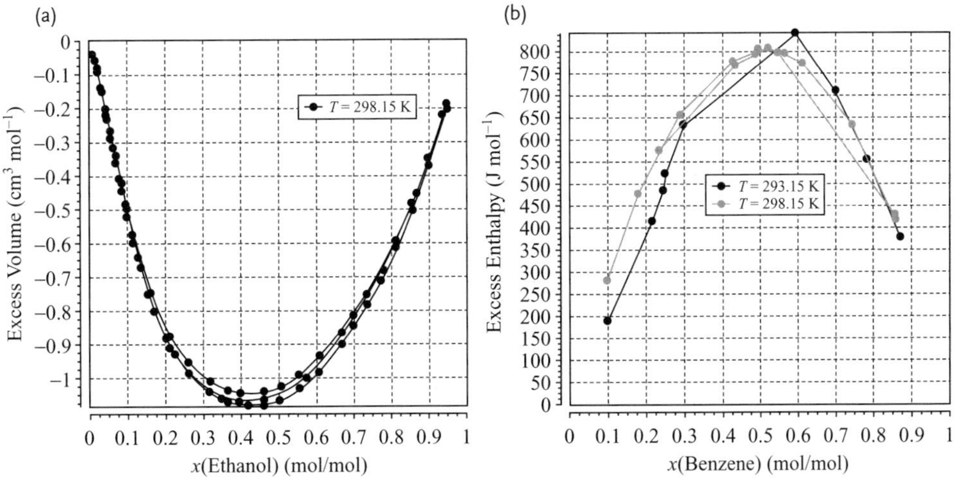

Figure 11.3 (a) Excess volume V^{E} for a mixture of H_2O–C_2H_5OH; (b) Excess enthalpy H^{E} for a mixture of benzene in cyclohexane. Images obtained from the Dortmund Data Base.

Problem 11.7

Verify Eqs (11.26) and (11.27).

11.3
Colligative Properties

So far, we have seen that the vapor pressure of solutions deviates from that for pure liquids. Solutions also show other deviations from pure liquids, for example, freezing point depression. To analyze this effect we assume that the solid solvent is a pure phase. Then, since at the freezing point the chemical potential of the solvent (component 1) is equal in the solid (s) and liquid (l) phases, we have

$$\mu_1(\mathrm{s}) = \mu_1(\mathrm{l}) + RT\ln a_1 \quad \text{or} \quad -\Delta_{\mathrm{fus}}G\,/\,RT = \ln a_1 \tag{11.30}$$

where $\Delta_{\mathrm{fus}}G \equiv \mu_1(\mathrm{l}) - \mu_1(\mathrm{s})$ is the Gibbs energy of fusion. Differentiation with respect to temperature and using the Gibbs–Helmholtz equation $\partial(\mu/T)/\partial(1/T) = H_{\mathrm{m}}$ yields

$$\frac{\partial \ln a_1}{\partial T} = \frac{\Delta_{\mathrm{fus}}H}{RT^2} \tag{11.31}$$

Upon integration, this equation from pure solvent conditions ($a_1 = 1$, $T = T^*$) to arbitrary values of a_1 and T, we obtain

$$\ln a_1 = \int_{T^*}^{T} \frac{\Delta_{\mathrm{fus}}H}{RT^2}\,\mathrm{d}T \tag{11.32}$$

which can be used to calculate the activity of the solvent in a solution. An approximation to Eq. (11.32) for sufficiently dilute solutions is obtained by using

$$\ln a_1 \cong \ln x_1 = \ln(1 - x_2) \cong -x_2 \tag{11.33}$$

Assuming that $\Delta_{\mathrm{fus}}H$ is independent of T over the range T^* to T, this results in

$$-x_2 = \frac{\Delta_{\mathrm{fus}}H}{R}\int_{T^*}^{T} \frac{\mathrm{d}T}{T^2} = -\frac{\Delta_{\mathrm{fus}}H}{R}\left(\frac{1}{T} - \frac{1}{T^*}\right) \tag{11.34}$$

or, since $T^{-1} - T^{*-1} = (T^* - T)/TT^* \cong \Delta T/T^{*2}$, in the freezing point depression

$$\Delta T = T^* - T = K_{\mathrm{fre}}x_2 \equiv \frac{RT^{*2}}{\Delta_{\mathrm{fus}}H}x_2 \tag{11.35}$$

A similar reasoning for boiling point elevation leads to a boiling point $T^* + \Delta T$, where T^* is the boiling point of the pure substance and $\Delta T = K_{\mathrm{boi}}x_2 \equiv R(T^*)^2x_2/\Delta_{\mathrm{vap}}H$. Although the temperature effects are small, the freezing point depression and boiling point elevation can be used to determine the molecular mass of the solute. Table 11.2 provides data of K_{fre} and K_{boi} for some solvents.

Table 11.2 Cryoscopic constants K_{fre} and ebullioscopic constants K_{boi} for some compounds.[a)]

Compound	K_{fre} (K kg mol^{-1})	K_{boi} (K kg mol^{-1})	Compound	K_{fre} (K kg mol^{-1})	K_{boi} (K kg mol^{-1})
Acetic acid	3.90	3.07	Naphthalene	6.94	5.8
Benzene	5.12	2.53	Phenol	7.27	3.04
CS_2	3.8	2.37	Water	1.86	0.51
CCl_4	30	4.95	Camphor	40	–

a) Data from Ref. [23].

Since in the dilute approximation the effects are only dependent on the number of solute molecules (and the type of solvent), these effects are denoted as *colligative properties*. Another colligative property is the *osmotic pressure*. This effect can be observed if we have a semipermeable membrane, that is, a membrane which is permeable to only one component (usually the solvent), with on one side the pure solvent at pressure P_0 and on the other side the solution at pressure P. This configuration leads to an excess pressure, normally addressed as osmotic pressure $\Pi = P - P_0$. The reasoning is similar as for the freezing point depression, and starts by equating the chemical potential of the pure solvent μ_1^* to that of the solution μ_1, so that we have

$$\mu_1(T,P) = \mu_1^*(T,P_0) \tag{11.36}$$

Since $\partial\mu_1/\partial P = V_1$, we may write

$$\mu_1(T,P) - \mu_1(T,P_0) = \int_{P_0}^{P} V_1\,\mathrm{d}P \text{ and thus} \tag{11.37}$$

$$\mu_1^*(T,P_0) - \mu_1(T,P_0) = \int_{P_0}^{P} V_1\,\mathrm{d}P \tag{11.38}$$

If we compare this expression with $\mu_1 = \mu_1^* + RT\ln a_1$, we have

$$RT\ln a_1 = -\int_{P_0}^{P} V_1\,\mathrm{d}P \cong -V_1^*\int_{P_0}^{P}\mathrm{d}P = -V_1^*\Pi \tag{11.39}$$

where the second step can be taken if we assume that the partial volume V_1 can be approximated by the molar volume of the pure solvent V_1^* and that the solution is incompressible. This yields an explicit expression for the activity of the solvent. Applying the volume and incompressibility approximation to Eq. (11.38), we obtain

$$\mu_1^*(T,P_0) - \mu_1(T,P_0) = V_1^*\Pi \tag{11.40}$$

which reduces, using $\mu_1 = \mu_1^* + gRT\ln x_1$ with g the osmotic coefficient, to

$$-gRT\ln x_1 = -gRT\ln(1-x_2) = -gRT\left(-x_2 - \frac{1}{2}x_2^2 - \ldots\right) = V_1^*\Pi \tag{11.41}$$

This expression explains why g is denoted as osmotic coefficient. For an ideal solution g = 1, and for $x_2 \to 0$ the expression reduces to the *van't Hoff law*

$$\blacktriangleright \quad \Pi \cong c_2RT \quad \text{with molarity} \quad c_2 = x_2/(x_1V_1^* + x_2V_2^*) \cong x_2/V_1^* \tag{11.42}$$

Alternatively, the use of ϕ leads to $\Pi = RTM_1m_2\phi/V_1^*$. For real solutions $g \neq 1$, and the easiest way to incorporate the effect is to write g as a power series in x_2 reading

$$g = 1 + C_2x_2 + \frac{1}{2}C_3x_2^2 + \ldots \quad \text{so that for a dilute solution } (c_2 \cong x_2/V_1^*)$$

$$\Pi = RTc_2\left[1 + \left(\frac{1}{2} + C_2\right)V_1^*c_2 + \ldots\right] \tag{11.43}$$

The coefficients C_2, C_3, . . . depend on the nature of the solute and solvent and, of course, on T and P. Note that the g-expansion (or *osmotic virial expansion*) starts with the term 1, because the limiting van't Hoff law should be recovered for $c_2 \to 0$. The osmotic effect is large, for example, for seawater with ~35 g l^{-1} NaCl or c_2(NaCl) ≅ 0.6 m, $\Pi \cong 27$ bar, to be compared with a freezing point depression of ~2 K for the same molarity. By using the osmotic pressure effect, the molar mass M_2 of the solute molecules can be determined by using a concentration series with decreasing c_2. The intercept of a plot of Π/w_2 versus w_2, where $w_2 = c_2/M_2$ is the weight fraction, provides M_2, independent of non-ideality. Normally, there is a distribution of molar masses, and since each component contributes equally to the sum, the mass so determined is the number average. For polymers, a molar mass up to 2×10 kg mol^{-1} can be determined since quite low molarities still yield measurable pressure differences. For increasing molar mass, however, the decreasing intercept RT/M_2 limits the accuracy.

Example 11.2

In the table below, the osmotic heights h at 298 K of various molarities c of poly(vinyl chloride) (PVC) in cyclohexanone ($\rho = 0.98$ g cm^{-3}) are given. The pressures are thus $\Pi = \rho gh$, where g is the acceleration due to gravity.

w_2 (g l^{-1})	1.0	2.0	4.0	7.0	9.0
h (cm)	0.28	0.71	2.01	5.10	8.00

From a graph of h/w_2 versus w_2, the intercept $I = RT/\rho gM_2$ is calculated as 0.21. Therefore, $M_2 = RT/\rho gI = 128$ kg mol^{-1}.

11.4
Ideal Behavior in Statistical Terms

Having reviewed some basics of ideal solutions in the previous sections, in the present section we reiterate some results for ideal behavior in statistical terms.

The starting point is to use a crystal lattice, with coordination number z, on which at each lattice cell a molecule is located. The size of the solvent and solute molecules is taken to be approximately the same, so that each cell is occupied by either a solvent or solute molecule. We also take only nearest-neighbor interactions into account. In this model, the internal energy U for a lattice of N similar molecules can be written as $U = \frac{1}{2}zNw_{11}$, where w_{11} is the "bond energy" between a pair of molecules of type 1. For a binary system of components 1 and 2, we then have to consider the 1–1, 2–2 and 1–2 pair interactions with an interaction[3] w_{11}, w_{22}, and w_{12}, respectively. If we then create two 1–2 bonds from a 1–1 and a 2–2 bond, the interaction change per bond is given by

$$\Delta U = w_{12} - \frac{1}{2}w_{11} - \frac{1}{2}w_{22} \equiv w \tag{11.44}$$

The parameter w so defined is often referred to as the *interchange energy*. Now, for such a mixed crystal to be ideal we assert that w should be zero. In this case any interchange of molecule 1 and 2 leaves the internal energy unchanged. Moreover, the distribution of molecules 1 and 2 will be random over the lattice cells. If we have a mixed crystal with $N = N_1 + N_2$ with $N_1 \equiv Nx_1$ molecules of type 1 (matrix) and $N_2 \equiv Nx_2$ molecules of type 2 (solute), the energy $U = \frac{1}{2}z(N_1w_{11} + N_2w_{22})$. Furthermore, the number of distinguishable ways W of arranging these molecules on the lattice is $W = N!/N_1!N_2! = N!/(Nx_1)!(Nx_2)!$, and therefore the entropy $S = k\ln W = -Nk\,(x_1\ln x_1 + x_2\ln x_2)$. The configurational Helmholtz energy $F = U - TS$ and hence the configurational partition function $Q = \exp(-\beta F)$ with, as before $\beta = 1/kT$, reads

$$Q = \frac{N!}{N_1!N_2!}\exp\left[-\frac{1}{2}\beta z\left(N_1w_{11} + \frac{1}{2}N_2w_{22}\right)\right] \tag{11.45}$$

The question is now: can we use this lattice model for liquid solutions? Obviously we can, but what are the limitations? The structure of liquids as described by the pair-correlation function $g(r)$ is far from static and continuously changing, although the average structure is constant and mainly determined by T and P. The average structure can be considered to be closer to an almost random packing of molecules with a certain coordination for the first and possibly second coordination shell than to lattice-like structure. A lattice-like structure is rare, although in some liquids much more order is present than in others, water (H_2O) being a good example. A lattice model is thus not a good reference structure for liquids. Nevertheless, for solutions lattice models were and still are frequently used, and we do so here. If we regard the liquid as quasi-crystalline (an image used previously in Chapter 8), the size of the molecules should be not too different[4]. Moreover, the internal degrees of freedom in the solution should be separable from the translational degrees of freedom and behave in the same way as in the pure liquid.

3) Note that for all i–j interactions the parameter $w_{ij} = -|w_{ij}| < 0$.

4) Bernal estimated that for more or less spherical molecules the ratio of molecular volumes must be between 1 and 2, equivalent to a diameter ratio of 1.26 (see Ref. [24]).

Overall, the concept of a lattice should not be taken too seriously: it is used to represent the average coordination in the first coordination shell, and we are dealing with nearest-neighbor interactions mainly.

If we accept this image, bearing in mind that so far we have $w = 0$, the configurational Helmholtz energy F, using Stirling's theorem, reads

$$F = -kT\ln Q = \frac{1}{2}N_1zw_{11} + \frac{1}{2}N_2zw_{22} + kT(N_1\ln N_1 + N_2\ln N_2) \tag{11.46}$$

Since in this model the Helmholtz energy of a pure liquid, labeled α, is[5)]

$$F_\alpha = -kT\ln Q_\alpha = \frac{1}{2}N_\alpha zw_{\alpha\alpha} \tag{11.47}$$

the Helmholtz energy of mixing $\Delta_{\text{mix}}F = F - x_1F_1 - x_2F_2$ is given by

$$\Delta_{\text{mix}}F = kT\left(N_1\ln\frac{N_1}{N} + N_2\ln\frac{N_2}{N}\right) = NkT(x_1\ln x_1 + x_2\ln x_2) \tag{11.48}$$

For the entropy of mixing $\Delta_{\text{mix}}S$ we thus regain, as expected,

$$\Delta_{\text{mix}}S = -\partial\Delta_{\text{mix}}F/\partial T = -Nk(x_1\ln x_1 + x_2\ln x_2) \tag{11.49}$$

The chemical potential $\mu_\alpha = \partial F/\partial N_\alpha$ becomes

$$\mu_\alpha = \partial F/\partial N_\alpha = \frac{1}{2}zw_{\alpha\alpha} + kT\ln(N_\alpha/N) = \frac{1}{2}zw_{\alpha\alpha} + kT\ln x_\alpha \tag{11.50}$$

and since for a pure component we have $\mu_\alpha^\circ = ½zw_{\alpha\alpha}$, we find

$$\mu_\alpha - \mu_\alpha^\circ = kT\ln(N_\alpha/N) = kT\ln x_\alpha \tag{11.51}$$

Finally, we recall that the (absolute) activity λ is given by $\mu = kT\ln\lambda$, and obtain

$$\lambda_\alpha / \lambda_\alpha^\circ = x_\alpha \tag{11.52}$$

so that, if we may regard the vapor phase as a perfect gas and use the partial pressure $P_\alpha = x_\alpha P$ for the activity (see Chapter 2), the final result becomes

$$P_\alpha / P_\alpha^\circ = x_\alpha \tag{11.53}$$

which we recognize as Raoult's law.

Problem 11.8

Using the lattice model with $N = N_1 + N_2$ sites, show that the entropy of mixing can be estimated, using the Boltzmann expression, as

$$\Delta_{\text{mix}}S = k\ln W = k\ln(N!/N_1!N_2!)$$

5) Note that in this model for the pure component $F_\alpha = U_\alpha - TS_\alpha = U_\alpha$.

with W the number of possible configurations and that this estimate leads to

$$\Delta_{\text{mix}}S = -kN(x_1 \ln x_1 + x_2 \ln x_2) \text{ with } x_1 = N_1/N \text{ and } x_2 = N_2/N.$$

11.5 The Regular Solution Model

Because deviations from the ideal solution behavior occur frequently, an improvement over the ideal solution model is required. These improvements can be classified according to whether they improve upon the enthalpy, the entropy, or both.

The nomenclature used is indicated in Table 11.3. However, it should be borne in mind that other authors frequently deviate from this terminology.

In this section we deal with the regular solution model of solute 2 in solvent 1. Hence, we discuss only energetic interactions because in this model the entropy is given by the perfect solution model, Eq. (11.18). With respect to the excess enthalpy there are two options. First, the enthalpy change is positive, that is, $H^{\text{E}} > 0$ which occurs if the 1–2 interaction is unfavorable. Second, the enthalpy change is negative, that is, $H^{\text{E}} < 0$ which is the case when the 1–2 interaction is favorable. Our task is thus to model H^{E}, which equals $\Delta_{\text{mix}}H$ because in the perfect solution model $\Delta_{\text{mix}}H = 0$.

We derive the expression for $\Delta_{\text{mix}}H$ from the lattice model with coordination number z. As described in Section 11.4, we assume a random distribution of molecules over the lattice, and consider nearest-neighbor interactions only. For a binary system of components 1 and 2 we must then consider the 1–1, 2–2, and 1–2 pair interactions with an interaction w_{11}, w_{22} and w_{12}, respectively[6]. If we create two 1–2 bonds from a 1–1 and a 2–2 bond, the interchange energy per bond w is given by

$$w \equiv w_{12} - \frac{1}{2}w_{11} - \frac{1}{2}w_{22} \tag{11.54}$$

Table 11.3 Nomenclature for solutions.

	ΔH	ΔS
Ideal solution	$\Delta H = 0$	$\Delta S = \Delta S^{\text{ideal}}$
Athermal solution	$\Delta H = 0$	$\Delta S \neq \Delta S^{\text{ideal}}$
Regular solution[a)]	$\Delta H \neq 0$ ($\Delta H = Ax_1x_2$)	$\Delta S = \Delta S^{\text{ideal}}$
Irregular solution[b)]	$\Delta H \neq 0$	$\Delta S \neq \Delta S^{\text{ideal}}$

a) Guggenheim refers to this type as a simple solution.
b) This might also be called a real solution.

6) We recall that for all i–j interactions the parameter $w_{ij} = -|w_{ij}| < 0$.

where w represents the interchange energy per bond, and zw represents the interchange energy for replacing on molecule 1 (2) by one molecule 2 (1) in the matrix of molecules 1 (2). If N_1 (N_2) denotes the number of type 1 (2) atoms and $N = N_1 + N_2$ is the total number of atoms, the probability of having an arbitrary site occupied by a molecule of type 1 (2) is $x_1 = N_1/N$ ($x_2 = N_2/N$). Hence, the probability for a (nearest-neighbor) pair 1–2 is x_1x_2, and similarly for a pair 2–1. The total probability for a pair is thus $2x_1x_2$ and consequently, because the total number of pairs is $\frac{1}{2}zN$, the number of 1–2 pairs is $\frac{1}{2}zN \cdot 2x_1x_2 = zNx_1x_2$. Similarly, the number of 1–1 pairs is $\frac{1}{2}zN \cdot x_1^2$ and the number of 2–2 pairs is $\frac{1}{2}zN \cdot x_2^2$.

The resulting interaction energy of mixing is then

$$\Delta_{\text{mix}}U = U\,(\text{after mixing}) - U\,(\text{before mixing})$$
$$= zN\left[\left(x_1x_2w_{12} + \frac{1}{2}x_1^2w_{11} + \frac{1}{2}x_2^2w_{22}\right) - \left(\frac{1}{2}x_1w_{11} + \frac{1}{2}x_2w_{22}\right)\right] \qquad (11.55)$$

This yields for one mole after a small calculation using $N = nN_A$

$$\blacktriangleright \quad \Delta_{\text{mix}}U_m = \Delta_{\text{mix}}U/n = \frac{1}{2}N_Azx_1x_2(2w_{12} - w_{11} - w_{22}) = N_Ax_1x_2zw \qquad (11.56)$$

Since for an ideal solution $U_m = x_1u_1 + x_2u_2 = \frac{1}{2}zw_{11}x_1 + \frac{1}{2}zw_{22}x_2$, we also have $U^E = \Delta_{\text{mix}}U$. For now, we interpret U as internal energy but come back to this interpretation later. Since in this model volume effects are absent (and in practice are often small), we thus also have for the enthalpy

$$\Delta_{\text{mix}}H_m = \Delta_{\text{mix}}U_m \quad \text{and} \quad H^E = U^E\,(\Delta V = 0) \qquad (11.57)$$

Figure 11.4 shows the behavior of $\Delta_{\text{mix}}H_m$ for various values of the parameter zw.

The excess Gibbs function $G^E = \Delta G - \Delta G^{\text{ideal}}$ can now easily be calculated. For ideal solutions, we calculated the Gibbs energy from the ideal mixing entropy only, as the enthalpy change was considered to be zero. In the case of regular solutions

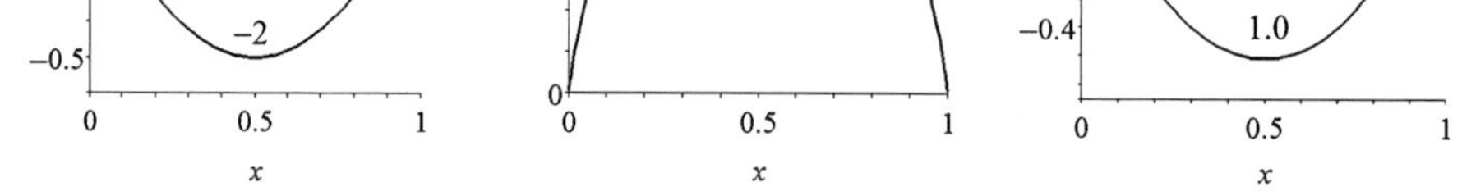

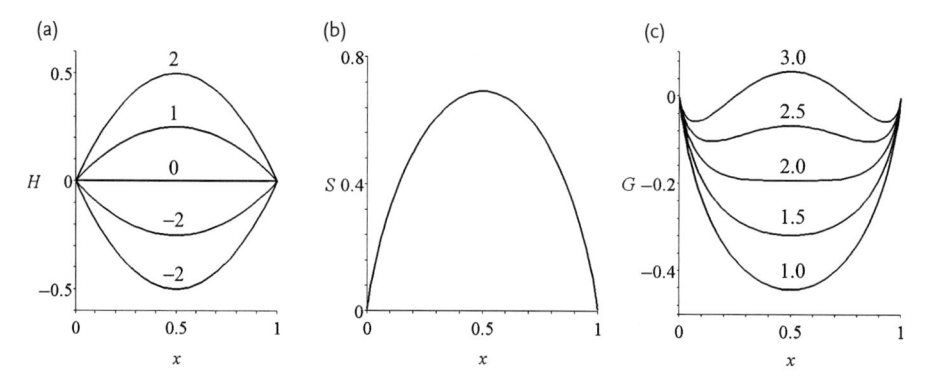

Figure 11.4 (a) The enthalpy of mixing $H = \Delta_{\text{mix}}H/RT$; (b) The entropy of mixing $S = \Delta_{\text{mix}}S/R$; (c) The Gibbs energy of mixing $G = \Delta_{\text{mix}}G/RT$ for various values of the interaction parameter zw, all as a function of mole fraction x.

we still use the ideal mixing entropy Eq. (11.18) (also plotted in Figure 11.4), in spite of the preferential interactions, and reading

$$\Delta_{\text{mix}}S_{\text{m}} = -R(x_1 \ln x_1 + x_2 \ln_2) \quad \text{or} \quad S^{\text{E}} = 0 \tag{11.58}$$

Hence, we also have $G^{\text{E}} = H^{\text{E}} = Nzwx_1x_2$. Finally, the molar Gibbs energy of mixing is given by

$$\Delta_{\text{mix}}G_{\text{m}} = N_{\text{A}}[zwx_1x_2 + kT(x_1 \ln x_1 + x_2 \ln x_2)] \tag{11.59}$$

In Figure 11.4 the behavior of $\Delta_{\text{mix}}G_{\text{m}}$ is sketched for various values of the parameter zw. We see that, depending on the value of the parameter zw, $\Delta_{\text{mix}}G_{\text{m}}$ shows a single or a double minimum; the latter situation leads to phase separation, to be discussed later.

Problem 11.9

Show that for the regular solution model volume changes are absent as long as the parameter zw is assumed to be pressure-independent.

11.5.1 The Activity Coefficient

The next step is to calculate the activity coefficients. To that purpose we recall that

$$G = \Sigma_\alpha n_\alpha \mu_\alpha \quad \text{with the mole number } n_\alpha = x_\alpha N/N_{\text{A}} \tag{11.60}$$

The chemical potential[7] is given by $\mu_\alpha = \mathring{\mu}_\alpha + RT \ln \gamma_\alpha x_\alpha$, so that G becomes

$$G = \Sigma_\alpha (n_\alpha \mathring{\mu}_\alpha + RTn_\alpha \ln x_\alpha + RTn_\alpha \ln \gamma_\alpha) \tag{11.61}$$

Since the Gibbs energy for the ideal solution is given by

$$G^{\text{ideal}} = \Sigma_\alpha (n_\alpha \mathring{\mu}_\alpha + RTn_\alpha \ln x_\alpha) \tag{11.62}$$

we thus have

$$G^{\text{E}} = \Delta_{\text{mix}}G - \Delta_{\text{mix}}G^{\text{ideal}} = (G - G_1 - G_2) - (G^{\text{ideal}} - G_1 - G_2) = RTn_\alpha \ln \gamma_\alpha \tag{11.63}$$

The activity coefficient γ_α (convention I) thus can be calculated from

$$\blacktriangleright \quad \partial G^{\text{E}}/\partial n_\alpha = RT \ln \gamma_\alpha \tag{11.64}$$

In the previous paragraph we derived using the regular solution model for the excess Gibbs energy

$$G^{\text{E}} = \Delta G - \Delta G^{\text{ideal}} = Nzwx_1x_2 = N_{\text{A}}zwn_1n_2/(n_1 + n_2) \tag{11.65}$$

7) The reference value $\mathring{\mu}_\alpha$ depends on the choice of the composition parameter used. A different value is thus used for the mole fraction, molarity or molality scale. As long as one adheres to a particular choice, this provides no problem.

Hence, the result for the activity coefficients, with $\beta = 1/kT$, is

$$\blacktriangleright \quad \ln\gamma_1 = \beta zwx_2^2 \quad \text{and} \quad \ln\gamma_2 = \beta zwx_1^2 \tag{11.66}$$

11.5.2
Phase Separation and Vapor Pressure

Mixtures are not always solutions, and from the above model we can assess the phase behavior of a mixture. The critical point for phase separation can be obtained from

$$\frac{\partial\mu_1}{\partial x_1} = \frac{\partial\mu_2}{\partial x_2} = \frac{\partial^2\mu_1}{\partial x_1^2} = \frac{\partial^2\mu_2}{\partial x_2^2} = 0 \tag{11.67}$$

Because $\mu = \mu° + RT\ln a = \mu° + RT\ln\gamma x$ and $\ln\gamma_1 = \beta zwx_2^2$ and $\ln\gamma_2 - \beta zwx_1^2$, we have

$$\frac{\partial\mu_1}{\partial x_1} = \frac{\partial\ln a_1}{\partial x_1} = \frac{1}{1-x_2} - 2x_2zw = 0 \qquad \frac{\partial^2\mu_1}{\partial x_1^2} = -\frac{1}{(1-x_2)^2} + 2zw = 0 \tag{11.68}$$

$$\frac{\partial\mu_2}{\partial x_2} = \frac{\partial\ln a_2}{\partial x_1} = \frac{1}{x_2} - 2(1-x_2)zw = 0 \qquad \frac{\partial^2\mu_2}{\partial x_2^2} = -\frac{1}{x_2^2} + 2zw = 0 \tag{11.69}$$

These equations are satisfied for $x_1 = x_2 = 0.5$ (as would be expected on symmetry grounds) and $(\beta zw)_{\text{cri}} = 2$. The critical temperature T_{cri} for phase separation is thus

$$\blacktriangleright \quad T_{\text{cri}} = zw/2k \tag{11.70}$$

At this point the activity coefficient is $\ln\gamma = 0.5$ or $\gamma = 1.649$, and the activity becomes $a = \gamma x = 0.824$. In fact, using the expression for $\Delta_{\text{mix}}G$, we can obtain the complete solution of the phase separation problem via $\partial\Delta_{\text{mix}}G/\partial x_2 = 0$. This leads to

$$\ln(x_2/x_1) = \beta zw(2x_2 - 1) \tag{11.71}$$

Using the transformation $s = 2x_2 - 1$ the final solution is

$$\beta zw = s^{-1}\ln\frac{1+s}{1-s} = 2s^{-1}\tanh^{-1}s \quad \text{or} \quad s = \tanh\left(\frac{1}{2}\beta zws\right) \tag{11.72}$$

representing the familiar implicit, *zeroth order* (or *mean field*) solution in x_2 for the order–disorder problem (see Appendix C).

The total pressure in the regular solution model becomes

$$P = P_1^*\gamma_1x_1 + P_2^*\gamma_2x_2 = P_1^*x_1\exp(\beta zwx_2^2) + P_2^*x_2\exp(\beta zwx_1^2) \tag{11.73}$$

and at T_{cri} becomes

$$P_{\text{cri}} = 0.824(P_1^* + P_2^*) \tag{11.74}$$

to be compared with $P = (P_1^* + P_2^*)$ for completely insoluble liquids and $P = 0.5(P_1^* + P_2^*)$ for completely soluble liquids. To check the agreement with

Table 11.4 Parameters $H^{E\prime} = H^E/Rx_1x_2$, $S^{E\prime} = S^E/Rx_1x_2$ and $C^{E\prime} = C^E/Rx_1x_2$ for some systems.

System	T (K)	$H^{E\prime}$ (–)	$S^{E\prime}$ (K)	T (K)	$H^{E\prime}$ (–)	$S^{E\prime}$ (K)	$C^{E\prime}$ (–)[a)]
CCl_4–C_6H_6	298	52.4	0.044	343	86.0	0.15	0.76
C_6H_6-*cyclo*-C_6H_{12}	293	400	0.84	343	324	0.60	−1.6
C_6H_6-1,2$C_2H_4Cl_2$	298	32	0.060	343	28	0.060	0.009

System	T (K)	$H^{E\prime}$ (–)	$S^{E\prime}$ (K)	System	T (K)	$H^{E\prime}$ (–)	$S^{E\prime}$ (K)
CCl_4-*cyclo*-C_6H_{12}	313	68	0.11	C_6H_6–$C_6H_6CH_3$	353	24	0.076
CS_2–$(CCH_3)_2CO$	308	700	0.64	*n*-C_5F_{12}-*n*-C_5H_{12}	277	744	0.56

Data from Pitzer [1].
a) The value for *C* is assumed constant, and applies to both temperatures.

experiment we use the experimental vapor pressure P_{exp} for equimolar solutions. From Eq. (11.73) we obtain $\beta zw = 4\ln[2P_{exp}/(P_1^* + P_2^*)]$. It appears that normally the value of zw decreases with increasing temperature, roughly according

$$zw = a + bT \tag{11.75}$$

although exceptions do occur. For example, for the system CS_2–$CHCl_3$, zw (J mol^{-1}) = 5305 − 13.2T, and for the system $(C_2H_5)_2O$–$CHCl_3$, zw (J mol^{-1}) = −7368 + 15.5T. These examples make it clear that zw cannot be related to the values for w_{ij} only, since the w_{ij} are assumed to be temperature-independent. Experimental thermodynamic values for several systems are given in Table 11.4. Problem 11.15 reveals that the description $zw = a + bT$ is generally insufficient.

Problem 11.10: Regular solution model

Figure 11.3 shows the excess enthalpy H^E for a solution of C_6H_6 in cyclo-C_6H_{12}.

a) Determine the interaction parameter w using the regular solution model assuming a coordination number of $z = 10$.
b) Calculate the activity coefficient for this solution for $x(C_6H_6) = 0.25$.

Problem 11.11

Verify Eqs (11.70) and (11.72).

Problem 11.12

A student suggests that the expression for the activity coefficient for one component is given as $\ln \gamma_1 = A(1 - x_1)^2$, and that the expression for the second component can be obtained right away by changing the indices. Is this correct?

Problem 11.13

Calculate the vapor pressure for component 2 in a regular solution of 2 in 1.

11.5.3
The Nature of w and Beyond

There remains to be discussed the physical interpretation of the parameter w and how a value for w can be obtained. Let us first discuss the interpretation. In the discussion of the lattice model, we (deliberately) often addressed the parameters w_{11}, w_{22} and w_{12} as interactions. Originally, $2w = 2w_{12} - w_{11} - w_{22}$ was interpreted as the energy (enthalpy) required to create two 1–2 pairs (bonds) from one 1–1 pair and one 2–2 pair. However, it was realized by Guggenheim [2] that w is averaged over all accessible states of the remainder of the system, each weighted with the relevant Boltzmann factor. In this sense, w represents – if taken at constant volume V – the Helmholtz energy, dependent on temperature T and the mole fraction x_1, or – if taken at constant pressure P – the Gibbs energy, again dependent on temperature T and the mole fraction x_1. Of course, the mixing entropy has still to be added. Hence, we can interpret

$$\Delta_{\text{mix}}G = Nzwx_1x_2 + NkT(x_1 \ln x_1 + x_2 \ln x_2) \tag{11.76}$$

or, equivalently, turning to molar excess quantities,

$$G_{\text{m}}^{\text{E}} = \Delta_{\text{mix}}G - \Delta_{\text{mix}}G^{\text{ideal}} = N_{\text{A}}zwx_1x_2 \tag{11.77}$$

as an integral expression for the excess Gibbs energy. Using for the moment $w_{\text{m}}(T,P) = N_{\text{A}}w\,(T,P)$ yields $G_{\text{m}}^{\text{E}} = w_{\text{m}}zx_1x_2$, from which we have

$$S_{\text{m}}^{\text{E}} = -\frac{\partial G_{\text{m}}^{\text{E}}}{\partial T} = -x_1x_2\frac{\partial zw_{\text{m}}}{\partial T},\ H_{\text{m}}^{\text{E}} = G_{\text{m}}^{\text{E}} - T\frac{\partial G_{\text{m}}^{\text{E}}}{\partial T} = x_1x_2\left(zw_{\text{m}} - T\frac{\partial zw_{\text{m}}}{\partial T}\right)_P \tag{11.78}$$

A nice illustration is presented by the system CCl_4–C_6H_{12} for which both G^{E} and H^{E} data are available [3], both plotted in Figure 11.5 as a function of x_1x_2. In fact, these data can be described well by the single expression

$$zw_{\text{m}}(\text{J mol}^{-1}) = 1176 + 1.96T\ln T - 14.18T \quad \text{so that} \tag{11.79}$$

$$(zw_{\text{m}} - T\partial zw_{\text{m}}/\partial T)(\text{J mol}^{-1}) = 1176 - 1.96T \tag{11.80}$$

Several other systems obey this approximate relationship (Table 11.4).

Let us now turn to a simple theoretical estimate for w. In the vdW EoS theory, where the parameter a represents the interactions, it was assumed that for mixtures $a_{12} = (a_{11}a_{22})^{1/2}$. Quantum considerations based on the vdW interaction model indicate indeed, that if some severe approximations are made (see Section 3.6), the interaction energy w_{12} between molecules 1 and 2 may be approximated[6)] by $w_{12} = -(w_{11}w_{22})^{1/2}$, where w_{11} and w_{22} refer to the interaction energies for the pairs 1–1 and 2–2, respectively. Using this approximation, denoted as *Berthelot's rule*, we write

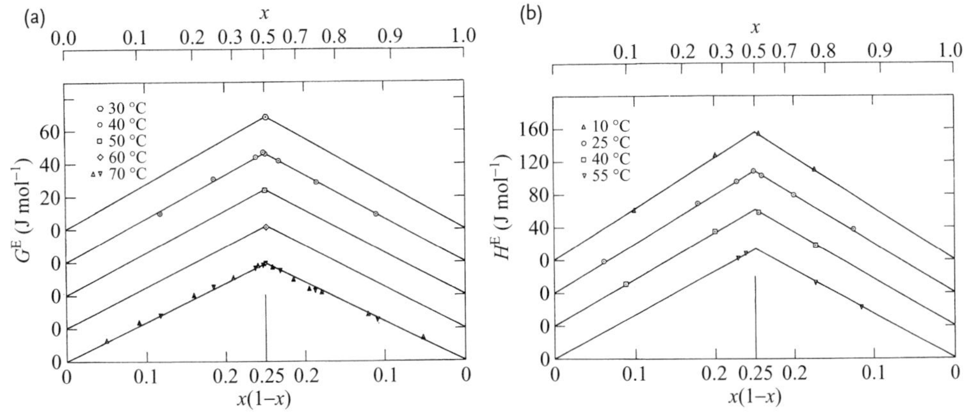

Figure 11.5 (a) Excess Gibbs energy function G^E and (b) excess enthalpy of mixing H^E for the system carbon tetrachloride and cyclohexane. For clarity, the various curves have been shifted vertically.

$$2w = 2w_{12} - w_{11} - w_{22} = -2(w_{11}w_{22})^{1/2} + |w_{11}| + |w_{22}| = (|w_{11}|^{1/2} - |w_{22}|^{1/2})^2 \tag{11.81}$$

This expression is always positive. In this approximation the internal energy change on vaporization for a pure liquid is $\Delta U_1 = \frac{1}{2}N_A z w_{11}$, and this provides a basis for estimating w for mixtures of (nonpolar) molecules. This so-called solubility parameter approach will be discussed in Section 11.6 and Chapter 13.

Having discussed the intermolecular interactions in Chapter 3, it seems logical to try to estimate values for w accordingly. We recall Eq. (3.18) reading

$$\phi_{\mathrm{vdW}} = -\mathcal{C}^2\left(\frac{\mu_1^2\mu_2^2}{3kT} + \mu_1^2\alpha_2 + \mu_2^2\alpha_1 + \frac{3}{2}\alpha_1\alpha_2\frac{I_1I_2}{I_1+I_2}\right)\frac{1}{r^6} \tag{11.82}$$

where the first term represents the dipolar energy (remember that for an interaction ~$1/T$, $U = 2F$), the second and third terms represent the induction energy, and the fourth term the dispersion energy. One easily obtains, with σ the molecular size, for the dipolar contribution w_{dip}, the induction contribution w_{ind}, and the dispersion contribution w_{dis},

$$w_{\mathrm{dip}} = -\beta\mathcal{C}^2 z(\mu_1^2 - \mu_2^2)/3kT\sigma^6, \tag{11.83}$$

$$w_{\mathrm{ind}} = -\mathcal{C}^2 z(\mu_1^2 - \mu_2^2)(\alpha_1 - \alpha_2)/\sigma^6 \quad \text{and} \tag{11.84}$$

$$w_{\mathrm{dis}} = -3\mathcal{C}^2 z[(I_1\alpha_1 - I_2\alpha_2)^2 + (\alpha_1 - \alpha_2)^2]/8(I_1 + I_2)\sigma^6, \tag{11.85}$$

respectively. Although the order of magnitude of the values obtained is given correctly, the agreement with the experimental values is not particularly striking, while the experimentally observed form $zw = a + bT$ $(+ cT\ln T)$ is not reproduced. The reason will be clear: w represents the Gibbs or Helmholtz energy, not the internal energy.

The regular solution model yields symmetric behavior with respect to the two components 1 and 2, although this is not so experimentally in many cases.

Moreover, although H^E, S^E and V^E may exhibit complex behavior, G^E is often still rather symmetric and described by $G^E = Nzwx_1x_2$. There are a number of deficiencies to remedy to go to beyond the regular solution model. First and foremost, there is an inconsistency in the coordination assumptions. We assumed that the coordination is random while there are preferred interactions leading to an interaction enthalpy, and the assumption of random mixing entropy is therefore not correct. However, the correction for this effect appears to be small. Second, the enthalpy estimate is based on nearest-neighbors on a lattice. However, the regular solution model can be freed of the lattice concept, as will be shown in Section 11.6. Further systematic improvement is possible, and some attempts are discussed in Section 11.8.

Although, today, lattice-like models are used less often for small-molecule liquids, they are still instructive because they are simple, and useful because they represent reasonably well the thermodynamics of solutions. Moreover, these models can be systematically improved, which implies that the behavior can be described by just a few parameters. For polymer solutions, lattice models are still extensively used and, of course, simulations can be made more easily nowadays then in the past. Nonetheless, the accuracy of simulations for solutions may still be insufficient.

Problem 11.14

Show that using $G^E_m = zw_m x_1 x_2$ the heat capacity is $C^E_m = -x_1x_2T(\partial^2 zw_m / \partial T^2)_P$.

Problem 11.15

Show that, if $zw = a + bT$, the parameters a and b relate to the enthalpy and entropy, respectively. Indicate why, considering data from Table 11.4, this description cannot be sufficient. Using $zw = a + bT + cT\ln T$, interpret and estimate a, b and c for the systems CCl_4–C_6H_6 and C_6H_6-*cyclo*-C_6H_{12}.

Problem 11.16

Verify the expressions for w_{dip}, w_{ind}, and w_{dis}.

11.6
A Slightly Different Approach

So far the model has been based on mole fractions, but here we will iterate the arguments for the regular solution model in slightly different terms, that is, using the pair-correlation function [4]. From that set of arguments we will see that volume fractions are to be preferred. Other arguments in favor of using volume fractions are given in Section 11.8. The outcome leads to the solubility parameter approach, based on the Berthelot rule, Eq. (11.81), and therefore we will assess the validity of that approximation once more. Finally, it leads to the so-called one-fluid and two-fluid models of mixtures.

The argumentation starts with the energy expression

$$U = \langle E_{\text{kin}} \rangle + \langle E_{\text{pot}} \rangle = \frac{3NkT}{2} + \frac{N^2}{2V} \int_0^\infty u(r) g(r) 4\pi r^2 \,\mathrm{d}r \tag{11.86}$$

in which the first term represents the kinetic energy and the second term the potential energy (denoting the integral by U_0 for later reference). Furthermore, $u(r)$ and $g(r)$ represent the pair energy[8] and pair-correlation function, respectively. We extend this expression to a binary mixture, considering the potential energy only. In a solution around a central molecule of component 1, molecules of type 1 and 2 are found. The probability of finding a molecule of component 1 is given by $(N_1/V)g_{11}(r)$, while the probability of finding a molecule of component 2 is given by $(N_2/V)g_{12}(r)$. Similar relationships hold for a central molecule of component 2. Since the probabilities that the central molecule is 1 or 2 are x_1 and x_2, respectively, we may write for the total volume V, using V_α and V_m for the partial and molar volume[9] of component α and the solution, respectively,

$$V = n_1V_1 + n_2V_2 = (n_1 + n_2)(x_1V_1 + x_2V_2) \equiv nV_\text{m} \tag{11.87}$$

$$\langle E_{\text{pot}} \rangle = \frac{2\pi N_\text{A}^2 n}{x_1V_1 + x_2V_2} \left[x_1 \left(x_1 \int_0^\infty u_{11} g_{11} r^2 \,\mathrm{d}r + x_2 \int_0^\infty u_{21} g_{21} r^2 \,\mathrm{d}r \right) \right. \\ \left. + x_2 \left(x_1 \int_0^\infty u_{12} g_{12} r^2 \,\mathrm{d}r + x_2 \int_0^\infty u_{22} g_{22} r^2 \,\mathrm{d}r \right) \right] \tag{11.88}$$

Remembering that $u_{12} = u_{21}$ and $g_{12} = g_{21}$, we may further simplify, meanwhile using volume fractions $\phi_i = (x_iV_i)/(x_1V_1 + x_2V_2)$, to obtain

$$\langle E_{\text{pot}} \rangle = (n_1V_1 + n_2V_2) 2\pi N_\text{A}^2 \left(\frac{\phi_1^2}{V_1^2} \int_0^\infty u_{11} g_{11} r^2 \,\mathrm{d}r \right. \\ \left. + \frac{2\phi_1\phi_2}{V_1V_2} \int_0^\infty u_{21} g_{21} r^2 \,\mathrm{d}r + \frac{\phi_2^2}{V_2^2} \int_0^\infty u_{22} g_{22} r^2 \,\mathrm{d}r \right) \tag{11.89}$$

Subtracting the energy of the separate components, given by

$$U_1 + U_2 = (n_1V_1 + n_2V_2) 2\pi N_\text{A}^2 \left(\frac{\phi_1}{V_1^2} \int_0^\infty u_{11} \mathring{g}_{11} r^2 \,\mathrm{d}r + \frac{\phi_2}{V_2^2} \int_0^\infty u_{22} \mathring{g}_{22} r^2 \,\mathrm{d}r \right) \tag{11.90}$$

we obtain for the mixing energy

$$\Delta_{\text{mix}} U = U - U_1 - U_2 = (n_1V_1 + n_2V_2) 2\pi N_\text{A}^2 \\ \times \left[\frac{\phi_1^2}{V_1^2} \int_0^\infty u_{11} (g_{11} - \mathring{g}_{11}) r^2 \,\mathrm{d}r + \frac{\phi_2^2}{V_2^2} \int_0^\infty u_{22} (g_{22} - \mathring{g}_{22}) r^2 \,\mathrm{d}r \right. \\ \left. + \phi_1\phi_2 \left(\frac{2}{V_1V_2} \int_0^\infty u_{21} g_{21} r^2 \,\mathrm{d}r - \frac{1}{V_1^2} \int_0^\infty u_{11} \mathring{g}_{11} r^2 \,\mathrm{d}r - \frac{1}{V_2^2} \int_0^\infty u_{22} \mathring{g}_{22} r^2 \,\mathrm{d}r \right) \right] \tag{11.91}$$

8) Here we use u for the pair potential energy since ϕ is (conventionally) used for the volume fraction.

9) The partial volume V_α is often approximated by the molar volume of the pure component V_α^*.

So far everything is exact, apart from the pair potential approximation. The next argument used is *scaling*. We use, based on the principle of corresponding states,

$$u_{11} = \varepsilon_{11} f(r/\sigma_{11}) \quad u_{12} = \varepsilon_{12} f(r/\sigma_{12}) \quad u_{22} = \varepsilon_{22} f(r/\varepsilon_{22}) \tag{11.92}$$

Moreover, we use the experimental similarity of g for simple molecules and write

$$g_{11} = \overset{\circ}{g}_{11} = g(r/\sigma_{11}) \quad g_{12} = g(r/\sigma_{12}) \quad g_{22} = \overset{\circ}{g}_{22} = g(r/\sigma_{22}) \tag{11.93}$$

in combination with $\sigma_{12} = ½(\sigma_{11} + \sigma_{22})$. Defining $y = r/\sigma$, we obtain for 1 mole

$$\Delta_{\text{mix}}U_{\text{m}} = V_{\text{m}}\phi_1\phi_2\left(\frac{2\varepsilon_{12}\sigma_{12}^3}{V_1V_2} - \frac{\varepsilon_{11}\sigma_{11}^3}{V_1^2} - \frac{\varepsilon_{22}\sigma_{22}^3}{V_2^2}\right)2\pi N_{\text{A}}^2\int_0^\infty f(y)g(y)y^2\,\text{d}y \tag{11.94}$$

We now make the identification

$$c_{\alpha\beta} = \frac{\varepsilon_{\alpha\beta}\sigma_{\alpha\beta}^3}{V_\alpha V_\beta}2\pi N_{\text{A}}^2\int_0^\infty f(y)g(y)y^2\,\text{d}y \tag{11.95}$$

which leads to

$$\Delta_{\text{mix}}U_{\text{m}} = V_{\text{m}}(2c_{12} - c_{11} - c_{22})\phi_1\phi_2 \tag{11.96}$$

Assuming that the Berthelot approximation $c_{12} = -(c_{11}c_{22})^{1/2}$ is valid, we get

$$\Delta_{\text{mix}}U_{\text{m}} = V_{\text{m}}(|c_{11}|^{1/2} - |c_{22}|^{1/2})^2\phi_1\phi_2 \tag{11.97}$$

or, defining $\delta_\alpha \equiv |c_{\alpha\alpha}|^{1/2} = |w_{\alpha\alpha}/V_\alpha|^{1/2}$,

$$\Delta_{\text{mix}}U_{\text{m}} = V_{\text{m}}(\delta_1 - \delta_2)^2\phi_1\phi_2 \tag{11.98}$$

This expression is essentially the same as Eq. (11.56) combined with Eq. (11.81), except that mole fractions have been replaced by volume fractions. The derivation shows that the regular solution model is not necessarily based on a lattice model, but can be based on the structural analogy between solvent and solution.

11.6.1
The Solubility Parameter Approach

Hildebrand [5] introduced the *solubility parameter* $\delta_\alpha = (\Delta U_\alpha/V_\alpha)^{1/2}$ so that

$$w = (\delta_1 - \delta_2)^2 V_{\text{m}} \quad \text{with} \quad V_{\text{m}} = x_1V_1 + x_2V_2 \quad \text{and}$$

$$\Delta_{\text{mix}}U_{\text{m}} = \Delta_{\text{mix}}H_{\text{m}} - RT = V_{\text{m}}(\delta_1 - \delta_2)^2 x_1x_2 \tag{11.99}$$

The term solubility parameter is based on the original application of the model in the solubility of compounds. Consider that miscibility occurs when $\Delta_{\text{mix}}G_{\text{m}} \le 0$. Since in this model $\Delta_{\text{mix}}S_{\text{m}} \ge 0$ and $\Delta_{\text{mix}}H_{\text{m}} \ge 0$ always, $\Delta_{\text{mix}}H_{\text{m}}$ must be not too large to result in miscibility. In other words, miscibility is predicted if w is small. Two ways to estimate δ-values exist. One way is to use the experimental enthalpy of vaporization $\Delta_{\text{vap}}H_\alpha$ for the pure compounds α to calculate $\Delta_{\text{vap}}U_\alpha = \Delta_{\text{vap}}H_\alpha - RT$ which, when combined with V_α, yields δ_α. Another approach employs group contribution methods which have been devised to estimate δ (see Chapter 13).

In order to estimate how well the geometric mean approximation for w is, we resort to the dispersion energy. Recall that in Chapter 3 an approximate expression for the dispersion interaction was given reading $\alpha\beta$

$$\varepsilon_{\alpha\beta} = -C_{\alpha\beta}/r^6 \quad \text{with} \quad C_{\alpha\beta} = \frac{3}{2}\mathcal{C}\frac{I_\alpha I_\beta}{I_\alpha + I_\beta}\alpha_1\alpha_2 \tag{11.100}$$

with α the polarizability and I the characteristic energy. It was indicated that estimating I as the ionization potential is doubtful, while an estimate for α is much more reliable.

Assuming $V_\alpha \sim \sigma_\alpha^3$, a reasonable assumption for spherical molecules, we may write from Eq. (11.95)

$$c_{12}^2 = c_{11}c_{22}\frac{\varepsilon_{12}^2\sigma_{12}^6}{\varepsilon_{11}\sigma_{11}^3\varepsilon_{22}\sigma_{22}^3} \tag{11.101}$$

Using Eq. (11.100) for $\varepsilon_{\alpha\beta}$ at $r = \sigma_{\alpha\beta}$ and the assumption $\sigma_{12} = (\sigma_{11} + \sigma_{22})/2$, we obtain

$$c_{12}^2 = c_{11}c_{22}\frac{I_1 I_2}{[(I_1+I_2)/2]^2}\frac{\sigma_{11}^3\sigma_{22}^3}{[(\sigma_{11}+\sigma_{22})/2]^6} \tag{11.102}$$

Because the geometric average is invariably less than the arithmetic average, the approximation $c_{12} = -(c_{11}c_{22})^{1/2}$ is only valid if $I_1 = I_2$ and $\sigma_{11} = \sigma_{22}$ (as discussed in Section 3.6); otherwise, $|c_{12}| < (c_{11}c_{22})^{1/2}$. Writing

$$w = w_{12} - \frac{1}{2}w_{11} - \frac{1}{2}w_{22} = \frac{1}{2}[(\delta_1 - \delta_2)^2 + 2\delta_1\delta_2(1 - f_I f_s)] \quad \text{where} \tag{11.103}$$

$$f_I = \frac{2I_1^{1/2}I_2^{1/2}}{I_1 + I_2} \quad \text{and} \quad f_\sigma = \left(\frac{2\sigma_{11}^{1/2}\sigma_{22}^{1/2}}{\sigma_{11}+\sigma_{22}}\right)^3 \tag{11.104}$$

Reed [6] estimated that for hydrocarbon + fluorocarbon systems $f_I \cong 0.97$ and $f_\sigma \cong 0.995$, rendering large deviations from the Berthelot rule using typical δ-values (this is consistent with our discussion in Section 3.6). Finally, we note that although the solubility parameter approach is empirically rather successful, its theoretical basis is flimsy.

11.6.2
The One- and Two-Fluid Model

The approach discussed provides a basis for the one-fluid and two-fluid model of mixtures. We saw that using $g_{\alpha\beta} = g\,(r/\sigma_{\alpha\beta})$ for all pairs of components α and β, $\rho = N/V$, and returning to mole fractions x_α, the configurational energy is given by

$$U_{\text{con}} = 2\pi N\rho\sum_{\alpha,\beta}\varepsilon_{\alpha\beta}\sigma_{\alpha\beta}^3 x_\alpha x_\beta\int_0^\infty f(y)g(y)y^2\,\mathrm{d}y \tag{11.105}$$

Using the symbols ε and σ for the mixture we can define

$$\varepsilon\sigma^3 = \sum\nolimits_{\alpha,\beta} x_\beta x_\beta \varepsilon_{\alpha\beta}\sigma_{\alpha\beta}^3 \quad \text{and} \quad \sigma^3 = \sum\nolimits_{\alpha,\beta} x_\alpha x_\beta \sigma_{\alpha\beta}^3 \tag{11.106}$$

In this way a mixture can be described by a hypothetical pure, model fluid with properties defined by ε and σ given the scaled interaction and correlation functions $f(y)$ and $g(y)$. In the literature this model is usually referred to as the *one-fluid van der Waals* (or *vdW1*) *model.*

Another "Ansatz" is to use $g_{\alpha\beta} = ½[g_{\alpha\alpha}(r/\sigma_{\alpha\alpha}) + g_{\beta\beta}(r/\sigma_{\beta\beta})]$. This leads to

$$U_{\text{con}} = 2\pi N\rho \sum\nolimits_\alpha \varepsilon_\alpha \sigma_\alpha^3 x_\alpha \int_0^\infty f(y) g_{\alpha\alpha}(y) y^2 \,\mathrm{d}y \quad \text{where} \tag{11.107}$$

$$\varepsilon_\alpha \sigma_\alpha^3 = \sum\nolimits_\beta x_j \varepsilon_{\alpha\beta}\sigma_{\alpha\beta}^3 \quad \text{and} \quad \sigma_\alpha^3 = \sum\nolimits_\beta x_\beta \sigma_{\alpha\beta}^3 \tag{11.108}$$

Here, $g_{\alpha\alpha}$ represents the correlation function for a pure, pseudo-component α with ε_α and σ_α and the mixture is described as an ideal mixture of two pseudo-components. It is referred to as the *two-fluid van der Waals* (or *vdW2*) *model.* Although it is difficult to discriminate between the vdW1 and vdW2 models on an experimental basis, simulations using the Lennard-Jones potential clearly show that the vdW1 model is better [7] that the vdW2 model (Figure 11.6). The expressions for F and G in the vdW1 model read

$$F(V,T) = \varepsilon F_0(V/\sigma^3, kT/\varepsilon) - 3NkT\ln\sigma \quad \text{and} \tag{11.109}$$

$$G(P,T) = \varepsilon G_0(P\sigma^3/\varepsilon, kT/\varepsilon) - 3NkT\ln\sigma \tag{11.110}$$

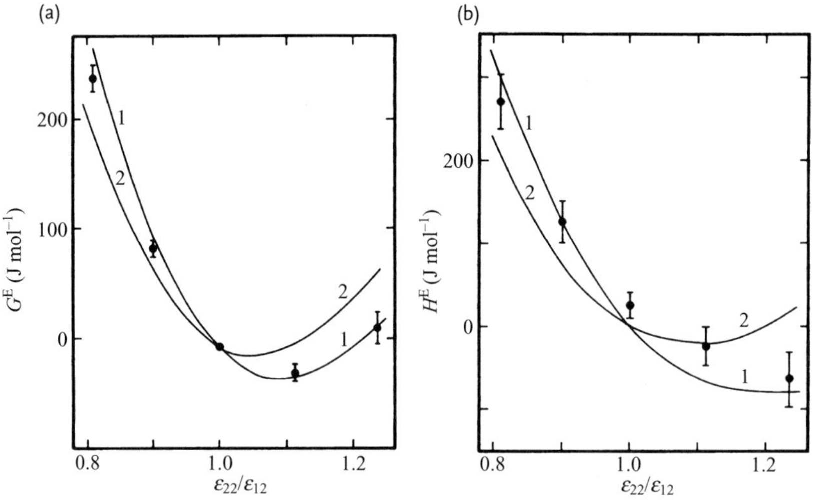

Figure 11.6 (a) Excess Gibbs energy of mixing and (b) excess enthalpy of mixing for an equimolar mixture of 6:12 molecules for which $\sigma_{22}/\sigma_{12} = 1.06$ at 97 K and $P = 0$ ($\varepsilon_{12} = 133.50$ K, $\sigma_{12} = 3.596$ Å). The points give the simulation results, and the curves marked 1 and 2 give the results of the vdWl and vdW2 theories, respectively.

respectively, where F_0 and G_0 correspond to U_0. For the vdW2 model, these expressions must be summed over the components using the appropriate values for ε and σ. In contrast to the vdW2 model, the vdW1 model introduces no ambiguities between F and G, which makes it also preferable. Since for mixtures, P and T usually are the normal variables, the use of G is more convenient.

11.7 The Activity Coefficient for Other Composition Measures*

So far, we have dealt with activity coefficients using the mole fraction x as a measure for composition. In practice, people use also molality m (m moles of solute per 1 kg of solvent) and molarity c (c moles per liter solution). Whichever measure is used, the difference $\mu - \mu^\infty$, where μ^∞ is the infinite dilution reference value, should be the same. Denoting with a_x and a_m the activities using the mole fraction and molality scale we must have

$$\mu - \mu^\infty = \left(\mathring{\mu}_x + RT\ln a_x\right) - \left(\mathring{\mu}_x + RT\ln a_x^\infty\right) = RT\ln\frac{a_x}{a_x^\infty} \quad \text{and}$$

$$\mu - \mu^\infty = \left(\mathring{\mu}_m + RT\ln a_m\right) - \left(\mathring{\mu}_m + RT\ln a_m^\infty\right) = RT\ln\frac{a_m}{a_m^\infty} \tag{11.111}$$

So we obtain

$$\frac{a_x}{a_x^\infty} = \frac{a_m}{a_m^\infty} = \frac{\gamma_x x}{\gamma_x^\infty x^\infty} = \frac{\gamma_m m}{\gamma_m^\infty m^\infty} \quad \text{or} \quad \frac{\gamma_m}{\gamma_x} = \frac{x}{m}\frac{m^\infty}{x^\infty}\frac{\gamma_m^\infty}{\gamma_x^\infty} \tag{11.112}$$

with γ the activity coefficient. For the molality scale we have (see Chapter 2)

$$\frac{m_2}{x_2} = \frac{n_1 + n_2}{n_1 M_1} = \frac{n}{n_1 M_1} \tag{11.113}$$

where M_1 is the molar mass of the solvent. For infinite dilution

$$n_2 \to 0 \text{ (actually } x_1 \to 1) \quad \text{and} \quad m_2^\infty / x_2^\infty = 1/M_1 \quad \text{or} \quad x_2^\infty / m_2^\infty = M_1$$

Moreover, for infinite dilution,

$$\gamma_m^\infty = \gamma_x^\infty \to 1 \quad \text{and thus} \quad \gamma_m^\infty / \gamma_x^\infty = 1$$

In total, we thus obtain

$$\blacktriangleright \quad \gamma_m / \gamma_x = x_2 / (m_2 M_1) \tag{11.114}$$

A similar calculation for the molarity scale yields $\gamma_c/\gamma_x = x_2\rho/c_2M_1$ with ρ the mass density and c_2 the molarity of the solute.

Problem 11.17

Calculate the relationship between the activity coefficients on the mole fraction scale and the molarity scale.

11.8 Empirical Improvements*

In this section we indicate briefly some empirical improvements that can be made on the regular solution model, largely based on an expansion of the excess Gibbs function. Theoretical improvements deal with a better description for the local configuration than being based on random mixing with consequences for entropy and energy, and we discuss these in the next section.

The excess Gibbs energy as used in the regular solution model can be systematically but largely empirically extended. As a first approximation in the regular solution model, we wrote $G_\mathrm{m}^\mathrm{E} = zw_\mathrm{m}x_1x_2$, where w_m is a (temperature-dependent) parameter (we omit the subscript m from now on). Corrections on this expression may be modeled as a power law in x_1 and x_2. Hence, we can write the *Redlich–Kister expansion* [8]

$$G^\mathrm{E}/x_1x_2 = A + B(x_1 - x_2) + C(x_1 - x_2)^2 + \cdots \tag{11.115}$$

The inclusion of these terms generates systematic improvements in the description.

When $A = B = C = \cdots = 0$, $G^\mathrm{E} = 0$, $RT\ln\gamma_1 = RT\ln\gamma_2 = 0$ and $\gamma_1 = \gamma_2 = 1$. Accordingly, we regain the ideal solution.

When $B = C = \cdots = 0$, $G^\mathrm{E} = Ax_1x_2$ and $RT\ln\gamma_1 = Ax_2^2$ and $RT\ln\gamma_2 = Ax_1^2$. Obviously, these results represent the regular solution and the expressions are symmetric in nature. For infinite dilution one obtains $\ln\gamma_1^\infty = \ln\gamma_2^\infty = A$.

When $C = \cdots = 0$, $G^\mathrm{E}/x_1x_2 = A + B\ (x_1 - x_2)$, implying a deviation from the regular solution model. Equivalently, since $x_1 + x_2 = 1$, we may write $G^\mathrm{E}/x_1x_2 = A\ (x_1 + x_2) + B\ (x_1 - x_2)$ or defining $A_{21} = A + B$ and $A_{12} = A - B$,

$$G^\mathrm{E}/x_1x_2 = A_{21}x_1 + A_{12}x_2 \tag{11.116}$$

a result known as the *Margules equation*. Obviously, this approach can be extended although, according to Guggenheim [9], an expansion further than third order is hardly ever required in view of the experimental accuracy. In Figure 11.7 and

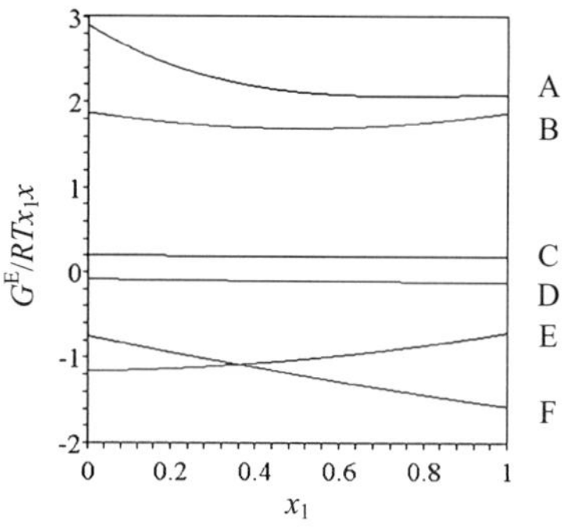

Figure 11.7 Excess Gibbs energies for the systems of Table 11.5.

Table 11.5 Excess Gibbs function parameters for various solutions.

System	*T* (K)	$RT \times A$	$RT \times B$	$RT \times C$
A: ethanol/methylcyclohexane	305	2.1180	−0.2390	0.3750[a)]
B: methylcyclohexane/acetone	318	1.6907	−0.0001	0.1832
C: pyridine/acetone	303	0.1919	0.0050	0.0075
D: chloroform/furan	303	−0.1083	−0.0177	0.0071
E: pyridine/chloroform	303	−1.0271	0.2270	0.0930
F: chloroform/1,4 dioxane	303	−1.2006	−0.4131	0.0318

Data from Pitzer [1].
a) Also using $RT \times D = -0.173$.
A, *B*, and *C* refer to the constants in Eq. (11.115).

Table 11.5 some examples are presented of G^{E} functions, which shows that this approach is flexible and can describe rather different type of behaviors, that is, continuously increasing G^{E}, continuously decreasing G^{E}, and a G^{E} showing a maximum or minimum.

One may also expand the reciprocal x_1x_2/G^{E} in the mole fractions to obtain the *van Laar expansion* [10]

$$x_1x_2/G^{\mathrm{E}} = A' + B'(x_1 - x_2) \tag{11.117}$$

Equivalently, we may write, since $(x_1 + x_2) = 1$,

$$x_1x_2/G^{\mathrm{E}} = A'(x_1 + x_2) + B'(x_1 - x_2) = (A' + B')x_1 + (A' - B')x_2 \tag{11.118}$$

Defining $1/A'_{21} = A' + B'$ and $1/A'_{12} = A' - B'$, we obtain

$$\frac{x_1x_2}{G^{\mathrm{E}}} = \frac{x_1}{A'_{21}} + \frac{x_2}{A'_{12}} \quad \text{or} \quad \frac{G^{\mathrm{E}}}{x_1x_2} = \frac{A'_{12}A'_{21}}{A'_{12}x_1 + A'_{21}x_2} \tag{11.119}$$

The expressions for the activity coefficients, known as the *van Laar equations*, are

$$RT\ln\gamma_1 = A'_{12}\left(1 + \frac{A'_{12}x_1}{A'_{21}x_2}\right)^{-2} \quad \text{and} \quad RT\ln\gamma_2 = A'_{21}\left(1 + \frac{A'_{21}x_2}{A'_{12}x_1}\right)^{-2} \tag{11.120}$$

In this case for $x_1 = 0$, $RT\ln\gamma_1^\infty = A'_{12}$ and for $x_2 = 0$, $RT\ln\gamma_2^\infty = A'_{21}$.

Some algebraic manipulation, meanwhile, identifying $A'_{21} = CV_1$ and $A'_{12} = CV_2$, where V_α is the partial volume of component α and C is a constant, shows that Eq. (11.119) is equivalent to an expansion of G^{E} in the volume fraction $\phi_1 = V_1x_1/(V_1x_1 + V_2x_2) = V_1x_1/V_{\mathrm{m}}$, where $V_{\mathrm{m}} \equiv V_1x_1 + V_2x_2$ is the molar volume of the mixture, that is, $G_{\mathrm{m}}^{\mathrm{E}} = C\phi_1\phi_2V_{\mathrm{m}}$. This relation fits data well for many binary systems of normal liquids. In the van Laar version the constant C was originally given in terms of the a- and b-parameters of the vdW equation, that is

$$C = \left(a_1^{1/2}/b_1 - a_2^{1/2}/b_2\right)^2 \tag{11.121}$$

To be preferred is the expression, rationalized by Scatchard [11],

$$C = \left[(\Delta U_1/V_1)^{1/2} - (\Delta U_2/V_2)^{1/2}\right]^2 \equiv (\delta_1 - \delta_2)^2 \tag{11.122}$$

where $\Delta U_\alpha = \Delta_{\text{vap}}H_\alpha - RT$ is the molar energy of vaporization, and the solubility parameter $\delta_\alpha = (\Delta U_\alpha/V_\alpha)^{1/2}$ is used. The latter expression is also given by the approximate statistical mechanical calculation in Section 11.6. Using A'_{12} and A'_{21} as parameters, the ratio A'_{12}/A'_{21} corresponds reasonably well to V_1/V_2, for example, for the mixture benzene and iso-octane [12] at 45 °C $A'_{12}/A'_{21} = 0.562$, while experimentally $V_1/V_2 = 0.536$.

Although these expressions show a relatively large flexibility in fitting the experimental data, their theoretical basis is flimsy and the equations are difficult to extend to multicomponent systems. Finally, the temperature dependence of the parameters involved is not explicitly included. Developments based on the concept of local composition have led to various models which remedy these defects, and of which the *Wilson model* [13] is the archetype. For a solution of components 1 and 2, the number of 1–1 and 2–2 interactions relative to the number of 1–2 interactions is given by the ratio of component 1 to 2 weighted by a Boltzmann factor. Hence

$$\frac{n_{11}}{n_{12}} = \frac{n_1 \exp(-\beta h_{11})}{n_2 \exp(-\beta h_{12})} \quad \text{and} \quad \frac{n_{22}}{n_{12}} = \frac{n_2 \exp(-\beta h_{22})}{n_1 \exp(-\beta h_{12})} \tag{11.123}$$

The volume fractions ϕ_α, using as pure component molar volume V_α, are then defined by

$$\phi_1 = n_1V_1\exp(-\beta h_{11})/[n_1V_1\exp(-\beta h_{11}) + n_2V_2\exp(-\beta h_{12})] \quad \text{and} \tag{11.124}$$

$$\phi_2 = n_2V_2\exp(-\beta h_{22})/[n_1V_1\exp(-\beta h_{12}) + n_2V_2\exp(-\beta h_{22})] \tag{11.125}$$

The Gibbs energy of mixing is assumed to be

$$\Delta_{\text{mix}}G_{\text{m}} = RT(x_1\ln\phi_1 + x_2\ln\phi_2) \quad \text{or for the excess Gibbs energy} \tag{11.126}$$

$$G_{\text{m}}^{\text{E}} = \Delta_{\text{mix}}G_{\text{m}} - \Delta_{\text{mix}}G_{\text{m}}^{\text{ideal}} = RT[x_1\ln(\phi_1/x_1) + x_2\ln(\phi_2/x_2)] \tag{11.127}$$

Defining

$$r_{21} = 1 - \frac{V_2\exp(-\beta h_{22})}{V_1\exp(-\beta h_{12})} \quad \text{and} \quad r_{12} = 1 - \frac{V_1\exp(-\beta h_{11})}{V_2\exp(-\beta h_{12})} \tag{11.128}$$

we can write

$$G_{\text{m}}^{\text{E}} = -RT[x_1\ln(1 - r_{21}x_2) + x_2\ln(1 - r_{12}x_2)] \tag{11.129}$$

After some calculation the result for the activity coefficients γ_1 and γ_2 reads

$$\ln\gamma_1 = \frac{1}{RT}\frac{\partial G^{\text{E}}}{\partial n_1} = -\ln(1 - r_{21}x_2) - \frac{x_1x_2r_{21}}{1 - r_{21}x_2} + \frac{x_2^2r_{12}}{1 - r_{12}x_1} \quad \text{and} \tag{11.130}$$

$$\ln\gamma_2 = \frac{1}{RT}\frac{\partial G^{\text{E}}}{\partial n_2} = -\ln(1 - r_{12}x_1) - \frac{x_1x_2r_{12}}{1 - r_{12}x_1} + \frac{x_1^2r_{21}}{1 - r_{21}x_2} \tag{11.131}$$

The quantities r_{12} and r_{21} are considered as empirical parameters and have been extensively tabulated [14]. For a fixed T, only the two parameters r_{12} and r_{21} are required to describe the system, but for a range of temperatures the parameters $\partial r_{12}/\partial T$ and $\partial r_{21}/\partial T$ are also required. The Wilson expressions can be easily generalized to a multicomponent system since, in principle, no new parameters are required. There are, however, several drawbacks. For example, when $r_{12} = r_{21} = 1$, $G_m^E = -\Delta_{mix}G_m^{ideal}$ or $\Delta_{mix}G_m = 0$, the model cannot describe phase separation, nor for other values of r_{12} and r_{21}. It also appears that the model cannot describe a minimum or maximum in $\gamma_\alpha(x)$.

In summary, Wilson-type models can be used effectively for binary and multicomponent systems, and contain an explicit temperature dependence. Consequently, the approach is rather extensively used. For further information, the reader is referred to the literature (e.g., Ref. [15]).

11.9 Theoretical Improvements*

Next, let us turn to attempts to improve the local configuration. In the ideal solution model it is assumed that the distribution of the components 1 and 2 is random. The same is assumed for the regular solution model, although preferential energetic interactions are also assumed. This state of affairs represents, at least, an inconsistency.

In order to improve, let us recapitulate the regular solution model in slightly different words. Consider the interchange 1–1 + 2–2 → 1–2 + 2–1 on a lattice with coordination number z, to which we attributed an interchange energy $2w$. We have a total of ½zN pairs of nearest-neighbor bonds, irrespective of their type. We now define zX to be the total number of 1–2 pairs for a particular distribution. Since the number of neighbors of type 1 molecules that are not of type 2 is $z(N_1 - X)$, the number of 1–1 pairs is ½$z(N_1 - X)$. Similarly, the total number of 2–2 pairs is ½$z(N_2 - X)$, respectively. The problem is now to determine the average value $\bar{X}$ of X.

Assuming a completely random distribution over sites, the relative probability of a 1–1 and a 2–2 pair is equal to the probability of a 1–2 and a 2–1 pair. Therefore

$$\frac{1}{2}z(N_1 - X)\cdot\frac{1}{2}z(N_2 - X) = \frac{1}{2}zX\cdot\frac{1}{2}zX \quad \text{or} \quad (N_1 - X)(N_2 - X) = X^2 \tag{11.132}$$

This immediately leads to the solution $\bar{X} = N_1N_2/(N_1 + N_2) = Nx_1x_2$, and from this all previous results follow. In particular, the configurational interaction energy

$$\begin{aligned} U_{con} &= \frac{1}{2}z(N_1 - \bar{X})w_{11} + \frac{1}{2}z(N_2 - \bar{X})w_{22} + \bar{X}zw_{12} \\ &= \frac{1}{2}N_1zw_{11} + \frac{1}{2}N_2zw_{22} + \bar{X}zw \quad \text{or} \quad \Delta_{mix}U = \bar{X}zw = Nx_1x_2zw \end{aligned} \tag{11.133}$$

This approximation has different names in different fields and we designate it here (in accordance with Guggenheim) as the *zeroth approximation*. It is also known as the *mean field* approximation.

The basic tenet here is the independent distribution of single molecules over sites, in spite of their nonzero interchange energy w. Systematic improvement is possible by considering the independent distribution of pairs of molecules (*first approximation*), of triplets of molecules (*second approximation*), and so on, taking into account the interchange energy within a multiplet. Of course, the independent pair model is also deficit because a certain local configuration cannot be realized by adding independent pairs. For example, in a four coordinated lattice the first three pairs can be chosen independently, but the fourth cannot. A similar statement can be made also for the higher approximations. However, we might expect to obtain a better description as with the independent distribution of single molecules. The independent distribution of pairs, as discussed in Appendix C [9, 15], provides some improvement (and is an interesting approach as such). The expression for $\Delta_{\text{mix}}G_{\text{m}}$ (or $\Delta_{\text{mix}}F_{\text{m}}$) is somewhat complex, and for details we refer to Appendix C. For the critical temperature the simple expression $zw/kT_{\text{cri}} = z\ln[z/(z-2)]$ results.

Before we discuss the improvement in numerical results, we first note that the scheme 1–1 + 2–2 → 1–2 + 2–1 strongly resembles the scheme for a chemical reaction. Even the equation $X^2 = (N_1 - X)\,(N_2 - X)$ resembles a reaction equilibrium expression. This led Guggenheim [16] to introduce the so-called *quasi-chemical approximation* by postulating that the equation $X^2 = (N_1 - X)\,(N_2 - X)$ must be modified heuristically to include the interchange energy so that it becomes $X^2 = (N_1 - X)\,(N_2 - X)\exp(-2\beta w)$. In this way, the analogy with a chemical reaction is more or less complete. Later, this approach was shown to be completely equivalent to the independent pair distribution approach, which we have labeled here as the *first approximation*, and this approach has been used ever since for many applications. Obviously, the use of triplets and quadruplets instead of pairs improves the estimate; however, the details become rapidly complex and we refrain from discussing details, referring the interested reader to Guggenheim [17, 18].

Let us now quote a few results of the zeroth, first, and higher approximations [17, 18] to see what the improvements are as compared to the zeroth approximation. From Table 11.6 it is clear that for the critical temperature T_{cri} and the Helmholtz energy of mixing $\Delta_{\text{mix}}F$, the first approximation is an improvement over the zeroth, but that higher approximations do not yield a significant improvement. A similar observation can be made for the vapor pressure (not shown). For the entropy of mixing $\Delta_{\text{mix}}S$, the values for the higher approximations are close to that of the zeroth. In conclusion, when using this combinatorial approach for molecular solutions, in all cases the first approximation is sufficiently accurate and, although this is a conceptual improvement, in many cases even the zeroth approximation is sufficient, in contrast to the case of polymeric solutions (see Chapter 13). Finally, we note also that the effect of next-nearest-neighbor interactions has been assessed. It appears that this effect is much smaller than the differences between the zeroth and first approximation, and therefore can be entirely neglected.

Table 11.6 Critical temperature, Helmholtz energy and entropy in several approximations for close-packed lattice with $z = 12$.

Property	Approximation			
	Zeroth	First	Second	Third
η_{cri}	–	1.2	1.20185	1.20409
zw/kT_{cri}	2	2.1878	2.2063	2.2288
$-\Delta_{mix}F/RT$; $x = 0.1$	0.1451	0.1317	0.1303	0.1279
$-\Delta_{mix}F/RT$; $x = 0.3$	0.1909	0.1693	0.1667	0.1635
$-\Delta_{mix}F/RT$; $x = 0.5$	0.1931	0.1711	0.1684	0.1652
$\Delta_{mix}S/R$; $x = 0.1$	0.3251	0.3214	0.3212	0.3202
$\Delta_{mix}S/R$; $x = 0.3$	0.6109	0.5927	0.5903	0.5873
$\Delta_{mix}S/R$; $x = 0.5$	0.6931	0.6683	0.6648	0.6605

In spite of the relative simplicity of the zeroth order model, an analytical solution is known so far only for 2D lattices. For a square 2D lattice with $z = 4$ this solution reads [19] $\sinh(zw/4kT_{cri}) = 1$, which gives $zw/kT_{cri} = 3.526$. For comparison, we recall that in the zeroth approximation $zw/kT_{cri} = 2.0$, whilst for the first approximation $zw/kT_{cri} = z\ln[z/(z-2)] = 2.773$ was derived. Hence, in conclusion, although the first approximation constitutes a conceptual and numerical improvement as compared to the zeroth approximation, the difference with the exact solution is still significant. One expects this also to be true in 3D, and a numerical evaluation of the 3D problem indeed showed this to be true (see Chapter 16 and Appendix C).

References

1 See Pitzer (1995).
2 Guggenheim, E.A. (1937) *Trans. Faraday Soc.*, **33**, 151.
3 Adcock, D.S. and McGlashan, M.L. (1954) *Proc. Roy. Soc.*, **A226**, 266. See also Guggenheim (1967), p. 199.
4 Hildebrand, J.L. and Wood, G.E. (1933) *J. Chem. Phys.*, **1**, 817. See also Hildebrand and Scott (1962).
5 Hildebrand, J.H. (1919) *J. Am. Chem. Soc.*, **41**, 1067. See also Hildebrand and Scott (1962).
6 Reed, T.M., III (1955) *J. Phys. Chem. Soc.*, **59**, 429.
7 (a) Henderson, D. and Leonard, P.J. (1971) *Proc. Natl Acad. Sci. USA*, **68**, 632; (b) Lee, L.L. (1988) *Molecular Thermodynamics of Nonideal Fluids*, Butterworth, Boston.
8 (a) Redlich, O., Kister, A.T., and Turnquist, C.E. (1952) *Chem. Eng. Progr. Symp. Ser. 2*, **48**, 49; (b) However, the expansion was already used by Guggenheim, E.A. (1937) *Trans. Faraday Soc.*, **33**, 151.
9 See Guggenheim (1952).
10 (a) Van Laar, J.J. (1910) *Z. Phys. Chem.*, **72**, 723; (b) Van Laar, J.J. (1913) *Z. Phys. Chem.*, **89**, 599; (c) See also: Van Laar, J.J. (1906) *Sechs Vorträge über das thermodynamische Potential*. Vieweg Verlag, Brunswick, Germany.
11 Scatchard, G. (1931) *Chem. Rev.*, **8**, 321.

12 Prausnitz, J.M., Lichtenthaler, R.N., and de Azevedo, E.G. (1986) *Molecular Thermodynamics of Fluid Phase Equilibria*, Prentice-Hall, Englewood Cliffs, NJ, Ch. 6.
13 Wilson, G.M. (1964) *J. Am. Chem. Soc.*, **86**, 127.
14 Ohe, S. (1989) *Vapor-liquid Equilibrium Data*, Elsevier, Amsterdam.
15 See Lucas (2007).
16 (a) Guggenheim, E.A. (1935) *Proc. Roy. Soc.*, **A148**, 304; (b) Rushbrooke, G.S. (1938) *Proc. Roy. Soc.*, **A166**, 296.
17 Guggenheim, E.A. (1952) *Applications of Statistical Mechanics*, Clarendon, Oxford.
18 Guggenheim, E.A. (1966) *Applications of Statistical Mechanics*, Clarendon, Oxford.
19 Onsager, L. (1944) *Phys. Rev.*, **65**, 117.
20 von Zawidski, J. (1900) *Z. Phys. Chem.*, **35**, 129.
21 See Smith *et al.* (2005).
22 Silbey, R.J. and Albert, R.A. (2001) *Physical Chemistry*, 3rd edn, J. Wiley & Sons Ltd, New York.
23 Atkins, P.W. (2002) *Physical Chemistry*, 7th edn, Oxford University Press.
24 Fowler, R.H. and Guggenheim, E.A. (1939) *Statistical Thermodynamics*, Cambridge University Press, London, p. 351.

Further Reading

Fawcett, W.R. (2004) *Liquids, Solutions and Interfaces*, Oxford University Press, Oxford.
Guggenheim, E.A. (1952) *Mixtures*, Clarendon, Oxford.
Guggenheim, E.A. (1967) *Thermodynamics*, 5th edn, North-Holland, Amsterdam.
Hildebrand, J.H. and Scott, R.L. (1962) *Regular Solutions*, Prentice-Hall, Englewood Cliffs, NJ.
Lucas, K. (2007) *Molecular Models of Fluids*, Cambridge University Press, Cambridge.
Pitzer, K.S. (1995) *Thermodynamics*, 3rd edn, McGraw-Hill, New York.
Smith, J.M., Van Ness, H.C., and Abbott, M.M. (2005) *Introduction to chemical engineering thermodynamics*, 7th ed., McGraw-Hill, New York.

12
Mixing Liquids: Ionic Solutions

In solutions in many cases ions are involved, which we discuss in this chapter. First, we deal with the solubility of salts and the thermodynamic reasons involved, followed by models and facts about hydration. Next, the Debye–Hückel theory is introduced leading to (approximate) thermodynamic expressions for ionic solutions. Thereafter, we discuss electrical conductivity, including the effect of ion pairing.

12.1
Ions in Solution

The existence of ions in solutions was not immediately clear. Following the demonstration of electrolysis by Michael Faraday (1791–1869), an important step was taken by Svante Arrhenius (1859–1927) who postulated the existence of ions in his thesis, for which he eventually received the Nobel Prize – though not after considerable debate. Other important contributions, in particular related to conductivity, were made by Friedrich Wilhelm Georg Kohlrausch (1846–1910) and Friedrich Wilhelm Ostwald (1853–1932). During this era, the distinction between strong and weak electrolytes was made, with a major breakthrough being the development of a theory for the activity coefficient of strong electrolytes by Erich Hückel (1896–1980) and Pieter Debye (1888–1966); this was later refined by Lars Onsager (1903–1976) and Hans Eduard Wilhelm Falkenhagen (1895–1971). Concepts to include association in strong electrolytes were introduced by Niels Bjerrum (1879–1958).

When dealing with ionic compounds, several questions come to mind. Why do certain salts dissolve easily but others do not? Do salts dissociate completely on dissolution of the compound? What is the configuration of the solvent around a dissolved ion? With regards to the solubility question, two approaches can be advanced – analytical and energetic, respectively – and these are dealt with in the next section. Aspects of hydration structure and dissociation are discussed in Sections 12.3 and 12.4, respectively.

Liquid-State Physical Chemistry, First Edition. Gijsbertus de With.

12.1.1 Solubility

From an analytical point of view, solubility is described by the solubility product K_{sol}, directly related to $\Delta_{sol}G°$ via

$$\Delta_{sol}G° = -RT\ln K_{sol} \tag{12.1}$$

For example, for AgCl we have the solubility product $K_{sol} = (a_{Ag})(a_{Cl})/(a_{AgCl}) = [a_{Ag}][a_{Cl}] \cong [Ag^+][Cl^-]$, with activity a_X and molarity [X] of component X. As a first reduction step we note that, according to the definition of activity of a solid, $a_{AgCl} = 1$, whilst for the second step we neglect the difference between activity and molarity. The larger the solubility product, the greater the solubility. In the example of AgCl, $K_{sol} = 1 \times 10^{-10}$ using concentration units (M or mol l^{-1}) so that, if there is no other source of either Ag^+ or Cl^- ions, we have, using the electroneutrality condition $[Ag^+] = [Cl^-]$, $[Ag^+] = 1 \times 10^{-5}$ M. For the solubility product data of some common salts, see Appendix E. Once a solubility product is determined experimentally, the rest follows.

The presence of other compounds – be it one with a common ion or an acid or base – influences the solubility to a large extent. The effect of the former is called the *common ion effect*, while the effect of the latter, though quite important, carries no specific name. Inert ions also influence solubility by changing the ionic strength of a solution. The influence of these ions on the solubility is conventionally addressed in basic or physical chemistry courses, and we refrain from further discussion here, apart from showing some illustrative examples.

Example 12.1: The common ion effect

Suppose that we want to dissolve AgCl in a solution of 1.0 M NaCl. The molarity of Cl^- ions is then $[Cl^-] = 1.0 + [Ag^+]$. The concentration of $[Ag^+]$ is negligible as compared to $[Cl^-]$, so that with $[Cl^-] \cong 1.0$ we obtain $K_{AgCl} = [Ag^+][Cl^-] \cong [Ag^+] \times 1.0 = 1 \times 10^{-10}$. In other words, the solubility of AgCl in 1.0 M NaCl solution is 1×10^{-10} M.

Example 12.2: The effect of acid–base equilibria

Suppose that we want to dissolve AgOH, a weakly soluble salt, in water. For the reaction $AgOH \leftrightarrow Ag^+ + OH^-$ the solubility product $K_{AgOH} = [Ag^+][OH^-] = 1 \times 10^{-8}$, while for the water dissociation $H_2O \leftrightarrow OH^- + H^+$ the equilibrium constant reads $K_{H2O} = [H^+][OH^-] = 1 \times 10^{-14}$. The overall reaction reads $AgOH + H^+ \leftrightarrow Ag^+ + H_2O$, with an equilibrium constant $K = K_{AgOH}/K_{H2O} = 1 \times 10^{-6}$.

The solubility of AgOH in a buffer of, say, pH 11 is calculated realizing that, for a buffer solution, $[H^+]$ and $[OH^-]$ are fixed. At pH 11, $[H^+] = 1 \times 10^{-11}$ and $[OH^-] = 1 \times 10^{-3}$. Since $K_{AgOH} = 1 \times 10^{-3}[Ag^+] = 1 \times 10^{-8}$, we obtain $[Ag^+] = 1 \times 10^{-5}$ M. Similarly, in a buffer of pH 7, $[H^+] = 1 \times 10^{-7}$ and $[OH^-] = 1 \times 10^{-7}$. Therefore, $K_{AgOH} = 1 \times 10^{-7}[Ag^+] = 1 \times 10^{-8}$ and $[Ag^+] = 1 \times 10^{-1}$ M.

In pure water, $[H^+]$ and $[OH^-]$ are not fixed so that we have to solve

$$[H^+][OH^-] = 1 \times 10^{-14} \quad \text{and} \quad [Ag^+][OH^-] = 1 \times 10^{-8}$$

simultaneously. As this involves three unknowns with only two equations, we need one more equation. The charge balance requires that

$$[H^+] + [Ag^+] = [OH^-]$$

Assuming that $[Ag^+] >> [H^+]$, this expression reduces to $[OH^-] \cong [Ag^+]$, so that $K_{AgOH} = [Ag^+]^2 = 1 \times 10^{-8}$ and $[Ag^+] = 1 \times 10^{-4}$ M. From $[OH^-] \cong [Ag^+] = 1 \times 10^{-4}$ M we calculate $[H^+] = 1 \times 10^{-10}$ M, so that the assumption $[Ag^+] >> [H^+]$ is warranted. Formally, from $K_{AgOH} = [Ag^+]\,[OH^-]$ and $K_{H2O} = [H^+]\,[OH^-]$

$$[Ag^+] = K_{AgOH}/[OH^-] = K_{AgOH}/([H^+] + [Ag^+]) \quad \text{and}$$

$$[H^+] = K_{H2O}/[OH^-] = K_{H2O}/([H^+] + [Ag^+])$$

Solving leads to

$$[Ag^+]^2 = K_{AgOH}/[1 + (K_{H2O}/K_{AgOH})]$$

In more complex situations, the mass balance must also be invoked.

From an energetic point of view, solubility is described by the configuration change of ions upon dissolution of a salt. During dissolution, the lattice of a salt becomes disrupted and the individual ions are hydrated. As for any process, dissolution is governed by the Gibbs energy change $\Delta_{sol}G$, combining the enthalpy change $\Delta_{sol}H$ with the entropy change $\Delta_{sol}S$. The question arises: What controls $\Delta_{sol}G$?

Let us first look at $\Delta_{sol}H$. The energetics of the dissolution process can represented by a Born diagram, similarly to the Born–Haber diagram used for lattice energies. Such a Born diagram for AgF is shown in Figure 12.1. This shows that the overall enthalpy of dissolution $\Delta_{sol}H = \Delta_{hyd}H - \Delta_{lat}H = -19\,\text{kJ mol}^{-1}$, and that the large lattice enthalpy $\Delta_{lat}H$ is compensated by an even larger hydration enthalpy $\Delta_{hyd}H$. The balance between $\Delta_{lat}H$ and $\Delta_{hyd}H$ determines whether $\Delta_{sol}H < 0$ (exothermic) or $\Delta_{sol}H > 0$ (endothermic). Since for AgF $\Delta_{sol}H < 0$, the liquid heats up during the process.

During dissolution, heat is usually absorbed but it can also be released (e.g., for anhydrous Na_2SO_4 and $CaSO_4$). There can even be a change in sign of $\Delta_{sol}H$ with

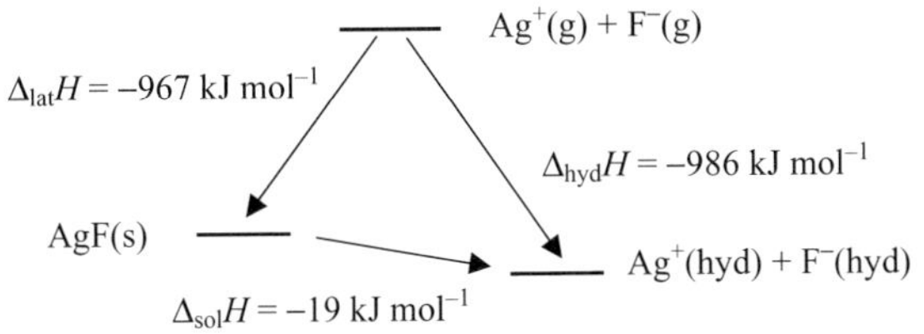

Figure 12.1 Born diagram for the dissolution of AgF in water, indicating the enthalpy changes.

temperature (e.g., for $CaCl_2 \cdot 2H_2O$ and LiCl). Within a series of compounds that are all very soluble in water, the trend may reverse: the dissolution of LiCl produces enough heat to boil the solvent, while the dissolution of NaCl produces very little temperature change, yet NH_4Cl may cool the solvent to below its freezing point.

Since both $\Delta_{lat}H$ and $\Delta_{hyd}H$ are large and comparable, we should also consider entropy change $\Delta_{hyd}S$. It appears that $\Delta_{hyd}H$ and $\Delta_{hyd}S$ show similar trends, for example, with ionic charge ze (z valency, e unit charge) and ionic radius a (see Section 12.2). Moreover, we should realize that ΔH and ΔS contribute oppositely to ΔG, and consequently the solubility of salts is difficult to predict. For a discussion of the dissolution trends for various salts, see Ref. [1].

The second question of whether a salt dissolves completely or incompletely into individual ions can be answered by taking measurements of the colligative properties and electrical conductivity. It appears that there are two types of electrolytes:

- *Strong electrolytes* showing (almost) complete dissociation, so that the colligative properties are proportional to the number of ions and the electrical conductivity is weakly dependent on the concentration.
- *Weak electrolytes,* with only partial dissociation, showing a strong dependence of the colligative properties and a strong dependence of the electrical conductivity on concentration.

Since we need to deal with the individual ions, we must ask also how the hydration enthalpy of one salt can be divided over the ions. It appears that the experimental values of $\Delta_{hyd}H$ of a series of salts with common ions show approximately constant differences. For example, $\Delta_{hyd}H(\text{NaCl}) = -774.6\,\text{kJ mol}^{-1}$ and $\Delta_{hyd}H(\text{KCl}) = -690.4\,\text{kJ mol}^{-1}$, so that the difference is $84.2\,\text{kJ mol}^{-1}$. For the corresponding bromides the data are $\Delta_{hyd}H(\text{NaBr}) = -740.7\,\text{kJ mol}^{-1}$ and $\Delta_{hyd}H(\text{KBr}) = -656.7\,\text{kJ mol}^{-1}$, resulting in a difference of $84.6\,\text{kJ mol}^{-1}$, almost the same as for the chlorides [1]. Hence, if the hydration enthalpy for a single ion could be known, from a series we could calculate the enthalpies for all other ions. For this reference, the H^+ ion is usually chosen. The choice $\Delta_{hyd}H(H^+) = 0$ represents the *conventional scale,* while taking for $\Delta_{hyd}H(H^+)$ an independently determined experimental value is designated as the *absolute scale.* Various experimental methods to estimate $\Delta_{hyd}H(H^+)$, for which we refer to the literature [2], resulted in $-1150.1 \pm 0.9\,\text{kJ mol}^{-1}$ at 25 °C[1)]. A value of $\Delta_{hyd}S(H^+) = -153\,\text{J K}^{-1}\text{mol}^{-1}$ is obtained for the entropy, leading to $\Delta_{hyd}G(H^+) = -1104.5 \pm 0.3\,\text{kJ mol}^{-1}$. In Section 12.2, we deal with a theoretical solution to the hydration question given by the Born hydration model, and with a few of its extensions in greater detail.

Problem 12.1

The solubility product of $BaCO_3$ is $K_{BaCO3} = 5 \times 10^{-9}$, while the equilibrium constants for H_2CO_3 are $K_1 = [HCO_3^-][H^+]/[H_2CO_3] = 4 \times 10^{-7}$ and $K_2 = [CO_3^{2-}][H^+]/[HCO_3^-] = 4 \times 10^{11}$. Calculate the solubility of $BaCO_3$ in water.

1) The value of $\Delta_{hyd}H(H^+)$ obtained is greater in magnitude than most older values. The ± sign indicates random errors only.

Problem 12.2

For NaCl, which is known to dissolve well in water, $\Delta_{\text{lat}}H = -788\,\text{kJ mol}^{-1}$, while $\Delta_{\text{hyd}}H = -784\,\text{kJ mol}^{-1}$. What is $\Delta_{\text{sol}}H$? Does the solvent cool down or heat up during dissolution? Estimate the minimum value required for $\Delta_{\text{sol}}S$.

12.2
The Born Model and Some Extensions

The first question that comes to the mind when modeling hydration is how to represent an ion. The conventional solution is to consider it as a sphere with a certain radius and charge. There are two size scales for ions in regular use:

- The Pauling[2)] scale [3], in which the effective nuclear charge is used to proportion the distance between ions in crystals into anionic and cationic radii. This analysis is based on $r_{\text{ion}}(O^{2-}) = 140\,\text{pm}$.
- The Shannon and Prewitt scale [4], which is based on a major review of crystallographic data. Different radii for different coordination numbers and for high and low spin states of the ions are given. The data are based on $r_{\text{ion}}(O^{2-}) = 126\,\text{pm}$, and are referred to as "crystal" ionic radii. Essentially, the cation radii are about 20 pm larger and the anion radii about 15 pm smaller than those of Pauling. Some data are given in Appendix E. Ionic radii in aqueous solutions have been reviewed by Marcus [5].

From electrostatics we know that the potential of a particle with charge q and radius a in a medium with relative permittivity ε_r is given by

$$\phi(q) = \frac{1}{4\pi\varepsilon}\frac{q}{a} = \frac{\mathcal{C}}{\varepsilon_\text{r}}\frac{q}{a} \quad \text{with} \quad \varepsilon \equiv \varepsilon_0\varepsilon_\text{r} \quad \text{and} \quad \mathcal{C} = (4\pi\varepsilon_0)^{-1} \tag{12.2}$$

If we assume that we may take a hard sphere for the ion and replace the solvent by a (continuous) dielectric, we can calculate from this expression the Gibbs energy difference $\Delta_{\text{hyd}}G$ necessary to transfer the ion with charge ze (with z the valency and e the unit charge) from the gas phase ($\varepsilon_\text{r} = 1$) to a hydrated state ($\varepsilon_\text{r} = \varepsilon_\text{r}$). In this model, due to Born [6], the process is achieved in three steps. First, we decharge the ion in the gas phase; second, we transfer the ion to the solution; and, third, we recharge the ions in the solvent.

For calculating $\Delta_{\text{hyd}}G$ we use the work w of charging given by

$$w = \int_0^{ze}\phi(q)\,\text{d}q = \int_0^{ze}\frac{q}{4\pi\varepsilon a}\text{d}q = \frac{1}{4\pi\varepsilon a}\int_0^{ze} q\,\text{d}q = \frac{1}{4\pi\varepsilon a}\frac{q^2}{2}\bigg|_0^{ze} = \frac{(ze)^2}{8\pi\varepsilon a} \tag{12.3}$$

2) To be consistent with Pauling's radii, Shannon also used a value of $r_{\text{ion}}(O^{2-}) = 140\,\text{pm}$; data using that value are referred to as "effective" ionic radii. Shannon states that "it is felt that crystal radii correspond more closely to the physical size of ions in a solid".

Table 12.1 Hydration Gibbs energy and entropy for various ions at 25 °C.[a]

Ion	$-\Delta_{hyd}G$ (kJ mol^{-1})	$-\Delta_{hyd}S$ (J mol^{-1} K^{-1})	Ion	$-\Delta_{hyd}G$ (kJ mol^{-1})	$-\Delta_{hyd}S$ (J mol^{-1} K^{-1})
H^+	1104	153	Ga^{3+}	4684	625
Li^+	529	164	In^{3+}	4134	451
Na^+	424	133	F^-	429	115
K^+	352	96	Cl^-	304	53
Be^{2+}	2498	354	Br^-	278	3761
Mg^{2+}	1931	375	I^-	243	14
Ca^{2+}	1608	296	OH^-	431	140
Al^{3+}	4676	604	S^{2-}	1238	79

a) Data from Ref. [32].

The total work done for a mole of ions is thus, assuming a zero energy change for step 2,

$$W = N_A w = N_A\left(\frac{(ze)^2}{8\pi\varepsilon a} - \frac{(ze)^2}{8\pi\varepsilon_0 a}\right) = \frac{N_A(ze)^2}{8\pi\varepsilon_0 a}\left(\frac{1}{\varepsilon_r} - 1\right) \tag{12.4}$$

and we see that it is dependent, apart from the permittivity ε_r, on the charge ze and the radius a of the ions. Since the work W is done at constant temperature and pressure, it represents the change in Gibbs energy $\Delta_{hyd}G°$ for the process.

As an example, take the Na^+ ion for which the Shannon–Prewitt (Pauling) radius is 116 (95) pm. We calculate that $\Delta_{hyd}G° = -591$ (-720) kJ mol^{-1}. Experimental hydration data for some ions are given in Table 12.1 (note that the entropy contributions are small), from which it can be seen that the experimental value for Na^+ is -424 kJ mol^{-1}. In this case – and this appears to be generally true – the Born equation overestimates $\Delta_{hyd}G°$. Moreover, the Born equation predicts equal values for $\Delta_{hyd}G$ for positive and negative ions of the same radius. Experimentally, this is not true when using the Pauling scale, as can be seen by comparing the values for the K^+ ion (a = 133 pm) and the F^- ion (a = 136 pm) in Table 12.1, although the estimates are much closer to each other when using the Shannon–Prewitt scale.

This type of discrepancy has led research groups to modify the Born expression, the simplest empirical modification being an adaptation of the effective ionic radius, recognizing that water coordinates differently with positive and negative ions. By adding 10 pm to the radius of a negative ion and 85 pm to the radius of a positive ion using the Pauling scale, a good correspondence with the experimental data [7] can be obtained.

Another consideration is that the value of ε_r close to the ion will be lower than the bulk value, in view of the high electric field strength close to the ion. As an example, we quote a simple empirical expression for this dependence due to Stiles [8] that reads

$$\varepsilon(r) = \varepsilon_a (\varepsilon_b / \varepsilon_a)^{f(r)} \quad \text{where} \quad f(r) = 1 - [(b-r)/(b-a)]^2 \tag{12.5}$$

In this way the permittivity varies from ε_a at $r = a$ to ε_b at $r = b$. Since ε is now position-dependent, we must use $\nabla \cdot \varepsilon(r)\boldsymbol{E}(r) = 0$ (instead of the usual $\varepsilon\nabla \cdot \boldsymbol{E}(r) = 0$), leading to $\boldsymbol{E}(\boldsymbol{r}) = \mathcal{C}ze\boldsymbol{r}/\varepsilon_{\mathrm{r}}(r)r^2$ with $\mathcal{C} = (4\pi\varepsilon_0)^{-1}$. The electrostatic energy, to be evaluated numerically, becomes

$$\Delta_{\mathrm{hyd}}G^\circ = \frac{z^2e^2}{8\pi\varepsilon_0}\int_a^\infty \frac{1}{\varepsilon(r)r^2} - \frac{1}{r^2}\mathrm{d}r = \frac{z^2e^2}{8\pi\varepsilon_0}\left[\frac{1}{b\varepsilon_b} - \frac{1}{a} + \frac{1}{\varepsilon_a}\int_a^b \left(\frac{\varepsilon_a}{\varepsilon_b}\right)^{f(r)} r^{-2}\,\mathrm{d}r\right] \tag{12.6}$$

It appears that $R = (\Delta_{\mathrm{hyd}}G)_{\mathrm{Stiles}}/(\Delta_{\mathrm{hyd}}G)_{\mathrm{Born}}$ is not particularly sensitive to the precise choice of the profile, but also that saturation effect is incapable of explaining the overestimate of the Born model completely. For example, using $\varepsilon_a = 2$ and $\varepsilon_b = 80$, $R = 0.85$, 0.92 and 0.96 for $a/b = 0.2$, 0.4 and 0.6, respectively.

While the above-described adaptations are based on continuum considerations, the mean spherical approximation (MSA) is based on the discrete nature of the solvent, as described by the integral equation approach. The ion is represented by a sphere with radius a and charge ze, while for the solvent a sphere with radius a and dipole moment μ is taken. Hydrogen bonding is neglected. We only quote here the results with some background and refer to the literature [9] for details. For ionic radii, the Shannon–Prewitt scale is used.

The MSA consists of the core condition $g(r) = 0$ for $r < 2a$ and the approximation $c(r) = -\beta\phi(r)$ for $r > 2a$, together with the OZ equation (see Chapter 6). For hard-spheres with $\phi(r) = \infty$ for $r < 2a$, this approach is identical to the PY approach and an analytical solution is possible (see Section 7.3). In the case of hydration, $\phi(r)$ represents either the ion–ion potential or the ion–dipole potential, as given by Eq. (3.7). The dipole–dipole interactions between the solvent molecules are taken into account by the factor f, which we take as $f = 1$ initially. The Gibbs energy of hydration in the limit of infinite dilution reads

$$\Delta_{\mathrm{hyd}}G = \frac{N_{\mathrm{A}}(ze)^2 f}{8\pi\varepsilon_0 a}\left(\frac{1}{\varepsilon_{\mathrm{r}}} - 1\right)\left(\frac{1}{1 + b/\lambda a}\right) \equiv \frac{N_{\mathrm{A}}(ze)^2 f}{8\pi\varepsilon_0}\left(\frac{1}{\varepsilon_{\mathrm{r}}} - 1\right)\left(\frac{1}{a+\delta}\right) \tag{12.7}$$

where the polarization parameter λ for the solvent is given by

$$\lambda^2(1+\lambda)^4 = 16\varepsilon_{\mathrm{r}} \tag{12.8}$$

If we take for ε_{r} the bulk value of the solvent, the effect of polarizability, softness and asymmetry of the solvent is taken effectively into account. For water at 25 °C, $\varepsilon_{\mathrm{r}} = 78.5$, resulting in $\lambda = 2.65$. The parameter $\delta \equiv b/\lambda$ depends only on the nature of the solvent. For $\delta = 0$ we obtain $\Delta_{\mathrm{hyd}}G = (\Delta_{\mathrm{hyd}}G)_{\mathrm{Born}}$, and hence δ acts as a radius correction. A reasonable hard-sphere radius for H_2O is $b = 142\,\mathrm{pm}$ [10], so that $\delta = 54\,\mathrm{pm}$, which leads to a significant change in $\Delta_{\mathrm{hyd}}G$. The entropy $\Delta_{\mathrm{hyd}}S$ is obtained by differentiating $\Delta_{hyd}G$ with respect to T, which leads (as δ is temperature-dependent) to

$$\Delta_{hyd}S = -\frac{\partial \Delta_{hyd}G}{\partial T} = \frac{N_A(ze)^2}{8\pi\varepsilon_0 a}\left\{\left(\frac{1}{\varepsilon_r^2}\frac{1}{a+\delta}\frac{\partial \varepsilon_r}{\partial T}\right) - \left[\frac{1}{(a+\delta)^2}\left(1-\frac{1}{\varepsilon_r}\right)\frac{\partial \delta}{\partial T}\right]\right\}$$

$$\frac{\partial \delta}{\partial T} = -\frac{a}{\lambda^2}\frac{\partial \lambda}{\partial T} \quad \text{and} \quad \frac{\partial \lambda}{\partial T} = \left[\frac{\lambda(\lambda+1)}{3\lambda+1}\right]\frac{1}{2\varepsilon_r}\frac{\partial \varepsilon_r}{\partial T} \tag{12.9}$$

and where $\partial\lambda/\partial T$ is obtained from differentiating Eq. (12.8). Using the same data as before, we obtain for Na^+ in H_2O, $\Delta_{hyd}G° = -403\,\text{kJ mol}^{-1}$, which is considerably closer to the experimental value of $-424\,\text{kJ mol}^{-1}$ than the Born estimate of $-591\,\text{kJ mol}^{-1}$. The value obtained obviously still depends on the choice of radius made. As the Shannon–Prewitt radii for positive ions are typically 20 pm larger than the Pauling radii, the agreement with experiment for these ions is quite good. However, for negative ions the size correction is overestimated. Although dipole–dipole interactions could be taken into account, it is felt that the formal treatment overestimates the effect. Hence, Eq. (12.7) can be rewritten as $-z^2/\Delta_{hyd}G = k(a + \delta)$ with $k = 8\pi\varepsilon_0\varepsilon_r/N_A e^2 f(\varepsilon_r - 1)$, where now the experimental data for $\Delta_{hyd}G$ are used and f and δ are determined separately for cations and anions. For alkali metal and earth alkali metal ions, it appears that the slope k obtained corresponds to $f = 1$ and $\delta^+ = 49$ pm. This suggests that for cations the dipole–dipole interactions are not important in hydration, while the value of δ^+ is close to the theoretical value $\delta = 54$ pm. For the anions used (halides and sulfide), the slope k obtained corresponds to $f = 0.74$ and $\delta^- = 0$ pm, showing that dipole–dipole interactions are important but also that the anions have a less disruptive effect on the water network than cations. It should be noted that the results depend heavily on the ionic radius used. A radius corresponding to coordination number six is used, whilst for smaller-sized ions (Li^+, Mg^{2+}) this may not be appropriate. Also, if Pauling radii are used, the agreement is not satisfactory. Nevertheless, overall a rather satisfying description is obtained.

For more complete reviews on hydration, we refer to Marcus [10] and Volkov *et al.* [11]. In principle, the above-described approach can also be used to calculate $\Delta_{tra}G$ for the transfer of an ion from one solvent to another. Several reviews on ion transfer are available in the literature (e.g., Ref. [12]).

Problem 12.3

Determine $T\Delta_{hyd}S$ using the Born model. Show that $T\Delta_{hyd}S$ is small as compared to $\Delta_{hyd}G$, and indicate the reason why. Estimate the numerical values of $T\Delta_{hyd}S$ and $\Delta_{hyd}G$ for Li^+ in water, given that $\varepsilon_r = 302.6\exp(-T/220.5)$.

Problem 12.4

As with any model, the Born model and its extensions contains assumptions. Indicate at least two aspects that deserve criticism, and include the reason(s) why.

Problem 12.5

A simple model for dielectric saturation is $\varepsilon_r = \varepsilon_{sat} \cong 4$ for $r < 0.2\,\text{nm}$, and $\varepsilon_r = \varepsilon_{bulk} \cong 78$ for $r > 0.4\,\text{nm}$ with a linear transition region in between. Using this model, calculate $\Delta_{hyd}G°$ for Na^+ and compare the answer with the Born result.

Problem 12.6

Calculate $\Delta_{hyd}G$ for Li^+ in H_2O using the MSA. Compare the result with the experimental value.

Problem 12.7

Indicate why a difference in effective radius for positive and negative ions, as determined from hydration energies, is expected.

12.3 Hydration Structure

Several methods exist to study solvation. We first address gas-phase hydration. After introducing a basic model for the liquid hydration structure, we then discuss some results from nuclear magnetic resonance, ion movement, and diffraction methods.

12.3.1 Gas-Phase Hydration

One experimental method of assessing the hydration of ions is based on the generation of ions $M^{\pm}$ by thermionic emission, from a suitable salt for a positive ion, or by reactive electron transfer by O_2^- for negative ions, in a water vapor-containing vessel. Equilibrium between the ions and the water molecules is established according to

$$M^{\pm}(H_2O)_{n-1} + H_2O \leftrightarrow M^{\pm}(H_2O)_n \qquad (12.10)$$

The concentrations are measured using mass spectroscopy, and by varying the temperature the enthalpy and entropy of the reactions can be determined. In this way, clusters containing up to six water molecules have been detected. One can then calculate the total enthalpy for the reaction

$$M^{\pm} + 6H_2O \leftrightarrow M^{\pm}(H_2O)_6 \qquad (12.11)$$

where, if necessary, extrapolation of the data to $n = 6$ is done. Table 12.2 provides data for several ions using this technique. The proton was also studied in this way. Adding all partial enthalpies for the H^+ ion results in $\Delta_{hyd}H = -1192\,\text{kJ mol}^{-1}$, which

Table 12.2 Enthalpy of hydration $-\Delta_{hyd}H$ (upper row) and $-\Delta_{hyd}G$ (lower row) in kJ mol^{-1} for several ions in the gas phase at 300 K.[a)]

N	H_3O^+	Li^+	Na^+	K^+	Rb^+	OH^-	F^-	Cl^-	Br^-	I^-
1	144.0[b)]	142.3	104.6	75.7	66.9	109.6	97.5	60.2	54.4	43.9
	100.7	113.8	78.5	49.5	41.0	74.7	79.8	35.9	30.6	22.0
2	230.5	250.2	187.4	143.1	123.8	188.3	172.4	113.8	104.6	84.6
	155.6	192.9	133.7	86.7	70.3	121.4	136.2	63.6	53.9	38.8
3	302.0	336.8	251.8	198.3	174.8	257.7	233.1	162.8	152.3	123.5
	155.6	248.5	171.4	113.1	91.2	153.6	169.9	84.2	72.3	51.6
4	356.8	405.4	307.0	247.7	221.7	317.1	290.4	208.4	198.3	162.0
	216.6	279.9	196.2	131.5	107.1	176.6	193.9	99.2	85.1	60.7
5	407.0	463.6	355.1	292.5	265.6	376.1	344.0	248.1	243.4	199.7
	233.3	298.7	211.3	144.9	118.8	194.2	211.9	111.9	94.9	67.2
6	454.3	514.2	399.9	334.3	–	–	389.6	284.9	286.5	–
	245.4	309.2	223.4	154.5	–	–	226.6	121.6	102.8	–

a) Data from Tissandier *et al.* [2] for clusters containing *N* molecules.
b) Starting value for H_3O^+ calculated as $\Delta_{hyd}H$ ($\Delta_{hyd}G$) = −1150(−1104) + 647 = −503 (−457) kJ mol^{-1}, where 647 kJ mol^{-1} is the value of the proton affinity of water. For H_3O^+ molecules 7 and 8 contribute still 49 and 43 kJ mol^{-1}, respectively.

is in reasonable agreement with liquid-phase data. For higher values of *n*, the contribution to $\Delta_{hyd}H$ is ~40 kJ mol^{-1}, to be compared with the evaporation enthalpy for H_2O, $\Delta_{eva}H$ = 44 kJ mol^{-1}. Note that $\Delta_{hyd}H$ for the H^+ ion is much larger than for any other ion.

Problem 12.8

Indicate why the proton behaves so differently from all other ions.

12.3.2
Liquid-Phase Hydration

The structure of water changes in the neighborhood of ions. It appears that ions are solvated and a general model, due to Frank [13], is sketched in Figure 12.2. In the first zone around an ion, the *inner hydration shell* (or *ordered zone*), the water molecules are more or less immobilized by the high field of the ion. This is sometimes described as an "iceberg," as the structure has some ice-like characteristics. Occasionally, the ordered zone is divided in a primary shell in direct contact with the ion, as well as a secondary region that is not in direct contact but is still influenced by the proximity of the central ion. In the second zone, *the structure-breaking zone* (or *disordered zone*), the structure is more disordered than in ordinary water, where there is considerable order due to hydrogen bonding. In this zone the

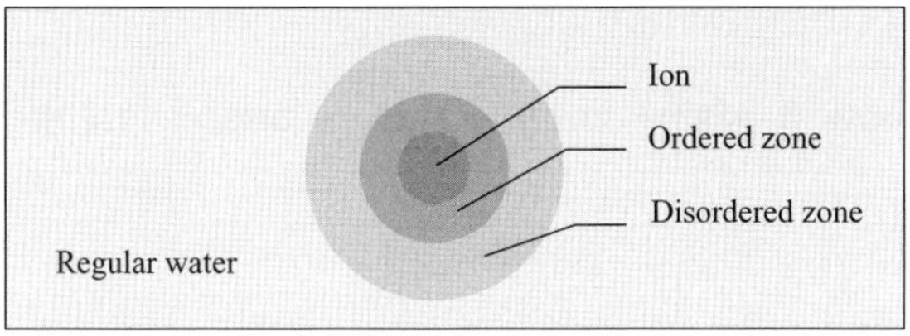

Figure 12.2 Frank model for the structure of water around an ion.

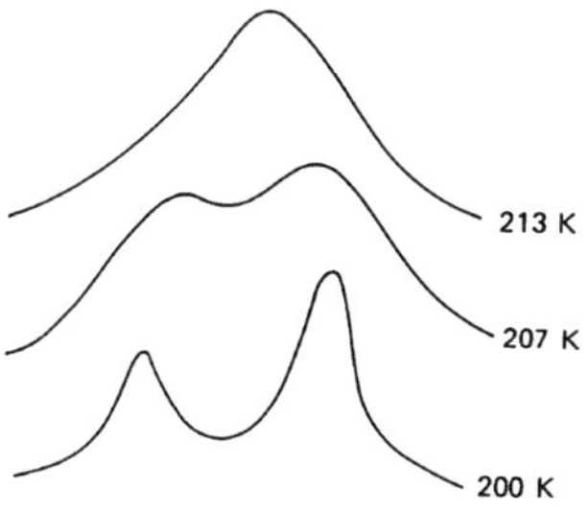

Figure 12.3 ^{1}H NMR spectra of magnesium perchlorate in aqueous acetone.

structure is due to the competing effect of the orientating influence of the ion or dipoles and the normal structural effect of the neighboring molecules. The third zone comprises normal water.

When using proton nuclear magnetic resonance (NMR) spectroscopy, a proton must remain for at least 10^{-4} s in a certain environment in order to be distinguished as a distinct entity, but this is generally a long time when compared to molecular diffusion times. If the residence time is less, only one resonance line will be observed for each type of proton, for example, one for H_2O and two for CH_3OH. If the resonance time is longer, because bonding of the solvent to the central ion is strong, then two lines may be observed for a given proton–one for the solvent molecules in the primary solvation shell, and one for the bulk including the secondary ordered region and disordered zone (Figure 12.3). The solvation number can be calculated from the peak areas if the overall composition is known. For example, a 1 M aqueous solution of Mg^{2+} ions with density $\rho = 1.115\,\mathrm{g\,cm^{-3}}$ will contain 985 g or 985/18 = 54.7 moles of H_2O. For a coordination number of six, of these 54.7 moles, six can be attributed to the primary hydration shell and 48.7 to "bulk" water. The ratio of the two peaks thus will be 6:48.7 or 1:8.1. Although these measurements can sometimes be made done at room temperature, it is often necessary to cool the solution in order to obtain sufficient "bonding" and thus produce two separate signals (Figure 12.3). This also provides the possibility of studying the solvent exchange rate and the associated activation energy.

Table 12.3 Solvation numbers as determined from NMR data in various solvents.[a)]

	Water	MeOH	TMP	MeCN	DMF	DMSO	Liq. NH_3
Be^{2+}	4	–	4	–	4	4	–
Mg^{2+}, Zn^{2+}	6, 6	6, 6	–	–	–	–	6, –
Al^{3+}, Ga^{3+}	6, 6	6, 6	4, 6	–, 6	6, 6	6, 6	6, –
In^{3+}, Sc^{3+}	–	–	4	–	–	–	–
V^{2+}, Mn^{2+}	6, 6	–, 6	–, –	–, 6	–, –	–, –	–, –
Fe^{2+}, Co^{2+}, Ni^{2+}	6	6	–	6	6	6	–
Ti^{3+}, V^{3+}	6	–	–	–	–	–	–
Cr^{3+}, Fe^{3+}	–	–	–	–	–	6	–
Pd^{2+}, Pt^{2+}	4	–	–	–	–	–	–
Ce^{3+}, Pr^{3+}, Nd^{3+}	–	–	–	–	9	–	–
$Tb^{3+} \rightarrow Yb^{3+}$	–	–	–	–	8	–	–
Th^{4+}	9	–	–	–	–	–	–

a) Data from Ref. [33].

Table 12.3 provides a summary of solvation numbers obtained in this way; the data show that the solvation number for positive ions is mostly six, in both aqueous and nonaqueous solutions. Only for very small ions (Be^{2+}), for square-planar coordination (Pt^{2+} and Pd^{2+}) and for very voluminous solvents, will a lower solvation number be observed. For the larger ions (lanthanides, actinides) a solvation number greater than six will result.

Overall, this is a simplified picture of the primary solvation of ions. It should be noted that the general applicability of this NMR method is somewhat limited, as only diamagnetic and some paramagnetic ions can be studied in this way.

Another approach to estimating solvation numbers is to use information from the motion of ions through the solvent, be it due to spontaneous motion (diffusion), mechanically stimulated motion (viscosity), or electrically stimulated motion (conductivity). The resistance of ions and their associated solvation shell reflects their effective size. For example, conductivity measurements indicate that conductivity is dependent on charge and size of the ion involved. From the conductivity, it is possible to estimate the mobility, and from the mobility the diffusivity. The latter then yields an estimate for the hydrodynamic radius or volume. Subtracting the volume of the ion results in the volume of the solvent molecules in the solvation shell, from which the number of molecules can be easily calculated. We refrain from further details of the measurements, and simply list some hydration results in Table 12.4. These data show that the solvation numbers so obtained are significantly larger than those obtained via NMR experiments. This is due to the fact that these methods include the primary as well as secondary solvation shell. Moreover, the results differ widely for different techniques. It is expected that similar trends occur for other solvents than water.

Table 12.4 Hydration numbers as determined by several methods.[a)]

Parameter	Li^+	Na^+	K^+	Cs^+	Mg^{2+}	Ca^{2+}	Ba^{2+}	Zn^{2+}	Fe^{2+}	Al^{3+}	Cr^{3+}
Transport number	13–22	7–13	4–6	4	12–14	8–12	3–5	10–13	–	–	–
Mobility	3–21	2–10	5–7	–	10–13	7–11	5–9	10–13	10–13	–	–
Conductivity	2–3	2–4	–	6	8	8	8	–	–	–	–
Diffusion	5	3	1	1	9	9	8	11	12	13	17
Entropy	5	4	3	3	13	10	8	12	12	21	–
Compressibility	3	4	3	–	–	–	–	–	–	31	–
Activity coefficient	–	–	–	–	5	4	3	–	12	12	–
NMR peak area	–	–	–	–	6	–	–	6	6	6	6

a) Data from Ref. [33].

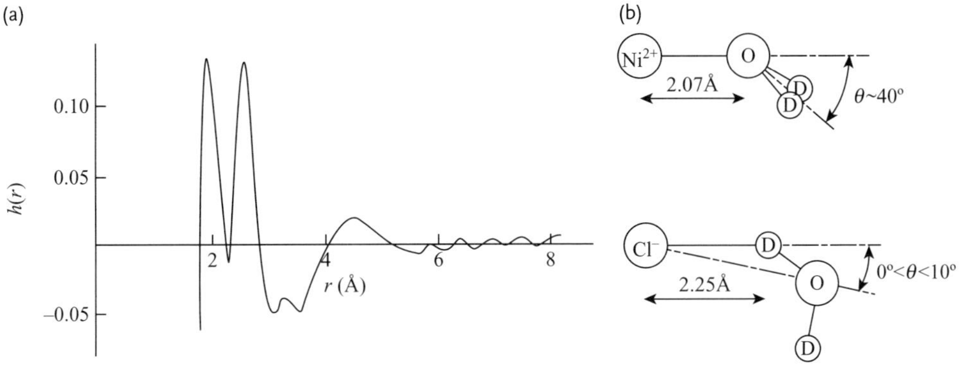

Figure 12.4 (a) Total-correlation function $h(r) = g(r) - 1$ of Ni^{2+} surrounded by D_2O; (b) The configuration of a D_2O molecule with respect to a Ni^{2+} and Cl^- ion as calculated from $g(r)$ data. Data from Ref. [34].

With scattering and diffraction experiments using X-rays and neutrons, research groups have also studied ionic solutions, a typical example being $NiCl_2$ in D_2O[3)]. In this case, the total correlation function $h(r) = g(r) - 1$, as measured with neutron ray diffraction (NRD), shows two peaks. If the area $r^2g(r)$ is calculated, the ratio of the first peak to the second peak is one-half. The first peak is due to the metal–oxygen distances of the (heavy) water surrounding the Ni^{2+} ion, while the second peak is due to the metal–deuterium distances, which explains the observed ratio. From the correlation function it can be inferred that the Ni–O distance is about 0.207 nm, and that the Ni–O axis makes an angle of about 40° with the D_2O plane (Figure 12.4). Calculating the coordination number using

3) Hydrogen atoms have a large so-called "incoherent" scattering cross-section for neutrons as compared to the coherent cross-section, which troubles the interpretation of these experimental data significantly. For deuterium, this ratio is much more favourable, rendering the interpretation much easier.

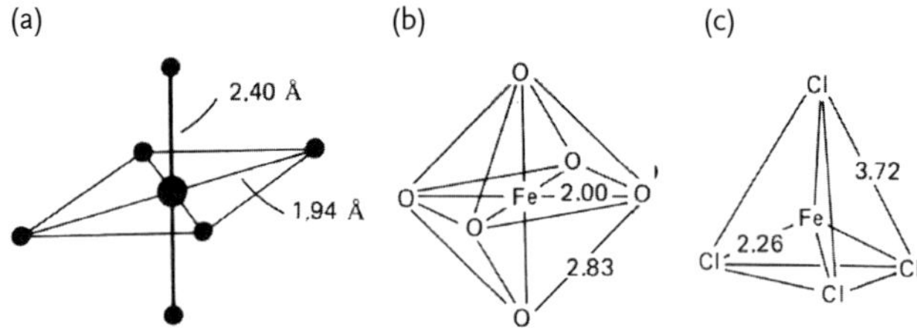

Figure 12.5 The structure of (a) $[Cu(OH_2)_6]^{2+}$; (b) $[Fe(OH_2)_6]^{2+}$; and (c) $[FeCl_4]^-$. Distances are indicated in Å.

$$N_1 = \int_0^{M_1} g(r) 4\pi r^2 \, dr \tag{12.12}$$

where M_1 denotes the first minimum in $g(r)$, one obtains $N_1 \cong 5.8$. This suggests a more or less octahedral arrangement of the D_2O molecules around the Ni^{2+} ion. In any case, the coordination is tight as the relevant distances are all below 0.4 nm. A similar result is shown for the Cl^- ion from NaCl and $CaCl_2$ solutions with $N_1 \cong 6$, but now with the Cl–D distance smaller than the Cl–O distance (Figure 12.4).

So, the most prominent feature in the pair-correlation function is the nearest-neighbor distance of metal ions to solvent. Table 12.5 shows an overview of the metal–oxygen distances as determined with X-ray and neutron techniques. Both, XRD and EXAFS[4] are capable of probing more complex geometries. The coordination of Cu^{2+} ions provides an example, and in this case a metal–oxygen distance at 0.194 nm and 0.240 nm is generally recognized. These distances correspond to the four equatorial and two axial water molecules in a (distorted) octahedral coordination (Figure 12.5). A similar result has been obtained for $[Cr(OH_2)]^{2+}$ with equatorial and axial distances of 0.207 nm and 0.230 nm, respectively.

A more complete picture can be obtained if the O–O distances can also be determined. For example, in octahedral $[Fe(OH_2)_6]^{2+}$ the O–O distance $\sqrt{2} = 1.41$ times the Fe–O distance, while in tetrahedral $[FeCl_4]^-$ the Cl–Cl distance is $\sqrt{3} = 1.73$ times the Fe–Cl distance (Figure 12.5). In general, the location and intensity of the peaks varies with the type of coordination, as summarized in Table 12.6. In principle, the secondary shell can also be studied in this way. The results of these studies suggest that about 14 to 18 water molecules are present in the complete hydration shell for such ions as Mg^{2+}, Al^{3+}, Cr^{3+} and Fe^{3+}, within a distance of 0.41 to 0.42 nm. Altogether, this is not an unreasonable estimate, although it should

4) On the one hand, X-ray diffraction (XRD) provides information on distances between all elements, while EXAFS (extended X-ray absorption fine structure) yields information specifically on the element whose absorption edge is selected. Unfortunately, EXAFS is limited to relatively nearby distances, while diffraction is informative to, say, 2 nm away.

Table 12.5 Metal to oxygen distances (Å) in metal–water complexes.[a)]

	Aqueous solution				
	XRD	EXAFS	NRD	Hydrates	Ionic radius[b)]
Li^+	–	–	1.90–1.95	1.93–1.98	0.88
Na^+	2.40	–	2.50	2.35–2.52	1.16
K^+	2.87–2.92	–	2.70	2.67–3.22	1.52
Ag^+	2.43	2.31–2.36	–	–	1.29
Mg^{2+}	2.10	–	–	2.01–2.14	0.86
Ca^{2+}	2.40	–	2.39–2.46	2.30–2.49	1.14
Cr^{2+}	2.07; 2.30	–	–	–	–
Mn^{2+}	2.20	2.18	–	2.00–2.18	0.96
Fe^{2+}	2.12	2.10	–	1.99–2.08	0.91
Co^{2+}	2.08	2.05	–	1.93–2.12	0.88
Ni^{2+}	2.04	2.05–2.07	2.07–2.10	2.02–2.11	0.84
Cu^{2+}[c)]	1.94; 2.43	1.95	1.95–2.05	1.93–2.00[d)]	0.87
Zn^{2+}	2.08–2.18	1.94	–	2.08–2.14	0.89
Cd^{2+}	2.31	–	–	2.24–2.31	1.09
Hg^{2+}	2.41	–	–	2.24–2.34	1.16
Al^{3+}	1.87–1.90	–	–	1.87	0.67
In^{3+}	2.15	–	–	2.23	0.93
Cr^{3+}	1.94	1.98	–	2.02	0.76
Fe^{3+}	2.05	1.99	–	2.09–2.20	0.79
Rh^{3+}	2.04–2.07	–	–	2.09–2.20	0.81
Ce^{3+}	2.55	–	–	2.48–2.60	1.19
Nd^{3+}	2.51	–	2.48	2.47–2.51	1.14
Dy^{3+}	2.40	–	2.37	2.38	1.05
U^{4+}	2.42	–	–	2.36	1.14

a) Data from Ref. [33].
b) For six coordination.
c) Cu^{2+} has tetragonal coordination in water.
d) Cu^{2+} complexes in hydrates are often less symmetrical then tetragonal.

be borne in mind that these experiments are usually performed using rather concentrated solutions.

Due to its small size, the H^+ ion associates itself strongly with a water molecule, $H^+ + H_2O \leftrightarrow H_3O^+$, and forms the *oxonium ion* H_3O^+. This (first) water molecule is tightly bound to the proton. It has been shown that the angle between the O–H bonds in H_3O^+ is about 115°, so that the molecule is almost, but not completely, flat. It is believed that three more water molecules are attracted by the H_3O^+-ion, forming a hydrogen-bonded structure in which the oxygen atoms of the surrounding water molecules are oriented to the hydrogen atoms of the oxonium ion. This structure has thus to be written as $H_3O^+–3H_2O$ or $H_9O_4^+$. Alternatively, it has been suggested that the H^+ ion is tetrahedrally surrounded by four water molecules. The hydrated oxonium ion is referred to as the *hydronium ion*, and the precise

Table 12.6 Peaks expected in the pair correlation function for various geometries.

Type		Peak	Intensity	Peak	Intensity	Peak	Intensity
Octahedral	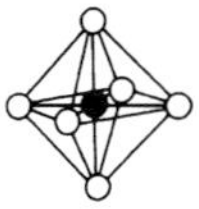	1	6	$\sqrt{2}$	10	2	3
Square-planar		1	4	$\sqrt{2}$	4	2	2
Tetrahedral	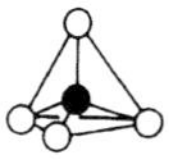	1	4	$\sqrt{3}$	6	–	–
Linear		1	2	2	1	–	–

structure of this is still the subject of debate, but in any case the proton is strongly hydrated. A similar situation arises for the OH^- ion; in this case, we have three hydrating H_2O molecules, so that one writes $H_7O_4^-$. The hydrogen atoms of the surrounding water molecules are closest to the O-atom of the central ion. The hydration enthalpy for the OH^- ion is large, but smaller than that for the H^+ ion.

Summarizing in one sentence – an ion in solution is far from being alone!

Problem 12.9

Calculate the angle between the water plane and the Li–O direction for a Li–D_2O complex, given that r(Li–O) = 0.195 nm and that r(Li–D) = 0.255 nm.

12.4 Strong and Weak Electrolytes

For strong electrolytes Bjerrum first made the proposition, now generally accepted, that complete dissociation is present [14]. NaCl in water provides an example. Generally, an electrolyte $Q = M_{\nu_+}X_{\nu_-}$ dissociates into $\nu = \nu_+ + \nu_-$ ions according to

$$Q = \nu_+ M^{z+} + \nu_- X^{z-} \tag{12.13}$$

where z_+ and z_- are the charges of M and X, respectively. Electroneutrality requires that

$$\nu_+ z_+ + \nu_- z_- = 0 \tag{12.14}$$

The chemical potential of the electrolyte with activities a_+ and a_- is

$$\mu_Q = \mu_Q^\circ + RT(\nu_+ \ln a_+ + \nu_- \ln a_-) = \mu_Q^\circ + RT \ln a_Q \tag{12.15}$$

with the activity a_Q defined by $a_Q \equiv a_+^{\nu_+} a_-^{\nu_-}$. As a reminder, the molality scale for component i is defined as m_i/m_0, with $m_0 = 1\,\text{kg}\,\text{mol}^{-1}$. Frequently, m_0 is omitted and we do so here (compare ln P, which actually means $\ln(P/P^\circ)$ with, say, the reference pressure P° = 1 bar). The activity coefficient[5] (on molality basis) is defined as $\gamma_i = a_i/m_i$, and for the salt the activity becomes

$$a_Q = a_+^{\nu_+} a_-^{\nu_-} = \gamma_+^{\nu_+} m_+^{\nu_+} \gamma_-^{\nu_-} m_-^{\nu_-} = \gamma_\pm^{\nu} m_+^{\nu_+} m_-^{\nu_-} = \nu_+^{\nu_+} \nu_-^{\nu_-} (\gamma_\pm m_Q)^{\nu} = (\nu_\pm \gamma_\pm m_Q)^{\nu} \tag{12.16}$$

using the mean activity coefficient $\gamma_\pm \equiv (\gamma_+^{\nu_+} \gamma_-^{\nu_-})^{1/\nu}$, the mean stoichiometric coefficient $\nu_\pm \equiv (\nu_+^{\nu_+} \nu_-^{\nu_-})^{1/\nu}$ and the fact that for an electrolyte of molality m_Q we have $m_+ = \nu_+ m_Q$ and $m_- = \nu_- m_Q$. For a symmetric electrolyte, the above expression reduces to $a_Q = (\gamma_\pm m_Q)^2$. For mixed electrolytes the full expression should be used. Similar equations can be written down using molarity c and mole fraction x instead of molality m. The mole fraction is, however, infrequently used in this connection.

For weak electrolytes, Arrhenius proposed in 1887 that dissociation is incomplete but that dissociation increases with increasing dilution. An example is a solution of acetic acid in water for which the dissociation equilibrium is

$$\mathrm{CH_3COOH} \leftrightarrow \mathrm{H^+} + \mathrm{CH_3COO^-} \quad \text{or} \quad \mathrm{HA} \leftrightarrow \mathrm{H^+} + \mathrm{A^-} \tag{12.17}$$

For the general electrolyte, Eq. (12.13), the molarity of positive ions for a fully dissociated solution is $\nu_+ m_Q$ with m_Q the molarity Q. With association this becomes $m_+ + \alpha\nu_+ m_Q$ with α the *degree of dissociation*. Because the total molarity of M^+ ions, whether dissociated or not, is $\nu_+ m_Q$, we have

$$\nu_+ m_Q = \alpha\nu_+ m_Q + (1-\alpha)\nu_+ m_Q \equiv \alpha\nu_+ m_Q + m_{IP} \tag{12.18}$$

where m_{IP} denotes the molarity of the ion pairs. The total molarity of negative ions, whether dissociated or not, is $\nu_- m_Q$ so that

$$\nu_- m_Q = m_- + m_{IP} = m_- + (1-\alpha)\nu_+ m_Q \quad \text{or} \quad m_- = \nu_- m_- - (1-\alpha)\nu_+ m_Q \tag{12.19}$$

Since $\gamma_\pm$ is given by $\gamma_\pm^{\nu} \equiv \gamma_+^{\nu_+} \gamma_-^{\nu_-}$ and the chemical potential reads

$$\mu = \mu^\circ + RT\ln(a_Q) = \mu^\circ + RT\ln(\gamma_\pm^{\nu} m_+^{\nu_+} m_-^{\nu_-}) \tag{12.20}$$

we find, using Eq. (12.15),

$$\mu = \mu^\circ + RT\ln\{\gamma_\pm^{\nu}(\alpha\nu_+)^{\nu_+}[\nu_- - (1-\alpha)\nu_+]^{\nu_-} m_+^{\nu_+} m_-^{\nu_-}\} \tag{12.21}$$

For $\nu_+ = \nu_- = 1$, the activity coefficient reduces to $(\alpha\gamma_\pm)^{\nu}$. For the acetic acid example mentioned before, the dissociation constant K is thus given by

$$K = a_{H^+} a_{A^-} / a_{HA} = \gamma_\pm^2 m_{H^+} m_{A^-} / \gamma_{HA} m_{HA} \tag{12.22}$$

5) In Chapter 2, we used $f_{x,X}$, $f_{m,X}$ and $f_{c,X}$ for component X for systematics. Here, we use $\gamma_X \equiv f_{m,X}$ for brevity. The coefficient $f_{m,X}$ is related to $f_{c,X}$ by $f_{m,X} = c_X f_{c,X}/\rho m_X$ with ρ the mass density of the solvent. Moreover, to avoid confusion with the degree of dissociation α, we use for the remainder of this chapter the label i, j (or +, −) for components instead of α, β.

Table 12.7 Constants for the pK_a expression values ($mol\,kg^{-1}$) for some acids in H_2O.

	pK_a at 25 °C	A_1 (K)	A_2 (–)	A_3 (K^{-1})
Acetic acid	4.756	1170.48	3.1649	0.013 399
Benzoic acid	4.203	819.63	1.287	0.009 19
n-Butyric acid	4.820	1033.39	2.6215	0.013 334
Formic acid	3.752	1342.85	5.2743	0.015 168
Propionic acid	4.874	1213.26	3.3860	0.014 055

For a solution of HA of molality m we have $m_{H^+} = m_{A^-} = \alpha m$, $m_{HA} = (1-\alpha)m$ and

$$K = \gamma_\pm^2 \alpha^2 m / \gamma_{HA}(1-\alpha) \cong \alpha^2 m / (1-\alpha) \tag{12.23}$$

where the last step only can be made if we take $\gamma_\pm = \gamma_{HA} = 1$.

Equation (12.23) is the *dilution law*, due to Ostwald in 1888, which expresses quantitatively that the lower the concentration, the larger the degree of dissociation. For acetic acid at 25 °C at infinite dilution (i.e., the limiting value), $K \cong 1.75 \times 10^{-5}\,mol\,kg^{-1}$, and does not deviate more than a few percent from this value up to $10^{-1}\,mol\,kg^{-1}$. From the overall reaction, Eq. (12.17), and the expression for K, Eq. (12.22), the chemical potential becomes $\mu_{HA} = \mathring{\mu}_{H+} + \mathring{\mu}_{A-} + RT\ln a_{H+}a_{A-}$, where $a_{HA} = a_{H+}a_{A-} = a^2\gamma_\pm^2 m^2$. An accurate empirical description for the dissociation constant K reads

$$-RT\ln K = \Delta G = A - CT + DT^2 \tag{12.24}$$

In practice, one uses pK_a values, defined by p$K_a \equiv -\log K = A_1/T - A_2 + A_3 T$. Values of the constants A_1, A_2 and A_3 for a few typical acids are listed in Table 12.7.

Van't Hoff found that the osmotic pressure of electrolyte solutions was always significantly higher than was predicted by colligative theory. For example, for the osmotic pressure equation Van't Hoff proposed $\Pi = icRT$, where i is now known as the *van't Hoff factor*. In principle this factor is equal to, but dependent on concentration, in practice usually less than the number of ions formed from one mole of compound. If, for a symmetric electrolyte 1 mole provides ν ions upon complete dissociation, the number of really dissociated molecules is $\nu\alpha$ and the number of undissociated molecules is $1 - \alpha$. Hence, the total number of ions in solution $i = 1 - \alpha + \nu\alpha$ or $\alpha = (i - 1)/(\nu - 1)$. The value for α obtained from colligative properties is only in approximate agreement with values determined via conductivity measurements.

Problem 12.10

Show that, if $-R \ln K_a = A/T - C + DT$, the K_a value shows a maximum given by $R \ln K_a = C - 2(AD)^{1/2}$ at temperature $T_{max} = (A/D)^{1/2}$.

Problem 12.11

Calculate ΔH, ΔS and ΔC_P from ΔG for the dissociation of a weak electrolyte.

12.5 Debye–Hückel Theory

The presence of charges has a significant influence on the structure and properties of solutions. The conventional theory to deal with these effects is due to Debye and Hückel, formulated in 1923, and attempts–common to all solution theories–to model the excess Gibbs (or Helmholtz) energy.

For the purpose of the model, several assumptions are made:

- It is assumed that, apart from a hard-sphere core, the ionic interactions form the only contribution to the excess Gibbs energy. This is the so-called *primitive model* (PM) for which we have $u(r) = \infty$ for $r < a$ with a often taken as the distance of closest approach $a = (\sigma_+ + \sigma_-)/2$ and $u(r) = e^2 z_+ z_- / 4\pi\varepsilon r$ for $r \geq a$, where ez_+ (ez_-) and σ_+ (σ_-) are the charge and diameter of the positive (negative) ions, respectively. If a is a constant, equal for all ions, the model is denoted as the *restricted primitive model* (RPM).
- The solvent is considered as a dielectric continuum in which the ions interact according to Coulomb's law. The permittivity ε is assumed to be equal to the permittivity of the pure liquid, thereby neglecting the influence of the ions on the permittivity. In the literature replacing the solvent by a dielectric continuum is referred to as the McMillan–Mayer picture.
- The ions are considered to be spherical, nonpolarizable charges producing spherically symmetric electric fields.
- The solution is sufficiently dilute so that at the average ionic distance the potential energy is small as compared with kT.
- The electrolytes are completely dissociated. In the model the structuring effect of the ionic interaction is counteracted by the thermal motion which is trying to disrupt structures. As a result any ion will surrounded by a group of ions, called *ionic atmosphere*, of which the net charge is of opposite sign to that of the reference ion.

To model this we recall that the electrical potential of a single ion j is given by

$$\Psi_j^{\text{ion}}(r) = ez_j / 4\pi\varepsilon r \quad \text{with} \quad \varepsilon \equiv \varepsilon_0 \varepsilon_r \tag{12.25}$$

For a solution we need $\Psi_j(r)$ representing the electrical potential due to the potential Ψ_j^{ion} of ion j itself *and* the potential Ψ_j^{atm} of its atmosphere. The charge density $\rho_j(r)$ at distance r from ion j is

$$\rho_j = \sum\nolimits_i ez_i n_{ij} \quad \text{(summation over all ions including type } j) \tag{12.26}$$

where n_{ij} denotes the number density of ion i at distance r from ion j. Charge density $\rho_j(r)$ and potential $\Psi_j(r)$ are self-consistently connected by a basic equation of electrostatics, *Poisson's equation* (see Appendix D), given by

$$\varepsilon\nabla^2\Psi_j = -\rho_j \tag{12.27}$$

where $\nabla^2 = \partial^2/\partial x^2 + \partial^2/\partial y^2 + \partial^2/\partial z^2$ is the Laplace operator in Cartesian coordinates. Since the charge distribution around an ion is spherically symmetric, it is convenient to use the Laplace operator in polar coordinates from which the angular coordinates are omitted. In this form it reads

$$\nabla^2 = \frac{1}{r^2}\frac{\partial}{\partial r}\left(r^2\frac{\partial}{\partial r}\right) \tag{12.28}$$

Combining Eqs (12.27) and (12.28) yields for the Poisson equation

$$\frac{1}{r^2}\frac{\partial}{\partial r}\left(r^2\frac{\partial\Psi_j}{\partial r}\right) = -\frac{\rho_j}{\varepsilon} \tag{12.29}$$

With the electrical potential at a certain point given by Ψ_j, the potential energy of an ion of charge ez_i is given by $ez_i\Psi_j$. This is also the work required in charging an ion up to charge ez_i in the potential Ψ_j. We assume that the number density of ions n_{ij} of ions i around a central ion j is given by the correlation function $g_{ij}(r)$ and the density at zero potential[6] n_i. For low ionic density the potential of mean force in $g_{ij}(r)$ is the potential energy $ez_i\Psi_j$ (see Chapter 5) or, equivalently, given by Boltzmann's law, so that

$$n_{ij} = n_i g_{ij}(r) \cong n_i \exp[-ez_i\Psi_j(r)/kT] \tag{12.30}$$

The charge density ρ_j is obtained by summing over all ions i and reads

$$\rho_j = \sum\nolimits_i ez_i n_{ij} = \sum\nolimits_i ez_i n_i \exp(-ez_i\Psi_j/kT) \tag{12.31}$$

Expanding the exponentials in Eq. (12.31) via $\exp(x) = 1 + x + x^2/2 + \ldots$, we obtain

$$\rho_j = \sum\nolimits_i ez_i n_i - \sum\nolimits_i \frac{e^2 z_i^2 n_i \Psi_j}{kT} + \sum\nolimits_i \frac{e^3 z_i^3 n_i \Psi_j^2}{2k^2T^2} - \ldots \tag{12.32}$$

The first order term of the expansion is zero because of the charge neutrality of the solution in total. For a symmetric binary salt the third order term is also zero. In any case we limit the expansion to the second order term only. Substituting Eq. 12.32 so truncated in Eq. (12.29), we obtain the (linearized) Poisson–Boltzmann expression

$$\frac{1}{r^2}\frac{\mathrm{d}}{\mathrm{d}r}\left(r^2\frac{\mathrm{d}\Psi_j}{\mathrm{d}r}\right) = \kappa^2\Psi_j \quad \text{with} \quad \kappa^2 = \frac{e^2}{\varepsilon kT}\sum\nolimits_i n_i z_i^2 \tag{12.33}$$

6) We use n for number density in order to avoid confusion with the charge density ρ.

Using $\Psi_j = Y(r)/r$, Eq. (12.33) is transformed to

$$\mathrm{d}^2Y/\mathrm{d}r^2 = \kappa^2 Y \tag{12.34}$$

for which the solution is

$$Y = A\exp(\kappa r) + B\exp(-\kappa r) \quad \text{or} \quad \Psi_j = A_j\frac{\exp(\kappa r)}{r} + B_j\frac{\exp(-\kappa r)}{r} \tag{12.35}$$

where A_j and B_j are constants to be determined from the boundary conditions. Since for $r \to \infty$, $\Psi \to 0$, the constant A_j must be zero and the general solution reads $\Psi_j^{\text{gen}} = B_j\exp(-\kappa r)/r$.

Before, we assumed that Ψ_j represents the electrical potential due to the potential Ψ_j^{ion} of ion j itself plus the potential Ψ_j^{atm} of its atmosphere. Let us further assume that there is a distance of closest approach a to the central ion for other ions (we take a the same for all pairs of ions[7], that is, the RPM model). Hence, for $r < a$ no other ions are present. The potential Ψ_j near an ion, that is, at $r = a$, of charge z_j is then given by

$$\Psi_j = \Psi_j^{\text{ion}} + \Psi_j^{\text{atm}}(a) \equiv (z_j e / 4\pi\varepsilon a) + C_j \tag{12.36}$$

where C_j as defined is a constant due to the presence of other ions for $r > a$. At $r = a$, both the potential Ψ_j and the electric field $E = -\partial\Psi_j/\partial r$ should be continuous. Since obviously Ψ_j^{gen} and Ψ_j should represent the same specific solution, we can from these two conditions evaluate both B_j and C_j and if we do so, we obtain

$$B_j = \frac{ez_j\exp(\kappa a)}{4\pi\varepsilon(1+\kappa a)} \quad \text{and} \quad C_j = -\frac{ez_j}{4\pi\varepsilon}\frac{\kappa}{(1+\kappa a)} \tag{12.37}$$

so that the total potential Ψ_j at a distance $r \geq a$ from a finite size ion j becomes

$$\Psi_j(r) = B_j\frac{\exp(-\kappa r)}{r} = \frac{ez_j}{4\pi\varepsilon}\frac{\exp[-\kappa(r-a)]}{(1+\kappa a)r} \tag{12.38}$$

The potential of the atmosphere Ψ_j^{atm} is the difference between the total potential Ψ_j and the potential of the single ion Ψ_j^{ion} and thus

$$\Psi_j^{\text{atm}}(r) = \Psi_j(r) - \Psi_j^{\text{ion}}(r) = \frac{ez_j}{4\pi\varepsilon}\frac{1}{r}\left\{\frac{\exp[-\kappa(r-a)]}{(1+\kappa a)} - 1\right\} \tag{12.39}$$

The value $C_j \equiv \Psi_j^{\text{atm}}(a)$, given by Eq. (12.37), represents the potential at the surface of a sphere with radius a due to the effect of all other ions.

To check the self-consistency, we calculate the charge of the atmosphere associated with the reference ion of charge ez_j. Substituting Eqs (12.36) and (12.37) in Eq. (12.32) we find the result

$$\rho_j(r) = -\frac{ez_j\kappa^2}{4\pi r_j}\frac{\exp(\kappa a)\exp(-\kappa r_j)}{(1+\kappa a)} \tag{12.40}$$

7) As said, typically one takes $a = (\sigma_+ + \sigma_-)/2$ (with σ_i the ionic diameter) because the probability for having pairs +/+ and −/− pairs is much smaller than for +/− pairs.

so that by integration over all space with $r > a$ the final result obtained is

$$\int_a^\infty \rho_j(r) 4\pi r^2 \,\mathrm{d}r = -\frac{ez_j}{1+\kappa a}\exp(\kappa a)\int_a^\infty \exp(-\kappa r_j)\kappa^2 r_j \,\mathrm{d}r_j = -ez_j \tag{12.41}$$

As expected, this charge equals the negative of the charge of the reference ion.

12.5.1
The Activity Coefficient and the Limiting Law

In the previous chapter we derived that the activity coefficient is related to the excess Gibbs energy. Here, G^E is calculated as the work required to introduce one extra ion to the solution containing a density n_j of each type of ion j, multiplied by N_A. The ion is introduced as uncharged, and this first step is assumed to require no electrical energy. The second step is to gradually charge the ion in solution from 0 to ez_j. For each increment the work at constant P and T is equal to the charge multiplied by the potential due to the other ions Ψ_j^atm. Hence [15], (using x as dummy variable)

$$kT\ln\gamma_j = \frac{\mu_j^\mathrm{E}}{N_\mathrm{A}} = \int_0^{ez_j} \Psi_j^\mathrm{atm}(x)\,\mathrm{d}x = \int_0^{ez_j} \frac{-\kappa x}{4\pi\varepsilon(1+\kappa a)}\mathrm{d}x = -\frac{\kappa e^2 z_j^2}{8\pi\varepsilon(1+\kappa a)} \tag{12.42}$$

For an ionic solution of ν_+ ions of charge z_+ and ν_- ions of charge z_-, the mean activity coefficient $\gamma_\pm$ is defined by

$$\nu\ln\gamma_\pm \equiv (\nu_+ + \nu_-)\ln\gamma_\pm \equiv \nu_+\ln\gamma_+ + \nu_-\ln\gamma_- \tag{12.43}$$

Moreover, we have charge neutrality expressed by

$$\nu_+ z_+ + \nu_- z_- = 0 \tag{12.44}$$

Substituting Eq. (12.42) in Eq. (12.43) meanwhile using Eq. (12.44) results in

$$\ln\gamma_\pm = -\frac{\kappa e^2 |z_+ z_-|}{8\pi\varepsilon kT(1+\kappa a)} \tag{12.45}$$

The next concept we need is the *ionic strength* I_m and I_c defined by, respectively,

$$I_m \equiv \frac{1}{2}\sum\nolimits_i m_i z_i^2 \quad \text{and} \quad I_c \equiv \frac{1}{2}\sum\nolimits_i c_i z_i^2 \tag{12.46}$$

for solutions of molality m and of concentration c, respectively. For a 1–1 electrolyte solution I_m is just the molality, while for other electrolytes it represents a type of "effective" molality. As is customary, we convert for κ from molarity c_i to molality m_i, which is important because this leads to extra terms upon evaluating the enthalpy, heat capacity, and so on. The general relationship between c_i and m_i is (see Section 2.1)

$$c_i = \rho' m_i \left(n_1 M_1 \Big/ \sum\nolimits_i n_i M_i \right) \cong \rho' m_i \quad \text{for small } c_i$$

where the subscript 1 refers to the solvent. Hence (remembering that $n_i = N_A c_i$ with N_A as Avogadro's number)

$$\sum_i n_i z_i^2 = N_A \sum_i c_i z_i^2 = 2N_A I_c \cong 2N_A \rho' I_m \tag{12.47}$$

with ρ' the mass density of the solution. For small m_i, $\rho' \cong M/V_m$, where M is the molar mass and V_m is the molar volume of the solvent. Introducing all of this into the expression for κ, we obtain

$$\kappa^2 = \frac{2e^2 N_A}{\varepsilon kT} I_c = \frac{2e^2 N_A \rho'}{\varepsilon kT} I_m \tag{12.48}$$

so that the final Debye–Hückel expression for the activity coefficient becomes

▶ $$\ln \gamma_\pm = \frac{-A|z_+ z_-| I_m^{1/2}}{1 + BaI_m^{1/2}} \quad \text{with} \tag{12.49}$$

$$A = (2\pi N_A \rho')^{1/2} \left(\frac{e^2}{4\pi\varepsilon kT} \right)^{3/2} \quad \text{and} \quad B = \left(\frac{2e^2 N_A \rho'}{\varepsilon kT} \right)^{1/2} \tag{12.50}$$

For water at 25 °C using $\rho' = 1.00\,\text{g cm}^{-3}$ and $\varepsilon_r = 78.5$, $A = 1.172\,\text{kg}^{1/2}\,\text{mol}^{-1/2}$ and $B = 3.29 \times 10^9\,\text{m}^{-1}\,\text{kg}^{1/2}\,\text{mol}^{-1/2}$. The expression is valid up to about $I_m \cong 0.1\,\text{m}$. In the original derivation of the Debye–Hückel model, point ions were assumed so that $a = 0$ leading to $\ln\gamma_\pm = -z_- z_+ A I_m^{1/2}$, the so-called Debye–Hückel *limiting law*. In this limit the activity coefficient is valid for low molality only, say $I_m < 0.01\,\text{m}$. The limiting Debye–Hückel expression corresponds with using

$$\Psi_j(r) = \frac{ez_j}{4\pi\varepsilon} \frac{\exp[-\kappa r]}{r} \quad \text{instead of} \quad \Psi_j(r) = \frac{ez_j}{4\pi\varepsilon} \frac{\exp[-\kappa(r-a)]}{(1+\kappa a)r} \tag{12.51}$$

The latter model is usually denoted as the *extended Debye–Hückel model.*

12.5.2 Extensions

For values of $I > 0.1\,\text{m}$, extensions for the extended Debye–Hückel model are required. An extensive literature on this topic exists, but we refer here only to one (semi-)empirical extension. According to Hückel, the alignment of the solvent molecules by the ionic atmosphere leads to a linear term in the ionic strength I_m, so that the expression for $\ln\gamma_\pm$ at high molality becomes

$$\ln \gamma_\pm = -\frac{A|z_+ z_-| I_m^{1/2}}{1 + aBI_m^{1/2}} + CI_m \tag{12.52}$$

In the literature this expression is often addressed as the *Davies equation* (1962), and considered to be a general expression for activity coefficients of solutions with $0.1 < I_m < 1.0$. Using a "universal" value for the ion radius of 0.3 nm and evaluating the constants at 25 °C, this reads[8] (see Ref. [16], p. 39).

8) Note the change from ln to log.

$$\log\gamma_\pm = -0.509|z_+z_-|\left(\frac{I_m^{1/2}}{1+I_m^{1/2}} - 0.30I_m\right) \tag{12.53}$$

The typical error is 5% at $I_m \sim 0.5$ m. Other authors use other values for C ranging from $C = 0.1$ to $C = 0.3$. The linear term is often addressed as the "salting-out" term, as it accounts for the lowered solubility of salts at high ionic strength.

Several other schemes exist. The most well-known and rather reliable semi-empirical scheme, which relies on the fitting of data and is valid up to $I_m \cong 6$, is probably due to Pitzer [17]. Generalization to multisolvent electrolytes is difficult, however. Another approach, based on the MSA using the (restrictive) primitive model, has been described by Lee [18] and Barthel *et al.* [19].

Problem 12.12

Verify Eq. (12.37).

Problem 12.13

Calculate κ for a 2–1 electrolyte in H_2O at 25 °C for $c = 0.01\,\text{mol}\,\text{l}^{-1}$.

Problem 12.14

For a 0.05 m solution of HCl in water, calculate the activity coefficient according to the limiting and extended Debye–Hückel models, as well as the Davies equation. Comment on the differences.

12.6 Structure and Thermodynamics

From the discussion on correlation functions in Section 6.2 we recall that the probability of finding two molecules at distance r is given by $n^2g(r)$ with $g(r)$ the (*pair*) *correlation function* and $n = N/V$ the number density of N particles in a volume V (note that in Section 6.2 we used $\rho = N/V$). For independent particles, that probability is given by n^2. If it is given that the reference molecule is at the origin, the probability of finding a molecule at distance r from the origin is thus $ng(r)$.

12.6.1 The Correlation Function and Screening

These concepts are applied here for ions in solution. We have disregarded the solvent (except for its permittivity) and consider only the correlation between the ions in solution. In the expression for the concentration of counterions i in the neighborhood of a central ion j, Eq. (12.30), we used $n_i g_{ij}(r)$ with the correlation function for the ions given by

$$g_{ij}(r) = \exp(-ez_i\Psi_j / kT) \tag{12.54}$$

Using Eq. (12.38) we obtain, with B_j given by Eq. (12.37),

$$\blacktriangleright \quad g_{ij}(r) = \exp[-q_{ij}(r)] \quad \text{with} \quad q_{ij}(r) = \frac{ez_i\Psi_j}{kT} = \frac{ez_iB_j}{kT}\frac{\exp(-\kappa r)}{r} \tag{12.55}$$

As expected, this expression is symmetric in z_i and z_j. Expanding g_{ij} we obtain

$$g_{ij} = 1 - q_{ij} + \frac{1}{2}(q_{ij})^2 - \ldots \tag{12.56}$$

The original Debye–Hückel approximation is given by the first two terms. It might be thought that keeping higher-order terms (or the exponential itself for that matter) is inconsistent with the Debye–Hückel approximation, since in the approximation of Eq. (12.33) the Debye–Hückel approach is self-consistent and obeys the superposition principle for potentials of electrostatics. Pitzer [20], however, has pointed out that for the charged hard-sphere the exponential expression represents the pair-correlation function well, as judged by comparison with Monte Carlo simulations for the same model (Figure 12.6).

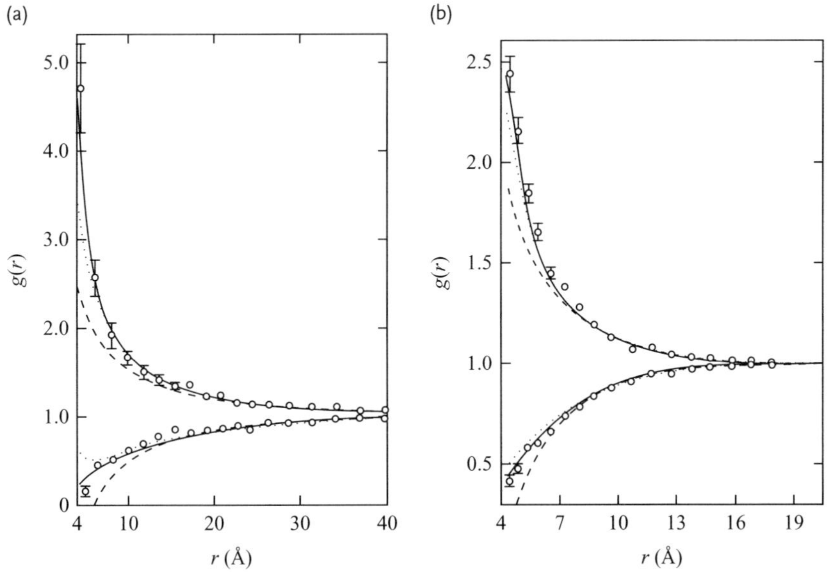

Figure 12.6 The pair correlation function at 25 °C for (a) a 0.00911 M and (b) a 0.425 M aqueous solution of a 1–1 electrolyte with a common hard core value of a = 0.425 nm. The points represent the Monte Carlo simulations, while the solid curve is the exponential Debye–Hückel expression. The dotted and dashed curves are three-term and two-term expansions of the exponential expression, respectively.

The work done by the system by the charging process leads to a lower potential energy as a result of the approaching of positive and negative charge to approximately a distance of κ^{-1}. In fact, κ^{-1} is the maximum in the charge distribution $4\pi r^2 g_{ij}(r)$, so that κ^{-1} is the *most probable thickness* of the ionic atmosphere. For a 1–1 electrolyte in H_2O at 25 °C we obtain $\kappa^{-1} = 3.04 \times 10^{-10} c^{-1/2}$ with c in mol l^{-1}. So, for $c = 0.01$ mol l^{-1}, the Debye length $\kappa^{-1} \cong 3.0$ nm. Another interpretation is that for a distance κ^{-1} the correlation between the ions is largely lost. Because of the formation of the oppositely charged atmosphere of thickness κ^{-1}, the long-range Coulomb potential is screened; this is why κ^{-1} is also called the Debye *screening length*.

From the fact that we neglected the structure of the solvent altogether, except for its permittivity, one might guess that the attempts to describe electrolyte liquids by virial expansion would be useful, and this indeed is the case. However, because $\Psi_j' \sim r^{-1}$ is of long range, the various integrals for the virial coefficients diverge. McMillan and Mayer showed that, by a clever combination of various expressions, a finite answer can be obtained, and today various advanced theories of ionic solutions employ the virial approach to model the properties of ionic solutions (see, e.g., Friedman (1962) or Barthel (1998)).

Another obvious starting point is to use one of the various integral equations, as they are rather successful in describing liquids, corresponding to highly concentrated solutions. Other approaches, such as the use of perturbation methods, are also available, and for these reference should be made to the specialized literature.

12.6.2
Thermodynamic Potentials*

From all of the above information, the Helmholtz and Gibbs energies of ionic solutions can be calculated. To obtain the required result we integrate $\mathrm{d}G_j = \Psi_j^{\mathrm{atm}}\mathrm{d}q_j$ with $q_j = ez_j$. Recall that all Ψ_j^{atm} are functions of all valencies z_j (see Eq. 12.39). The most convenient way to integrate is to increase the charges of all ions in the same ratio. If we then denote by λ the fraction of their final charges which the ions have at any stage during integration, we have

$$G_{\mathrm{ele}} = \int_0^1 \mathrm{d}G(\lambda) = \int_0^1 \sum\nolimits_j N_j \Psi_j^{\mathrm{atm}}(\lambda) e z_j \,\mathrm{d}\lambda \quad \text{with} \quad 0 \le \lambda \le 1 \tag{12.57}$$

Using Eq. (12.37) we obtain

$$G_{\mathrm{ele}} = -\int_0^1 \sum\nolimits_j N_j \frac{\lambda e z_j}{4\pi\varepsilon} \frac{\lambda\kappa}{1+\lambda\kappa a} e z_j \,\mathrm{d}\lambda \quad \text{leading to}$$

$$G_{\mathrm{ele}} = -\sum\nolimits_j \frac{e^2 z_j^2 N_j \kappa}{4\pi\varepsilon} \int_0^1 \frac{\lambda^2}{1+\lambda\kappa a} \mathrm{d}\lambda \quad \text{and performing the integration to}$$

$$G_{\mathrm{ele}} = -\sum\nolimits_j \frac{e^2 z_j^2 N_j \kappa}{12\pi\varepsilon} \tau(\kappa a) \quad \text{with} \tag{12.58}$$

$$\tau(x) = 3x^{-3}\left[\ln(1+x) - x + \frac{1}{2}x^2\right]$$
$$= 1 - \frac{3}{4}x + \frac{3}{5}x^2 - \frac{3}{6}x^3 + \frac{3}{7}x^4 - \ldots \cong 1 \quad \text{for} \quad x \to 0 \tag{12.59}$$

From Eq. (12.58) all thermodynamic properties can be calculated in the usual way.

Problem 12.15

Show that the activity coefficient $\gamma_1 = \dfrac{\partial G_{\text{ele}}}{\partial N_1} = \dfrac{\partial G_{\text{ele}}}{\partial \kappa}\dfrac{\partial \kappa}{\partial V}\dfrac{\partial V}{\partial N_1}$ for the solvent is

$$\gamma_1 = \sum_j \frac{e^2 z_j^2 N_j}{12\pi\varepsilon}\frac{\kappa}{2V}V_1\sigma(\kappa a) \quad \text{with} \quad \sigma(x) = \frac{3}{x^3}\left[(1+x) - \frac{1}{(1+x)} - 2\ln(1+x)\right]$$

Problem 12.16*

To show once more the consistency of the Debye–Hückel expressions, derive the activity coefficient for the solute, Eq. (12.42), from G_{ele}.

12.7 Conductivity

In 1833/1834, Faraday published his *law of electrolysis* which, in modern terms, states that if two electrodes with a potential difference V are placed in an ionic solution, then material at the electrodes will be liberated, the amount of which is proportional to the amount of charge $Q = \int I\,\mathrm{d}t$, where I is the electric current and t the time. Moreover, for a given amount of charge, two elements deposit an amount of material proportional to the equivalent mass, that is, M/z where M is the molar mass and z the charge of the ion. The macroscopic measure of charge $F = eN_A$ is nowadays called the *Faraday constant* where, as usual, e denotes the unit charge and N_A is Avogadro's constant. Electrolysis can only take place because ionic solutions conduct electricity. The resistance R in *Ohm's law* $R = V/I$ of a solution in a container of cross-section A and length L is given by

$$R = \rho L / A \tag{12.60}$$

where ρ is the *resistivity*. One often uses the reciprocal $\kappa = \rho^{-1}$, which is addressed as the *conductivity*[9]. The latter is usually measured in a parallel plate set-up, that is, using a homogeneous system, so that we can also write $\kappa = i/E$ with $i = I/A$ the current density and $E = V/L$ the electric field strength[10]. In order to avoid electrode

9) Unfortunately the conventional symbol for conductivity κ is the same as used for the Debye length κ_{-1}.

10) We use here a linear relationship between the "generalized force" (here, the field E) and the "generalized displacement" (here, the current density i). Irreversible thermodynamics shows that this linear relationship is often applicable. Anyway, nonlinear response is beyond the scope of this book.

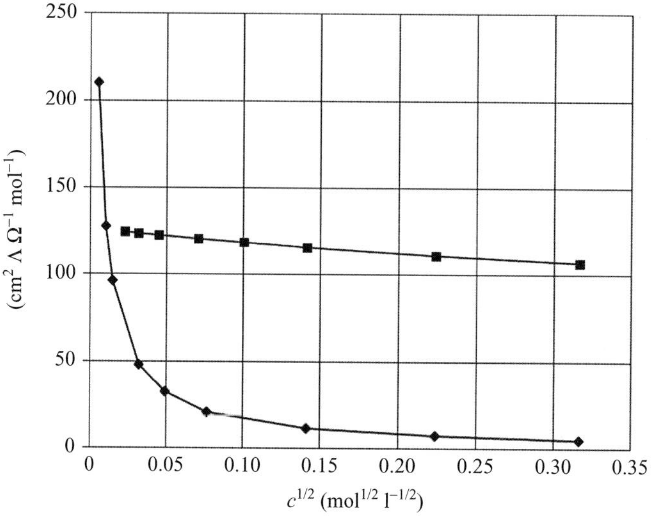

Figure 12.7 Molar conductivities of CH_3COOH (◆) and NaCl (■) as a function of concentration at 298 K.

reactions, changing the concentration of the electrolyte in the neighborhood of the electrode, and because the electrolysis products provide an opposing potential (an effect often addressed as *polarization*), one measures with an alternating potential difference, typically with a frequency in the order of kilohertz. For pure water at 25 °C, $\kappa \cong 5.5 \times 10^{-10}\,\Omega^{-1}\,m^{-1}$. For solutions, it is convenient to define the *molar conductivity* $\Lambda_m = \kappa/c$, where c is the concentration and the *equivalent conductivity* $\Lambda = \Lambda_m/n_{equ}$, where the equivalent $n_{equ} = \nu_+ z_+ = \nu_-|z_-|$ represents the amount of charges released .

Experimental examination of the conductivity revealed that there are two types of ionic solution, as illustrated in Figure 12.7. One type , known as a *strong electrolytic solution,* has a high and (almost) constant value of Λ that slowly decreases with increasing concentration; examples are NaCl or LiCl in H_2O. An empirical relationship to describe Λ given by Kohlrausch in 1885 [36] is

$$\Lambda = \Lambda° - Ac^{1/2} \tag{12.61}$$

It appears that all 1 : 1 electrolytes have approximately the same value for A.

The other type of ionic solution, known as a *weak electrolytic solution,* has a much smaller value of Λ which greatly increases with dilution; examples are CH_3COOH and $CuSO_4$ in H_2O. It should be noted that the labels "strong" and "weak" refer to the combination of solute and solvent. Strong electrolytes in H_2O (e.g., KBr) may be weak in other solvents (e.g., acetic acid). There are relatively few intermediate cases between strong and weak electrolytes. In both cases the limiting value of Λ for $c \to 0$, the limiting conductivity $\Lambda°$, plays an important role in theory.

Although it would be beneficial to know the contributions $\lambda_{\pm}^{\circ}$ of the individual ions to Λ, it is impossible to determine these contributions from Λ° values alone. Earlier, in 1875, Kohlrausch [37] had noted that the differences in Λ° for pairs of salts having a common ion are approximately independent of that ion. He attributed this effect to independent and additive contributions of the positive and negative ions to Λ°, that is, $\Lambda^{\circ} = \lambda_{+}^{\circ} + \lambda_{-}^{\circ}$, usually referred to as the law of *independent migration of ions*. So, if the *single ion conductivity* λ_{j}° for a particular ion j is known, the rest can be determined.

To obtain these data we consider the speed of ions. The drift speed of an ion $s_{\pm}$ is proportional to the electric field E with, as the proportionality constant, the *mobility* $u_{\pm}$. Suppose that we have a container with field E containing an electrolyte of concentration c. For a salt $Q = M_{\nu+}X_{\nu-}$ dissociating into $\nu_{+}M^{z+}$ and $\nu_{-}X^{z-}$ ions, the concentration of ion j is $c_j = \nu_j c$. The drift speed is then (for positive ions) $s_{+} = u_{+}E$, and all ions within distance $u_{+}E$ of the negative electrode will reach that electrode in unit time. The number of such ions is $\nu_{+}cu_{+}E$, and these carry a charge $z_{+}F\nu_{+}cu_{+}E = n_{\text{equ}}Fcu_{+}E$. Hence, the conductivity κ_{+} due to the positive ions is

$$\kappa_{+} = i_{+} / E = n_{\text{equ}}Fcu_{+}E / E = n_{\text{equ}}Fcu_{+} \quad \text{or} \quad \lambda_{+} = \kappa_{+} / n_{\text{equ}}c = Fu_{+} \tag{12.62}$$

We further introduce the *transport number*, t, which reflects the relative contribution of positive and negative charge carriers, that is, not necessarily the molecules or ions as such, to the total current density i and which reads

$$t_j = i_j / i = |z_j| c_j \lambda_j \Big/ \sum\nolimits_j |z_j| c_j \lambda_j \tag{12.63}$$

Since electroneutrality requires $\nu_{+}z_{+} + \nu_{-}z_{-} = 0$, the expression for t_{+} for a binary electrolyte reduces to

$$t_{+} = u_{+} / (u_{+} + u_{-}) \tag{12.64}$$

If t_{+} can be measured over a range of concentrations, it is possible to extrapolate their values to zero concentration t_{+}° and thus calculate $\lambda_{+}^{\circ} = t_{+}^{\circ}\Lambda^{\circ}/\nu_{+}$. Note that for weak electrolytes the degree of dissociation α is involved in the total conductivity Λ since $c_{+} = \alpha\nu_{+}c$, but not in the transport numbers.

Transport numbers were first measured in 1853 by Johann Wilhelm Hittorf (1824–1914), using an ingenious set-up as sketched in Figure 12.8. In this set-up the amounts of material deposited in the cathode and anode compartments are determined. From the amount of charge and material deposited after a certain time, the number of ions transported can be calculated, leading directly to the transport numbers. It should be noted that during the measurement the concentration changes yielding the result an effective transport number over the changing concentration range during the experiment. Another method for determining the transport numbers is the moving boundary method, which is based on the motion of the interface between two layered electrolyte solutions, visible by differences in either color or refractive index, as a function of the amount of charge passing through the system. Another method which involves measurement of the electromotive force for two combined cells, in one of which the concentration is varied,

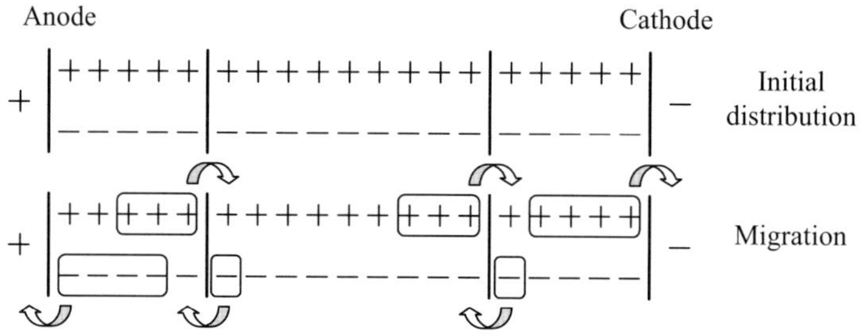

Figure 12.8 Hittorf set-up for measuring ionic velocities. In this case, the ratio of the mobilities of the ions, as determined from the concentrations of the ions in the anode and cathode department, is 1 to 3. Hence, $t^-/t^+ = 1/3$ or $t^- = t/4$ and $t^+ = 3t/4$.

Table 12.8 Limiting ionic conductivities $\lambda^\circ_\pm/\Omega^{-1}\,\text{mol}^{-1}\,\text{cm}^2$ in water at 298 K and for NaCl at various other temperatures. Data mainly from Ref. [19]. The parameter a is the (Pauling) ionic radius or the Kapustinski radius [35], if labeled with an asterisk.

	λ°_+	*a* (pm)		λ°_-	*a* (pm)		$\lambda^\circ_\pm$	*a* (pm)
H^+	349.6	–	OH^-	199.2	133*	NH_4^+	73.5	137*
Li^+	38.78	60	F^-	55.4	136	Me_4N^+	44.4	201*
Na^+	51.10	95	Cl^-	76.3	181	Et_4N^+	32.2	–
K^+	73.50	133	Br^-	78.1	195	Pr_4N^+	23.2	–
Ca^{2+}	59.5	99	I^-	77.0	216	nBu_4N^+	19.3	–
Ba^{2+}	63.6	135	Ac^-	40.8	162*	nAm_4N^+	17.4	–
T/K	273	298	323	348	373	NO_3^-	71.4	179*
Na^+	26	51	82	116	155	ClO_4^-	67.3	240*
Cl^-	41	76	116	160	207	SO_4^{2-}	80.0	258*

is also used. We refrain from further details here and refer to the literature [21]. Table 12.8 provides some experimental data from which it can be observed that for increasing ionic radius a, the (equivalent) conductivity λ° also increases, contrary to the expected decrease with ionic radius. This is due to the fact that not the bare but rather the hydrated ion moves. As $\Delta_{\text{hyd}}G$ decreases with increasing ionic radius, the effective radius decreases. The data for NaCl indicate a strong increase with temperature, and with increasing temperature the average hydration number will decrease so that the effective size becomes smaller, leading to an increase in λ°.

For weak electrolytes, which are typically of 1–1 type, we derived Eq. (12.23), $K = \alpha^2 m/(1-\alpha)$, assuming that the activity coefficient $\gamma = 1$. Normally, $\alpha = \Lambda/\Lambda_{\text{equ}}$ is used as a measure of dissociation, where Λ_{equ} is the sum of the molar conductivities of the positive and negative ions at the ionic concentration at equilibrium. As the mobility is concentration-dependent, but the ionic concentrations at equilibrium are only known if α is known, an iterative procedure should be used to obtain α. Data obtained in this way, although reasonably constant, show a definite

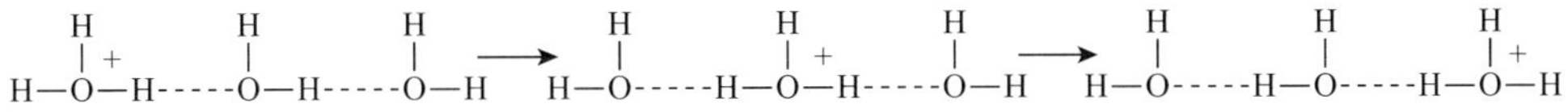

Figure 12.9 Schematic of the Grotthuss mechanism.

concentration dependence which is the result of neglecting the activity coefficients. By plotting the $\ln K$ values obtained versus square-root concentration $(\alpha c)^{1/2}$ for a series of diluted solutions, one can extrapolate to zero concentration to obtain the thermodynamic equilibrium constant.

Note from Table 12.8 that the values of $\lambda°$ for the H^+ and OH^- ions are much larger than for most other ions. Moreover, while hydration data suggest $\lambda°(H^+) < \lambda°(Li^+)$, in reality $\lambda°(H^+) >> \lambda°(Li^+)$. In fact, it is the network structure of water and not the hydration shell that determines the anomalous conductance[11] of the H^+ and OH^- ions, as explained by the Grotthuss mechanism [22], sketched in Figure 12.9. The protons can easily "debond" from one H_2O to "bond" again to the neighboring H_2O molecule (tunneling), as the hydrogen bonding makes the two configurations already much more similar as otherwise would be the case. It is noteworthy that this process, although faster than translational diffusion, proves to be slower than might be expected from its mechanism. This relative sluggishness may be due to the rotation of molecules required for trains of sequential proton movement and the consequential necessity for the breakage of hydrogen bonds. The strange effect [23] of degassing increasing proton motion over tenfold indicates, however, that nonpolar dissolved gas molecules (which are naturally present) disrupt the linear chains of water molecules necessary for the Grotthuss mechanism and so slow down the proton movement. Other mechanisms have also been proposed.

Although, in the past, a mechanism similar to that for the H^+ ion was assumed to operate for the OH^- ion, today it is thought that the movement of the OH^- ion is accompanied by a fourth hydrogen-bonded donor water molecule. The hydrated hydroxide ion is coordinated to four electron-accepting water molecules such that when an incoming electron-donating hydrogen bond forms (necessitating the breakage of one of the original hydrogen bonds) a fully tetrahedrally coordinated water molecule may be easily formed by hydrogen ion transfer.

12.7.1 Mobility and Diffusion

The mobility as determined from electrical conductivity is related to the diffusivity, and this can be seen as follows. The driving force for diffusion is given $F_{\text{dri}} = -(kT/c)\,\mathrm{d}c/\mathrm{d}x$ (ignoring the difference between activity a and concentration c), while the frictional force is proportional to the drift speed s, that is, $F_{\text{fri}} = fs$, where f is a friction coefficient and $1/f$ a mobility coefficient. Under steady-state conditions

11) Water has been intensively studied: See http://www1.lsbu.ac.uk/water/index2.html (31-08-2012).

these two forces are equal and so $fs = -(kT/c)\mathrm{d}c/\mathrm{d}x$. Defining the flux j by $j \equiv cs$ we obtain

$$j = -(kT/f)(\mathrm{d}c/\mathrm{d}x) \tag{12.65}$$

Comparing this expression with *Fick's first law*, $j = -D(\mathrm{d}c/\mathrm{d}x)$, we obtain the equation, first derived in 1905 by Albert Einstein (1879–1955),

$$D = kT/f = kT/6\pi\eta a \tag{12.66}$$

where the last step can be made if the frictional force is given by Stokes' law $F_{\text{fri}} = 6\pi\eta as$, where η is the viscosity and a is the radius of the charge carrier (ion plus hydration shell). In this form it is referred to as the *Stokes–Einstein equation*. In the electric case the driving force F_{dri} for a positive ion with charge z_+e in an electric field E is given by $F_{\text{dri}} = z_+eE$, while its frictional force F_{fri} is proportional to its speed, $F_{\text{fri}} = f_+s_+$. Under steady-state conditions the forces F_{dri} and F_{fri} are equal and opposite, and thus $z_+eE = f_+s_+$. Because the speed is also given by $s_+ = u_+E$, where u_+ is the mobility, we obtain $f_+ = z_+e/u_+$. Inserting this result into Eq. (12.66) leads to the *Einstein equation*

▶ $$D_+ = (kT/z_+e)u_+ \tag{12.67}$$

Writing $\lambda_+^\circ = Fu_+$ and $z_+e = z_+F/N_A$, we can also write $D_+ = RT\lambda_+^\circ/z_+^2F^2$, known as the *Nernst–Einstein equation*, first derived in 1888 by Walther Nernst (1864–1941). A similar equation can be written down for D_-. To obtain the diffusion coefficient for ambipolar diffusion of a symmetrical electrolyte in a solution, the diffusion coefficients of the positive ions D_+ and negative ions D_- must be combined via

$$D = 2D_+D_-/(D_+ + D_-) \tag{12.68}$$

Problem 12.17: Conductivity of water

Make an estimate of the lifetime t_H of hydrogen bonds in water from its ionic conductance, based on the Grotthuss mechanism.

Problem 12.18: Ionic conductivity

Consider a Teflon tube with Pt electrodes that close the ends of the tube (tube cross-section 5 cm², tube length 10 cm). The tube is filled with an aqueous solution of 0.01 M CaI_2 and a voltage of 10 V is set across the tube.

a) Calculate the rise in temperature ΔT for preparing this solution.

b) What is the molar conductivity Λ_m of CaI_2?

c) Calculate the electrical current I on the basis of Λ_m° data.

d) Calculate I using for the ions the Stokes–Einstein equation $F = 6\pi\eta as$ (F friction force, η viscosity, a radius and s velocity) and the Pauling ionic radii, and compare the result with that obtained in (c).

e) What is the transport number t of Ca^{2+} in this case?

f) Suppose that the tube and electrodes have a negligible thermal capacity and are thermally perfectly isolated from the surroundings. What is the temperature rise per second of the solution in the tube, given that $C_P(\text{solution}) \cong 4180\,\text{J}\,\text{kg}^{-1}\,\text{K}^{-1}$?

Problem 12.19: Diffusion coefficient*

Derive Eq. (12.68) for electrolyte $Q = M_{\nu+}X_{\nu-}$ using a 1D configuration. Do so by equating the flux of the positive and negative ions in terms of the electrochemical potential and solving this equation for $d\phi/dx$. Note that the electrochemical potential for component α is defined as $\mu_\alpha^* = \mu_\alpha + z_\alpha e\phi$, and that in equilibrium $\mu_Q^* = \mu_+^* + \mu_-^*$ for every point in the solution. By substituting $d\phi/dx$ in $d\mu_Q^*/dx$, calculate the flux j_Q and identify the proportionality constant as the diffusion constant D.

12.8 Conductivity Continued*

Let us continue with conductivity and try to explain the Kohlrausch expression Eq. (12.61). One can ask two questions with respect to this equation: why is the dependence $c^{1/2}$; and what is the value of the constants $\Lambda°$ and A. To explain this we must consider that when an ion in solution is dragged through the solution by an applied electric field, two extra effects are present. The first is the electrophoretic effect, and the second is the relaxation effect. Thereafter, we discuss ion association.

Let us consider first the *electrophoretic effect*, first described by Onsager and Fuoss in 1932. The total net charge of an ion atmosphere is equal and opposite to that of the central ion. If, therefore, under the influence of an applied electric field, the central ion moves to the right (positive direction) the atmosphere will move to the left (negative direction). The central ion therefore finds itself in a stream of solution moving opposite to the direction of the central ion, diminishing its net velocity. If the ionic strength becomes smaller, the atmosphere becomes more extended and the effect smaller.

Since bulk flow of the solution is negligible, it follows that the force f_j acting on ion j must be balanced by the force f_s acting on the solvent molecules, and thus we have $\Sigma_j n_j f_j + n_s f_s = 0$, where the number density is given by n_j. The local density of ions i around a central ion j in a volume element dV reads $n_{ij} = n_i g_{ij}(r)$, determined by the distribution function $g_{ij}(r)$ as given by Eq. (12.56). Therefore, the force near ion j will be $\Sigma_i n_{ij} f_i\, dV$ which is different from the volume-averaged force $\Sigma_i n_i f_i dV$. Since the solvent is neutral, the local density is the average density and the force on the solvent is given by $n_s f_s dV = -\Sigma_i n_i f_i\, dV$. The net force dF_j acting on ion j is therefore

$$dF_j = \left[\sum\nolimits_i (n_{ij} f_i) + n_s f_s\right] dV = \sum\nolimits_i (n_{ij} - n_i) f_i 4\pi r^2 \, dr \tag{12.69}$$

where the last step can be made if we take for dV a spherical volume element with radius r and thickness dr. Evaluating the dF_j on this volume element by approximating g_{ij} by Eq. (12.56), the result becomes

$$dF_j = \left[-A_1 \exp(-\kappa r)/r + \frac{1}{2} A_2 \exp(-2\kappa r)\Big/ r^2\right] r^2 dr \quad \text{with} \tag{12.70}$$

$$A_1 = 4\pi \sum\nolimits_i n_i f_i (z_i e B_j / kT) \quad \text{and} \quad A_2 = 4\pi \sum\nolimits_i n_i f_i (z_i e B_j / kT)^2 \tag{12.71}$$

This force is counteracted by the friction experienced by ion j and its atmosphere, as described by Stokes' law. So, for the drift speed change we have $\Delta s_j = dF_j / 6\pi\eta r$, where η is the viscosity of the medium and r the radius of each volume element. This leads to

$$\Delta s_j = (6\pi\eta)^{-1} \int_a^\infty \left[-A_1 \exp(-\kappa r) + \frac{1}{2} A_2 \exp(-2\kappa r)\Big/ r\right] dr \tag{12.72}$$

Let us now turn our attention to the *relaxation effect*, already considered by Debye and Hückel in 1923, but more successfully by Onsager in 1927 and extended in 1954 by Falkenhagen [21]. The basic idea is that there will be a disturbance of the spherical symmetry of the atmosphere if the central ion moves under the influence of the applied field. If, for example, a specific type of ion moves to the right, each ion will constantly have to build up its ionic atmosphere to the right, while the charge density to the left gradually decays. The rate at which this occurs is related to the relaxation time τ of the atmosphere. The time required for the atmosphere to fall essentially to zero is $4h\tau$, where

$$h = \frac{|z_+ z_-|}{|z_+| + |z_-|} \frac{\lambda_+^\circ + \lambda_-^\circ}{|z_+|\lambda_+^\circ + |z_-|\lambda_-^\circ} \tag{12.73}$$

and where z and λ have the same meaning as before. The relaxation time appears to be

$$\tau = f / \kappa^2 kT \tag{12.74}$$

where f is the friction force constant for a single ion. Due to this finite relaxation time, the atmosphere will lag behind the moving central ion by approximately τs_{cen} and deviate from the spherical symmetry. As a consequence, there will a backwards pull on the central ion, which for a solvent with permittivity ε has been calculated by Onsager and Falkenhagen *et al.* as

$$\frac{\Delta E}{E} = \frac{z_1 z_2 e^2}{12\pi\varepsilon kT} \frac{h}{1 + h^{1/2}} \frac{\kappa}{1 + \kappa a} \equiv \frac{z_1 z_2 e^2}{12\pi\varepsilon kT} H \frac{\kappa}{1 + \kappa a} \tag{12.75}$$

where ΔE is the relaxation electric field acting in the opposite direction of E and the constant $H \equiv h/(1 + h^{1/2})$ is introduced. Adding the field of the relaxation effect to the applied field leads to the total force f_j on an ion j given by

$$f_j = z_j eE[1+(\Delta E/E)] = z_j e(E+\Delta E) \tag{12.76}$$

The force f_j produces a velocity s'_j relative to the solvent that becomes

$$s'_j = u_j^{\circ} E[1+(\Delta E/E)] = s_j^{\circ}[1+(\Delta E/E)] \tag{12.77}$$

with u_j° the limiting mobility, as defined by the limiting velocity at infinite dilution $s_j^{\circ} \equiv u_j^{\circ} E$. The force f_j is also the force that has to be used in Eq. (12.71) for evaluating the electrophoretic velocity drag Δs_j. Evaluating Δs_j with this force leads to

$$\Delta s_j = \left[-(6\pi\eta)^{-1} z_j e\kappa/(1+\kappa a) + \frac{1}{2}(6\pi\eta)^{-1} A_2 \mathrm{Ei}(2\kappa a)\right](E+\Delta E) \tag{12.78}$$

where $\mathrm{Ei}(x) = \int_x^{\infty} \exp(-y) y^{-1}\,\mathrm{d}y$ is the exponential integral and where $\kappa^2 = e^2\Sigma_i n_i z_i^2/\varepsilon kT$, Eq. (12.33), is used. For symmetrical electrolytes $z_1 = -z_2$ and, since f_i in Eq. (12.71) is proportional to ez_i, A_2 will vanish.

Although we have included the second-order electrophoretic term, we recall that a mathematically consistent solution of the Poisson equation in combination with the Boltzmann distribution can be taken only as far as first order for unsymmetrical electrolytes and to a second-order term for symmetrical electrolytes. However, since for symmetrical electrolytes the second-order term with A_2 vanishes anyway, we need to consider only the first-order term. The final drift speed (to first order) then becomes

$$s_j = s'_j + \Delta s_j = u_j^{\circ}(E+\Delta E) - \frac{ez_j\kappa}{6\pi\eta(1+\kappa a)}(E+\Delta E) \tag{12.79}$$

The ratio $s_j/s_j^{\circ} = u_j E/u_j^{\circ} E$ equals $\lambda_j/\lambda_j^{\circ}$, and the conductivity λ_j of ion j becomes

$$\lambda_j = \left(\lambda_j^{\circ} - \frac{e^2 N_A}{6\pi\eta}\frac{|z_j|\kappa}{1+\kappa a}\right)\left(1+\frac{\Delta E}{E}\right) \tag{12.80}$$

using $u_j^{\circ} = \lambda_j^{\circ}/F$ with F the Faraday constant. Combining the expressions for the positive and negative ions in the usual way via $\Lambda = \lambda_+ + \lambda_-$, introducing $\kappa = (2N_A e^2 I_m \rho'/\varepsilon kT)^{1/2} \equiv \kappa' I_m^{1/2}$, Eq. (12.48), and neglecting cross terms results in the equivalent conductivity, we obtain

$$\Lambda = \Lambda^{\circ} - S\sqrt{I_m}/(1+a\kappa'\sqrt{I_m}) \tag{12.81}$$

Here, S is a complex constant, dependent on temperature, charge and (limiting) conductivity of the ions and relative permittivity and viscosity of the solvent, given by

$$S = P + Q\Lambda^{\circ} \equiv \frac{(|z_1|+|z_2|)e^2 N_A \kappa'}{6\pi\eta} + \frac{z_1 z_2 e^2 H\kappa'}{12\pi\varepsilon kT}\Lambda^{\circ} \tag{12.82}$$

For a symmetrical electrolyte we have $z_1 = z_2 = z$. The relevant data for water at 25 °C are: $\varepsilon_r = 78.30$ and $\eta = 0.8903$ mPa s. Expressing the value for Λ in $\mathrm{cm^2\,\Omega^{-1}\,mol^{-1}}$, the values of P and Q reduce to

$$P = 30.32(|z_1|+|z_2|) \quad \text{and} \quad Q = 0.7852|z_1 z_2|H \tag{12.83}$$

This model explains why the conductivity of strong electrolytes is described to first order by $\Lambda = \Lambda° - Ac^{1/2}$ and provides a value[12] for A. Moreover, it gives a fair account of the conductivity of aqueous solutions of 1–1 electrolytes up to 0.05 or 0.1 m. For other electrolytes the situation is less favorable. Robinson and Stokes [21] showed that the convergence of the model using the full expansion of Eq. (12.54) depends mainly on the factor $(z_1^n + z_2^n)^2/a^n(|z_1|+|z_2|)$, where n is the order of the expansion. Because most ions have a radius of, say, 4 Å, it will be clear that convergence for 1–1 electrolytes is reasonable and the behavior can be described by the truncated expression Eq. (12.56). Although for 2–2 electrolytes the second-order term vanishes, the third-order term is about as large as the first-order term, and hence a more extensive theory is called for. Also, for higher concentrations a more extensive theory is required. These theories are beyond the scope of this book.

12.8.1 Association

A regularly occurring deviation in plots of Λ versus $c^{1/2}$ is that the (negative) slopes are larger than predicted by the Debye–Hückel–Onsager model, which means that the experimental conductivities are smaller than predicted. Estimating ionic radii from the fitted parameters often yielded unphysically small radii. In essence, there are two ways to remedy the situation. The first approach is to extend the Debye–Hückel–Onsager solution of the Poisson–Boltzmann equation to a higher order (a brief discussion can found in Fowler and Guggenheim [24]), but this solution does not have the property of self-consistency, in contrast to the first-order solution. This implies, for example, that if the activity is calculated from the electrostatic energy (see Problem 12.15), the result deviates from the result as obtained via the Güntelberg procedure (see Section 12.5). Here, we address the alternative, second approach that takes ion association into account.

Recall that in the Debye–Hückel–Onsager model, the complete dissociation of ion pairs is assumed. This is not necessarily true, however, and in 1926 Niels Bjerrum [25] introduced a theory in which ions of opposite charge would temporarily form an associated pair via Coulomb forces. A reference ion, say a positive ion at the origin, and a counterion, say a negative ion at distance R, are considered to be an ion pair if no other ions can be found in the sphere with radius R centered at the reference ion. For ease of presentation, we limit the discussion to symmetric ionic solutions ($z_1 = z_2$), so that an ion pair (IP) in solution is neutral. We denote the ionic radii by a_1 and a_2. It appears to be convenient for later to introduce the parameter $q = z_1z_2e^2/8\pi\varepsilon kT$, nowadays called the *Bjerrum length*.

12) It does not, however, provide a value for $\Lambda°$, since this parameter is related to ion–solvent interaction. Equating Stokes' frictional force to the electrical driving force provides an estimate, of course, but an estimate for the radius a must still be made. This value can be estimated by employing a concentration series and fitting Eq. (12.81), using a as a parameter.

We first need to discuss the expression for the energy difference W between two (solvated) ions and the ion pair. To zeroth order this is given by the Coulomb expression $W_C(r) = e^2z_1z_2/4\pi\varepsilon r$. However, from the Debye–Hückel theory we know that in an ionic solution shielding of the Coulomb potential occurs and Eq. (12.38) applies. However, it should be noted that the shielded Coulomb potential applies only outside the sphere of the bound ion pair as there are no other ions inside the sphere. Moreover, apart from the Coulomb forces, we also have the induction, dispersion, and repulsion forces. Finally, we have also to consider the effect of the solvation shells. Each ion has a solvation shell, and if ions come close together these shells interpenetrate and this leads to an interaction force, often termed the *Gurney force*. Hence, these interactions together can be represented as the sum of the Coulomb interaction $W_C(r)$ and a short-range part $W_{SR}(r)$, the latter being the sum of the induction, dispersion, Gurney, and repulsion forces. A simple but effective representation of $W_{SR}(r)$ is given by

$$\begin{aligned}
&W_{SR}^{(0)}(r) = \infty && \text{for } r \le a \text{ with } a = a_1 + a_2 \text{ (reference ion)} \\
&W_{SR}^{(1)}(r) = W^* && \text{for } a \le r \le R \text{ (region 1, ion pair)} \\
&W_{SR}^{(2)}(r) = 0 && \text{for } r \ge R \text{ (region 2, solution outside ion pair)}
\end{aligned} \tag{12.84}$$

Now, the Debye–Hückel model has to be solved for the potential $W = W_C(r) + W_{SR}(r)$ in the three regions indicated. Similar to the original Debye–Hückel model, the potential as well as its derivative must be continuous at $r = R$ (as well as $r = a$). This leads for region 2 right away, in the same way as in Section 12.5, to

$$W_{SR}^{(2)} = \frac{z_1z_2e^2}{4\pi\varepsilon kT}\frac{\exp[\kappa(R-r)]}{1+\kappa R}\frac{1}{r} \equiv 2qkT\frac{\exp[\kappa(R-r)]}{1+\kappa R}\frac{1}{r} \tag{12.85}$$

For region 1, no ions are present between the reference ion and the counterion. In principle, the Coulomb potentials $W_C(r) = z_1z_2e^2/4\pi\varepsilon r = 2qkT/r$ acts, but since the potential and derivative should match with $W_{SR}^{(2)}$ at $r = R$, the potential is modified to

$$W_{SR}^{(1)} = \frac{2qkT}{r} - \frac{2qkT}{1+\kappa R} + W^* \tag{12.86}$$

Now, we are ready to deal with association reaction. In Chapter 6 we learned that the number of molecules in a sphere with radius R is given by the integral over the radial distribution function $g(r)$. We apply the same expression here, but modified for an ionic solution with concentration of ions c and degree of dissociation α. Approximating $g(r)$ by $\exp(-\beta W_{SR}^{(1)})$ with $\beta = 1/kT$, the number N of dissociated ions in a sphere of radius R is given by

$$N = 4\pi N_A \alpha c \int_a^R g(r) r^2 \,dr = 4\pi N_A \alpha c \int_a^R \exp(-\beta W_{SR}^{(1)}) r^2 \,dr \tag{12.87}$$

The lower limit of the integral is taken as $a = a_+ + a_-$, since this is the distance of closest approach. Inserting Eq. (12.86) the result becomes

$$N = 4\pi N_A \alpha c \exp\left(\frac{-2\kappa q}{1+\kappa R}\right)\int_a^R \exp\left(\frac{2q}{r} - \frac{W^*}{kT}\right) r^2 \,\mathrm{d}r \equiv \alpha c N' \tag{12.88}$$

For the association reaction $M^{z+} + X^{z-} \leftrightarrow IP$ in the sphere we have

$$\alpha c + \alpha c \leftrightarrow (1-\alpha)c \quad \text{or} \quad 1+1 \leftrightarrow (1-\alpha)/\alpha \tag{12.89}$$

The fraction bound ion pairs with respect to the number of ions is thus $(1-\alpha)/\alpha$. Since the actual number of ions in the sphere is given by N, we obtain

$$(1-\alpha)/\alpha = N \quad \text{or} \quad (1-\alpha)/\alpha^2 c = N' \tag{12.90}$$

Now consider the association constant K'_{ass} of the reaction in terms of activities. Let us assume that the activity coefficient of the bound ion pair $\gamma_{\mathrm{IP}} = 1$. For the ions, the activity coefficient $\gamma'_{\pm} = \exp[-q\kappa/(1+\kappa R)]$ as given by Eq. (12.45), but with the distance of closest approach a replaced by R. In that case, K'_{ass} is given by

$$K'_{\mathrm{ass}} = (1-\alpha)c/(\alpha\gamma'_{\pm}c)^2 = (1-\alpha)/(\alpha\gamma'_{\pm})^2 c \equiv K_{\mathrm{ass}}/(\gamma'_{\pm})^2 \tag{12.91}$$

with K_{ass} the association constant in terms of concentrations. Using Eqs (12.88) and (12.90), combined with the expression for $(\gamma'_{\pm})^2$, we find

$$K_{\mathrm{ass}} = 4\pi N_A \int_a^R \exp\left(\frac{2q}{r} - \frac{W^*}{kT}\right) r^2 \,\mathrm{d}r \tag{12.92}$$

The activity coefficient with respect to the total amount of ions dissolved becomes $\gamma_{\pm} = \alpha\gamma'_{\pm}$. Since the final set of expressions is coupled, an iterative solution is required.

However, before solving the equations, a choice for R and W^* must be made. It appears that with increasing distance r from the reference ion, there is a decreasing probability of finding a counterion per unit volume. The volume of the shell increases, however, and these effects when combined yield a minimum probability of finding a counterion on the sphere around the reference ion for $R = q$.

So, Bjerrum [25] selected $R = q$ and ignored W^*, while much later–taking a somewhat different point of view–Fuoss [26] chose $R = 4a/3$ and $W^* = 0$. Fuoss [27] also presented a more elaborate theory which showed that the precise choice of R is immaterial since by using a value of ½q or $2q$ instead of q, the result would differ only insignificantly. For water at 25 °C with $\varepsilon_r = 78.3$, one calculates for $z_1 = z_2 = 1$, $q = e^2/8\pi\varepsilon kT = 0.358$ nm, and at this distance the electrostatic energy is $2kT$. This energy, being fourfold the kinetic energy, is sufficient to have a significant ion association, albeit of a dynamic nature with rapid exchange with the surrounding ions. Hence, we can say that if ions are closer than 0.358 nm, they are associated. For multivalent ions with charge ez, $q = |z_1z_2|e^2/8\pi\varepsilon kT$, implying that in water association occurs at a distance $|z_1z_2|\cdot 0.358$ nm. While for univalent ions it is difficult to approach each other closer than 0.358 nm and associate, the q expression for ions of higher charge shows that, for these ions, association is much more important. From the above it also follows that ion association is much more important for solvents with a low relative permittivity.

There remains only a discussion of the agreement with experiment, which we limit here to just two examples. Bjerrum [25] showed that values for the distance of closest approach *a*, as calculated from the osmotic coefficient g, yielded much more reasonable values than with the Debye–Hückel–Onsager theory, especially for bivalent ions as well as for univalent ions in low permittivity solution, such as alcohols. A decisive set of experiments on the conductivity of solutions of tetra-iso-ammonium nitrate in mixtures of water and dioxane, covering a wide range of relative permittivities from 2.9 to 78, was presented by Fuoss and Krauss [28]. On applying the Debye–Hückel–Onsager model, marked deviations for the limiting conductivity from experiment were observed for low-permittivity solutions. Using Bjerrum's equations it appeared that, assuming a distance of closest approach of ~0.64 nm , K_{ass} was indeed constant over the range of ε_r used. This is a severe test in view of the range of ε_r used.

12.9 Final Remarks

In this chapter we have discussed only a few basic aspects of ionic solutions. The field of ionic solutions is extensive yet, despite its old age, its practical importance means that much research is still being conducted on electrolytes and monographs written on the subject, for example by Barthel *et al.* [19] and Wright [29]. In spite of all these developments, the Debye–Hückel–Onsager theory remains a cornerstone of the topic. Whilst the approximations involved have been discussed [30], perhaps the most appealing approach for higher concentrations is a combined use of the Debye–Hückel–Onsager model and Bjerrum's ion-association concepts. The relationship between the ion-pair concept and solution of the nonlinear Poisson–Boltzmann equation has been discussed [31]. For the many other effects that occur, for example by using strong or time-dependent electric fields, we refer to the literature.

References

1 Ketelaar, J.A.A. (1958) *Chemical Constitution*, Elsevier, Amsterdam.
2 Tissandier, M.D.J., *et al.* (1998) *Phys. Chem.*, **A102**, 7787.
3 Pauling, L. (1960) *The Nature of the Chemical Bond*, 3rd edn, Cornell University Press, Ithaca.
4 (a) Shannon, R.D. and Prewitt, C.T. (1969) *Acta Crystallogr.*, **B25**, 925; (b) Shannon, R.D. (1976) *Acta Crystallogr.* **A32**, 751.
5 Marcus, Y. (1988) *Chem. Rev.*, **88**, 1475.
6 Born, M. (1920) *Z. Phys.*, **1**, 45.
7 Latimer, W.M., Pitzer, K.S., and Slansky, C.M. (1939) *J. Chem. Phys.*, **7**, 108.
8 Stiles, P.J. (1980) *Aust. J. Chem.*, **33**, 1389.
9 Blum, L. and Fawcett, W.R. (1992) *J. Phys. Chem.*, **96**, 408 and references cited therein.
10 Marcus, Y. (1985) *Ion Solvation*, Wiley-Interscience, New York.
11 See Volkov *et al.* (1998).
12 Marcus, Y. (1996) *Pure Appl. Chem.*, **89**, 1495.
13 (a) Frank, H.S. and Evans, M.W. (1945) *J. Chem. Phys.*, **13**, 507; (b) Frank, H.S.

and Wen, W.Y. (1957) *Disc. Faraday Soc.* **24**, 133.

14 Bjerrum, N. (1909) Proceedings, 7th International Congress of Pure and Applied Chemistry, London, section X, p. 58; See also Bjerrum, N. (1909) *Z. Elektrochem.*, **23**, 321.

15 Güntelberg, E., page 155 in Bjerrum, N. (1926) *Z. Phys. Chem.*, **119**, 145.

16 See Davies (1962).

17 Pitzer, K.S. (1995) *Thermodynamics*, 3rd edn, McGraw-Hill, New York and references cited therein.

18 See Lee (2008).

19 See Barthel *et al.* (1998).

20 (a) Pitzer, K.S. (1977) *Acc. Chem. Res.*, **10**, 371; (b) Pitzer, K.S. (1973) *J. Phys. Chem.*, **77**, 268.

21 See Robinson and Stokes (1970).

22 (a) de Grotthuss, C.J.T. (1806) *Ann. Chim.*, **LVIII**, 54; (b) Cukierman, S.L. (2006) Et tu, Grotthuss! and other unfinished stories. *Biochim. Biophys. Acta Bioenerg.*, **1757**, 876.

23 Pashley, R.M., Francis, M.J., and Rzechowicz, M. (2008) *Curr. Opin. Coll. Interface Sci.*, **13**, 236.

24 Fowler, R.H. and Guggenheim, E.A. (1939) *Statistical Thermodynamics*, Cambridge University Press, London.

25 (a) Bjerrum, N. (1926) *Kgl. Danske Vid. Selsk., Math.-fys. Medd.*, **7**, 9; (b) Fuoss, R.M. (1934) *Trans. Faraday Soc.* **30**, 967. See also Davies (1962), Chapter 15.

26 Fuoss, R.M. (1958) *J. Am. Chem. Soc.*, **80**, 5059.

27 Fuoss, R.M. (1934) *Trans. Far. Soc.*, **30**, 967.

28 Fuoss, R.M. and Kraus, C.A. (1933) *J. Am. Chem. Soc.*, **55**, 1019 and 2387.

29 See Wright (2007).

30 Kirkwood, J.G. (1934) *J. Chem. Phys.*, **2**, 767.

31 (a) Guggenheim, E.A. (1957) *Disc. Faraday Soc.*, **24**, 53; (b) Skinner, J.F. and Fuoss, R.M. (1964) *J. Am. Chem. Soc.* **86**, 3423.

32 Fawcett, W.R. (1999) *J. Phys. Chem.*, **B103**, 11181.

33 Burgess, J. (1988) *Ions in solution*. Ellis Horwood, Chichester.

34 Enderby, J.E. and Neilson, G.W. (1979) *Water* (ed. F. Franks), Plenum, New York, Vol. 7, Ch. 1.

35 Jenkins, H.D.B. and Thakur, K.P. (1979) *J. Chem. Ed.*, **56**, 576.

36 Kohlrausch, F. (1885) *Ann. Phys. Chem.* **26**, 161.

37 Kohlrausch, F. and Grotrian, F. (1875) *Ann. Phys.* **154**, 215.

Further Reading

Barthel, J.M.G., Krienke, H., and Kunz, W. (1998) *Physical Chemistry of Electrolyte Solutions*, Springer, Berlin.

Davies, C.W. (1962) *Ion Association*, Butterworths, London.

Falkenhagen, H. (1971) *Theorie der Elektrolyte*, Hirzel, Stuttgart.

Fowler, R.H. and Guggenheim, E.A. (1939) *Statistical Thermodynamics*, Cambridge University Press, London.

Friedman, H.L. (1962) *Ionic Solution Theory*, Interscience, London.

Harned, H.S. and Owens, B.B. (1950) *The Physical Chemistry of Electrolyte Solutions*, Reinhold, New York.

Lee, L.L. (2008) *Molecular Thermodynamics of Electrolyte Solutions*, World Scientific, Singapore.

McInnes, D.A. (1939) *The Principles of Electrochemistry*, Reinhold, New York.

Robinson, R.A. and Stokes, R.H. (1970) *Electrolyte Solutions*, 2nd rev. ed, Butterworth, London. See also Dover Publishers reprint, 2002.

Volkov, A.G., Deamer, D.W., Tanelian, D.L., and Markin, V.S. (1998) *Liquid Interfaces in Chemistry and Biology*, John Wiley & Sons, Inc., New York.

Wright, M.R. (2007) *An Introduction to Aqueous Electrolyte Solutions*, John Wiley & Sons, Ltd, Chichester.

13
Mixing Liquids: Polymeric Solutions

In discussing the behavior of molecular solutions, although we kept track of the differences in the size of the molecules, the entropy was (nearly) always assumed to behave ideally. This is, however, no longer the case for polymeric solutions, in which the difference in size between the components becomes appreciable. In this chapter we discuss the extension leading to the famous Flory–Huggins theory.[1] We first discuss entropy and then deal with energy modifications; thereafter, the essence of some alternative theories is indicated. A brief review of polymer structural characteristics precedes all of this.

13.1
Polymer Configurations

Polymers consist of long molecular chains of covalently bonded atoms. Typically, the molecule is constructed from a set of repeating units, the *monomers*. To a good approximation the energy of a single molecule can be estimated by the adding bond energies (Table 13.1).

The chemical structure of the chains is complicated somewhat by *isomerism*, a phenomenon whereby one overall composition can have different geometric structures. A simple example of chemical isomerism is provided by the vinyl polymers for which one may have either *head-to-head* (–CH_2–CHX–CHX–CH_2–) or *head-to-tail* (–CH_2–CHX–CH_2–CHX–) addition. A somewhat more complex case involves steric isomerism. Consider again the case of vinyl polymers in which a side group is added to every alternate carbon atom. If the groups are all added in an identical way, we obtain an *isotactic* polymer (Figure 13.1). However, if there is an inversion for each monomer unit, we obtain a *syndiotactic* polymer. Finally, an irregular addition sequence leads to an *atactic* polymer. Of course, a chain can be branched and, generally, the chains are entangled.

Each sample of polymer will consist of molecular chains of varying length and, consequently, of varying molecular mass. The molecular mass distribution

1) Flory suggested that Huggins' name should be listed first as he had published several months earlier. See Current Contents, Citation Classic 18, May 6 (1965).

Liquid-State Physical Chemistry, First Edition. Gijsbertus de With.

Table 13.1 Bond energy U_{bon} and bond length d for various bonds.[a)]

Bond	U_{bon} (eV)	d (Å)	Bond	U_{bon} (eV)	d (Å)
C–H	4.3	1.08	C–Cl	2.8	1.76
C–C	3.6	1.54	C–Si	3.1	1.93
C=C	6.3	1.35	Si–H	3.0	1.45
C≡C	8.7	1.21	Si–Si	1.8	2.34
C–O	3.6	1.43	Si–F	5.6	1.81
C=O	7.6	1.22	Si–Cl	3.7	2.16
C–F	5.0	1.36	Si–O	3.8	1.83

a) Data from Ref. [36].
1 eV = 96.48 kJ mol^{-1}.

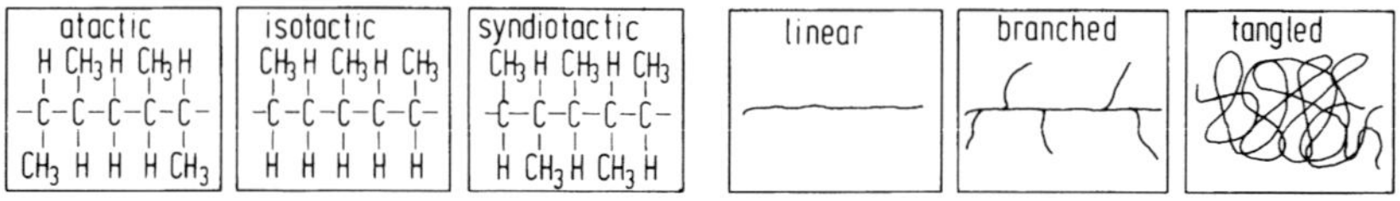

Figure 13.1 Tacticity and chain structure of polymers.

is important for many properties. One can distinguish between the number average M_n and weight average M_w, defined by

$$M_n = \frac{\sum_i N_i M_i}{\sum_i N_i} \quad \text{and} \quad M_w = \frac{\sum_i (N_i M_i) M_i}{\sum_i N_i M_i} \tag{13.1}$$

respectively, where N_i is the number of molecules with molecular mass M_i and the summation is over all molecular masses. Since the degree of polymerization (DP) involved is generally very high, the difference between a discrete and a continuous distribution is usually negligible, and we can write

$$M_n = \frac{\int p(M) M \, dM}{\int p(M) \, dM} = \int p(M) M \, dM \quad \text{and} \quad M_w = \frac{\int p(M) M^2 \, dM}{\int p(M) M \, dM}$$

where the second step for M_n can be made if $p(M)$ is assumed to be normalized. Since the second central moment or *variance* of the distribution function is given by

$$\langle \Delta M^2 \rangle = \int p(M)[M - M_n]^2 \, dM = \int p(M) M^2 \, dM - M_n^2 = M_w M_n - M_n^2$$

we have

$$\langle \Delta M^2 \rangle / M_n^2 = (M_w / M_n) - 1$$

and the *polydispersity index* (PDI) M_w/M_n describes the width of the distribution. The shape of $p(M)$ can vary widely, dependent on the polymerization process.

Mixtures of various types of polymers are possible. A *blend* is a mixture of two or more polymers. In a *graft*, a chain of a second polymer is attached to the base polymer. If in the main chain a chemical combination exists between two monomers [A] and [B], then the material is a *copolymer*. In the latter case, we can distinguish between: (i) a *block* copolymer, where the monomer A is followed first by a sequence of other monomers A such as AAA and subsequently by a series of B monomers; and (ii) *random* copolymers, where a monomer A is followed randomly by either monomer A or B. This results in the absence of long sequences of A and B monomers.

We now turn to the molecular conformations, the most important of which are *gauche* and *trans* conformations. Consider first the central bond between two C atoms as in ethane, C_2H_6. Figure 13.2 shows the two extremes in conformation, namely *cis* and *trans*, in a view along the C–C bond axis. In ethane, three equivalent minimum energy or *trans* conformations are present. To rotate the two CH_3 groups with respect to each other, energy must be spent and an energy barrier exists between the two *trans* states. Substituting on each C atom one H atom by a CH_3 group, to produce butane, C_4H_{10}, the equivalence between the *trans* states is lost and we obtain one *trans* (t) conformation and two equivalent (g^+, g^-) *gauche* conformations with dihedral angles $\phi = 0°$ and $\phi = +120°$ and $\phi = -120°$, respectively, for the minimum energy conformations (Figure 13.2). Continuing with the substitution of end H atoms with CH_3 groups results in polyethylene (PE). Although the details for each C–C bond for this molecule may slightly differ, one *trans* and two *gauche* conformations are present for each C–C bond. They must all be specified for a complete description of the molecule. For PE, the lowest energy conformation is the all-*trans* conformation with a zig-zag structure of the C–C bonds.

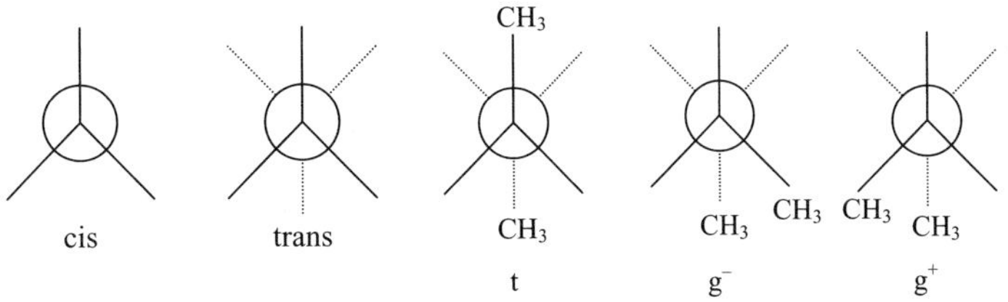

Figure 13.2 The *cis* and *trans* conformations in ethane and the t, g^+, and g^- conformations in butane, as shown by the Newman projection.

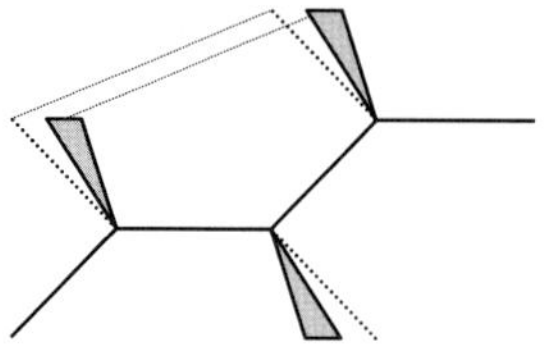

Figure 13.3 Nonbonded interactions between CX_2 groups of the second-nearest C atoms.

This is no longer true for other polymers, where the H atoms have been replaced by other atoms or groups. Consider for example polytetrafluorethylene (PTFE), where all H atoms have been replaced by F atoms. As the F atoms are larger than the H atoms, the nonbonded repulsive interactions between CF_2 groups of the second-nearest carbon atoms become much more important (Figure 13.3), and repulsive energy can be decreased by rotating slightly along the C–C axis of each bond. Of course, this increases the bond rotation energy, and by balancing the two contributions an equilibrium is reached. In the case of PTFE, an optimum dihedral angle of $\phi \cong 16.5°$ is obtained. The result of all this is that the molecule forms a helix along its axis in which the positions of the side groups (the F atoms in the case of the PTFE) rotate along the molecular axis. After n screws along the axis the position of the mth monomer regains the position of the first monomer, apart from a shift along the axis. When described in this way, we refer to these as m/n helices. For example, PE has a 2/1 helix, while PTFE has a 13/6 helix below 19 °C and a 15/7 helix above 19 °C. This description for the (semi-)crystalline state is not as exact as it appears however, as the "periodicity" along the chain may vary slightly [1].

Focusing on the chains themselves, a first estimate of the end-to-end distance[2)] X is made via the *freely jointed chain* model: n bonds, each of length l, connected without any restriction. The probability distribution of the end-to-end vectors for long chain molecules is described by the random walk model resulting in

$$P(\mathbf{r})\mathrm{d}\mathbf{r} = \left(\frac{3}{2\pi nl^2}\right)^{3/2} \exp\left[-\frac{3r^2}{2nl^2}\right]\mathrm{d}\mathbf{r} \tag{13.2}$$

For such a model chain one obtains, in the limit of a large number of atoms,

$$\langle r^2\rangle = \int r^2 P(\mathbf{r})\mathrm{d}\mathbf{r} = 4\pi\int_0^\infty r^4 P(\mathbf{r})\mathrm{d}r = nl^2 \tag{13.3}$$

where $\langle r^2\rangle = \langle x^2\rangle + \langle y^2\rangle + \langle z^2\rangle$ is the mean square end-to-end distance of the chains. The *end-to-end distance* $X = \langle r^2\rangle^{1/2}$ is thus proportional to $n^{1/2}$.

2) For derivations we refer to the literature, see, e.g. Gedde [6], which we have taken as guide.

However, we know that the bonds are not freely connected but have a certain bond angle τ. Leaving the bonds otherwise unrestricted, we obtain the *freely rotating chain* model for which it holds in the limit of a large number of bonds that

$$\langle r^2\rangle = nl^2\left[\frac{1-\cos\tau}{1+\cos\tau}\right] \tag{13.4}$$

As expected, the square-root dependence on n is preserved but the proportionality factor is changed. In the case of sp^3-hybridized carbon atoms, for example, in a PE chain, with a bond angle of $\tau = 109.5\,°$, we have approximately $\langle r^2\rangle = 2.0nl^2$.

A further improvement is obtained by using the *independent hindered rotation model*, that is, a rotating chain but with preferential orientation for the dihedral (bond rotation) angle ϕ between two groups connected by a bond. In this model one has

$$\langle r^2\rangle = nl^2\left[\frac{1-\cos\tau}{1+\cos\tau}\right]\left[\frac{1+\langle\cos\phi\rangle}{1-\langle\cos\phi\rangle}\right] \tag{13.5}$$

Again, the square-root dependence on n is preserved and the proportionality factor changes. For the PE chain we have one *trans* (t) configuration with a dihedral angle $\phi = 0\,°$ and two equivalent *gauche* (g^+, g^-) configurations with a dihedral angle $\phi = 120\,°$ and $\phi = -120\,°$, respectively (Figure 13.2). The latter have a higher energy by an amount E_{gau}. Denoting the Boltzmann factor by $\sigma = \exp(-E_{gau}/RT)$, we obtain for the average dihedral angle

$$\langle\cos\phi\rangle = \frac{1+\sigma\cos(120°)+\sigma\cos(-120°)}{1+\sigma+\sigma} = \frac{1-\sigma}{1+2\sigma} \tag{13.6}$$

For the end-to-end distance we thus have

$$\langle r^2\rangle = nl^2\left[\frac{1-\cos\tau}{1+\cos\tau}\right]\left[\frac{2+\sigma}{3\sigma}\right] \tag{13.7}$$

For PE at 140 °C, using $E_{gau} = 2.1\,\text{kJ mol}^{-1}$, we find $\sigma = 0.54$ leading to $\langle r^2\rangle \cong 3.4nl^2$.

Finally, we recognize that the hindered rotation around a bond is correlated, and this is taken into account in the *correlated hindered rotation model*. The final expression becomes

► $$\langle r^2\rangle = Cnl^2 \tag{13.8}$$

where the *characteristic ratio* C is a function of the correlation of the rotations along the chain and is therefore a measure of the stiffness of the chain. For PE, Flory calculated, taking into account the correlation up to two bonds away, that $C = 6.7 \pm 0.2$, in good agreement with experiment; for other polymers, different values of C are obtained (see Table 13.2). Actually the solvent used should be indicated if non-theta conditions are present (see Section 13.2).

The above-described considerations led to the introduction of the *equivalent chain* (Figure 13.4), in which a real chain, containing n correlated and rotation-hindered bonds of length l, is described as a freely jointed chain of n_K segments of length b.

Table 13.2 Values for characteristic ratio *C* for various polymers.

Material	*C*	Material	*C*
PEO	4.0/4.1	i-PMMA	10
PE	6.7/6.8	s-PMMA	7
a-PS	10.0	PVC	13
i-PS	10.7	a-PVAc	8.9/9.4
a-PP	5.5	PDMS	6.2
i-PP	5.8	a-PiB	6.6
s-PP	5.9	PC	2.4
a-PMMA	8.4		

PE, polyethylene; PEO, polyoxyethylene; PS, polystyrene; PP, polypropylene; PMMA, poly(methyl methacrylate); PVC, poly(vinyl chloride); PVAc, poly(vinyl acetate); PDMS, poly(dimethylsiloxane); PiB, poly(isobutylene); PC, poly(carbonate); a, atactic; i, isotactic; s, syndiotactic.

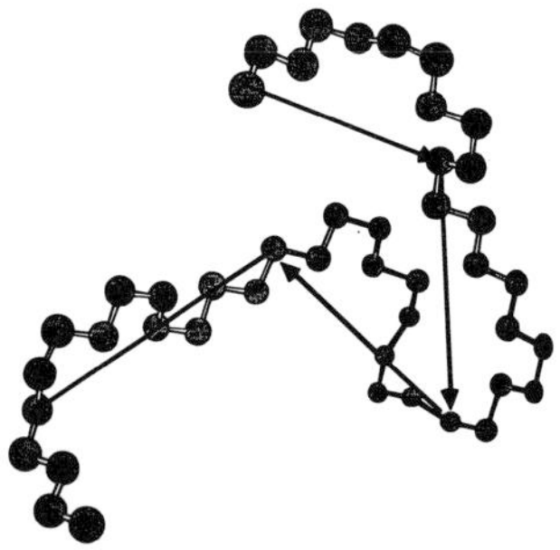

Figure 13.4 The equivalent chain for polyethylene with segments containing approximately one-tenth of the bonds of the real polyethylene chain, and each segment having a length approximately eightfold that of a C–C bond length. This results in the same end-to-end distance *X* and contour length *L* as the real polyethylene chain.

Each of the *segments* thus represents a number of real bonds; however, as the correlation along the chain is limited to a few bonds, these segments can be considered as freely jointed. For this description we use the end-to-end distance *X* and the *contour length L*, which is the length of the fully extended real chain with all valence angles and bond lengths at their equilibrium value. In fact, we match $X = \langle r^2 \rangle = Cnl^2$ with $n_K b^2$ and L with $n_K b$. This can be done in a unique way, leading to $b = \langle r^2 \rangle / L$ and $n_K = L^2 / \langle r^2 \rangle$. Let us take again PE as an example. For the PE chain with a bond angle $\tau = 109\,°$, $L = nl \sin(\tau/2) \cong 0.83\, nl$, while $\langle r^2 \rangle = 6.7nl^2$. This leads to $b \cong 8l$ and $n_K \cong 0.1n$.

The segment thus contains about 10 (real) bonds, and its length b is often addressed as the *Kuhn length*. For other polymers, of course, various Kuhn lengths are obtained, with higher values reflecting a greater stiffness of the molecular chain. In discussing the properties of polymers, frequent use is made of the equivalent chain model since, for this model many mathematical results are known. In the polymer literature, mention is often made of Kuhn monomers or even monomers when addressing the entities that we have labeled here as segments. In order to avoid confusion, we systematically use the term "monomer" for a repetitive chemical unit, and the term "segment" when speaking about a unit for the equivalent chain. The results obtained for the equivalent chain are then translated to real chain via the characteristic ratio C.

Justification 13.1

A simple justification for the distribution function $P(n,\mathbf{r})$ runs as follows. We consider a freely jointed (ideal) chain of n bonds or segments of length l. If we add to the chain an extra segment with vector $\boldsymbol{l}$ (components l_i) we have $P(n+1,\mathbf{r}) = P(n,\mathbf{r}+\boldsymbol{l})$. Assuming that $l \equiv |\boldsymbol{l}| \ll r \equiv |\mathbf{r}|$, we may expand $P(n+1,\mathbf{r})$ in a Taylor series in $\boldsymbol{l}$ obtaining

$$P(n,\mathbf{r}+\boldsymbol{l}) = P(n,\mathbf{r}) + \nabla P(n,\mathbf{r})\cdot\boldsymbol{l} + \frac{1}{2}\boldsymbol{l}\cdot\nabla^2 P(n,\mathbf{r})\cdot\boldsymbol{l} + \cdots$$

Averaging over all possible orientations of $\boldsymbol{l}$, taking into account that $\langle\boldsymbol{l}\rangle = 0$, the result is

$$\langle P(n,\mathbf{r}+\boldsymbol{l})\rangle = \langle P(n,\mathbf{r})\rangle + (l^2/6)\langle\nabla^2 P(n,\mathbf{r})\,|_{l=0}\rangle + \cdots$$

where $\langle l_i l_j\rangle = \tfrac{1}{3}l^2\delta_{ij}$ has been used. We further write $\langle P(n,\mathbf{r})\rangle = P(n,r)$ for the average over all orientations of $\mathbf{r}$. Using $\langle P(n+1,\mathbf{r})\rangle = P(n+1,r)$ we also have for $n >> 1$

$$P(n+1,r) = P(n,r) + \langle\partial P(n,\mathbf{r})/\partial n\rangle + \cdots$$

and equating $P(n+1,r)$ with $\langle P(n,\mathbf{r}+\boldsymbol{l})\rangle$ we have approximately

$$\partial P(n,\mathbf{r})/\partial n = (l^2/6)\nabla^2 P(n,\mathbf{r})$$

The relevant solution depends only on r, since the undisturbed molecule has a spherical shape and reads $P \sim n^{-3/2}\exp(-3r^2/2nl^2)$. Using the normalization condition $\int P(n,\mathbf{r})4\pi r^2\,\mathrm{d}r = 1$, we have for the complete solution

$$P(n,\mathbf{r}) = (3/2\pi nl^2)^{3/2}\exp(-3r^2/2nl^2)$$

Note that for $r > nl$, $P(n,r) \neq 0$, though this condition should be obeyed for a realistic solution. However, for $n >> 1$, $P(n,r) \cong 0$ when $r > nl$.

The possibility of having trans or gauche states at each bond renders polymer molecules relatively flexible, so that for polymers in good solvents (and in polymer melts as well as in glassy amorphous polymers) the chains exhibit a random configuration and the polymer molecule appears as a coil. The configuration of a polymer molecule, whether in solution or in the melt, can be (partially) characterized by the end-to-end distance[3] X. For the freely jointed (and therefore the equivalent) chain the probability analysis shows that the molecules have a Gaussian end-to-end distribution, also indicative of the fact that polymer molecules normally are coiled, and one often speaks of a *Gaussian chain*. In a good solvent, polymer–solvent attractions prevail, the coil expands, and X increases. In a poor solvent, polymer–polymer attraction prevails, irrespective of whether they are due to parts from the same or from a different chain. The coil shrinks and X decreases until the effective monomer–monomer (segment–segment) repulsion due to excluded volume forces sets in. Under certain conditions the intramolecular interactions are similar in magnitude to the intermolecular interactions. In other words, the enthalpy and entropy contributions from solvent–monomer (solvent–segment) and monomer–monomer (segment–segment) interactions to the Helmholtz energy of the assembly of molecules under consideration compensate each other, and one part of the molecule seems not to "notice" other parts of the same molecule, nor other molecules. The molecules behave like "phantoms," and are indeed sometimes referred to as phantom chains. The temperature at which this occurs is termed the *Flory temperature* θ, and one speaks of *theta conditions*, under which the coil neither shrinks nor expands, and has unperturbed dimensions. The influence of the solvent can be described by

$$\chi = \alpha \chi_\theta \tag{13.9}$$

where the subscript denotes the theta conditions and α a parameter dependent on solvent, temperature, and molecular mass. At the Flory temperature $T = \theta$, theta conditions hold and $\alpha = 1$. We will discuss the behavior of α somewhat further in the next paragraph. In addition to the main topic, we note here that for solids the *Flory theorem* is important: in a dense polymeric system theta conditions prevail. Describing theta conditions as the configuration where intramolecular and intermolecular interactions compensate, and since the "solvent" is the polymer melt itself, the theorem is highly plausible. Rephrasing, on the one hand, the monomers of a certain reference chain is subject to a repulsive potential due to the excluded volume effect of its own monomers and this leads to an expansion of the coil. On the other hand, the other chains, which interpenetrate the reference chain, generate a counteracting attractive potential acting inwards on the reference chain such that, under theta conditions, the two effects cancel leading to (pseudo)-unperturbed chains. The results of small-angle neutron-scattering experiments have supported this theorem.

Finally, we note it is not difficult to calculate the Helmholtz energy F for a Gaussian chain of n bonds with length l. We use Boltzmann's estimate for the entropy $S = k\ln W$, where W is the number of configurations (to be precise, with the same

3) Occasionally, the root mean square distance of the atoms from the center of gravity, the *radius of gyration s*, is also used. It holds that $\langle s^2\rangle = \langle r^2\rangle/6$.

energy). The value of W can be estimated from $P(n,r)$, realizing that $P(n,r) = W(n,r)/\int W(n,r)\mathrm{d}r$. By using Eq. (13.2) we find

$$S = -\frac{3kr^2}{2nl^2} + 3k\ln\left(\frac{3}{2\pi nl^2}\right) + k\ln\int W(n,r)\mathrm{d}r = -\frac{3kr^2}{2nl^2} + S(n,0) \qquad (13.10)$$

Since we assume that all conformations have the energy, $F = U - TS$ becomes $F = -TS = 3kTr^2/2nl^2$, so that for the force to extend this chain we obtain $f = \partial F/\partial r = 3kTr/nl^2$.

Problem 13.1

Consider a polymer with the main chain carbon atoms in sp^2 conformation (bond length $l = 0.1$ nm) and with a strong preference for a rotation angle of 45° deviating from the zig-zag conformation. Calculate the contour length L and end-to-end distance X for the conditions, as given in the accompanying table. Compare the results and explain the differences qualitatively.

Case	n	Bond angle	Rotation angle
A	100	Free	Free
B	1000	Free	Free
C	1000	Fixed	Free
D	1000	Fixed	Fixed by repulsion

Problem 13.2

Suppose that two adjacent bonds have a correlated orientation due to the (fixed) bond angle τ, but that no correlation exists between bonds with at least one bond in between. Calculate the end-to-end distance for a polymer with 1000 bonds in the main chain and a bond length of 0.1 nm.

13.2 Real Chains in Solution

In the previous paragraph we discussed mainly chains in the absence intermolecular interactions, that is, ideal chains. Furthermore, we indicated that the balance between segment–segment interactions on the one hand and segment–solvent interactions on the other hand determine the behavior of polymer in solution. In this section, we elaborate on the latter topic using the equivalent chain with n segments of length b, inspired by the treatment as given by Rubinstein and Colby [2].

Let us thus be somewhat more precise about the interactions. The effective segment–segment interactions are obviously dependent on the solvent used, and are often called *excluded volume interactions*. If the effective interaction energy between segments is $\Phi(r)$, the statistics are determined by the Boltzmann factor $\exp(-\beta\Phi)$; however, if we have no effective interaction, then $\Phi(r) = 0$ so that $\exp(-\beta\Phi) = 1$. The net result is thus conveniently described, using the Mayer factor $f(r) = \exp(-\beta\Phi) - 1$, by the integral

$$\upsilon = -\int_0^\infty f(r)\mathrm{d}\mathbf{r} = 4\pi\int_0^\infty [1-\exp(-\beta\Phi)]r^2\,\mathrm{d}r \tag{13.11}$$

representing the *excluded volume* υ. If $\upsilon < 0$, the net interaction is attractive, while for $\upsilon > 0$ the net interaction is repulsive. If there is hardly any difference between segment–segment and segment–solvent interactions, the solvent is denoted as an *athermal solvent*. At low concentration, the Helmholtz energy F can be described by a virial expression in terms of the segment concentration. We recall that such an expansion for F containing an ideal and an interaction part reads

$$F/V = F_{\text{ide}}/V + F_{\text{int}}/V = kTc + \frac{1}{2}kT\left[\upsilon c^2 + \omega c^3 + \ldots\right] \tag{13.12}$$

where υ represents the two-segment contribution (the excluded volume) and ω the three-segment interaction.

The two-segment and three-segment interactions also depend on the shape of the segments. So far, we have assumed spherical segments, but it is somewhat more realistic to use cylindrical segments with length b and (somewhat) smaller diameter d. Flexible polymers typically have a b/d ratio in the range 2 to 3, whereas for stiffer polymers the value is somewhat larger. For athermal spherical segments with diameter d, we have $\upsilon_s \sim d^3$ and $\omega_s \sim d^6$. However, the interaction must not change if we redefine our segment. A chain with n_s spherical segments of diameter d can also be considered as a chain of $n_c = n_s d/b$ cylindrical segments of diameter d and length b. So, $\upsilon_s n_s^2 = \upsilon_c n_c^2$ and $\omega_s n_s^3 = \omega_c n_c^3$. For the cylindrical segments we therefore obtain[4)]

$$\upsilon_c \sim \upsilon_s\left(\frac{n_s}{n_c}\right)^2 \sim \upsilon_s\left(\frac{b}{d}\right)^2 \sim b^2 d \quad \text{and} \quad \omega_c \sim \omega_s\left(\frac{n_s}{n_c}\right)^3 \sim \omega_s\left(\frac{b}{d}\right)^3 \sim b^3 d^3 \tag{13.13}$$

The ratio of excluded volume $\upsilon_c \sim b^2 d$ over occupied volume $\upsilon_0 \sim bd^2$ is b/d, and it will be clear that the excluded volume depends strongly on the shape of the segment. Note that by taking $b = d$ we regain the expression appropriate for spherical segments.

Using this image we can distinguish between several types of solvent:

- *Athermal solvents.* For these solvents the segment–segment interactions equal the segment–solvent interactions, so that only the core repulsions between the segments control the interaction. Hence, $\upsilon \cong b^2 d$.
- *Good solvents.* In these solvents the segment–segment attraction is somewhat larger than the solvent–segment attraction. Consequently, $0 < \upsilon < b^2 d$.

4) We use scaling arguments here and leave out all proportionality factors.

- *Theta solvents.* At a certain temperature (the θ-temperature) the segment attractions cancel the core repulsions. This leads to $\upsilon \cong 0$. At $T = \theta$ the chains have near-ideal conformations.

- *Poor solvents.* For $T < \theta$, the attractive interactions between the segments are relatively strong so that the segments clog together. The excluded volume is negative, and we have $-b^2d < \upsilon < 0$.

- *Non-solvents.* For strong segment–segment attraction we obtain the limit of a poor solvent, that is, the non-solvent. In this case, $\upsilon \cong -b^2d$.

Our task is now to provide a model for this behavior.

Flory developed a relatively simple theory with energetic and entropic contributions to describe a polymer solution using a good solvent. For his model, Flory assumed that the n segments are homogeneously distributed without any correlation in a sphere of radius $R > R_0 = n^{1/2}b$. The probability of finding a second segment in an excluded volume υ of a given segment equals the product of υ and the number density n/R^3, that is, $\upsilon n/R^3$. Because the energy increase of an excluded volume interaction is kT per exclusion, we have for the n segments in the chain $U \sim kT\upsilon n^2/R^3$. The entropy is estimated by the Gaussian chain estimate $-TS \sim kTR^2/nb^2$. In total, we thus have

$$F = U - TS \sim kT\left[\left(\upsilon n^2 / R^3\right) + \left(R^2 / nb^2\right)\right] \tag{13.14}$$

From $\partial F/\partial R = 0$ we obtain for the Flory chain $R_F = \upsilon^{1/5}n^{3/5}b^{2/5}$, to be compared with the Gaussian chain result $R = R_0 = n^{1/2}b$. The relative size of the Flory chain with respect to the Gaussian chain is therefore $R_F/R_0 \sim \upsilon^{1/5}n^{3/5}b^{2/5}/n^{1/2}b \sim (\upsilon n^{1/2}/b^3)^{1/5}$, and chains only swell when the *chain interaction parameter* $z = \upsilon n^{1/2}/b^3$ is positive and sufficiently large. For $z < 0$, the chain remains nearly ideal.

It appears that the Flory theory describes experiments rather well, but this is the result of some canceling errors (see, for example, Ref. [2]). The description as given here is akin to the self-avoiding random walk (SAW) model in which one stipulates that the freely jointed chain cannot intersect itself, contrary to the Gaussian chain where all conformations are possible. More sophisticated theory yields for the exponent $\nu = 0.588$ instead of $\nu = 3/5$.

The differences between the Gaussian and Flory chain become rather clear if we compare their behaviors under tension. To that purpose, we note that most of the entropy is the result of conformational freedom at the smallest length scale. Hence, we use the image that a chain containing n segments is constructed from smaller sections of size ξ, each containing g segments. Upon applying tension to the chain, each section – often referred to as a *tension blob* – is essentially unperturbed, but the blob more or less moves as a whole under the applied tension. Therefore, one degree of freedom is restricted, increasing the Helmholtz energy by kT. The result for F thus becomes $F \sim kTn/g \sim kTL^2/nb^2$, in agreement with Eq. (13.10). The physical interpretation of a blob is that ξ is the length for which the external tension changes the overall conformation from almost undeformed for length scales $r < \xi$ to extended for length scales $r > \xi$.

For the size of the blob ξ and the contour length L we have, respectively,

$$\xi^2 \sim gb^2 \quad \text{and} \quad L \sim \xi n / g \sim nb^2 / \xi \tag{13.15}$$

Solving for ξ and g we obtain

$$\xi \sim nb^2 / L \quad \text{and} \quad g \sim n^2b^2 / L^2 \tag{13.16}$$

For the Gaussian and Flory chain we have, respectively,

$$R_0 \sim bn^{1/2} \quad \text{and} \quad R_F \sim bn^{3/5} \tag{13.17}$$

Furthermore, we assume that a subsection of size r of the chain scales similarly, that is

$$r_0 \sim bn^{1/2} \quad \text{and} \quad r_F \sim bn^{3/5} \tag{13.18}$$

so that for the blob size we obtain

$$\xi_0 \sim bg^{1/2} \quad \text{and} \quad \xi_F \sim bg^{3/5} \tag{13.19}$$

The end-to-end distances for the two chains, here again labeled by X, are

$$X_0 \sim \frac{\xi n}{g} \sim \frac{nb^2}{\xi} \sim \frac{R_0^2}{\xi} \quad \text{and} \quad X_F \sim \frac{\xi n}{g} \sim \frac{nb^{5/3}}{\xi^{2/3}} \sim \frac{R_F^{5/3}}{\xi^{2/3}} \tag{13.20}$$

Solving for ξ we obtain

$$\xi_0 \sim R_0^2 / X_0 \quad \text{and} \quad \xi_F \sim R_F^{5/3} / X^{3/2} \tag{13.21}$$

and calculating the Helmholtz energy, recalling that the energy for stretching is kT per blob, the result becomes

$$F_0 \sim \frac{kTn}{g} \sim \frac{kTX_0}{\xi} \sim kT\left(\frac{X_0}{R_0}\right)^2 \quad \text{and} \quad F_F \sim \frac{kTn}{g} \sim \frac{kTX_F}{\xi} \sim kT\left(\frac{X_F}{R_F}\right)^{5/2} \tag{13.22}$$

The forces for stretching are, as usual, given by $f = \partial F/\partial X$ and become

$$f_0 \sim kT\frac{X_0}{R_0^2} \quad \text{and} \quad f_F \sim \frac{kT}{R_F}\left(\frac{X_F}{R_F}\right)^{3/2} \tag{13.23}$$

or, equivalently,

$$\frac{f_0 b}{kT} \sim \frac{X_0}{nb} \quad \text{and} \quad \frac{f_F b}{kT} \sim \left(\frac{X_F}{nb}\right)^{3/2} \tag{13.24}$$

For both types of chains the stretching energy is kT per segment when the chain is nearly fully stretched, but, although the force increases faster with X for the Flory chain, it is always smaller than for the ideal chain. For both chains the number of available conformations decreases when stretched, but the Flory chain has fewer conformations to lose, which results in a smaller force.

13.2.1
Temperature Effects

We will consider thermal effects also using a scaling approach. Since we encounter both attractive and repulsive excluded volume interactions, we use $|\upsilon|$ in definitions and indicate the sign explicitly where needed. Similar to stretching, we assume that a thermal length scale exists, the *thermal blob* size ξ_T, for which the excluded volume interactions are smaller than the thermal energy kT, if the length scale $r < \xi_T$. The conformations of the thermal blob are then nearly ideal and if the blob contains g_T segments, we have

$$\xi_T \sim b g_T^{1/2} \tag{13.25}$$

The size of the blob ξ_T can be estimated by equating the excluded volume interaction $kT|\upsilon|g_T^2/\xi_T^3$ for a single blob with the thermal energy kT, that is

$$kT|\upsilon|g_T^2 / \xi_T^3 \sim kT \tag{13.26}$$

Combining Eqs (13.25) and (13.26) we obtain

$$g_T \sim b^6 / \upsilon^2 \quad \text{and} \quad \xi_T \sim b^4 / |\upsilon| \tag{13.27}$$

The blob size is the value where excluded volume interactions become important. For $\upsilon \sim b^3$, $\xi_T \sim b$, the blob has the size of a segment and is fully swollen in an athermal solvent. For $\upsilon \sim -b^3$, again $\xi_T \sim b$ and the chain is fully collapsed in a non-solvent. For $|\upsilon| < b^3 n^{-1/2}$, $\xi_T > R_0$, and the chain behaves as nearly ideal. Finally, for $b^3 n^{-1/2} < |\upsilon| < b^3$, ξ_T is between segment and chain size with moderate swelling in a good solvent ($\upsilon > 0$) or moderate collapse in a poor solvent ($\upsilon < 0$).

For athermal and good solvents ($\upsilon > 0$) on a length scale $r > \xi_T$, the excluded volume repulsion is larger than kT. The polymer is then swollen with n/g_T thermal blobs and an end-to-end distance $X \sim \xi_T(n/g_T)^\nu \sim b(\upsilon/b^3)^{2\nu-1} n^\nu$ with $\nu = 3/5$ (or $\nu = 0.588$ using more sophisticated theory). For poor solvents ($\upsilon < 0$) on a length scale $r > \xi_T$, the exclusive volume attraction is larger than kT. In this case, the blobs adhere to each other to form a dense globule. Assuming a dense packing of blobs, the size of this globule becomes $R_{glo} \sim \xi_T(n/g_T)^{1/3} \sim b^2 n^{1/3}/|\upsilon|^{1/3}$. The globule is approximately spherical and has a volume fraction $\phi \sim nb^3/R_{glo}^3 \sim |\upsilon|/b^3$, independent of the number of segments n. In Figure 13.5 the behavior for the various cases is sketched. Clearly, in reality the transitions indicated are less sharp than suggested by the graph in the figures.

For poor solvents, the Flory theory requires some extension because for $\upsilon < 0$ we find $R = 0$ as the optimum value. Two effects are relevant. The first is the effect of confinement in the blobs, while the second effect appears to be the excluded volume effect for two segments, to be extended even to three segments.

Consider first the confinement effect. The number of segments in a blob is given by $g \sim (R/b)^2$. The energy cost per blob is kT, and because we have n/g blobs per chain that fully overlap we have the contribution $F_{con} \sim kTn/g \sim kTnb^2/R^2$. Unfortunately, obtaining the optimum in the usual way, the result is still $R = 0$, and we therefore need to invoke other stabilization mechanisms in order to be able to describe the behavior of a polymer in a poor solvent.

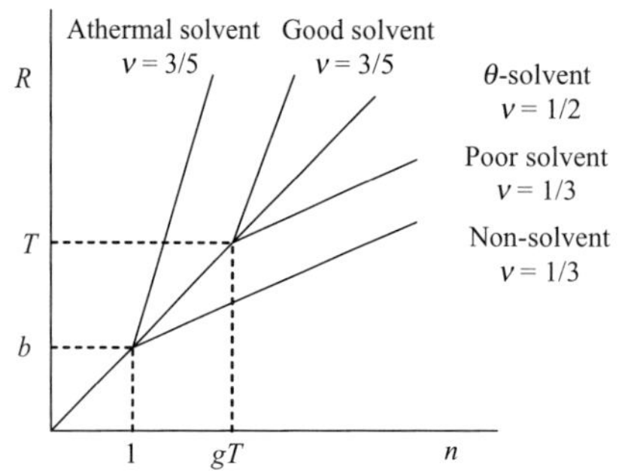

Figure 13.5 Schematic of the relationship between the size R of a chain as a function of number of segments n where for each linear part $R \sim n^{\nu}$.

For the excluded volume effect, we need the virial expansion of the interaction as expressed by Eq. (13.12), or actually its interaction part,

$$F_{\text{int}} / V \sim kT\left[\upsilon c^2 + \omega c^3 + \ldots\right] \tag{13.28}$$

Recall that υ and ω represent the two-segment and three-segment contributions, respectively. The number density or concentration of segments inside a coil is given by $c = n/R^3$. This implies that we have

$$F_{\text{int}} / V \sim F_{\text{int}} / R^3 \sim kT\left[\upsilon(n^2 / R^6) + \omega(n^3 / R^9) + \ldots\right] \quad \text{or}$$

$$F_{\text{int}} \sim kT\left[\upsilon(n^2 / R^3) + \omega(n^3 / R^6) + \ldots\right] \tag{13.29}$$

Adding this contribution to the terms that we already had, that is, $kT\upsilon n^2/R^3$ and kTR^2/nb^2 from the original Flory treatment for good solvents and $kTnb^2/R^2$ from the confinement contribution, we obtain

$$F \sim kT\left[(R^2 / nb^2) + (nb^2 / R^2) + \upsilon(n^2 / R^3) + \omega(n^3 / R^6) + \ldots\right] \quad \text{or} \tag{13.30}$$

$$F \sim kT\left[\upsilon(n^2 / R^3) + \omega(n^3 / R^6) + \ldots\right] \quad \text{if} \quad R \ll R_0 \tag{13.31}$$

For the case that $\upsilon < 0$ and $\omega > 0$, the globule size R_{glo} becomes $R_{\text{glo}} \sim (\omega n/|\upsilon|)^{1/3}$, so that for spherical segments with $\omega \sim b^6$, we regain $R_{\text{glo}} \sim b^2 n^{1/3}/\upsilon^{1/3}$. For cylindrical segments $\omega \sim b^3 d^3$ and the globule size becomes $R_{\text{glo}} \sim bdn^{1/3}/|\upsilon|^{1/3}$. The volume fraction in the globule is then $\phi \sim nbd^2/R_{\text{glo}}^3 \sim |\upsilon|/b^2 d$. For the case of a non-solvent, $\upsilon \sim -b^2 d$, $\phi \sim 1$, and $R_{\text{glo}} \sim (bd^2 n)^{1/3}$. Hence, the globule is fully collapsed, having a dense packing of segments, each of volume bd^2.

An approximate temperature dependence for υ can be obtained if we reconsider

$$\upsilon = -\int_0^\infty f(r)\,\mathrm{d}\mathbf{r} = 4\pi\int_0^\infty [1 - \exp(-\beta\Phi)]r^2\,\mathrm{d}r \tag{13.32}$$

and separate the integration in two parts. For $r < b$, $\beta\Phi(r) \gg 1$ and $f = -1$, while for $r > b$ we have $\exp[-\beta\Phi(r)] - 1 \cong -\beta\Phi(r)$ if $\beta\Phi(r) < 1$. Hence, the integral becomes approximately

$$\upsilon = -\int_0^\infty f(r)\,\mathrm{d}r \cong 4\pi\int_0^b r^2\,\mathrm{d}r + 4\pi\int_b^\infty \beta\Phi r^2\,\mathrm{d}r = 4\pi b^3/3 + 4\pi\int_b^\infty \beta\Phi r^2\,\mathrm{d}r \quad (13.33)$$

Defining now a characteristic temperature θ via

$$k\theta = -b^{-3}\int_b^\infty \Phi r^2\,\mathrm{d}r \quad (13.34)$$

we obtain

$$\upsilon \sim [(4\pi b^3/3) + 4\pi(\theta/T)b^3] \sim (T-\theta)b^3/T \quad (13.35)$$

This implies that the chain interaction parameter z becomes $z \sim \upsilon n^{1/2}/b^3 \sim (T-\theta)n^{1/2}/T$ or $z^2 \sim \upsilon^2 n/b^6 \sim n/g_T$. Finally, we obtain for the number of segments in a thermal blob $g_T \sim b^6/\upsilon^2 \sim n/z^2 \sim [T/(T-\theta)]^2$. Hence, for $T < \theta$ we have $R_F/R_0 \sim R_F/n^{1/2}b \sim b/\upsilon^{1/3}n^{1/6} \sim z^{-1/3}$ and the blob contracts. Oppositely for $T > \theta$, $R_F/R_0 \sim R_F/n^{1/2}b \sim z^{2\nu-1} \sim z^{2\cdot(3/5)-1} \sim z^{0.2}$ and the blob swells. Overall, a fairly consistent description is obtained.

13.3 The Flory–Huggins Model

Let us now turn to the thermodynamics of polymer solutions. In 1941, Flory [3] and Huggins [4] published their (in the meantime) classic papers on the solutions of polymers. The principle of dealing with the problem remains the same as for molecular solutions: calculate the entropy of mixing $\Delta_{\text{mix}}S$, the enthalpy of mixing $\Delta_{\text{mix}}H$, and from these two quantities the Gibbs energy of mixing $\Delta_{\text{mix}}G$. If $\Delta_{\text{mix}}G < 0$, the polymer is soluble in the solvent. In this section we deal with this approach, using again the lattice model with coordination number z and having n sites, based on the original discussion of Flory [5] and the more recent version provided by Gedde [6]. As usual, we refer to component 1 as the solvent, and component 2 as the polymer.

13.3.1 The Entropy

The basis of the thermodynamic estimates for polymeric solutions is similar to that of molecular solutions (Figure 13.6). For molecular solutions, we have seen that the entropy of mixing was given by (see Section 11.4)

$$\Delta_{\text{mix}}S = -R(n_1 \ln x_1 + n_2 \ln x_2) \quad (13.36)$$

where, as before, n_i and x_i represent the number of moles and the mole fraction of component i, respectively, while R represents the gas constant. This expression can also be written (because it is assumed that for the molar volumes we have $V_1^* = V_2^*$) as:

$$\Delta_{\text{mix}}S = -R(n_1 \ln \phi_1 + n_2 \ln \phi_2) \quad (13.37)$$

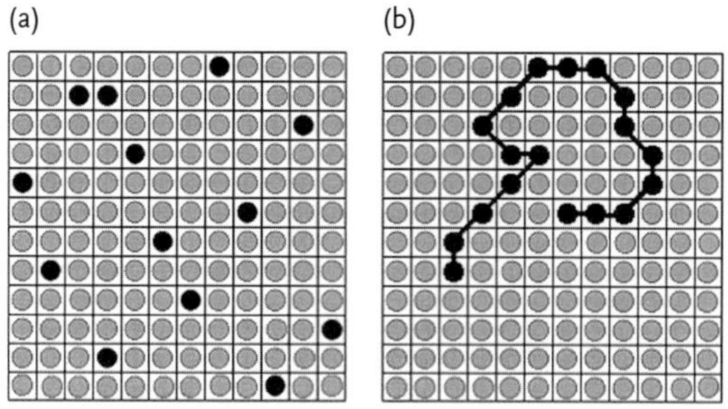

Figure 13.6 Lattice models for (a) molecular and (b) polymeric solutions.

where ϕ_i represents the volume fraction of component i. In Chapter 11, arguments were advanced to show that the proper expression is indeed in volume fractions.

Here, we derive the entropy expression for a polymer solution. In fact, in one set of considerations we can derive the relevant expressions not only for solutions but also for polymer blends and molecular solutions. To that purpose, we deal with solvent and solute on a symmetrical basis. We introduce first a reference volume V_0 that represents the molar volume of a lattice site which can be chosen arbitrarily, for example, $V_0 = V_1$, V_2 or $(V_1V_2)^{1/2}$. Often, one takes for V_0 the molar volume of the pure solvent V_1^*. In the sequel we will also use the molecular reference volume $v_0 = V_0/N_A$ where, as usual, N_A represents Avogadro's constant. Assuming no volume change, for a volume V_j of component j, the total volume of the solution is $V = V_1 + V_2$, and the number of sites[5] is $n = V/v_0$. Of these sites a volume fraction $\phi_j = V_j/V$ is occupied by component j, and therefore $V_j/v_{\text{ref}} = n\phi_j$.

Now, consider a molecule which in terms of the equivalent chain consists of r segments. If the molecular volume v_j of component j is given by $v_j = V_j/N_A$, the *degree of polymerization* is defined by $r_j = v_j/v_0 = V_j/V_0 = M_j/\rho_jV_0$, where M_j is the molar mass of the molecule that has density ρ_j (volume V_j) in the amorphous state at the solution temperature. Because $v_j = r_jv$, the number of molecules n_j of component j is $n_j = V_j/v_j = V_j/r_jv_0 = n\phi_j/r_jv_0$. Hence, on the lattice r_1n_1 sites are occupied by solvent molecules and r_2n_2 sites by polymer molecules, and the total number of lattice sites is $n = r_1n_1 + r_2n_2$.

We now turn again to solutions and use $r_1 = 1$ and $r_2 = r$. The analysis starts with having already i polymer molecules on the lattice, so that the number of vacant sites is then $n - ir$. This number equals the number of different ways of placing the $(i + 1)$th molecule on the lattice, which equals the product of the coordination number z and the fraction of remaining vacancies $(1 - f_i)$, that is, $z(1 - f_i)$.

5) We consequently redefine $N = nN_A$ to n in order to avoid the repetitious use of the factor N_A.

The next segment can only occupy $(z-1)(1-f_i)$ positions. The number of different ways of arranging the $(i+1)$th molecule is then

$$v_{i+1} = (n-ir)z(z-1)^{r-2}(1-f_i)^{r-1} \tag{13.38}$$

The number of ways to arrange all polymer molecules is accordingly

$$W_2 = v_1 v_2 \ldots v_{n2} / n_2! \tag{13.39}$$

where the denominator $n_2!$ takes into account that the solute molecules are indistinguishable. The fraction vacant sites is $(1-f_i) = (n-ir)/n$ which, upon substitution in Eq. (13.38), meanwhile replacing once z by $z-1$, results in

$$v_{i+1} = (n-ir)^r[(z-1)/n]^{r-1} \tag{13.40}$$

Since we consider only diluted solutions, we have $n >> r$ and therefore we can approximate Eq. (13.40) by

$$v_{i+1} = \frac{(n-ir)!}{[n-r(i+1)]!}\left(\frac{z-1}{n}\right)^{r-1} \tag{13.41}$$

Substitution of Eq. (13.41) in Eq. (13.39) results in

$$\begin{aligned} W_2 &= \left(\frac{z-1}{n}\right)^{n_2(r-1)} \frac{1}{n_2!} \frac{(n-r)!(n-2r)!\ldots(n-n_2r)!}{(n-2r)!(n-3r)!\ldots[n-(n_2+1)r]!} \\ &= \left(\frac{z-1}{n}\right)^{n_2(r-1)} \frac{1}{n_2!} \frac{(n-r)!}{[n-(n_2+1)r]!} \cong \left(\frac{z-1}{n}\right)^{n_2(r-1)} \frac{n!}{(n-rn_2)!n_2!} \end{aligned} \tag{13.42}$$

where the last step can be made only for dilute solutions with $n >> r$.

The entropy becomes $S = k\ln W_2$ which, upon some straightforward evaluation, using Stirling's approximation ($\ln x! = x\ln x - x$) and the fact that $n = n_1 + n_2 r$, becomes

$$S(n_1,n_2) = k\left[-n_1\ln\left(\frac{n_1}{n_1+rn_2}\right) - n_2\ln\left(\frac{n_2}{n_1+rn_2}\right) + n_2(r-1)\ln\left(\frac{z-1}{\mathrm{e}}\right)\right] \tag{13.43}$$

The entropy difference upon dissolution is calculated as $\Delta S = S(n_1,n_2) - S(0,n_2)$ because $S(0,n_2) = kn_2\ln r + kn_2(z-1)\ln[(z-1)/\mathrm{e}]$ represents the entropy of the amorphous polymer. Considering that the volume fractions of the solvent and solute are $\phi_1 = n_1/(n_1 + rn_2)$ and $\phi_1 = rn_2/(n_1 + rn_2)$, respectively, the result is (after some algebra)

$$\Delta_{\text{mix}}S(n_1,n_2) = -k(n_1\ln\phi_1 + n_2\ln\phi_2) \tag{13.44}$$

Using $n = n_1 + rn_2$, the entropy change per site becomes $\Delta_{\text{mix}}S/n$ (note that $\Delta_{\text{mix}}S/n$ is an intensive quantity), so that $\Delta_{\text{mix}}S/nv_0$ represents the entropy change per unit volume

$$\Delta_{\text{mix}}S/nv_0 = -k[\phi_1\ln\phi_1 + (\phi_2/r)\ln\phi_2] \tag{13.45}$$

The entropy expression also applies if both components are polymers, that is, for a blend. In this case we use again $r_j = M_j/\rho_j V_0$ leading to volume fractions given by

$$\phi_j = n_j r_j \Big/ \sum_j n_j r_j \tag{13.46}$$

The entropy change per unit volume then becomes

$$\Delta_{\text{mix}} S / n v_0 = -k[(\phi_1/r_1)\ln\phi_1 + (\phi_2/r_2)\ln\phi_2] \tag{13.47}$$

and this quantity, since $\phi \leq 1$, is always positive. From these equations we conclude that, in the case of a polymer blend, the entropy gain per unit volume is relatively small as compared to that of the polymer solution because, in the former case $n_1 = (\phi_1 \Sigma_j n_j r_j / r_1)$ is much smaller. Note that, if $r_1 = r_2 = 1$, we regain the ideal solution entropy. We included here only the mixing entropy, due to the translational degrees of freedom, and assumed that the configurational contribution would remain the same as for the pure polymer. Although this assumption is appropriate for blends, we discussed that excluded volume interactions for solutions have a significant influence. Moreover, we neglected volume changes which, although small, are actually present.

Problem 13.3

Estimate for which size ratio the volume fraction expression for the entropy deviates by more than 5% from the mole fraction expression.

13.3.2
The Energy

For the energy we must calculate the interaction between the solvent and solute segments, and we do this in a rather similar fashion to that used for molecular solutions. In particular, we neglect (again) volume changes and assume that the segments are placed at random over the lattice, thereby neglecting all correlation between segments. We again use a lattice of n sites with coordination number z and n_1 solvent molecules and n_2 polymer molecules, so that $n = n_1 + rn_2$. Each solute segment is surrounded by $z\phi_1$ solvent molecules and $z\phi_2$ solute segments, and therefore the interaction energy for that segment reads $z\phi_1 w_{12} + z\phi_2 w_{22}$. Because the lattice contains $n\phi_2$ segments, each interacting in this way, the energy U_2 for the solute becomes

$$U_2 = \frac{1}{2} n z \phi_2 (\phi_1 w_{12} + \phi_2 w_{22}) \tag{13.48}$$

where the factor ½ is introduced to avoid counting interactions twice. Similarly, for the solvent molecules we obtain the energy

$$U_1 = \frac{1}{2} n z \phi_1 (\phi_1 w_{11} + \phi_2 w_{12}) \tag{13.49}$$

The energy for the pure solvent and solute are $U_{01} = \frac{1}{2} n\phi_1 z w_{11}$ and $U_{02} = \frac{1}{2} n\phi_2 z w_{22}$, respectively, so that the energy difference upon dissolution is calculated as

$$\begin{aligned}\Delta_{\text{mix}}U &= (U_1 + U_2) - (U_{01} + U_{02}) \\ &= \frac{1}{2}nz[\phi_2(\phi_2 w_{22} + \phi_1 w_{12}) + \phi_1(\phi_1 w_{11} + \phi_2 w_{12}) - \phi_1 w_{11} - \phi_2 w_{22}] \\ &= \frac{1}{2}nz(2w_{12} - w_{11} - w_{22})\phi_1\phi_2 \equiv nzw\phi_1\phi_2 = (n_1 + rn_2)zw\phi_1\phi_2 \end{aligned} \tag{13.50}$$

where $w \equiv w_{12} - ½(w_{11} + w_{22})$ is the *interchange energy*, the increase in energy when a solvent–polymer contact is formed from molecules that were originally contacting molecules of the same type.

In Flory–Huggins theory it is conventional to render the interaction parameter w dimensionless according to $\chi = zw/kT$, where χ (chi) is the dimensionless interaction parameter. Values for many polymers are listed in handbooks [7]. Since $\phi_1 = n_1/(n_1 + n_2 r_2)$, we can also write

$$\Delta_{\text{mix}}U = kT\chi n_1\phi_2 \tag{13.51}$$

The total volume of the solution is $V = nv_0 = (n_1 + n_2 r)v_0$, where v_0 is the reference volume. The energy of mixing per unit volume then becomes

$$\frac{\Delta_{\text{mix}}U}{nv_0} = \frac{kT\chi n_1\phi_2}{(n_1 + n_2 r)v_0} = \frac{kT\chi\phi_1\phi_2}{v_0} \tag{13.52}$$

13.3.3 The Helmholtz Energy

From the entropy and energy, we obtain the Helmholtz energy

$$\Delta_{\text{mix}}F = \Delta_{\text{mix}}U - T\Delta_{\text{mix}}S = kT(\chi n_1\phi_2 + n_1 \ln\phi_1 + n_2 \ln\phi_2) \tag{13.53}$$

For the case of mixture of two polymers we obtain the result

$$\Delta_{\text{mix}}F = \Delta_{\text{mix}}U - T\Delta_{\text{mix}}S = nkT[\chi\phi_1\phi_2 + (\phi_1 / r_1)\ln\phi_1 + (\phi_2 / r_2)\ln\phi_2] \tag{13.54}$$

with, as before, the DP $r_j = M_j/\rho_j V_0$. Note that the value of χ depends on the choice of V_0. Remembering that χ contributes to the Helmholtz energy, one should not be surprised that χ is temperature-dependent. Often, one writes

$$\chi = \beta zw = A + B / T \tag{13.55}$$

using the "entropic" and "enthalpic" parameters A and B, respectively, comparable to $zw = a + bT$, Eq. (11.75), for regular solutions. A typical example is the system polystyrene (PS)/polymethyl methacrylate (PMMA) with $A = 0.0129$ and $B = 1.96$ K using $v_0 = 100$ Å^3, resulting in $\chi \cong 0.01$, so that only low-molar mass components form a miscible blend. Empirically, this relationship is not generally observed, that is, the graph of χ versus $1/T$ may show curvature, local minima or even change sign, and is somewhat dependent on composition and chain length. Finally, we note that, as expected, for $r_1 = r_2 = 1$, we regain the regular solution description.

13.3.4 Phase Behavior

From the Helmholtz energy expression one can calculate the stability of the polymer solution via the conventional route, as used before for molecular (regular) solutions (see Eqs 11.67 to 11.70) which led in that case to $kT_{\rm cri} = ½zw$. Here, the main result for a binary polymer mixture with $\phi_1 = \phi$ and $\phi_2 = 1 - \phi$, is

$$\partial^2\Delta_{\rm mix}F / \partial\phi^2 = nkT\{(n_1\phi)^{-1} + [n_1(1-\phi)]^{-1} - 2\chi\} \qquad (13.56)$$

Equating this result to zero, one obtains the expression for the inflection points of the Helmholtz curve, dependent on composition and separating unstable and metastable regions. It is conventionally labeled as the *spinodal (line)* (see Sections 16.1 and 16.2), and reads for this case

$$\chi_{\rm spi} = \frac{1}{2}\{(n_1\phi)^{-1} + [n_1(1-\phi)]^{-1}\} \qquad (13.57)$$

Using $\chi = A + B/T$ one obtains

$$T_{\rm spi} = \frac{2B}{\{(n_1\phi)^{-1} + [n_1(1-\phi)]^{-1}\} - A} \qquad (13.58)$$

The critical point corresponds to the minimum in this curve and is obtained from

$$\partial\chi_{\rm spi} / \partial\phi = \frac{1}{2}\{-(n_1\phi)^{-2} + [n_1(1-\phi)^2]^{-1}\} = 0 \qquad (13.59)$$

which corresponds to the critical composition

$$\phi_{\rm cri} = n_1^{1/2} / (n_1^{1/2} + n_2^{1/2}) \qquad (13.60)$$

and upon substitution this result in the spinodal to

$$\chi_{\rm cri} = \frac{1}{2}(n_1^{1/2} + n_2^{1/2})/n_1^{1/2}n_2^{1/2} = \frac{1}{2}(n_1^{-1/2} + n_2^{-1/2})^2 \qquad (13.61)$$

Since n_2 is usually large this implies for solutions ($n_1 = 1$) that, in the limit of infinite molar mass, $\chi_{\rm cri} = ½$ or $kT_{\rm cri} = 2zw$, in contrast to the result for regular solutions $kT_{\rm cri} = ½zw$. For the most common case where $B > 0$, χ decreases with increasing T and the solubility increases. The highest temperature for a two-phase system to exist is the *upper critical solution temperature* (UCST) and for $T > T_{\rm cri}$ the homogeneous solution is stable. However, $B < 0$ also occurs and χ increases with increasing T so that the solubility decreases. The lowest temperature for a homogeneous solution to exist is the *lower critical solution temperature* (LCST). If B changes sign as a function of T, the system will show both an UCST and a LCST.

In order to be able to describe the UCST, both the "entropic" part A and "enthalpic" part B can be considered to be temperature-independent. From Problem 13.5 we have

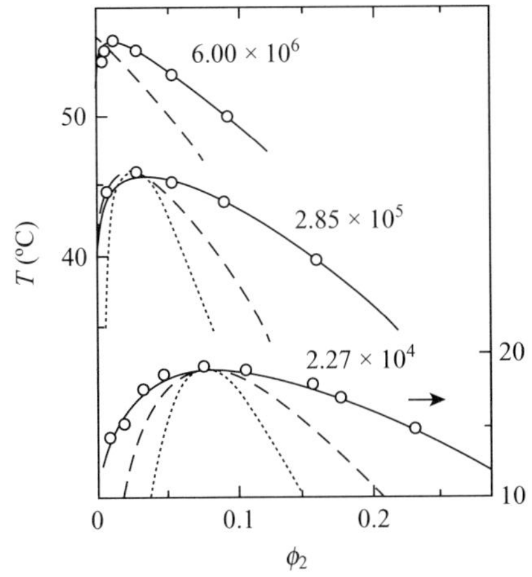

Figure 13.7 Experimental and calculated phase diagrams for three polyisobutylene fractions in diisobutyl ketone of the molecular weights indicated. The dashed curves are the calculated Flory–Huggins binodals, while the dotted curves are the spinodals. The model description is due to Casassa [37], using the data of Schultz and Flory [38].

$$\chi_{\text{cri}} = (1/2) + (1/2r) + \left(1/\sqrt{r}\right) \tag{13.62}$$

so that for the DP $r_2 \equiv r \to \infty$, $\chi_{\text{cri}} = ½$. Combining with $\chi = A + B/T$ and solving for $T_{\text{cri}}(r = \infty) \equiv \theta$, we obtain $½ = A + B/\theta$ and therefore

$$1/T_{\text{cri}} = 1/\theta + B^{-1}\left[(1/2r) + \left(1/\sqrt{r}\right)\right] \tag{13.63}$$

describing $1/T_{\text{cri}}$ as a function of θ and r. Hence, given a set of experimental values of T_{cri} as a function of r, the slope of Eq. (13.63) determines B and the intercept determines θ, and therefore A. An example is shown in Figure 13.7.

The temperature θ can be shown to be indeed the theta temperature. From Problem 13.6 we have, remembering that $a_1 = \gamma_1 x_1$ and that $\phi_1 = 1 - \phi_2$,

$$\begin{aligned} -\ln a_1 &= -\ln \gamma_1 x_1 = -\ln \phi_1 - (1 - r^{-1})\phi_2 - \chi\phi_2^2 = -\ln(1 - \phi_2) - (1 - r^{-1})\phi_2 - \chi\phi_2^2 \ldots \\ &\cong \phi_2 + \frac{1}{2}\phi_2^2 + \ldots - (1 - r^{-1})\phi_2 - \chi\phi_2^2 \\ &= \phi_2 + \left(\frac{1}{2} - \chi\right)\phi_2^2 + \ldots \equiv V_1^*[A_1 c_2 + A_2 c_2^2 + \ldots] \end{aligned} \tag{13.64}$$

where in the third line $r \to \infty$ is used. In the last step we apply an expansion similar to that used for the osmotic pressure (see Section 13.2), but now for $\ln a_1$ instead of the osmotic factor g. The concentration of the polymer is given by $c_2 = n_2 M_2/V$, where n_2 is the number of moles of polymer of molecular mass M_2 in a volume V. If we assume that we have no volume change upon mixing, the volume of the polymer V_2 is given by $V_2 = v_2^* n_2 M_2$, with v_2^* the specific volume of the pure polymer as defined by $r \cong V_2^*/V_1^* = v_2^* M_2/V_1^*$, so that $\phi_2 = v_2^* c_2$. The

Flory–Huggins first and second osmotic virial coefficients A_1 and A_2 become, respectively,

$$A_1 = M_2^{-1} \quad \text{and} \quad A_2 = \left(\frac{1}{2} - \chi\right)(v_2^*)^2 / V_1^* \tag{13.65}$$

Similarly to what is discussed in Section 11.3, the first osmotic virial coefficient A_1 is determined entirely by the molecular mass M_2. From Eq. (13.65) we see further that A_2 vanishes if χ = ½ and the solution behaves to third order as an ideal solution, that is, the chains are "unperturbed." This behavior is comparable to the pseudo-ideal pressure behavior of gases at the Boyle temperature, where the second virial coefficient vanishes. We have already indicated that for $r \to \infty$, χ_{cri} = ½, and therefore it is appropriate to consider θ as the theta temperature.

In the spirit of the osmotic expansion, we expect that the second osmotic virial coefficient A_2 characterizes the onset of intermolecular interactions between the polymer solute molecules, and therefore depends on M_2. However, we predict that A_2 is independent of M_2, which experimentally is not the case. This shows one of the deficiencies of the Flory–Huggins model, namely that it is not valid for very dilute systems because the assumption that a segment experiences the interaction for a random distribution of segments and solvent molecules cannot be true for such a solution.

Let us now consider the LCST. Whereas, for the UCST the parameters A and B could be considered as temperature-independent, when describing the LCST they should be considered as temperature-dependent. This can be done, for example, by writing

$$B(T) = B(\theta) + \int_\theta^T \Delta C_P \, \mathrm{d}T \tag{13.66}$$

with ΔC_P the exchange heat capacity. Using the Gibbs–Helmholtz equation, we obtain

$$\chi = \frac{1}{2} + \int_\theta^T B(T)\mathrm{d}\beta = \frac{1}{2} + k^{-1}\int_\theta^T B(T)\mathrm{d}(1/T) \tag{13.67}$$

If ΔC_P is considered as constant, we easily evaluate this expression to

$$\chi = \frac{1}{2} - [B(\theta)/k](\theta^{-1} - T^{-1}) + (\Delta C_P/k)[1 - (\theta/T) + \ln(\theta/T)] \tag{13.68}$$

If we accept that it is possible that $\Delta C_P < 0$, χ as a function of T shows a minimum and two critical temperatures exist: $T = \theta$, representing the UCST, and another temperature representing the LCST. Typical values of ΔC_P required are $\Delta C_P \cong -k$ to $-2k$. One interpretation for $\Delta C_P < 0$ is that the interchange energy between solute segments and solvent molecules becomes lower as the system expands with increasing temperature.

Some further aspects of the FH theory are discussed by Boyd and Phillips [1].

Problem 13.4: Asymmetry

Plot $\Delta_{mix}F$ as resulting from the Flory–Huggins theory with $\chi = 1$, 2, and 3 for:

a) $r_1 = 1$ and $r_2 = 1$, 10, 10^2, and 10^4.
b) $r_1 = 10^2$ and $r_2 = 10^2$ and 10^4.
c) $r_1 = 10^4$ and $r_2 = 10^4$.

Problem 13.5: Phase separation

Using the Flory–Huggins expression Eq. (13.54) as well as $r_1 = 1$ and $r_2 = r$:

a) Show that the critical values of ϕ_2 and χ for phase separation to occur in polymer solutions are given by, respectively,

$$\ln(\phi_{2,\text{cri}}) = \left(1+\sqrt{r}\right)^{-1} \quad \text{and} \quad \chi_{\text{cri}} = (1/2)+(1/2r)+\left(1/\sqrt{r}\right)$$

Check that $\chi_{\text{cri}} = ½$ for infinite molecular mass of the polymer.

b) Show that the critical temperature T_{cri} can be described by

$$T_{\text{cri}} = B/R\chi_{\text{cri}} \quad \text{with} \quad B = \left(\sqrt{\Delta_{\text{vap}}E_1} - \sqrt{\Delta_{\text{vap}}E_2}\right)^2$$

Problem 13.6: Activity coefficient

Show by using Eq. (11.9) that the activity coefficients of the solvent (1) and the polymer (2) containing r monomers per polymer in the Flory–Huggins model are given by, respectively,

$$\ln\gamma_1 = \ln\frac{\phi_1}{x_1} + \left(1-\frac{1}{r}\right)\phi_2 + \chi\phi_2^2 \quad \text{and} \quad \ln\gamma_2 = \ln\frac{\phi_2}{x_2} - (r-1)\phi_1 + r\chi\phi_1^2$$

Problem 13.7: LCST

Derive Eq. (13.68).

13.4 Solubility Theory

The χ parameter can be estimated using the solubility parameter approach. To that purpose, compare the regular solution model expression for the interchange energy in terms of solubility parameters, (Eq. 11.98), $\Delta_{\text{mix}}U_{\text{m}} = V_{\text{m}}(\delta_1 - \delta_2)^2\phi_1\phi_2$ with the Flory–Huggins expression, Eq. (13.52), $\Delta_{\text{mix}}U = nkT\chi\phi_1\phi_2$. Equating both expressions using meanwhile $V_{\text{m}} = nv_0$, we easily obtain

$$\chi = v_0(\delta_1 - \delta_2)^2/kT \tag{13.69}$$

Therefore, χ relates directly to the solution while the δs refer to each component of the solution. While for χ we have either $\chi < 0$ or $\chi > 0$, approximating χ using the solubility parameter concept we have $\chi > 0$ always.

Although in these models the difference between U and H is neglected, in practice the difference exists. Experimentally, the evaporation enthalpy $\Delta_{\text{vap}}H$ is measured. To obtain the energy $\Delta_{\text{vap}}U$ one has to subtract the product PV, to a good approximation equal to RT. This implies that for the solubility parameter δ, since this parameter refers to energy values, we should use $\Delta_{\text{mix}}H - RT$ if experimental values are used. Since $\delta^2 = \Delta_{\text{vap}}U/V^*$, we have $\delta^2 = (\Delta_{\text{vap}}H - RT)/V^* = \rho(\Delta_{\text{vap}}H - RT)/M$. For polymers, however, the evaporation enthalpy is usually unavailable. Although other experimental methods to estimate χ are available – that is, osmotic pressure and viscosity measurements – research teams have nevertheless devised a group contribution scheme to estimate the δ-values. Originally in this approach, $\Delta_{\text{vap}}U$ was considered as an additive property of various chemical groups in the molecule for low-molecular-mass molecules. Later, it was shown [8] that $F = (\Delta_{\text{vap}}U)^{1/2}$ is actually a more useful additive quantity, both for low-molecular-weight as well as high-molecular-weight compounds. The value of δ is then supposed to be given by

$$\delta = \rho \sum\nolimits_i F_i \big/ M_0 \tag{13.70}$$

where ρ is the (mass) density of the amorphous polymer and M_0 the molar mass of the repeating unit. The contributions F_i are tabulated for various chemical groups in the repeating unit, and a selection is given in Table 13.3 using the parameterization of Hoy [9]. A similar approach is used for random copolymers. One then takes $\delta_c = \Sigma_i\delta_i w_i$, where δ_i and w_i represent the delta and weight fraction of the repeating unit i in the copolymer. For alternating copolymers one takes the copolymer repeating unit as the monomer, but for block copolymers and grafted copolymers no satisfactory methods are available. For solvents, the group contribution approach is also applicable, but a small correction for end effects is usually applied (Table 13.3).

For mixtures, one usually uses the *rule-of-mixtures* given by

$$\delta_{\text{mix}} = \delta_1\phi_1 + \delta_2\phi_2 \tag{13.71}$$

where ϕ_1 (δ_1) and ϕ_2 (δ_2) represent the volume fraction (solubility parameter) of components 1 and 2. Further, one usually assumes that the temperature dependence of δ can be neglected, with most tables providing data for δ at 25 °C.

If the difference in solubility parameters $\Delta\delta = |\delta_2 - \delta_1|$ between two components is sufficiently small, the components are soluble in each other; otherwise a two-phase system results. For "normal" systems – that is, for systems without hydrogen bonding – a value of about $\Delta\delta = 4\ (\text{J m}^3)^{1/2}$ is often used.

Table 13.3 Group contribution for the calculation of the solubility parameter.[a)]

Group	F_i (J m^3)$^{1/2}$ mol^{-1}	Group	F_i (J m^3)$^{1/2}$ mol^{-1}
$-CH_3$	0.3032	–H acidic dimer	−0.1032
$-CH_2-$	0.2689	–OH aromatic	0.3496
>CH–	0.1758	$-NH_2$	0.4632
–C– with no H	0.0655	>NH	0.3681
$CH_2=$ olefin	0.2587	>N–	0.1249
>CH= olefin	0.2485	–C≡N	0.7249
>C= olefin	0.1728	–NCO	0.7333
–CH= aromatic	0.2395	–S–	0.4282
>C= aromatic	0.2006	$>Cl_2$	0.7006
–O– ether, acetal	0.2351	–Cl primary	0.4192
–O– epoxide	0.3602	–Cl secondary	0.4258
–COO–	0.6677	–Cl aromatic	0.3292
>C=O	0.5376	–Br	0.5272
–CHO	0.5983	–Br aromatic	0.4204
–(CO)–O–(CO)– (anhydride)	1.1598	F	0.0845
–OH	0.4617		
Structural feature			
Conjugation	0.0476	Six-membered ring	−0.0479
Cis	−0.0146	*Ortho* substitution	0.0198
Trans	−0.0276	*Meta* substitution	0.0135
Four-membered ring	0.1590	*Para* substitution	0.0825
Five-membered ring	0.0429	Non-polymeric solvent[b)]	0.2770

a) Data from Ref. [10].
b) A "base" value of 0.277, essentially a correction for end groups, must be included in the summation when the system is used for small molecules, for example, monomers or solvent molecules.

Example 13.1

Calculate the δ-value for polystyrene (PS), $-(CH_2-CHC_6H_5)_n$.

Group	F_i	Number	$\Sigma_i F_i$
$-CH_2-$	0.2689	1	0.2689
>CH–	0.1758	1	0.1758
–CH= (aromatic)	0.2395	5	1.1975
>C= (aromatic)	0.2006	1	0.2006
6-membered ring	−0.0479	1	−0.0479
		Total	1.795 (J m^3)$^{1/2}$ mol^{-1}

The density $\rho = 1050\,\text{kg m}^{-3}$ and $M_0 = 0.104\,\text{kg mol}^{-1}$. Hence $\delta = \rho\Sigma_i F_i/M_0 = 1050 \times 1.795/0.104 = 18.1\,\text{MPa}^{1/2}$. Testing whether PS dissolves in n-hexane, one needs

$\rho = 654\,\text{kg m}^{-3}$ and $M_0 = 0.0861\,\text{kg mol}^{-1}$, while $\Sigma_i F_i = 1.957$. Therefore $\delta = 14.7\,\text{MPa}^{1/2}$. Hence, PS should dissolve in *n*-hexane, as in fact it does.

As might be expected, the above-described method – often denoted as the *Hildebrand solubility parameter* approach – is not suitable for use outside its original area, which is nonpolar, non-hydrogen-bonding solvents. Several attempts have been reported to extend and improve upon this scheme (see, e.g., van Krevelen [10]). One frequently used approach is that of Hansen [11]; whereas, in the original approach essentially only the dispersion forces were considered, Hansen divided the cohesive energy δ^2 into three parts, namely

$$\delta^2 = \delta_d^2 + \delta_p^2 + \delta_H^2 \tag{13.72}$$

where the subscripts d, p, and H refer to the contributions due to dispersion forces, polar forces, and hydrogen bonding, respectively. From the experimental solubility data for many solvents, the parameters δ_d, δ_p and δ_H are determined using a least-squares approach. Data for a number of solvents are given in Table 13.4.

Using the values so obtained, one can again estimate whether two solvents are miscible, this being the case if the resulting value for $\delta_2 - \delta_1$ is sufficiently small. One can also "compose" in this way a solvent mixture with desired properties. For polymers, there is an extra parameter R_0 that describes the radius of the "solubility sphere" in $2\delta_d$-δ_p-δ_H space. Hence, if $R^2 = (2\delta_{d,1} - 2\delta_{d,2})^2 + (\delta_{d,1} - \delta_{d,2})^2 + (\delta_{H,1} - \delta_{H,2})^2 < R_0^2$, component 1 is a solvent for component 2 (the polymer), whereas

Table 13.4 The δ-parameters for several solvents in $(\text{MPa})^{1/2}$.

Solvent	δ_d	δ_p	δ_H
n-Hexane	14.7	0	0
Benzene	18.8	0	0
Chloroform	18.0	3.1	5.7
Nitrobenzene	18.6	12.3	4.1
Diethyl ether	14.5	2.9	5.1
Iso-amylacetate	15.3	3.1	7.0
Di-octylphthalate	16.6	7.0	3.1
Methyl isobutyl ketone	15.3	6.1	4.1
Tetrahydrofuran	16.8	5.7	8.0
Methylethylketone	16.0	9.0	5.1
Acetone	15.5	10.4	7.0
Dimethylsulfoxide	18.4	16.4	10.2
Acetic acid	14.5	8.0	13.5
m-Cresol	18.0	5.1	12.9
1-Butanol	16.0	5.7	15.7
Methylene glycol	16.2	14.7	20.5
Methanol	15.1	12.3	22.3

Table 13.5 The δ-parameters for several polymers in $(MPa)^{1/2}$ [10].

Polymer	δ	δ_d	δ_p	δ_H
Polyisobutylene	17.6	16.0	2.0	7.2
Polystyrene	20.1	17.6	6.1	4.1
Poly(vinyl chloride)	22.5	19.2	9.2	7.2
Poly(vinyl acetate)	23.1	19.0	10.2	8.2
Poly(methyl methacrylate)	23.1	18.8	10.2	8.6
Poly(ethyl methacrylate)	22.1	18.8	10.8	4.3
Polybutadiene	18.8	18.0	5.1	2.5
Polyisoprene	18.0	17.4	3.1	3.1

when $R^2 > R_0^2$, component 1 is a non-solvent for the polymer. The factor of 2 in front of the dispersion term has been the subject of debate. Whilst there is some theoretical basis [11, 12] for this factor, it was mainly introduced to match the experimental data. However, there are clearly systems [13] where the regions of solubility are far more eccentric than predicted by the standard Hansen theory. Values of δ_d, δ_p and δ_H for a number of polymers are reported in Table 13.5.

Although the Hildebrand parameter for non-polar solvents is usually close to the Hansen value, the latter has a wider applicability area. A typical example is provided by the fact that two solvents, butanol and nitroethane, which have the same Hildebrand parameter, are each incapable of dissolving typical epoxy polymers. Yet, a 50 : 50 mix is a good solvent for epoxy polymers. This can be easily explained by the fact that the Hansen parameters for this mix are close to the Hansen parameter of epoxy polymers.

The need to use a "matching factor" for the dispersion component clearly shows the semi-empirical nature of the whole scheme, in which many approximations are made. Yet, in spite of this, the approach is – practically speaking – rather useful, and estimates of the χ parameters often are reasonably accurate.

Problem 13.8

Verify the dimension of the molar attraction constant F in Table 13.3.

Problem 13.9

Calculate the solubility parameter of toluene from its vaporization enthalpy $\Delta_{vap}H$. To estimate $\Delta_{vap}H$, use the empirical rule that relates $\Delta_{vap}H$ of a liquid to its boiling temperature T_n according to $\Delta_{vap}H$ (at 298 K in J mol^{-1}) $= 99.3T_n + 0.084T_n^2 - 12\,330$. For toluene: $M = 92$, $\rho = 867\,\text{kg m}^{-3}$, $T_n = 383.7\,\text{K}$ at 1 atm.

Problem 13.10

Calculate the solubility parameter δ of toluene from the group molar attraction constants F_i.

Problem 13.11

Calculate the composition by volume of a blend of *n*-hexane, *t*-butanol and dioctyl phthalate (use indices 1, 2, and 3 for these liquids for convenience) that would have the same solvent properties as tetrahydrofuran (use Table 13.4 and match δ_p and δ_H values). Use an equation analogous to Eq. (12.22) to assess the d, p, and H parts of the solubility parameter for solvent mixtures.

13.5 EoS Theories*

In the FH approach it is assumed that the Helmholtz energy F is fully determined by the temperature (and composition), and that volume is not an explicit variable. In reality, the volume does change, and this leads to Equation of State (EoS) theories in which both temperature and volume are taken explicitly into account. In this section, we outline a few of these theories in which use will be made of various aspects of lattices and holes, as discussed in Chapter 8. When using a lattice theory, one has three options to include "free volume":

- The first option is to use temperature-dependent cell volumes, which is addressed as *cell theory*.
- The second option is to introduce empty cells or holes (while keeping the cell volume constant); in Chapter 8 for simple and normal liquids this was referred to as hole theory, but in relation to polymer solutions it is often termed the *lattice fluid* (or *lattice gas*) theory.
- The third option is to combine these approaches, in relation with polymers often denoted as *hole theory*, in obvious contradiction to the terminology as used for simple fluids.

An example of each of these options is discussed below, but first we briefly review basic cell theory and its extension to chain molecules.

13.5.1 A Simple Cell Model

In Section 8.2, we learned that for the cell model with N molecules in a volume V on a lattice with coordination number z, using a Lennard-Jones potential with hard-core parameter σ, or alternatively an equilibrium distance $r_0 = 2^{1/6}\sigma$, an energy well depth ε, and expanding the lattice from σ to a to mimic free volume for the liquid, the configurational partition function Q is given by

$$Q=[Q(1)]^N=\exp[-\beta E_0](v_f)^N=\exp\left[-\frac{1}{2}\beta zN\phi(0)\right](v_f)^N \qquad (13.73)$$

where $E_0 = \frac{1}{2}zN\phi(0)$ represents the lattice energy of the cell model and $\phi(0)$ represents the pair energy for all molecules (segments) occupying the origin of their cell. In connection with polymer mixtures, this quantity is sometimes referred to as *contact energy*. Using $v_0 = V_0/N \equiv \gamma^{-1}\sigma^3 = \gamma^{-1}(2^{-1/6}r_0)^3$ or[6] $v^* \equiv \gamma^{-1}r_0^3 = \gamma^{-1}(2^{1/6}\sigma)^3 = 2^{1/2}v_0$ and $v_a = V/N \equiv \gamma^{-1}a^3$ with $\gamma = 2^{1/2}$ for a FCC lattice, $\phi(0)$ is given by

$$\phi(0)=\varepsilon[A(v^*/v_a)^4-2B(v^*/v_a)^2] \qquad (13.74)$$

Here, $A = 1$ and $B = 1$ if one includes only nearest-neighbor interactions, and $A = 1.011$ and $B = 1.2045$ using the lattice sum for the FCC lattice. If we have for the total potential $\phi(r) = \phi(0) + \psi(r)$ with $\psi(r)$ the vibrational part, the free volume is given by

$$v_f=\int_0^\infty \exp[-\psi(r)/kT]4\pi r^2\,dr \qquad (13.75)$$

Approximating the potential $\psi(r)$ by a square well with radius $(a - \sigma)$, the free volume v_f becomes

$$v_f=4\pi\gamma(v_a^{1/3}-v_0^{1/3})^3/3=4\pi\gamma[v_a^{1/3}-(2^{-1/2}v^*)^{1/3}]^3/3 \qquad (13.76)$$

For the configurational Helmholtz energy $F_{con} = -kT\ln Q$, we thus have

$$F_{con}=\frac{1}{2}zN\phi(0)-NkT\ln v_f \qquad (13.77)$$

Because $P = -\partial F/\partial V$ and in $F = F_{kin} + F_{con} \equiv -kT\ln\Lambda^{-3N} - kT\ln[Q(1)^N]$, with as usual $\Lambda = (h^2/2\pi mkT)^{1/2}$, only F_{con} depends on V, we have

$$PV/NkT=v_a^{1/3}\left[v_a^{1/3}-(2^{-1/2}v^*)^{1/3}\right]^{-1}+(2\beta z\varepsilon)[A(v^*/v_a)^4-B(v^*/v_a)^2] \qquad (13.78)$$

Introducing the reduced variables $v = v_a/v^*$ (with $v^* = \gamma^{-1}r_0^3 = \gamma^{-1}(2^{1/6}\sigma)^3$), $t = T/T^*$ where $T^* = z\varepsilon/k$ and $p = P/P^*$ where $P^* = z\varepsilon/v^* = kT^*/v^*$), this expression becomes

$$pv/t=(v)^{1/3}\left[v^{1/3}-2^{-1/6}\right]^{-1}+(2/t)[Av^{-4}-Bv^{-2}] \qquad (13.79)$$

For a small molecule, the external degrees of freedom (DoF)–that is, translation and rotation–are fully available. However, for a chain each of the r segments of the N polymer molecules cannot move completely freely because they are bonded covalently to their neighbors. Prigogine *et al.* [14] solved this issue empirically by proposing that only a fraction c of the external DoF, in particular the "lower" frequency modes that are available for "completely free segments," are available as external DoF for chain-like molecules. By accepting this, the exponent 3 in the

6) In the sequel, we use v^* instead of v_0 in order to be consistent with most literature. Moreover, it avoids the use of double subscripts for solutions later on.

expression for v_f changes to $3c$, and the volume $v = V/rN$ now refers to the segment volume. The lattice energy does not change, except that the coordination number might be altered. Hence, instead of Eq. (13.77) we have

$$F_{\text{con}} = \frac{1}{2} rzN\phi(0) - rNckT \ln v_f \tag{13.80}$$

The reduced EoS remains the same but with reduced variables $v = v_a/v^*$ (with $v^* = \gamma^{-1} r_0^3 = \gamma^{-1}(2^{1/6}\sigma)^3$), $t = T/T^*$ where $T^* = z\varepsilon/ck$ and $p = P/P^*$ where $P^* = z\varepsilon/v^* = ckT^*/v^*$). In total, we thus have three parameters: $z\varepsilon$, v^*, and c (or alternatively, P^*, v^*, and T^*).

On extending this simple cell (SC) model to solutions, one assumes that the EoS expression obtained for the pure liquid is also valid for solutions. Since this model has three parameters, in order to describe solutions one also requires three combining rules. It appears – as might be expected – that a geometric combining rule for ε is not sufficiently accurate, and therefore ε is often determined experimentally, for example, from the heat of mixing. One also needs to take into account various ways in which the molecules can be distributed over the lattice, and this is done by adding, for example, a FH-type entropy term. In general, this leads to an extra combinatorial factor $g(Y)$ in Eq. (13.73), where Y indicates the set of relevant variables. Using a cell model for a pure component $g = 1$, but for a (binary) solution with N_j molecules of type j, we have the set $Y = (N_1, N_2)$. For a lattice theory for a pure component of N molecules we have $Y = (N, \phi)$, with ϕ the volume fraction occupied cells, whereas for a solution we have $Y = (N_1, N_2, \phi)$. The latter set also applies to a hole theory.

13.5.2
The FOVE Theory

Apart from contributing to the above-discussed FH theory, Flory also contributed the EoS theories and, together with some collaborators, constructed what presently is known as the Flory–Orwoll–Vrij–Eichinger (FOVE) theory [15]. In this theory, a lattice is again assumed on which the solute segments and solvent molecules are distributed with no additional holes –that is, it is a cell model.

For larger volumes, the attractive part in Eq. (13.74) proportional to v^{-2} dominates, and Flory *et al.* argued that, in the spirit of the vdW theory, $\phi(0)$ should be replaced by $\phi(0) = -s\eta/v$. Here, η is the energy parameter associated with the segment–segment contact (with dimension J m^3 instead of J only, as for ε), s is the number of such contacts per segment, and v is the segment volume. Furthermore, Eq. (13.76) is accepted for the free volume so that F becomes

$$F_{\text{con}} = \frac{1}{2} rNs\eta / v_a - 3rNckT \ln\{4\pi\gamma[v_a^{1/3} - (2^{-1/2}v^*)^{1/3}]^3/3\} \tag{13.81}$$

Following the usual route of calculating $P = -\partial F/\partial V$, the reduced EoS obtained reads

$$pv/t = v^{1/3}[v^{1/3} - 2^{-1/6}]^{-1} - 2^{-1/6} / tv \tag{13.82}$$

The more usual form of the FOVE EoS is obtained by replacing $2^{-1/6}$ by 1, resulting from the use of the LJ potential. This does not influence the EoS if the reducing parameters are defined accordingly. The reducing parameters are usually obtained by fitting to data for pure liquids. This model gives a reasonable description for pure polymer liquids, taking reasonable ranges of temperature and pressure.

On changing to solutions, we now need to state how we obtain the combining rules for the reduced variables which we can take to be v^*, T^* and c, recalling that $P^* = ckT^*/v^*$. If we use the same segment volume for both components, we obtain from the EoS of the pure polymers $V_j^* = r_j v^*$, and therefore $r_1/r_2 = V_1^*/V_2^*$. Hence, we need only two combining rules, either for T^* and P^* or for c and r. From the energy term for pure liquids in Eq. (13.81), $U = -½rNs\eta/v = rNP^*(v^*)^2/v$. Using the same reasoning as for Eq. (8.37), but with possibly different coordination numbers s_1 and s_2 for components 1 and 2, respectively, we have

$$s_1N_1 = 2N_{11} + N_{12} \quad \text{and} \quad s_2N_2 = 2N_{22} + N_{12} \tag{13.83}$$

We approximate now the number of unlike pairs by $N_{12} = (s_1r_1N_1)(s_2r_2N_2)/srN$, where we use $r = (r_1N_1 + r_2N_2)/N$ and $s = (s_1r_1N_1 + s_2r_2N_2)/rN$. The energy of the mixture (in the notation of the present paragraph and using η instead of ε) becomes

$$\begin{aligned} U &= -(N_{11}\eta_{11} + N_{22}\eta_{22} + N_{12}\eta_{12})/v \\ &= -[s_1r_1N_1\eta_{11} + s_2r_2N_2\eta_{22} - (s_1r_1N_1)(s_2r_2N_2)\Delta\eta/srN]/2v \end{aligned} \tag{13.84}$$

where $\Delta\eta = \eta_{11} + \eta_{11} - 2\eta_{12}$. The volume fractions for the segments read $\phi_j = r_jN_j/rN$, and therefore the energy becomes

$$U = -rN[s_1\phi_1\eta_{11} + s_2\phi_2\eta_{22} - (s_1s_2/s)\Delta\eta\phi_1\phi_2]/2v \tag{13.85}$$

Using $P_j^* = s_j\eta_{jj}/2(v^*)^2$ for the pure liquids, one easily transforms this expression to

$$U = -rN[\phi_1P_1^* + \phi_2P_2^* - (s_1s_2/s)\Delta\eta\phi_1\phi_2/(2v^*)^2](v^*)^2/v \tag{13.86}$$

Comparing with $U = -rNP^*(v^*)^2/v$, one obtains

$$P^* = \phi_1P_1^* + \phi_2P_2^* - \Delta P_{12}^*\phi_1\phi_2 \tag{13.87}$$

with $\Delta P_{12}^* = s_1s_2\Delta\eta/2s(v^*)^2$. Writing for the number of external DoF $c = (c_1r_1N_1 + c_2r_2N_2)/rN = \phi_1N_1 + \phi_2N_2$, the use of the reduced variables $T^* = P^*v^*/ck$ and $T_j^* = P_j^*v^*/c_jk$ yield

$$1/T^* = (\phi_1P_1^*/T_1^* + \phi_2P_2^*/T_1^*)/P^* \tag{13.88}$$

Finally, the reduced Gibbs energy $G_{\text{red}} = G/rN\varepsilon^*$ (with $G = F + PV$ and $\varepsilon^* = -½z\varepsilon$) reads

$$G_{\text{red}} = -v^{-1} - 3t\ln[(v^*)^{1/3}(v^{1/3} - 1)] + pv + tc^{-1}(\phi_1r_1^{-1}\ln\phi_1 + \phi_2r_2^{-1}\ln\phi_2) \tag{13.89}$$

where, obviously, the reduced variables and their combining rules must be used. From this expression all other properties can be calculated in the usual way.

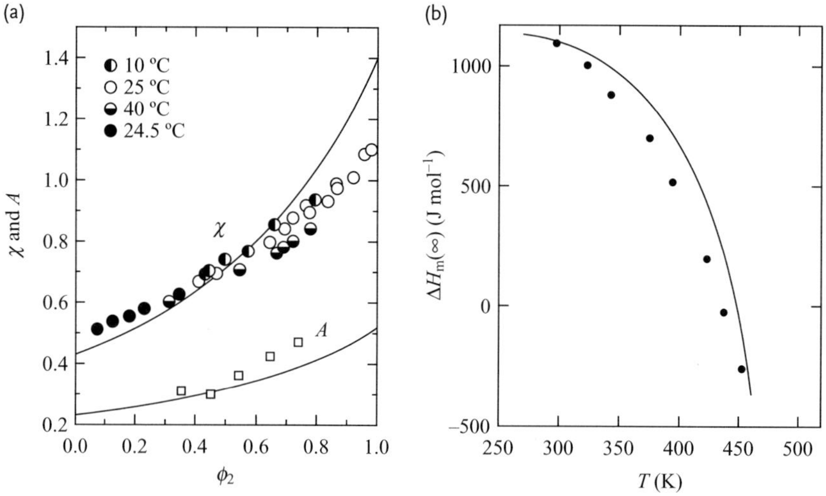

Figure 13.8 (a) Apparent χ and enthalpic A parameters (Eq. 13.55) versus segment volume fraction ϕ_2 for polyisobutylene in benzene as calculated from the FOVE EoS theory. The entropic B parameter is the difference between χ and A. The symbols indicate the experimental data points, while the curves are calculated from the FOVE theory; (b) A comparison of experimental (solid circles) and theoretical (solid line) heats of mixing for dilute solutions of polyisobutylene in benzene. The theoretical curve was calculated from the LF EoS theory, using temperature-independent pure-component parameters.

We provide only one example for comparison with experiment. Figure 13.8a shows the apparent χ parameter and its enthalpic part A [Eq. (13.55)] for a solution of polyisobutylene in benzene. Although evidently not perfect the agreement is, nevertheless, quite good.

13.5.3
The LF Theory

Sanchez and Lacombe [16] proposed another means of discussing polymer liquids and solutions. In essence, this theory–labeled as the Lattice Fluid Theory (LFT)–is a straightforward extension of the hole theory of molecular solutions to polymer molecules. A fixed cell size of volume v^* is chosen, and the overall volume change occurs via the introduction of extra holes. For a polymer solution with r segments for the N molecules, the volume is $V = N_0 + rN$ if N_0 holes are present. If $\phi = N/(N_0 + rN)$ is the fraction of cells occupied, then we obtain for the energy (similar to Section 8.3), $U = ½rN\phi z\varepsilon$, where ε refers to the segment–segment interaction. Hence, Eq. (13.54) becomes

$$\Delta_{\text{mix}}F = kT(N_0 + rN)[(\phi^2 z\varepsilon / 2kT) + (1-\phi)\ln(1-\phi) + (\phi / r)\ln\phi] \tag{13.90}$$

Again, following the usual route of calculating $P = -\partial F/\partial V = -\partial F/\partial(V/v^*) = -\partial F/\partial N_0$, the EoS is obtained which in this case reads

$$Pv^* = -kT[\ln(1-\phi) + (1-1/r)\phi] + \frac{1}{2}\phi^2 z\varepsilon \tag{13.91}$$

or in reduced form

$$p/\rho t = -\rho^{-1}[\ln(1-\rho) + (1-1/r)\rho] - \rho/t \tag{13.92}$$

with reduced variables $v = v^*/(V/rN)$, $\rho = 1/v = \phi rNv^*/V$ and $t = T/T^*$, where $T^* = \varepsilon^*/k$, $\varepsilon^* = -\frac{1}{2}z\varepsilon$ and $p = P/P^*$, with $P^* = \varepsilon^*/v^* = kT^*/v^*$. Here, we have also three parameters: two reduced parameters, v^* and ε^* (or alternatively v^* and T^*) and the DP, r. Again, these parameters are obtained by fitting to data for pure liquids. This model also gives a reasonable description for pure polymer liquids, taking reasonable ranges of temperature and pressure. The reduced Gibbs energy $g = G/rN\varepsilon^*$ with $G = F + PV$ reads

$$g = -\rho + tv[(1-\rho)\ln(1-\rho) + (\rho/r)\ln\rho] + pv \tag{13.93}$$

For solutions, Sanchez and Lacombe estimated the combining rules for v^* and r as follows. The first assumption is that the close-packed volume is additive. Denoting the DP for pure components by r_j°, and for the mixture by r_j, the volume of the mixture becomes $V^* = r_1^\circ N_1 v_1^* + r_2^\circ N_2 v_2^* = (r_1 N_1 + r_2 N_2)v^*$, which implies $V_j^* = r_j^\circ v_j^* = r_j v^*$. Accordingly, $r_1/r_2 = V_1^*/V_2^*$, as was also obtained for the FOVE theory. The second assumption is that the number of pair interactions in the mixture is the same as the sum of those interactions in the pure liquids. Hence, $\frac{1}{2}(r_1^\circ N_1 + r_2^\circ N_2)z = \frac{1}{2}(r_1 N_1 + r_2 N_2)z \equiv \frac{1}{2}rNz$, where $N = N_1 + N_2$. These two assumptions lead to

$$r = x_1 r_1^\circ + x_2 r_2^\circ \quad \text{and} \quad v^* = (x_1 r_1^\circ v_1^* + x_2 r_2^\circ v_2^*)/r \tag{13.94}$$

where, as usual, x_j denotes the mole fraction. This further leads to

$$\phi_1 = r_1 N_1/rN = x_1 r_1/r \quad \text{and} \quad \phi_2 = 1 - \phi_1 \tag{13.95}$$

The combining rule for P^* follows the FOVE approach, but s_1 and s_2 are assumed to be equal and thus

$$P^* = \phi_1 P_1^* + \phi_2 P_2^* - \phi_1\phi_2 \Delta P_{12}^* \tag{13.96}$$

where ΔP_{12}^* now can be taken as a parameter independent of composition since $s_1 = s_2$. From these results it follows that $\varepsilon^* = kT^* = P^*V^*$. Adding the entropy term, one finally obtains for the reduced Gibbs energy $g = G/Nr\varepsilon^*$

$$g = -\rho + tv[(1-\rho)\ln(1-\rho) + \rho r^{-1}\ln\rho] + pv + t[\phi_1 r^{-1}\ln\phi_1 + \phi_2 r_2^{-1}\ln\phi_2] \tag{13.97}$$

using the same reduced variables as for Eq. (13.89). From this expression all other properties can be calculated in the usual way.

Also here we provide only one comparison with experiment. Figure 13.8b shows the heats of mixing for dilute solutions of polyisobutylene in benzene with a quite good agreement with experiment.

13.5.4
The SS Theory

Another approach is due to Simha and his collaborators. In particular, Simha and Somcynski [17] (SS) described a pure polymer liquid, which was extended later by Jain and Simha [18] to solutions. This, in polymer solution jargon, is a hole theory. A recent perspective has been provided by Moulinié and Utracki [19].

Employing, as before, for each of the N molecules the equivalent chain with r segments and denoting the fraction occupied cells with ϕ, we have $\phi = rN/(rN + N_0)$. The configurational partition function Q is given by Eq. (13.73), with the additional combinatorial factor $g(X)$ for solutions. In the original SS theory for $g(X)$ the FH-form was chosen which, keeping only the pertinent part, reads $g(N,\phi) = \phi^{-N}(1-\phi)^{-rN(1-\phi)/\phi}$ for pure compounds. Instead of using $\frac{1}{2}zN\phi(0)$, one uses the expression $\frac{1}{2}q_c zN\phi\varepsilon_c$, where the number of nearest-neighbor sites per chain $q_c z = r(z-2)+2$ replaces the coordination number z, and the introduction of the volume fraction ϕ corrects for the presence of holes. The contact energy for a pair of segments ε_c replaces $\phi(0)$, as given by Eq. (13.74) while the volume $v_a = \phi V/rN = \phi v$ again refers to the average volume per segment. Similarly, as in the Significant Liquid Structures (SLS) theory (see Section 8.4), the free volume is estimated as an average of a solid-like contribution $v_{f,s}$ and a gas-like contribution $v_{f,g}$, so that

$$v_f = \left[\phi v_{f,s} + (1-\phi)v_{f,g}\right]^3 = \left\{\phi\left[v_a^{1/3} - (2^{-1/2}v^*)^{1/3}\right] + (1-\phi)v_a^{1/3}\right\}^3 \tag{13.98}$$

Finally for the number of external DoF $3c$ is used so that the partition function $Q(N,\phi)$, now a function of N and ϕ, reads

$$Q(N,\phi) = g(N,\phi)\exp\left[-\frac{1}{2}\beta qzN\phi\varepsilon_c)\right](v_f)^{crN} \tag{13.99}$$

The various properties follow from $F(N,\phi) = -kT\ln Q(N,\phi)$, where the optimum value for ϕ is obtained from $\partial F/\partial\phi = 0$, leading to

$$(r/3c)\{[(1-r)/r] + \phi^{-1}\ln(1-\phi)\} = L(\phi,v) + (\phi/6t)(\phi v)^{-2}[2B - 3A(\phi v)^{-2}] \tag{13.100}$$

with $L(\phi,v) = [2^{-1/6}\phi(\phi v)^{-1/3} - 1/3]/[1 - 2^{-1/6}\phi(\phi v)^{-1/3}]$. As usual, the EoS is obtained from $P = -\partial F/\partial V$. Solving the coupled equations, Eq. (13.100) and the pressure equation, leads to the SS description. Taking $r = c = q = 1$, one obtains the expressions for simple, spherical molecules. The theory was tested on Ar, for which it appeared that the volume of the cell at zero pressure is remarkably constant at about $0.96v^*$ to $0.98v^*$, while as a function of volume and temperature the volume varied between 1.02 for $1/v = 0.05$ and $t = 1.20$, and 0.84 for $1/v = 0.70$ and $t = 3.0$. The (reduced) critical constants were predicted to be $v_{cri} = 3.9$ (3.16), $t_{cri} = 1.27$ (1.26) and $p_{cri} = 0.115$ (0.116), showing fair agreement with the experimental data (shown in brackets). For chain-like molecules the cell volume depends weakly on the ratio $r/3c$, so that often a universal value of $r/3c = 1$ is chosen. Figure 13.9 shows the comparison with experiment for the specific volume for polystyrene–cyclohexane mixtures, showing quite reasonable agreement.

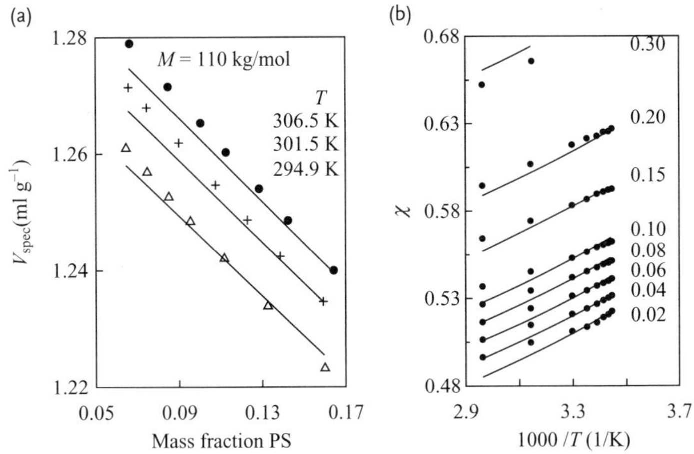

Figure 13.9 (a) Comparison of the specific volume V_{spec} of polystyrene–cyclohexane mixtures as a function of composition at different temperatures. The points show the experimental data, while the lines represent the result of the SS theory; (b) Comparing the apparent χ parameter for polystyrene ($M_w = 520\,\text{kg mol}^{-1}$) in cyclohexane as a function of temperature. The points show the experimental data, while the lines represent the result of the HH theory.

Several extensions for the SS theory have been proposed [20], the first of which is the Holey Huggins (HH) theory.[7] Whereas, in the original SS theory the external contact parameter q was evaluated via $q_c z = r(z - 2) + 2$, in the HH theory the refinement is made to take the occupied fraction ϕ into account via $q = Nq_c z/(Nq_c z + zN_0) = (1 - \alpha)\phi/(1 - \alpha\phi)$, where $\alpha = 2(1 - r^{-1})/z$. This leads essentially to the Huggins [4] expression for the entropy. Obviously, for $2/z \to \infty$, $\alpha \to 0$ and the HH theory reduces to the SS theory.

This modification leads to

$$(\ln g)/Nr = -r^{-1}\ln\phi - \phi^{-1}(1-\phi)\ln(1-\phi) + \frac{1}{2}z(1-\alpha\phi)\phi^{-1}\ln(1-\alpha\phi), \qquad (13.101)$$

$$E_0 = \frac{1}{2}q_c q z N\varepsilon_c \quad \text{and} \quad v_f = \left[q v_{f,s} + (1-q)v_{f,g}\right]^3$$

The Huggins expression for the combinatorial function g is in considerable better agreement with simulation results then the FH-expression.

However, the random nature of the entropy term remains. Using the quasi-chemical approximation, an improvement labeled as non-random mixing (NRM) HH theory results. To that purpose, a parameter X is introduced which takes in to

7) The name Holey Huggins was suggested by Dr Walter Stockmayer, in his capacity as editor of the journal *Macromolecules*, to indicate the addition of holes to the original Huggins theory (personal information from Prof. Erik Nies, October 2012).

account the fraction of segment–hole contacts. Denoting the total number of contacts by $Q = \frac{1}{2}z(Nq_c + N_0) = \frac{1}{2}zNq_c/q$, the product QX is the number of segment–hole contacts, while the number of segment–segment contacts is $Q(q - X)$ and the number of hole–hole-contacts is $Q(1 - q - X)$. The combinatorial function g is now described by Eq. (C.24), while E_0 and v_f become

$$E_0 = Q(q - X)\varepsilon_c \quad \text{and} \quad v_f = \left[(1 - Xq^{-1})v_{f,s} + Xq^{-1}v_{f,g}\right]^3 \tag{13.102}$$

Here, X/q represents the fraction of segment–hole contacts with respect to the total number of external contacts, and $1 - X/q$ the remaining fraction of segment–segment contacts. As usual, the optimum parameters are determined by minimization with respect to ϕ and X, that is, from $\partial F/\partial\phi = 0$ and $\partial F/\partial X = 0$. Figure 13.9 shows a comparison of the specific volume of polystyrene–cyclohexane mixtures as function of composition at different temperatures, as well as the apparent χ parameter for polystyrene in cyclohexane as a function of temperature. The agreement is quite fair.

Simplifications for the SS theory have also been proposed. These attempts have been focused on providing an explicit expression for ϕ, so that solving coupled equations becomes unnecessary. A relatively simple approach that, nevertheless, provides results as good as the original SS theory is to replace [21] the expression for ϕ by $1 - \phi = \exp(-c/2st)$, as inspired by the expression for the fraction of vacancies in a solid. This decouples the original SS equations, Eq. (13.100) and the pressure equation $P = -\partial F/\partial V$. Subsequently (for one reason or another), the commonly used FCC lattice with $z = 12$ was replaced by the BCC lattice with $z = 8$. The authors refer to this model as the simplified hole theory (SHT). When comparing 67 polymers and 61 solvents, the conclusion was that SHT performed for the polymers as well as the SS theory – that is, there was a grand average deviation D for the density over all compounds and temperature ranges available of $D = 0.066$ and $D = 0.065$, respectively. For solvents, the SHT was only slightly worse, with $D = 0.097$ for the original SS theory and $D = 0.125$ for SHT. Simha *et al.* [22] also proposed a simplification by using a fitted expression for the fraction of holes, so that a decoupled set of equations also resulted.

Several comparisons of EoS theories have been made [20, 23]. It appears that, in general, the SS theory is superior but is closely followed by the much simpler simple cell model. Wang *et al.* showed that D, as defined above, for the simple cell model was 0.084 for polymers and 0.154 for solvents. The unexpected success of the simple cell theory was improved even more by a slight modification [24], whereby multiplication of the hard-core "length" $V^{*1/3}$ by a factor f was introduced, hopefully to identify a hard-core volume factor describing the local geometry of close-packed flexible segments in a better way. This yielded an average value of $f = 1.07$, and led to a better description of the EoS than with the SS theory for the seven polymers studied. In conclusion, whilst the SS theory or its simplification has done rather well, it is followed surprisingly closely by the SC model.

13.6 The SAFT Approach*

After having investigated hard-sphere fluids (see Chapter 7), Wertheim focused his attention on "sticky" hard spheres in order to be able to discuss the association of molecules – that is, dimerization and polymerization [25]. Subsequently, Wertheim developed a Thermodynamic Perturbation Theory (TPT) that uses a (hard-sphere) repulsive core and one or more directional short-range attractions. In this theory, the Helmholtz energy is calculated from a graphical summation of interactions between different species (the derivation is complex and is omitted here). Wertheim's research led to the development of the Statistical Associating Fluid Theory (SAFT), as developed by Chapman and Gubbins [26] on the one hand and Huang and Radosz [27] on the other hand. Although only minor differences exist between these two developments, the Huang–Radosz development has gained more acceptance, and the present discussions are mainly limited to this approach. Although conventional EoS theories are capable of describing simple and normal liquids rather satisfactorily, their use for complex liquids often results in a description with unphysical parameter values and/or mixing rules. The reason for this is that electrostatic associative interactions (as occur in hydrogen bonding, highly polar liquids and electrolytes) are not taken explicitly into account. The basic idea of SAFT is to take association explicitly into account, where association refers to both covalent bonding (as for polymeric chains) as well as noncovalent interactions (as for hydrogen bonding). Covalent bonding is considered as a limiting case of noncovalent bonding, so that for both type of interaction Wertheim's results can be used.

In a perturbation approach, a reference fluid is chosen to which desirable properties are added. In SAFT it is assumed that the excess Helmholtz energy F^{E} can be composed from three major contributions to a (pair-wise additive) potential. These are: the segment contribution F^{seg}, representing the interaction between the individual segments; the chain contribution F^{chain}, due to the fact that the segments form a chain; and the association contribution F^{ass}, due to the possibility that some segments associate with other segments. SAFT is thus more of a framework for an EoS than a specific theory and has, therefore, led to many variants, some of which are indicated here. For the many other variants, we refer to the (review) literature [28].

Consequently, the excess (or residual) molar Helmholtz energy is given by

$$F^{\mathrm{E}}(T,\rho) = F^{\mathrm{seg}}(T,\rho) + F^{\mathrm{chain}}(T,\rho) + F^{\mathrm{ass}}(T,\rho) + \ldots \tag{13.103}$$

where T and ρ represent the temperature and density of the system, and the excess is defined with respect to the Helmholtz energy of the ideal gas. The segment contribution contains a repulsion and dispersion contribution, $F^{\mathrm{seg}}(T,\rho) = F^{\mathrm{rep}}(T,\rho) + F^{\mathrm{dis}}(T,\rho) = \Sigma_\alpha x_\alpha m_\alpha F^{\mathrm{mon}}$, where F^{mon} represents the Helmholtz energy of a fluid in which the segments are not connected. The summation is over all species α with mole fraction x_α and number of segments m_α. Originally, the repulsion term was approximated by a hard-sphere (HS) contribution, expressed by the Carnahan–Starling expression, so that

$$\beta F^{\text{rep}} / N_{\text{A}} = \beta F^{\text{HS}} / N_{\text{A}} = m\eta(4-3\eta)/(1-\eta)^2 \tag{13.104}$$

where m is the number of spherical segments per molecule and η is the reduced density $\eta = 0.74048\ \rho m v^\circ$ with v° the close-packed hard-core volume of the fluid. The latter is approximated by

$$v^\circ = v^{\circ\circ}[1 - C\exp(-3\beta u^\circ)]^3 \tag{13.105}$$

in which u° is a dispersion energy parameter per segment and $C = 0.12$ is a constant (except for hydrogen for which it is $C = 0.241$). The parameters m, $v^{\circ\circ}$ and u° are three parameters which are normally fitted to pure-component vapor pressure and liquid density data. In the original formulation the dispersion term F^{disp} is based on molecular simulation data for the square-well (SW) fluid, and reads

$$\beta F^{\text{disp}} / N_{\text{A}} = m \sum\nolimits_{i=1}^{4} \sum\nolimits_{j=1}^{9} D_{ij}(\beta u)^i (\eta / 0.74048)^j \tag{13.106}$$

with $u/k = (u^\circ/k)\ (1 + \beta e)$, where $e/k = 10\,\text{K}$ for all molecules except for a few small molecules [27]. Alternatively, one uses other potentials such as the Lennard-Jones (LJ) potential, for which a closed-form EoS expression [29] is available (soft-SAFT), if necessary adding dipole–dipole interactions, possibly approximated by a Padé approximant for the perturbation expression (SAFT-LJ). As long as the EoS for the reference fluid used is known, it can be used directly.

The chain term F^{chain} is based on the Wertheim TPT expression to first order for association in the limit of infinitely strong bonding between infinitely small association sites. In this way, polymerization is accounted for. By using one association site per segment, one obtains a description of the thermodynamics of hard-sphere dumbbell fluids which is almost as accurate as the exact solution [25d]. Using two diametrically positioned association sites, one obtains a linear chain for which holds that

$$\beta F^{\text{chain}} / N_{\text{A}} = \sum\nolimits_{\alpha} x_\alpha (1 - m_\alpha) \ln y_\alpha^{\text{seg}}(l) \tag{13.107}$$

where y^{seg} is the cavity function evaluated at (bond) length l. For tangent LJ-spheres or hard spheres, the cavity function y^{seg} reduces to $y^{\text{seg}}(\sigma) = g^{\text{seg}}(\sigma)$, where g^{seg} denotes the pair-correlation function and σ either the LJ hard-core or HS diameter. In the latter case, for a single component, Eq. (13.107) reduces to

$$\beta F^{\text{chain}} / N_{\text{A}} = (1-m)\ln[(1-0.5\eta)/(1-\eta)^3] \tag{13.108}$$

The first-order theory provides a good approximation for linear chains. Bond angles are not taken in to account, although this could be done using second-order TPT though at the cost of considerable complexity. In any case, for accurate results a high-quality pair-correlation function must be used.

If we add regular association sites to the molecule, characterized as before by a non-central potential located at the edge of the molecule, one can account for association. Each of these sites can bond to only one other site located at another molecule with a single bond.[8] The detailed calculation is complex, and Justification 13.2 provides a heuristic derivation from which the association energy is given by

$$\beta F^{\text{ass}} / N_{\text{A}} = \sum_{\alpha} x_{\alpha} \sum_{i=1}^{i=M_{\alpha}} \left[\ln X_{\alpha,j} - \frac{1}{2} X_{\alpha,j} \right] + \frac{1}{2} M_{\alpha} \quad (13.109)$$

with M_α the number of association sites per species α and $X_{\alpha,j}$ is the mole fraction of species α not bonded at site j. The latter quantity is calculated from

$$X_{\alpha,i} = \left(1 + \rho \sum_{\beta} \sum_{j} x_{\beta,j} \Delta_{ij}\right)^{-1} \quad (13.110)$$

where Δ_{ij} is the association strength evaluated from

$$\Delta_{ij} = \int g^{\text{seg}}(\boldsymbol{r}_{ij}) f_{ij}(\boldsymbol{r}_{ij}) \text{d}\boldsymbol{r}_{ij} = \sqrt{2} v^{\circ\circ} \frac{(1-0.5\eta)}{(1-\eta)^3} [\exp(-\beta\varepsilon_{ij}) - 1] \kappa_{ij} \quad (13.111)$$

where in the first part on the right-hand side, as before, g^{seg} is the segment pair correlation function and $f_{ij} = \exp[-\beta\phi(\boldsymbol{r}_{ij})] - 1$, which is the Mayer function with $\phi(\boldsymbol{r}_{ij})$ the association potential. In the second part on the right-hand side, the SW potential is introduced with two pure-component parameters, namely the energy of association ε_{ij} and volume of association κ_{ij}. Although these parameters can be determined from spectroscopic data of the pure components, in practice they are fitted to experimental liquid density and vapor pressure data of the mixture, as are the other parameters.

For a dimerizing fluid with only one associative site per molecule, Eq. (13.109) reduces to

$$\beta F^{\text{ass}} / N_{\text{A}} = \ln X - \frac{1}{2} X + \frac{1}{2} \quad \text{with} \quad X = (1 + \rho N_{\text{A}} X \Delta_{ij})^{-1} \quad (13.112)$$

from which we obtain for the fraction monomers

$$X = 2 \Big/ \left(1 + \sqrt{1 + 4 N_{\text{A}} \Delta_{ij}}\right) \quad (13.113)$$

For two association sites i and j per molecule, for example in alcohols, the oxygen lone pair and the hydroxyl proton for creating a hydrogen bond, we have

$$\beta F^{\text{ass}} / N_{\text{A}} = \ln X_i - \frac{1}{2} X_i + \ln X_j - \frac{1}{2} X_j + 1 \quad (13.114)$$

with still for the fraction monomers $X = X_i = X_j$ the analytical expression as given by Eq. (13.113). For more complex models or other parameters, implicit expressions often result.

For mixtures, for the repulsion an expression for hard-spheres mixtures [30] is used. The chain and association terms can still be evaluated by the exact TPT expressions, while the dispersion terms are evaluated using the van der Waals

8) These restrictions can be remedied, for which we refer to the literature as given in Ref. [28].

one-fluid approximation (vdW1, see Section 11.6) reading, using x_i as the mole fraction for component i,

$$m = \sum_i x_i m_i \quad u = \sum_i \sum_j x_i x_j m_i m_j u_{ij} v_{ij}^\circ \Big/ \sum_i \sum_j x_i x_j m_i m_j v_{ij}^\circ,$$

$$v_{ij}^\circ = \left\{ \frac{1}{2}[(v_{ii}^\circ)^{1/3} + (v_{jj}^\circ)^{1/3}] \right\}^3 \quad \text{and} \quad u_{ij} = (u_{ii} u_{jj})^{1/2}(1 - k_{ij}) \qquad (13.115)$$

where k_{ij} is a binary adjustable parameter for the interaction energy. Sometimes, a similar parameter is used for the hard-core volume or, equivalently, the diameter.

Summarizing so far, in SAFT a minimum of two parameters is used for each segment, namely a characteristic energy and characteristic size, and this suffices for simple conformal fluids. For non-associating chain molecules, however, a third additional parameter m characterizing the DP is required. For association, another two parameters are required, characterizing the association energy and the volume available for association for each of the interaction sites defined. All of these parameters are usually obtained from experimental data fitting, but they can also be obtained from *ab initio* calculations or spectroscopic information.

SAFT has been successfully applied to many systems, including low-pressure polymer phase equilibria. As an example, Figure 13.10 shows the solubility of CH_4 and N_2 in polyethylene, and the weight fraction Henry's law constant for various small molecules in low-density polyethylene at 1 atm, both over a wide temperature range. SAFT using a single temperature-independent binary interaction parameter provides an accurate correlation of the experimental data.

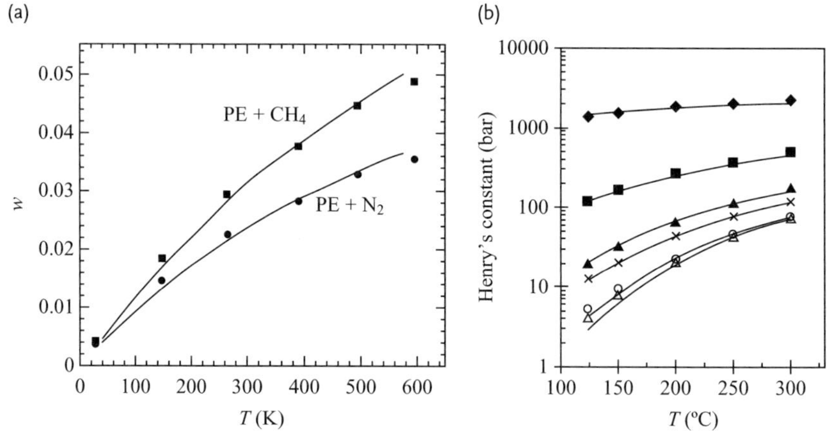

Figure 13.10 (a) Solubilities (w weight fraction of gas) of methane (upper curve) and nitrogen (lower curve) in polyethylene, as calculated from SAFT. Data from Ref. [39].; (b) Weight fraction Henry's constant of ethylene (◆), *n*-butane (■), *n*-hexane (◆), *n*-octane (△), benzene (×), and toluene (○) in low-density polyethylene at 1 atm. Experimental data (points) and SAFT correlation (lines) using a single temperature-independent binary interaction parameter. Data from Ref. [40].

Several adjustments have been made to the original SAFT model. For example, in the hard-sphere SAFT (SAFT-HS) the dispersion term is replaced by a simple van der Waals term reading $\beta F^{\text{disp}}/N_{\text{A}} = m\beta u^{\circ}\eta$, meanwhile using the vdW1 rules for mixtures. This version was used (among others) to predict generalized phase diagrams for model water–*n*-alkane mixtures. An empirical modification is provided by replacing the hard-sphere, chain and dispersion terms by a cubic EoS (see Chapter 4) and retaining the original TPT association term (Cubic-Plus-Association, CPA). In spite of this simplification, this approach provides an accurate description of vapor–liquid equilibria (VLE) and liquid–liquid equilibria (LLE) [31]. Obviously, polarity is not explicitly included and, in spite of some semi-empirical options, a "polar CPA" is not available. The SAFT-LJ model was originally applied to water, and was further applied to pure alkanes and alcohols and to binary VLE mixtures of water, alkanes, and alcohols. The soft-SAFT model was applied successfully to pure *n*-alkanes, 1-alkenes, and 1-alcohols and to binary and ternary *n*-alkane mixtures, including the critical region. Many other modifications are discussed in the literature [24], for example, taking in account the effect of intramolecular interactions [32] and interfaces in liquid–liquid and liquid–solid systems, and its use for ionic liquids, electrolytes, liquid crystals biomaterials, and oil reservoir fluids. For these modifications, we refer to the literature [28].

While many forms of SAFT use the hard-sphere as a reference fluid, add the dispersion term and then form chains (for polymeric solutions), in perturbed chain-SAFT (PC-SAFT) the chains are formed first and a dispersive interaction for chains is added thereafter. This implies that $F^{\text{seg}}(T,\rho) = F^{\text{HS}}(T,\rho) + F^{\text{dis}}(T,\rho)$, later adding $F^{\text{chain}}(T,\rho)$, is replaced by $F^{\text{seg}}(T,\rho) = F^{\text{HS}}(T,\rho) + F^{\text{chain}}(T,\rho)$ and adding $F^{\text{dis}}(T,\rho)$ later. Obviously, this requires a (hard-)sphere chain reference fluid. The original [33] and a simplified [34] form yield very comparable results (Figure 13.11), and in both cases only one extra energy interaction parameter k_{ij} is used.

SAFT accounts for chain connectivity rigorously, and is therefore accurate for polymers. Pure polymer EoS parameters are usually calculated by fitting melt *PVT* data over a wide temperature and pressure range (typically 0–200 MPa), similar to the CM, FOVE, LFT, and other models. In the case of SAFT, polymer parameters regressed from *PVT* data do not provide a good description of mixture phase equilibria, even with the use of large binary interaction parameters. This drawback is attributed to the use of parameter values obtained from *PVT* data, with parameters probably not accounting fully for the intramolecular structure, which becomes important as the molecular size increases. Similarly, intramolecular interactions become more important when the density decreases and SAFT leads to unrealistic, predicted low-density limits. On the other hand, SAFT pure-component parameters for small molecules in a homologous series vary smoothly with molecular mass and, as a result, they can be extrapolated to high-molecular-mass values and used for mixtures containing heavy oil fractions or polymers.

In conclusion, SAFT-based models have received an impressive acceptance in both academia and industry as the leading models for phase equilibrium calculations of polymer mixtures and, to a lesser extent, of aqueous systems. For polymers one limitation (especially in the case of polar polymers) is the method currently

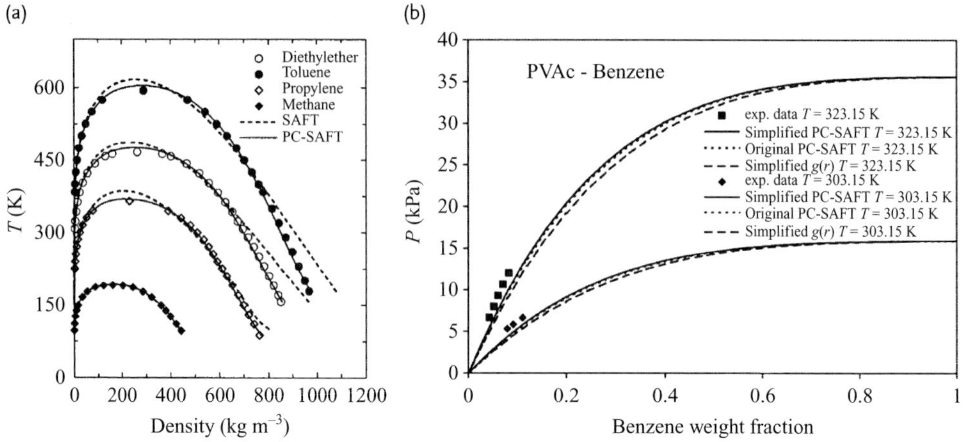

Figure 13.11 (a) Saturated liquid and vapor densities for methane, propylene, diethyl ether and toluene comparing PC-SAFT (—) and SAFT (- - -) results to experimental data; (b) Comparing the vapor pressure of polyvinyl acetate (PVAc) in benzene at 303 and 323 K using PC-SAFT (···) and simplified PC-SAFT (—).

used to estimate polymer model parameters, that is, by fitting to *PVT* data. It is possible that a simultaneous fitting to other thermodynamic data, as well including those from solutions, might be an option. Nonetheless, the use of five parameters may end in non-unique values – that is, in several data sets correctly describing the pure liquids. The use of other reference fluids, such as that employed for chains in PC-SAFT, is another option for improvement.

Justification 13.2: Heuristic derivation*

Wertheim's derivation is complex, and we provide here an heuristic derivation [35] of the SAFT association equations. The pair potential ϕ is split into a reference part ϕ^0 and a perturbation part ϕ^1, summed over components α, β, . . . and interaction sites $i, j, \ldots$,

$$\phi_{\alpha\beta}(12) = \phi^0_{\alpha\beta}(12) + \phi^1_{\alpha\beta}(12) = \phi^0_{\alpha\beta}(12) + \lambda \sum\nolimits_{i,j} \phi_{\alpha i,\beta j}(r_{ij}) \tag{13.116}$$

where (12) indicates the set $(\boldsymbol{r}_{12},\boldsymbol{\omega}_1,\boldsymbol{\omega}_2)$ with $\boldsymbol{r}_{ij}$ the distance between sites i and j and $\boldsymbol{\omega}_j$ the orientation of site j. As usual, the Helmholtz energy F is given by $F = F^{\mathrm{IG}} + F^{\mathrm{E}}$. Here, $F^{\mathrm{IG}} = -kT\ln Z$ is the ideal gas contribution originating from the partition function Z and given by $\beta F^{\mathrm{IG}}/V = \Sigma_\alpha \rho_\alpha \ln(\rho \Lambda_\alpha) - \rho_\alpha$, where ρ_α is the number density of component α and Λ_α is the kinetic contribution from the translation, rotation and vibration parts of the molecular partition function. The excess part is labeled F^{E}. For $\lambda = 0$ we have the reference potential, while for $\lambda = 1$ the full potential is obtained. In Chapter 7, we used first-order perturbation theory, and generalizing to several components, we have that

$$F_\lambda^{\mathrm{E}} = \frac{1}{2}\sum_\alpha \sum_\beta \rho_\alpha \rho_\beta \langle g_{\alpha\beta}^\lambda(12)[\partial\phi_{\alpha\beta}^\lambda(12)/\partial\lambda]_{\lambda=0}\rangle_{\omega_1\omega_2} \tag{13.117}$$

Using the Mayer expansion with $f_{\alpha\beta} = \exp(-\beta\phi_{\alpha\beta}) - 1$, we have $\phi_{\alpha\beta} = \phi_{\alpha\beta}^0 - kT \ln[1+f_{\alpha\beta}^\lambda]$, and thus $\partial\phi_{\alpha\beta}/\partial\lambda = -kTf_{\alpha\beta}$. The general first-order expression thus reads

$$\beta[F^{\mathrm{E}} - F^{\mathrm{E},0}]/V = \frac{1}{2}\sum_\alpha \sum_\beta \rho_\alpha \rho_\beta \int \langle g_{\alpha\beta}^0(12) f_{\alpha\beta}(12)\rangle_{\omega_1\omega_2}\,\mathrm{d}\mathbf{r}_{12} \tag{13.118}$$

In the sequel, the integral will be denoted by $\Delta_{\alpha\beta}$. Usually, the reference potential is chosen in such a way that the first-order contribution vanishes, but here another choice will be made. For strongly associating fluids the first-order term is large, and the perturbation expansion is of limited value. Wertheim proposed using different entities (monomers, dimers, trimers, . . .) as distinct, and by limiting ourselves here to monomers and dimers we have $\rho_\alpha = \rho_{\mathrm{m}\alpha} + \rho_{\mathrm{d}\alpha}$, where $\rho_{\mathrm{m}\alpha}$ and $\rho_{\mathrm{d}\alpha}$ denote the density contribution from monomers and dimers, respectively. For a weak association $\rho_\alpha \cong \rho_{\mathrm{m}\alpha}$, but for a strong association we have $\rho_\alpha >> \rho_{\mathrm{m}\alpha}$. This suggests a renormalization of the splitting of F in $F = (F^{\mathrm{IG}})' + (F^{\mathrm{E}})'$ with

$$\beta(F^{\mathrm{IG}})'/V = \sum_\alpha [\rho_\alpha \ln(\rho_{\mathrm{m}\alpha}\Lambda_\alpha) - \rho_{\mathrm{m}\alpha}] \quad \text{and} \tag{13.119}$$

$$\beta[(F^{\mathrm{E}})' - (F^{\mathrm{E},0})']/V = -\frac{1}{2}\sum_\alpha \sum_\beta \rho_{\mathrm{m}\alpha}\rho_{\mathrm{m}\beta}\Delta_{12} \tag{13.120}$$

implying that we only include the interactions between monomers at density ρ_α and neglect interactions between monomers and bonded molecules. The interactions from bonded molecules are implicitly contained in $(F^{\mathrm{E}})'$. Now, the usual procedure is applied to obtain the optimum value for $\rho_{\mathrm{m}\alpha}$, that is, we put $\partial(F^{\mathrm{E}})'/\partial\rho_{\mathrm{m}\alpha} = 0$. Some calculation yields

$$\rho_\alpha = \rho_{\mathrm{m}\alpha} + \rho_{\mathrm{m}\alpha}\sum_\beta \rho_\beta \Delta_{\alpha\beta} \tag{13.121}$$

To simplify, we note that for system where bonding interactions are absent, we have $\rho_\alpha = \rho_{\mathrm{m}\alpha}$ and that

$$\beta(F^{\mathrm{IG}})'/V = \sum_\alpha [\rho_\alpha \ln(\rho_\alpha\Lambda_\alpha) - \rho_\alpha] \quad \text{and thus} \tag{13.122}$$

$$\beta[(F^{\mathrm{IG}})' - (F^{\mathrm{IG},0})']/V = \sum_\alpha \rho_\alpha(\ln X_\alpha - X_\alpha + 1) \tag{13.123}$$

where for the fraction nonbonded molecules $X_\alpha = \rho_{\mathrm{m}\alpha}/\rho_\alpha$ is used. From Eqs (13.120) and (13.123) one easily obtains

$$\beta(F - F^0)'/V = \sum_\alpha \rho_\alpha(\ln X_\alpha - X_\alpha + 1) - \frac{1}{2}\sum_\alpha \sum_\beta \rho_{\mathrm{m}\alpha}\rho_{\mathrm{m}\beta}\Delta_{\alpha\beta} \tag{13.124}$$

From Eq. (13.121) the last term on the right in Eq. (13.124) equals $-½\Sigma_\alpha\rho_\alpha(1 - X_\alpha)$, so that Eq. (13.124) finally becomes

$$\beta(F - F^0)'/V = \sum_\alpha \rho_\alpha\left(\ln X_\alpha - \frac{1}{2}X_\alpha + \frac{1}{2}\right) \tag{13.125}$$

which is the first part of Eq. (13.112). Dividing Eq. (13.121) by ρ_α, we obtain for the fraction nonbonded molecules

$$X_\alpha = \left(1 + \sum\nolimits_\beta \rho_\beta X_\beta \Delta_{\alpha\beta}\right)^{-1} \tag{13.126}$$

which corresponds to the second part of Eq. (13.112).

References

1 See Boyd and Phillips (1993).
2 See Rubinstein and Colby (2003).
3 (a) Flory, P.J. (1941) *J. Chem. Phys.*, **9**, 660; (b) Flory, P.J. (1942) *J. Chem. Phys.*, **10**, 51; (c) Flory, P.J. (1953) *Principles of Polymer Chemistry*. Cornell University Press, Ithaca, NY.
4 (a) Huggins, M.L. (1941) *J. Chem. Phys.*, **9**, 440; (b) Huggins, M.L. (1942) *Ann. N. Y. Acad. Sci.*, **43**, 1.
5 See Flory (1953).
6 See Gedde (1995).
7 Balsara, N.P. (1996) Thermodynamics of polymer blends, in *Physical Properties of Polymers Handbook* (ed. J.E. Mark), AIP Press.
8 (a) Dunkel, M. (1928) *Z. Physik. Chem.*, **A138**, 42; (b) Small, P.A. (1953) *J. Appl. Chem.* **3**, 71.
9 (a) Hoy, K.L. (1970) *J. Paint Technol.*, **42**, 76; (b) Hoy, K.L. (1989) *J. Coated Fabrics*, **19**, 53.
10 See van Krevelen (1990).
11 See Hansen (2007).
12 Patterson, D. (1968) *J. Polym. Sci. C*, **16**, 3379.
13 Bottino, A., Capannelli, G., Munari, S., and Tarturo, A. (1988) *J. Polym. Sci. B Polym. Phys.*, **26**, 785.
14 Prigogine, I. (1957) *The Molecular Theory of Solutions*, North-Holland, Amsterdam.
15 (a) Flory, P.J., Orwoll, R.A., and Vrij, A. (1964) *J. Am. Chem. Soc.*, **86**, 3507 and 3515; (b) Eichinger, B.E. and Flory, P.J. (1968) *Trans. Faraday Soc.* **64**, 2035, 2053, 2061 and 2066.
16 (a) Sanchez, I.C. and Lacombe, R.H. (1976) *J. Phys. Chem.*, **80**, 2352 and 2588; (b) Sanchez, I.C. and Lacombe, R.H. (1978) *Macromolecules*, **11**, 1145.
17 Simha, R. and Somcynski, T. (1969) *Macromolecules*, **2**, 342.
18 Jain, R.K. and Simha, R. (1980) *Macromolecules*, **13**, 1501.
19 Moulinié, P., and Utracki, L.E. (2010) in *Polymer Physics* (eds L.A. Utracki and A.M. Jamieson), Wiley, Hoboken, NJ, Chapter 6, pp. 277–322.
20 Xie, H., Nies, E., Stroeks, A., and Simha, R. (1992) *Polym. Eng. Sci.*, **32**, 1654.
21 Wang, W., Liu, X., Zhong, C., Twu, C.H., and Coon, J.E. (1997) *Ind. Eng. Chem. Res.*, **36**, 2390.
22 Olabisi, O. and Simha, R. (1977) *J. Appl. Polym. Sci.*, **21**, 149.
23 (a) Dee, G.T. and Walsh, D.J. (1988) *Macromolecules*, **21**, 811; (b) Rodgers, P.A. (1993) *J. Appl. Polym. Sci.*, **48**, 1061.
24 Dee, G.T. and Walsh, D.J. (1988) *Macromolecules*, **21**, 815.
25 (a) Wertheim, M.S. (1984) *J. Stat. Phys.*, **35**, 19, 35; (b) Wertheim, M.S. (1984) *J. Stat. Phys.*, **42**, 459, 477; (c) Wertheim, M.S. (1986) *J. Chem. Phys.*, **85**, 2929; (d) Wertheim, M.S. (1987) *J. Chem. Phys.*, **87**, 7323.
26 (a) Chapman, W.G., Jackson, G., and Gubbins, K.E. (1988) *Mol. Phys.*, **65**, 1 and 1057; (b) Chapman, W.G., Gubbins, K.E., and Jackson, G. (1990) *Ind. Eng. Chem. Res.* **29**, 1709.
27 (a) Huang, S. and Radosz, M. (1990) *Ind. Eng. Chem. Res.*, **29**, 2284; (b) Huang, S. and Radosz, M. (1991) *Ind. Eng. Chem. Res.*, **30**, 1994.
28 (a) For reviews, see, Müller, E.A., and Gubbins, K.E. (2001) *Ind. Eng. Chem. Res.*, **40**, 2193; (b) Economou, I.G. (2002) *Ind. Eng. Chem. Res.* **41**, 953; (c) Tan, S.P., Adidharma, H., and Radosz, M. (2008) *Ind. Eng. Chem. Res.* **47**, 8063; (d) Kontogeorgis, G.M. and Economou,

I.G. (2010) *J. Supercrit. Fluids* **55**, 421. The latter reference also provides a review of conventional EoS expressions.
29 Tang, Y.P., and Lu, B.C.-Y. (1998) *Fluid Phase Eq.*, **146**, 73.
30 Mansoori, G.A., Carnahan, N.F., Starling, K.E., and Leland, T.W. (1971) *J. Chem. Phys.*, **54**, 1523.
31 (a) For a review, see, Kontogeorgis, G.M., Michelsen, M.L., Folas, G., Derawi, S., von Solms, N., and Stenby, E.H. (2006) *Ind. Eng. Chem. Res.*, **45**, 4855; (b) Kontogeorgis, G.M., Michelsen, M.L., Folas, G., Derawi, S., von Solms, N., and Stenby, E.H. (2006) *Ind. Eng. Chem. Res.*, **45**, 4869; (c) Kontogeorgis, G.M. and Folas, G. (2010) *Thermodynamics for Industrial Applications. From Classical and Advanced Mixing Rules to Association Theories*, John Wiley & Sons, Inc., New York.
32 Avlund, A.S., Kontogeorgis, G.M., and Chapman, W.G. (2011) *Mol. Phys.*, **109**, 1759.
33 Gross, J., and Sadowski, G. (2001) *Ind. Eng. Chem. Res.*, **40**, 1244.
34 von Solms, N., Kouskoumvekaki, I.A., Michelsen, M.L., and Kontogeorgis, G.M. (2006) *Fluid Phase Eq.*, **241**, 344.
35 Joslin, C.G., Gray, C.G., Chapman, W.G., and Gubbins, K. (1987) *Mol. Phys.*, **62**, 843.
36 Guy, A.G. (1976) *Essentials of Materials Science*, McGraw-Hill, New York.
37 Casassa, E.F. (1976) *J. Polym. Sci. Symp.*, **54**, 53.
38 Schultz, A.R. and Flory, P.J. (1952) *J. Am. Chem. Soc.*, **74**, 4760.
39 Wu, C.-S. and Chen, Y.P. (1994) *Fluid Phase Equilib.*, **100**, 103.
40 Maloney, D.P. and Prausnitz, J.M. (1976) *AIChE J.*, **22**, 74.

Further Reading

Boyd, R.H. and Phillips, P.J. (1993) *The Science of Polymer Molecules*, Cambridge University Press, Cambridge.
Flory, P.J. (1953) *Principles of Polymer Chemistry*, Cornell University Press, Ithaca.
Gedde, U.W. (1995) *Polymer Physics*, Chapman & Hall, London.
Hansen, C.M. (2007) *Hansen Solubility Parameters: A User's Handbook*, 2nd edn, CRC Press, Boca Raton, London.
van Krevelen, D.W. (1990) *Properties of Polymers*, 3rd edn, Elsevier, Amsterdam.
Rubinstein, M. and Colby, R.H. (2003) *Polymer Physics*, Oxford University Press, Oxford.
Rudin, A. (1999) *The Elements of Polymer Science and Engineering*, 2nd edn, Academic Press, London.
Strobl, G. (2007) *The Physics of Polymers*, 3rd edn, Springer, Berlin.
Utracki, L.A., and Jamieson, A.M. (eds) (2010) *Polymer Physics*, Wiley, Hoboken, NJ.

14
Some Special Topics: Reactions in Solutions

An important driving force for discussing liquids, apart from their intrinsic scientific interest, is their use as solvent for reactions. In this chapter reactions in solution are dealt with. First, we briefly review the basics of kinetics, and thereafter deal with transition state theory for gas-phase reactions. In liquid media, the transport of reactants is important, and thus we briefly review the effects of diffusion and viscosity. The transition from diffusion-controlled to chemically controlled reactions is indicated. Finally, the physical effect of solvents on reactions is discussed.

14.1
Kinetics Basics

For the description of reactions some terms and concepts are required which we briefly iterate here. A typical reaction may be written as

$$\nu_A A + \nu_B B \leftrightarrow \nu_X X + \nu_Y Y \tag{14.1}$$

where A and B represents the reactants (R), and X and Y represents the products (P). The *stoichiometric sum* is defined by $\Sigma_j \nu_j = \Sigma_P \nu_P - \Sigma_R \nu_R$. Thus, for Eq. (14.1) $\Sigma_j \nu_j$ becomes $(\nu_X + \nu_Y) - (\nu_A + \nu_B)$. Progress of the reaction as a function of time t is given by the *extent of reaction* ξ (t), defined by

$$\xi(t) \equiv [n_j(t) - n_j(0)]/\nu_j \tag{14.2}$$

where n_j and ν_j denote the number of moles and the reaction coefficient for component j, respectively. The *rate of reaction* is defined by

$$r \equiv \frac{1}{V}\frac{d\xi}{dt} = \frac{1}{\nu_j V}\frac{dn_j}{dt} \quad \text{or} \quad r = \frac{1}{\nu_j}\frac{dc_j}{dt} \quad \text{for the volume } V \text{ constant} \tag{14.3}$$

So, for Eq. (14.1) we have (using the notation [Z] for the molarity of Z)

$$r = -\frac{1}{\nu_A}\frac{d[A]}{dt} = -\frac{1}{\nu_B}\frac{d[B]}{dt} = \frac{1}{\nu_X}\frac{d[X]}{dt} = \frac{1}{\nu_Y}\frac{d[Y]}{dt} \tag{14.4}$$

Liquid-State Physical Chemistry, First Edition. Gijsbertus de With.

Often, the reaction rate r can be described by

$$r = k_{\text{for}}[\text{A}]^{\alpha}[\text{B}]^{\beta} \tag{14.5}$$

with the (forward) *rate constant* k_{for}. The *reaction order* n is given by $n = \alpha + \beta$, in which α and β are the *partial reaction order*. It must be stated that r is not necessarily described by a power law. For example, it may also be described by (with P and Q constants)

$$r = \frac{P[\text{S}]}{Q + [S]} \tag{14.6}$$

If $[\text{S}] \ll Q$ we obtain $r = P[\text{S}]/Q$ and the apparent order $n = 1$. On the other hand, if $Q \ll [\text{S}]$ we have $r = P$ and the apparent order $n = 0$. For an expression like Eq. (14.6) the reaction order is defined by $\alpha = \partial \ln r/\partial \ln[S]$. In many cases a reaction is discussed in terms of *elementary reactions* such as A + BC $\leftrightarrow$ AB + C. More complex reaction schemes, involving two or more elementary reactions, are accordingly denoted as *composite reactions*. Finally, with *molecularity* the number of molecules participating in the reaction is indicated. Clearly, for the reaction as given above the molecularity is two. Molecularity and reaction order generally have different values, but for elementary reactions (with certain exceptions[1]) they have the same value.

The total reaction rate r is the difference between the rate r_{for} for the forward reaction $\nu_A\text{A} + \nu_B\text{B} \rightarrow \nu_X\text{X} + \nu_Y\text{Y}$ and the rate r_{rev} for the reverse reaction $\nu_A\text{A} + \nu_B\text{B} \leftarrow \nu_X\text{X} + \nu_Y\text{Y}$; that is,

$$r = r_{\text{for}} - r_{\text{rev}} = k_{\text{for}}[\text{A}]^{\nu_A}[\text{B}]^{\nu_B} - k_{\text{rev}}[\text{X}]^{\nu_X}[\text{Y}]^{\nu_Y} \tag{14.7}$$

At equilibrium $r = 0$ or $k_{\text{rev}}/k_{\text{for}} = [\text{X}]^{\nu_X}[\text{Y}]^{\nu_Y}/[\text{A}]^{\nu_A}[\text{B}]^{\nu_B} \equiv K_c$ with K_c the *equilibrium constant* (in terms of molarities[2]). For gas-phase reactions often the equilibrium constant K_P in terms of pressure P is used. The partial pressure is $P_j = (N_j/V)kT = c_j kT$, where k is Boltzmann's constant and c_j is the molarity of component j. Hence, we have

$$(kT)^{\Sigma_j \nu_j} K_c = (kT)^{\Sigma_P \nu_P - \Sigma_R \nu_R} K_c \equiv K_P \tag{14.8}$$

The rate equations can be integrated, and this yields the concentration as a function of time explicitly. We have, defining $c_{j,0} \equiv c_j(t = 0)$ for component j,

$$\text{0th-order: } -\mathrm{d}c_A/\mathrm{d}t = k_{\text{rea}} \quad \rightarrow \quad k_{\text{rea}}t = c_{A,0} - c_A$$

$$\text{1st-order: } -\mathrm{d}c_A/\mathrm{d}t = k_{\text{rea}}c_A \quad \rightarrow \quad k_{\text{rea}}t = \ln(c_{A,0}/c_A)$$

$$\text{2nd-order: } -\mathrm{d}c_A/\mathrm{d}t = k_{\text{rea}}c_A^2 \quad \rightarrow \quad k_{\text{rea}}t = (1/c_A) - (1/c_{A,0})$$

$$\text{2nd-order: } -\mathrm{d}c_A/\mathrm{d}t = k_{\text{rea}}c_A c_B \quad \rightarrow \quad k_{\text{rea}}t = (c_{B,0} - c_{A,0})^{-1}\ln(c_{A,0}c_B/c_{B,0}c_A)$$

1) Participation of a solvent in a reaction is not easily discerned since its molarity changes very little. Also participation of a catalyst is cumbersome since after reaction its concentration is unchanged.

2) This argument is based on the existence of order, as defined by Eq. (14.5), but the result is generally valid.

Various methods exist to determine the order of the reaction from experimental data (see, e.g., Laidler [1], Connors [2], or Arnaut *et al.* [3]).

The rate of a reaction is influenced by several variables. Apart from the molarity [X] of component X, the temperature T is important. The change of equilibrium constants K_P or K_c with temperature is given by the *van't Hoff equation*

$$\frac{\partial \ln K_P}{\partial T} = \frac{\Delta H^\circ}{RT^2} \quad \text{or} \quad \frac{\partial \ln K_c}{\partial T} = \frac{\Delta U^\circ}{RT^2} \tag{14.9}$$

where ΔH° and ΔU° are the enthalpy and energy under standard conditions, respectively. Empirically, the temperature dependence of the rate constant k_{rea} for limited temperature range can be described by the *Arrhenius equation*

$$k_{\text{rea}} = A\exp(-E_{\text{act}}/RT) \tag{14.10}$$

where A is the *pre-exponential factor*, dependent on temperature T, and E_{act} is the *activation energy*. For a wider temperature range one often uses the modified Arrhenius equation

$$k_{\text{rea}} = A'T^m \exp(-E_{\text{act}}/RT) \tag{14.11}$$

where A' is a temperature-independent factor. The range of E_{act} for reactions in liquids is typically 40 to 120 kJ mol^{-1}, which implies an increase in rate by a factor of 2 to 3 for every 10 K. The value of the parameter m can be positive as well as negative, but typically $|m| < 2$.

Furthermore, the reaction medium is relevant – that is, the type of solvent and ionic strength. We discuss these effects in Section 14.3. Finally, light and catalysts are important, but for the effect of these we refer to the literature [3].

14.2 Transition State Theory

Transition state theory (TST), also known as "activated complex theory", is an important follow-up from statistical mechanics, used throughout in chemistry and physics, which has been developed mainly by Eyring and coworkers. In this section we will illustrate this theory for chemical reactions in the gas phase.

14.2.1 The Equilibrium Constant

The first result we need is the equilibrium condition. Associating (as in Chapter 5) the macroscopic internal energy U with the statistical mechanics expression for the average energy $\langle E \rangle$, the macroscopic number of molecules N with the average number of molecules $\langle N \rangle$ and recalling that the grand canonical partition function reads $kT \ln \Xi = PV$, we conclude from the expression for the grand canonical entropy

$$TS = -kT\sum_i p_i \ln p_i = -\mu\langle N\rangle + \langle E\rangle + kT\ln\Xi \tag{14.12}$$

that $\partial S/\partial N = -\mu/T$. The equilibrium condition at constant U and V for a system of several components thus becomes $\mathrm{d}S = \Sigma_j(\partial S_j/\partial N_j)\,\mathrm{d}N_j = -\Sigma_j(\mu_j/T)\,\mathrm{d}N_j = 0$, where the index j denotes the various reactants and products. If we have a reaction

$$\nu_\mathrm{A}\mathrm{A} + \nu_\mathrm{B}\mathrm{B} \leftrightarrow \nu_\mathrm{X}\mathrm{X} + \nu_\mathrm{Y}\mathrm{Y} \quad \text{or equivalently} \quad \sum_\mathrm{R}\nu_\mathrm{R}\mathrm{R} \leftrightarrow \sum_\mathrm{P}\nu_\mathrm{P}\mathrm{P}$$

where R and P denote reactants and products, respectively, we obtain

$$\frac{1}{T}\sum_\mathrm{R}\nu_\mathrm{R}\mu_\mathrm{R} = \frac{1}{T}\sum_\mathrm{P}\nu_\mathrm{P}\mu_\mathrm{P}$$

Using the relation $\mu_j = kT\ln(N_j/z_j)$, obtained from $\mu_j = \partial F/\partial N_j$ with z_j denoting the partition function of component j, the result is

$$\sum_\mathrm{R}\nu_\mathrm{R}\ln(N_\mathrm{R}/z_\mathrm{R}) = \sum_\mathrm{P}\nu_\mathrm{P}\ln(N_\mathrm{P}/z_\mathrm{P}) \quad \text{or, using } N_j/V = c_j = [j],$$

$$\blacktriangleright \quad \frac{\prod_\mathrm{P} c_\mathrm{P}^{\nu_\mathrm{P}}}{\prod_\mathrm{R} c_\mathrm{R}^{\nu_\mathrm{R}}} = K_c(T) \quad \text{with} \quad K_c(T) = \frac{\prod_\mathrm{P}(z_\mathrm{P}/V)^{\nu_\mathrm{P}}}{\prod_\mathrm{R}(z_\mathrm{R}/V)^{\nu_\mathrm{R}}} \tag{14.13}$$

where $K_c(T)$ is the *equilibrium constant*. This relationship is known as the *law of mass action*. The reference level of energy for each factor in Eq. (14.13) is the same. However, for the evaluation of the various partition functions it is more convenient to use the ground state of each species as a reference. Let us take the gas-phase chemical reaction AB + C ↔ A + BC as a simple example with $K(T) = z_\mathrm{A}z_\mathrm{BC}/z_\mathrm{AB}z_\mathrm{C}$. Shifting the reference level of all species to an arbitrary level and denoting the partition function with respect to the ground state for each species by z', we obtain

$$\blacktriangleright \quad K(T) = \frac{(z'_\mathrm{A}\mathrm{e}^{-E_\mathrm{A}/kT}/V)(z'_\mathrm{BC}\mathrm{e}^{-E_\mathrm{BC}/kT}/V)}{(z'_\mathrm{AB}\mathrm{e}^{-E_\mathrm{AB}/kT}/V)(z'_\mathrm{C}\mathrm{e}^{-E_\mathrm{C}/kT}/V)} = \frac{z'_\mathrm{A}z'_\mathrm{BC}}{z'_\mathrm{AB}z'_\mathrm{C}}\mathrm{e}^{-(E_\mathrm{A}+E_\mathrm{BC}-E_\mathrm{AB}-E_\mathrm{C})/kT}$$

$$\equiv \frac{z_\mathrm{A}z_\mathrm{BC}}{z_\mathrm{AB}z_\mathrm{C}}\exp[-\Delta E/kT] \tag{14.14}$$

where in the last step the primes have been removed (using, from now on, as reference level for each of the species the ground state) and $\Delta E \equiv (E_\mathrm{A} + E_\mathrm{BC} - E_\mathrm{AB} - E_\mathrm{C})$ represents the difference in ground-state energies of the reactants and products.

14.2.2
Potential Energy Surfaces

The second concept we need is the potential energy surface. We note that the potential energy of a system containing atoms or molecules can be written as a function of special combinations of the nuclear spatial coordinates, usually referred as *generalized* or *normal coordinates*. Generally, the pictorial representation of the potential energy hypersurface is difficult. The concept can be grasped from a simple example, for which we take the collinear reaction between three atoms A,

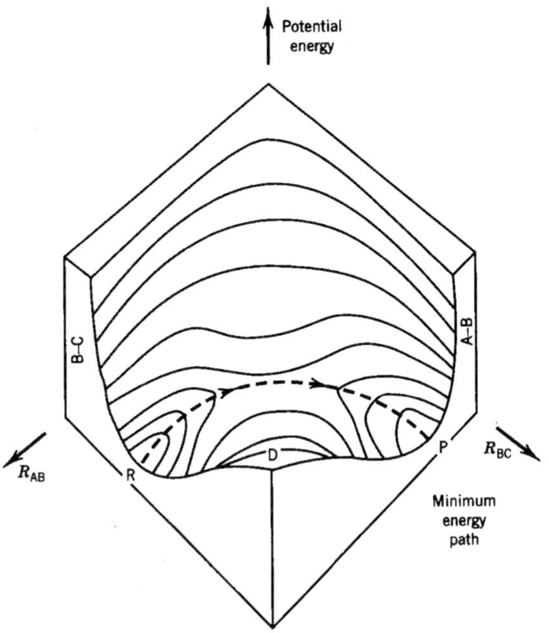

Figure 14.1 The potential energy surface for the reaction A + BC ↔ AB + C, showing a col between two valleys.

B, and C. In Figure 14.1 a map is shown for this reaction with, as axes, the distances AB and BC, respectively. The map shows two valleys separated by a col, and in general, thermal fluctuation creates continuous attempts to pass from one valley to another. In order to calculate the rate constant for the chemical reaction A + BC ↔ AB + C, we must calculate all trajectories on the potential energy surface. With this in mind we need the concept of a *dividing surface,* defined as a surface which cannot be passed without passing a barrier. In Figure 14.1, for example, one of the dividing surfaces is given by $R_{AB} = R_{BC}$. When calculating the rate constant, we must take into account only those trajectories that do pass the dividing surface. An upper limit to the reaction rate $r(T)$ is then given by

$$r(T) \leq j(T) \tag{14.15}$$

where $j(T)$ represents the total amount of reactant systems crossing a dividing surface per unit volume and per unit time at a temperature T – that is, the flux. Equation (14.15) is known as the *Wigner variational theorem.* All statistical methods for computing reaction rates are based on this principle. The calculation of all allowed trajectories to estimate the flux $j(T)$ provides an enormous task if performed in a rigorous fashion, and therefore approximate methods are normally introduced. The variation theorem provides the opportunity to make an optimum choice for the dividing surface – that is, to select the surface that provides the lowest flux $j(T)$. This may be achieved by using a surface depending on several parameters and choosing these parameters in such a way that a minimum $j(T)$ is obtained.

14.2.3
The Activated Complex

The simplest approach along the above lines is to take into account only the trajectory following the minimum energy path from one valley to the other valley. In the case of the chemical reaction A + BC ↔ AB + C, this means the path from the configuration represented by A + BC (point R in Figure 14.1) to the configuration represented by AB + C (point P in Figure 14.1). The coordinate along this path is called the *reaction coordinate*. The top of the col between the two valleys, which actually is a saddle point of the potential energy surface, is known as the *transition state*. A configuration in the neighborhood of the transition state is addressed as an *activated complex*. Transition state theory is based on a number of assumptions. The first assumption is that the optimum dividing surface passes through the transition state and is perpendicular to the reaction coordinate. The second assumption is that activated complexes are at all times in equilibrium with both the reactants and the products. For the forward reaction A + BC ↔ $(ABC)^{\ddagger}$, where $(ABC)^{\ddagger}$ denotes the activated complex, this implies that the equilibrium constant K_{for} is given by

$$K_{\text{for}} = [(\text{ABC})^{\ddagger}]_{\text{for}} / [\text{A}][\text{BC}] \tag{14.16}$$

where, as usual, [X] denotes the concentration of species X. The equilibrium constant K_{rev} for the reverse reaction AB + C ↔ $(ABC)^{\ddagger}$ is

$$K_{\text{rev}} = [(\text{ABC})^{\ddagger}]_{\text{rev}} / [\text{AB}][\text{C}] \tag{14.17}$$

Since in Eqs (14.16) and (14.17) $[(ABC)^{\ddagger}]_{for}$ and $[(ABC)^{\ddagger}]_{rev}$ are equal, we have $[(ABC)^{\ddagger}]_{for} = [(ABC)^{\ddagger}]_{rev} = ½[(ABC)^{\ddagger}]$. The combination of these two conditions together is called the *quasi-equilibrium assumption*. Finally, it is assumed that once the transition state is reached the reaction completes – that is, the reactants do not return to the nonreacted state.

Here, we consider the forward reaction in some detail. Equation (14.14) shows that the equilibrium constant for the reaction A + BC ↔ $(ABC)^{\ddagger}$ can be written as

$$K_{\text{for}} = \frac{z_{(\text{ABC})^{\ddagger}}}{z_{\text{A}} z_{\text{BC}}} \exp(-\Delta E / kT) \tag{14.18}$$

We also assume that the partition function of each species X can be (approximately) factorized as

$$z = z_{\text{ele}} z_{\text{tra}} z_{\text{vib}} z_{\text{rot}}$$

where z_{ele}, z_{tra}, z_{vib}, and z_{rot} represent the electronic, translation, vibration, and rotation partition functions, respectively. The contribution of z_{ele} reduces to the degeneracy number of the ground state, usually 1, since we assume the reaction to proceed on a potential energy surface, which exists only by virtue of the adiabatic assumption. Therefore, no electronic transitions are allowed.

If we have sufficient information on the transition state – that is, we know the structure and relevant force constants – we can construct its partition function. The translation and rotation partition function do not provide a problem in principle, as they can be constructed as for normal molecules. However, the vibration partition function requires some care. The usual normal coordinate analysis of an N-atom molecule can be made, and from that we obtain $3N - 6$ normal coordinates for a nonlinear molecule or $3N - 5$ normal coordinates for a linear molecule. Of these vibration coordinates, all but one has a positive coefficient in the second-order terms in the potential energy expansion. The last, negative coefficient corresponds to an imaginary frequency for this vibration coordinate, and represents the reaction coordinate. This implies that, in the transition state, a small fluctuation in the reaction coordinate will lead to an unstable configuration with respect to this coordinate. It is customary [4][3] to treat this coordinate as a translation coordinate over a small length δ at the top of the potential energy barrier with a partition function $z = (\delta/h)(2\pi m^{\ddagger}kT)^{1/2}$. The complete partition function $z_{(\mathrm{ABC})}{}^{\ddagger}$ for the activated complex is then

$$z^{\ddagger}_{(\mathrm{ABC})} = (\delta/h)(2\pi m^{\ddagger}kT)^{1/2}z^{\ddagger} \tag{14.19}$$

where $z^{\ddagger}$ represents the partition function for the remaining coordinates – that is, the true vibration coordinates, the translation and rotation coordinates. Further, $m^{\ddagger}$ denotes the mass of the activated complex and k, T and h have their usual meanings.

The concentration of the activated complexes due to the forward reaction is given by Eq. (14.16), where the forward equilibrium constant is given by Eq. (14.18). Since there are $[(\mathrm{ABC})^{\ddagger}]_{\mathrm{for}}$ complexes per unit volume which populate the length δ and which are moving forward, the forward reaction rate r_{for} is $r_{\mathrm{for}} = [(\mathrm{ABC})^{\ddagger}]_{\mathrm{for}}/\tau$, where τ is the average time to traverse the length δ, given by $\tau = \delta/v_{\mathrm{ave}}$. Thus, we need the average velocity v_{ave} in one direction over the length δ. Using the same approximation of a free translatory motion again, we borrow from kinetic gas theory [5]

$$v_{\mathrm{ave}} = \int_0^{\infty} v\exp(-m^{\ddagger}v^2/2kT)\mathrm{d}v \Big/ \int_0^{\infty}\exp(-m^{\ddagger}v^2/2kT)\mathrm{d}v = (2kT/\pi m^{\ddagger})^{1/2} \tag{14.20}$$

Combining Eqs (14.16), (14.18), (14.19) and (14.20) with $r_{\mathrm{for}} = [(\mathrm{ABC})^{\ddagger}]v_{\mathrm{ave}}/\delta$, we obtain

$$r_{\mathrm{for}} = k_{\mathrm{for}}[\mathrm{A}][\mathrm{BC}] \quad \text{with} \quad k_{\mathrm{for}} \equiv \frac{kT}{h}\frac{z^{\ddagger}}{z_{\mathrm{A}}z_{\mathrm{BC}}}\exp(-\Delta E/kT) \equiv \frac{kT}{h}K^{\ddagger} \tag{14.21}$$

3) Alternatively, we consider the reaction coordinate as the coordinate for which $\omega \to 0$. In that limit the vibrational contribution to the partition function reads $kT/\hbar\omega$. For an ordinary vibration of a bond, the circular frequency ω would be the reciprocal of the time τ to make a complete vibration; that is, traversing the length δ twice. Here, we need the frequency for traversing δ only once, that is, 2ω. Hence, the rate $r = ½[(\mathrm{ABC})^{\ddagger}]\ 2\omega$, leading to the same final result.

A similar expression is obtained for the reverse reaction, and it is easily verified that the forward and reverse reactions are in equilibrium. The forward rate constant k_{for} can thus be calculated given the relevant information. However, for a first-principles calculation many pieces of information are required: the energy barrier for the reaction; the structure; and the force constants associated with the dynamics of the reactants, products, and activated complex. Typically, this information is incompletely available. This is also true for other mechanisms, although of course for mechanisms other than gas reactions different arguments apply. The exponential dependence on the barrier energy is generally valid, however, and rationalizes the generally observed *Arrhenius-type behavior.*

It remains to be discussed how to connect the results to experimental data. Experimentally, it is often observed that, if the logarithm of the rate constant is plotted against the reciprocal temperature, a straight line is obtained. The gradient of this line is used to define the *(empirical) activation energy* E_{act} by

$$\frac{\mathrm{d}\ln k_{for}}{\mathrm{d}(1/T)} \equiv -\frac{E_{act}}{R} \quad \text{or} \quad \frac{\mathrm{d}\ln k_{for}}{\mathrm{d}T} \equiv \frac{E_{act}}{RT^2} \tag{14.22}$$

Substituting Eq. (14.21), taking logarithms, and differentiating with respect to T yields

$$\frac{\mathrm{d}\ln k_{for}}{\mathrm{d}T} = \frac{\mathrm{d}\ln K^{\ddagger}}{\mathrm{d}T} + \frac{1}{T}$$

Using Eq. (14.22) and the van't Hoff relation $\mathrm{d}\ln K_c^{\ddagger}/\mathrm{d}T = \Delta U^{\ddagger}/RT^2$, where $\Delta U^{\ddagger}$ is the change in energy, we obtain

$$\frac{E_{act}}{RT^2} = \frac{\Delta U^{\ddagger}}{RT^2} + \frac{1}{T} \quad \text{or} \quad E_{act} = \Delta U^{\ddagger} + RT$$

This equation thus provides a link between the experimentally observed E_{act} and $\Delta U^{\ddagger}$. It should be noted that neither E_{act} nor $\Delta U^{\ddagger}$ is identical to ΔE, which appears in Eq. (14.18) because of the temperature dependence of the various partition functions. For the various partition functions, the temperature dependencies read

$$z_{tra} \approx T^{3/2}, \quad z_{rot,lin} \approx T, \quad z_{rot,nonlin} \approx T^{3/2} \quad \text{and} \quad z_{ele} \approx T^0$$

For the vibrations, however, the behavior is less simple. For $kT \ll \hbar\omega$, $Z_{vib} \approx T^0$, while for $kT >> \hbar\omega$, $Z_{vib} \approx T$, where n is the number of vibrational modes of the molecule. At intermediate temperature, $Z_{vib} \approx T^a$ with $0 \le a \le n$. For a restricted (intermediate) temperature range a is approximately constant. In total, this yields

$$k_{for} \approx T^m \exp(-\Delta E / RT)$$

explaining the modified Arrhenius equation. As noted above, typically $|m| < 2$.

Finally, we note that formalism can also be interpreted thermodynamically, and to this purpose we write Eq. (14.21) in terms of molar quantities

$$k_{for} \equiv \frac{kT}{h}\frac{z^{\ddagger}}{z_A z_{BC}} \mathrm{e}^{-\Delta E/RT} \equiv \frac{kT}{h} K^{\ddagger} \quad \text{as} \tag{14.23}$$

$$k_{\text{for}} = \frac{kT}{h}\text{e}^{-\Delta G^{\ddagger}/RT} = \frac{kT}{h}\text{e}^{\Delta S^{\ddagger}/R}\text{e}^{-\Delta H^{\ddagger}/RT} = \frac{kT}{h}\text{e}^{\Delta S^{\ddagger}/R}\text{e}^{-\Delta U^{\ddagger}/RT}\text{e}^{P\Delta V^{\ddagger}/RT} \tag{14.24}$$

with $\Delta S^{\ddagger}$, $\Delta H^{\ddagger}$, $\Delta U^{\ddagger}$ and $\Delta V^{\ddagger}$ the molar entropy, enthalpy, energy, and volume of activation, respectively. In a gas reaction, the term $P\Delta V^{\ddagger}$ may be put equal to $RT\Sigma_j\nu_j$, where $\Sigma_j\nu_j$ is the stoichiometric sum for activated complex formation. For a unimolecular reaction the stoichiometric sum $\Sigma_j\nu_j = 0$ and $\Delta H^{\ddagger} = \Delta U^{\ddagger} = E_{\text{act}} - RT$, or $E_{\text{act}} = \Delta H^{\ddagger} + RT$. Consequently, $A = \text{e}(kT/h)\exp(\Delta S^{\ddagger}/R)$. For a bimolecular reaction $\Sigma_j\nu_j = -1$ and $\Delta H^{\ddagger} = \Delta U^{\ddagger} - RT = E_{\text{act}} - 2RT$, or $E_{\text{act}} = \Delta H^{\ddagger} + 2RT$. Consequently, $A = \text{e}^2(kT/h)\exp(\Delta S^{\ddagger}/R)$. In solution, the volume term is nearly always negligible, and therefore $\Delta H^{\ddagger} \cong \Delta U^{\ddagger}$. We see that the pre-exponential factor in the empirical expression $k_{\text{rea}} = A\exp(-E_{\text{act}}/RT)$ corresponds approximately to the entropy.

Here, our brief overview on gas-phase TST ends, and in the next sections we apply TST to reactions in liquids. It must be re-emphasized that, for a detailed calculation, a considerable amount of information is required.

Problem 14.1: The dissociation of I_2

Consider the reaction $I_2 \leftrightarrow 2I$ at 300 K. Note that the ground state for the I atom is $^2P_{2/3}$ (hence fourfold degenerate), while that for the I_2 molecule is $^1\Sigma_g^+$ (hence nondegenerate). For I_2 the rotational temperature $\theta_{\text{rot}} = 0.054$ K, the vibrational temperature $\theta_{\text{vib}} = 308$ K, the dissociation energy from the ground state is $D = 1.5417$ eV, and the molecular mass $m_I = 127\,\text{g mol}^{-1}$.

a) Show that the translational and electronic partition function for the I atom are $z_{\text{tra}} = \Lambda^{-3}V$ and $z_{\text{ele}} = 4\exp[-\tfrac{1}{2}(D/kT + \theta_{\text{vib}}/2T)]$, respectively.

b) Show that the translational, rotational, vibrational and electronic partition function for the I_2 molecule are $z_{\text{tra}} = 2^{-3/2}\Lambda^{-3}V$, $z_{\text{rot}} = T/2\theta_{\text{rot}}$, $z_{\text{vib}} = [\exp(\theta_{\text{vib}}/2T) - \exp(-\theta_{\text{vib}}/2T)]^{-1}$, and $z_{\text{ele}} = 1$, respectively.

c) Show that $K_c = 32(\pi mkT/h^2)^{3/2}(\theta_{\text{rot}}/T)[1 - \exp(-\theta_{\text{vib}}/T)]\exp(-D/kT)$, and that $K_P = kTK_c$.

d) Calculate the numerical value for K_P at 300 K.

14.3 Solvent Effects

The description of reactions given in Section 14.2 is for gas-phase reactions. Although the study of gas-phase reactions has led to a great deal of insight into reaction mechanisms, in practice most reactions are carried out in liquids. The difference between a gas-phase reaction and a liquid-phase reaction is related to the presence of a solvent. In solution reactants are usually *solvated*, that is, bound to one or more solvent molecules. The transport of reactants in liquids, depending on the diffusion coefficients, is more hampered then in gases; moreover, the solvent

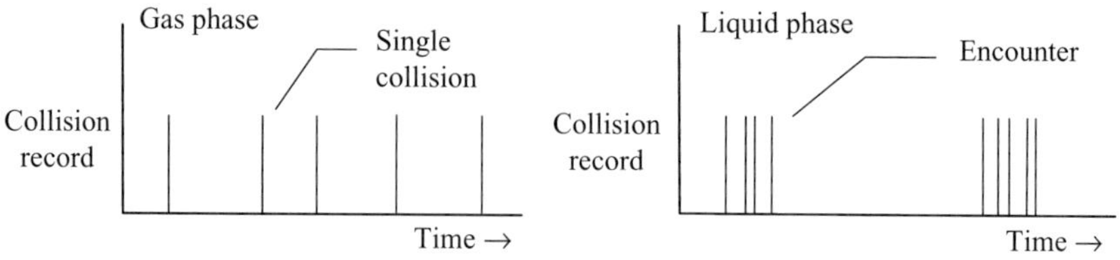

Figure 14.2 Collision pattern between reactants in the gas phase and in the liquid phase.

may catalyze the reaction and thus change the mechanism. The reaction rate will thus be influenced not only by temperature and concentrations but also by the type of solvent. Whereas, reactions involving ions are rare in gas-phase reactions, they are abundant in liquid-phase reactions. Therefore, electrostatic forces, dependent on permittivity and ionic strength, are also important. Finally, pressure will influence the rate of reaction. Before discussing these effects, we should briefly sketch the situation of a solute in a solvent and some associated preliminaries. We note upfront that the analysis is a mixture of molecular and macroscopic arguments in which macroscopic physical properties of solvents are used.

In the discussion of simple liquids in Chapter 8, we indicated that there is only some justification for the lattice model in which a molecule occupies for a certain time a cell formed by its neighbors. Such a cell exists only for about 10 collisions. However, in molecular solutions the cell for a solute, in this connection denoted as *cage*, exists somewhat longer. Indeed, in time, two reactants will meet in the same cage and collide many times, perhaps 20 to 200 times [6], before they either diffuse further or react in that period. Such a set of events is called an *encounter*. The time between encounters is large when compared to the time between collisions within an encounter. In the gas phase, the time interval between collisions of reactants is much more homogeneously distributed (Figure 14.2). With respect to reactions, the lower frequency of encounters is compensated by the higher frequency within an encounter, as evidenced by the near constancy of rate constants for certain reactions.

There is some experimental evidence for the cage effect from, for example, the photochemical decomposition [8] of CH_3NNCH_3 and CD_3NNCD_3. The absorbed light dissociates the molecules in N_2 molecules and CH_3 and CD_3 radicals. In the gas phase the latter recombine to ethane in proportions that indicate a random mixing before recombination, whereas in *i*-octane only C_2H_6 and C_2D_6 are formed. This absence of CH_3CD_3 indicates that the radicals are kept together by the cage until recombination occurs.

With this cage model in mind, we can distinguish three stages for a reaction: 1) the diffusion of reactants A and B towards each other to form the encounter pair (A···B) in a cage; 2) the reaction of the reactants (A···B) in the cage to products (X···Y); and 3) the separation of products (X···Y) away from each other. These processes can be represented by

$$\mathrm{A}+\mathrm{B} \underset{k_{-\mathrm{D}}}{\overset{k_\mathrm{D}}{\rightleftharpoons}} (\mathrm{A}\cdots\mathrm{B}) \xrightarrow{k_\mathrm{R}} (\mathrm{X}\cdots\mathrm{Y}) \longrightarrow \mathrm{X}+\mathrm{Y} \tag{14.25}$$

The rate of reaction, as measured by d[A]/dt, is given by

$$-\frac{d[A]}{dt} = k_D[A][B] - k_{-D}[(A\cdots B)] \tag{14.26}$$

while the rate of formation of the encounter pair (A$\cdots$B) is given by

$$\frac{d[(A\cdots B)]}{dt} = k_D[A][B] - k_{-D}[(A\cdots B)] - k_R[(A\cdots B)] \tag{14.27}$$

Now, we assume steady-state conditions for (A$\cdots$B), that is, d[(A$\cdots$B)]/dt = 0, so that

$$[(A\cdots B)] = \frac{k_D}{k_{-D} + k_R}[A][B] \tag{14.28}$$

Substitution of Eq. (14.28) in Eq. (14.26) results in

► $$-\frac{d[A]}{dt} = k[A][B] \quad \text{with} \quad k = \frac{k_D k_R}{k_{-D} + k_R} \tag{14.29}$$

There are two limiting conditions. First, for $k_R \ll k_{-D}$, $k \cong k_R(k_D/k_{-D})$ and the rate becomes

$$-\frac{d[A]}{dt} \cong k_R \frac{k_D}{k_{-D}}[A][B] \tag{14.30}$$

In this case we have *reaction control*. Second, for $k_R >> k_{-D}$, $k \cong k_D$ and the rate becomes

$$-\frac{d[A]}{dt} \cong k_D[A][B] \tag{14.31}$$

and we have *diffusion control*. The behavior for these limiting behaviors is discussed in the next two sections.

We limit the discussion to the physical effects of solvents, where the properties of the solvents influence the rate but the solvent does not participate as a reactant. For these aspects we refer to, for example, Buncel *et al.* [7] and Reichardt [8].

Problem 14.2

Verify Eqs (14.29) to (14.31).

14.4
Diffusion Control

The main task in this section is to estimate the rate constants for diffusion k_D and k_{-D}. While the reactants may be either neutral or ionic, we start with ionic reactions.

The *flux* j_A – that is, the number of moles of ions crossing a unit area per unit time – of ions A with molarity c_A across a surface normal to the mass velocity v_A is given by

$$j_A = v_A c_A = \frac{z_A}{|z_A|} u_A E c_A \tag{14.32}$$

with the mobility u_A and electric field E. The factor $z_A/|z_A|$ is introduced to fix the sign of the charge $z_A e$ of the ion A, since a positive ion has a positive flux in the direction of E. For this expression it is assumed that the flux is parallel to the electric field. With an additional concentration gradient, also assumed to be parallel to the electric field, the total flux in the direction of decreasing concentration becomes

$$j_A = -\left(D_A \frac{dc_A}{dz} - \frac{z_A}{|z_A|} u_A E c_A \right) \tag{14.33}$$

with D_A the diffusion coefficient relative to the solvent. For the rate with which the reactant ions A and B approach each other we must replace D_A by $D_A + D_B$.

Consider now the first step of the scheme in Eq. (14.25), where the labels A and B represent ions with charge z_A and z_B, respectively. We assume that a gradient in c_B is present around each A ion, and *vice versa*. Using the *Nernst–Einstein relation* (Eq. 12.67)

$$\frac{u_X}{D_X} = \frac{|z_X e|}{kT} \tag{14.34}$$

and the relation between electric field $E_z = -\partial\phi/\partial z$, with ϕ the electric potential, and potential energy $\Phi(r)$

$$\Phi_X(r) = z_X e \phi_X(r) \tag{14.35}$$

we obtain for the flux of B ions across a sphere of radius r around each ion A

$$j_B = (D_A + D_B)\left[\frac{dc_B(r)}{dr} + \frac{c_B(r)}{kT}\frac{d\Phi_B(r)}{dr}\right] \tag{14.36}$$

where the overall negative sign disappears because the flux is now taken in the direction of increasing concentration. To illustrate the following argument more clearly, we ignore the $d\Phi_B(r)/dr$ contribution for the moment and add it again later.

The *flow* is $J_B = 4\pi r^2 j_B$, and we obtain for the molarity $c_B(r)$

$$\int_{c_B(r)}^{c_B(\infty)} dc_B(r) = c_B(\infty) - c_B(r) = \frac{J_B}{4\pi(D_A + D_B)} \int_r^\infty \frac{1}{r^2} dr = \frac{J_B}{4\pi(D_A + D_B)} \frac{1}{r} \tag{14.37}$$

The flow J_B can be taken outside the integral because under steady-state conditions the value of the flow is independent of the radius of the sphere. Because $c_B(\infty)$ represents the bulk molarity [B], the total result is

$$c_B(r) = [B] - \frac{J_B}{4\pi(D_A + D_B)} \frac{1}{r} \tag{14.38}$$

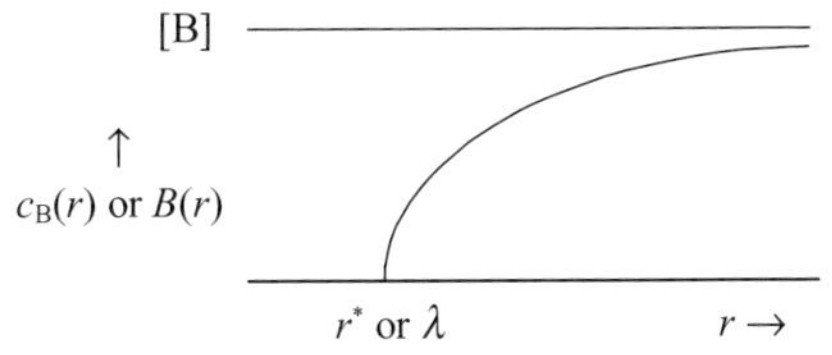

Figure 14.3 The behavior of $c_B(r)$ and $B(r)$ as a function of r.

We now assume that, if ion A approaches ion B within a critical radius r^*, a reaction occurs and $c_B(r) = 0$ for $r < r^*$ (Figure 14.3). This implies that the concentration gradient does not extend to $r = 0$ since all ions are lost reaching r^*. The flow thus becomes

$$J_B = 4\pi(D_A + D_B)r^*[B] \tag{14.39}$$

so that the rate of reaction reads

$$v = 4\pi(D_A + D_B)r^*[A][B] \equiv k_D[A][B] \quad \text{with} \quad k_D = 4\pi(D_A + D_B)r^* \tag{14.40}$$

The rate constant k_D is thus proportional to the diffusion constants, as expected, and a critical length r^*.

Let us now introduce the $d\Phi_B(r)/dr$ contribution again. To solve the complete Eq. (14.36) we use a somewhat unexpected "Ansatz," and note that if we differentiate the function

$$B(r) = c_B(r)\exp[\Phi_B(r)/kT] \tag{14.41}$$

we obtain

$$\frac{dB(r)}{dr} = \left[\frac{dc_B(r)}{dr} + \frac{c_B(r)}{kT}\frac{d\Phi_B(r)}{dr}\right]\exp[\Phi_B(r)/kT] \tag{14.42}$$

From Eq. (14.36) we thus have

$$j_B = (D_A + D_B)\exp[-\Phi_B(r)/kT]\frac{dB(r)}{dr} \tag{14.43}$$

The integration corresponding to Eq. (14.37) becomes

$$B(\infty) - B(r) = \frac{J_B}{4\pi(D_A + D_B)}\frac{1}{\lambda} \quad \text{with} \quad \lambda^{-1} = \int_r^\infty \exp[\Phi_B(r)/kT]r^{-2}\,dr \tag{14.44}$$

where $B(\infty)$ is simply $c_B(\infty)$ = [B] and λ has the dimensions of length. Making the same assumption of a critical radius r^* where $c_B(r) = 0$ and hence $B(r) = 0$, the total result is

$$J_B = 4\pi(D_A + D_B)\lambda[B] \quad \text{and} \tag{14.45}$$

$$v = k_D[A][B] \quad \text{with} \quad k_D = 4\pi(D_A + D_B)\lambda \tag{14.46}$$

Finally, we must link λ with r^*, and this is done by assuming that the potential energy between two ions is given by

$$\Phi(r) = \frac{z_A z_B e^2}{4\pi\varepsilon_0\varepsilon_r}\frac{1}{r} \equiv W\frac{1}{r} \tag{14.47}$$

where the subscripts are omitted since the expression is symmetric in A and B. As usual, ε_0 and ε_r are the permittivity of vacuum and the relative permittivity, respectively. Substituting in Eq. (14.43) for λ and integrating results in

$$\lambda = \frac{\beta W}{\exp(\beta W/r^*) - 1} \tag{14.48}$$

which, as it should, results in $\lambda = r^*$ for $\Phi(r) = 0$ for $r > r^*$. The behavior of the concentration profile $c_B(r)$ and function $B(r)$ as a function of r is sketched in Figure 14.3.

For the dimension of k_D we have to consider the dimension of D which is $m^2\,s^{-1}$; hence, k_D will have dimension $m^3\,s^{-1}$. Conversion to molar quantities is achieved by multiplying with Avogadro's number N_A.

The rate constant for the reverse diffusion k_{-D} in the scheme of Eq. (14.25) is the reciprocal of the time in which A and B remain nearest neighbors. If specific interactions are absent, this quantity can be estimated from random walk diffusion $d = (6Dt)^{1/2}$, where d is the diffusion distance in time t. Hence, in this case k_{-D} is given by

$$k_{-D} = 6(D_A + D_B)/(r^*)^2 \tag{14.49}$$

Clearly, the controlling parameters for the controlled reactions are the charges z_A and z_B, the permittivity ε_r, and the diffusion coefficients D_A and D_B. Moreover, we need an estimate for r^*, which is essentially an empirical parameter. We discuss the agreement with experiment after we have dealt with reaction control.

14.5 Reaction Control

In TST, the reaction rate is given in terms of the molarity of the components. The conventional way of introducing medium effects is to realize that the molarity is the activity a_X divided by the activity coefficient γ_X, and to replace the molarities c_X in the TST expression accordingly.

In Section 14.2 we derived for the reaction $A + BC \leftrightarrow (ABC)^\ddagger$

$$k_{for} \equiv \frac{kT}{h}\frac{z^\ddagger}{z_A z_{BC}}\exp(-\Delta E/kT) \equiv \frac{kT}{h}K^\ddagger \tag{14.50}$$

Here, the equilibrium constant $K^\ddagger$ is not the true equilibrium constant $K_0^\ddagger$ using activities a_X, but the corresponding expression using molarities c_X. In many cases we have an elementary reaction of the type

$$A \rightarrow M^\ddagger \rightarrow X \text{ (unimolecular)} \quad \text{or} \quad A + B \rightarrow M^\ddagger \rightarrow X + Y \text{ (bimolecular)}$$

Table 14.1 Pre-exponential factor and activation energy (Eq. 14.10) for the decomposition of N_2O_5 at 20 °C.[a)]

Solvent	lnA (l mol^{-1} s^{-1})	E_{act} (J mol^{-1})	Solvent	lnA (l mol^{-1} s^{-1})	E_{act} (J mol^{-1})
Gas phase	31.4	103.4	Chloroform	31.8	103.0
Nitromethane	31.0	102.6	Ethylidine chloride	32.5	104.3
Bromine	30.5	100.5	Nitrogen tetraoxide	32.7	104.7
Pentachlorethane	32.2	104.7	Propylene dichloride	34.8[b)]	117.2
Carbon tetrachloride	30.8	100.9	Nitric acid	34.0[b)]	118.5
Ethylene chloride	31.3	102.2			

a) Data from Ref. [11].
b) Data at 45 °C.

For a bimolecular reaction the true equilibrium constant and rate constant are then

$$K_0^{\ddagger} = a_{\ddagger}/a_A a_B \quad \text{and} \quad k_{AB,0} = k_{AB}(\gamma_{\ddagger}/\gamma_A\gamma_B) \tag{14.51}$$

while we used

$$K^{\ddagger} = c_{\ddagger}/c_A c_B = K_0^{\ddagger}(\gamma_A\gamma_B/\gamma_{\ddagger}) \quad \text{and} \quad k_{AB} = k_{AB,0}(\gamma_A\gamma_B/\gamma_{\ddagger}) \tag{14.52}$$

For a unimolecular reaction, similarly $k_A/k_{A,0} = \gamma_A/\gamma_{\ddagger}$, and if the structure of the activated complex is not too different from the reactant we have approximately $k_A/k_{A,0} = 1$. This is, for example, the case for the decomposition of N_2O_5 in various solvents. Data for the decomposition of N_2O_5 at 20 °C are listed in Table 14.1; these data show that the pre-exponential factor and activation energy are approximately constant over the range of solvents indicated.

Similar measurements in propylene dichloride and nitric acid showed relatively low rate constants, while the activation energies were relatively high. This suggests the formation of a complex between the reactant and solvent. Another example is the dimerization of cyclopentadiene (C_5H_6 to $C_{10}H_{12}$). At 50 °C the rate constant is 6×10^{-6} l mol^{-1} s^{-1} for the gas phase, while the rate constants were 6×10^{-6}, 10×10^{-6} and 20×10^{-6} l mol^{-1} s^{-1} for the solvents CS_2, C_6H_6, and C_2H_5OH, respectively. However, in many cases there is a significant influence of the solvent. For different types of solution we have discussed expressions for the activity coefficients, and we will discuss their effect on the reaction rates in the next sections.

14.6 Neutral Molecules

In Chapter 11 we discussed regular solutions and obtained for the activity coefficients of apolar compounds

$$\ln\gamma_1 = (wz/kT)x_2^2 \quad \text{and} \quad \ln\gamma_2 = (wz/kT)x_1^2 \tag{14.53}$$

where w is the interaction parameter, z the coordination number, and x_1 and x_2 are the mole fractions. The conversion of activity coefficients for mole fractions γ_x to activity coefficients for molarities γ_c was also provided.

Using regular solution theory (see Chapter 11) the final result is

$$RT\ln\gamma_1 = V_1\left(\frac{x_2V_2}{x_1V_1+x_2V_2}\right)^2\left[\frac{w_{11}}{V_1}+\frac{w_{22}}{V_2}-\frac{2w_{12}}{(V_1V_2)^{1/2}}\right] \tag{14.54}$$

where V is the molar volume and x the mole fraction of the species, designated by the subscripts 1 and 2. This type of result can be plugged directly into the rate equations.

In solution, solvation effects can play a role, and the entropy of activation can help to study this effect. From the kinetic analysis it will be clear that in $k_{\text{rea}} = A\exp(-E_{\text{act}}/RT)$ for unimolecular reactions in solution the factor $A = \text{e}(kT/h)\exp(\Delta S^{\ddagger}/R)$, so that, if the specific reaction rate and energy of activation are known, it is possible to estimate the entropy of activation. For the reaction of triethylamine with ethyl bromide in mixtures of acetone and benzene, the data are given in Table 14.2. It can be seen that the entropy of activation remains almost constant over the range from 20 to 100% acetone in the mixture, implying that the activated state is solvated as long as the solvent contains at least one-fifth of acetone. In benzene, the activated state is probably hardly solvated, which leads to a lower entropy. The accompanying small decrease in activation energy is insufficient to compensate for the increase in entropy.

For polar molecules, orientation effects will also play a role. Kirkwood [7] derived, on the basis of electrostatic considerations and neglecting short-range van der Waals interactions, the Gibbs energy of transferring a molecule with a strong dipole from a solvent with permittivity $\varepsilon_r = 1$ to a solvent with $\varepsilon_r = \varepsilon_r$. This quantity is related to the activity coefficient γ for which the dilute gas with $\varepsilon_r = 1$ is the standard state. The result is

$$\Delta_{\text{sol}}G = kT\ln\gamma = -\frac{\mu^2}{4\pi\varepsilon_0 r^3}\frac{\varepsilon_r-1}{2\varepsilon_r+1} \tag{14.55}$$

with r the molecular radius.

For the reaction $A + B \rightarrow M^{\ddagger} \rightarrow X + Y$, the rate constant becomes

$$\ln k_{\text{AB}} = \ln k_{\text{AB},0} - \frac{1}{4\pi\varepsilon_0 kT}\frac{\varepsilon_r-1}{2\varepsilon_r+1}\left(\frac{\mu_{\text{A}}^2}{r_{\text{A}}^3}+\frac{\mu_{\text{B}}^2}{r_{\text{B}}^3}-\frac{\mu_{\ddagger}^2}{r_{\ddagger}^3}\right) \tag{14.56}$$

Table 14.2 Entropies and activation energies in acetone–benzene mixtures for the reaction between ethyl bromide and triethylamine.[a)]

% Acetone	E_{act} (J mol^{-1})	$\Delta S^{\ddagger}$ (J K^{-1}mol^{-1})	% Acetone	E_{act} (J mol^{-1})	$\Delta S^{\ddagger}$ (J K^{-1}mol^{-1})
100	49.0	−181	20	51.1	−186
80	50.7	−177	0	46.9	−210
50	50.2	−183			

a) Data from Ref. [11].

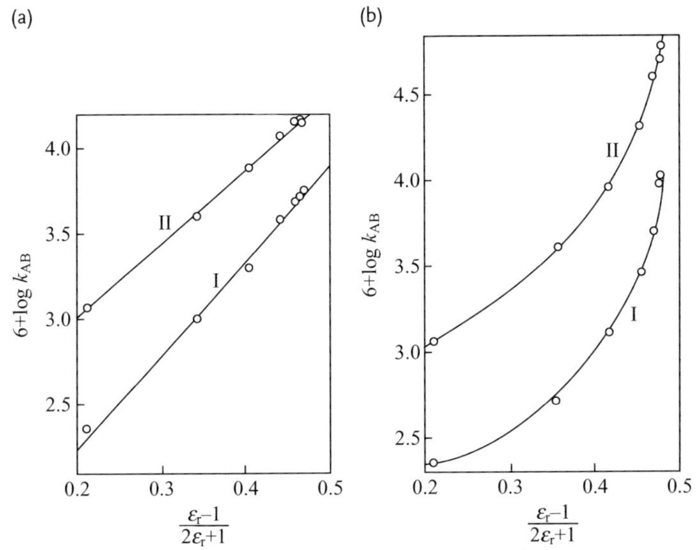

Figure 14.4 Specific rate constants for the reactions between pyridine and benzylbromide (I) and triethylamine and benzyl bromide (II) in (a) alcohol–benzene mixtures and (b) benzene–nitrobenzene mixtures. Data from Ref. [11].

A plot of $\ln k_{AB}$ versus $(\varepsilon_r - 1)/(2\varepsilon_r + 1)$ should yield a straight line for the same reaction carried in different solvents. Figure 14.4 shows an example of the specific rate constant for the reactions between pyridine and benzylbromide (I) and triethylamine and benzyl bromide (II) in alcohol–benzene mixtures (Figure 14.4a) and benzene–nitrobenzene mixtures (Figure 14.4b) as a function of $(\varepsilon_r - 1)/(2\varepsilon_r + 1)$. The graph for the reactions in alcohol–benzene mixtures is almost linear, in agreement with theory, whereas the graph for the benzene–nitrobenzene mixtures showed considerable deviations. This implies that nonelectrostatic forces play a considerable role and must not be neglected.

Problem 14.3

Discuss why, for reactions between triethylamine and benzyl bromide in benzene–nitrobenzene mixtures, deviations from a linear plot of $\ln k_{AB}$ versus $(\varepsilon_r - 1)/(2\varepsilon_r + 1)$ are to be expected.

14.7
Ionic Solutions

In Chapter 11 we derived the activity coefficients for ionic solutions using the Debye–Hückel theory, which can be used immediately if the structure of the activated complex is known. For this structure, two extremes have been proposed: (i)

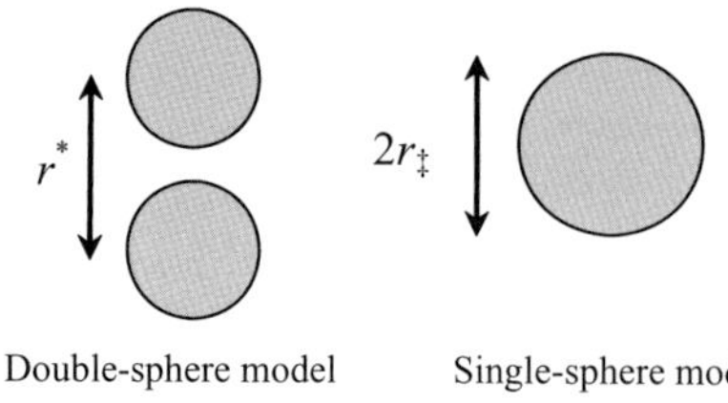

Figure 14.5 The double-sphere and single-sphere models for the activated complex in an ionic reaction.

a *double-sphere model*; and (ii) a *single-sphere model*. Each of these models will be discussed in the following sections (Figure 14.5).

14.7.1
The Double-Sphere Model

For the double-sphere model consider the reaction A + B → $M^{\ddagger}$, where the reactants have charges $z_A e$, $z_B e$ and $(z_A + z_B)e$, respectively. The activated complex is considered to be a double-sphere with r^* the distance between A and B. The concentration of the activated complex is taken as proportional to the bulk concentration of A multiplied by the average concentration of B at a distance r^*. The latter is given by

$$c_B(r^*) = c_B \exp(-ez_B\psi / kT) \tag{14.57}$$

where c_B is the bulk concentration of B, ψ is the mean electrostatic potential at a distance r^* from A, and e is the unit charge. Hence, the concentration of activated complex is

$$c_{\ddagger} = k' c_A c_B \exp(-ez_B\psi / kT) \tag{14.58}$$

with k' the proportionality constant. Using the Debye–Hückel expression for ψ (Eq. 12.48), the expression becomes

$$c_{\ddagger} = k' c_A c_B \exp\left[-\frac{e^2 z_A z_B}{4\pi\varepsilon_0\varepsilon_r kT}\frac{\exp(-\kappa r^*)}{r^*}\frac{\exp(\kappa a)}{1+\kappa a}\right] \tag{14.59}$$

where a is the mean distance of closest approach of A and B. The parameter κ is defined by the Debye length $\kappa^2 = 2N_A e^2 I_c/\varepsilon_0\varepsilon_r kT$, where N_A is Avogadro's constant and the ionic strength I_c is defined by $I_c = \frac{1}{2}\Sigma_i c_i z_i^2$. At high dilution, Eq. (14.59) becomes

$$c_{\ddagger}^0 = k' c_A^0 c_B^0 \exp\left(-\frac{e^2 z_A z_B}{4\pi\varepsilon_0\varepsilon_r kTr^*}\right) \tag{14.60}$$

and since the activity coefficient γ is equal to c^0/c, we obtain

$$\ln\frac{\gamma_A\gamma_B}{\gamma_{\ddagger}} = \ln\frac{c_A^0 c_B^0}{c_{\ddagger}^0} - \ln\frac{c_A c_B}{c_{\ddagger}} \tag{14.61}$$

or, using Eqs (14.59) and (14.60)

$$\ln\frac{\gamma_A\gamma_B}{\gamma_\ddagger} = \frac{e^2 z_A z_B}{4\pi\varepsilon_0\varepsilon_r kTr^*}\left(1-\frac{e^{-\kappa r^*}e^{\kappa a}}{1+\kappa a}\right) \cong \frac{e^2 z_A z_B}{4\pi\varepsilon_0\varepsilon_r kT}\frac{\kappa}{1+\kappa a} \tag{14.62}$$

The activity coefficient is with reference to the infinitely diluted solution, but we need this with respect to the gas phase. Here, we use α for the activity coefficient of the solution with respect to the gas phase, β for the activity coefficient of an infinitely diluted solution with respect to the gas phase, and γ for the activity coefficient of the solution with respect to the infinitely diluted solution. So

$$\alpha = \beta\gamma \quad \text{and} \quad \frac{\alpha_A\alpha_B}{\alpha_\ddagger} = \frac{\gamma_A\gamma_B}{\gamma_\ddagger}\frac{\beta_A\beta_B}{\beta_\ddagger} \tag{14.63}$$

where the quantity involving the β-terms is given by

$$\ln\frac{\beta_A\beta_A}{\beta_\ddagger} = \ln\frac{c_\ddagger^0}{c_A^0 c_B^0} - \ln\left(\frac{c_\ddagger^0}{c_A^0 c_B^0}\right)_{\text{gas}} \tag{14.64}$$

and the subscript "gas" indicates the ideal gas state with permittivity $\varepsilon_r = 1$. Applying the Debye–Hückel expression, this reads

$$\ln\frac{\beta_A\beta_B}{\beta_\ddagger} = \frac{e^2 z_A z_B}{4\pi\varepsilon_0 kTr^*}\left(1-\frac{1}{\varepsilon_r}\right) \tag{14.65}$$

Eliminating the activity coefficients α using Eqs (14.62), (14.63) and (14.65), the final result for the double-sphere model employing $k_{AB} = k_{AB,0}(\alpha_A\alpha_B/\alpha_\ddagger)$ is

$$\blacktriangleright \quad \ln k_{AB} = \ln k_{AB,0} + \frac{e^2 z_A z_B}{4\pi\varepsilon_0 kTr^*}\left(1-\frac{1}{\varepsilon_r}\right) + \frac{e^2 z_A z_B}{kT 4\pi\varepsilon_0\varepsilon_r}\frac{\kappa}{1+\kappa a} \tag{14.66}$$

14.7.2
The Single-Sphere Model

For the single-sphere model, one uses essentially the Born hydration model (see Section 12.2). The activated complex is considered to be a single-sphere of radius r, which is transferred from the gas phase to the liquid phase. For this process we showed in Chapter 12 that

$$\Delta G = \frac{e^2 z^2}{8\pi\varepsilon_0 r}\left(\frac{1}{\varepsilon_r}-1\right) \tag{14.67}$$

and because for this process $\Delta G = kT\ln\beta$ (with β as the activity coefficient defined in the previous section), we obtain

$$\ln\beta_X = \frac{e^2 z_X^2}{8\pi\varepsilon_0 kTr_X}\left(\frac{1}{\varepsilon_r}-1\right) \tag{14.68}$$

Hence, it follows that

$$\ln\frac{\beta_A\beta_B}{\beta_\ddagger} = \frac{e^2 z_A z_B}{8\pi\varepsilon_0\varepsilon_r kT}\left(\frac{1}{\varepsilon_r}-1\right)\left[\frac{z_A^2}{r_A}+\frac{z_B^2}{r_B}-\frac{(z_A+z_B)^2}{r_\ddagger}\right] \tag{14.69}$$

where $r_\ddagger$ is the radius of the (spherical) activated complex with charge $(z_A + z_B)e$. The activity coefficient γ is as given by the Debye–Hückel model (see Section 12.5)

$$\ln\gamma_X = -\frac{e^2 z_X^2}{8\pi\varepsilon_0\varepsilon_r kT}\frac{\kappa}{1+\kappa a_X} \tag{14.70}$$

and, if a mean distance a is assumed for the distance of closest approach of the ions, it follows that

$$\ln\frac{\gamma_A\gamma_B}{\gamma_\ddagger} = \frac{e^2 z_A z_B}{4\pi\varepsilon_0\varepsilon_r kT}\frac{\kappa}{1+\kappa a} \tag{14.71}$$

Now, employing again $k_{AB} = k_{AB,0}(\alpha_A\alpha_B/\alpha_\ddagger)$ and introducing Eqs (14.69) and (14.71) in the α-terms of Eq. (14.63), the final result becomes

$$\ln k_{AB} = \ln k_{AB,0} + \frac{e^2}{8\pi\varepsilon_0 kT}\left(\frac{1}{\varepsilon_r}-1\right)\left[\frac{z_A^2}{r_A}+\frac{z_B^2}{r_B}-\frac{(z_A+z_B)^2}{r_\ddagger}\right]+\frac{e^2 z_A z_B}{4\pi\varepsilon_0\varepsilon_r kT}\frac{\kappa}{1+\kappa a} \tag{14.72}$$

The expressions for the single-sphere and double-sphere models become identical when $r_A = r_B = r_\ddagger$. The expressions derived in this and the previous section can be tested in two ways: first, via the dependence on ionic strength while keeping the permittivity constant; and, second, via the dependence on permittivity of the solvent using data for infinite dilution, that is, extrapolated to zero ionic strength. This will be discussed in the next two sections.

Problem 14.4

Verify Eqs (14.66) and (14.72).

Problem 14.5

Discuss the difference in dependence on ionic radii between the single-sphere and double-sphere models for the rate constant. To that purpose, express r_A, r_B and $r_\ddagger$ for the single-sphere model in r^*, and consider when the resulting expression differs significantly from the corresponding expression for the double-sphere model. Do you expect that experimental data on the rate constant can help to distinguish between the two models?

14.7.3 Influence of Ionic Strength

For both Eqs (14.66) and (14.72) the influence of the ionic strength is given by

$$\ln k_{AB} = \ln k_{AB,0}^{sol} + \frac{e^2 z_A z_B}{4\pi\varepsilon_0\varepsilon_r kT}\frac{\kappa}{1+\kappa a} \tag{14.73}$$

where $k_{AB,0}^{sol}$ is the rate constant for infinite dilution. Inserting $\kappa^2 = (e^2/\varepsilon kT)\Sigma_i n_i z_i^2 = 2e^2 N_A I_c/\varepsilon kT$ (Eq. 12.48) and neglecting κa in the term $1 + \kappa a$ in comparison with unity, this expression reduces to

$$\ln k_{AB} = \ln k_{AB,0}^{sol} + \frac{2^{1/2}e^3N_A^2}{4\pi(\varepsilon_0\varepsilon_r RT)^{3/2}} z_A z_B \sqrt{I_c} \equiv \ln k_{AB,0}^{sol} + A z_A z_B \sqrt{I_c} \tag{14.74}$$

Introducing all numerical factors it appears that $A = 1.02$ (using mol l^{-1} for molarity) for aqueous solutions at 25 °C. In a plot of k_{AB} versus $\sqrt{I_c}$, one should thus find a straight line with slope $1.02z_Az_B$. The influence of the ionic strength for various reactions is shown in Figure 14.6a, where the experimental slopes nicely match the expected $1.02z_Az_B$, including the zero slope if either z_A or z_B is zero.

Of course, deviations occur when the solution is too concentrated, that is, if I_c is too high. This occurs first, due to the further (mathematical) approximation made in the Debye–Hückel expression and, second, due to the approximate nature of the Debye–Hückel model itself. Moreover, for solvents with a low permittivity, the association of the ions may cause discrepancies, as has been shown for the reaction of sodium bromoacetate (Na^+BrAc^-) with sodium thiosulfate $Na_2^{2+}S_2O_3^{2-}$. At high dilutions $BrAc^-$ reacts with $S_2O_3^{2-}$ so that $z_Az_B = +2$. However, at higher concentrations the complexes $Na^+\cdots BrAc^-$, $Na^+\cdots S_2O_3^{2-}$ and $2Na^+\cdots S_2O_3^{2-}$ are formed, and the effective value of z_Az_B changes to zero. The repulsion between

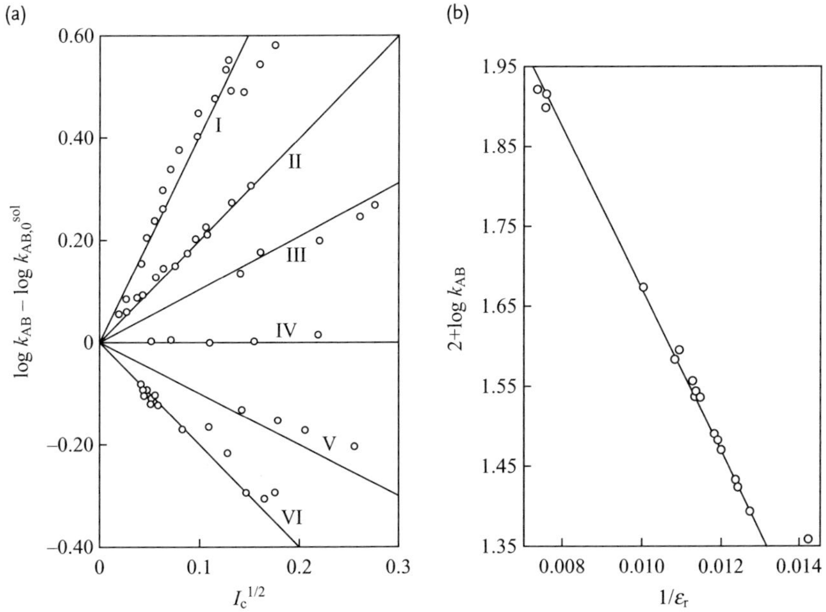

Figure 14.6 Dependence of the specific rate constants on (a) ionic strength I_c and (b) permittivity ε_r. Panel (a): I. $Co(NH_3)_5Br^{2+}$ + Hg^{2+}, $z_Az_B = +4$; II. $S_2O_8^{2-} + I^-$, $z_Az_B = +2$; III. $NO_2{:}N\cdot CO_2\cdot C_2H_5^- + OH^-$, $z_Az_B = +1$; IV. $CH_3CO_2C_2H_5 + OH^-$, $z_Az_B = 0$; V. $H_2O_2 + H^+ + Br^-$, $z_Az_B = -1$; VI. $Co(NH_3)_5Br^{2+} + OH^-$, $z_Az_B = -2$. Panel (b): The reaction between bromoacetate and thiosulfate in various media. Data from Ref. [11].

the negatively charged ions $BrAc^-$ and $S_2O_3^{2-}$ is decreased by the formation of complex ions, such that the specific reaction rate increases.

14.7.4
Influence of Permittivity

For a solution with ionic strength equal to zero, the terms in $\sqrt{I_c}$ in Eqs (14.66) and (14.72) disappear and we are left with only the effect of the permittivity. Plotting $\ln k_{AB}$, extrapolated to zero ionic strength, versus $1/\varepsilon_r$ results for Eq. (14.66) in

$$\frac{d\ln k_{AB,0}}{d(1/\varepsilon_r)} = -\frac{e^2 z_A z_B}{4\pi\varepsilon_0 kTr^*} \tag{14.75}$$

while for Eq. (14.72) one obtains

$$\frac{d\ln k_{AB,0}}{d(1/\varepsilon_r)} = \frac{e^2}{8\pi\varepsilon_0 kT}\left[\frac{z_A^2}{r_A} + \frac{z_B^2}{r_B} - \frac{(z_A+z_B)^2}{r_\ddagger}\right] \tag{14.76}$$

Both expressions give the same order of magnitude answer because the values of r^*, r_A, r_B and $r_\ddagger$ are not too different. From Eq. (14.75) it is immediately clear that the slope has a negative sign if the ions are similarly charged, and positive if they are oppositely charged. This result can also be obtained from Eq. (14.76).

The influence of the permittivity on the rate constants extrapolated to zero ionic strength for the reaction between bromoacetate and thiosulfate in various media is shown in Figure 14.6b. Here, the line corresponds to a value of $r^* = 0.51$ nm, but a good fit can also be obtained by using $r_A = 0.33$ nm (bromoacetate), $r_B = 0.17$ nm (thiosulfate) and $r_\ddagger = 0.50$ nm (activated complex), respectively. The use of mole fractions instead of volume fractions (which are actually used in the single-sphere model) leads also to a good fit with $r^* = 0.56$ nm. Yet, deviations can also occur, in particular for low-permittivity solvents.

Problem 14.6: The dissociation of $AgNO_3$

The dissociation constants K for $AgNO_3$ in H_2O and C_2H_5OH are 1170×10^3 mol l^{-1} and 4.42×10^3 mol l^{-1}, respectively. Estimate the value for K in CH_3OH using the double-sphere model, and compare the result with the experimental value $K = 20.5 \times 10^3$ mol l^{-1}. Estimate the value for $2r^*$ and compare the result with the sum of the Stokes radii for Ag^+ and NO_3^-. Use, where necessary, data from Appendix E.

14.8
Final Remarks

In this chapter, following an introduction to TST, we discussed the relevance of reaction- and diffusion-limited reactions. Moreover, the influence of dipole moment, permittivity and ionic strength of the solvent was treated. It is clear that trends can be predicted reasonably well, but that precise estimates cannot be expected.

References

1 See Laidler (1987).
2 See Connors (1990).
3 See Arnaut *et al.* (2007).
4 This approximation can be rationalized, see, e.g., (a) Bernasconi, C.F. (ed.) (1986) *Investigation of Rates and Mechanisms of Reactions, Part 1*, John Wiley & Sons, New York; (b) Steinfeld, J.I., Francisco, J.S., and Hase, W.L. (1998) *Chemical Kinetics and Dynamics*, 2nd edn, Prentice-Hall.
5 (a) Raff, L.M. (2001) *Principles of Physical Chemistry*, Prentice-Hall, Upper Saddle River, NJ; (b) Levine, I.N. (2002) *Physical Chemistry*, 5th edn, McGraw-Hill, Boston; (c) Kirkwood, J.G. (1934), *J. Chem. Phys.*, **2**, 351.
6 Benesi, A.J. (1982) *J. Phys. Chem.*, **86**, 4926.
7 Lyon, R.K. and Levy, D.H. (1961) *J. Am. Chem. Soc.*, **83**, 4290.
8 See Buncel *et al.* (2003).
9 See Reichardt (2003).
10 Kirkwood, J.G. (1934) *J. Chem. Phys.*, **2**, 351.
11 See Glasstone *et al.* (1941).

Further Reading

Arnaut, L., Formosinho, S., and Burrows, H. (2007) *Chemical Kinetics*, Elsevier, Amsterdam.

Buncel, E., Stairs, R., and Wilson, H. (2003) *The Role of Solvents in Chemical Reactions*, Oxford University Press, Oxford.

Connors, K.A. (1990) *Chemical Kinetics*, Wiley-VCH Verlag GmbH, Weinheim, Germany.

Glasstone, S., Laidler, K.J., and Eyring, H. (1941) *The Theory of Rate Processes*, McGraw-Hill, New York.

Laidler, K.J. (1987) *Chemical Kinetics*, 3rd edn, Harper and Row, London.

Moore, J.W. and Pearson, R.G. (1981) *Kinetics and Mechanism*, 3rd edn, John Wiley & Sons, Inc., New York.

Reichardt, R. (2003) *Solvents and Solvent Effects in Organic Chemistry*, 3rd edn, Wiley-VCH Verlag GmbH, Weinheim, Germany; See also 2nd edn (1988), Wiley-VCH, Weinheim, Germany.

Soustelle, M. (2011) *An Introduction to Chemical Kinetics*, John Wiley & Sons, Ltd, Chichester.

Wright, M.R. (2004) *Introduction to Chemical Kinetics*, John Wiley & Sons, Ltd, Chichester.

15
Some Special Topics: Surfaces of Liquids and Solutions

So far, in discussing the behavior of liquids and solutions we have limited ourselves to bulk fluids. However, in many cases interfaces are important, and in this chapter we deal with some aspects of these. In view of the wide range of phenomena associated with liquid interfaces, we limit ourselves to the thermodynamics and structure of planar (nonelectrified) surfaces of single-component liquids and binary solutions, the latter in relation to adsorption. For all other aspects, we refer to the literature.

15.1
Thermodynamics of Surfaces

To be able to discuss a multicomponent, multiphase system with interfaces, we need a concise notation. As before, we indicate the summation over the components, here labeled with the subscript i, by Σ_i. It is useful to refer to a particular component as the reference compound; this is usually taken to be the solvent and is labeled as 1. In considering two phases we have also to indicate the phase, and we do so by using a superscript (α) and (β) for the two bulk phases and (σ) for the interface, where the brackets are used to avoid confusion with exponentiation.

There are two approaches possible to describe the interfacial behavior of, say, a liquid and a vapor. First, the approach considering the interface as a real phase with volume, entropy, and so on (Figure 15.1a), as originated by van der Waals and Bakker and propagated in particular by Guggenheim. In this approach there are three phases to consider: the liquid; the vapor; and the interfacial phase (also called the *interphase*). For each phase the normal thermodynamic relations apply. There is only an additional work term for the interphase, to which we will come shortly. Second, we can associate the interface between two phases with a *dividing* (or *geometric*) *plane*, positioned in such a way that the total amount of a particular component, usually the solvent, is the same as if both bulk phases remained homogeneous up to that dividing plane (Figure 15.1a). This approach is due to Gibbs. Obviously, the dividing plane has volume zero but nevertheless it has other excess properties such as entropy and energy, which consequently can be positive or negative. The amount of component i in excess over the amount when both

Liquid-State Physical Chemistry, First Edition. Gijsbertus de With.

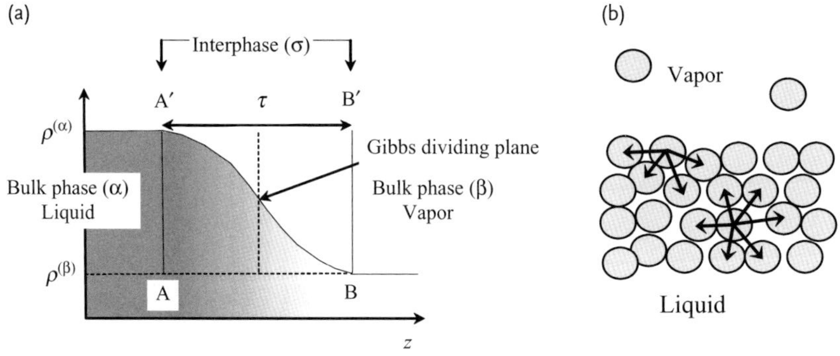

Figure 15.1 (a) Schematic representation of the density profile over a liquid surface; (b) Schematic representation of the liquid–vapor interface, showing the difference in environment for a molecule at the surface and in the bulk.

phases remain homogeneous up to the dividing plane as defined by component 1, is indicated by $n_i^{(1)}$ or per unit area as $\Gamma_i^{(1)} = n_i^{(1)}/A$, where A is the surface area. The quantity $\Gamma_i^{(1)}$ is addressed as *surface excess*. Obviously, $\Gamma_1^{(1)} = 0$. We will use both approaches.

Simple considerations reveal that a surface must have an associated energy. If we consider the surface in comparison with the bulk, we note that there is a significant difference in surroundings of the molecules resulting in a lesser number of intermolecular interactions for the surface, even if the average structure in the surface (apart from the cut) remains the same as in the bulk. After surface creation, relaxation takes place resulting in an average structure in the surface different from the one in the bulk, but an excess energy remains (Figure 15.1b). Generally, the thickness τ of the interfacial region is extremely small, say, a few molecular diameters or about 1 nm for normal liquids and molecular solutions. A larger thickness might have to be considered if ions are involved in view of the associated Debye length (see Chapter 12).

So far, we have treated liquids as homogeneous and obeying Euler's condition, so that total energy and entropy are additive if two volumes of liquid are added together. Moreover, we considered so far mainly volume work. However, surfaces have surface energy, so that if we create additional surface the work involved changes the various thermodynamic potentials, such as the Helmholtz energy. An elementary (though not so easy to carry out) experiment is the well-known surface extension experiment (Figure 15.2), which shows that to extend a surface by $\mathrm{d}A$, an amount of work $\delta W = \gamma\,\mathrm{d}A$ is needed, where γ is called the *interfacial tension*[1]. Stability requires that $\gamma > 0$, consistent with the missing interactions in the molecular picture. This implies that, without other constraints present, for a liquid drop minimization of the surface (Gibbs or Helmholtz) energy leads to a spherical surface.

1) For a one-component system, such as a liquid–vapor system, usually addressed as surface tension.

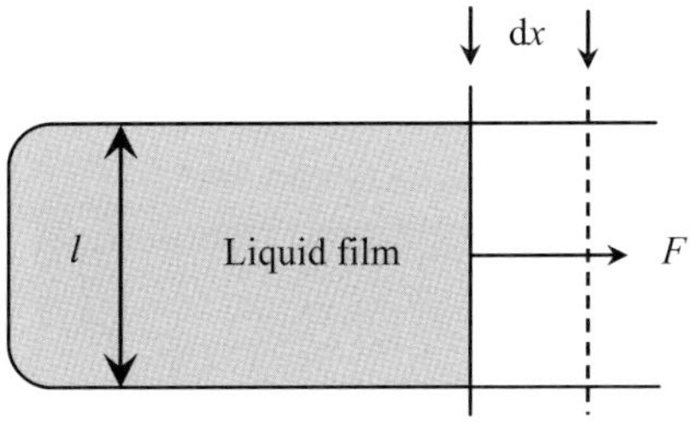

Figure 15.2 Schematic representation of a sliding-wire experiment to extend the area A of a liquid film at constant temperature T by $dA = 2l dx$, requiring a force F. Because the interface region of the film has a smaller density than the bulk of the film, during the process the volume changes and the process occurs actually at constant pressure P. Since the interfacial thickness τ is extremely small, say a few molecular diameters or about 1 nm, and therefore the interface volume $V^{(\sigma)}$ is small as compared to the bulk volume $V^{(L)}$, to a high degree of precision the process still occurs at constant volume V. In principle, the volume V could be kept really constant by carrying the experiment out in a closed set-up with cylinder and piston to keep the total volume constant. For a single-component liquid, because the work done is $F dx = \gamma\, dA$, the surface tension γ represents both the surface stress per unit length and the Helmholtz (Gibbs) energy per unit area.

It is useful to consider the work γdA somewhat more carefully. Consider to that purpose a system (Figure 15.1a), where the planes AA′ and BB′ are positioned in the bulk phases α and β parallel to the interface, so that their distance is τ and the volume of the interphase is $V^{(\sigma)} = A\tau$. Recall that the pressure $P = P^{(k)} + P^{(c)}$, with $P^{(k)}$ the kinetic and $P^{(c)}$ the configurational contribution (see Chapter 6), is the perpendicular force per unit area on an infinitesimal virtual test plane in the bulk phase, resulting from the collisions of all molecules on one side of this plane with all molecules at the other side. Consider now for a static liquid an infinitesimal cube with the xy plane parallel to the interface. The pressure on the yz-, xz-, and xy-planes is P_{xx}, P_{yy} and P_{zz}, respectively, and equilibrium of forces[2] requires that $\partial P_{xx}/\partial x = \partial P_{yy}/\partial y = \partial P_{zz}/\partial z = 0$. Because we have a planar interface, there is isotropy in the xy-plane, implying that $P_{xx} = P_{yy} \equiv P_t$ with P_t the *tangential pressure*. This also means that P_{xx}, P_{yy} and P_{zz} are functions of z only. From $\partial P_{zz}/\partial z = 0$, we have $P_{zz} \equiv P_n$ with P_n the *normal pressure*, constant everywhere in the fluid. Actually, the bulk is fully isotropic and thus $P_{xx} = P_{yy} = P_{zz} = P$, where P is the hydrostatic pressure and thus $P_n = P$. For the interphase we have only planar isotropy so that $P_t = P_t(z)$ varies with the height of the test plane within the interphase (Figure 15.3a). Summarizing so far, $P_n = P$ everywhere and equal to P_t in the bulk phases, but P_t is possibly different from P_n in the interphase.

Experimentally indeed $P_t \neq P_n$, as can be rationalized as follows. We recall that repulsive forces between molecules are much shorter ranged than attractive forces, so that we can neglect them and take the range of the attractive forces as finite.

2) The subscript xx refers to the force in the x-direction on a plane with normal in the x-direction, and is the standard notation for more complete discussions about mechanical stability. For full details see, e.g. Ref. [51].

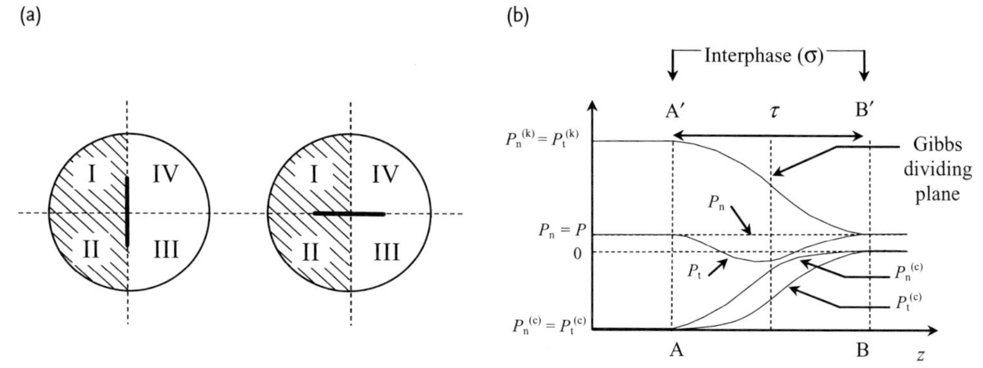

Figure 15.3 (a) The difference in attractive interactions between a test plane parallel and perpendicular to the interface. The hatched area represents the liquid, say phase (α), while the non-hatched area represents the vapor say phase, say phase (β). The range of interaction is indicated by the circles; (b) The contributions of the kinetic part of the pressure $P_n^{(k)} = P_t^{(k)}$ and the configurational part of the pressure $P_n^{(c)}$ and $P_t^{(c)}$ as function of z, that is, over the interphase (σ). While $P_n^{(c)} = P_t^{(c)}$ for the two bulk phases, $P_n^{(c)} \neq P_t^{(c)}$ for the interphase. The integral over the difference $P_n = P$ and P_t in the interphase represents the interfacial tension γ.

We take the test plane in the interphase (Figure 15.3), and note that the kinetic contribution to the pressure $P^{(k)}$ depends only on density (see Chapter 6). This is not true for the configurational part $P^{(c)}$. The decrease of attraction between molecules in quadrants I+II and III+IV for the orientation of the test plane parallel to the interface will be larger than the decrease in attraction between quadrants I+IV and II+III for the perpendicular orientation, because the density in both quadrants III and IV is small while quadrants I and II are densely populated [1]. The difference between $P = P_n$ and $P_t(z)$ integrated over the thickness τ of the interphase represents the interfacial tension

$$\gamma = \int_{-\frac{1}{2}\tau}^{+\frac{1}{2}\tau} [P - P_t(z)]\mathrm{d}z = \int_{-\infty}^{+\infty} [P - P_t(z)]\mathrm{d}z \tag{15.1}$$

where the last step can be made because in the bulk phases $P = P_t$. Summarizing, the surface energy corresponds to the surface stress parallel to the surface.

Consider now a change in area $\mathrm{d}A$, associated with a change in thickness $\mathrm{d}\tau$ and a change in volume to $V^{(\sigma)} + \mathrm{d}V^{(\sigma)}$ for a constant amount of material. For the work in the direction perpendicular to the interface we have $-PA\mathrm{d}\tau$. The work in the direction parallel to the interface is $-(P\tau-\gamma)\mathrm{d}A$, where the difference term, γ represents the interfacial tension. The total work done is then

$$\blacktriangleright \quad \delta W = -PA\mathrm{d}\tau - (P\tau - \gamma)\mathrm{d}A = -P(A\mathrm{d}\tau + \tau\mathrm{d}A) + \gamma\mathrm{d}A = -P\mathrm{d}V^{(\sigma)} + \gamma\mathrm{d}A \tag{15.2}$$

and this expression replaces $-P\mathrm{d}V^{(\alpha)}$ as used for the bulk phase.

In the Guggenheim picture, the Helmholtz energy of the system F is defined by

$$F \equiv F^{(\alpha)} + F^{(\beta)} + F^{(\sigma)} \tag{15.3}$$

where $F^{(\alpha)}$, $F^{(\beta)}$ and $F^{(\sigma)}$ denote the Helmholtz energies of the two homogeneous phases and the interphase, respectively. A similar expression is used for U, V, S and n_i. For the bulk phase we have

$$\mathrm{d}F^{(\alpha)} = -S^{(\alpha)}\mathrm{d}T - P\mathrm{d}V^{(\alpha)} + \sum\nolimits_i \mu_i \mathrm{d}n_i^{(\alpha)} \tag{15.4}$$

Clearly, the intensive properties T, P and μ_i do not require a phase superscript, since in equilibrium they are constant throughout the phases. Equation (15.4) is homogeneous of the first degree in the extensive properties $V^{(\alpha)}$ and $n_i^{(\alpha)}$. Therefore, from Euler's theorem (see Appendix B) we have

$$F^{(\alpha)} = -PV^{(\alpha)} + \sum\nolimits_i \mu_i n_i^{(\alpha)} \tag{15.5}$$

Similarly, for the surface phase we have

$$\mathrm{d}F^{(\sigma)} = -S^{(\sigma)}\mathrm{d}T - P\mathrm{d}V^{(\sigma)} + \sum\nolimits_i \mu_i \mathrm{d}n_i^{(\sigma)} + \gamma \mathrm{d}A \tag{15.6}$$

where γ is the surface tension. Equation (15.6) yields the Maxwell relation $(\partial S^{(\sigma)}/\partial A)_{T,V^{(\sigma)},n_i^{(\sigma)}} = -(\partial \gamma/\partial T)_{A,V^{(\sigma)},n_i^{(\sigma)}}$. Using Euler's theorem we find

$$F^{(\sigma)} = -PV^{(\sigma)} + \sum\nolimits_i \mu_i n_i^{(\sigma)} + \gamma A \tag{15.7}$$

From Eqs (15.6) and (15.3) we easily obtain

$$S^{(\sigma)} = -\left(\frac{\partial F^{(\sigma)}}{\partial T}\right)_{V^{(\sigma)},n_i^{(\sigma)},A} \quad \text{and} \quad \gamma = \left(\frac{\partial F^{(\sigma)}}{\partial A}\right)_{T,V^{(\sigma)},n_i^{(\sigma)}} = \left(\frac{\partial F}{\partial A}\right)_{T,V,n_i} \tag{15.8}$$

For the Gibbs energy[3] we use

$$G^{(\alpha)} \equiv F^{(\alpha)} + PV^{(\alpha)} \quad \text{and} \quad G^{(\sigma)} \equiv F^{(\sigma)} + PV^{(\sigma)} - \gamma A \tag{15.9}$$

so that

$$\mathrm{d}G^{(\sigma)} = -S^{(\sigma)}\mathrm{d}T + V^{(\sigma)}\mathrm{d}P + \sum\nolimits_i \mu_i \mathrm{d}n_i^{(\sigma)} - A\mathrm{d}\gamma \quad \text{and} \quad G^{(\sigma)} = \sum\nolimits_i \mu_i n_i^{(\sigma)} \tag{15.10}$$

the last expression in correspondence with the same result for the bulk phase. Because $F = F^{(\alpha)} + F^{(\beta)} + F^{(\sigma)} = -PV + \Sigma_i\mu_i n_i + \gamma A$, this requires that $G = F + PV$ is taken as $G = G^{(\alpha)} + G^{(\beta)} + G^{(\sigma)} + \gamma A$ [deviating from the form of Eq. (15.3)]. Calculating $\mathrm{d}G$ we find

► $$\mathrm{d}G = -S\mathrm{d}T + V\mathrm{d}P + \sum\nolimits_i \mu_i \mathrm{d}n_i + \gamma \mathrm{d}A \quad \text{and} \quad \gamma = \left(\frac{\partial G}{\partial A}\right)_{T,P,n_i} \tag{15.11}$$

consistent with the sliding-wire experiment (Figure 15.2).

Differentiating Eq. (15.7) and subtracting the result from Eq. (15.6), we obtain

$$\mathrm{d}\gamma = -s^{(\sigma)}\mathrm{d}T + \tau \mathrm{d}P - \sum\nolimits_i \Gamma_i \mathrm{d}\mu_i \quad \text{with} \tag{15.12}$$

3) Other definitions for the Gibbs energy of the interface exist because one can decide to transform with respect to all work terms, as in Eq. (15.9), or just using the volume work, i.e. using $G'^{(\sigma)} \equiv F^{(\sigma)} + PV^{(\sigma)}$. The latter choice results in $G'^{(\sigma)} = \Sigma_i \mu_i n_i^{(\sigma)} + \gamma A$ (deviating from the form for the bulk) and $\mathrm{d}G'^{(\sigma)} = -S^{(\sigma)}\mathrm{d}T - V^{(\sigma)}\mathrm{d}P + \Sigma_i\mu_i \mathrm{d}n_i^{(\sigma)} + \gamma \mathrm{d}A$ so that $\gamma = (\partial G'^{(\sigma)}/\partial A)_{T,P,ni}$ and $(\partial S^{(\sigma)}/\partial A)_{T,P,ni} = (\partial \gamma/\partial T)_{A,P,ni}$. In this case we have $G = G^{(\alpha)} + G^{(\beta)} + G'^{(\sigma)}$ (consistent with Eq. (15.3)), leading to the same result $\gamma = (\partial G/\partial A)_{T,P,n}$.

$$s^{(\sigma)} = S^{(\sigma)}/A, \quad \Gamma_i = n_i^{(\sigma)}/A \quad \text{and} \quad \tau = V^{(\sigma)}/A \tag{15.13}$$

Equation (15.12) is the surface analog of the Gibbs–Duhem equation, known as the *Gibbs adsorption equation*. Approximating the activity a_i in the expression for the chemical potential by the concentration c_i, we have

$$\mu_i = \mu_i^\circ + RT\ln a_i \cong \mu_i^\circ + RT\ln c_i \quad \text{or} \quad \mathrm{d}\mu_i = (RT/c_i)\mathrm{d}c_i \tag{15.14}$$

This results in

$$\mathrm{d}\gamma = -s^{(\sigma)}\mathrm{d}T + \tau\mathrm{d}P - \sum\nolimits_i (RT\Gamma_i/c_i)\mathrm{d}c_i \tag{15.15}$$

It can be shown that this change is independent of the precise choice for the position of the planes [2] AA′ and BB′. Using the Gibbs approach with $V^{(\sigma)} = 0$ is equivalent to taking the limit $\tau \to 0$ and referring the surface excess with respect to the resulting dividing plane determined by component 1. The result is

▶ $$\mathrm{d}\gamma = -s^{(\sigma)}\mathrm{d}T - \sum\nolimits_i (RT\Gamma_i^{(1)}/c_i)\mathrm{d}c_i \tag{15.16}$$

which is the form of the adsorption equation usually employed. Here, $\Gamma_i^{(1)}$ is the surface excess (or *relative adsorption*) of component i with respect to component 1. It can be shown that also this quantity is independent of the precise choice for the position of the dividing plane [3]. For a two-component system at constant T and P, we have $\Gamma_2^{(1)} = -(c_2/RT)(\partial\gamma/\partial c_2)_T = -RT^{-1}(\partial\gamma/\partial\ln c_2)_T$. Evidently, $\Gamma_2^{(1)}$ can be obtained by measuring the change in surface tension γ with concentration c.

Let us focus for the moment on a single-component system, for example, the liquid–vapor system. In this case, Eq. (15.12) reduces to

$$\mathrm{d}\gamma = -s^{(\sigma)}\mathrm{d}T + \tau\mathrm{d}P - \Gamma\mathrm{d}\mu \tag{15.17}$$

Employing $\mathrm{d}\mu = -S_\mathrm{m}^{(\alpha)}\mathrm{d}T + V_\mathrm{m}^{(\alpha)}\mathrm{d}P = -S_\mathrm{m}^{(\beta)}\mathrm{d}T + V_\mathrm{m}^{(\beta)}\mathrm{d}P$, where $S_\mathrm{m}^{(\alpha)}$ indicates the molar entropy and $V_\mathrm{m}^{(\alpha)}$ the molar volume, and eliminating $\mathrm{d}\mu$ and $\mathrm{d}P$, we obtain

$$-\frac{\mathrm{d}\gamma}{\mathrm{d}T} = \left(s^{(\sigma)} - \Gamma S_\mathrm{m}^{(\alpha)}\right) - \left[\left(\tau - \Gamma V_\mathrm{m}^{(\alpha)}\right)\frac{S_\mathrm{m}^{(\beta)} - S_\mathrm{m}^{(\alpha)}}{V_\mathrm{m}^{(\beta)} - V_\mathrm{m}^{(\alpha)}}\right] \cong s^{(\sigma)} - \Gamma S_\mathrm{m}^{(\alpha)} \tag{15.18}$$

The last step can be made for surfaces between a liquid (α) and a vapor (β). In the interphase the density is comparable to that in the liquid so that $\tau/\Gamma \cong V_\mathrm{m}^{(\alpha)}$, but $\tau/\Gamma << V_\mathrm{m}^{(\beta)}$. Therefore, remote from the critical temperature, $(\tau/\Gamma) - V_\mathrm{m}^{(\alpha)} << V_\mathrm{m}^{(\beta)} - V_\mathrm{m}^{(\alpha)}$. Similarly, employing $\mu = G_\mathrm{m}^{(\alpha)} = U_\mathrm{m}^{(\alpha)} - TS_\mathrm{m}^{(\alpha)} + PV_\mathrm{m}^{(\alpha)} = G_\mathrm{m}^{(\beta)} = U_\mathrm{m}^{(\beta)} - TS_\mathrm{m}^{(\beta)} + PV_\mathrm{m}^{(\beta)}$ and $g^{(\sigma)} = \Gamma\mu = u^{(\sigma)} - Ts^{(\sigma)} + P\tau - \gamma$ with $u^{(\sigma)} = U^{(\sigma)}/A$ and $g^{(\sigma)} = G^{(\sigma)}/A$, and eliminating $S_\mathrm{m}^{(\alpha)}$, $S_\mathrm{m}^{(\beta)}$ and $s^{(\sigma)}$, the result is

$$\gamma - T\frac{\mathrm{d}\gamma}{\mathrm{d}T} = \left(u^{(\sigma)} - \Gamma U_\mathrm{m}^{(\alpha)}\right) - \left[\left(\tau - \Gamma V_\mathrm{m}^{(\alpha)}\right)\frac{U_\mathrm{m}^{(\beta)} - U_\mathrm{m}^{(\alpha)}}{V_\mathrm{m}^{(\beta)} - V_\mathrm{m}^{(\alpha)}}\right] \cong u^{(\sigma)} - \Gamma U_\mathrm{m}^{(\alpha)} \tag{15.19}$$

where the last step can be made for the reason indicated before. For a created surface of unit area we may thus interpret γ as the work done, $-T(\partial\gamma/\partial T)$ as the heat absorbed (the difference in entropy of surface and bulk for the same amount of matter, multiplied by T), and $\gamma - T(\partial\gamma/\partial T)$ as the increase in internal energy (the difference in internal energy of surface and bulk for the same amount of matter).

If we use the Gibbs dividing surface for a single component system so that $V^{(\alpha)} = 0$, and position the dividing plane in such a way that $n_1^{(1)} = 0$, we obtain

$$dF^{(\sigma)} = -S^{(\sigma)}dT + \gamma dA \quad \text{and} \quad F^{(\sigma)} = \gamma A \tag{15.20}$$

and we see again that that the surface tension is equal to the surface Helmholtz energy per unit area[4]. For the (*specific*) *surface internal energy* one obtains

▶ $$u^{(\sigma)} = U^{(\sigma)} / A = (F^{(\sigma)} + TS^{(\sigma)}) / A = \gamma - T(\partial\gamma / \partial T) \tag{15.21}$$

as expected, consistent with Eqs. (15.18) and (15.19).

Finally, as an aside, let us briefly consider the equilibrium of an infinitesimal, not necessarily planar, interfacial area with pressure $P^{(\alpha)}$ on the one side and $P^{(\beta)}$ on the other side (at constant temperature and for constant amount of material). In the Gibbs approach, we have from Eqs (15.3), (15.4) and (15.6)

$$dF = dF^{(\alpha)} + dF^{(\beta)} + dF^{(\sigma)} = -P^{(\alpha)}dV^{(\alpha)} - P^{(\beta)}dV^{(\beta)} + \gamma dA = 0 \tag{15.22}$$

4) The terms "surface tension," "surface stress", and "surface energy" are often used indiscriminately in the literature. Although not that important for liquids, it is still useful to make the distinction. Since all work is given by the product of a "generalized displacement," for surface work the increase in area dA, and a "generalized force," for surface work called *surface stress* Ψ, we have for a single-component, single-phase system $\delta W = \Psi\, dA = d(\gamma A)$. As all work terms in the Gibbs equation, Ψ is dependent on the conditions being kept constant. For example, at constant T and V, $\Psi\, dA$ contributes to the Helmholtz energy. For liquid surfaces we have only a plastic increase of the surface area, that is, the average distance between molecules does not change upon increasing the area. For solid surfaces also an elastic increase of the surface area is possible, that is, the average distance between molecules does change upon increasing the area. Hence, in general the increase in area is given by $dA = dA_{ela} + dA_{pla}$. For a single-component, single-phase system we easily obtain:

$$\begin{aligned}\delta W &= d(\gamma A) = \gamma dA + A d\gamma \\ &= [\gamma + A(\partial\gamma / \partial A)]dA \\ &= [\gamma(dA_{pla} + dA_{ela}) / dA + A(\partial\gamma / \partial A_{ela})(\partial A_{ela} / \partial A)]dA \\ &= \{\gamma(dA_{pla} / dA) + [\gamma + A(\partial\gamma / \partial A_{ela})](dA_{ela} / dA)\}dA\end{aligned}$$

The above expression is conventionally designated as the *Shuttleworth equation*. For liquids, we regain the usual description $\delta W = \gamma\, dA$ so that *surface tension* γ equals the surface stress Ψ, that is the force per unit length in the liquid that tends to contract and minimize the surface. For solids, the full term in square brackets describes the surface stress Ψ. In general, measuring Ψ for solids is a difficult task as different methods lead to different values for Ψ and solid surfaces are usually not in equilibrium, as the relaxation might be kinetically hampered. *Surface energy* should refer to surface internal energy, similarly as energy refers to internal energy. Surface tension and surface Helmholtz energy have the same dimension and for liquids are numerically equal, as long as relaxation is fast. Note that surface internal energy has also the same dimension as surface tension but is not numerically equal. We use *surface tension* to avoid the somewhat longer *surface Helmholtz energy* and use, whenever required, the full designation.

Since $V = V^{(\alpha)} + V^{(\beta)}$ is constant, we have $dV^{(\alpha)} = -dV^{(\beta)}$, resulting in

$$(P^{(\alpha)} - P^{(\beta)}) = \gamma(dA / dV^{(\alpha)}) \tag{15.23}$$

From differential geometry we learn that $dA/dV = r_1^{-1} + r_2^{-2}$, where r_1 and r_2 are the radii of curvature in two orthogonal directions in the infinitesimal plane. It can be shown that $r_1^{-1} + r_2^{-1} = R_1^{-1} + R_2^{-1}$, where R_1 and R_2 are respectively the minimum and maximum radii of curvature in two orthogonal directions, that is, the principal curvatures. Using $dA/dV = R_1^{-1} + R_2^{-1}$ in Eq. (15.23), one usually refers to this as the *Laplace equation*, relating the surface tension to the pressure difference over a curved surface. It is here derived in an almost completely thermodynamic way. For a sphere, $R_1 = R_2 = R$ and

$$(P^{(\alpha)} - P^{(\beta)}) = 2\gamma / R \tag{15.24}$$

The above result can also be obtained using $V = 4\pi R^3/3$ and $A = 4\pi R^2$. It will be clear that for a planar surface $R_1 = R_2 = \infty$, resulting in $P^{(\alpha)} - P^{(\beta)} = 0$.

Problem 15.1

Calculate the surface energy of water at 25 °C, given that the surface tension $\gamma(0\,°C) = 75.6\,mJ\,m^{-2}$, $\gamma(25\,°C) = 72.0\,mJ\,m^{-2}$, and $\gamma(50\,°C) = 67.9\,mJ\,m^{-2}$.

Problem 15.2

Discuss the advantages and disadvantages of using either Eq. (15.17) or (15.20) as a basis for experimental work.

Problem 15.3

Show for a planar system containing bulk phases α and β and an interphase σ that the intensive quantities chemical potential μ, pressure P and temperature T are constant throughout the system.

Problem 15.4

Show for a planar system containing bulk phases α and β and an interface σ that the surface excess $\Gamma_i^{(1)}$ is independent of the position of the dividing plane. Hint: express $n_i^{(\sigma)}$ in terms of n_i, $c_i^{(\alpha)}$, $c_i^{(\beta)}$ and $V^{(\beta)}$, subtract $n_1^{(\sigma)}(c_i^{(\alpha)} - c_i^{(\beta)})/(c_1^{(\alpha)} - c_1^{(\beta)})$ from the result, and note that one side is independent of the position of the dividing plane.

15.2 One-Component Liquid Surfaces

The surface tension of a liquid decreases with temperature T and must vanish at the critical temperature T_{cri}. One of the first empirical expressions for the surface

Table 15.1 Various data for liquids.[a)]

Liquid	$K \times 10^7$ (J K^{-1} mol$^{-2/3}$)	T_{cri} (K)	$U^{(\sigma)}$ (cal mol^{-1})	L (cal mol^{-1})	$U^{(\sigma)}/L$
Ar	2.02	150.7	745	2 278	0.431
N_2	2.00	126.0	612	1 711	0.358
CO	2.00	132.9	647	1 981	0.326
CS_2	2.022	546.2	2 784	8 444	0.330
CCl_4	2.105	556.3	2 951	11 220	0.263
C_6H_6	2.104	556.6	2 952	11 934	0.258

a) Data from Ref. [52].
K, Eötvös constant; T_{cri}, critical temperature; $U^{(\sigma)}$, surface internal energy; *L*, energy of vaporization.

tension γ as a function of T, applicable to simple and normal liquids, is *Eötvös' rule* [4] which states that, if V_m is the molar volume, γ for a normal liquid is given by

$$\gamma V_m^{2/3} = K(T_{cri} - T) \quad \text{with} \quad K = 2.1 \times 10^{-7}\ \text{J K}^{-1}\ \text{mol}^{-2/3} \tag{15.25}$$

The constant K is supposed to be a universal constant valid for all liquids (Table 15.1), which appears to be approximately true for many nonpolar and nonhydrogen-bonding liquids[5)]. In practice, the value for hydrogen-bonding liquids varies with the type of compound, typically 0.7–1.5 × 10^{-7} J K^{-1}mol$^{-2/3}$ for alcohols and 0.9–1.7 × 10^{-7} J K^{-1}mol$^{-2/3}$ for organic acids. The lower values indicate association in these liquids, and one has tried to estimate the degree of association from these data. Although qualitatively sound, accurate values for the degree of association cannot be obtained.

Estimating the molecular area $a \equiv A/N_A$ as $a = \kappa(V_m/N_A)^{2/3}$, $F^{(\sigma)} = \gamma A$ becomes $F^{(\sigma)} = \kappa N_A^{1/3}\gamma V_m^{2/3}$. Here, κ is a numerical factor close to one, for example, using a lattice model with FCC arrangement for the molecules, one has $\kappa = 2^{1/3} = 1.26$ for the (100) plane, $2^{5/6} = 1.78$ for (110) plane, and $2^{-2/3}3^{1/2} = 1.09$ for the (111) plane. To obtain the surface internal energy $U^{(\sigma)}$ we use the Gibbs–Helmholtz relationship $U^{(\sigma)} = \partial(F^{(\sigma)}/T)/\partial(1/T)$, and from Eq. (15.25) easily obtain $U^{(\sigma)} = \kappa N_A^{1/3}KT_{cri}$. To obtain the surface entropy $S^{(\sigma)}$, we use $S^{(\sigma)} = -\partial F^{(\sigma)}/\partial T$ and obtain $S^{(\sigma)} = \kappa N_A^{1/3}K$. Inserting the value of K, it appears that $S^{(\sigma)} \cong 2.0k$. For water, $S^{(\sigma)} \cong 1.1k$.

In a simple lattice picture of liquids with N molecules in the bulk with coordination number z and nearest-neighbor interaction w, the potential energy Φ and heat of vaporization L read, respectively,

$$\Phi = \frac{1}{2}Nzw \quad \text{and} \quad L = -\Phi \tag{15.26}$$

5) Eötvös used $K = 2.27 \times 10^{-7}$ J K^{-1}mol$^{-2/3}$, but the present value represents the state of affairs somewhat better. Slightly more precise estimates are obtained using the *Eötvös–Ramsey–Shields rule* $\gamma V_m^{2/3} = K'(T_{cri} - T - \delta)$, with $K' \cong 2.12 \times 10^{-7}$ J K^{-1}mol$^{-2/3}$ and $\delta \cong 6$. There is no theoretical argument whatsoever for this modification [53].

If we cleave a liquid into two parts, we transfer molecules to the surface with surface coordination number $z^{(\sigma)}$ and the work expended $W^{(\sigma)}$ is

$$W^{(\sigma)} = \frac{1}{2}N(z^{(\sigma)} - z)w \tag{15.27}$$

since the number of molecules N which in the bulk had a potential energy $\frac{1}{2}zw$ have now a potential energy $\frac{1}{2}z^{(\sigma)}w$. At 0 K, where the argument applies, we have for the molar energy associated with surface $U^{(\sigma)} = W^{(\sigma)}$. Note that $w < 0$ and $z^{(\sigma)} < z$, so that $U^{(\sigma)} > 0$. We obtain

$$U^{(\sigma)} / L = 1 - (z^{(\sigma)} / z) \tag{15.28}$$

Using the FCC (111) plane for the surface, we have $z^{(\sigma)} = 9$ and $z = 12$, resulting in $U^{(\sigma)}/L = 0.25$. While experimentally for simple liquids values of ~0.4 are observed, for normal liquids the estimate $U^{(\sigma)}/L = 0.25$ is reasonably good (Table 15.1)[6].

Empirically extending the dependence of γ on temperature T, still using the fact that γ vanishes at the critical temperature T_{cri}, one can write [5]

▶ $$\gamma = \gamma_0 (1 - T/T_{\text{cri}})^{1+r} \tag{15.29}$$

with γ_0 and r constants. It appears that for simple liquids [2] excellent agreement is obtained with $r = 2/9$, while for a range of organic compounds the average value $r = 0.210 \pm 0.015$ describes the experimental data well [6a, 6b][7]. Katayama [7] argued that the surface tension is determined by the difference in interactions between the liquid and vapor phase and therefore related to the density difference. Defining $y = (\rho_{\text{L}} - \rho_{\text{V}})/M$ or $yV_{\text{cri}} = (\rho_{\text{L}} - \rho_{\text{V}})/\rho_{\text{cri}}$, with M the molecular mass and ρ_{L} and ρ_{V} the density of the liquid and vapor, respectively, Katayama proposed to use

$$\gamma y^{-2/3} = a(T_{\text{cri}} - T) \quad \text{with} \quad a = 2.04 \times 10^{-7}\ \text{J K}^{-1}\ \text{mol}^{-2/3} \tag{15.30}$$

and where a is given as the average value for six liquids. Guggenheim [8] showed that for simple liquids, the density behavior of liquids and gases can be accurately represented by (see Chapter 4)

$$(\rho_{\text{L}} - \rho_{\text{V}})/\rho_{\text{cri}} = (7/2)(1 - T/T_{\text{cri}})^{1/3} \tag{15.31}$$

Consequently, using Eq. (15.31), we can write

$$y = (7\rho_{\text{cri}} / 2M)(1 - T/T_{\text{cri}})^{1/3} \tag{15.32}$$

and eliminating y between Eqs (15.30) and (15.32), we obtain

$$\gamma = aT_{\text{cri}}(7\rho_{\text{cri}} / 2M)^{2/3}(1 - T/T_{\text{cri}})^{11/9} \equiv \gamma_0 (1 - T/T_{\text{cri}})^{11/9} \tag{15.33}$$

explaining nicely the empirically observed relation with $r = 2/9$.

We can also eliminate T/T_{cri} between Eqs (15.30) and (15.31) with as the result

$$\gamma = aT_{\text{cri}} \left(\frac{2}{7}\right)^3 \left(\frac{\rho_{\text{cri}}}{M}\right)^{2/3} \left(\frac{\rho_{\text{L}} - \rho_{\text{V}}}{\rho_{\text{cri}}}\right)^{11/3} \sim (\rho_{\text{L}} - \rho_{\text{V}})^{11/3} \tag{15.34}$$

6) Note that $U^{(\sigma)} = \gamma A$ where A is the molar area, which here is estimated with $\kappa = 2^{1/3}$.

7) A recent compilation of data [6c] yields for 85 normal liquids, $r = 0.249$ with a sample standard deviation of 0.054.

In 1923, McLeod [9], unaware of Katayama's results, also suggested a relationship between the surface tension γ and liquid and vapor densities ρ_L and ρ_V given by

$$\gamma = c(\rho_L - \rho_V)^4 \quad \text{with} \quad c = \text{molecule-dependent constant} \tag{15.35}$$

to be compared with $\gamma \sim (\rho_L - \rho_V)^{11/3}$. In fact, even the data McLeod used–that is, the data of Ramsay and Shields[5]–showed the exponent to be somewhat smaller than 4.

Sugden [10], ignoring the fact that an exponent 11/3 is more suitable than 4, and hence using $r = 1/5$ in Eq. (15.29), rewrote the McLeod expression to

$$P = \gamma^{1/4} M / (\rho_L - \rho_V) \tag{15.36}$$

with M the molecular mass (for polymers the molecular mass of the repeating unit).

The parameter P, which appears experimentally to be independent of temperature over a reasonable temperature range, is denoted as *parachor P*, and Sugden related it to the molecular structure. In essence, he used a group contribution method for which the necessary data were refined by Quayle [11]. Table 15.2 provides data as given by Poling, Prausnitz, and O'Connell [12], who state that they are often as accurate as those from Quayle (fitted to surface tension data), except for compounds such as benzonitrile, carbon disulfide, carbon tetrachloride and pyridine, where the errors are unacceptably high. Originally, the parachor was used to study molecular structure, but nowadays spectroscopic and diffraction methods have superseded this approach, although it is still used for estimating surface tension. Nevertheless, there are some serious flaws in this correlator, as demonstrated clearly by Exner[8] [13], but unfortunately his critique has been largely neglected in the literature.

It appears that the surface tension also obeys the principle of corresponding states (PoCS). Guggenheim [2] noticed that a dimensionless parameter is obtained if we take $\gamma_0 V_{cri}^{2/3} T_{cri}$. For the set Ne, Ar, Xe, N_2, O_2 and CH_4, its average value is (4.30 ± 0.16), where $\pm$ indicates the sample standard deviation (see Table 4.3). An improved correlation is obtained by using the acentric factor ω (see Chapter 4). As a first step we use the reduced density $\rho_{red} = \rho/\rho_{cri}$, which we take as a function of $T_{red} = T/T_{cri}$ and ω. According to Riedel[9] [14], we can take this correlation as

$$\rho / \rho_{cri} = 1 + 0.85(1 - T_{red}) + (1 - T_{red})^{1/3}(1.89 + 0.91\omega) \tag{15.37}$$

At $T_{red} = 0$ we obtain ρ_0 representing the (hypothetical) density at 0 K. For $T_{red} < 0.8$ it appears that the value of ρ/ρ_0 is more or less constant, independent of the acentric factor ω. Using this correlation and knowing the ω value, a single measurement below $T_{red} = 0.8$ is sufficient to predict the density variation over the full temperature range. As a side product we are also able to estimate the thermal expansion coefficient or expansivity α.

The second step is using Eq. (15.33), where γ_0 is interpreted as the (hypothetical) surface tension at 0 K. It appears that the correlation

$$\gamma_0 V_0^{2/3} T_{cri}^{-1} = 1 + 1.18\omega \tag{15.38}$$

8) His paper has the inspiring title: *How to get wrong results from good experimental data.*
9) In fact, Riedel used a parameter $\alpha = 5.808 + 4.93\omega$, which is obviously fully equivalent to ω.

Table 15.2 Structural contributions for the parachor P.[a)]

Carbon-hydrogen		**R–[–CO–]–R′ for CO in ketones**	
C	9.0	R + R′ = 2	51.3
H	15.5	R + R′ = 3	49.0
CH_3	55.5	R + R′ = 4	47.5
CH_2 in $-CH_2-$		R + R′ = 5	46.3
$n < 12$	40.0	R + R′ = 6	45.3
$n > 12$	40.3	R + R′ = 7	44.1
Alkyl groups		–CHO	66
1-methylethyl	133.3	O (if not noted above)	20
1-methylpropyl	171.9	N (if not noted above)	17.5
1-methylbutyl	211.7	S	49.1
2-methylpropyl	173.3	P	40.5
1-ethylpropyl	209.5	F	26.1
1,1-dimethylcthyl	170.4	Cl	55.2
1,1-dimethylpropyl	207.5	Br	68.0
1,2-dimethylpropyl	207.9	I	90.3
1,1,2-trimethylpropyl	243.5		
C_6H_5	189.6	Ethylenic bonds =C<	19.1
Special groups		**Terminal**	
–COO– (esters)	63.8	2,3 position	17.7
–COOH (acids)	73.8	3,4 position	16.3
–OH	29.8	Triple bond	40.6
$-NH_2$	42.5	**Ring closure**	
–O–	20.0	Three-membered	12
$-NO_2$	74	Four-membered	6.0
$-NO_3$ (nitrate)	93	Five-membered	3.0
$-CO(NH_2)$	91.7	Six-membered	0.8

a) Data from Ref. [12], as modified from Ref. [11], to be used with γ[dyn cm = 10^{-3} N m^{-2}] and M/ρ [cm^3 mol^{-1}].

with V_0 the (hypothetical) volume at 0 K as calculated from ρ_0, describes the behavior rather well (Figure 15.4). Using the density ρ at a given T_{red}, we obtain ρ/ρ_0 from Eq. (15.37) and the reference volume V_0 is then $V_0 = (m/\rho)(\rho/\rho_0)$. Together with T_{cri}, Eq. (15.38) provides an estimate for γ_0. The fit of Eq. (15.38) is rather good–typically the deviations for normal fluids are less than 5%–and this led Pitzer [15] to suggest that γ_0 could be used to test whether liquids behave as "normal" (see Chapter 4).

Another approach [16] to estimate the surface tension is based on the comparison of lattice considerations (for a liquid and a surface taking interactions up to the fifth shell into account) and conventional thermodynamics. Assuming a stepwise change in density for the interface, usually denoted as the *Fowler (–Kirkwood–Buff) approximation*, the analysis leads to

$$\gamma = (V_{\text{m}} / N_{\text{A}})^{1/3} (\partial U / \partial V)_T / 8 \tag{15.39}$$

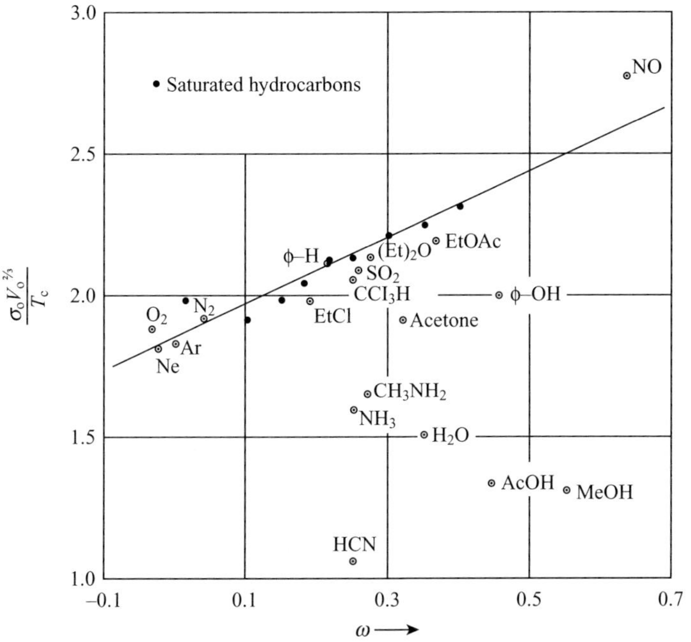

Figure 15.4 The reduced surface tension $\gamma_0 V_0^{2/3}/T_{\text{cri}}$ as a function of the acentric factor ω. The data points for normal fluids are all within 5% of the line given by Eq. (15.38). Data from Ref. [54].

The derivative $(\partial U/\partial V)_T$ is the internal pressure of the liquid which can be calculated from (the experimentally accessible) $\alpha T/\kappa_T - P$, where α and κ_T are the expansivity and compressibility, respectively, and P is the ambient pressure. The agreement of the predicted values with respect to the experimental values for eight liquids at various temperatures was ~17%, and quite good considering the simplified nature of the model. In fact, it was shown [17] for a variety of liquids that $\gamma\kappa_T = l$ was relatively constant, and it was suggested that l represents an intrinsic length scale. A first-principles statistical mechanical derivation of such a relation for monoatomic liquids by using density functional theory (DFT) arguments for the density $\rho(z)$ through the liquid–vapor interface is available [18]. Furthermore, it was shown that a good empirical correlation [19] exists between γ and κ_T for 24 organic liquids near 298 K that reads $\gamma\kappa_T = 27.4 \pm 3.4\,\text{pm}$, where $\pm$ indicates the sample standard deviation. Sanchez [20] provides $\gamma(\kappa_T/\rho)^{1/2} = A^{1/2}$ with $A^{1/2} = 2.78 \pm 0.13\ (\text{erg}\,\text{cm}^2/\text{g})^{1/2}$ as a temperature-independent constant, and where about half of the variation was attributed to experimental errors. Although the constant $A^{1/2}$ was shown to be proportional to the second moment of the direct correlation function, and was empirically related to the Lennard-Jones energy parameter ε and size parameter σ via $A^{1/2} = 0.26(\varepsilon\sigma^2/M)^{1/2}$ with M the molecular mass, the physical implications of this correlation are unclear.

Table 15.3 Surface tension as estimated for various liquids using the Kirkwood–Buff approach.

Liquid	T (K)	σ (Å)	ε/k (K)	γ_{cal} (mJ m^{-2})	γ_{exp} (mJ m^{-2})	$u^{(\sigma)}_{cal}$ (mJ m^{-2})	$u^{(\sigma)}_{exp}$ (mJ m^{-2})
Ar	84.3	3.255	147.23	15.1	13.2	27.08	35.0
Kr	117	3.599	168.51	17.1	16.1	33.44	40.1
Xe	161.5	3.750	296.40	24.5	19.3	41.93	50.0
Ne	33.1	2.761	33.44	4.5	2.7	8.05	14.3
N_2	64	3.341	146.43	12.7	12.0	22.93	27.5
O_2	77	3.032	199.30	17.6	16.5	30.85	37.1
CH_4	96	3.579	181.25	15.8	16.0	30.56	35.8

σ and ε, Lennard-Jones parameters; γ, surface tension; $u^{(\sigma)}$, internal surface energy.

Yet another approach is based on the pair correlation function $g(r)$. Accepting again Fowler's approximation for the interface and using Eq. (6.39) in combination with Eq. (15.1), Kirkwood and Buff derived, denoting as before the potential energy and its derivative by $\phi(r)$ and $d\phi(r)/dr = \phi'(r)$, respectively, the expression for the surface tension γ and surface internal energy $u^{(\sigma)}$ as

$$\gamma = \left(\frac{\pi}{8}\right)\rho^2\int_0^\infty g(r)\phi'(r)r^4\,dr \quad \text{and} \quad u^{(\sigma)} = -\left(\frac{\pi}{2}\right)\rho^2\int_0^\infty g(r)\phi(r)r^3\,dr \tag{15.40}$$

In their approach [21] $\rho = \rho_L$ denotes the number density of molecules in the liquid. Using the Lennard-Jones potential for $\phi(r)$ as determined from the internal energy U and vapor pressure P, the surface tension γ and surface internal energy $u^{(\sigma)}$ were calculated [22] for several simple molecules (Table 15.3). The agreement for γ with experiment is good. Evidently, the agreement for γ is much less as for $u^{(\sigma)}$, and this was attributed to the use of the Fowler approximation and neglect of the vapor phase. It should be noted that the Lennard-Jones parameters used should be chosen properly, that is, from liquid state data U and P, as indicated. These parameters are normally estimated from the second virial coefficient $B(T)$, or from viscosity data $\eta(T)$. Generally, $\sigma\{U,P\} < \sigma\{B(T),\eta(T)\}$ and $\varepsilon\{U,P\} > \varepsilon\{B(T),\eta(T)\}$. Values for γ and $u^{(\sigma)}$ calculated from either $B(T)$ or $\eta(T)$ resulted, in most cases, in physically impossible values. The drawback of the approach is similar as for the bulk, namely that it is largely applicable only to molecules that are more or less spherical, since otherwise the orientational angles must be introduced and this complicates the formalism tremendously.

Later [23], it was suggested that $\rho = \rho_L - \rho_V$ be used instead of just $\rho = \rho_L$. This extended the applicability of the Kirkwood–Buff predictions from only close to the triple point to close to the critical point. Further, one could wonder about the effect of the use of the Fowler approximation for Lennard-Jones fluids. Its effect has been evaluated [24], and it appears that any agreement of theoretical results using this approximation with simulations and experiments is fortuitous. It accounts for ~70% of the value for γ as obtained from simulations for Lennard-Jones liquids.

Finally, we note that the surface tension itself becomes size-dependent for drops with very small radius r, and that this dependency [25] is expressed by $\gamma = \gamma_0(1 - 2\delta/r)$, where γ_0 is the surface tension for a flat surface and δ is the *Tolman length* with a magnitude on the order of a few tenths of the molecular diameter.

In conclusion, in spite of being important, the surface structure of pure liquids is still incompletely known and the debate continues. While the basic theory, as exposed in several reviews [26], is essentially clear, development and experimentation are continuing, for example using optical and X-ray scattering techniques [27], to provide continuously new insights.

Problem 15.5

Show that for the FCC lattice, $\kappa = 2^{1/3}$ for the (100) plane, $\kappa = 2^{5/6}$ for (110) plane, and $\kappa = 2^{-2/3}3^{1/2}$ for the (111) plane.

Problem 15.6

Show that for the Eötvös–Ramsey–Shields model, $\gamma V_m^{2/3} = K'(T_{cri} - T - \delta)$, $U^{(\sigma)} = \kappa N_A^{1/3} K'(T_{cri} - \delta)$, and $S^{(\sigma)} = \kappa N_A^{1/3} K'$. Indicate why this expression for γ cannot be generally valid.

Problem 15.7

Verify Eqs (15.33) and (15.34).

Problem 15.8

Estimate the surface tension for cyclohexane, using a vaporization energy $L = 30.5\,\text{kJ mol}^{-1}$, a density $\rho = 773\,\text{kg m}^{-3}$, and molecular mass $M = 84.16\,\text{g mol}^{-1}$, by assuming a cubic structure. Compare your result with the experimental value of $0.0247\,\text{N m}^{-1}$, and comment.

15.3 Gradient Theory

While the theory of fluids is somewhat complex, the theory for their surfaces is even more complex. We limit ourselves here to the classical picture initiated by van der Waals, often called (square) gradient theory, along the line as presented by Widom [28]. For more sophisticated treatments we refer to the literature, for example, Rowlinson and Widom [29], Davis [30], Safran [31], and Hansen and McDonald [32]. The basic idea of gradient theory is that, locally in the interphase, equilibrium prevails and therefore the total Helmholtz energy can be written as an integral over the local Helmholtz energy density, the latter being a function of

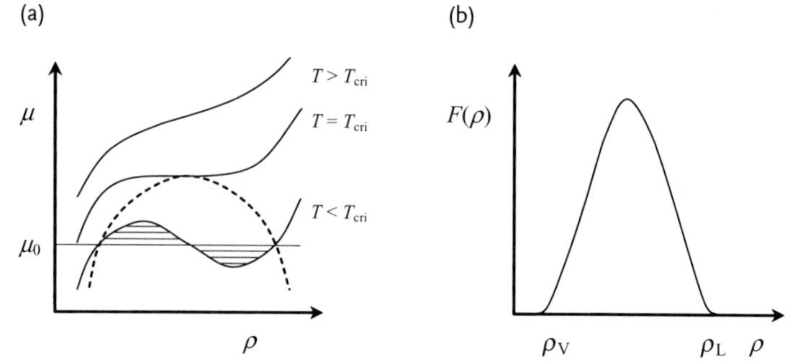

Figure 15.5 (a) Schematic representation of the change of chemical potential μ with density ρ for $T < T_{cri}$, $T = T_{cri}$ and $T > T_{cri}$; (b) The excess Helmholtz energy $F(\rho)$ between ρ_L and ρ_V.

density ρ only. We take the surface as the x–y plane and assume that the liquid is isotropic so that ρ is a well-defined function of z only, that is, $\rho = \rho(z)$. For the interphase, the gradient in density is indicated by $\rho'(z)$, where (here and in the sequel) the prime indicates the derivative with respect to the argument. The change in chemical potential μ of a liquid as a function of ρ is shown in Figure 15.5. In the interphase the density takes values between ρ_L and ρ_V which are excluded for the bulk phase by the equal-area rule. At any temperature in the two-phase region, the chemical potential in the interphase follows the metastable line $f(\rho)$ between ρ_V and ρ_L, so that the "excess" Helmholtz energy $F(\rho)$ is the difference between $f(\rho)$ and the tie line indicated by μ_0. Hence, it reads

$$F(\rho) = \int_{\rho_L \text{ or } \rho_V}^{\rho} [f(\rho) - \mu_0]\mathrm{d}\rho \tag{15.41}$$

The lower limit can be taken as either ρ_V or ρ_L since, in view of the equal-area rule

$$\int_{\rho_V}^{\rho_L} [f(\rho) - \mu_0]\mathrm{d}\rho = 0 \tag{15.42}$$

with as consequence that $F(\rho_V) = F(\rho_L) = 0$. Figure 15.5 shows a schematic of $F(\rho)$ indicating that, at any density between ρ_V and ρ_L, a positive excess energy is present.

The Helmholtz energy of the system would be minimized by an infinitely sharp transition between regions with density ρ_L and density ρ_V if a gradient in density in the interphase did not exist. However, a gradient $\rho'(z)$ does exist and this gradient must stabilize the liquid at the densities in the interphase, since keeping the liquid at a density different from ρ_L and ρ_V will cost a significant amount of Helmholtz energy. In fact, the transition between ρ_L and ρ_V contributes to the excess Helmholtz energy, as can be seen in the following (this is in fact the same argument as used in Section 15.1). Consider two levels in the interphase with density $\rho(z)$ and $\rho(z + \xi)$, respectively. With r as the distance between a fixed point at level

z and an arbitrary volume element $\mathrm{d}\mathbf{r}$ at level $z + \xi$, we have an extra contribution to the Helmholtz (actually internal) energy density given approximately by

$$\frac{1}{2}\rho(z)\int \phi(r)[\rho(z+\xi)-\rho(z)]\mathrm{d}\mathbf{r} \tag{15.43}$$

where $\phi(r)$ is the potential energy between molecules a distance r apart. The integration is over all space with r and ξ varying in the integration, while z and the point chosen at height z are fixed. The total excess Helmholtz energy is according to the above given by

$$\gamma = \int_{-\infty}^{+\infty}\left\{F[\rho(z)] + \frac{1}{2}\rho(z)\int \phi(r)[\rho(z+\xi)-\rho(z)]\mathrm{d}\mathbf{r}\right\}\mathrm{d}z \tag{15.44}$$

The integral is function of a function, that is, a *functional*. The minimum of this functional of the density profile $\rho(z)$, subject to the conditions $\rho(-\infty) = \rho_L$ and $\rho(+\infty) = \rho_V$, provides the surface tension.

For small gradients we can expand $\rho(z + \xi)$ in powers of ξ to second order and obtain

$$\rho(z+\xi)-\rho(z) = \rho'(z)\xi + \frac{1}{2}\rho''(z)\xi^2 + \ldots \tag{15.45}$$

The first-order term does not contribute to the integral in Eq. (15.44), since the integration is over all space and $\phi(r)$ is a spherically symmetric function while ξ is odd. Using the same symmetry, the second-order term contributes one-third from what it would contribute if ξ^2 were to be replaced by r^2 (see also Justification 13.1). Therefore

$$\gamma = \int_{-\infty}^{+\infty}\left\{F[\rho(z)] - \frac{1}{2}m\rho(z)\rho''(z)\right\}\mathrm{d}z \quad \text{with} \quad m = -6^{-1}\int \phi(r)r^2\,\mathrm{d}\mathbf{r} \tag{15.46}$$

The *influence parameter m* is proportional to the second moment of $\phi(r)$, and therefore measures the range of $\phi(r)$. A more elaborate analysis shows that m is related to the direct correlation function $c(r)$. Approximating by $c(r) \cong c_0(r) - \beta\phi(r)$, with $c_0(r)$ the correlation function of a reference system and $\beta = 1/kT$, results in

$$m = 6^{-1}kT\int c(r)r^2\,\mathrm{d}\mathbf{r} \cong 6^{-1}kT\int [c_0 - \beta\phi(r)]r^2\,\mathrm{d}\mathbf{r} \cong -6^{-1}\int_0^{\infty}\phi(r)r^2\,\mathrm{d}\mathbf{r} \tag{15.47}$$

The last step can be made in view of the large weight given to $\phi(r)$ by the factor r^2. The term $-\frac{1}{2}m\rho(z)\rho''(z)$ in Eq. (15.46) can be transformed by partial integration and, because the boundary terms vanish identically, this results in $\frac{1}{2}m\rho'(z)^2$. The final result is thus

► $$\gamma = \int_{-\infty}^{+\infty}\left\{F[\rho(z)] + \frac{1}{2}m\rho'(z)^2\right\}\mathrm{d}z \tag{15.48}$$

Because the potential is attractive, we have $\phi(r) < 0$, $m > 0$ and, consequently, $\gamma > 0$.

The minimum energy profile $\rho(z)$ of Eq. (15.46) can be obtained by using the Euler condition (see Appendix B, Eq. B.61), which in this case becomes[10)]

$$F'(\rho) - m\rho''(z) = 0 \tag{15.49}$$

Integrating, we obtain (note that $F(\rho_L) = F(\rho_V) = 0$):

$$\int F'(\rho)\mathrm{d}\rho - \frac{1}{2}m\int \rho''(z)\mathrm{d}z = 0 \quad \text{or} \quad F(\rho) - \frac{1}{2}m\rho'(z)^2 = 0 \tag{15.50}$$

from which we easily derive that

$$\rho'(z) = [2F(\rho)/m]^{1/2} \tag{15.51}$$

Using Eq. (15.48), the expression for the surface tension thus becomes

$$\blacktriangleright \quad \gamma = m\int_{-\infty}^{+\infty} \rho'(z)^2\,\mathrm{d}z = m\int_{\rho_V}^{\rho_L} \rho'(z)\mathrm{d}\rho = (2m)^{1/2}\int_{\rho_V}^{\rho_L} F(\rho)^{1/2}\,\mathrm{d}\rho \tag{15.52}$$

A second integration yields a parametric representation of the density profile

$$z = (m/2)^{1/2}\int_{\rho(0)}^{\rho(z)} [F(\rho)]^{-1/2}\,\mathrm{d}\rho \tag{15.53}$$

At liquid–gas coexistence, $F(\rho)$ has two minima of equal depth. One is located at $\rho = \rho_L$ and another at $\rho = \rho_V$. A simple parameterization, valid near the critical point, is

$$F(\rho) = C(\rho_L - \rho)^2(\rho - \rho_V)^2 \quad \text{with} \quad C = C(T) \tag{15.54}$$

a temperature-dependent parameter. Substituting Eq. (15.54) in Eq. (15.53), we obtain

$$z = (m/2C)^{1/2}\int_{\rho(0)}^{\rho(z)} [(\rho_L - \rho)(\rho - \rho_V)]^{-1}\,\mathrm{d}\rho = \lambda \ln\left(\frac{\rho - \rho_V}{\rho_L - \rho}\right) \tag{15.55}$$

Here, $\lambda = (m/2C)^{1/2}/(\rho_L - \rho_V)$ is a characteristic length for the interfacial width. Solving for ρ we find

$$\rho(z) = \frac{1}{2}(\rho_L + \rho_V) - \frac{1}{2}(\rho_L - \rho_V)\tanh(z/2\lambda) \tag{15.56}$$

For this profile the density ρ changes continuously from ρ_L to ρ_V, is anti-symmetric with respect to the mid-point z_0 as a consequence of the assumed form of Eq.

10) An interesting analogy exists between the gradient theory and classical mechanics (see Section 2.2). Denoting association with ↔, we have $-F(\rho)$ ↔ potential energy Φ, m ↔ mass m, z ↔ time t and $f(\rho)-\mu_0$ ↔ force F, so that Eq. (15.49) corresponds with Newton's law. The kinetic energy $T \leftrightarrow \frac{1}{2}m\rho'(z)^2$ and thus for the Lagrange function $L = T - V \leftrightarrow \frac{1}{2}m\rho'(z)^2 + F(\rho)$, while for the total energy (Hamilton function) $H = T + V \leftrightarrow \frac{1}{2}m\rho'(z)^2 - F(\rho) = 0$. Minimizing γ as an integral over $\frac{1}{2}m\rho'(z)^2 + F(\rho)$ corresponds with Hamilton's principle, which minimizes the action, that is, the integral over L.

(15.54), and diverges at the critical point. Within mean field theory (see Chapter 16) it appears that $\rho_L - \rho_V \sim (T_{cri} - T)^{1/2}$ so that λ diverges as $(T_{cri} - T)^{-1/2}$.

Substituting Eq. (15.54) in Eq. (15.52), we find for γ close to T_{cri}

$$\gamma = (2mC)^{1/2} \int_{\rho_V}^{\rho_L} (\rho_L - \rho)(\rho - \rho_V)\,d\rho = 6^{-1}(2mC)^{1/2}(\rho_L - \rho_V)^3 \tag{15.57}$$

Hence, close to T_{cri}, $\gamma \sim (\rho_L - \rho_V)^3 \sim (T_{cri} - T)^{3/2}$. Experiment and renormalization theory (see Chapter 16) show that the exponent really is about 1.26.

One of the better aspects of gradient theory is that, in principle, any equation of state can be used, whether experimental or theoretical. Moreover, although in principle derived for conditions near the critical point, it appears that its validity covers a much wider range [33]. Although rather successful qualitatively, for gradient theory a number of conceptual problems do exist. The density over the interphase is considered to be a well-defined function of the height in the interphase. Gravity is taken to localize the interface and to stabilize it against long-wavelength, thermally excited waves, the so-called *capillary waves* [34], which would smear out the profile even for, in principle, an infinitely sharp interface. For these aspects and other aspects we neglected, we refer to the literature.

Problem 15.9

Show that $-\frac{1}{2}m\rho(z)\rho''(z)$ in Eq. (15.46) can be transformed by partial integration to $\frac{1}{2}m\rho'(z)^2$.

Problem 15.10

Verify Eq. (15.49).

15.4 Two-Component Liquid Surfaces

For solutions, the preferential adsorption of one of the components at the surface becomes important. For a single-component liquid we described the thermodynamics of the surface first exactly, and thereafter simplified the expression by neglecting the effects resulting from the interphase thickness. This can also be achieved for a two-component solution, but the resulting equations are cumbersome [2], and therefore we neglect this contribution from the start. Rather, we limit ourselves to surfaces – that is, liquid–vapor interfaces. The adsorption behavior is described by the Gibbs adsorption expression Eq. (15.12), and in the indicated approximation reads

$$d\gamma = -s^{(\sigma)}dT + \tau dP - \Gamma_1 d\mu_1 - \Gamma_2 d\mu_2 \cong -s^{(\sigma)}dT - \Gamma_1 d\mu_1 - \Gamma_2 d\mu_2 \tag{15.58}$$

Substituting the expression for $d\mu_1$ reading

$$d\mu_1 = -S_1 dT + V_1 dP + \frac{\partial \mu_1}{\partial x} dx \cong -S_1 dT + \frac{\partial \mu_1}{\partial x} dx \tag{15.59}$$

and the corresponding expression for $d\mu_2$, we have

$$-d\gamma = \left(s^{(\sigma)} - \Gamma_1 S_1 - \Gamma_2 S_2\right) dT + \left(\Gamma_1 \frac{\partial \mu_1}{\partial x} + \Gamma_2 \frac{\partial \mu_2}{\partial x}\right) dx \tag{15.60}$$

Employing the Gibbs–Duhem equation at constant T and P, reading

$$(1-x)\frac{\partial \mu_1}{\partial x} + x \frac{\partial \mu_2}{\partial x} = 0 \tag{15.61}$$

where $1 - x = x_1$ and $x = x_2$, the result is

$$-d\gamma = \left(s^{(\sigma)} - \Gamma_1 S_1 - \Gamma_2 S_2\right) dT + \Gamma_2^{(1)} dx \quad \text{with} \quad \Gamma_2^{(1)} = \Gamma_2 - \left(\frac{x\Gamma_1}{1-x}\right)\frac{\partial \mu_2}{\partial x} \tag{15.62}$$

From this expression it is easy to calculate the temperature derivative

$$\blacktriangleright \quad -\left(\frac{\partial \gamma}{\partial T}\right)_x = s^{(\sigma)} - \Gamma_1 S_1 - \Gamma_2 S_2 \tag{15.63}$$

From the chemical potential and Helmholtz energy expressions reading, respectively,

$$\mu_i = G_i = F_i + PV_i \cong F_i = U_i - TS_i \quad \text{and} \tag{15.64}$$

$$F^{(\sigma)} + PV^{(\sigma)} - \gamma A \cong F^{(\sigma)} - \gamma A = n_1^{(\sigma)}\mu_1 + n_2^{(\sigma)}\mu_2 \quad \text{or}$$

$$u^{(\sigma)} - Ts^{(\sigma)} - \gamma = \Gamma_1 \mu_1 + \Gamma_2 \mu_2 \tag{15.65}$$

we obtain

$$\gamma - T\left(\frac{\partial \gamma}{\partial T}\right)_x = u^{(\sigma)} - \Gamma_1 U_1 - \Gamma_2 U_2 \tag{15.66}$$

From Eqs (15.60) or (15.62) we can calculate similarly the composition derivative as

$$\blacktriangleright \quad -\left(\frac{\partial \gamma}{\partial x}\right)_T = \Gamma_1 \frac{\partial \mu_1}{\partial x} + \Gamma_2 \frac{\partial \mu_2}{\partial x} = RT\left(\Gamma_1 \frac{\partial \ln a_1}{\partial x} + \Gamma_2 \frac{\partial \ln a_2}{\partial x}\right) \quad \text{or} \tag{15.67}$$

$$-\left(\frac{\partial \gamma}{\partial x}\right)_T = RT\Gamma_2^{(1)} \frac{\partial \ln a_2}{\partial x} \tag{15.68}$$

respectively. Since, for an ideal mixture, $\mu_i = \mu_i^{\circ} + RT \ln x_i$, Eq. (15.68) reduces to

$$-\left(\frac{\partial \gamma}{\partial x}\right)_T = RT\frac{\Gamma_2^{(1)}}{x} \tag{15.69}$$

For solutes with a measurable vapor pressure, we can also replace $\ln a_2$ with $\ln P_2$. The theory discussed applies to all types of solutes. Nevertheless, it is useful to distinguish between various types of solutes, and we do so in Section 15.6.

15.5 Statistics of Adsorption

Apart from a thermodynamic description of liquid surfaces, another route is to use a statistical mechanical approach. We do so briefly in this section, limiting ourselves again to surfaces, and to that purpose consider a binary system for which we denote the bulk composition with $A_{1-x}B_x$ and the surface composition $A_{1-y}B_y$. We again use a lattice-type model, and assume that only the first molecular layer has a composition different from the bulk. As before, the components are indicated by subscript i, while the surface and bulk phase are indicated by superscripts (σ) and (α), respectively. We start with an ideal solution and thereafter consider the influence of nonideality.

We recall that the chemical potential of component i is given by

$$\mu_i^{(\alpha)} = \mu_i^{(\circ\alpha)} + kT\ln x \quad \text{and} \quad \mu_i^{(\sigma)} = \mu_i^{(\circ\sigma)} + kT\ln y \tag{15.70}$$

where $\mu_i^{(\circ\alpha)}$ and $\mu_i^{(\circ\sigma)}$ are the standard chemical potentials of the pure components in the bulk and the surface, respectively, while x and y denote the mole fractions. Further, at equilibrium we have $\mu_i^{(\alpha)} = \mu_i^{(\sigma)}$. From these equations we easily obtain

$$\mu_A^{(\circ\sigma)} = \mu_A^{(\sigma)} + kT\ln[(1-x)/(1-y)] \quad \text{and} \quad \mu_B^{(\circ\sigma)} = \mu_B^{(\sigma)} + kT\ln(x/y) \tag{15.71}$$

The Helmholtz energies are then given by

$$F_i^{(\alpha)} = N_i\mu_i^{(\circ\alpha)} + N_ikT \quad \text{and} \quad F_i^{(\sigma)} = N_i\mu_i^{(\sigma)} + N_ikT \tag{15.72}$$

For a one-component system the surface tension is given by $\gamma A = F_i^{(\alpha)} - F_i^{(\sigma)}$, so that

$$\mu_i^{(\sigma)} - \mu_i^{(\circ\alpha)} = a_i\gamma \tag{15.73}$$

where $a_i = A/N_i$ denotes the area per molecule. If we subtract Eq. (15.71-2) from Eq. (15.71-1) and Eq. (15.72-2) from Eq. (15.72-1) and insert Eq. (15.73) for both components, we obtain

$$\blacktriangleright \quad kT\ln\frac{y}{1-y} = kT\ln\frac{x}{1-x} - (a_B\gamma_B - a_A\gamma_A) \tag{15.74}$$

This is the adsorption equation for ideal solutions.

We now introduce interactions and "change gear" to statistics, employing the grand (canonical) partition function Ξ in connection with the lattice model. Note that Ξ is the proper partition function to use, since we keep μ_i, N and V constant (see Chapter 5). The configurational partition function of the surface layer $Q^{(\sigma)}$ is given by

$$Q^{(\sigma)} = \frac{N^{(\sigma)}!}{N_A^{(\sigma)}!N_B^{(\sigma)}!}\exp(-U^{(\sigma)}/kT) \tag{15.75}$$

where

$$N_A^{(\sigma)} = N^{(\sigma)}(1-y), \quad N_B^{(\sigma)} = N^{(\sigma)}y \quad \text{and} \quad N^{(\sigma)} = N_A^{(\sigma)} + N_B^{(\sigma)} \tag{15.76}$$

The energy $U^{(\sigma)}$ is evaluated as for a regular solution (in the zeroth approximation) and reads, using as before the interaction energy $2w = 2w_{AB} - w_{AA} - w_{BB}$,

$$U^{(\sigma)} = \frac{1}{2}z^{(\sigma)}(N_A^{(\sigma)}w_{AA} + N_B^{(\sigma)}w_{BB}) + w\left(z_t^{(\sigma)}\frac{N_A^{(\sigma)}N_B^{(\sigma)}}{N^{(\sigma)}} + z_n^{(\sigma)}\frac{N_A^{(\sigma)}N_B + N_A N_B^{(\sigma)}}{N}\right) \tag{15.77}$$

Here, $z_t^{(\sigma)}$ is the coordination number for "bonds" in the surface layer with composition $A_{1-y}B_y$, and $z_n^{(\sigma)}$ is the coordination number for "bonds" between the surface layer and the first bulk layer, so that the total surface coordination number reads $z^{(\sigma)} = z_t^{(\sigma)} + z_n^{(\sigma)}$ (e.g., for a FCC (111) plane $z_t^{(\sigma)} = 6$ and $z_n^{(\sigma)} = 3$). The grand partition function becomes

$$\Xi^{(\sigma)} = \sum_{N_A^{(\sigma)}}\sum_{N_B^{(\sigma)}} \exp[(N_A^{(\sigma)}\mu_A + N_B^{(\sigma)}\mu_B)/kT]Q^{(\sigma)} \tag{15.78}$$

We now use the maximum-term method (see Justification 5.4), replacing the sum by its largest term and obtain

$$\begin{aligned} kT\ln\Xi^{(\sigma)} &= -N^{(\sigma)}kT[y\ln y + (1-y)\ln(1-y)] \\ &\quad + yN^{(\sigma)}\left(\mu_A - \frac{1}{2}z^{(\sigma)}w_{AA}\right) + (1-y)N^{(\sigma)}\left(\mu_B - \frac{1}{2}z^{(\sigma)}w_{BB}\right) \\ &\quad - N^{(\sigma)}w\{z_t^{(\sigma)}y(1-y) + z_n^{(\sigma)}[y(1-x) + x(1-y)]\} \end{aligned} \tag{15.79}$$

The equilibrium condition is

$$\partial \ln \Xi^{(\sigma)}/\partial y = 0 \tag{15.80}$$

and leads to

$$\mu_B - \mu_A = \frac{1}{2}z^{(\sigma)}(w_{BB} - w_{AA}) + w[z_t^{(\sigma)}(1-2y) + z_n^{(\sigma)}(1-2x) + kT[\ln[y/(1-y)] \tag{15.81}$$

From $\mu_i^{(\alpha)} = \mu_i^{(\sigma)}$ and Eq. (15.71) we obtain

$$\mu_B - \mu_A = \frac{1}{2}z(w_{BB} - w_{AA}) + zw(1-2x) + kT[\ln[x/(1-x)] \tag{15.82}$$

so that combining Eqs (15.81) and (15.82) gives

$$\blacktriangleright \quad \begin{aligned} kT\ln\frac{y}{1-y} &= kT\ln\frac{x}{1-x} + \frac{1}{2}(z - z^{(\sigma)})(w_{BB} - w_{AA}) \\ &\quad - w[z_t^{(\sigma)}(1-2y) - (z - z_n^{(\sigma)})(1-2x)] \end{aligned} \tag{15.83}$$

This is the equivalent of Eq. (15.74) in the zeroth approximation. We see that the surface tension term $-(a_B\gamma_B - a_A\gamma_A)$ is replaced by $\frac{1}{2}(z - z^{(\sigma)})(w_{BB} - w_{AA})$, while the interaction term with w is obviously absent for the ideal solution.

Estimating w_{ii} from the heat of vaporization L_i in the nearest-neighbor approximation $L_i = -\frac{1}{2}zN_A w_{ii}$ (with N_A = Avogadro's constant) and w from the heat of solution L_{AB} of component A in component B (or *vice versa* from L_{BA}), we have

$$\Delta L_{AB} = -zw = \Delta L_{BA} \quad \text{with} \quad 2w = 2w_{AB} - w_{AA} - w_{BB} \tag{15.84}$$

Note that the heat of vaporization is counted positive, while the bond energy is counted negative. In the zeroth approximation of the regular solution theory $\Delta L_{AB} = \Delta L_{BA}$. Experimentally, this condition is only exceptionally fulfilled, which indicates a need for further improvement. However, the approach does provide a clear basis for a picture in which the surface tension – that is, the surface Helmholtz energy – is estimated as, with again N_A as Avogadro's number,

$$\gamma_i A = -\frac{1}{2}(z - z^{(\sigma)})w_{ii}N_i^{(\sigma)} = \frac{z - z^{(\sigma)}}{z}\frac{N_i^{(\sigma)}}{N_A}L_i \tag{15.85}$$

Using this approximation we obviously neglect entropy terms and have identified $\frac{1}{2}zw_{ii}$ with μ_i° and $\frac{1}{2}z^{(\sigma)}w_{ii}$ with $\mu_i^{(\circ\sigma)}$. The discussion in Section 15.2 showed that, numerically, the difference is far from negligible. The fraction of missing "bonds" $(z-z^{(\sigma)})/z$, as also indicated in Section 15.2, is ~1/4.

15.6 Characteristic Adsorption Behavior

Generally, three types of behavior of γ for the various types of solutes with concentration can be distinguished, as illustrated schematically in Figure 15.6. Type I behavior includes *hydrophilic* compounds which prefer to be in water and therefore deplete at the interface, resulting in a, generally small, increase in γ. The most common examples are salts such as NaCl. As an illustration we quote that, at room temperature, γ rises from ~72.6 mJ m^{-2} for pure water to ~76.0 mJ m^{-2} for a 2 M aqueous solution of NaCl. Type II behavior comprises the *hydrophobic* compounds; these do not prefer to be in the water and therefore segregate at the interface, resulting in a decrease in γ. Aliphatic alcohols with a somewhat longer chain such as hexanol provide an example. Type III behavior includes *amphiphilic* compounds or *surfactants* that generally contain strongly hydrophobic "tails" and strongly hydrophilic "heads." One can distinguish between *anionic* surfactants generating negative molecules (and a positive counterion) and *cationic* surfactants generating positive molecules (and a negative counterion). Typical examples are sodium dodecylsulfate (SDS; $C_{12}H_{25}OSO_3Na$) and hexadecyl trimethylammonium bromide (CTAB; $C_{12}H_{25}N(CH_3)_3Br$). Moreover, one has *nonionic* surfactants containing highly polar groups, such as polyethylene oxide, and *amphoteric* surfactants carrying a positive and negative charge while being on the whole neutral. The cationic and anionic compounds strongly enrich at the interface because their tails prefer

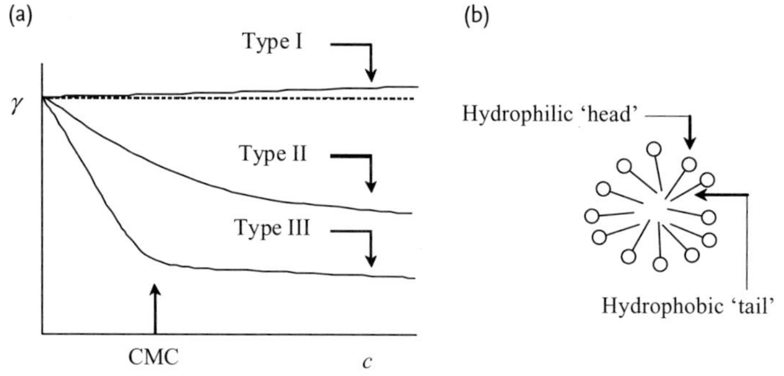

Figure 15.6 (a) Schematic representation of the interfacial tension with concentration for hydrophilic (type I), hydrophobic (type II), and amphiphilic solutes (type III, surfactants). The CMC is indicated; (b) Schematic representation of the structure of micelles occurring at $c > c_{\text{CMC}}$ for surfactants, where the hydrophilic head is indicated by O and the hydrophobic tail by —.

to be out of the water while their heads prefer to be in the water. The polar groups of the nonionic surfactants prefer the water phase with a similar effect. Most surfactants are anionic, followed by nonionic surfactants. Cationic surfactants often pose environmental problems, while amphoteric surfactants are generally expensive and therefore are only used for special applications. Surfactants generally enrich strongly at the interface, so that γ is decreased considerably at small concentrations. Above a certain concentration they do not further enrich at the surface (the surface is "occupied") but rather form *micelles*, clusters of typically 50 surfactant molecules in which the tails stick together in order to minimize the contact with water, and of which the structure is shown schematically in Figure 15.6. This concentration is called the critical micelle concentration (CMC). The solubility of surfactants is generally low but increases with temperature. The temperature at which the CMC equals the solubility is normally addressed as the *Kraft temperature*. Above this temperature, micelle formation is possible and surfactants are much more active above this temperature than below.

15.6.1
Amphiphilic Solutes

For amphiphilic solutes (surfactants), we may suppose that the concentration in the bulk is small and that segregation occurs essentially at the first surface layer (a higher concentration leads to micelles). Thermodynamically, this means that we assume that $n_i^{(\alpha)} = n_i^{(\beta)} \cong 0$, so that $n_i^{(\sigma)} \cong n_i$ and $\Gamma_i^{(1)} \cong n_i/A$. We discuss a kinetic and a statistical–mechanical approach, both leading to the Langmuir isotherm.

For the kinetic approach we consider the surface as a plane with a certain density of surface sites. The process in the surface layer can be represented as a chemical reaction where the solute dissolved in the bulk (B) "reacts" with an empty surface site (S) to an occupied adsorbed site (A); that is, B + S $\leftrightarrow$ A. Obviously, we assume that the molecules at the surface do not interact. Since the solubility is low, we approximate the activity a by concentration[11] $c = N_B/V$, where N_B is the number of dissolved molecules in the volume V. We denote the fraction occupied surface sites by θ, that is, $\theta = N_A/N_{max}$, where N_{max} is the number of sites at monolayer coverage. Hence, $N_S/N_{max} = 1 - \theta$. The equilibrium constant K for this adsorption "reaction" is then

$$K = \frac{\theta}{(1-\theta)c} \quad \text{or} \quad \theta = \frac{Kc}{1+Kc} \tag{15.86}$$

usually called the *Langmuir adsorption isotherm*. The amount of segregated material $\Gamma^{(\sigma)}$ is then, with Γ_{max} a proportionality constant representing monolayer coverage,

$$\Gamma^{(\sigma)} = \Gamma_{max}\theta \quad \text{or} \quad 1/\Gamma^{(\sigma)} = 1/\Gamma_{max} + (1/\Gamma_{max}K)\cdot 1/c \tag{15.87}$$

The latter form can be used for regression analysis of experimental data.

Before we discuss the behavior and some consequences of the Langmuir isotherm, we first derive it via statistical mechanics. In this approach, we consider the solvent and the surface phase and equate the chemical potential of the solute in the bulk phase $\mu^{(\alpha)}$ with that of the surface phase $\mu^{(\sigma)}$. In view of the limited solubility, we assume ideal behavior so that the N solute molecules can move freely through the solvent with volume V with potential energy ϕ_α. Hence, we have for the partition function $z^{(\alpha)}$ of a single molecule (see Chapter 5)

$$z^{(\alpha)} = z_{int}^{(\alpha)} z_{ext}^{(\alpha)} \exp(-\phi_\alpha / kT) \tag{15.88}$$

where $z_{int}^{(\alpha)}$ is the internal partition function and $z_{ext}^{(\alpha)} = V/\Lambda^3$ the external partition function, representing the translation motion of the molecule as a whole, with $\Lambda \equiv (h^2/2\pi mkT)^{1/2}$ the thermal length. For the total partition function $Z^{(\alpha)}$ and Helmholtz energy $F^{(\alpha)}$ in the bulk we have, respectively,

$$Z^{(\alpha)} = (N!)^{-1}(z^{(\alpha)})^N \quad \text{and} \quad F^{(\alpha)} = -kT\ln Z^{(\alpha)} \tag{15.89}$$

so that

$$\mu^{(\alpha)} = (\partial F^{(\alpha)} / \partial N)_{V,T} = -kT\ln(Vz_{int}^{(\alpha)} e^{-\phi_\alpha/kT} / \Lambda^3 N) \tag{15.90}$$

For the surface we suppose that we have M sites of which S are occupied. Hence, the fraction adsorbed molecules $\theta = S/M$ and the fraction of nonoccupied sites is $1 - \theta = (M - S)/M$. The partition function for a single molecule becomes

$$z^{(\sigma)} = z_{int}^{(\sigma)} z_{ext}^{(\sigma)} \exp(-\phi_\sigma / kT) \tag{15.91}$$

where similar labels are used as for the bulk. The total partition function $Z^{(\sigma)}$ and Helmholtz energy $F^{(\sigma)}$ for the surface then read, respectively,

$$Z^{(\sigma)} = \frac{M!}{S!(M-S)!}(z^{(\sigma)})^S \quad \text{and} \quad F^{(\sigma)} = -kT\ln Z^{(\sigma)} \tag{15.92}$$

11) This derivation also applies if we replace the concentration in the liquid phase $c = N/V$ with P/kT (= N/V), where P is the pressure in the gas phase, representing the adsorption of a gas at the surface.

where the factor $M!/S!(M-S)!$ is introduced in view of the indistinguishability. Employing the Stirling approximation for the factorials, we have $F^{(\sigma)} = -kT\ln[M^M(z_{\text{int}}^{(\sigma)}z_{\text{ext}}^{(\sigma)})^S \exp(-\phi_\sigma/kT)/S^S(M-S)^{M-S}]$. The chemical potential $\mu^{(\sigma)}$ accordingly becomes

$$\mu^{(\sigma)} = \left(\frac{\partial F^{(\sigma)}}{\partial S}\right)_{M,T} = -kT\ln\left(\frac{1-\theta}{\theta}z_{\text{int}}^{(\sigma)}z_{\text{ext}}^{(\sigma)}\text{e}^{-\phi_\sigma/kT}\right) \quad (15.93)$$

Equating $\mu^{(\alpha)}$ with $\mu^{(\sigma)}$ we obtain, using for the bulk concentration $c = N/V$,

$$\blacktriangleright \quad \frac{\theta}{1-\theta} = b(T)c \quad \text{with} \quad b(T) = \left(\frac{z_{\text{int}}^{(\sigma)}z_{\text{ext}}^{(\sigma)}}{z_{\text{int}}^{(\alpha)}}\right)\Lambda^3\text{e}^{-(\phi_\sigma-\phi_\alpha)/kT} \quad (15.94)$$

equivalent to Eq. (15.86) if we take[12] $K = b(T)$.

Let us now evaluate the Langmuir behavior. For small c, the increase in $\theta \cong Kc$, that is, the occupied surface site fraction is linear in c, while for large c, we have $\theta \to 1$, that is, the occupied surface site fraction saturates at 1. This behavior is displayed in Figure 15.7 for various values of K. Taking $K = 10^4c_0$ as an example, we note that $\theta \cong 1$ at $c/c_0 \cong 0.002$, so that the CMC in this model will be about $0.002c_0$. A few more remarks can be made. First, recall that the equilibrium constant for adsorption K is related to the Gibbs energy of adsorption G_{ads} by $K = \exp(-G_{\text{ads}}/kT)$, to be compared with Eq. (15.94). Having determined K as a function of T, all other thermodynamic properties, such as the heat of adsorption, can be determined in the usual way. Second, substituting the Langmuir isotherm $\Gamma^{(\sigma)} = \Gamma_{\max}\theta = \Gamma_{\max}Kc/(1+Kc)$ in the Gibbs adsorption expression $\Gamma^{(\sigma)} = -(RT)^{-1}\partial\gamma/\partial\ln c$ and solving for γ, one obtains the *Szyszkowski equation* [35] reading

$$\gamma = \gamma_0 - RT\Gamma_{\max}\ln(1+Kc) \quad (15.95)$$

which appears to describe experimentally the amount of adsorbed material often rather well.

Obviously, other assumptions lead to different adsorption isotherms. In fact, if we assume that the molecules at the surface do attract each other so that the energy changes by $2zw\theta$, where z denotes the surface coordination number and $2w$ the energy when a new pair of neighbors is formed, but that the surface entropy is not affected[13], we have

$$\mu^{(\sigma)} = \left(\frac{\partial F^{(\sigma)}}{\partial S}\right)_{M,T} = -kT\ln\left(\frac{1-\theta}{\theta}z_{\text{int}}^{(\sigma)}z_{\text{ext}}^{(\sigma)}\text{e}^{-(\phi_\sigma+2zw\theta)/kT}\right) \quad (15.96)$$

This leads to the *Fowler–Guggenheim adsorption isotherm* reading

$$\frac{\theta}{1-\theta} = b'(T)c \quad \text{with} \quad b'(T) = \left(\frac{z_{\text{int}}^{(\sigma)}z_{\text{ext}}^{(\sigma)}}{z_{\text{int}}^{(\alpha)}}\right)\Lambda^3\text{e}^{-(\phi_\sigma+2zw\theta-\phi_\alpha)/kT} = b(T)\text{e}^{-2zw\theta/kT} \quad (15.97)$$

12) Often, $z_{\text{ext}}^{(\sigma)} = 1$ is taken on the argument that the sites are localized but the molecules still vibrate with respect to the surface, leading to $z_{\text{ext}}^{(\sigma)} \neq 1$.

13) That is, use the zeroth order approximation for a regular solution model for the surface with w the average interaction energy per molecule (see Sections 8.3 and 15.5). Note that $w < 0$ since otherwise no attraction between the adsorbed molecules occurs.

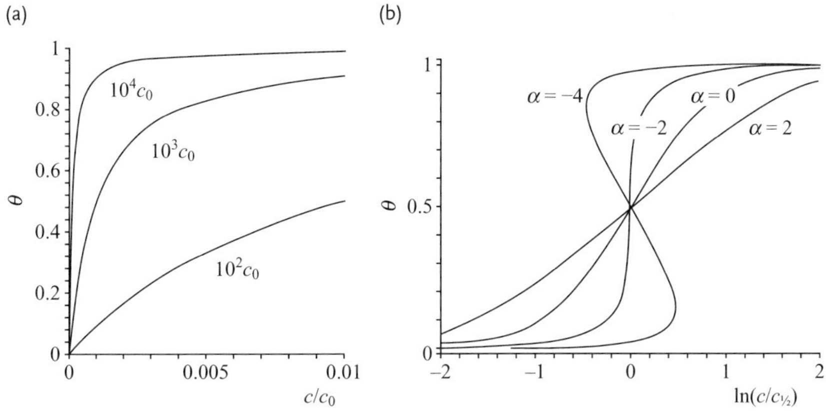

Figure 15.7 (a) The Langmuir isotherm $\theta(c) = K(c/c_0)/[1 + K(c/c_0)]$ for $K = 10^2c_0$, 10^3c_0, and 10^4c_0; (b) The Fowler–Guggenheim adsorption isotherm for various values of $\alpha = zw/kT$. The critical value is $\alpha = -2$, while $\alpha = 0$ corresponds to the Langmuir isotherm.

Plotting θ as a function of $\ln(c/c_{½})$, where $c_{½}$ is the concentration where $\theta = ½$, we see that this isotherm produces a van der Waals shape with two loops for $\alpha = zw/kT < -2$, characteristic of a phase transformation. The critical temperature is thus determined by $zw/kT_{\text{cri}}^{(\sigma)} = -2$, occurring at $\theta_{\text{cri}} = ½$. Figure 15.7 shows the behavior for the various values of α. In view of the fact that the solution is obtained via the zeroth approximation, more sophisticated solutions lead to different values for $T_{\text{cri}}^{(\sigma)}$. For example, the first-order approximation leads for a square lattice ($z = 4$) to $zw/kT_{\text{cri}}^{(\sigma)} = -2.773$ while an exact solution yields $zw/kT_{\text{cri}}^{(\sigma)} = -3.526$ (see Appendix 3).

Another extreme for a liquid surface would be that the adsorbed molecules can move freely, much like a 2D-gas. This is suggested by the fact that the *surface pressure* $\pi = \gamma_0 - \gamma$, where γ_0 is the surface tension of the pure liquid and γ is the surface tension of the solution, for low concentration c is linearly related to c, that is, $\pi = \lambda c$, with λ a constant. The surface pressure is usually measured with a *surface pressure balance*, often denoted as *Langmuir–Blodgett trough*, of which a schematic is shown in Figure 15.8. On one side of a movable barrier, surfactant is added to the surface while the other side has a pure water surface. By measuring the surface pressure as function of surface area A and using the Gibbs adsorption expression $-\mathrm{d}\gamma = \mathrm{d}\pi = kT\Gamma\, \mathrm{d}\ln c$ with $\Gamma \equiv \Gamma_2^{(1)} = n_2^{(1)}/A$ in conjunction with the experimental relation $\pi = \lambda c$, we obtain

$$\pi = \Gamma kT \tag{15.98}$$

This expression resembles that of the ideal gas law, and suggests similar improvements for the surface as has been used for bulk gases. By employing the van der Waals approximation, in analogy with the 3D expression $(P + a\rho^2)(1 - b\rho) = \rho RT$ (see Chapter 4), for 2D reading

$$(\pi + a\theta^2)(1 - b\theta) = \theta kT \quad \text{or} \quad (\pi + a\sigma^{-2})(\sigma - b) = kT \tag{15.99}$$

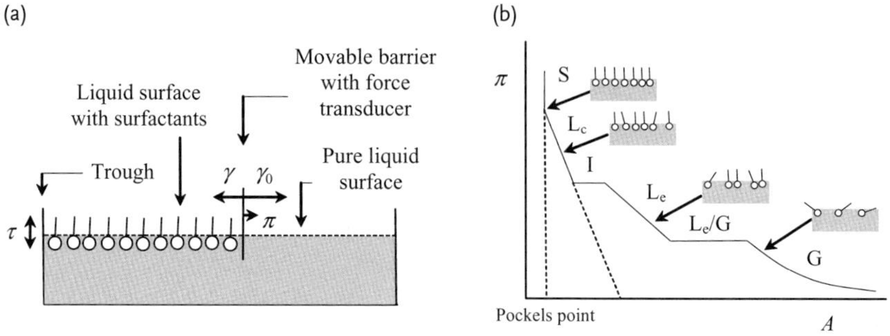

Figure 15.8 (a) Schematic representation of a Langmuir–Blodgett trough; (b) Schematic representation of the surface pressure $\pi = \gamma_0 - \gamma$ versus surface area A and the associated structure, showing the regimes of gas-like (G), liquid-expanded (L_e), liquid-condensed (L_c), and solid (S).

with a and b parameters and where θ (or $\sigma = 1/\theta$) in 2D corresponds to ρ (or $V_m = 1/\rho$) in 3D, the *van der Waals adsorption isotherm* becomes, with the constant $\varepsilon = 2a/b$,

$$\frac{\theta}{1-\theta} = b''(T)c \quad \text{with} \quad b''(T) = \left(\frac{z_{\text{int}}^{(\sigma)}}{z_{\text{int}}^{(\alpha)}}\right)\Lambda e^{-\left[\frac{\theta}{1-\theta} + \frac{(\phi_\sigma - \phi_\alpha) - \varepsilon\theta}{kT}\right]} \tag{15.100}$$

This isotherm predicts a phase change, similarly as in the case of localized adsorption with attractive interaction. Characteristics are $\theta_{\text{cri}} = 1/3$ and $kT_{\text{cri}}^{(\sigma)}/\varepsilon = -4/27 = -0.148$. This isotherm is not symmetric around $\theta = 1/2$. While for $\theta > 1/2$, the vdW isotherm resembles the Langmuir isotherm, for $\theta < 1/2$ it decreases much more rapidly and resembles the Fowler–Guggenheim isotherm.

In fact, the surface pressure $\pi = \gamma_0 - \gamma$ of a film of amphiphilic molecules as a function of area A shows several features akin to gas compression. A schematic is provided in Figure 15.8, which shows the regimes labeled solid (S), liquid-condensed (L_c), intermediate (I), liquid-expanded (L_e), and gas (G). The structure changes during compression and this change is also schematically indicated. We have already mentioned the gas-like region in which the tails still make a considerable amount of contact with the liquid surface. In the liquid-expanded region the surface contains still a disordered, homogeneous distribution of molecules, but the tails lift off the surface. There is an intermediate region where gas-like and liquid-like areas coexist. The liquid-condensed phase is actually more solid-like because the tails are already aligned. Here, the molecules are relatively (surface) close-packed but are still rather mobile. In the solid-like region, the molecules are really close-packed and compressed to each other. Occasionally, a phase intermediate between the L_c and L_e phase is observed. The transition point between the S and L_c phase, the *Pockels point*, is a measure for the area occupied by the molecules. By knowing the amount of molecules on the surface, one can estimate the molecular area, since at that point they are just close-packed.

A pioneering experiment was that of Langmuir [36], who studied amphiphilic molecules with different chain lengths. For example, he used $CH_3(CH_2)_nCOOH$

with $n = 14$, 16 and 24, which resulted in 21, 22 and 25 Å^2, respectively. From this, Langmuir concluded that the "head" area or COOH area is about 23 Å^2, independent of the chain length. This field has undergone tremendous development since then and now contains many interesting structures and phenomena, detailed discussions of which are available elsewhere [37, 38].

It should be noted that, although the interpretation of the above transition between solid and liquid compressed phase of the adsorption isotherm is straightforward, and has generally been accepted as correct, debate still persists in the literature [39][14].

15.6.2
Hydrophobic Solutes

Although, for hydrophobic solutes the change in γ is relatively small, even with fully soluble solutes segregation and orientation still occur. To illustrate such behavior we use the example of water and ethanol. Estimating the surface energy of hydroxyl groups by extrapolation from water as ~190 mJ m^{-2}, and that of hydrocarbon groups as ~50 mJ m^{-2}, one can easily reach the conclusion that the hydrocarbon tails should "stick out" of the solution. Supporting this suggestion is the fact that the actual surface tension of ethanol is 22 mJ m^{-2}, which is quite similar to the value for hydrocarbons. To illustrate such behavior further, we can use the data on water–ethanol solutions [40]. From measurements of the vapor pressure of water P_1, of ethanol P_2, and the surface tension γ, the quantity $-\partial\gamma/\partial \ln P_2$ can be calculated, and thus $\Gamma_2^{(1)}$. Table 15.4 provides the relevant data. A useful intermediate is $I = -(RT)^{-1}(1-x)\mathrm{d}\gamma/\mathrm{d}\ln P_2 = (1-x)\Gamma_2 - x\Gamma_1 = (1-x)\Gamma_2^{(1)}$. Estimates for

Table 15.4 Mixture of water and ethanol at 25 °C.[a]

x	P_1 (mmHg)	P_2 (mmHg)	γ (mJ m^{-2})	$-\partial\gamma/\partial\ln P_2$ (mJ m^{-2})	$10^{10}I$ (mol cm^{-2})	$10^{10}\Gamma_2$ (mol cm^{-2})	$10^{10}\Gamma_1$ (mol cm^{-2})	$\Gamma_2/(\Gamma_1+\Gamma_2)$
0.0	23.75	0.0	72.2	0.0	0.0	0.0	25.0	0.00
0.1	21.7	17.8	36.4	15.6	5.6	6.8	4.6	0.60
0.2	20.4	26.8	29.7	16.0	5.1	7.25	3.25	0.69
0.3	19.4	31.2	27.6	14.6	4.1	7.25	3.25	0.69
0.4	18.35	34.2	26.35	12.6	3.0	7.25	3.25	0.69
0.5	17.3	36.9	25.4	10.5	2.1	7.3	3.1	0.70
0.6	15.8	40.1	24.6	8.45	1.4	7.45	2.65	0.74
0.7	13.3	43.9	23.85	7.15	0.8	7.65	2.0	0.79
0.8	10.0	48.3	23.2	6.2	0.5	7.9	1.3	0.86
0.9	5.5	53.3	22.6	5.45	0.2	8.1	0.7	0.94
1.0	0.0	59.0	22.0	5.2	0.0	8.35	0.0	1.00

a) Data from Ref. [40a].

14) The title of this paper is the rhetorical question: *Should the Gibbs analysis be revised?*

the individual values of Γ_1 and Γ_2 can be made only if extra assumptions are made. If we assume that the surface is covered with a monolayer, and that each molecule contributes a constant area A_i to the interface, we have $A_1\Gamma_1 + A_2\Gamma_2 = 1$. Estimating the molecular areas for H_2O and CH_3CH_2OH as $A_1 = 0.4 \times 10^{10}\,\text{cm}^2\,\text{mol}^{-1}$ ($A_1/N_A \cong 7$ Å^2) and $A_2 = 1.2 \times 10^{10}\,\text{cm}^2\,\text{mol}^{-1}$ ($A_2/N_A \cong 20$ Å^2), one can estimate Γ_1 and Γ_2 from the above two equations, as given in Table 15.4. Further calculating $y = \Gamma_2/(\Gamma_1 + \Gamma_2)$, which can be considered as the mole fraction in the surface, we see that y increases rapidly with the bulk mole fraction x and reaches already $y \cong 0.6$ at $x = 0.1$. Later analysis supports this image, including the sticking out of the surface of the hydrocarbon tail [41] (suggesting an average angle of ~40° with respect to the surface with a distribution width of about 30°). Even for this relatively simple system the structure of the surface layer appears to be not completely known.

The molecular structure of liquid solution surfaces is still intensively investigated. Today, the literature is extended and many–often opposing–views have been expressed. Important experimental methods, sum-frequency spectroscopy [42] and neutron reflection [43], as well as theoretical methods [44], have been reviewed.

15.6.3
Hydrophilic Solutes

Contrary to hydrophobic and amphiphilic solutes, hydrophilic solutes show depletion at the interface, leading to an increase in surface tension, albeit of a much smaller magnitude as compared to the decrease in surface tension for the other two types of solute. Theories fall into three groups: (i) direct calculation of the work of separation for a surface originally in the interior of the solution; (ii) calculation of the depletion, followed by the use of the Gibbs adsorption equation; and (iii) calculation by statistical mechanics using the pair correlation function using an extended Kirkwood–Buff approach. Here, we briefly limit the discussion to theories of the second type, since those of the first type are used only to a limited degree, while those of the third type are limited to more or less spherical molecules.

Within the theories of type two, two basic pictures have been used. In the oldest case, the increase in γ has been modeled by Wagner, Onsager, and Samaras [45], invoking the image charges that repel the ions from the surface. In this picture, polarizability is entirely neglected, but Stairs [46] has modified the model so as to include this effect. The second picture uses the fact that the hydration shell of ions cannot stick out of the surface. Disruption of the hydration shell would involve a large amount of energy as the hydration energies are large; hence, to keep the shell intact ions cannot approach the surface too closely [47]. Again, the field is wide, but has been reviewed [48].

It should be mentioned that an alternative picture for the surface structure of salt solutions has been proposed [49] on the basis of MD simulation in which the polarizability of the ions plays a significant role. Whilst for the small and hardly polarizable ions, such as Na^+ and F^-, depletion does indeed occur, for the larger and more polarizable ions an enrichment was actually calculated. This was attributed to an anisotropic solvation, leading to a substantial dipole on the ion. The positive ions were indeed depleted from the surface, whereas the negative ions were enriched. The solvated-ion-dipole/water-dipole interactions were shown to compensate for the loss of pure-ion-charge/water-dipole interactions. This picture was consistent with, for example, surface potential measurements.

Problem 15.11: Langmuir isotherm

Show that, taking $\theta = n/n_{max} = Kc/(1 + Kc)$, with n the number of moles adsorbed and n_{max} the number of moles adsorbed at full monolayer coverage, the Langmuir isotherm can also be linearized as $c/n = (1/n_{max}K) + (c/n_{max})$.

Problem 15.12

Derive Eq. (15.95).

Problem 15.13: van der Waals isotherm

Using the 2D vdW equation $(\pi + a\sigma^{-2})(\sigma - b) = kT$ and the Gibbs adsorption expression $\Gamma = -(RT)^{-1}\partial\gamma/\partial \ln c$, derive the vdW adsorption expression Eq. (15.100). Hint: write d ln$c$ as a function of dπ and integrate the resulting expression term by term.

15.7 Final Remarks

It seems that the eminent physicist Wolfgang Pauli used to say [50], "God made the bulk; surfaces were invented by the devil", which was a succinct way of saying that surfaces are bestowed with many subtleties. In this chapter, we have only touched briefly on a few of these. In particular, the effects of curved and electrified surfaces were avoided, in spite of their obvious importance. For these aspects and other details we refer to the references in the section "Further Reading". Even for those aspects that were touched, many more considerations can be made, just indicating the richness of the field, though further discussion of the various aspects of liquid surfaces is beyond the scope of the present text.

References

1 (a) Orowan, E. (1970) *Proc. R. Soc.*, **A316**, 473; (b) Goodman, F.O. (1990) *Surf. Sci. Lett.*, **232**, 224.

2 Guggenheim, E.A. (1967) *Thermodynamics*, 5th edn, North-Holland, Amsterdam.

3 See Butt *et al.* (2006).
4 Eötvös, R. (1886) *Ann. Phys.*, **263**, 448.
5 Ferguson, A. (1916) *Philos. Mag.*, **31**, 37.
6 (a) Ferguson, A. (1923) *Trans. Faraday Soc.*, **19**, 407; (b) Ferguson, A. (1940) *Proc. Phys. Soc. London*, **52**, 759; (c) Lin, H., Duan, Y.-Y., and Min, Q. (2007) *Fluid Phase Eq.*, **254**, 75.
7 Katayama, M. (1916) *Rep. Tôhoku Univ.*, **4**, 373.
8 Guggenheim, E.A. (1945) *J. Chem. Phys.*, **13**, 253.
9 McLeod, D.B. (1923) *Trans. Faraday Soc.*, **19**, 38.
10 (a) Sugden, S. (1924) *J. Chem. Soc.*, **125**, 32 and 1177; (b) Sugden, S. (1930) *The Parachor and Valency*, Routledge, London.
11 Quayle, O.R. (1953) *Chem. Rev.*, **53**, 439.
12 Poling, B.E., Prausnitz, J.M., and O'Connell, J.P. (2001) *The Properties of Gases and Liquids*, 5th edn, McGraw-Hill, London.
13 (a) Exner, O. (1962) *Nature*, **196**, 890; (b) Exner, O. (1997) *J. Phys. Org. Chem.*, **10**, 797.
14 Riedel, L. (1954) *Chem. Ing. Tech.*, **26**, 679.
15 Pitzer, K.S. (1995) *Thermodynamics*, 3rd edn, McGraw-Hill, New York.
16 Davis, H.T. and Scriven, L.E. (1976) *J. Phys. Chem.*, **80**, 2805.
17 Egelstaff, P.A. and Widom, B. (1970) *J. Chem. Phys.*, **53**, 2667.
18 Bhatia, A.B. and March, N.H. (1978) *J. Chem. Phys.*, **68**, 19999.
19 Freeman, G.R. and March, N.H. (1998) *J. Chem. Phys.*, **109**, 10521.
20 Sanchez, I.C. (1983) *J. Chem. Phys.*, **79**, 405.
21 Kirkwood, J.G. and Buff, F.P. (1949) *J. Chem. Phys.*, **17**, 338.
22 Shoemaker, P.D., Paul, G.W., and Marc de Chazal, L.E. (1970) *J. Chem. Phys.*, **52**, 491.
23 Lekner, J. and Henderson, J.R. (1977) *Mol. Phys.*, **34**, 333.
24 Mulero, A., Galan, C., and Cuadros, F. (2003) *J. Phys. Condens. Matter*, **15**, 2285.
25 Blokhuis, E.M. and Kuipers, J. (2006) *J. Chem. Phys.*, **124**, 074701.
26 (a) Toxvaerd, S. (1972) *Progr. Surf. Sci.*, **3**, 189; (b) Evans, R. (1979) *Adv. Phys.*, **28**, 143.
27 Penfold, J. (2001) *Rep. Prog. Phys.*, **64**, 777.
28 Widom, B. (1996) *J. Phys. Chem.*, **100**, 13190.
29 See Rowlinson and Widom (1982).
30 See Ted Davis (1995).
31 See Safran (1994).
32 See Hansen and McDonald (2006).
33 See Ted Davis and Scriven (1982).
34 Buff, F.P., Lovett, R.A., and Stillinger, F.H. Jr (1965) *Phys. Rev. Lett.*, **15**, 621.
35 (a) von Szyszkowski, B. (1908) *Z. Phys. Chem.*, **64**, 385; (b) Meissner, H.P. and Michaels, A.S. (1949) *Ind. Eng. Chem.*, **41**, 2782.
36 (a) Langmuir, I. (1917) *J. Am. Chem. Soc.*, **39**, 1848; this is the second paper of a series of three pioneering surface chemistry papers. The other two papers are: (b) Langmuir, I. (1916) *J. Am. Chem. Soc.*, **38**, 1848; and (c) Langmuir, I. (1918) *J. Am. Chem. Soc.*, **40**, 1361.
37 See Hiemenz and Rajagopalan (1997).
38 See Adamson and Gast (1997).
39 Laven, J. and de With, G. (2012) *Langmuir*, **27**, 7958.
40 (a) Guggenheim, E.A. and Adam, N.K. (1933) *Proc. Roy. Soc.*, **A139**, 218; (b) Guggenheim, E.A. (1967) *Thermodynamics*, 5th edn, North-Holland, Amsterdam.
41 (a) Surface tension data: Yano, Y.F. (2005) *J. Colloid Interface Chem.*, **284**, 255; (b) Surface structure via sum-frequency spectroscopy: Sung, J.S., Park, K., and Kim, D. (2005) *J. Phys. Chem.*, **B109**, 18507.
42 Richmond, G.L. (2001) *Annu. Rev. Phys. Chem.*, **52**, 357.
43 (a) Penfold, J. and Thomas, R.K. (1990) *J. Phys. Condens. Matter*, **2**, 1369; (b) Thomas, R.K. (2004) *Annu. Rev. Phys. Chem.*, **55**, 391.
44 Tarazona, P., Chaćon, E., and Bresme, F. (2012) *J. Phys. Condens. Matter*, **24**, 284123.
45 (a) Wagner, C. (1924) *Phys. Z.*, **25**, 474; (b) Onsager, L. and Samaras, N.N.T. (1934) *J. Chem. Phys.*, **2**, 528; (c) Belton, J.W. (1937) *Trans. Faraday Soc.*, **33**, 1449.

46 Stairs, R.A. (1994) *Can. J. Phys.*, **73**, 781.
47 Schäfer, K.L., Pérez Masiá, A., and Jüntgen, H. (1955) *Z. Elektrochem.*, **59**, 425.
48 (a) Petersen, P.B. and Saykally, R.J. (2006) *Annu. Rev. Phys. Chem.*, **57**, 333; (b) Jungwirth, P. and Winter B. (2008) *Annu. Rev. Phys. Chem.*, **59**, 343.
49 (a) Jungwirth, P. and Tobias, D.J. (2001) *J. Phys. Chem.*, **B105**, 10468; (b) Jungwirth, P. and Tobias, D.J. (2001) *J. Phys. Chem.*, **B105**, 6361.
50 As quoted in Jamtveit, B. and Meakin, P. (1999) *Growth, Dissolution, and Pattern Formation in Geosystems*, Kluwer Academic Publishers, Dordrecht.
51 de With, G. (2006), *Structure, Deformation, and Integrity of Materials*, Wiley-VCH Verlag GmbH, Weinheim.
52 Moelwyn-Hughes, E.A. (1961) *Physical Chemistry*, 2nd edn, Pergamon, Oxford.
53 Ramsay, W. and Shields, J. (1893) *Philos. Trans.* **A184**, 647.
54 Curl, R.F. and Pitzer, K.S. (1958) *Ind. Eng. Chem.*, **50**, 265.

Further Reading

Adamson, A.W. and Gast, A.P. (1997) *Physical Chemistry of Surfaces*, 6th edn, John Wiley & Sons, Inc., New York.

Butt, H.-J., Graf, K., and Kappl, M. (2006) *Physics and Chemistry of Interfaces*, 2nd edn, Wiley-VCH Verlag GmbH, Weinheim.

Croxton, C.A. (1980) *Statistical Mechanics of the Liquid Surface*, John Wiley & Sons, Ltd, Chichester.

Erbil, H.Y. (2006) *Surface Chemistry of Solid and Liquid Interfaces*, Blackwell, Oxford.

Hansen, J.-P. and McDonald, I.R. (2006) *Theory of Simple Liquids*, 3rd edn, Academic, London (1st edn 1976, 2nd edn 1986).

Hiemenz, P.C. and Rajagopalan, R. (1997) *Principles of Colloid and Interface Chemistry*, 3rd edn, Marcel Dekker, New York.

Hunter, R.J. (1987) *Foundations of Colloid Science*, vol. I, Oxford University Press, Oxford.

Rowlinson, J.S. and Widom, B. (1982) *Molecular Theory of Capillarity*, Oxford University Press, Oxford. See also Dover Publishers reprint, 1988.

Safran, S.A. (1994) *Statistical Thermodynamics of Surfaces, Interfaces and Membranes*, Westview Press, Boulder, CO.

Stokes, R.J. and Evans, D.F. (1997) *Fundamentals of Interfacial Engineering*, John Wiley & Sons, Inc., New York.

Ted Davis, H. (1995) *Statistical Mechanics of Phases, Interfaces and Thin Films*, John Wiley & Sons, Inc., New York.

16
Some Special Topics: Phase Transitions

In Chapter 2 we reviewed general thermodynamics but avoided phase transitions. In this chapter, we discuss some aspects of these phenomena, dealing first with some general aspects and thereafter with discontinuous and continuous transitions, in particular the latter providing a rich part of physical chemistry.

16.1
Some General Considerations

By changing the conditions – such as P or T – for many materials, a transition from one phase to another can be induced. Under certain conditions even two phases of the same material may coexist. As each of the two phases has its own Gibbs energy expression, under these conditions the chemical potentials of the phases are equal. The Gibbs energy G itself is always continuous over the transition, but the partial derivatives $\partial G/\partial T$ and $\partial G/\partial P$ may be discontinuous (Figure 16.1). In that case, the phase transformation is denoted *discontinuous*[1] (or *first-order*), while for the situation where the first derivative is continuous, but the higher derivatives are either zero or infinite, one speaks of a *continuous* (or *second-order*) phase transition.

The angle of intersection of the G_1 and G_2 curve for phases 1 and 2, respectively, determines the entropy and volume change associated with the phase transformations, and hence the type of phase transition. Experimentally, it appears that by moving along the liquid–vapor (L-V) coexistence line over the critical point (CP), the differences in properties, in particular the density, between the liquid and gas phase vanish in a continuous way and the transition is continuous (Figure 16.2).

1) In the past, the transitions were often labeled as first and second order, according to the discontinuity of their first- or second-order derivatives of the Gibbs energy. However, the second-order "class" appeared to be more complex than anticipated, and therefore these transitions are nowadays often labeled as continuous, due to the fact that in all cases a continuous transition from a one-phase state to a two-phase state occurs with a continuous change in order parameter ($\Delta\rho$ for fluids) over the transition. Although the label "first order" stuck, for consistency, we refer to this transition as discontinuous, the more so since the density behavior for fluids is discontinuous over the transition.

Liquid-State Physical Chemistry, First Edition. Gijsbertus de With.

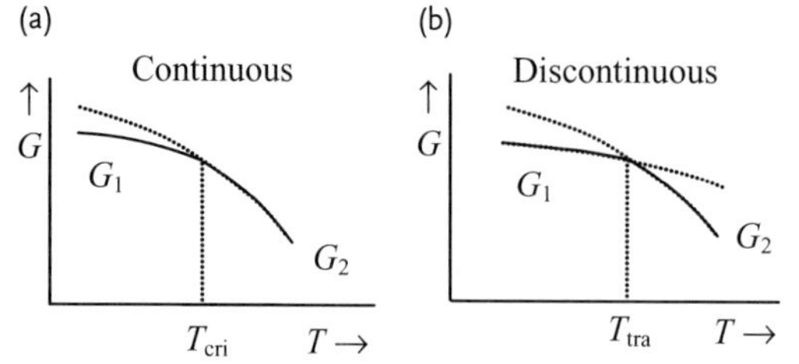

Figure 16.1 Schematic of the behavior of the Gibbs energy G for two phases around (a) continuous phase transition and (b) discontinuous phase transformation. In both cases the stable states below the transition temperature have a Gibbs energy G_1, while above the transition temperature the Gibbs energy is G_2. The continuous transition occurs at the critical temperature T_{cri} with a continuous change in G, that is, $\partial \Delta G/\partial T = 0$, where $\Delta G = G_2 - G_1$. The dotted line indicates the metastable continuation of the high temperature G below T_{cri}. The discontinuous transition occurs at a certain transition temperature T_{tra} with a discontinuous change in G ($\partial \Delta G/\partial T \neq 0$).

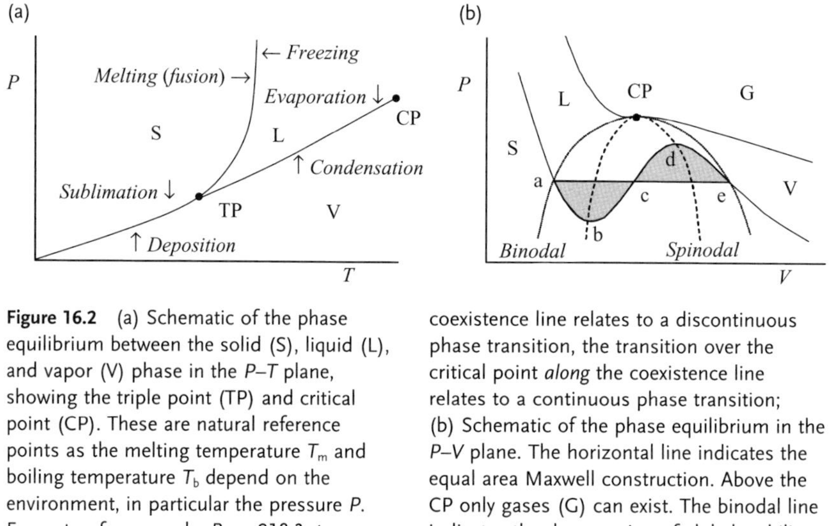

Figure 16.2 (a) Schematic of the phase equilibrium between the solid (S), liquid (L), and vapor (V) phase in the P–T plane, showing the triple point (TP) and critical point (CP). These are natural reference points as the melting temperature T_m and boiling temperature T_b depend on the environment, in particular the pressure P. For water, for example, $P_{cri} = 218.3$ atm, $T_{cri} = 374.15\,°C$, $\rho_{cri} = 320\,kg\,m^{-3}$, and $T_{tri} = 0.01\,°C$. While the transition *over* a coexistence line relates to a discontinuous phase transition, the transition over the critical point *along* the coexistence line relates to a continuous phase transition; (b) Schematic of the phase equilibrium in the P–V plane. The horizontal line indicates the equal area Maxwell construction. Above the CP only gases (G) can exist. The binodal line indicates the demarcation of global stability, while the spinodal line indicates the limits of local stability.

Moving across the L-V curve from the liquid to the gas phase, and vice versa, leads to a discontinuous transition.

Following the coexistence (vapor pressure) line between liquid and vapor in the P–T diagram, we end at the critical point with temperature T_{cri}. In this process the density of the liquid decreases, while the density of the gas increases. At T_{cri}, the gas density ρ_{gas} and liquid density ρ_{liq} become identical. Moreover, for $T < T_{cri}$ a meniscus – that is, a sharp transition region between liquid and vapor – is present

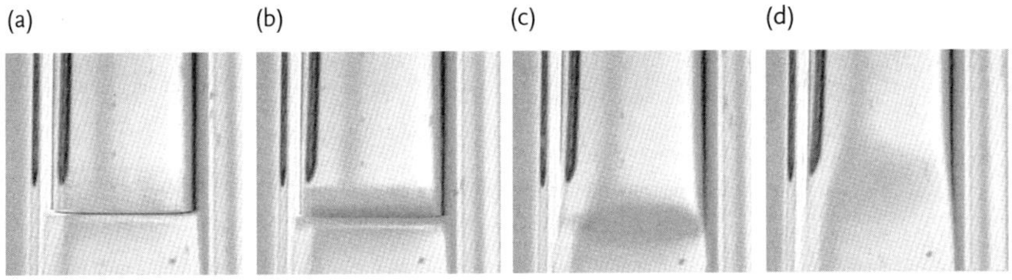

Figure 16.3 The disappearing of the meniscus of benzene along the coexistence curve from far below (a) to close to (b) and just below (c) to just above T_{cri} (d).

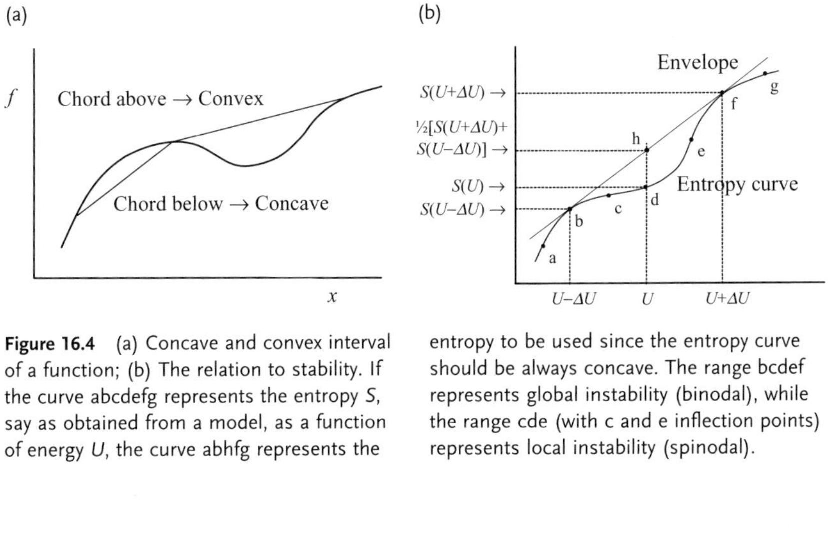

Figure 16.4 (a) Concave and convex interval of a function; (b) The relation to stability. If the curve abcdefg represents the entropy S, say as obtained from a model, as a function of energy U, the curve abhfg represents the entropy to be used since the entropy curve should be always concave. The range bcdef represents global instability (binodal), while the range cde (with c and e inflection points) represents local instability (spinodal).

except for temperatures close to T_{cri} (say within one degree), where the meniscus widens and suddenly disappears at T_{cri}. Figure 16.3 illustrates this behavior[2)].

Before discussing some ideas on how to describe transitions, let us first discuss some stability considerations. For this we need the concept of *convexity*. A curve is *convex* (or convex up) if the chord is above the curve, whereas a curve is *concave* (or convex down) is the chord is below the curve (Figure 16.4a). Thermodynamics requires that for changes in entropy we always have $(\Delta S)_U \geq 0$, which implies that entropy S is concave over its entire domain. If we obtain from a model or from experimental data a curve such as abcdefg in Figure 16.4b, the above implies that the envelope abhfg must be considered as the acting entropy function. Considering for the moment S as function energy U and volume V, concavity of S results in

2) Actually, the transition process is strongly influenced by gravity, and the density profile over the meniscus near the critical temperature can be described by an expression akin to the barometric formula (see Problem 16.1).

$$S(U+\Delta U,V)+S(U-\Delta U,V)\le 2S(U,V) \quad \text{or} \quad \left(\frac{\partial^2 S}{\partial U^2}\right)_V \le 0 \tag{16.1}$$

where the reduction to the differential form only follows if $\Delta U \to 0$. Similarly,

$$S(U,V+\Delta V)+S(U,V-\Delta V)\le 2S(U,V) \quad \text{or} \quad \left(\frac{\partial^2 S}{\partial V^2}\right)_U \le 0 \tag{16.2}$$

where again the differential form only follows if $\Delta V \to 0$. In fact, these considerations also apply for a combined change of U and V,

$$S(U+\Delta U,V+\Delta V)+S(U-\Delta U,V-\Delta V)\le 2S(U,V) \tag{16.3}$$

leading again to Eqs (16.1) and (16.2) as well as to (Problem 16.3)

$$\frac{\partial^2 S}{\partial U^2}\frac{\partial^2 S}{\partial V^2}-\left(\frac{\partial^2 S}{\partial V\partial U}\right)^2 \ge 0 \tag{16.4}$$

So, the concavity leads to conditions for global stability while the differential forms lead to conditions for local stability. For example, Eq. (16.1) leads to

$$\left(\frac{\partial^2 S}{\partial U^2}\right)_V = -\frac{1}{T^2}\left(\frac{\partial T}{\partial U}\right)_V = -\frac{1}{NT^2C_V} \le 0 \quad \text{or} \quad C_V \ge 0 \tag{16.5}$$

indicating that the molar heat capacity C_V must be positive for a system to be thermally stable. Although Eqs (16.2) and (16.4) can also be used to establish further stability requirements, it is easier to employ thermodynamic potentials instead of the entropy. To that purpose, we first recall that the principle $(\Delta S)_U \ge 0$ is always equivalent to $(\Delta U)_S \le 0$, stating that the energy U should always be convex. This leads to

$$U(S+\Delta S,V+\Delta V)+U(S-\Delta S,V-\Delta V)\ge 2U(S,V), \tag{16.6}$$

$$\frac{\partial^2 U}{\partial S^2}\ge 0, \quad \frac{\partial^2 U}{\partial V^2}\ge 0 \quad \text{and} \quad \frac{\partial^2 U}{\partial S^2}\frac{\partial^2 U}{\partial V^2}-\left(\frac{\partial^2 U}{\partial V\partial S}\right)^2 \ge 0 \tag{16.7}$$

which is fully analogous to the entropy case. It is convenient to consider also the Helmholtz and Gibbs energy. We first note that, if we have $U(S,V)$ and $F(T,V)$, then $T = \partial U/\partial S$ and $S = -\partial F/\partial T$, respectively. Therefore

$$\frac{\partial S}{\partial T} = -\frac{\partial^2 F}{\partial T^2} \quad \text{and} \quad \left(\frac{\partial S}{\partial T}\right)^{-1} = \left(\frac{\partial^2 U}{\partial S^2}\right)^{-1} \quad \text{or} \quad -\frac{\partial^2 F}{\partial T^2} = \left(\frac{\partial^2 U}{\partial S^2}\right)^{-1} \tag{16.8}$$

Hence, the sign of $\partial^2 F/\partial T^2$ is the negative of the sign of $\partial^2 U/\partial S^2$, implying that if U is a convex function of S, then F is a concave function of T. This type of result holds for all transforms, so that we obtain for the Helmholtz function and Gibbs function, respectively,

$$\left(\frac{\partial^2 F}{\partial T^2}\right)_V \le 0, \quad \left(\frac{\partial^2 F}{\partial V^2}\right)_T \ge 0, \quad \left(\frac{\partial^2 G}{\partial T^2}\right)_P \le 0 \quad \text{and} \quad \left(\frac{\partial^2 G}{\partial P^2}\right)_T \le 0 \tag{16.9}$$

Summarizing, the thermodynamic potentials – that is, the energy and its Legendre transforms – are convex functions of their extensive variables and concave functions of their intensive variables[3].

From Eq. (16.9) we easily obtain

$$\left(\frac{\partial^2 F}{\partial V^2}\right)_T = -\left(\frac{\partial P}{\partial V}\right)_T = \frac{1}{V\kappa_T} \geq 0 \quad \text{or} \quad \kappa_T \geq 0 \tag{16.10}$$

indicating that the isothermal compressibility κ_T must be positive for a system to be mechanically stable[4]. By combining $C_V \geq 0$ and $\kappa_T \geq 0$ with $\kappa_T - \kappa_S = TV\alpha^2/C_P$ and $C_P - C_V = TV\alpha^2/\kappa_T$ (see Chapter 2, Eq. 2.28), one can further infer that

$$C_P \geq C_V \geq 0 \quad \text{and} \quad \kappa_T \geq \kappa_S \geq 0 \tag{16.11}$$

The first of these equations states that, upon the addition of heat to a system, the temperature rises but more at constant volume than at constant pressure. The second equation states that, upon decreasing the volume of the system, the pressure increases, but more at constant temperature than at constant entropy.

Problem 16.1

Near the critical point the density profile of molecules with molar mass M over the meniscus is significantly influenced by gravity. If we describe the gravitational potential (approximately) by $\phi = gh$, with g the acceleration due to gravity and h the height, the equilibrium condition reads $\mathrm{d}(\mu + M\phi) = 0$.

a) What is the expression for the molar volume V_m?

b) Calculate the pressure P at height h with respect to the meniscus level h_0.

c) Calculate the density $\rho = M/V_m$ as a function of height h with respect to h_0.

d) In many cases properties are scaled with respect to their value at critical point, for example, $T_{red} = T/T_{cri}$, $P_{red} = P/P_{cri}$ and $V_{red} = V/V_{cri}$. What is the appropriate expression for μ_{red} in terms of P_{cri}, V_{cri} and T_{cri}?

e) Scaling the equilibrium condition $\mathrm{d}(\mu + M\phi) = \mathrm{d}(\mu + Mgh) = 0$, we obtain $\mathrm{d}\mu_{red} = -\mathrm{d}h_{red}$. Calculate h_{cri} in $h_{red} = h/h_{cri}$.

f) Consider Ne and H_2O. For which compound is the width of the transition zone between liquid and gas near the critical point as characterized by h_{cri} larger? What does the order of magnitude of h_{cri} indicate to you? Use data as given in Appendix E.

3) A similar exercise for the Massieu functions, i.e. the entropy and its Legendre transforms, leads to concave functions of their extensive variables and convex functions of their intensive variables.

4) Although the derivation of $\kappa_T \geq 0$ is easier from F than from S, F refers to isothermal conditions so that the thermal stability condition $C_V \geq 0$ cannot be derived from F.

Problem 16.2

Using the relation $\partial^2 U/\partial V^2 \geq 0$, show that $\kappa_S \geq 0$.

Problem 16.3

Show by expanding the left-hand side of Eq. (16.3) in a Taylor series to second order that:

$$S_{UU}(\Delta U)^2 + 2S_{UV}\Delta U\Delta V + S_{VV}(\Delta V)^2 \leq 0$$

where $S_{UU} \equiv \partial^2 S/\partial U^2$, $S_{UV} \equiv \partial^2 S/\partial U\partial V$, and $S_{VV} \equiv \partial^2 S/\partial V^2$. Recalling that $S_{UU} \leq 0$, show that this expression can be written as

$$[S_{UU}\Delta U + S_{UV}\Delta V]^2 + [S_{UU}S_{VV} - S_{UV}^2](\Delta V)^2 \geq 0$$

and that this subsequently leads to Eq. (16.4).

16.2 Discontinuous Transitions

The most important discontinuous phase transitions for liquids are melting (fusion) and evaporation. For equilibrium between phases their chemical potential μ should be equal, otherwise transfer of matter occurs until $\mathrm{d}G = (\mu_2 - \mu_1)\mathrm{d}n = 0$ is fulfilled. If we consider that each phase has its own Gibbs energy function, the crossover temperatures between solid and liquid and liquid and vapor determine the melting and boiling temperatures, respectively. The effect of pressure is illustrated in Figure 16.5; this shows that, upon increasing the pressure, the melting point rises if the molar volume of the solid $V_\mathrm{m}(\mathrm{s})$ < the molar volume of the liquid $V_\mathrm{m}(\mathrm{l})$, while the melting point decreases if $V_\mathrm{m}(\mathrm{s}) > V_\mathrm{m}(\mathrm{l})$.

The coexistence curves for two phases in the P–T plane (Figure 16.2a) can be obtained from the *Clapeyron equation*, which follows from $\Delta G = \Delta H - T\Delta S = 0$ with the Maxwell relation

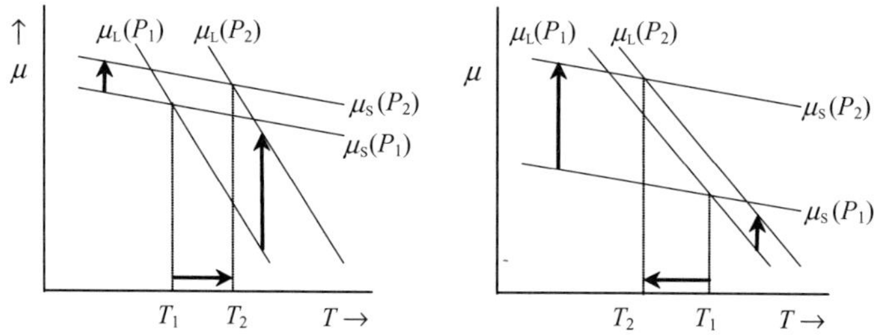

Figure 16.5 The dependency of melting point on pressure. (a) $V_\mathrm{m}(\mathrm{S}) < V_\mathrm{m}(\mathrm{L})$ leading to melting point rising ($T_2 > T_1$); (b) $V_\mathrm{m}(\mathrm{S}) > V_\mathrm{m}(\mathrm{L})$, leading to melting point lowering ($T_2 < T_1$).

$$\blacktriangleright \quad \frac{dP}{dT}=\frac{dS}{dV}\cong\frac{\Delta S}{\Delta V}=\frac{\Delta_{vap}H}{T\Delta V} \tag{16.12}$$

When liquid and vapor are both present in equilibrium we have $G_L = G_V$. Hence, on the one hand, we have (Figure 16.2b), since $G = F + PV$, the relation $F_a - F_e = -P(V_a - V_e)$. On the other hand, the work required to go from vapor to liquid is $F_a - F_e = -\int_e^a P\,dV$. From this we conclude that the (gray) area (Figure 16.2b) described by the curve abc must equal the (gray) area described by the curve cde. Hence, phase equilibrium is determined by the horizontal line for which these two areas are equal[5]. This is *Maxwell's equal area rule*. From a stability point of view, for the range abcde we have global instability (binodal), while for the range bcd we have local instability (spinodal).

16.2.1
Evaporation

We note that with a transition from the liquid to gas state the volume changes and the evaporation enthalpy is involved, the latter largely determining the vapor pressure. We denote the evaporation enthalpy, normally indicated by $\Delta_{vap}H$, here for brevity by L so that *Kirchhoff's equation* reads

$$L = L_0 + \int_0^T [C_P(\text{vapor}) - C_P(\text{liquid})]dT \equiv L_0 + \int_0^T \Delta C_P\,dT \tag{16.13}$$

We can write for the volume change $\Delta V = RT\Delta Z/P$, where Z is the compression factor $Z = PV/RT$ and thus ΔZ represents the compression factor change upon vaporization. This results in

$$\frac{d\ln P}{dT}=\frac{L}{RT^2\Delta Z} \quad \text{or} \quad \frac{d\ln P}{d(1/T)}=-\frac{L}{R\Delta Z} \tag{16.14}$$

The *Clausius–Clapeyron equation* results if the gas behaves perfectly (hence $\Delta Z = 1$). Applying this equation and integrating we obtain from Eq. (16.13)

$$\ln P = -L_0/RT + \int_0^T (RT^2)^{-1}\left(\int_0^T \Delta C_P\,dT\right)dT + \text{constant} \tag{16.15}$$

Assuming that $\Delta C_P = \Delta C_{P,0}[1 - (T/T_{cri})]$, the result is (Problem 16.5)

$$\ln P = -\frac{L_0}{RT}+\frac{\Delta C_{P,0}}{R}\left(\ln T-\frac{T}{2T_{cri}}\right)+\text{constant} \quad \text{or} \tag{16.16}$$

$$\ln P = a - b\ln T + (bT/2T_{cri}) - L_0/RT \cong a - b\ln T - L_0/RT \tag{16.17}$$

with a and b positive constants. The last step can be made if $\Delta C_P = \Delta C_{P,0}$, that is, the temperature dependence of ΔC_P can be neglected. In this approximation evaporation enthalpy is thus $L = L_0 + \Delta C_{P,0}T$. The experimental vapor pressure

5) Note, though, that in practice doing reversible work along the curve bcd is impossible.

indeed can be described within at least less than 2% between boiling and melting point by

$$\log P = a - b\log T - c/T \tag{16.18}$$

where a, b and c are constants[6]. Evaluating the vaporization enthalpy L via Eq. (16.14) leads to

$$L = RT^2(\mathrm{d}\ln P/\mathrm{d}T) = R(2.303c - bT) \quad \text{and therefore} \tag{16.19}$$

$$L_0 = 2.303Rc \quad \text{and} \quad \Delta C_{P,0} = -Rb \tag{16.20}$$

Values for a, b and c for several liquids are given in Appendix E. One step further is to assume that $\Delta C_{P,0} = 0$; that is, L is constant, resulting in $\log P = a - b \log T$. As this expression obviously can only describe P over a limited temperature range, one often uses the empirical *Antoine equation* [1] reading

$$\log P(T) = A - [B/(T + C)] \tag{16.21}$$

For a series of molecules the values for L_0 and b vary systematically with structure. Analyzing the data one obtains for aliphatic hydrocarbons

$$L_0(C_nH_{2n+2}) = 1.1 + 1.7n \quad \text{and} \quad b(C_nH_{2n+2}) = 0.5 + 0.94n \tag{16.22}$$

where n denotes the number of carbon atoms and the estimate is in kcal mol^{-1}. For aliphatic alcohols, similarly,

$$L_0(C_nH_{2n+1}OH) = 10.0 + 2.2n \quad \text{and} \quad b(C_nH_{2n+1}OH) = 2.5 + 1.88n \tag{16.23}$$

These types of estimate can be used to estimate unknown data. For example, by estimating data for $C_{12}H_{26}$ at 400 K and $C_{18}H_{38}$ at 500 K we obtain 12.0 kcal mol^{-1} and 15.3 kcal mol^{-1}, to be compared with the experimental data of 12.3 ± 1.3 kcal mol^{-1} and 15.3 ± 0.7 kcal mol^{-1}, respectively.

Empirically, the evaporation enthalpy L is often described by *Trouton's rule*

$$L/RT_\mathrm{n} = C, \quad \text{with} \quad C \cong 10.2 \tag{16.24}$$

where the subscript n indicates the normal boiling temperature (at 1 atm). The correlation is not particularly accurate in general (see Appendix E). Compounds that form dimers in the liquid state have a large Trouton constant, for example, NO with $C = 13.7$. Furthermore, compounds that form dimers in the gaseous state have a small Trouton constant, for example, CH_3COOH with $C = 7.5$ and HF with $C = 3.1$. Other differences are less easily rationalized. A much more accurate rule is *Hildebrand's rule*, which states that L divided by the vaporization temperature T_{vap} for all liquids is the same when vaporization is achieved at a temperature where the vapor density is the same. The value of $L/T_{vap} = 93$ J K^{-1} mol^{-1} when the vapor density is taken as 2.02×10^{-2} mol l^{-1}. Both, the Trouton and the Hildebrand rule are typical examples of empirical estimators for thermodynamic data.

6) Often, experimental data are given as in terms of $\log P$ instead of $\ln P$.

Table 16.1 Molar volumes at the melting point.[a)]

Liquid	T_m or T_{tri} (K)	V_S (cm^3 mol^{-1})	V_L (cm^3 mol^{-1})	$\Delta V/V_S$ (%)
H_2O	273.1	19.82	18.18	−8.3
CH_4	90.67	30.94	33.63	+8.6
CD_4	89.78	29.2	31.7	+8.6
CCl_4	250.4	87.9	91.87	+4.5
C_2H_4	103.97	39.06	43.63	+11.7
C_6H_6	278.5	77.28	88.28	+11.4
$C_{10}H_8$	353.2	112.2	130.9	+11.7

a) Data from Ref. [13].

16.2.2 Melting

As has been alluded to several times before, the molar volume increases upon melting, and Table 16.1 provides some typical data. A simple argument [2] indicates the reason why. Consider an arbitrary plane in the liquid. For such a plane the arrangement of molecules must be such as to allow them to pass to another plane while still being in contact with (a number of) neighbors. If we take a close-packed configuration for this plane, the molecule is sixfold coordinated in this plane and the molecular area is $o_L = \frac{1}{2}\sqrt{3}\sigma^2$, where σ is the diameter of the molecule. It is not unreasonable then to suppose that the volume available to the molecule in the liquid is $v_L = o_L^{3/2} = 3^{3/4}2^{-3/2}\sigma^3$. If we compare v_L with the volume available in the BCC lattice, $v_{BCC} = 2^2 3^{-3/2}\sigma^3$, we obtain $v_L/v_{BCC} = 1.05$, in good correspondence with the value for CCl_4 (Table 16.1). Similarly, for the FCC structure $v_{FCC} = 2^{-1/2}\sigma^3$ we obtain $v_L/v_{FCC} = 1.14$, which is in good correspondence with the value for inert gases (indeed, crystallizing in the FCC or HCP structure).

Problem 16.4

Derive the Clapeyron equation from $\mu_{liq} = \mu_{vap}$ at the coexistence curve.

Problem 16.5

Verify Eq. (16.17).

16.3 Continuous Transitions and the Critical Point

For continuous phase transitions, considerable effort has been paid to adequately describe and understand the process, and this has led to an image that captures the essentials well. We recall that thermodynamic equilibrium is determined by minimization of the appropriate potential with respect to the internal variable or variables. Because $\eta = \Delta\rho = \rho - \rho_{cri}$ is single-valued above T_{cri} and has two values below T_{cri}, this difference is conventionally used as the internal variable (see Chapter 2) characterizing the liquid–gas transition. In connection with transitions, the internal variable used is normally called the *order parameter*. In the sequel of this section we first discuss some experimental facts, and thereafter mean field theory.

16.3.1
Limiting Behavior

Both, single compounds and mixtures exhibit continuous transitions. In Figure 16.6 the relative density $(\rho_L - \rho_V)/2\rho_{cri}$ of CO_2 versus the relative temperature $t \equiv (T - T_{cri})/T_{cri}$ is given, while Figure 16.7 shows the difference in volume fraction $(\phi^{(1)} - \phi^{(2)})$ for CCl_4 in C_7F_{14} versus t. In both cases, the behavior is described by a power law [3].

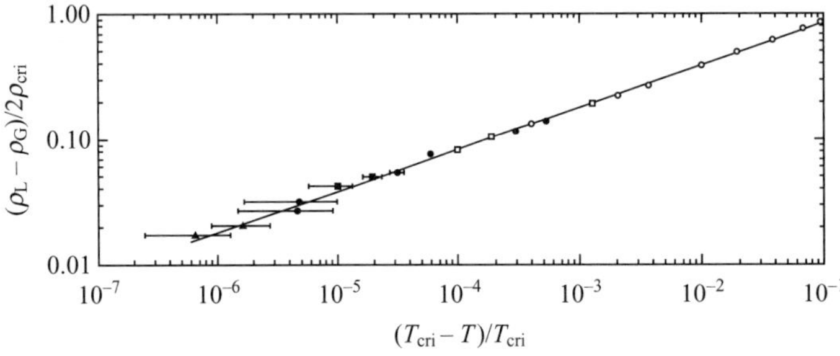

Figure 16.6 The relative density $(\rho_L - \rho_G)/2\rho_{cri}$ of CO_2 versus the relative temperature $-t = (T_{cri} - T)/T_{cri}$ with $T_{cri} = 304.18$ K, as described by $(\rho_L - \rho_G)/2\rho_{cri} = B(-t)^\beta$ with $B = 1.85$ and $\beta = 0.340 \pm 0.015$.

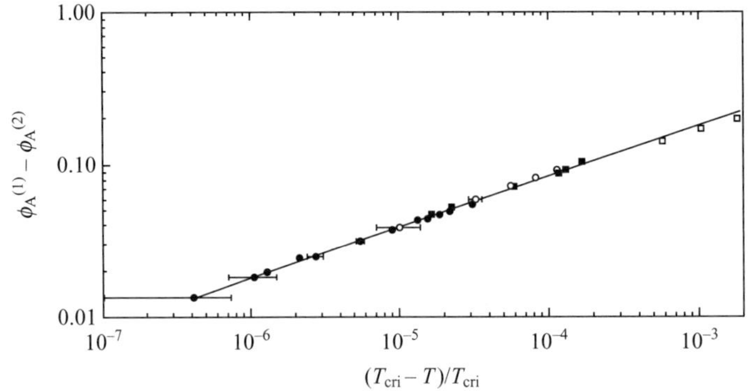

Figure 16.7 The difference in volume fraction $\phi^{(1)}$ in the upper phase and $\phi^{(2)}$ in the lower phase for CCl_4 in C_7F_{14} versus the relative temperature $-t = (T_{cri} - T)/T_{cri}$ with $T_{cri} = 301.786$ K, as described by $\phi^{(1)} - \phi^{(2)} = B(-t)^\beta$ with $B = 1.81$ and $\beta = 0.335 \pm 0.020$.

This type of experiment has led to the definition of various critical exponents. A critical exponent of a function, say a for a function $f(x)$ of x, is defined by

$$a \equiv \lim_{x \to 0} \frac{\mathrm{d} \ln f(x)}{\mathrm{d} \ln x} \tag{16.25}$$

and we write $f(x) \sim x^a$. This does not imply that $f(x) = Ax^a$, but rather that $f(x) = Ax^a(1 + Bx^b + \ldots)$ with $b > 0$, and where Ax^a is called the *singular part* of $f(x)$ and $(1 + Bx^b + \ldots)$ the *regular part* or *background function*. The exponent[7] a defines the rate of approach of $f(x)$ to zero for $a > 0$ or to infinity for $a < 0$. If $a = 0$, the result is ambiguous. In this case $f(x)$ has either a discontinuity or else a logarithmic singularity, that is, $f(x)$ behaves like $\ln(T - T_{\text{cri}})$.

The order parameter, that is, $\Delta\rho$ along the saturation curve for $T < T_{\text{cri}}$, is described by

$$\frac{\rho_{\text{liq}}(T) - \rho_{\text{gas}}(T)}{2\rho_{\text{cri}}(T)} = B(-t)^{\beta}(1 + \ldots) \tag{16.26}$$

where B and β are parameters and $(1 + \ldots)$ is the background function to which the critical exponent β is insensitive. Experimentally, the range for β is $0.3 < \beta < 0.4$.

The response function–that is, the compressibility κ_T–behaves similarly but becomes infinite at $T = T_{\text{cri}}$. For $T < T_{\text{cri}}$ (for $\rho = \rho_{\text{liq}}$ or $\rho = \rho_{\text{gas}}$ on the saturation curve) and $T > T_{\text{cri}}$ (for $\rho = \rho_{\text{cri}}$), κ_T is described by, respectively,

$$\frac{\kappa_T}{\kappa_T^{\circ}} = C'(-t)^{-\gamma'}(1 + \ldots) \quad \text{and} \quad \frac{\kappa_T}{\kappa_T^{\circ}} = C(t)^{-\gamma}(1 + \ldots) \tag{16.27}$$

where C, C', γ, and γ', are parameters and $\kappa_T^{\circ} = 1/P_{\text{cri}}^{\circ} = m/kT\rho_{\text{cri}}$ is the compressibility of an ideal gas with molecules of mass m at of density ρ_{cri} at T_{cri}. Experimentally, the range for γ and γ' is $1.2 < \gamma < 1.4$.

Such a power-law relation also holds for the heat capacity C_V. This is given in a two-phase system for $T < T_{\text{cri}}$ (for $\rho_{\text{liq}} + \rho_{\text{gas}} = 2\rho_{\text{cri}}$) and $T > T_{\text{cri}}$ (for $\rho = \rho_{\text{cri}}$) by, respectively,

$$\frac{C_V}{Nk} = A'(-t)^{-\alpha'}(1 + \ldots) \quad \text{and} \quad \frac{C_V}{Nk} = A(t)^{-\alpha}(1 + \ldots) \tag{16.28}$$

where A, A', α and α' are parameters. Experimentally, the range for α and α' is $-0.2 < \alpha < 0.2$.

For the pressure P along the critical isotherm $T = T_{\text{cri}}$, data are described by

$$\frac{P - P_{\text{cri}}}{P_{\text{cri}}^{\circ}} = D\left|\frac{\rho - \rho_{\text{cri}}}{\rho_{\text{cri}}}\right|^{\delta} \operatorname{sgn}(\rho - \rho_{\text{cri}}) \tag{16.29}$$

where $\operatorname{sgn}(I) = 1$ if $I > 0$ and $\operatorname{sgn}(I) = -1$ if $I < 0$. Experimentally, we have $4 < \delta < 5$.

Finally, we examine the (total) pair correlation function $h(r) = g^{(2)}(r) - 1$ (see Section 7.2). For large r, $h \to 0$ and we expect, and also from Ornstein–Zernike theory it appears, that $h(r)$ for $T > T_{\text{cri}}$ behaves like

7) In the next paragraphs we use a labeling for the various critical exponents which is largely standardized. Unfortunately several labels, such as β and α, are also used for other properties.

$$h(r) = f(r)\exp(-r/\xi)/r^{D-2} \tag{16.30}$$

where $f(r)$ is a weakly varying function[8] of r, D is the dimension of space, and where the *coherence* (or *correlation*) *length* ξ measures the range of order in the liquid. In essence, ξ is a measure of the greatest distance from a given particle at which this particle can still cause nonrandom effects on the particle density. Under normal conditions the order of magnitude for ξ is a few molecular diameters, but for $T \to T_{cri}$ and $\rho = \rho_{cri}$, the value of ξ diverges to macroscopic size. In this case, high- and low-density regions appear on an increasingly larger scale until macroscopic regions of different density in the fluid lead to the two-phase region for $T \le T_{cri}$. However, the small(er) scale fluctuations do not disappear and within high-density regions there will be smaller regions of lower density, and vice versa. As soon as ξ approaches the wavelength of visible light, say 0.5–0.8 μm, these fluctuations lead to the phenomenon of *critical opalescence*. Using experimental data it appears that ξ also scales with T and for $T < T_{cri}$ (for $\rho = \rho_{liq}$ or $\rho = \rho_{gas}$ on the saturation curve) and $T > T_{cri}$ (for $\rho = \rho_{cri}$) is given by, respectively,

$$\xi = \xi_0'(-t)^{-\nu'} \quad \text{and} \quad \xi = \xi_0(t)^{-\nu} \tag{16.31}$$

Using the result from Eq. (16.30), Fisher [4] showed that for $D = 2$, $h(r) \sim \ln r$, that is, increases with increasing r, which is physically impossible. He therefore introduced the additional exponent η (obeying $0 < \eta < 2$) for $r \to \infty$ at ($\rho = \rho_{cri}$) and ($\rho = \rho_{cri}$ and $T = T_{cri}$), respectively, by

$$h(r) \to H_D \exp(-r/\xi)/r^{D-2+\eta} \quad \text{and} \quad h(r) \to H_D/r^{D-2+\eta} \tag{16.32}$$

It appears that in 3D the value of η is small, typically about 0.03. Moreover, η is related to ν and γ (Example 16.1).

Example 16.1: The Fisher relation

In Chapter 6 we derived for the compressibility

$$\rho\kappa_T = \beta\left[1 + \rho\int_0^\infty h(r)4\pi r^2\,\mathrm{d}r\right]$$

Using $h(r) \to H_D\exp(-r/\xi)/r^{D-2+\eta}$ we obtain

$$\rho\int_0^\infty h(r)4\pi r^2\,\mathrm{d}r \to 4\pi\rho\int_0^\infty \frac{H_3\mathrm{e}^{-r/\xi}}{r^{1-\eta}}r^2\,\mathrm{d}r = 4\pi\rho H_3\xi^{2-\eta}\int_0^\infty x^{1-\eta}\mathrm{e}^{-x}\,\mathrm{d}x$$

with $x = r/\xi$. If $\eta < 1$, the integral converges and it has value 1 for $\eta = 0$. Because $\kappa_T \sim t^{-\gamma}$, $1 + \rho\int_0^\infty h(r)4\pi r^2\,\mathrm{d}r \sim \xi^{2-\eta}$ and $\xi \sim t^{-\nu}$, we have

$$t^{-\gamma} = \xi^{2-\eta} = (t^{-\nu})^{2-\eta} \quad \text{or} \quad \gamma = (2-\eta)\nu$$

---

8) For example, in 3D $h(r)$ can be approximately described by $h(r) = (a/r)\cos(br + c)\exp(-r/\xi)$.

Table 16.2 Critical points and their order parameters.[a)]

Critical point	Order parameter	Example	T_{cri} (K)
Liquid-gas	Density difference	H_2O	647.05
Fluid mixture	Fraction of components	CCl_4–C_7F_{14}	301.78
Binary alloy	Fraction species on sublattice	Cu–Zn	739
Ferromagnetic	Magnetic moment	Fe	1044.0
Antiferromagnetic	Sublattice magnetic moment	FeF_2	78.26
Ferroelectric	Electric moment	$(NH_2CH_2COOH)_3\cdot H_2SO_4$	322.5

a) Data from Ref. [5].

There appear to be more relations between the critical exponents. They were originally derived as (thermodynamic) inequalities between the various critical exponents, and usually named after their discoverer. We have for example,

$$\text{Rushbrooke's inequality} \quad \alpha' + 2\beta + \gamma' \geq 2, \tag{16.33}$$

$$\text{Griffiths' inequality} \quad \alpha' + \beta(1+\delta) \geq 2 \quad \text{and} \tag{16.34}$$

$$\text{Widom's inequality} \quad \gamma' \geq \beta(\delta - 1) \tag{16.35}$$

Later, we will see that they actually are equalities and, moreover, that $\alpha' = \alpha$ and $\gamma' = \gamma$. The derivation of some of them is easy, but for others it is quite involved. As an example we derive the Rushbrooke inequality. To do so we need *Stanley's lemma* that states that for $f(x) \sim x^a$ and $g(x) \sim x^b$ with $f(x) \leq g(x)$ and for $0 < x < 1$, so that ln $x < 0$, we have $a \geq b$. From thermodynamics, we know that $C_P - C_V = TV\alpha^2/\kappa_T$ (all quantities positive, see Section 2.1), so that $C_P \geq TV\alpha^2/\kappa_T$. Because we have for the heat capacity $C_P \sim (-t)^{-\alpha'}$, for the compressibility $\kappa_T \sim (-t)^{-\gamma'}$, and for the thermal expansion coefficient $\alpha \sim (-t)^{\beta-1}$, we obtain at once $-\alpha' \leq \gamma' + 2(\beta - 1)$ or $\alpha' + 2\beta + \gamma' \geq 2$ (for $0 < \alpha' < 1$, while for $\alpha' \leq 0$, the most we can say is that $2\beta + \gamma' \geq 2$).

Furthermore, it appears that the critical exponents for a number of phenomena are essentially the same (Table 16.2). This happens to be the case if, for two phenomena, both the spatial dimensionality, for example, 2D or 3D, and the dimensionality of the order parameter, for example, a scalar or vector, are the same. These phenomena are said to belong to the same *universality class*. For the phenomena listed in Table 16.2 the lattice gas model (see Appendix C), with a scalar order parameter, forms the prototype. The liquid–gas transition also belongs to this class (Problem 16.9). We limit ourselves entirely to this class, and some experimental results are given in Table 16.3.

16.3.2
Mean Field Theory: Continuous Transitions

Remember that in thermodynamics the equilibrium situation is conveniently obtained by minimizing the proper thermodynamic potential, say the "generalized" Helmholtz energy $F(\eta;T,V)$, dependent on temperature T, volume V with

Table 16.3 Experimental values of critical exponents for various fluids.

Compound[a)]	α	β	γ	δ	ν	η
CO_2	0.10	0.321	1.24	4.85	0.57	~0
Xe	0.11	0.329	1.23	4.74	–	0.05
N_2	–	0.327 ± 0.002	1.23 ± 0.01	–		
Ne	–	0.327 ± 0.002	1.25 ± 0.01	–		

a) CO_2 and Xe Data from Ref. [14].

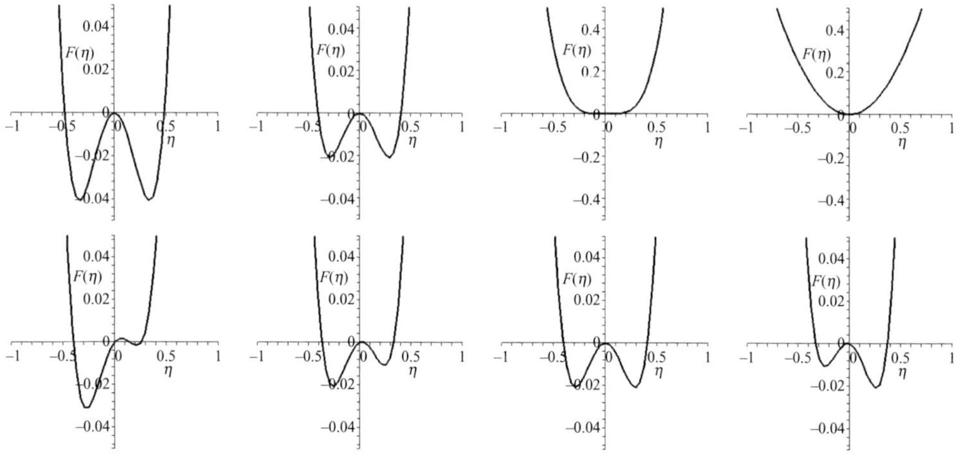

Figure 16.8 The shape of the Helmholtz energy curve F. Upper row: Continuous transition along the coexistence curve where (from left to right) the images represent the shape changing to T_{cri} (mid-right) and above T_{cri} (outer-right); Lower row: Discontinuous transition across the coexistence curve where images (from left to right) represent the shape change from liquid via coexistence line (mid-right) to gas (outer-right).

respect to the order parameter (internal variable) η. We used the designation "generalized" because the "generalized" F only becomes the thermodynamic F after minimization with respect to η. In general, η varies with position, that is, is a field quantity, and we have to use the energy density. For a uniform distribution of η we can use the energy itself, but we will still use the energy density, here denoted by $\mathcal{L}(\eta; T, V)$ or by just $\mathcal{L}(\eta)$ or $\mathcal{L}$. The order parameter η can be a local average of an underlying microscopic parameter describing the physics of the system. For example, for a ferromagnetic material the order parameter is the magnetization m which in the lattice model (see Appendix C) results as the average of the magnetic spins σ with value ± 1. Above T_{cri}, $m = 0$, while below T_{cri} the magnetization is $\pm m$. The order parameter η can also be purely macroscopic. For example, for a fluid the order parameter is conveniently taken as the density difference $\Delta\rho$ between liquid and gas. In Figure 16.8 the qualitative shape of the

$F(\eta;T,V)$ curves as a function of temperature for various conditions of V and T for a fluid are given. The simplest approach to describe these curves is to use the so-called mean field approach, originally due to Landau.

In this approach, $\mathcal{L}(\eta)$ typically represents the grand potential density (see Chapter 2), and we assume that the Helmholtz part $F(\eta)$ of $\mathcal{L}(\eta)$ near T_{cri} can be expanded in a power series in η, reading

$$\mathcal{L}(\eta) = F(\eta) - h\eta = a_0 + a_1\eta + a_2\eta^2 + a_3\eta^3 + a_4\eta^4 + \ldots - h\eta \qquad (16.36)$$

The driving force h is the conjugate variable to the order parameter η. For a fluid the driving force is the chemical potential μ (with order parameter $\Delta\rho$), while for a ferromagnetic material mentioned above it is the external magnetic field H (with order parameter m). In principle, all coefficients a_n depend on temperature T. The parameter a_0 is a reference energy and can be omitted. Equilibrium requires that $\partial\mathcal{L}/\partial\eta = 0$, and in the absence of an external driving force, this requires that a_1 is absent. Further note that a_4 should be > 0 if the system is to be stable. In cases the order parameter η can have only the two values $\pm\eta$, as for the magnetization case, the third-order terms must also be absent because $\mathcal{L}(\eta)$ should equal $\mathcal{L}(-\eta)$. For the fluid case this is obviously not the case if we use $\eta = \Delta\rho = (\rho - \rho_{\text{cri}})$. Nevertheless, we assume this to be true for the moment, that is, we essentially discuss the magnetic problem, and come back later to the case $a_3 \neq 0$. For the moment we also consider the case when $h = 0$. So, we are left with $\mathcal{L}(\eta) = a_2\eta^2 + a_4\eta^4$. As usual, equilibrium is reached when $\partial\mathcal{L}(\eta)/\partial\eta = 0$ and the solutions are simply found to be

$$\eta^* = 0 \quad (T > T_{\text{cri}}) \quad \text{and} \quad \eta^* = \pm(-a_2/2a_4)^{1/2} \quad (T < T_{\text{cri}}) \qquad (16.37)$$

The coefficient a_2 can be expanded with respect to temperature reading $a_2 = a_{2,0} + a_{2,1}\Delta T + \mathcal{O}(\Delta T^2)$, where $a_{2,1} = \partial^3\mathcal{L}/\partial\eta^2\partial T$ (generally $a_{i,j} = \partial^{i+j}\mathcal{L}/\partial\eta^i\partial T^j$). Since for $T < T_{\text{cri}}$, η should be non-zero for η arbitrarily close to T_{cri}, we must have $a_{2,0} = 0$ and hence $a_2 = a_{2,1}\Delta T$. A similar expansion could be used for a_4, but as this will lead to second-order temperature effects, we take a_4 as a constant. This implies that for $T < T_{\text{cri}}$, the solution is $\eta^* = \pm(-a_{2,1}\Delta T/2a_4)^{1/2}$. To ease the notation slightly, we write

$$\mathcal{L}(\eta) = at\eta^2 + b\eta^4, \quad \text{so that} \quad \eta = \pm(-at/2b)^{1/2} \quad (T < T_{\text{cri}}) \qquad (16.38)$$

with a and b as the parameters, t as before, and omitting the asterisk from now on.

We now calculate the critical exponents. For $T < T_{\text{cri}}$ the solution is $\eta = (-at/2b)^{1/2}$, so comparing with $\eta \sim (T_{\text{cri}} - T)^\beta$ results in $\beta = ½$.

For $T > T_{\text{cri}}$, $\eta = 0$ and thus $\mathcal{L}(\eta) = 0$. For $T < T_{\text{cri}}$, $\eta = (-at/2b)^{1/2}$ and thus $\mathcal{L}(\eta) = at(-at/2b) + b(-at/2b)^2 = -a^2t^2/4b$. Because $C_V = -T\partial^2\mathcal{L}/\partial T^2$, we easily obtain

$$C_V = 0 \quad (T > T_{\text{cri}}) \quad \text{and} \quad C_V = Ta^2/2bT_{\text{cri}}^2 \quad (T < T_{\text{cri}}) \qquad (16.39)$$

The heat capacity thus changes discontinuously at T_{cri}, which can be represented by taking the exponent $\alpha = 0$.

To compute the other exponents we need to let $h \neq 0$. Equilibrium is obtained from $\partial \mathcal{L}/\partial \eta = 0$ and, using $\mathcal{L}(\eta) = at\eta^2 + b\eta^4 - h\eta$, reads $h = 2at\eta + 4b\eta^3$. On the critical isotherm $t = 0$ and therefore $h \sim \eta^3$ so that $\delta = 3$.

The response function, the isothermal susceptibility X_T in this case, is given by

$$X_T(h) \equiv \left.\frac{\partial \eta}{\partial h}\right|_T = \left(\frac{\partial^2 \mathcal{L}(\eta)}{\partial \eta^2}\right)^{-1} = \left[2at + 12b\eta(h)^2\right]^{-1} \tag{16.40}$$

where $\eta(h)$ is a solution of $h = 2at\eta + 4b\eta^3$. Since for $T > T_{\text{cri}}$ we have $\eta = 0$, $X_T = (2at)^{-1}$. Similarly, for $T < T_{\text{cri}}$ at $h = 0$, we have $\eta^2 = -at/b$, $X_T = (-4at)^{-1}$, and thus $\gamma = \gamma' = 1$. Note that $X_T(T < T_{\text{cri}}) = -\frac{1}{2}X_T(T > T_{\text{cri}})$ always in this model.

16.3.3
Mean Field Theory: Discontinuous Transitions

Let us consider briefly whether the Landau approach can be applied to discontinuous transitions. So far, we have used $\mathcal{L}(\eta) = at\eta^2 + b\eta^4 - h\eta$. We omitted a linear term because we required that $\eta = 0$ for $T > T_{\text{cri}}$, and a cubic term on the basis of symmetry. We now allow the cubic term so that we have

$$\mathcal{L}(\eta) \cong at\eta^2 + c'\eta^3 + b\eta^4 - h\eta \tag{16.41}$$

Considering only the case that $h = 0$, we obtain equilibrium from $\partial \mathcal{L}(\eta)/\partial \eta = 0$ and the solutions are found to be

$$\eta = 0 \quad (T > T_{\text{cri}}) \quad \text{and} \quad \eta = -c \pm \sqrt{c^2 - 2at/b} \quad (T < T_{\text{cri}}) \tag{16.42}$$

with $c \equiv 3c'/8b$. The solution for $T < T_{\text{cri}}$ is acceptable (i.e., real) if $c^2 - 2at/b > 0$, that is, for a temperature $t < t^* \equiv bc^2/2a$. Because $t^* > 0$, the temperature $T^* > T_{\text{cri}}$, the latter being the temperature where the coefficient of η^2 vanishes. Recall that for the continuous transition an acceptable solution becomes available only for $t < 0$ or $T > T_{\text{cri}}$. Figure 16.8 (lower row) shows a schematic of the behavior for the discontinuous transition. For $t > t^*$, $\mathcal{L}(\eta)$ shows only one minimum. For $t < t^*$ a secondary minimum appears which at $t = t_1$ is equal to the one at $\eta = 0$. For $t < t_1$ this minimum becomes the global minimum and the value of η jumps discontinuously to a nonzero value, representing the discontinuous transition. Note that $\eta(t_1)$ is not arbitrarily small, as required by Landau theory. Generally – but not always – the presence of a cubic term in η will give rise to a discontinuous transition.

16.3.4
Mean Field Theory: Fluid Transitions

Let us now return to our case of the liquid–gas transition for which $a_1, a_3 \neq 0$. In this case, $\mathcal{L}(\eta)$ reads (using the chemical potential μ as the proper driving force)

$$\mathcal{L}(\eta) = a_1\eta + a_2\eta^2 + a_3\eta^3 + a_4\eta^4 - \mu\eta \tag{16.43}$$

For the general case a solution of Eq. (16.43) is rather difficult to obtain. For the fluid case, however, we apply a trick by writing $\eta = \eta_0 + \Delta\eta$. If we do so we obtain

$$\mathcal{L}(\eta_0 + \Delta\eta) = (a_1 - \mu)(\eta_0 + \Delta\eta) + a_2(\eta_0 + \Delta\eta)^2 + a_3(\eta_0 + \Delta\eta)^3 + a_4(\eta_0 + \Delta\eta)^4 \quad \text{or} \tag{16.44}$$

$$\mathcal{L} = A(\eta_0) + B(\eta_0)\Delta\eta + C(\eta_0)\Delta\eta^2 + D(\eta_0)\Delta\eta^3 + E(\eta_0)\Delta\eta^4 \quad \text{with} \tag{16.45}$$

$$A(\eta_0, \mu) = (a_1 - \mu)\eta_0 + a_2\eta_0^2 + a_3\eta_0^3 + a_4\eta_0^4, \tag{16.46}$$

$$B(\eta_0) = (a_1 - \mu) + 2a_2\eta_0 + 3a_3\eta_0^2 + 4a_4\eta_0^3, \tag{16.47}$$

$$C(\eta_0) = a_2 + 3a_3\eta_0 + 4a_4\eta_0^2, \quad D(\eta_0) = a_3 + 4a_4\eta_0 \quad \text{and} \quad E(\eta_0) = a_4 \tag{16.48}$$

Each of the functions $B(\eta_0), \ldots, E(\eta_0)$ can be considered as a function of temperature, which we expand to first order in T. The function $A(\eta_0,\mu)$ depends on η_0 and μ, so that a first-order expansion in T and μ with respect to T_{cri} and μ_{cri} reads $A(\eta_0, m) = a_{1,1}\Delta T + \Delta\mu + \mathcal{O}(\Delta^2 T)$ with $\Delta T = T - T_{\text{cri}}$ and $\Delta\mu = \mu - \mu_{\text{cri}}$ using the notation $a_{i,j} = \partial^{i+j}\mathcal{L}/\partial\eta^i\partial T^j$. We can eliminate the linear and cubic terms in $\Delta\eta$ by setting them equal to zero. First, solving the cubic term $D(\eta_0)$ for η_0 leads to $\eta_0 = -a_3/4a_4 = -a_{3,1}\Delta T/4a_4$. Substitution in the linear term $B(\eta_0)$ results, to first order in ΔT, in

$$\begin{aligned} B(\eta_0) &\cong (a_1 - \mu) + 2a_2\eta_0 + \ldots \\ &= (a_{1,1}\Delta T - \Delta\mu) + 2a_{3,1}\Delta T\eta_0 \cong (a_{1,1}\Delta T - \Delta\mu) \end{aligned} \tag{16.49}$$

where the last step can be made because $\Delta T\eta_0$ is of second order in ΔT. Since $B(\eta_0) = 0$ requires $\Delta\mu = a_{1,1}\Delta T$, we see that this condition is already satisfied. Therefore, along the line determined by $\eta_0 = -a_{3,1}\Delta T/4a_4$, the Landau function $\mathcal{L}(\eta)$ will read

$$\mathcal{L} = A(\eta_0) + C(\eta_0)\Delta\eta^2 + E(\eta_0)\Delta\eta^4 = A(\eta_0) + a_{2,1}\Delta T\Delta\eta^2 + a_4\Delta\eta^4 \tag{16.50}$$

which has the same form as $\mathcal{L}(\eta)$ for the magnetic problem. The solution accordingly is also the same and reads for $T < T_{\text{cri}}$

$$\Delta\eta = \pm(-a_{2,1}\Delta T\,/\,2a_4)^{1/2} \tag{16.51}$$

All critical exponents are thus also the same as before. The complete solution is obtained by superposing the $\Delta\eta$ solution on the η_0 behavior. To interpret η_0 we recall that $\eta = \eta_0 + \Delta\eta = \rho - \rho_{\text{cri}}$. Using $\eta_{\text{liq}} = \eta_0 + \Delta\eta$ and $\eta_{\text{gas}} = \eta_0 - \Delta\eta$, we see that $\rho_{\text{liq}} + \rho_{\text{gas}} = 2\eta_0$, so that η_0 equals the average density and represents the *law of rectilinear diameters* (see Chapter 4). Because $\eta_0 = -a_{2,1}\Delta T/4a_4$, the average density is a linear function of temperature, as approximately confirmed by experimental evidence. However, expanding all functions to second order leads to nonlinear behavior [6].

In conclusion to this part, we note that the order of magnitude of the critical exponents as obtained from the mean field approach is correct, but that quantitative agreement is missing. The Landau expression is approximate and we neglected fluctuations, and further steps are therefore required. This includes scaling, using the idea of generalized homogeneous functions applied to thermodynamic potentials, from which it becomes clear that the thermodynamic inequalities become equalities, and renormalization, a theory that states that phenomena behave the same at different length scales near the critical point and provides accurate numerical values for the critical exponents.

Problem 16.6

Show that $f(x) \sim x^a \leq g(x) \sim x^b$ for $0 < x < 1$ and $a \geq b$.

Problem 16.7

Show, by evaluating $d(F/V)$ and using the Euler expression $F = G - PV$, that the differential of the Helmholtz density reads $\mu d\rho - s dT$, with $s = S/V$.

Problem 16.8

Verify that the inequalities for the critical exponents as given by Eqs (16.33)–(16.35) are satisfied as equalities for the mean field critical exponent values.

Problem 16.9: The van der Waals critical exponents

The vdW expression, $[P + a/V^2](V - b) = NkT$, rewritten as polynomial in V reads $V^3 - [b + (NkT/P)]V^2 + (a/P)V - (ab/P) = 0$.

a) Comparing with the expression $(V - V_{\text{cri}})^3 = V^3 - 3V_{\text{cri}}V^2 + 3V_{\text{cri}}^2 V - V_{\text{cri}}^3$, show that $T_{\text{cri}} = 8a/27Nkb$, $P_{\text{cri}} = a/27b^2$, and $V_{\text{cri}} = 3b$.

b) Show that $P_{\text{cri}}V_{\text{cri}}/RT_{\text{cri}} = 3/8$.

c) Defining $T_{\text{red}} \equiv T/T_{\text{cri}}$, $P_{\text{red}} \equiv P/P_{\text{cri}}$ and $V_{\text{red}} \equiv V/V_{\text{cri}}$, show that vdW expression reduces to $(P_{\text{red}} + 3/V_{\text{red}}^2)(3V_{\text{red}} - 1) = 8T_{\text{red}}$.

d) Further defining $\omega \equiv (\rho - \rho_{\text{cri}})/\rho_{\text{cri}}$, $p \equiv (P - P_{\text{cri}})/P_{\text{cri}}$ and $t \equiv (T - T_{\text{cri}})/T_{\text{cri}}$, show that the vdW expression reads $p(2 - \omega) = 8t(1 + \omega) + 3\omega^3$.

e) To calculate the exponent β, consider the behavior near the critical point. Show that the vdW expression reduces to $\omega(3\omega^2 + 12t) = 0$, using that for $\omega = 0$, $p = 4t$.

f) Show that this solution leads to $\omega = \pm 2(-t)^{1/2} = \pm(1 - T_{\text{red}})^{1/2}$ or $\beta = ½$.

g) To calculate the exponent δ, consider the behavior at T_{cri}, that is, $t = 0$. Show that the vdW expression becomes $p(2 - \omega) = 3\omega^3$.

h) Show that this solution leads to $p \cong (3\omega^3/2)[1 + (\omega/2) + \ldots]$ or $\delta = 3$.

i) To calculate the critical exponent γ, show that the compressibility $\kappa_T \sim (\partial p/\partial \omega)_T$ is given by $\kappa_T \sim (24t + 18\omega^2 - 6\omega^3)/(2 - \omega)^2$. Show that this solution, for $t > 0$, leads to $\kappa_T \sim 6t$ or $\gamma = 1$.

16.4 Scaling

In the previous paragraphs we have seen that the mean field theory delivers values for the critical coefficients, satisfying the thermodynamic inequalities as equalities, but that the agreement with experiment is only approximate. In this section we show that the inequalities are indeed equalities using the concept of scaling. To that purpose, we first discuss (generalized) homogeneous functions and thereafter consider thermodynamic potentials as homogeneous functions and lattice scaling.

16.4.1 Homogeneous Functions

We recall that a homogeneous function is defined by $f(\lambda r) = g(\lambda)f(r)$, in which $g(\lambda)$ is an unspecified function of the scale factor λ. For example, if $f(r) = Br^2$, we have $f(\lambda r) = \lambda^2 f(r)$ so that $g(\lambda) = \lambda^2$.

The general expression for $g(\lambda)$ appears to be λ^p, which can be derived as follows. Suppose we have the function $f[\lambda(\mu r)]$ depending on λ and μ. Then, we can write

$$f[\lambda(\mu r)] = g(\lambda) f(\mu r) = g(\lambda) g(\mu) f(r) \quad \text{and} \tag{16.52}$$

$$f[\lambda(\mu r)] = g(\lambda\mu) f(r) \quad \rightarrow \quad g(\lambda\mu) = g(\lambda) g(\mu) \tag{16.53}$$

Now, differentiating both sides with respect to μ, denoting $\mathrm{d}g(x)/\mathrm{d}x$ by g', we obtain

$$\frac{\partial}{\partial \mu} g(\lambda\mu) = \lambda g'(\lambda\mu) = g(\lambda) g'(\mu) \quad \text{and} \quad \frac{\partial}{\partial \mu} g(\lambda)g(\mu) = g(\lambda)g'(\mu) \tag{16.54}$$

Let $\mu = 1$ and set $g'(\mu = 1) \equiv p$. This leads to

$$\frac{g'(\lambda)}{g(\lambda)} = \frac{\mathrm{d}}{\mathrm{d}\lambda} \ln g(\lambda) = \frac{p}{\lambda} \quad \text{or} \quad \ln g(\lambda) = p \ln \lambda + c \quad \text{or} \quad g(\lambda) = \mathrm{e}^c \lambda^p \tag{16.55}$$

Consequently

$$g'(\lambda) = p\mathrm{e}^c \lambda^{p-1} \quad \text{but} \quad g'(1) = p \quad \rightarrow \quad c = 0 \quad \text{and} \quad g(\lambda) = \lambda^p \tag{16.56}$$

If a homogeneous function is dependent on two variables we have $f(\lambda x,\lambda y) = \lambda^p f(x,y)$. Let us take $\lambda = 1/y$, resulting in $f(\lambda x,\lambda y) = f(x/y,1) = y^{-p} f(x,y)$. If we denote $f(z,1)$ by $F(z)$, we have $f(x,y) = y^p F(x/y)$. So, a homogeneous function $f(x,y)$ may be written as y^p times a function of x/y. Similarly, $f(x,y) = x^p G(y/x)$ with $G(z) = f(1,z)$. This implies that, by a suitable choice of variables, say $f(x,y)/x^p$ and y/x, all data collapse on a single curve. We now generalize the concept by writing $f(\lambda^a x,\lambda^b y) = \lambda f(x,y)$. One could think that writing $f(\lambda^a x,\lambda^b y) = \lambda^p f(x,y)$ provides an even more general form, but that is not the case because this expression is equivalent to $f(\lambda^{a/p} x,\lambda^{b/p} y) = \lambda f(x,y)$ and merely constitutes a rescaling of a and b. For such a *generalized homogeneous function* of two variables, there are at most two undetermined parameters.

16.4.2
Scaled Potentials

One of the main problems of the Landau mean field approach is that the various thermodynamic potentials are not analytical and therefore cannot be expanded, strictly speaking, with respect to the critical temperature. It was Widom [7] who considered thermodynamic potentials as generalized homogeneous functions. So, let us write the Gibbs energy $G(t,P)$ where, as before, $t = (T - T_{\text{cri}})/T_{\text{cri}}$. We assume that the singular part of G is a generalized homogeneous function of t and P. So we write

$$G(\lambda^p t, \lambda^q P) = \lambda G(t, P) \tag{16.57}$$

Differentiating with respect to P yields

$$\lambda^q \frac{\partial G(\lambda^p t, \lambda^q P)}{\partial \lambda^q P} = \lambda \frac{\partial G(t, P)}{\partial P} \quad \text{or} \quad -\lambda^q V(\lambda^p t, \lambda^q P) = -\lambda V(t, P) \tag{16.58}$$

Now for the moment we take $P = 0$ to obtain

$$V(t, 0) = \lambda^{q-1} V(\lambda^p t, 0) \tag{16.59}$$

In the spirit of the procedure used before for the homogeneous functions, we use

$$\lambda = (-t)^{-1/p} \quad \text{leading to} \quad V(t, 0) = (-t)^{(1-q)/p} V(-1, 0) \tag{16.60}$$

and approaching T_{cri} via $t \to 0^-$, the latter notation implying that t approaches zero from the negative side, the result is

$$V(t, 0) \sim (-t)^{\beta} \quad \text{with} \quad \beta = (1-q)/p \tag{16.61}$$

In a similar vein we take $t = 0$ and let the pressure P go to zero. The first step yields

$$V(0, P) = \lambda^{q-1} V(0, \lambda^q P) \tag{16.62}$$

and taking

$$\lambda = P^{-1/q} \quad \text{leads to} \quad V(0, P) = P^{(1-q)/q} V(0, 1) \tag{16.63}$$

Using $P \to 0$ leads to

$$V(0, P) \sim P^{1/\delta} \quad \text{with} \quad \delta = q/(1-q) \tag{16.64}$$

Solving the results $\beta = (1 - q)/p$ and $\delta = q/(1 - q)$ for p and q leads to

$$p = \frac{1}{\beta} \frac{1}{\delta + 1} \quad \text{and} \quad q = \frac{\delta}{\delta + 1} \tag{16.65}$$

The next step is to differentiate the Gibbs energy G twice with respect to P, leading to

$$\lambda^{2q} \kappa_T(\lambda^p t, \lambda^q P) = \lambda \kappa_T(t, P) \tag{16.66}$$

Using the same procedure as before, that is, taking again $P = 0$ and assuming

$$\lambda = (-t)^{-1/p} \quad \text{leads to} \quad \kappa_T(t,0) = (-t)^{-2(1-q)/p}\kappa_T(-1,0) \tag{16.67}$$

Approaching T_{cri} via $t \to 0^-$ the result is

$$\kappa_T(t,0) \sim (-t)^{-\gamma'} \quad \to \quad \gamma' = (2q-1)/p \tag{16.68}$$

Substituting Eq. (16.65) the final result is

$$\gamma' = \beta(2\delta - 1) \tag{16.69}$$

which we recognize as Widom's (in)equality. We conclude for now there are only two independent critical exponents, p and q, instead of the original three, β, δ and γ'.

Repeating the last procedure but now with using $t \to 0^+$, we obtain

$$\lambda = t^{-1/p}, \quad \kappa_T(t,0) = t^{-2(1-q)/p}\kappa_T(-1,0) \quad \text{and} \tag{16.70}$$

$$\kappa_T(t,0) \sim t^{-\gamma} \quad \text{with} \quad \gamma = (2q-1)/p \tag{16.71}$$

Hence $\gamma' = \gamma$ stating that the critical exponents below and above T_{cri} are equal.

Continuing with differentiating G twice with respect to T leads to

$$\lambda^{2p}C_P(\lambda^p t, \lambda^q P) = \lambda C_P(t,P) \tag{16.72}$$

Taking $P = 0$ and assuming

$$\lambda = (-t)^{-1/p} \quad \text{leads to} \quad \alpha' = 2 - (1/p) \tag{16.73}$$

From a similar calculation we obtain $\alpha = \alpha'$. From the last result and $p = [\beta(\delta - 1)]^{-1}$, we conclude that $\alpha + \beta(\delta + 1) = 2$, which we recognize as Griffiths' (in)equality. Finally, combining $\gamma = \beta(\delta - 1)$ with $\alpha + \beta(\delta + 1) = 2$, we rediscover Rushbrooke's (in)equality, $\alpha + 2\beta + \gamma = 2$.

In conclusion to this part, we note that the thermodynamic inequalities have become equalities. Moreover, we showed that the values of the critical exponents are equal above and below the critical temperature, and that the number of independent critical exponents has been reduced to two. The remaining questions are: (i) why is a power law describing the critical behavior; and (ii) what are the values of the critical exponents? The first question will answered in the next section, while a sketch of an answer to the second question will be given Section 16.5.

16.4.3 Scaling Lattices

The next step in the elucidation-of-transitions process was provided by Kadanoff [8]. Essentially, the idea of scaling is applied to a lattice. One can think of the lattice gas model for which a Hamilton function describes the equilibrium distribution of occupied and nonoccupied lattice sites (see Appendix C). The correlation between the sites is given by the (total) correlation function $h(r) \equiv \langle\rho(r)\rho(0)\rangle - \langle\rho^2\rangle$,

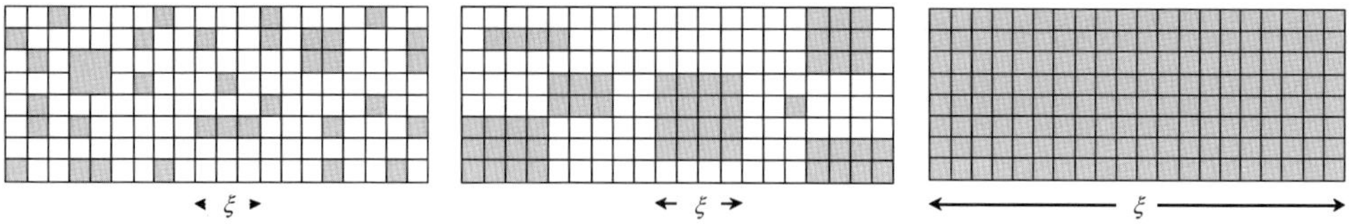

Figure 16.9 The coherence length ξ for the lattice gas model for (from left to right) $T \ll T_{cri}$, $T < T_{cri}$, and $T = T_{cri}$.

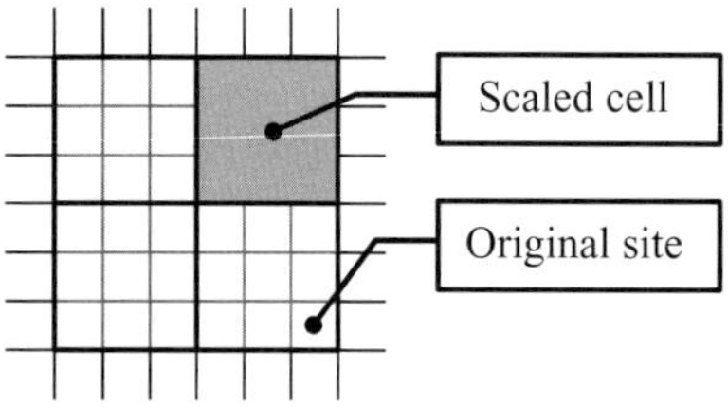

Figure 16.10 Lattice scaling illustrated for a 2D square lattice.

and we use the *coherence length* ξ as a measure of the range of the correlation function (Figure 16.9).

Consider therefore a D-dimensional lattice with N sites and lattice constant a and partition this lattice in cells (or blocks) of size La with $L >> 1$ (Figure 16.10). Hence, there are $n = N/L^D$ cells, each containing L^D sites. Further consider that we are close to T_{cri} so that the correlation length $\xi >> La$. Next, we associate with each cell a parameter that states whether the cell as a whole is considered to be occupied, or not. For this process we use a definite rule. For example, in the case of averaging over three lattice sites in a triangular lattice, we obtain a cell for which we apply the *majority rule*, that is, a value of 1 is taken if at least two of three original sites have also the value 1. Hence, 000 $\rightarrow$ 0 (occurrence 1), 001 $\rightarrow$ 0 (occurrence 3), 011 $\rightarrow$ 1 (occurrence 3), and 111 $\rightarrow$ 1 (occurrence 1). Finally, we assume that for the cells the same Hamiltonian applies as for the sites. If that is the case then one might expect that the thermodynamics of the site model would be the same as that for the cell model. Denoting the relative temperature difference (with respect to T_{cri}) as before by t and the pressure by P, we have for a cell (site) t_{cell} (t_{site}) and P_{cell} (P_{site}). The Gibbs energy g per site or per cell of the site and cell model, respectively, then relate as

$$g(t_{cell}, P_{cell}) = L^D g(t_{site}, P_{site}) \tag{16.74}$$

Now, we need relations between t_{cell} and t_{site} on the one hand, and P_{cell} and P_{site} on the other hand. We expect that the pressure of the cell model is proportional to the pressure of the site model, but the proportionality constant B will be dependent on L. The same assumption can be made for t. So we have

$$t_{\text{cell}} = A(L)t_{\text{site}} \quad \text{and} \quad P_{\text{cell}} = B(L)P_{\text{site}} \tag{16.75}$$

Although it is possible to continue without further assumptions [9], we assume that

$$A(L) = L^x \quad \text{and} \quad B(L) = L^y \tag{16.76}$$

with x and y arbitrary numbers (see Justification 16.1). This leads right away to

$$g(L^x t_{\text{site}}, L^y P_{\text{site}}) = L^D g(t_{\text{site}}, P_{\text{site}}) \tag{16.77}$$

This expression, stating that g is a generalized homogeneous function of t and P, leads essentially to scaling with exponents $p = x/D$ and $q = y/D$.

Justification 16.1: Power laws as scale-free functions

Since any dimensionless number can be raised to a certain power, one might ask why a power relation $(x/x_0)^b$ is more scale-free than, say, an exponential relation $\exp(x/x_0)$, since for both a scale length x_0 is involved. Suppose that we have a power relation reading $y/y_0 = a(x/x_0)^b$, and that we have data available for the range $x = x_0/2$ to $x = 2x_0$. This implies that the ratio between maximum and minimum measured values is 4^b. If we scale x_0 to cx_0 and use the same range of data, we still have a factor of 4^b between maximum and minimum measured values. This implies that a superposition of these data ranges can be realized by a simple linear change of scale, and appears the same no matter on what scale it is probed. No other function is capable of doing this.

16.5 Renormalization

The plot thickens, as they say, and the next step is considerably more complex than the previous ones. That is why only a sketch of an answer to the remaining questions will be given. Full details can be found in the literature listed in the section labeled "Further Reading".

To continue, we need fluctuations for which in Figure 16.11 snapshots from an early Monte Carlo (MC) simulation of the 2D lattice gas model using 64×64 cells are shown. From these pictures and other evidence it became clear that, near T_{cri}, *fluctuations occur on every length scale*, and that the structure of the pictures is rather independent of the density, suggesting *self-similar behavior*. The *coherence length* ξ, which measures the range of the correlation function $g(r)$, scales as $\xi/\xi_0 \sim |t|^{-\nu}$ above as well as below T_{cri}. In order to have self-consistency, fluctuations in $g(r)$ should be much smaller than $\langle\rho^2\rangle$. We have $\langle\rho^2\rangle \sim t^{2\beta}$ and $g(r) \sim \exp(-r/\xi)/r^{D-2+\eta}$. Hence, for $r \cong \xi$, we obtain $t^{2\beta} > (t^{-\nu})^{-(D-2+\eta)}$ or, using Stanley's lemma, $2\beta < \nu\,(D-2+\eta)$. Using scaling theory, this inequality also appears to be an equality. Using the mean field values $\beta = ½$, $\nu = ½$ and $\eta = 0$, results in $D \geq 4$. This result

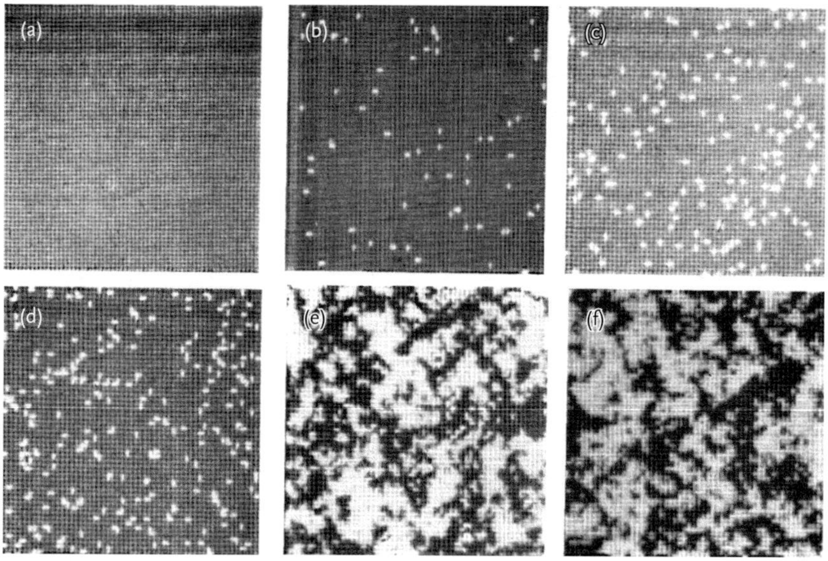

Figure 16.11 (a–c) The lattice gas model for $T = 0$, $T = T_{cri}/4$, $T = T_{cri}/2$; (d–f): The lattice gas model for $T = 3T_{cri}/4$, $T \cong T_{cri}$, and $T > T_{cri}$. Images taken from Ref. [15].

states that for fluctuations to be unimportant, the dimensionality should be at least four, which implies that in 3D, fluctuations do matter. One might think that adding fluctuations to the mean field result would remedy the situation, but since the fluctuations are more divergent than the mean values, this approach is not fruitful.

The answers to the question of what are the numerical values of the critical exponents can be found in the approach, normally called the *renormalization group* (RG) theory [10]. In the simplest form of this theory we choose a transformation which transforms the original Hamiltonian $\mathcal{H}$ of the site lattice with lattice constant a to a new Hamiltonian $\mathcal{H}_1$ of the cell lattice with lattice constant La, without changing the expression for the partition function, that is, $Z(\mathcal{H}, N) = Z(\mathcal{H}_1, N/L^D)$. Further, we use the usual lattice model approximation in which a site (cell) is occupied or not, representing repulsion, and only nearest-neighbor interactions with parameter J, representing attraction, are considered. In fact, this idea was used in the previous paragraph with respect to the Gibbs energy of the lattice, but here we employ the Hamiltonian. The basic idea is that this transformation can be described by $r_1 = f(r)$ so that the parameters r_1 of the Hamiltonian $\mathcal{H}_1$ of the cell lattice can be obtained from the parameters r of the Hamiltonian $\mathcal{H}$ of the site lattice. This process can be iterated, that is, $r_2 = f(r_1) = f[f(r)]$ and so on, for increasingly larger blocks until La reaches ξ. We can describe this process also by $r_{i+1} = f(r_i)$. It may happen that for $i \to \infty$ we obtain $r^* = f(r^*)$, where r^* is called a *fixed point*. As an example we use the function

$$r_{i+1} = cr_i(1 - r_i) \quad \text{with} \quad 0 < r < 1 \quad \text{and} \quad c > 1 \tag{16.78}$$

The fixed points are then the solutions of $r^* = cr^*(1 - r^*)$, which are $r^* = 0$ and $r^* = 1 - 1/c$. Stability can be probed by adding a small perturbation ε_i to r^* so that $x_i = r^* + \varepsilon_i$. From substitution in Eq. (16.78) and expanding $f(r_i)$ in a Taylor series while keeping only the linear term, we obtain

$$\varepsilon_{i+1} = \left(\frac{\mathrm{d}f}{\mathrm{d}r}\right)_{r=r^*} \varepsilon_i \tag{16.79}$$

If $(\mathrm{d}f/\mathrm{d}r)_{r=r^*} < 1$, then during each iteration the distance to r^* will decrease and the fixed point will be called *stable*. If, on the other hand, $(\mathrm{d}f/\mathrm{d}r)_{r=r^*} > 1$, the distance will increase and the fixed point is called *unstable*. For our example, Eq. (16.78), $\mathrm{d}f/\mathrm{d}r = c(1 - 2r)$ and since $c > 1$, $r^* = 0$ is an unstable fixed point and $r^* = 1 - 1/c$ is a stable fixed point.

From the above procedure we can determine not only the fixed points but also the way in which they are approached. Using $r_1 = f(r)$, one obtains

$$r_1 - r^* = f(r) - f(r^*) \cong \left(\frac{\mathrm{d}f}{\mathrm{d}r}\right)_{r=r^*} (r - r^*) + \ldots \tag{16.80}$$

Now, during the transformation the Gibbs energy per particle $g(r)$ must be unaltered, that is, $g(a_1) = L^D g(a)$ or, equivalently, for the partition function $Z(\mathcal{H}, N) = Z(\mathcal{H}_1, N/L^D)$. This scaling can only be done if the cell size is less than the coherence length ξ. Since for $T \to T_{\mathrm{cri}}$, $\xi \to \infty$, at T_{cri} the scaling can be done for any cell size. Hence, we identify the critical point of the phase transition T_{cri} with the fixed point T^*. Taking $r = J/kT$, with J the interaction energy and k Boltzmann's constant, we can rewrite Eq. (16.80), for r not too far removed from r^*, as

$$T' - T^* \cong \left(\frac{\mathrm{d}f}{\mathrm{d}r}\right)_{r=r^*} (T - T^*) \tag{16.81}$$

On comparison with the general expression $t_1 \equiv (T - T_{\mathrm{cri}})/T_{\mathrm{cri}} = Lt$, we obtain

$$x = \frac{\ln(\mathrm{d}f/\mathrm{d}r)_{r=r^*}}{\ln(L)} \tag{16.82}$$

If there are more parameters, say $r = r(t)$ as before and $s = s(P)$ (actually $s = P/kT$ for the pressure ensemble), we have $f = f(r,s)$ and we have two equations, namely

$$r^* = f_1(r^*, s^*) \quad \text{and} \quad s^* = f_2(r^*, s^*) \tag{16.83}$$

and using the same procedure we obtain

$$\begin{pmatrix} (r_1 - r^*) \\ (s_1 - s^*) \end{pmatrix} = \begin{pmatrix} \left(\frac{\mathrm{d}f_1}{\mathrm{d}r}\right)_{r=r^*} & \left(\frac{\mathrm{d}f_1}{\mathrm{d}s}\right)_{r=r^*} \\ \left(\frac{\mathrm{d}f_2}{\mathrm{d}r}\right)_{s=s^*} & \left(\frac{\mathrm{d}f_2}{\mathrm{d}s}\right)_{s=s^*} \end{pmatrix} \begin{pmatrix} (r - r^*) \\ (s - s^*) \end{pmatrix} \tag{16.84}$$

Because r_1 and s_1 are coupled, we transform to normal coordinates to obtain the scaling behavior, and for this we must find the eigenvalues of the derivative matrix. For convenience, we label the derivative matrix as $\mathbf{T}$ and denote the set

$(r - r^*)$ and $(s - s^*)$ by the column $\delta\mathbf{k}$, so that Eq. (16.84) reads $\delta\mathbf{k}_1 = \mathbf{T}\delta\mathbf{k}$. However, as $\mathbf{T}$ is non-symmetric[9], generally its left and right eigenvectors will be different. Using $\boldsymbol{\varphi}$ as the column for the left eigenvectors, we write $(\boldsymbol{\varphi}^{(\alpha)})^{\mathrm{T}}\mathbf{T} = \lambda^{(\alpha)}(\boldsymbol{\varphi}^{(\alpha)})^{\mathrm{T}}$ or $\Sigma_j \varphi_j^{(\alpha)} T_{ji} = \lambda^{(\alpha)}\varphi_i^{(\alpha)}$, where the superscript α labels the eigenvalues λ. In spite of $\mathbf{T}$ being a non-symmetric matrix, the eigenvectors do not mix with one another under a RG transformation. This is readily shown by introducing the normal coordinates $u^{(\alpha)} = (\boldsymbol{\varphi}^{(\alpha)})^{\mathrm{T}}\delta\mathbf{k}$, from which one obtains

$$u_1^{(\alpha)} = (\boldsymbol{\varphi}^{(\alpha)})^{\mathrm{T}}\delta\mathbf{k}_1 = (\boldsymbol{\varphi}^{(\alpha)})^{\mathrm{T}}\mathbf{T}\delta\mathbf{k} = \lambda^{(\alpha)}(\boldsymbol{\varphi}^{(\alpha)})^{\mathrm{T}}\delta\mathbf{k} = \lambda^{(\alpha)}u^{(\alpha)} \tag{16.85}$$

If we assume that all eigenvalues are real, the local geometry of the Hamiltonian surface around a fixed point is thus described by the normal coordinates $u^{(\alpha)}$, which are moreover scale-invariant. This implies that the Hamiltonian and therefore the Gibbs energy per particle $g(\mathbf{k})$ can be written in terms of the normal coordinates $u^{(\alpha)}$. We write $g(u^{(1)},u^{(2)}, \ldots) = b^{-D}g(\lambda^{(1)}u^{(1)}, \lambda^{(2)}u^{(2)}, \ldots)$, and we identify the coordinates $u^{(\alpha)}$ as the scaling parameters of the problem. Since in general at T_{cri} all $u^{(\alpha)}$ are zero, we can determine for a given coordinate, say number 1, its behavior close to T_{cri}, by setting $u^{(2)}$, $u^{(3)}$, ... equal to zero so that we have $g(u^{(1)},0, \ldots) = L^{-D}g(\lambda^{(1)}u^{(1)}, 0, \ldots)$. If only $r = r(t)$ and $s = s(P)$ are the variables of interest, we have $g(r,s) = L^{-D}g(\lambda^{(1)}r, \lambda^{(2)}s)$ and comparing this expression with the Widom scaling function $g(\lambda^p t,\lambda^q P) = \lambda g(t,P)$ the result becomes

$$\lambda = L^D, \quad p = \frac{1}{D}\frac{\ln\lambda^{(1)}}{\ln L} \quad \text{and} \quad q = \frac{1}{D}\frac{\ln\lambda^{(2)}}{\ln L} \tag{16.86}$$

In Section 16.4 we have shown that all exponents follow from p and q. Hence, all critical exponents are determined by the scaling behavior of the eigenvalues of the linearized transformation matrix $\mathbf{T}$ near the critical point.

Applying this process, the critical exponents of the lattice model have been obtained to high accuracy (Table 16.4). As an illustration of the accuracy of some approximate results, the values for the 2D lattice gas using different lattices and blocks are also given. Note the good agreement of the 3D values as calculated using renormalization theory with the experimental values (Table 16.3). To conclude this part, we note that the above-outlined procedure is the simplest possible way of renormalizing real space. Moreover, we note that the inverse of these transformations does not exist, so that the designation "renormalization group" actually should read "renormalization semi-group." However, the nomenclature is conventional.

9) A square matrix $\mathbf{A}$ of order m operating on a column $\mathbf{u}$ yields a new column $\mathbf{u}'$. If this operation yields a multiple of $\mathbf{u}$, say $\lambda\mathbf{u}$, the resulting equation $(\mathbf{A} - \lambda\mathbf{I})\mathbf{u} = 0$ is an *eigenvalue equation* with λ the *eigenvalue*, $\mathbf{I}$ the unit matrix, and $\mathbf{u}$ the *eigenvector*. This equation can be solved by $\det(\mathbf{A} - \lambda\mathbf{I}) = 0$, yielding m eigenvalues and associated eigenvectors. Collecting the set of columns $\mathbf{u}$ in the matrix $\mathbf{U}$, the complete set of equations can be written as $\boldsymbol{\Lambda} = \mathbf{U}^{\mathrm{T}}\mathbf{A}\mathbf{U}$, where $\boldsymbol{\Lambda}$ is the diagonal matrix containing all eigenvalues. For a symmetric matrix $\mathbf{A}$, $\mathbf{U}$ is an orthogonal matrix, i.e. $\mathbf{U}^{-1} = \mathbf{U}^{\mathrm{T}}$. For a non-symmetric matrix $\mathbf{A}$ this is no longer the case, and we have to keep track of whether the eigenvector is positioned on the left or right side of the matrix $\mathbf{A}$, i.e. whether we solve $\mathbf{A}\mathbf{u} = \lambda\mathbf{u}$ or $\mathbf{u}^{\mathrm{T}}\mathbf{A} = \lambda\mathbf{u}^{\mathrm{T}}$.

Table 16.4 Critical exponents according to various models.

Model	α	β	γ	δ	ν	η
Mean field	0	1/2	1	3	1/2	0
2D lattice gas[a)]	0	1/8	7/4	15	1	1/4
2D lattice gas[b)]	−0.27	0.72	0.83	2.2	–	–
2D lattice gas[c)]	–	–	–	–	1.07	–
2D lattice gas[d)]	–	–	–	–	0.901	–
3D lattice gas[e)]	0.110(1)	0.3265(3)	1.2372(5)	4.789(2)	0.6301(4)	0.0364(5)

a) Details of exact solution from Ref. [16].
b) Triangular lattice with (three sites → one cell) renormalization, leading to $\lambda_1 = 1.623$ and $\lambda_1 = 2.121$, so that $p = \ln\lambda_1/2 \ln 3^{1/2} = 0.441$ and $q = \ln\lambda_2/2 \ln 3^{1/2} = 0.684$.
c) Square lattice with removing half the sites [11].
d) Square lattice with (five sites → one cell) renormalization [11].
e) Data from Ref. [17].

This approach is apparently abstract and we refrain from detailing one or another model. It may be useful, however, to illustrate the methods by a much simpler example, namely that of percolation [12]. If we mix electrically conducting and insulating particles at random, using an increasing fraction of conducting particles, the mixture will at first be insulating; however, at a certain fraction a transition to a conductive mixture will occur. This change occurs over a small volume fraction change of conductive particles, and thereafter the conductivity still increases, albeit slowly. This phenomenon is called *percolation*, and the transition is termed the *percolation threshold*. The phenomenon can be considered as a geometric phase transition where, at the percolation threshold infinite, connected clusters of conductive particles will appear. For 2D (i.e., for disks) the transition occurs at 50 vol.%, whereas for 3D (i.e., for spheres), it occurs at about 16 vol.%. The development of these clusters in the percolation model is similar to that of clusters of molecules in phase transformations. Each realization of a packing at a certain volume fraction can be considered as a snapshot of a configuration of two types of molecule. Of course, the analog is static in time, but the basic idea is the same.

The physical problem is now to derive when and how electrical conduction across a macroscopic region occurs, in terms of the microdetails of the packing of the particles. As a concrete example, we use a 2D structure with a renormalization by three particles (disks). For the original packing the microdescription is in terms of sites that are either occupied or unoccupied with a conductive particle. We denote the probability that any site is occupied by p, independent of whether any other site is occupied, or not. The occupancy distribution of sites provides the exact microdescription of the system for which we only know the single site probability p.

In the spirit of the renormalization theory, we increase the length scale by b, that is, we increase the area by $b^2 = 3$. At the higher level of description, in this

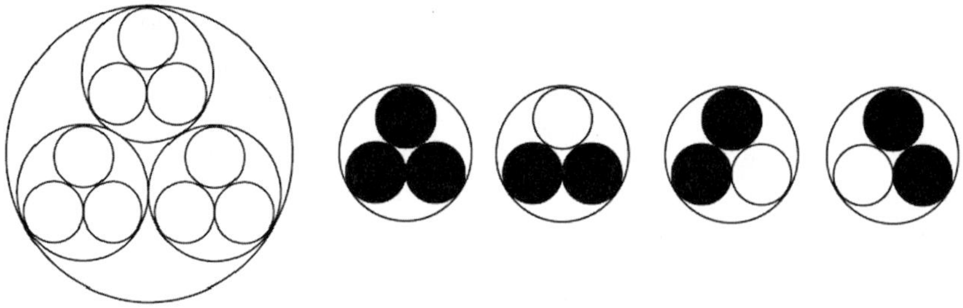

Figure 16.12 Percolation and renormalization by taking three spheres together, to which the majority rule is applied. The left schematic shows the original packing; the smaller schematics (at right) show the four possible "renormalized" sites.

case a cell of three sites, and using the majority rule, a cell will be occupied if, and only if, two or more of the sites are occupied (Figure 16.12). Let us call p_1 the probability that a cell of three sites is occupied. The probability p_1 is then the sum of probabilities of four mutually exclusive states. Using p for occupied sites and $(1 - p)$ for unoccupied sites, we have once p^3 and three times $p^2(1 - p)$.

The percolation threshold (critical point) occurs if the original probability p equals the renormalized probability p_1. We thus have

$$p_1 = p^3 + 3p^2(1-p) \tag{16.87}$$

The solution of this equation does not require iteration but yields directly as solutions $p^* = 0$ and $p^* = 1$ which represent the stable fixed points, and the solution $p^* = 0.5$ representing the unstable fixed point. The unstable fixed point solution thus appears to be the exact solution for this 2D percolation problem, which however, is fortuitous.

Let us now calculate the critical exponent ν which describes the onset of the percolation phenomenon. We take the coherence length ξ for the original packing to be given by $\xi = C|p - p_{\text{cri}}|^{-\nu}$, while for the renormalized packing we have $\xi_1 = C|p_1 - p_{\text{cri}}|^{-\nu}$ with $\xi_1 = \xi/b$. Combining results in

$$b|p_1 - p_{\text{cri}}|^{-\nu} = |p - p_{\text{cri}}|^{-\nu} \quad \text{or} \tag{16.88}$$

$$\frac{1}{\nu} = \frac{\ln[(p_1 - p_{\text{cri}})/(p - p_{\text{cri}})]}{\ln b} = \frac{\ln \lambda}{\ln b} \quad \text{with} \quad \lambda = \frac{p_1 - p_{\text{cri}}}{p - p_{\text{cri}}} \cong \frac{\mathrm{d}p_1}{\mathrm{d}p} \tag{16.89}$$

Expanding p_1 as $p_1 = p^* + \lambda^*(p - p^*) + \ldots$ with $\lambda^* = (\mathrm{d}p_1/\mathrm{d}p)_{p=p^*}$ and substitution in the original equation $p_1 = p^3 + 3p^2(1 - p)$ yields

$$\frac{\mathrm{d}p_1}{\mathrm{d}p} = 3p^2 + [6p(1-p) - 3p^2] = 6p(1-p) \tag{16.90}$$

so that $\lambda^* = 3/2$. Because $b^2 = 3$, we obtain

$$\nu = \frac{\ln b}{\ln \lambda} = \frac{\ln 3^{1/2}}{\ln(3/2)} = 1.355 \tag{16.91}$$

The exact value is $\nu = 4/3$. Approaching the threshold, the coherence length ξ of the clusters increases until ξ diverges at the threshold itself. This implies that conductive clusters span the complete system and therefore become macroscopically conductive.

Problem 16.10

Show that for $T' - T^* \ll T^*$, Eq. (16.80) is equivalent with Eq. (16.81).

Problem 16.11*

Show that the scaling relation for conduction percolation on a BCC lattice reads $p_1 = p^9 + 9p^8(1-p) + 36p^7(1-p)^2 + 84p^6(1-p)^3 + 126p^5(1-p)^4$. If you feel up to it, obtain the percolation threshold p^* and the critical exponent ν.

16.6 Final Remarks

We have come a long way: from a phenomenological description recognizing universality classes, via mean field theory, generalized homogenous functions, lattice gas and scaling to renormalization. These seven pillars (of wisdom?) constitute the modern basis of phase transition theory. The overall picture is complex, in particular the renormalization part, but is capable of catching most of the essentials of phase transformations quite well.

References

1 (a) Antoine, C. (1888) *Compt. Rend.*, **107**, 681 and 728; (b) Thomas, G.W. (1940) *Chem. Rev.*, **38**, 1; (c) For data, see Reid R.C., Prausnitz J.M., and Poling B.E. (1988) *The Properties of Gases and Liquids*, 4th edn, McGraw-Hill.
2 Clusius, K. and Weigand, K. (1940) *Z. Phys. Chem.*, **B46**, 1.
3 Heller, P. (1967) *Rep. Progr. Phys.*, **30**, 731. For an extensive review, see Sengers and Levelt-Sengers (1978).
4 (a) Fisher, M.E. (1964) *J. Math. Phys.*, **5**, 944; (b) Fisher, M.E. (1967) *Rep. Prog. Phys.*, **30**, 615.
5 See Ma (1976).
6 Barieau, R. (1966) *J. Chem. Phys.*, **45**, 3175.
7 Widom, B. (1965) *J. Chem. Phys.*, **43**, 3892 and 3898.
8 (a) Kadanoff, L.P. (1966) *Physics*, **2**, 263; (b) Kadanoff, L.P. *et al.* (1967) *Rev. Mod. Phys.*. **39**, 395.
9 Cooper, M.J. (1968) *Phys. Rev.*, **168**, 183.
10 Wilson, K.G. (1971) *Phys. Rev.*, **B4**, 3174 and 3184.
11 See Gitterman and Halpern (2004).
12 This illustration was first given by Chowdhury, D. and Stauffer, D. (2000) *Principles of Equilibrium Statistical Mechanics*, Wiley-VCH Verlag GmbH, Weinheim.
13 Moelwyn-Hughes, E.A. (1961) *Physical Chemistry*, 2nd edn, Pergamon, Oxford.
14 Hocken, R. and Moldover, M.R. (1976) *Phys. Rev. Lett.*, **37**, 29.

15 Ogita, N., Ueda, A., Matsubara, T., Matsuda, H., and Yonezawa, F. (1969) *J. Phys. Soc. Jpn*, **26** (Suppl.), 145.

16 Onsager, L. (1944) *Phys. Rev.*, **65**, 117.

17 Pelissetto, A. and Vicari, E. (2002) *Phys. Rep.*, **368**, 549.

Further Reading

Binney, J.J., Dowrich, N.J., Fisher, A.J., and Newman, M.E.J. (1992) *The Theory of Critical Phenomena*, Clarendon Press, Oxford.

Gitterman, M. and Halpern, V.H. (2004) *Phase Transitions*, World Scientific, Singapore.

Goldenfeld, N. (1993) *Lectures on Phase Transitions and the Renormalization Group*, Addison-Wesley.

Herbut, I. (2007) *A Modern Approach to Critical Phenomena*, Cambridge University Press, New York.

Ma, S.-K. (1976) *Modern Theory of Critical Phenomena*, Benjamin, Reading.

Sengers, J.V. and Levelt Sengers, J.M.H. (1978) Critical phenomena in classical fluids, in *Progress in Liquid Physics* (ed. C.A. Croxton), John Wiley & Sons, Ltd, Chichester, Ch. 4, pp. 103–174.

Stanley, H.E. (1971) *Introduction to Phase Transitions and Critical Phenomena*, Oxford University Press, New York.

Uzunov, D.I. (1993) *Introduction to the Theory of Critical Phenomena*, World Scientific, Singapore.

Appendix A
Units, Physical Constants, and Conversion Factors

Basic and Derived SI Units

Quantity	Unit	Symbol
Length	meter	m
Mass	kilogram	kg
Time	second	s
Electric current	ampere	A
Temperature	kelvin	K
	°C	t/°C = T/K −273.15
Amount of substance	mole	mol
Force	newton	N = kg m s^{-2}
Work, energy, heat	joule	J = N m
Power	watt	W = J s^{-1}
Pressure	pascal	Pa = N m^{-2}
Frequency	hertz	Hz = s^{-1}
Electrical charge	coulomb	C = A s
Electrical potential	volt	V = J C^{-1}
Electrical resistance	ohm	Ω = V A^{-1}

Liquid-State Physical Chemistry, First Edition. Gijsbertus de With.

Physical Constants

Constant	Symbol	Value
Avogadro's number	N_A	$6.022 \times 10^{23}\,mol^{-1}$
Elementary charge	e	$1.602 \times 10^{-19}\,C$
Electron rest mass	m_e	$9.109 \times 10^{-31}\,kg$
Proton rest mass	m_p	$1.673 \times 10^{-27}\,kg$
Neutron rest mass	m_n	$1.675 \times 10^{-27}\,kg$
Atomic mass unit (dalton)	amu (Da)	$1.661 \times 10^{-27}\,kg$
Gas constant	R	$8.315\,J\,mol^{-1}\,K^{-1}$
Boltzmann's constant	$k = R/N_A$	$1.381 \times 10^{-23}\,J\,K^{-1}$
Planck's constant	h	$6.626 \times 10^{-34}\,J\,s$
	$\hbar = h/2\pi$	$1.055 \times 10^{-34}\,J\,s$
Standard acceleration of gravity	g	$9.807\,m\,s^{-2}$
Speed of light	c_0	$2.998 \times 10^{8}\,m\,s^{-1}$
Faraday constant	$F = eN_A$	$9.649 \times 10^{4}\,C\,mol^{-1}$
Permeability of the vacuum	μ_0	$4\pi \times 10^{-7}\,N\,A^{-2}$ (exact)
Permittivity of the vacuum	$\varepsilon_0 = 1/\mu_0\, c_0^2$	$8.854 \times 10^{-12}\,C^2\,N^{-1}\,m^{-2}$

Conversion Factors for Non-SI Units

$1\,dyne = 10^{-5}\,N$
$1\,bar = 10^{5}\,Pa$
$1\,atm = 1.013\,bar$
$1\,torr = 1/760\,atm$
$1\,psi = 6.895 \times 10^{3}\,Pa$
$1\,int.\,cal = 4.187\,J$
$1\,erg = 10^{-7}\,J$

$1\,eV = 1.602 \times 10^{-19}\,J$
$1\,eV/particle = 96.48\,kJ\,mol^{-1}$
$1\,D = 3.336 \times 10^{-30}\,C\,m$
$1\,l = 1\,dm^3 = 10^{-3}\,m^3$
$1\,cm^{-1} = 1.986 \times 10^{-23}\,J$
$hc/k = 1.438\,cm\,K$

Prefixes

pico p = 10^{-12}
nano n = 10^{-9}
micro μ = 10^{-6}
milli m = 10^{-3}

kilo k = 10^{3}
mega M = 10^{6}
giga G = 10^{9}
tera T = 10^{12}

Greek Alphabet

Α, α	alpha	Ν, ν	nu
Β, β	beta	Ξ, ξ	xi
Γ, γ	gamma	Ο, ο	omicron
Δ, δ	delta	Π, π	pi
Ε, ε	epsilon	Ρ, ρ	rho
Ζ, ς	zeta	Σ, σ	sigma
Η, η	eta	Τ, τ	tau
Θ, θ, ϑ	theta	Υ, υ	upsilon
Ι, ι	iota	Φ, φ, ϕ	phi
Κ, κ	kappa	Χ, χ	chi
Λ, λ	lambda	Ψ, ψ	psi
Μ, μ	mu	Ω, ω	omega

Standard Values

kT at 298 K = 207.2 cm^{-1} = 4.116 × 10^{-21} J = 2.569 × 10^{-2} eV
Standard molar volume $V_m{}^\circ = RT/P^\circ$ = 2.479 × 10^{-4} m^3 mol^{-1}
Standard pressure P° = 1 bar = 100 kN = 10^5 N m^{-2}

Appendix B
Some Useful Mathematics

In the theoretical description of liquids, we will encounter functions, matrices and determinants. Moreover, we will encounter scalars, vectors, and tensors. We will also need occasionally coordinate transformations, some transforms and calculus of variations. In the following we will briefly review these concepts, as well as some of the operations between them and some useful general results.

B.1
Symbols and Conventions

Often, we will use quantities with subscripts, for example, A_{ij}. With respect to these quantities a convenient symbol is the *Kronecker delta*, denoted by δ_{ij} and defined by

$$\delta_{ij} = 1 \quad \text{if} \quad i = j \quad \text{and} \quad \delta_{ij} = 0 \quad \text{if} \quad i \neq j \tag{B.1}$$

Note that we can write the identity $\Sigma_i \delta_{ij} a_i = a_j$, and therefore δ_{ij} is also called the *substitution operator*. Applying δ_{ij} to A_{ij} results in $\Sigma_{ij} A_{ij} \delta_{ij} = \Sigma_j A_{jj}$.

We also introduce the *alternator* e_{ijk} for which it holds that

$$\begin{aligned} e_{ijk} &= 1 \text{ for } ijk = 123, 231, 312 \text{ (even permutations[1])}, \\ e_{ijk} &= -1 \text{ for } ijk = 132, 213, 321 \text{ (odd permutations) and} \\ e_{ijk} &= 0 \text{ otherwise, that is, if any of the three indices are equal.} \end{aligned} \tag{B.2}$$

An alternative expression is given by $e_{ijk} = \tfrac{1}{2}(i - j)(j - k)(k - i)$.

B.2
Partial Derivatives

A function f may be dependent on *variables* x_i and *parameters* p_j, denoted by $f(x_i;p_j)$. Reference to the parameters is often omitted by writing $f(x_i)$. In practice, reference

1) An even (odd) permutation is the result of an even (odd) number of binary interchanges. The character (even or odd) of a permutation is independent of the order of the binary interchanges.

Liquid-State Physical Chemistry, First Edition. Gijsbertus de With.

to the variable is also often omitted by writing just f. For a function f, several derivatives exist. If all variables but one, say x_1, are kept constant during differentiation, the derivative of f with respect to x_1 is called the *partial derivative* and is denoted by $(\partial f(x_i)/\partial x_1)_{x_i \neq x_1}$. Once a choice of independent variables is made, there is no need to indicate, as frequently done, which variables are kept constant. Therefore $(\partial f(x_i)/\partial x_1)_{x_i \neq x_1}$ can be indicated without confusion by $\partial f(x_i)/\partial x_1$. The function $\partial f/\partial x_i$ generally is a function of all variables x_i and, if continuous, may be differentiated again to yield the *second partial derivatives* $\partial^2 f/\partial x_i \partial x_j$ $(= \partial^2 f/\partial x_j \partial x_i)$.

Example B.1

For $f(x,y) = x^2y^3$ one simply calculates

$$\partial f/\partial x = 2xy^3 \quad \partial^2 f/\partial x^2 = 2y^3 \quad \partial^2 f/\partial x \partial y = 6xy^2$$

$$\partial f/\partial y = 3x^2y^2 \quad \partial^2 f/\partial y^2 = 6x^2y \quad \partial^2 f/\partial y \partial x = 6xy^2$$

In case the independent variables x_i increase by $\mathrm{d}x_i$, the value of the function f at $x_i + \mathrm{d}x_i$ is given by Taylor's expansion

$$f(x_i + \mathrm{d}x_i) = f(x_i) + \sum_i \frac{\partial f}{\partial x_i}\mathrm{d}x_i + \frac{1}{2!}\sum_{i,j} \frac{\partial^2 f}{\partial x_i \partial x_j}\mathrm{d}x_i \mathrm{d}x_j + \cdots \tag{B.3}$$

One can also write symbolically

$$f(x_i + \mathrm{d}x_i) = \exp\left(\sum_j \mathrm{d}x_j \frac{\partial}{\partial x_j}\right) f(x_i) \tag{B.4}$$

Expansion of the exponential yields Eq. (B.3). Another way is to write

$$f(x_i + \mathrm{d}x_i) = f(x_i) + \mathrm{d}f(x_i) + \frac{1}{2!}\mathrm{d}^2 f(x_i) + \cdots + \frac{1}{n!}\mathrm{d}^n f(x_i) \tag{B.5}$$

where the (*nth order*) *differential* is given by

$$\mathrm{d}^n f(x_i) = \left(\mathrm{d}x_j \frac{\partial}{\partial x_j}\right)^n f(x_i) \tag{B.6}$$

Example B.2

Consider again the function $f(x,y) = x^2y^3$. A first-order estimate for $f(2.1,2.1)$ is

$$f(2.1, 2.1) = f(2, 2) + (\partial f/\partial x)\mathrm{d}x + (\partial f/\partial y)\mathrm{d}y$$
$$= 32 + 2xy^3\mathrm{d}x + 3x^2y^2\mathrm{d}x = 32 + 2\cdot 2\cdot 8\cdot 0.1 + 3\cdot 4\cdot 4\cdot 0.1 = 40.00$$

using $f(2,2)$ as the reference value. The actual value is 40.84.

A function $f(x)$ is *analytic* at $x = c$ if $f(x)$ can be written as (a sum of) Taylor series (with a positive convergence radius). If $f(x)$ is analytic at each point on the open interval I, $f(x)$ is analytic on the interval I. For a function $w(z)$ of a complex variable $z = x + \mathrm{i}y$ ($\mathrm{i} = \sqrt{-1}$) to be analytic, it must satisfy the *Cauchy–Riemann conditions*

$$\frac{\partial u}{\partial x} = \frac{\partial v}{\partial y} \quad \text{and} \quad \frac{\partial v}{\partial x} = -\frac{\partial u}{\partial y} \tag{B.7}$$

where $u(x,y)$ = Re $w(z)$ and $v(x,y)$ = Im $w(z)$ denote the real and imaginary parts of w, respectively. Moreover, if $u(x,y)$ and $v(x,y)$ have continuous second derivatives the function $w(z)$ obeys the *Laplace equation* $\partial^2u/\partial^2x + \partial^2u/\partial^2y = \partial^2v/\partial^2x + \partial^2v/\partial^2y = 0$, and is said to be *harmonic*.

Example B.3

Consider the function $w(z) = \mathrm{e}^x \cos y + \mathrm{i}\, \mathrm{e}^x \sin y = \exp(z)$. Then it holds that

$$u = \mathrm{Re}\, w = \mathrm{e}^x \cos y \quad v = \mathrm{Im}\, w = \mathrm{e}^x \sin y$$

The derivatives are given by

$$\partial u/\partial x = \mathrm{e}^x \cos y \qquad \partial v/\partial x = \mathrm{e}^x \sin y$$

$$\partial v/\partial y = \mathrm{e}^x \cos y \qquad \partial u/\partial y = -\mathrm{e}^x \sin y$$

$$\partial^2 u/\partial x^2 = \mathrm{e}^x \cos y \qquad \partial^2 v/\partial y^2 = -\mathrm{e}^x \sin y$$

Hence, Cauchy–Riemann conditions are satisfied and the function is harmonic.

B.3 Composite, Implicit, and Homogeneous Functions

If for a function f the variables x_i are themselves a function of y_j, it is called a *composite function*. For the first-order differentials the *chain rule* applies so that

$$\mathrm{d}x_i = \sum\nolimits_j \frac{\partial x_i}{\partial y_j}\mathrm{d}y_j \quad \text{and} \quad \mathrm{d}f = \sum\nolimits_{i,j} \frac{\partial f}{\partial x_i}\frac{\partial x_i}{\partial y_j}\mathrm{d}y_j \tag{B.8}$$

In many cases the variables x_i are not independent, that is, a relation exists between them meaning that an arbitrary member, say x_1, can be expressed as a function of $x_2, \ldots, x_n$. Often, this relation is given in the form of an *implicit function*, that is, $f = f(x_i)$ = constant. Of course, if the equation can be solved, the relevant differentials can be obtained from the solution. The appropriate relations between the differentials can also be obtained by observing that $\mathrm{d}f = \Sigma_i(\partial f/\partial x_i) = 0$. If x_1 is the dependent variable, putting $\mathrm{d}x_1 = 0$ and division by $\mathrm{d}x_i$ ($i{\neq}1$) yields

$$\left(\frac{\partial f}{\partial x_i}\right)_{x_j,x_1} + \left(\frac{\partial f}{\partial x_j}\right)_{x_i,x_1}\left(\frac{\partial x_j}{\partial x_i}\right)_{f,x_1} = 0 \quad \text{or} \quad \left(\frac{\partial x_j}{\partial x_i}\right)_{f,x_1} = -\frac{(\partial f/\partial x_i)_{x_j,x_1}}{(\partial f/\partial x_j)_{x_i,x_1}} \tag{B.9}$$

Example B.4

Consider explicitly a function f of three variables x, y and z, where z is the dependent variable. If we take $x_1 = z$, $x_i = x$ and $x_j = y$, Eq. (B.9) reads

$$(\partial f/\partial x)_{y,z} + (\partial f/\partial y)_{x,z}(\partial y/\partial x)_{f,z} = 0 \quad \text{or} \quad (\partial y/\partial x)_{f,z} = -(\partial f/\partial x)_{y,z}/(\partial f/\partial y)_{x,z}$$

On the other hand, taking $x_i = y$ and $x_j = x$ for $x_1 = z$ results in

$$(\partial f/\partial y)_{x,z} + (\partial f/\partial x)_{y,z}(\partial x/\partial y)_{f,z} = 0 \quad \text{or} \quad (\partial x/\partial y)_{f,z} = -(\partial f/\partial y)_{x,z}/(\partial f/\partial x)_{y,z}$$

Hence, it easily follows that

$$\bullet\, (\partial x/\partial y)_{f,z} = 1/(\partial y/\partial x)_{f,z} \tag{B.10}$$

By cyclic permutation of the variables we obtain

$$(\partial f/\partial x)_{y,z} + (\partial f/\partial y)_{z,x}(\partial y/\partial x)_{f,z} = 0$$

$$(\partial f/\partial y)_{z,x} + (\partial f/\partial z)_{x,y}(\partial z/\partial y)_{f,x} = 0$$

$$(\partial f/\partial z)_{x,y} + (\partial f/\partial x)_{y,z}(\partial x/\partial z)_{f,y} = 0$$

resulting, after substitution in each other, in

$$\bullet\, (\partial x/\partial y)_{f,z}(\partial y/\partial z)_{f,x}(\partial z/\partial x)_{f,y} = -1 \tag{B.11}$$

Now consider x, y and z to be composite functions of another variable u. If f is constant, there is a relation between x, y and z and thus also between $\partial x/\partial u$, $\partial y/\partial u$ and $\partial z/\partial u$. Moreover, $\mathrm{d}f = 0$ and Eq. (B.9) explicitly reads

$$\mathrm{d}f = [(\partial f/\partial x)_{y,z}(\partial x/\partial u)_f + (\partial f/\partial y)_{z,x}(\partial y/\partial u)_f + (\partial f/\partial z)_{x,y}(\partial z/\partial u)_f]\mathrm{d}u = 0$$

Further taking z as constant, independent of u, results in

$$(\partial f/\partial x)_{y,z}(\partial x/\partial u)_{f,z} + (\partial f/\partial y)_{z,x}(\partial y/\partial u)_{f,z} = 0 \quad \text{or}$$

$$(\partial y/\partial u)_{f,z}/(\partial x/\partial u)_{f,z} = -(\partial f/\partial x)_{y,z}/(\partial f/\partial y)_{z,x}$$

Comparing with $(\partial y/\partial x)_{f,z} = -(\partial f/\partial x)_{y,z}(\partial f/\partial y)_{z,x}$ one obtains

$$\bullet\, (\partial y/\partial x)_{f,z} = (\partial y/\partial u)_{f,z}/(\partial x/\partial u)_{f,z} \tag{B.12}$$

The equations, indicated by •, are frequently employed in thermodynamics.

A function $f(x_i)$ is said to be positively *homogeneous* of degree n if for every value of x_i and for every $\lambda > 0$ we have

$$f(\lambda x_i) = \lambda^n f(x_i) \tag{B.13}$$

For such a function we have *Euler's theorem*

$$\sum\nolimits_i x_i(\partial f/\partial x_i) = nf(x_i)$$

to be proven by differentiation with respect to λ first and taking $\lambda = 1$ afterwards.

Example B.5

Consider the function $f(x,y) = x^2 + xy - y^2$. One easily finds

$$\partial f/\partial x = 2x + y \quad \text{and} \quad \partial f/\partial y = x - 2y$$

Consequently, $x(\partial f/\partial x) + y(\partial f/\partial y) = x(2x + y) + y(x - 2y) = 2(x^2 + xy - y^2) = 2f$. Hence, f is homogeneous of degree 2.

B.4
Extremes and Lagrange Multipliers

For obtaining an extreme of a function $f(x_i)$ of n independent variables x_i the first variation δf has to vanish (see Section B.10). This leads to

$$\delta f = \sum\nolimits_i \frac{\partial f}{\partial x_i}\delta x_i = 0 \tag{B.14}$$

and, since the variables x_i are independent and the variations δx_i are arbitrary, to $\partial f/\partial x_i = 0$ for $i = 1, \ldots, n$. If, however, the extreme of f has to be found when the x_i are dependent and satisfy r constraint functions

$$c_j(x_i) = C_j \quad (j = 1, \ldots, r \quad \text{and} \quad r < n) \tag{B.15}$$

where the parameters C_j are constants, the variables x_i must also obey

$$\sum\nolimits_i \frac{\partial c_j(x_i)}{\partial x_i}\delta x_i = 0 \quad (j = 1, \ldots, r) \tag{B.16}$$

Of course, the system can be solved in principle by solving Eq. (B.15) for the independent $n - r$ variables x_i as functions of the others, but the procedure is often complex. It can be shown that finding the extreme of f subject to the constraint of Eq. (B.15) is equivalent to finding the extreme of a function g defined by

$$g(x_i, \lambda_j) = f(x_i) - \sum\nolimits_{j=1}^{r} \lambda_j [c_j(x_i) - C_j] \tag{B.17}$$

where now the original variables x_i and the additional variables λ_j, called the *Lagrange (undetermined) multipliers*, are to be considered independent. Variation of λ_j leads to Eq. (B.15) and variation of x_i to

$$\sum\nolimits_i \left[\frac{\partial f(x_i)}{\partial x_i} - \lambda_j \frac{\partial c_j(x_i)}{\partial x_i}\right] = 0 \tag{B.18}$$

From Eq. (B.18) the values for x_i can be determined. These values are still functions of λ_j but they can be eliminated using Eq. (B.15). In physics, chemistry and materials science, the Lagrange multiplier often can be physically interpreted.

Example B.6

One can ask what is the minimum circumference L of a rectangle given the area A. Denoting the edges by x and y, the circumference is given by $L = 2(x + y)$, while the area is given by $A = xy$. The equations to be solved are

$$\partial L/\partial x - \lambda(\partial A/\partial x) = 2 - \lambda y = 0 \quad \rightarrow \quad y = 2/\lambda$$

$$\partial L/\partial y - \lambda(\partial A/\partial y) = 2 - \lambda x = 0 \quad \rightarrow \quad x = 2/\lambda$$

Hence, the solution is $x = y$, $\lambda = 2/\sqrt{A}$ and $\min(L) = 4\sqrt{A}$.

B.5 Legendre Transforms

In many problems we meet the demand to interchange between dependent and independent variables. If $f(x_i)$ denotes a function of n variables x_i, we have

$$\mathrm{d}f = \sum_i \frac{\partial f}{\partial x_i}\mathrm{d}x_i \equiv \sum_i X_i \mathrm{d}x_i \quad (i = 1, \ldots, n) \tag{B.19}$$

Elimination of x_i from $X_i(x_i)$ and $f(x_i)$ leads to $f = f(X_i)$. However, from $f(X_i)$ it is impossible to uniquely recover $f(x_i)$ by repeating this procedure for $f(X_i)$. Now consider the function $g = f - X_1x_1$. For the differential we obtain

$$\mathrm{d}g = \mathrm{d}f - \mathrm{d}(X_1x_1) = -x_1\mathrm{d}X_1 + \sum_j X_j \mathrm{d}x_j \quad (j = 2, \ldots, n) \tag{B.20}$$

and we see that the roles of x_1 and X_1 have been interchanged. This transformation can be applied to only one variable, to several variables or to all variables. In the last case we use $g = f - \Sigma_i X_i x_i$ and obtain $\mathrm{d}g = -\Sigma_j x_j \mathrm{d}X_j$ ($j = 1, \ldots, n$). This so-called *Legendre transformation*, if applied to the transform, results in the complete original expression and is often used in thermodynamics. For example, the Gibbs energy $G(T,P)$ with pressure P and temperature T as independent variables results from the internal energy $U(S,V)$ with entropy S and volume V as independent variables using $G = U - TS + PV$.

Example B.7

Consider the function $f(x) = \frac{1}{2}x^2$. The dependent variable X is given by

$$X = \partial f/\partial x = x$$

which can be solved to yield $x = X$. Therefore, the function expressed in the variable X reads $f(X) = \frac{1}{2}X^2$. For the transform $g(X)$ one thus obtains

$$g(X) = f(X) - Xx = \frac{1}{2}X^2 - XX = -\frac{1}{2}X^2$$

B.6 Matrices and Determinants

A *matrix* is an array of numbers (or functions), represented by a roman boldface uppercase symbol, for example, **A**, or by an italic uppercase symbol with indices, for example, A_{ij}. In full we write

$$\mathbf{A} = A_{ij} = \begin{pmatrix} A_{11} & A_{12} & . & A_{1n} \\ A_{21} & & & . \\ . & & & . \\ A_{m1} & . & . & A_{mn} \end{pmatrix} \tag{B.21}$$

The numbers A_{ij} are called the *elements*. The matrix with m rows and n columns is called an $m \times n$ matrix or a matrix of order (m,n). The *transpose* of a matrix, indicated by a superscript T, is formed by interchanging rows and columns. Hence

$$\mathbf{A}^{\mathrm{T}} = A_{ji} \tag{B.22}$$

Often, we will use square matrices A_{ij} for which $m = n$ (order n). A *column matrix* (or *column* for short) is a matrix for which $n = 1$ and is denoted by a lowercase italic standard symbol with an index, for example, by a_i, or by a lowercase roman bold symbol, for example, **a**. A row matrix is a matrix for which $m = 1$ and is the transpose of a column matrix and thus denoted by $(a_i)^{\mathrm{T}}$ or $\mathbf{a}^{\mathrm{T}}$.

Two matrices of the same order are *equal* if all their corresponding elements are equal. The *sum* of two matrices **A** and **B** of the same order is given by the matrix **C** whose corresponding elements are the sums of the elements of **A** and **B** or

$$\mathbf{C} = \mathbf{A} + \mathbf{B} \quad \text{or} \quad C_{ij} = A_{ij} + B_{ij} \tag{B.23}$$

The *product* of two matrices **A** and **B** results in matrix **C** whose elements are given by

$$\mathbf{C} = \mathbf{AB} \quad \text{or} \quad C_{ij} = \sum\nolimits_k A_{ik} B_{kj} \tag{B.24}$$

representing the *row-into-column* rule. Note that, if **A** represents a matrix of order (k,l) and **B** a matrix of order (m,n), the product **BA** is not defined unless $k = n$. For square matrices, we generally have $\mathbf{AB} \neq \mathbf{BA}$, so that the order must be maintained in any multiplication process. The *transpose* of a *product* $(\mathbf{ABC}..)^{\mathrm{T}}$ is given by $(\mathbf{ABC}..)^{\mathrm{T}} = ..^{\mathrm{T}}\mathbf{C}^{\mathrm{T}}\mathbf{B}^{\mathrm{T}}\mathbf{A}^{\mathrm{T}}$.

A *real* matrix is a matrix with real elements only while a *complex* matrix is a matrix with complex elements. The *complex conjugate* of a matrix **A** is the matrix $\mathbf{A}^*$ formed by the complex conjugate elements of **A** or

$$\mathbf{A}^* = A_{ij}^* \tag{B.25}$$

If a real (complex), square matrix **A** is equal to its transpose (complex conjugate)

$$\mathbf{A} = \mathbf{A}^{\mathrm{T}} \; (\mathbf{A} = (\mathbf{A}^*)^{\mathrm{T}}) \tag{B.26}$$

then **A** is a *symmetric* (*Hermitian*) matrix. For an *antisymmetric* matrix it holds that

$$\mathbf{A} = -(\mathbf{A}^{\mathrm{T}}) \tag{B.27}$$

and thus **A** has the form

$$\mathbf{A} = \begin{pmatrix} 0 & A_{12} & . & A_{1n} \\ -A_{12} & 0 & & . \\ . & & & . \\ -A_{1n} & . & . & 0 \end{pmatrix} \tag{B.28}$$

A *diagonal* matrix has only non-zero entries along the diagonal:

$$\mathbf{A} = \begin{pmatrix} A_{11} & 0 & . & 0 \\ 0 & A_{22} & & . \\ . & & . & . \\ 0 & . & . & A_{nn} \end{pmatrix} \tag{B.29}$$

The *unit* matrix **I** is a diagonal matrix with unit elements:

$$\mathbf{I} = \delta_{ij} = \begin{pmatrix} 1 & 0 & . & 0 \\ 0 & 1 & & . \\ . & & . & . \\ 0 & . & . & 1 \end{pmatrix} \tag{B.30}$$

Obviously, **IA** = **AI** = **A**, where **A** is any square matrix of the same order as the unit matrix.

The *determinant* of a square matrix of order n is defined by

$$\det \mathbf{A} = \sum (\pm A_{1i} A_{2j} A_{3k} \cdots A_{np}) \tag{B.31}$$

where the summation is over all permutations of the indices $i, j, k, \cdots, p$. The sign in brackets is positive (negative) when the permutation involves an even (odd) number of permutations from the initial term $A_{11}A_{22}A_{33}..A_{nn}$.

Example B.8

For a matrix **A** of order 3, Eq. (B.31) yields

$$\det \mathbf{A} = A_{11}A_{22}A_{33} + A_{12}A_{23}A_{31} + A_{13}A_{21}A_{32} - A_{12}A_{21}A_{33} - A_{11}A_{23}A_{32} - A_{13}A_{22}A_{31}$$

Alternatively, it can be written as $\det \mathbf{A} = \Sigma_{r,s,t}\, e_{rst}\, A_{1r}\, A_{2s}\, A_{3t}$.

The determinant of the product **AB** is given by

$$\det \mathbf{AB} = (\det \mathbf{A})(\det \mathbf{B}) \tag{B.32}$$

Further, the determinant of a matrix equals the determinant of its transpose, that is

$$\det \mathbf{A} = \det \mathbf{A}^{\mathrm{T}} \tag{B.33}$$

The *inverse* of a square matrix **A** is denoted by $\mathbf{A}^{-1}$ and it holds that

$$\mathbf{A}\mathbf{A}^{-1} = \mathbf{A}^{-1}\mathbf{A} = \mathbf{I} \tag{B.34}$$

where **I** is the unit matrix of the same order as **A**. Hence, a square matrix *commutes* with its inverse. The inverse only exists if det $\mathbf{A} \neq 0$. The inverse of the product $(\mathbf{ABC}..)^{-1}$ is given by $(\mathbf{ABC}..)^{-1} = ..^{-1}\mathbf{C}^{-1}\mathbf{B}^{-1}\mathbf{A}^{-1}$. The inverse of a transpose is equal to the transpose of the inverse, that is, $(\mathbf{A}^{\mathrm{T}})^{-1} = (\mathbf{A}^{-1})^{\mathrm{T}}$, often written as $\mathbf{A}^{-\mathrm{T}}$.

The *cofactor* α_{ij} of the element A_{ij} is $(-1)^{i+j}$ times the *minor* θ_{ij}. The latter is the determinant of a matrix obtained by removing row i and column j from the original matrix. The inverse of **A** is then found from *Cramers' rule*

$$\left(\mathbf{A}^{-1}\right)_{ij} = \alpha_{ji} / \det \mathbf{A} \tag{B.35}$$

Note the reversal of the element and cofactor indices.

Example B.9

Consider the matrix $\mathbf{A} = A_{ij} = \begin{pmatrix} 1 & 2 \\ 3 & 4 \end{pmatrix}$. The determinant is det $\mathbf{A} = -2$. The cofactors are given by

$$\alpha_{11} = (-1)^{1+1}\theta_{11} = (-1)^{1+1}4 = 4 \qquad \alpha_{12} = (-1)^{1+2}\theta_{12} = (-1)^{1+2}3 = -3$$

$$\alpha_{22} = (-1)^{2+2}\theta_{22} = (-1)^{2+2}1 = 1 \qquad \alpha_{21} = (-1)^{2+1}\theta_{21} = (-1)^{2+1}2 = -2$$

The elements of the inverse $\mathbf{A}^{-1}$ are thus given by

$$[A_{ij}^{-1}] = \begin{pmatrix} \alpha_{11} & \alpha_{21} \\ \alpha_{12} & \alpha_{22} \end{pmatrix} \Big/ \det \mathbf{A} = \begin{pmatrix} 4 & -2 \\ -3 & 1 \end{pmatrix} \Big/ -2 = \begin{pmatrix} -2 & 1 \\ 1.5 & -0.5 \end{pmatrix}$$

For a *diagonal* matrix **A** the inverse is particularly simple and given by

$$\mathbf{A}^{-1} = \begin{pmatrix} A_{11}^{-1} & 0 & . & 0 \\ 0 & A_{22}^{-1} & & . \\ . & & . & . \\ 0 & . & . & A_{nn}^{-1} \end{pmatrix} \tag{B.36}$$

For an *orthogonal* (*unitary*) matrix it holds that

$$\mathbf{A}^{\mathrm{T}} = \mathbf{A}^{-1} \quad \text{or} \quad \mathbf{A}^{\mathrm{T}}\mathbf{A} = \mathbf{I} \quad (\mathbf{A}^{*} = \mathbf{A}^{-1} \quad \text{or} \quad \mathbf{A}^{*}\mathbf{A} = \mathbf{I}) \tag{B.37}$$

implying det $\mathbf{A} = \pm 1$. With det $\mathbf{A} = 1$, the matrix **A** is a *proper* orthogonal matrix.

B.7
Change of Variables

It is also often required to use different independent variables, in particular in integrals. For definiteness consider the case of three "old" variables x, y and z and three "new" variables u, v and w. In this case, we have

$$u = u(x, y, z) v = v(x, y, z) \quad \text{and} \quad w = w(x, y, z)$$

where the functions u, v and w are continuous and have continuous first derivatives in some region R^*. The transformations $u(x,y,z)$, $v(x,y,z)$ and $w(x,y,z)$ are such that a point (x,y,z) corresponding to (u,v,w) in R^* lies in a region R and that there is a one-to-one correspondence between the points (u,v,w) and (x,y,z). The Jacobian matrix $J = \partial(x,y,z)/\partial(u,v,w)$ is defined by

$$J = \frac{\partial(x, y, z)}{\partial(u, v, w)} = \begin{pmatrix} \frac{\partial x}{\partial u} & \frac{\partial x}{\partial v} & \frac{\partial x}{\partial w} \\ \frac{\partial y}{\partial u} & \frac{\partial y}{\partial v} & \frac{\partial y}{\partial w} \\ \frac{\partial z}{\partial u} & \frac{\partial z}{\partial v} & \frac{\partial z}{\partial w} \end{pmatrix} \tag{B.38}$$

The determinant, det J, should be either positive or negative throughout the region R^*. Consider now the integral

$$I = \int_R F(x, y, z) \mathrm{d}x\mathrm{d}y\mathrm{d}z \tag{B.39}$$

over the region R. If the function F is now expressed in u, v and w instead of x, y and z, the integral has to be evaluated over the region R^* as

$$I = \int_{R^*} F(u, v, w) |\det J| \mathrm{d}u\mathrm{d}v\mathrm{d}w \tag{B.40}$$

where $|\det J|$ denotes the absolute value of the determinant of the Jacobian matrix[2)] J. The expression is easily generalized to more variables than 3.

Example B.10

In many cases the use of cylindrical coordinates is convenient. Here, we consider the Cartesian coordinates x_1, x_2 and x_3 as "new" variables and the cylindrical coordinates r, θ and z as "old" variables. The relations between the Cartesian coordinates and the cylindrical coordinates (Figure B.1) are

$$x_1 = r\cos\theta \qquad x_2 = r\sin\theta \qquad x_3 = z$$

while the inverse equations are given by

$$r = (x_1^2 + x_2^2)^{1/2} \qquad \theta = \tan^{-1}(x_2/x_1) \qquad z = x_3$$

The Jacobian determinant is easily calculated as $|\det J| = r$. Similarly, for spherical coordinates[3)]

2) In the literature the name Jacobian sometimes indicates the Jacobian determinant instead of the matrix of derivatives. To avoid confusion, we use Jacobian matrix and Jacobian determinant explicitly.

3) Unfortunately in the usual convention for spherical coordinates the angle φ corresponds to the angle θ in cylindrical coordinates.

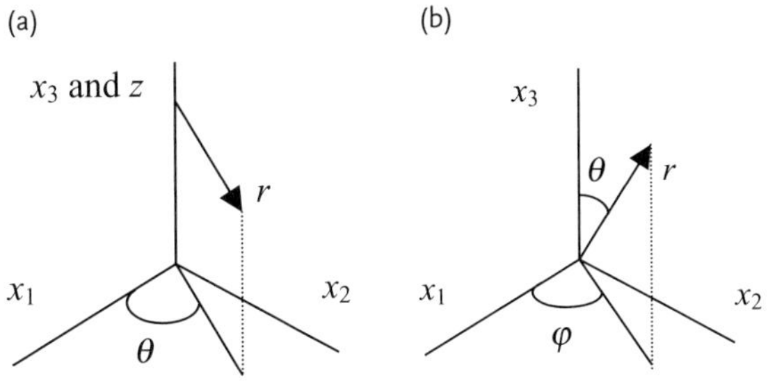

Figure B.1 (a) Cylindrical and (b) spherical coordinates.

$$x_1 = r\cos\varphi\sin\theta \qquad x_2 = r\sin\varphi\sin\theta \qquad x_3 = r\cos\theta$$

and the corresponding inverse equations

$$r = (x_1^2 + x_2^2 + x_3^2)^{1/2} \qquad \varphi = \tan^{-1}(x_2/x_1) \qquad \theta = \tan^{-1}[(x_1^2 + x_2^2)^{1/2}/x_3]$$

In this case the Jacobian determinant becomes $|\det J| = r^2 \sin\theta$.

B.8
Scalars, Vectors, and Tensors

A *scalar* is an entity with a magnitude. It is denoted by an italic, lowercase or uppercase, roman or greek letter, for example, a, A, γ, or Γ.

A *vector* is an entity with a magnitude and direction. It is denoted by a lowercase boldface, italic letter, for example, $\boldsymbol{a}$. It can be interpreted as an arrow from a point O (origin) to a point P. Let $\boldsymbol{a}$ be this arrow. Its *magnitude* (length), equal to the distance OP, is denoted by $a = |\boldsymbol{a}|$. A unit vector in the same direction as the vector $\boldsymbol{a}$, here denoted by $\boldsymbol{e}$, has a length of 1. Vectors obey the following rules (Figure B.2):

- $\boldsymbol{c} = \boldsymbol{a} + \boldsymbol{b} = \boldsymbol{b} + \boldsymbol{a}$ (commutative rule)
- $\boldsymbol{a} + (\boldsymbol{b} + \boldsymbol{d}) = (\boldsymbol{a} + \boldsymbol{b}) + \boldsymbol{d}$ (associative rule)

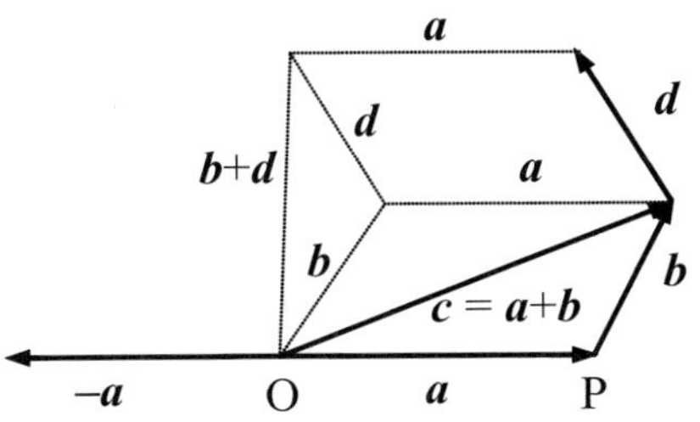

Figure B.2 Vector properties.

- $(\boldsymbol{a} + \boldsymbol{b}) \cdot \boldsymbol{c} = \boldsymbol{a} \cdot \boldsymbol{c} + \boldsymbol{b} \cdot \boldsymbol{c}$ (distributive law)
- $\boldsymbol{a} + (-\boldsymbol{a}) = \boldsymbol{0}$ (zero vector definition)
- $\boldsymbol{a} = |\boldsymbol{a}|\boldsymbol{e}, \ |\boldsymbol{e}| = 1$ (unit vector definition)
- $\boldsymbol{0} \cdot \boldsymbol{u} = 0$
- $\alpha\boldsymbol{a} = \alpha|\boldsymbol{a}|\boldsymbol{e}, \ |\boldsymbol{e}| = 1$

Various products can be formed using vectors. The *scalar* (or *dot* or *inner*) *product* of two vectors $\boldsymbol{a}$ and $\boldsymbol{b}$ yields a scalar and is defined as $\boldsymbol{a} \cdot \boldsymbol{b} = \boldsymbol{b} \cdot \boldsymbol{a} = |\boldsymbol{a}| \ |\boldsymbol{b}|\cos(\phi)$, where ϕ is the enclosed angle between $\boldsymbol{a}$ and $\boldsymbol{b}$. From this definition it follows that $a = |\boldsymbol{a}| = (\boldsymbol{a} \cdot \boldsymbol{a})^{1/2}$. Two vectors $\boldsymbol{a}$ and $\boldsymbol{b}$ are orthogonal if $\boldsymbol{a} \cdot \boldsymbol{b} = 0$. The scalar product is commutative ($\boldsymbol{a} \cdot \boldsymbol{b} = \boldsymbol{b} \cdot \boldsymbol{a}$) and distributive ($\boldsymbol{a} \cdot (\boldsymbol{b} + \boldsymbol{c}) = \boldsymbol{a} \cdot \boldsymbol{b} + \boldsymbol{a} \cdot \boldsymbol{c}$).

The *vector* (or *cross* or *outer*) *product* of $\boldsymbol{a}$ and $\boldsymbol{b}$ denotes a vector $\boldsymbol{c} = \boldsymbol{a} \times \boldsymbol{b}$. We define a unit vector $\boldsymbol{n}$ perpendicular to the plane spanned by $\boldsymbol{a}$ and $\boldsymbol{b}$. The sense of $\boldsymbol{n}$ is right-handed: rotate from $\boldsymbol{a}$ to $\boldsymbol{b}$ along the smallest angle and the direction of $\boldsymbol{n}$ is given by a right-hand screw. It holds that $\boldsymbol{a} \cdot \boldsymbol{n} = \boldsymbol{b} \cdot \boldsymbol{n} = 0$ and $|\boldsymbol{n}| = 1$. Explicitly, $\boldsymbol{n} = \boldsymbol{a} \times \boldsymbol{b}/|\boldsymbol{a}| \ |\boldsymbol{b}|$. The vector product $\boldsymbol{c}$ is equal to $\boldsymbol{c} = \boldsymbol{a} \times \boldsymbol{b} = -\boldsymbol{b} \times \boldsymbol{a} = |\boldsymbol{a}| \ |\boldsymbol{b}| \sin(\phi)\boldsymbol{n}$. The length of $|\boldsymbol{c}| = |\boldsymbol{a}| \ |\boldsymbol{b}|\sin(\phi)$ is numerically equal to the area of the parallelogram whose sides are given by $\boldsymbol{a}$ and $\boldsymbol{b}$. The vector product is anti-commutative ($\boldsymbol{a} \times \boldsymbol{b} = -\boldsymbol{b} \times \boldsymbol{a}$) and distributive ($\boldsymbol{a} \times (\boldsymbol{b} + \boldsymbol{c}) = \boldsymbol{a} \times \boldsymbol{b} + \boldsymbol{a} \times \boldsymbol{c}$), but not associative ($\boldsymbol{a} \times (\boldsymbol{b} \times \boldsymbol{c}) \neq (\boldsymbol{a} \times \boldsymbol{b}) \times \boldsymbol{c}$).

The *triple product* is a scalar and given by $d = \boldsymbol{a} \cdot \boldsymbol{b} \times \boldsymbol{c} = \boldsymbol{a} \times \boldsymbol{b} \cdot \boldsymbol{c}$. It yields the volume of the block (or parallelepepid) the edges of which are $\boldsymbol{a}$, $\boldsymbol{b}$, and $\boldsymbol{c}$. Three vectors $\boldsymbol{a}$, $\boldsymbol{b}$, and $\boldsymbol{c}$ are independent if from $\alpha\boldsymbol{a} + \beta\boldsymbol{b} + \gamma\boldsymbol{c} = 0$ it follows that $\alpha = \beta = \gamma = 0$. This is only the case if $\boldsymbol{a}$, $\boldsymbol{b}$ and $\boldsymbol{c}$ are noncoplanar or, equivalently, the product $\boldsymbol{a} \cdot \boldsymbol{b} \times \boldsymbol{c} \neq 0$.

Finally, we need the *tensor* (or *dyadic*) *product* $\boldsymbol{ab}$. Operating on a vector $\boldsymbol{c}$, it associates with $\boldsymbol{c}$ a new vector according to $\boldsymbol{ab} \cdot \boldsymbol{c} = \boldsymbol{a}(\boldsymbol{b} \cdot \boldsymbol{c}) = (\boldsymbol{b} \cdot \boldsymbol{c})\boldsymbol{a}$. Note that $\boldsymbol{ba}$ operating on $\boldsymbol{c}$ yields $\boldsymbol{ba} \cdot \boldsymbol{c} = \boldsymbol{b}(\boldsymbol{a} \cdot \boldsymbol{c}) = (\boldsymbol{a} \cdot \boldsymbol{c})\boldsymbol{b}$.

A *tensor* (of rank 2), denoted by an uppercase boldface, italic letter, for example, $\boldsymbol{A}$, is a linear mapping that associates with a vector $\boldsymbol{a}$ another vector $\boldsymbol{b}$ according to $\boldsymbol{b} = \boldsymbol{A} \cdot \boldsymbol{a}$. Tensors obey the following rules:

- $\boldsymbol{C} = \boldsymbol{A} + \boldsymbol{B} = \boldsymbol{B} + \boldsymbol{A}$ (commutative law)
- $\boldsymbol{A} + (\boldsymbol{B} + \boldsymbol{C}) = (\boldsymbol{A} + \boldsymbol{B}) + \boldsymbol{C}$ (associative law)
- $(\boldsymbol{A} + \boldsymbol{B}) \cdot \boldsymbol{u} = \boldsymbol{A} \cdot \boldsymbol{u} + \boldsymbol{B} \cdot \boldsymbol{u}$ (distributive law)
- $\boldsymbol{A} + (-\boldsymbol{A}) = \boldsymbol{O}$ (zero tensor definition)
- $\boldsymbol{I} \cdot \boldsymbol{u} = \boldsymbol{u}$ ($\boldsymbol{I}$ unit tensor definition)
- $\boldsymbol{O} \cdot \boldsymbol{u} = \boldsymbol{0}$ ($\boldsymbol{O}$ zero tensor, $\boldsymbol{0}$ zero vector)
- $\boldsymbol{A} \cdot (\alpha\boldsymbol{u}) = (\alpha\boldsymbol{A}) \cdot \boldsymbol{u} = \alpha(\boldsymbol{A} \cdot \boldsymbol{u})$

where α is an arbitrary scalar and $\boldsymbol{u}$ is an arbitrary vector. The simplest example of a tensor is the tensor product of two vectors, for example, if $\boldsymbol{A} = \boldsymbol{bc}$, the vector associated with $\boldsymbol{a}$ is given by $\boldsymbol{A} \cdot \boldsymbol{a} = \boldsymbol{bc} \cdot \boldsymbol{a} = (\boldsymbol{c} \cdot \boldsymbol{a})\boldsymbol{b}$.

So far we have discussed vectors and tensors using the *direct notation* only, that is using a symbolism, which represents the quantity without referring to a coordinate system. It is convenient though to introduce a coordinate system. In this book we will make use primarily of Cartesian coordinates, which are a rectangular and rectilinear coordinate system with origin O and unit vectors $\boldsymbol{e}_1$, $\boldsymbol{e}_2$ and $\boldsymbol{e}_3$ along the axes. The set $\boldsymbol{e}_i = \{\boldsymbol{e}_1, \boldsymbol{e}_2, \boldsymbol{e}_3\}$ is called an *orthonormal basis*. It holds that $\boldsymbol{e}_i\boldsymbol{e}_j = \delta_{ij}$. The vector OP = $\boldsymbol{x}$ is called the position of point P. The real numbers x_1, x_2 and x_3, defined uniquely by the relation $\boldsymbol{x} = x_1\boldsymbol{e}_1 + x_2\boldsymbol{e}_2 + x_3\boldsymbol{e}_3$, are called the (Cartesian) *components* of the vector $\boldsymbol{x}$. It follows that $x_i = \boldsymbol{x}\cdot\boldsymbol{e}_i$ for $i = 1, 2, 3$. Using the components x_i in equations, we use the *index notation*. Using the index notation the scalar product $\boldsymbol{u}\cdot\boldsymbol{v}$ can be written as $\boldsymbol{u}\cdot\boldsymbol{v} = u_1v_1 + u_2v_2 + u_3v_3 = \Sigma_i u_i v_i$. The length of a vector $\boldsymbol{x}$, $|\boldsymbol{x}| = (\boldsymbol{x}\cdot\boldsymbol{x})^{1/2}$, is thus also equal to $(x_1^2 + x_2^2 + x_3^2)^{1/2} = (\Sigma_i x_i x_i)^{1/2}$. Sometimes, it is also convenient to use *matrix notation*; in this case the components x_i are written collectively as a column matrix $\mathbf{x}$. In matrix notation the scalar product $\boldsymbol{u}\cdot\boldsymbol{v}$ is written as $\mathbf{u}^{\mathrm{T}}\mathbf{v}$. The tensor product $\boldsymbol{ab}$ in matrix notation is given by $\mathbf{ab}^{\mathrm{T}}$.

Using the alternator e_{ijk} the relations between the unit vectors can be written as $\boldsymbol{e}_i \times \boldsymbol{e}_j = \Sigma_k e_{ijk}\boldsymbol{e}_k$. Similarly, the vector product $\boldsymbol{c} = \boldsymbol{a}\times\boldsymbol{b}$ can alternatively be written as $\boldsymbol{c} = \boldsymbol{a}\times\boldsymbol{b} = \Sigma_{j,k}\, e_{ijk}\boldsymbol{e}_i a_j b_k$. In components this leads to the following expressions:

$$c_1 = a_2b_3 - a_3b_2 \quad c_2 = a_3b_1 - a_1b_3 \quad c_3 = a_1b_2 - a_2b_1 \tag{B.41}$$

The triple product $\boldsymbol{a}\cdot\boldsymbol{b}\times\boldsymbol{c}$ in components is given by $e_{ijk}a_ib_jc_k$, while the tensor product $\boldsymbol{ab}$ is represented by a_ib_j. A useful relation involving three vectors using the tensor product is $\boldsymbol{a}\times(\boldsymbol{b}\times\boldsymbol{c}) = (\boldsymbol{ba} - (\boldsymbol{a}\cdot\boldsymbol{b})\boldsymbol{I})\cdot\boldsymbol{c}$.

If $\boldsymbol{e}_i = \{\boldsymbol{e}_1, \boldsymbol{e}_2, \boldsymbol{e}_3\}$ is a basis, the tensor products $\boldsymbol{e}_i\boldsymbol{e}_j$, $i, j = 1, 2, 3$, form a basis for representing a tensor, and we can write $\boldsymbol{A} = A_{kl}\boldsymbol{e}_k\boldsymbol{e}_l$. The nine real numbers A_{kl} are the (Cartesian) components of the tensor $\boldsymbol{A}$, and are conveniently arranged in a square matrix. It follows that $A_{kl} = \boldsymbol{e}_k\cdot(\boldsymbol{A}\cdot\boldsymbol{e}_l)$, which can be taken as the definition of the components. Applying this definition to the unit tensor, it follows that δ_{kl} are the components of the unit tensor, that is, $\boldsymbol{I} = \delta_{kl}\boldsymbol{e}_k\boldsymbol{e}_l$. If $\boldsymbol{v} = \boldsymbol{A}\cdot\boldsymbol{u}$, we also have $\boldsymbol{v} = (A_{kl}\boldsymbol{e}_k\boldsymbol{e}_l)\cdot\boldsymbol{u} = \boldsymbol{e}_k A_{kl} u_l$. Tensors, like vectors, can form different products. The inner product $\boldsymbol{A}\cdot\boldsymbol{B}$ of two tensors (of rank 2) $\boldsymbol{A}$ and $\boldsymbol{B}$ yields another tensor of rank 2 and is defined by $(\boldsymbol{A}\cdot\boldsymbol{B})\cdot\boldsymbol{u} = \boldsymbol{A}\cdot(\boldsymbol{B}\cdot\boldsymbol{u}) = \Sigma_{p,m}A_{kp}B_{pm}u_m$ wherefrom it follows that $(\boldsymbol{A}\cdot\boldsymbol{B})_{km} = \Sigma_p A_{kp}B_{pm}$, representing conventional matrix multiplication. The expression $\boldsymbol{A}{:}\boldsymbol{B}$ denotes the double inner product, yields a scalar and is given in index notation by $\Sigma_{i,j}A_{ij}B_{ij}$. Equivalently, $\boldsymbol{A}{:}\boldsymbol{B} = \mathrm{tr}\,\mathbf{AB}^{\mathrm{T}} = \mathrm{tr}\,\mathbf{A}^{\mathrm{T}}\mathbf{B}$.

Recall that the components of a vector $\boldsymbol{a}$ can be transformed to another Cartesian frame by $a'_p = \Sigma_i C_{pi}a_i$ in index notation, or $\mathbf{a}' = \mathbf{Ca}$ in matrix notation. Since a tensor $\boldsymbol{A}$ of rank 2 can be interpreted as the tensor product of two vectors $\boldsymbol{b}$ and $\boldsymbol{c}$, that is,

$$\boldsymbol{A} = \boldsymbol{bc} \quad \text{or} \quad A_{ij} = b_ic_j \quad \text{or} \quad \mathbf{A} = \mathbf{bc}^{\mathrm{T}}$$

the transformation rule for the components of a tensor $\boldsymbol{A}$ obviously is

$$A'_{pq} = b'_pc'_q = \Sigma_{i,j}C_{pi}b_iC_{qj}c_j = \Sigma_{i,j}C_{pi}C_{qj}b_ic_j = \Sigma_{i,j}C_{pi}C_{qj}A_{ij}$$

or, in matrix notation[4)]

4) Obviously if the transformation is interpreted as a rotation of the tensor instead of the frame, we obtain $\mathbf{A}' = \mathbf{C}^{\mathrm{T}}\mathbf{AC}$. This is the conventional definition of an orthogonal transformation.

$$\mathbf{A}' = \mathbf{b}'\mathbf{c}'^{\mathrm{T}} = (\mathbf{C}\mathbf{b})(\mathbf{c}^{\mathrm{T}}\mathbf{C}^{\mathrm{T}}) = \mathbf{C}\mathbf{A}\mathbf{C}^{\mathrm{T}}$$

If $\boldsymbol{A}' = \boldsymbol{A}$ and thus $\mathbf{A}' = \mathbf{A}$, then $\boldsymbol{A}$ is an *isotropic* (or *spherical*) *tensor*. Further, if the component matrix of a tensor has a property which is not changed by a coordinate axes rotation that property is shared by $\boldsymbol{A}'$ and $\boldsymbol{A}$. Such a property is called an *invariant*. An example is the transpose of a tensor of rank 2: If $\mathbf{A}' = \mathbf{C}\mathbf{A}\mathbf{C}^{\mathrm{T}}$, then $\mathbf{A}'^{\mathrm{T}} = \mathbf{C}\mathbf{A}^{\mathrm{T}}\mathbf{C}^{\mathrm{T}}$. Consequently, we may speak of the transpose $\boldsymbol{A}^{\mathrm{T}}$ of the tensor $\boldsymbol{A}$, and we may define the symmetric parts $\boldsymbol{A}^{(\mathrm{s})}$ and antisymmetric $\boldsymbol{A}^{(\mathrm{a})}$ parts by

$$\boldsymbol{A}^{(\mathrm{s})} = (\boldsymbol{A} + \boldsymbol{A}^{\mathrm{T}})/2 \quad \text{and} \quad \boldsymbol{A}^{(\mathrm{a})} = (\boldsymbol{A} - \boldsymbol{A}^{\mathrm{T}})/2$$

While originally a distinction in terminology is made for a scalar, a vector and a tensor, it is clear that they all transform similarly under a coordinate transformation. Therefore, a scalar is sometimes denoted as a tensor of rank 0 and a vector as a tensor of rank 1. Expressed in components, all tensors obey the same type of transformation rules, for example, $A_{i..j} = \Sigma_{p,..,q} C_{ip} \ldots C_{jq} A_{p..q}$, where the transformation matrix $\mathbf{C}$ represents the rotation of the coordinate system. Scalars have no index, a vector has one index, a tensor of rank 2 has two indices, while a tensor of rank 4 has four indices. Their total transformation matrix contains a product of respectively 0, 1, 2, and 4 individual transformation matrices C_{ij}. Obviously, if we define (Cartesian) tensors as quantities obeying the above transformation rules[5)], extension to any order is immediate.

Finally, we have to mention that, like scalars, vectors and tensors can be a function of one or more variables x_k. The appropriate notation is $f(x_k)$, $a_i(x_k)$ and $A_{ij}(x_k)$ or, equivalently, $f(\boldsymbol{x})$, $\boldsymbol{a}(\boldsymbol{x})$ and $\boldsymbol{A}(\boldsymbol{x})$. If x_k represent the coordinates, $f(\boldsymbol{x})$, $\boldsymbol{a}(\boldsymbol{x})$ and $\boldsymbol{A}(\boldsymbol{x})$ are referred to as a *scalar, vector,* and *tensor field*, respectively.

Example B.11

Consider the vectors $\boldsymbol{a}$, $\boldsymbol{b}$ and $\boldsymbol{c}$ the matrix representations of which are

$$\mathbf{a}^{\mathrm{T}} = (1, 0, 0), \quad \mathbf{b}^{\mathrm{T}} = (0, 1, 1) \quad \text{and} \quad \mathbf{c}^{\mathrm{T}} = (1, 1, 2). \text{ Then}$$

$$\boldsymbol{a} + \boldsymbol{b} = \mathbf{a} + \mathbf{b} = (1, 1, 1)^{\mathrm{T}}$$

$$(\boldsymbol{b} \times \boldsymbol{c})^{\mathrm{T}} = (b_2c_3 - b_3c_2, \quad b_3c_1 - b_1c_3, \quad b_1c_2 - b_2c_1)$$
$$= (1\cdot2 - 1\cdot1, \quad 1\cdot1 - 0\cdot2, \quad 0\cdot1 - 1\cdot1) = (1 \quad 1 \quad -1)$$

5) This transformation rule is only appropriate for proper rotations of the axes. For improper rotations, which involve a reflection and change of handedness of the coordinate system, there are two possibilities. If the rule applies we call the tensor *polar*; if an additional change of sign occurs for an improper rotation we call the tensor *axial*. Hence, generally $L_{i..j} = (\det \mathbf{C})^p C_{ik} \ldots C_{jl} L_{k..l}$, where $p = 0$ for a polar tensor and $p = 1$ for an axial tensor. Since $\det \mathbf{C} = 1$ for a proper rotation and $\det \mathbf{C} = -1$ for an improper rotation, this results in an extra change of sign for an axial tensor under improper rotation. It follows that the inner and outer product of two polar or two axial tensors is polar, while the product of a polar and an axial tensor is axial. The permutation tensor e_{ijk} is axial since $e_{123} = 1$ for both right-handed and left-handed systems. Hence, the vector product of two polar vectors is axial. If one restricts oneself to right-handed systems, the distinction is irrelevant.

$$|c| = \left(\mathbf{c}^{\mathrm{T}}\mathbf{c}\right)^{1/2} = \left[(1 \quad 1 \quad 2)\begin{pmatrix}1\\1\\2\end{pmatrix}\right]^{1/2} = \sqrt{6} \qquad \boldsymbol{b}\cdot\boldsymbol{c} = \mathbf{b}^{\mathrm{T}}\mathbf{c} = (0 \quad 1 \quad 1)\begin{pmatrix}1\\1\\2\end{pmatrix} = 3$$

$$\boldsymbol{ab} = \mathbf{ab}^{\mathrm{T}} = \begin{pmatrix}1\\0\\0\end{pmatrix}(0 \quad 1 \quad 1) = \begin{pmatrix}0 & 1 & 1\\0 & 0 & 0\\0 & 0 & 0\end{pmatrix} \qquad \boldsymbol{a}\cdot\boldsymbol{b}\times\boldsymbol{c} = (1 \quad 0 \quad 0)\begin{pmatrix}1\\1\\-1\end{pmatrix} = 1$$

$$\boldsymbol{ab}\cdot\boldsymbol{c} = \begin{pmatrix}0 & 1 & 1\\0 & 0 & 0\\0 & 0 & 0\end{pmatrix}\begin{pmatrix}1\\1\\2\end{pmatrix} = \begin{pmatrix}3\\0\\0\end{pmatrix} = 3\boldsymbol{a}$$

B.9
Tensor Analysis

In this section we consider various differential operators, the divergence and some other theorems and their representation in cylindrical and spherical coordinates. Consider a_i as a typical representative of tensors of rank 1 and take the partial derivatives $\partial a_i/\partial x_j$. Such a derivative transforms like a tensor of rank 2, is called the *gradient* and is denoted by grad a_i in index notation or as $\nabla\boldsymbol{a}$ in direct notation. The gradient can operate on any tensor thereby increasing its rank by 1.

Summing over an index, known as *contraction*, decreases the rank of a tensor by 2. If we apply contraction to a tensor A_{ij} we calculate $\Sigma_i A_{ii}$ and the result is known as the *trace*, written in direct notation as tr $\boldsymbol{A}$. Contraction of a gradient $\Sigma_i(\partial a_i/\partial x_i)$ yields a scalar, called *divergence* and denoted in direct notation by div $\boldsymbol{a}$ or $\nabla\cdot\boldsymbol{a}$.

Another operator is the *curl* (or *rot* as abbreviation of rotation) of a vector $\boldsymbol{a}$, in direct notation written as $\nabla\times\boldsymbol{a}$, which is a vector with components $\partial_{k,j}e_{ijk}\partial a_k/\partial x_j$. It is defined for 3D space only.

Finally, we have the *Laplace operator*. This can act on a scalar a or on a vector $\boldsymbol{a}$, is denoted by ∇^2 or Δ, and is defined by $\nabla^2 a = \Delta a = \partial a^2/\partial^2 x_i$ or $\nabla^2\boldsymbol{a} = \Delta\boldsymbol{a} = \partial a_j^2/\partial^2 x_i$.

Example B.12

If a vector field $\boldsymbol{a}(\boldsymbol{x})$ represented by $\mathrm{a}^{\mathrm{T}}(\mathbf{x}) = (3x^2 + 2y,\ x + z^2,\ x + y^2)$, then

$$\frac{\partial a_i}{\partial x_j} = \nabla\boldsymbol{a} = \begin{pmatrix}\partial/\partial x\\\partial/\partial y\\\partial/\partial z\end{pmatrix}\left(3x^2+2y, \quad x+z^2, \quad x+y^2\right) = \begin{pmatrix}6x & 1 & 1\\2 & 0 & 2y\\0 & 2z & 0\end{pmatrix}$$

$$\nabla\times\boldsymbol{a} = \sum\nolimits_{j,k} \mathrm{e}_{ijk}\mathbf{e}_i\frac{\partial a_k}{\partial x_j} = \begin{pmatrix}2y-2z\\0-1\\1-2\end{pmatrix} = \begin{pmatrix}2(y-z)\\-1\\-1\end{pmatrix}, \qquad \Delta\boldsymbol{a} = \frac{\partial^2 a_j}{\partial x_i^2} = \begin{pmatrix}6\\2\\2\end{pmatrix}$$

and $\nabla\cdot\boldsymbol{a} = \sum\nolimits_i(\partial a_i/\partial x_i) = 6x + 0 + 0 = 6x$.

Introducing now some general theorems, let us first recall that if $\boldsymbol{X} = \nabla x$ with x a scalar, we have

$$\int_A^B \nabla x \,\mathrm{d}V = x_B - x_A \tag{B.42}$$

Second, without proof we introduce the divergence theorem. Therefore, we consider a region of volume V with a piecewise smooth surface S on which a single-valued tensor field $\boldsymbol{A}$ or A_{ij} is defined. The body may be either convex or nonconvex. The components of the exterior normal vector $\boldsymbol{n}$ of S are denoted by n_i. The *divergence theorem* or the *theorem of Gauss* states that

$$\int \frac{\partial A_{ij}}{\partial x_j}\,\mathrm{d}V = \int \sum\nolimits_j A_{ij} n_j \,\mathrm{d}S \quad \text{or equivalently} \quad \int \nabla \cdot \boldsymbol{A}\,\mathrm{d}V = \int \boldsymbol{A}\cdot\boldsymbol{n}\,\mathrm{d}S \tag{B.43}$$

The divergence theorem connects a volume integral (integrating over $\mathrm{d}V$) to a surface integral (integrating over $\boldsymbol{n}\mathrm{d}S$) and is mainly used in theoretical work. Applying the divergence theorem to a scalar a, a vector $\boldsymbol{a}$ or a tensor $\Sigma_{k,j}\varepsilon_{ijk}\partial a_k/\partial x_j$ we obtain in direct notation

$$\int \nabla a \,\mathrm{d}V = \int \boldsymbol{n} a \,\mathrm{d}S, \qquad \int \nabla \cdot \boldsymbol{a}\,\mathrm{d}V = \int \boldsymbol{n}\cdot\boldsymbol{a}\,\mathrm{d}S \qquad \text{and}$$
$$\int \nabla \times \boldsymbol{a}\,\mathrm{d}V = \int \boldsymbol{n}\times\boldsymbol{a}\,\mathrm{d}S \tag{B.44}$$

Third, one can also derive *Stokes' theorem*

$$\int \sum\nolimits_{i,j,k} n_i \varepsilon_{ijk} \frac{\partial a_k}{\partial r_j}\,\mathrm{d}S = \int \sum\nolimits_i a_i \,\mathrm{d}r_i \qquad \text{or} \qquad \int (\nabla \times \boldsymbol{a})\cdot\boldsymbol{n}\,\mathrm{d}S = \int \boldsymbol{a}\cdot\mathrm{d}\boldsymbol{r} \tag{B.45}$$

The theorem connects a surface (integrating over $\boldsymbol{n}\mathrm{d}S$) and line integral (integrating over $\mathrm{d}\boldsymbol{r}$) and implies that the surface integral is the same for all surfaces bounded by the same curve. By the way, the surface element $\boldsymbol{n}\mathrm{d}S$ is also often written as $\mathrm{d}\boldsymbol{S}$.

From Eqs (B.42), (B.43) and (B.45) one can derive many transformations of integrals. We only mention *Green's first identity* (using $\boldsymbol{n}\mathrm{d}S = \mathrm{d}\boldsymbol{S}$)

$$\int (\nabla\phi\cdot\nabla\psi)\mathrm{d}V = \int \phi(\nabla\psi\cdot\mathrm{d}\boldsymbol{S}) - \int \phi\nabla^2\psi\cdot\mathrm{d}V = \int \psi(\nabla\phi\cdot\mathrm{d}\boldsymbol{S}) - \int \psi\nabla^2\phi\,\mathrm{d}V \tag{B.46}$$

and *Green's second identity*

$$\int (\phi\nabla^2\psi - \psi\nabla^2\phi)\mathrm{d}V = \int (\phi\nabla\psi - \psi\nabla\phi)\cdot\mathrm{d}\boldsymbol{S} \tag{B.47}$$

The above operations can also be performed in other coordinate systems. Often, one considers systems where the base vectors *locally* still form an orthogonal basis, although the orientation may differ through space. These systems are normally addressed as *orthogonal curvilinear coordinates*. Cylindrical and spherical coordinates form examples with practical importance. Using the relations of Example B.10 one can show that the unit vectors for cylindrical coordinates are

$$e_r = e_1\cos\theta + e_2\sin\theta \qquad e_\theta = -e_1\sin\theta + e_2\cos\theta \qquad e_z = e_3$$

so that the only non-zero derivatives are

$$\mathrm{d}e_r/\mathrm{d}\theta = e_\theta \qquad \mathrm{d}e_\theta/\mathrm{d}\theta = -e_r$$

Using the chain rule for partial derivatives, one may show that the gradient becomes

$$\nabla = e_r\frac{\partial}{\partial r} + e_\theta\frac{1}{r}\frac{\partial}{\partial\theta} + e_z\frac{\partial}{\partial z} \tag{B.48}$$

The divergence of a vector $\boldsymbol{a}$ becomes

$$\nabla\cdot\boldsymbol{a} = \frac{\partial a_r}{\partial r} + \frac{1}{r}\left(a_r + \frac{\partial a_\theta}{\partial\theta}\right) + \frac{\partial a_z}{\partial z} \tag{B.49}$$

while the Laplace operator acting on a scalar a is expressed by

$$\nabla^2 a = \frac{\partial^2 a}{\partial r^2} + \frac{1}{r}\frac{\partial a}{\partial r} + \frac{1}{r^2}\frac{\partial^2 a}{\partial\theta^2} + \frac{\partial^2 a}{\partial z^2} \tag{B.50}$$

Using again the relations of Example B.10, one can show that the unit vectors for spherical coordinates are

$$e_r = (e_1\cos\varphi + e_2\sin\varphi)\sin\theta + e_3\cos\theta$$

$$e_\theta = -e_1\sin\varphi + e_2\cos\varphi$$

$$e_\varphi = (e_1\cos\varphi + e_2\sin\varphi)\cos\theta - e_3\sin\theta$$

so that the only non-zero derivatives are

$$\mathrm{d}e_r/\mathrm{d}\theta = e_\theta \qquad \mathrm{d}e_\theta/\mathrm{d}\theta = -e_r$$

$$\mathrm{d}e_r/\mathrm{d}\varphi = e_\varphi\sin\theta \qquad \mathrm{d}e_\theta/\mathrm{d}\varphi = e_\varphi\cos\theta \qquad \mathrm{d}e_\varphi/\mathrm{d}\varphi = -e_r\sin\theta - e_\varphi\cos\theta$$

The gradient operator becomes

$$\nabla = e_r\frac{\partial}{\partial r} + e_\theta\frac{1}{r}\frac{\partial}{\partial\theta} + e_\varphi\frac{1}{r\sin\theta}\frac{\partial}{\partial\varphi} \tag{B.51}$$

The divergence of a vector $\boldsymbol{a}$ becomes

$$\nabla\cdot\boldsymbol{a} = \frac{\partial a_r}{\partial r} + \frac{2a_r}{r} + \frac{1}{r}\frac{\partial a_\theta}{\partial\theta} + \frac{\cot\theta}{r}a_\theta + \frac{1}{r\sin\theta}\frac{\partial a_\varphi}{\partial\varphi} \tag{B.52}$$

while the Laplace operator acting on a scalar a is expressed by

$$\nabla^2 a = \frac{\partial^2 a}{\partial r^2} + \frac{2}{r}\frac{\partial a}{\partial r} + \frac{1}{r^2}\frac{\partial^2 a}{\partial\theta^2} + \frac{\cot\theta}{r^2}\frac{\partial a}{\partial\theta} + \frac{1}{r^2\sin^2\theta}\frac{\partial^2 a}{\partial\varphi^2} \tag{B.53}$$

B.10
Calculus of Variations

One of the chief applications of the calculus of variations is to find a function for which some given integral has an extreme. We treat the problem essentially as one-dimensional, but extension to more than one dimension is straightforward.

Suppose we wish to find a path $x = x(t)$ between two given values $x(t_1)$ and $x(t_2)$ such that the *functional*[6)] $J = \int_{t_1}^{t_2} f(x, \dot{x}, t)\mathrm{d}t$ of some function $f(x, \dot{x}, t)$ with $\dot{x} = \mathrm{d}x/\mathrm{d}t$ is an extremum. Let us assume that $x_0(t)$ is the solution we are looking for. Other possible curves close to $x_0(t)$ are written as $x(t,\alpha) = x_0(t) + \alpha\eta(t)$, where $\eta(t)$ is any function that satisfies $\eta(t_1) = \eta(t_2) = 0$. Using such a representation, the integral J becomes a function of α,

$$J(\alpha) = \int_{t_1}^{t_2} f\left[x(t,\alpha), \dot{x}(t,\alpha), t\right]\mathrm{d}t \tag{B.54}$$

and the condition for obtaining the extremum is $(\mathrm{d}J/\mathrm{d}\alpha)_{\alpha=0} = 0$. We obtain

$$\frac{\partial J}{\partial \alpha} = \int_{t_1}^{t_2}\left(\frac{\partial f}{\partial x}\frac{\partial x}{\partial \alpha} + \frac{\partial f}{\partial \dot{x}}\frac{\partial \dot{x}}{\partial \alpha}\right)\mathrm{d}t = \int_{t_1}^{t_2}\left(\frac{\partial f}{\partial x}\eta + \frac{\partial f}{\partial \dot{x}}\dot{\eta}\right)\mathrm{d}t \tag{B.55}$$

Through integration by parts the second term of the integral evaluates to

$$\int_{t_1}^{t_2}\frac{\partial f}{\partial \dot{x}}\dot{\eta}\,\mathrm{d}t = \frac{\partial f}{\partial \dot{x}}\eta\bigg|_{t_1}^{t_2} - \int_{x_1}^{x_2}\eta\frac{\mathrm{d}}{\mathrm{d}t}\frac{\partial f}{\partial \dot{x}}\mathrm{d}t \tag{B.56}$$

At t_1 and t_2, $\eta(t) = \partial x/\partial\alpha$ vanishes and we obtain for Eq. (B.55)

$$\frac{\partial J}{\partial \alpha} = \int_{t_1}^{t_2}\left(\frac{\partial f}{\partial x} - \frac{\mathrm{d}}{\mathrm{d}t}\frac{\partial f}{\partial \dot{x}}\right)\eta\,\mathrm{d}t$$

If we define the *variations* $\delta J = (\mathrm{d}J/\mathrm{d}\alpha)_{\alpha=0}\mathrm{d}\alpha$ and $\delta x = (\mathrm{d}x/\mathrm{d}\alpha)_{\alpha=0}\mathrm{d}\alpha$, we find

$$\delta J = \int_{t_1}^{t_2}\left(\frac{\partial f}{\partial x} - \frac{\mathrm{d}}{\mathrm{d}t}\frac{\partial f}{\partial \dot{x}}\right)\delta x\,\mathrm{d}t = \left[\int_{t_1}^{t_2}\left(\frac{\partial f}{\partial x} - \frac{\mathrm{d}}{\mathrm{d}t}\frac{\partial f}{\partial \dot{x}}\right)\eta\,\mathrm{d}t\right]\mathrm{d}\alpha = 0 \tag{B.57}$$

and since η must be arbitrary

$$\frac{\partial f}{\partial x} - \frac{\mathrm{d}}{\mathrm{d}t}\frac{\partial f}{\partial \dot{x}} = 0 \tag{B.58}$$

Once this so-called *Euler condition* is fulfilled an extremum is obtained. It should be noted that this extremum is not necessarily a minimum. Finally, we note that in case the variations at the boundaries do not vanish, that is, the values of η are not prescribed, the boundary term evaluates, instead of to zero, to

$$\left[\frac{\partial f}{\partial \dot{x}}\eta\right]_{t_1}^{t_2} \tag{B.59}$$

6) A function maps a number on a number. A functional maps a function on a number.

If we now require $\delta J = 0$ we obtain in addition to Eq. (B.58) also the boundary condition $\partial f/\partial \dot{x} = 0$ at $t = t_1$ and $t = t_2$.

Example B.13

Let us calculate the shortest distance between two points in a plane. An element of an arc length in a plane is

$$ds = \sqrt{dx^2 + dy^2}$$

and the total length of any curve between two points 1 and 2 is

$$I = \int_1^2 ds = \int_{x_1}^{x_2} f(\dot{y}, x)\, dx = \int_{x_1}^{x_2} \sqrt{1 + \left(\frac{dy}{dx}\right)^2}\, dx \quad \text{where} \quad \dot{y} = \frac{\partial y}{\partial x}$$

The condition that the curve is the shortest path is

$$\frac{\partial f}{\partial y} - \frac{d}{dx}\frac{\partial f}{\partial \dot{y}} = 0$$

Since $\dfrac{\partial f}{\partial y} = 0$ and $\dfrac{\partial f}{\partial \dot{y}} = \dot{y}/\sqrt{1+\dot{y}^2}$, we have $\dfrac{d}{dx}\left(\dfrac{\dot{y}}{\sqrt{1+\dot{y}^2}}\right) = 0$ or $\dfrac{\dot{y}}{\sqrt{1+\dot{y}^2}} = c$, where c is a constant. This solution holds if $\dot{y} = a$ where a is given by $a = c/(1 + c^2)^{1/2}$. Obviously, this is the equation for a straight line $y = ax + b$, where b is another constant of integration. The constants a and b are determined by the condition that the curve should go through (x_1, y_1) and (x_2, y_2).

B.11 Gamma Function

The *gamma function* is defined by

$$\Gamma(t) = \int_0^\infty x^{t-1} e^{-x}\, dx$$

For this function it generally holds that $\Gamma(t+1) = t\Gamma(t)$. For integer n it is connected to the more familiar *factorial function* $n!$ by $\Gamma(n+1) = n! = n(n-1)(n-2)\cdots(2)(1)$. Using $x = y^2$, we obtain

$$\Gamma(t) = 2\int_0^\infty y^{2t-1} e^{-y^2}\, dy$$

which results by setting $t = ½$ in $\Gamma\left(\frac{1}{2}\right) = 2\int_0^\infty e^{-y^2}\, dy = \sqrt{\pi}$. Consequently,

$$(-3/2)! = -2\cdot\pi^{1/2}, \quad \left(-\frac{1}{2}\right)! = \pi^{1/2}, \quad \left(\frac{1}{2}\right)! = \frac{1}{2}\cdot\pi^{1/2}, \quad \text{and} \quad (3/2)! = \frac{3}{4}\cdot\pi^{1/2}$$

B.12
Dirac and Heaviside Function

The *Dirac (delta) function* $\delta(x)$ is in one dimension defined by

$$\int_{-a}^{+a} \delta(x) f(x) \mathrm{d}x = f(0) \qquad \text{or} \qquad \int_{-a}^{+a} \delta(x-t) f(x) \mathrm{d}t = f(t) \tag{B.60}$$

where $a > 0$ and $a = \infty$ is included and which selects the value of a function f at the value of variable t from an integral expression. Alternatively, $\delta(x)$ is defined by

$$\delta(x-t) = \infty \quad \text{if } x = t \quad \text{and} \quad 0 \quad \text{otherwise,} \qquad \int_{-a}^{a} \delta(x-t) \mathrm{d}t = 1 \tag{B.61}$$

Some properties of the delta function for $a > 0$ are

$$\delta(-x) = \delta(x), \qquad x\delta(x) = 0, \qquad \delta(ax) = a^{-1}\delta(x) \qquad \text{and}$$

$$\delta(x^2 - a^2) = \frac{1}{2} a^{-1} [\delta(x-a) + \delta(x+a)]$$

The derivative[7] $\delta'(x)$ is related to $f'(x)$ as follows

$$\int_{-a}^{+a} \delta'(x-t) f(t) \mathrm{d}x = -f'(t) \equiv -\int_{-a}^{+a} \delta(x-t) f'(t) \mathrm{d}x$$

This leads further to

$$\delta'(x) = -\delta(-x) \qquad \text{and} \qquad x\delta'(x) = -\delta(x)$$

Related is the *Heaviside (step) function* $h(x)$, defined by

$$h(x-t) = 0 \quad \text{if } x < t \qquad \text{and} \qquad h(x-t) = 1 \quad \text{if } x > t \tag{B.62}$$

For a, $b > 0$ we have

$$\int_{-a}^{+b} h'(x) f(x) \mathrm{d}x = h(x) f(x)|_{-a}^{+b} - \int_{-a}^{+b} h(x) f'(x) \mathrm{d}x = f(b) - \int_{0}^{+b} f'(x) \mathrm{d}x = f(0)$$

so that the step function can be considered as the integral of the delta function.

B.13
Laplace and Fourier Transforms

The *Laplace transform* of a function $f(t)$, defined by

$$L[f(t)] = \hat{f}(s) = \int_{0}^{\infty} f(t) \exp(-st) \mathrm{d}t \tag{B.63}$$

transforms $f(t)$ into $\hat{f}(s)$ where s may be real or complex. The operation is linear, that is,

7) A prime denotes differentiation with respect to its argument.

$$L[c_1 f_1(t) + c_2 f_2(t)] = c_1 \hat{f}_1(s) + c_2 \hat{f}_2(s)$$

The *convolution theorem* states that the product of two Laplace transforms $L[f(t)]$ and $L[g(t)]$ equals the transform of the convolution of the functions $f(t)$ and $g(t)$

$$L[f(t)]L[g(t)] = L\left[\int_0^t f(t-\lambda)g(\lambda)\mathrm{d}\lambda\right] = L\left[\int_0^t f(\lambda)g(t-\lambda)\mathrm{d}\lambda\right] \tag{B.64}$$

Since the Laplace transform has the property

$$L\left[\frac{\mathrm{d}f(t)}{\mathrm{d}t}\right] = s\hat{f}(s) - f(0)$$

it can transform differential equations in t to algebraic equations in s. Generalization to higher derivatives is straightforward and reads

$$L\left[\frac{\mathrm{d}^n f(t)}{\mathrm{d}t^n}\right] = s^n \hat{f}(s) - s^{n-1} f(0) - \cdots - s f^{n-2}(0) - f^{n-1}(0)$$

Similarly, for integration it is found that

$$L\left[\int_0^t f(u)\mathrm{d}u\right] = \frac{1}{s}\hat{f}(s)$$

Some useful transforms are given in Table B.1.

The *Fourier transform* of a function $f(t)$ and its inverse are defined by

$$F[f(t)] = \tilde{f}(\omega) = N^{(-)}\int_{-\infty}^{+\infty} f(t)\exp(-\mathrm{i}\omega t)\mathrm{d}t \quad \text{and}$$

$$F^{-1}[\tilde{f}(\omega)] = f(t) = N^{(+)}\int_{-\infty}^{+\infty} \tilde{f}(\omega)\exp(\mathrm{i}\omega t)\mathrm{d}\omega \tag{B.65}$$

The normalization constants $N^{(-)}$ and $N^{(+)}$ can take any value as long as their product is $(2\pi)^{-1}$. If $N^{(-)} = N^{(+)} = (2\pi)^{-1/2}$ is taken, the transform is called symmetric. The Fourier transform is a linear operation for which the convolution theorem holds.

Table B.1 Laplace transform pairs.

$f(t)$	$\hat{f}(s)$	$f(t)$	$\hat{f}(s)$
1	$1/s$	t^n, $n > -1$	$\Gamma(n+1)/s^{n+1}$
a	a/s	$\exp(-at)$	$1/(s+a)$
$h(t)$	$1/s$	$t^n \exp(-at)$, $n = 0,1,\cdots$	$n!/(s+a)^{n+1}$
$h(t-a)$	$\exp(-as)/s$	$\sin at$	$a/(s^2+a^2)$
$\delta(t)$	1	$\cos at$	$s/(s^2+a^2)$
$\delta(t-a)$	$\exp(-as)$	$\sinh at$	$a/(s^2-a^2)$
t	$1/s^2$	$\cosh at$	$s/(s^2-a^2)$

Since for the delta function $\delta(t)$ it holds that

$$\tilde{\delta}(\omega) = \int_{-\infty}^{+\infty} \delta(t)\exp(-\mathrm{i}\omega t)\,\mathrm{d}t = 1$$

we have as a representation of the delta function

$$\delta(t) = (2\pi)^{-1}\int_{-\infty}^{+\infty} \exp(\mathrm{i}\omega t)\,\mathrm{d}\omega$$

Using

$$\int_{-a}^{+a} \exp(\mathrm{i}\omega t)\,\mathrm{d}\omega = (\mathrm{i}\omega)^{-1}\left[\exp(\mathrm{i}\omega a) - \exp(-\mathrm{i}\omega a)\right] = (2/\omega)\sin(\omega a)$$

we can represent $\delta(t)$ as

$$\delta(t) = \lim_{a\to\infty}\left[\sin(a\omega)/\pi\omega\right]$$

Similarly, for the 3D delta function $\delta(\boldsymbol{t})$ for a vector $\boldsymbol{t}$ we have

$$\tilde{\delta}(\boldsymbol{\omega}) = \int_{-\infty}^{+\infty} \delta(\boldsymbol{t})\exp(-\mathrm{i}\boldsymbol{\omega}\cdot\boldsymbol{t})\,\mathrm{d}\boldsymbol{t} = 1 \qquad \text{and} \qquad \delta(\boldsymbol{t}) = (2\pi)^{-3}\int_{-\infty}^{+\infty} \exp(\mathrm{i}\boldsymbol{\omega}\cdot\boldsymbol{t})\,\mathrm{d}\boldsymbol{\omega}$$

Finally, we note that by the Gauss theorem applied to a sphere with radius r

$$\int \nabla^2\left(\frac{1}{r}\right)\mathrm{d}V = \int \nabla\left(\frac{1}{r}\right)\cdot\boldsymbol{n}\,\mathrm{d}S = -\int \frac{\boldsymbol{r}\cdot\boldsymbol{n}}{r^3}\,\mathrm{d}S = -4\pi \qquad \text{or} \qquad \nabla^2\left(\frac{1}{r}\right) = -4\pi\delta(\boldsymbol{r})$$

since $\nabla^2(1/r) = 0$ for $r \neq 0$ and $\nabla^2(1/r) = \infty$ for $r = 0$. Therefore, we have

$$\begin{aligned}
\nabla^2 t^{-1} &= \nabla^2 F^{-1}\left[F[t^{-1}]\right] = \nabla^2(2\pi)^{-3}\int F[t^{-1}]\exp(\mathrm{i}\boldsymbol{\omega}\cdot\boldsymbol{t})\,\mathrm{d}\boldsymbol{\omega} \\
&= \frac{1}{(2\pi)^3}\int F[t^{-1}](-\omega^2)\exp(\mathrm{i}\boldsymbol{\omega}\cdot\boldsymbol{t})\,\mathrm{d}\boldsymbol{\omega} \\
&= \frac{-4\pi}{(2\pi)^3}\int \exp(\mathrm{i}\boldsymbol{\omega}\cdot\boldsymbol{t})\,\mathrm{d}\boldsymbol{\omega} \qquad \text{or} \qquad F[t^{-1}] = \frac{4\pi}{\omega^2}
\end{aligned}$$

Applying the inverse transform we obtain $F^{-1}\left[F[t^{-1}]\right] = F^{-1}\left[4\pi/\omega^2\right]$ or

$$\frac{1}{t} = \frac{1}{2\pi^2}\int_{-\infty}^{+\infty} \frac{1}{\omega^2}\exp(\mathrm{i}\boldsymbol{\omega}\cdot\boldsymbol{t})\,\mathrm{d}\boldsymbol{\omega}$$

B.14
Some Useful Integrals and Expansions

Several integrals and expansions, given without further comment, are useful throughout.

Integrals

$$n=0:\quad \int_0^\infty \exp(-ax^2)x^0\,dx = \frac{1}{2}\sqrt{\frac{\pi}{a}}$$

$$n=2:\quad \int_0^\infty \exp(-ax^2)x^2\,dx = \frac{1}{4}\sqrt{\frac{\pi}{a^3}}$$

$$n=4:\quad \int_0^\infty \exp(-ax^2)x^4\,dx = \frac{3}{8}\sqrt{\frac{\pi}{a^5}}$$

$$n \text{ even:}\quad \int_0^\infty \exp(-ax^2)x^n\,dx = 1\cdot 3\cdot 5\cdot\cdot(n-1)\frac{(\pi a)^{1/2}}{(a)^{\frac{1}{2}n+1}}$$

$$n=1:\quad \int_0^\infty \exp(-ax^2)x^1\,dx = \frac{1}{2a}$$

$$n=3:\quad \int_0^\infty \exp(-ax^2)x^3\,dx = \frac{1}{2a^2}$$

$$n=5:\quad \int_0^\infty \exp(-ax^2)x^5\,dx = \frac{1}{a^3}$$

$$n \text{ odd:}\quad \int_0^\infty \exp(-ax^2)x^n\,dx = \frac{\left[\frac{1}{2}(n-1)\right]!}{(2a)^{\frac{1}{2}(n+1)}}$$

Bi- and Multinomial Expansion

$$(1+x)^{-1} = 1 - x + x - x^3 + x^4 - \cdots$$

$$(1+x)^{-2} = 1 - 2x + 3x^2 - 4x^3 + 5x^4 - \cdots$$

$$(x+y)^n = x^n + nx^{n-1}y + n(n-1)x^{n-2}y^2/2! + n(n-1)(n-2)x^{n-3}y^3/3! + \cdots,$$
valid for $y^2 < x^2$

$$(x_1 + x_2 + \cdots + x_k)^n = \sum\nolimits_n \frac{k!}{n_1!n_2!\cdots n_k!} x_1^{n1} x_2^{n2} x_k^{nk} \quad \text{with} \quad n = \sum\nolimits_i n_i$$

Sine Cosine and Tangent

$$\sin x = x - x^3/3! + x^5/5! - x^7/7! + \cdots$$

$$\cos x = 1 - x^2/2! + x^4/4! - x^6/6! + \cdots$$

$$\tan x = x + x^3/3 + 2x^5/15 + 17x^7/315 + \cdots$$

Exponential and Logarithm

$$\exp x = 1 + x + x^2/2! + x^3/3! + x^4/4! + \cdots$$

$$\ln(1+x) = x - x^2/2 + x^3/3 - x^4/4 + \cdots$$

Reversion of series

$$y = b_1x + b_2x^2 + b_3x^3 + \cdots$$

$$x = c_1y + c_2y^2 + c_3y^3 + \cdots$$

$$b_1c_1 = 1, b_1^3c_2 = -b_2, b_1^5c_3 = 2b_2^2 - b_1b_3, \cdots$$

Euler McLaurin formula

Denoting the derivative $d^nf(x)/dx^n|_{x=a}$ by $f_n(a)$, the Euler–McLaurin expression reads

$$\sum_{i=a}^{i=b} f(i) = \int_a^b f(x)dx + \frac{1}{2}[f(a) + f(b)] + \sum_{n\geq 1}(-1)^n \frac{B_n}{(2n)!}[f_{2n-1}(a) - f_{2n-1}(b)]$$

with Bernoulli numbers $B_1 = 1/6$, $B_2 = 1/30$, $B_3 = 1/42$, $B_4 = 1/30, \ldots$.

Further Reading

Adams, R.A. (1995) *Calculus*, 3rd edn, Addison-Wesley, Don Mills, ON.

Jeffreys, H. and Jeffreys, B.S. (1972) *Methods of Mathematical Physics*, Cambridge University Press, Cambridge.

Kreyszig, E. (1988) *Advanced Engineering Mathematics*, 6th edn, John Wiley & Sons, Inc., New York.

Margenau, H. and Murphy, G.M. (1956) *The Mathematics of Physics and Chemistry*, 2nd edn, van Nostrand, Princeton, NJ.

Appendix C
The Lattice Gas Model

In this appendix we review briefly the lattice gas model [1], which is widely used in physical chemistry and known under different names in various fields, for example, as the regular solution model for liquid solutions, as the order–disorder model for alloys [2], or as the Ising model [3] for magnetic problems [4]. This broad range of applications explains its importance. For pure liquids the model was introduced by Frenkel [5] in 1932, and somewhat later independently by Eyring (and Cernuschi) [6]. It is one of the simplest models, if not the simplest model, showing a phase transition[1], but we should note here that, although the model looks deceptively simple, its solution is complex.

C.1
The Lattice Gas Model

The basic idea of the *lattice gas model* is that the volume available to the fluid is divided into *cells* of molecular size. Usually, for simplicity, the cells are arranged in a regular lattice with coordination number z, for example, a simple cubic (SC) lattice with $z = 6$, a body-centered cubic (BCC) lattice with $z = 8$, or a face- centered cubic (FCC) lattice with $z = 12$. Normally, these cells are occupied by one molecule at most, representing repulsion, and only nearest-neighbor cell attractive interactions are considered.

The lattice gas model can be solved exactly for 1D, and in this case the model does not show a phase transition. It can also be solved exactly for 2D without an applied field, as was achieved for the first time by Onsager [7]; in this case, the model does show a phase transition. A complete exact solution (i.e., with applied field) is neither known for 2D, nor for 3D, so that one has to approximate. We will not deal with the exact 2D solution [2] but discuss here first the (conventional) *zeroth approximation* or *mean field solution*[2] in terms of a model for liquid solutions, and thereafter the *first approximation* or *quasi-chemical solution*. In fact we have used these approximations in Chapters 8 and 11 already, although expressed in slightly different terms.

1) For a very readable introduction to phase transitions, see Ref. [8].
2) Also denoted as the Bragg-Williams approximation.

Liquid-State Physical Chemistry, First Edition. Gijsbertus de With.

C.2
The Zeroth Approximation or Mean Field Solution

For a lattice model of a liquid solution we distribute molecules of type 1 and type 2 over the lattice (without any further holes). In Section 8.3, we showed that if we have N_1 molecules of type 1 with N_2 molecules of type 2, so that the total number becomes $N = N_1 + N_2$, we obtain the general relations

$$N_{11} = \frac{1}{2}zN_1 - \frac{1}{2}N_{12} \quad \text{and} \quad N_{22} = \frac{1}{2}zN_2 - \frac{1}{2}N_{12} \tag{C.1}$$

while the energy is given by

$$E = \varepsilon_{11}N_{11} + \varepsilon_{22}N_{22} + \varepsilon_{12}N_{12} = \left(\varepsilon_{12} - \frac{1}{2}\varepsilon_{11} - \frac{1}{2}\varepsilon_{22}\right)N_{12} + \frac{1}{2}z\varepsilon_{11}N_1 + \frac{1}{2}z\varepsilon_{22}N_2 \tag{C.2}$$

Here N_{11} (N_{22}, N_{12}) represent the number of type 1–1 (2–2, 1–2) pairs, ε_{11} (ε_{22}, ε_{12}) the energy of the pair 1–1 (2–2, 1–2), and the quantity $w \equiv \varepsilon_{12} - ½\varepsilon_{11} - ½\varepsilon_{22}$ is denoted as *interchange energy*. To be consistent with Section 8.3 and Chapters 8 and 11, we use the notation $zX \equiv N_{12}$, so that

$$E = zwX + \frac{1}{2}z\varepsilon_{11}N_1 + \frac{1}{2}z\varepsilon_{22}N_2 \tag{C.3}$$

The canonical partition function is easily constructed and we obtain

$$Q(T,N,V) = q_1^{N_1}q_2^{N_2}\sum\nolimits_{\{X\}}\exp\left[-\beta\left(zwX + \frac{1}{2}z\varepsilon_{11}N_1 + \frac{1}{2}z\varepsilon_{22}N_2\right)\right] \tag{C.4}$$

Recognizing that the term $B \equiv ½(z\varepsilon_{11}N_1 + z\varepsilon_{22}N_2)$ is constant, this reduces to

$$Q(T,N,V) = q_1^{N_1}q_2^{N_2}\exp(-\beta B)\sum\nolimits_{\{X\}}\exp(-\beta zwX) \tag{C.5}$$

where the sum $\Sigma_{\{X\}}$ is over all configurations having X type 1–2 pairs and q_j is the internal partition function of molecule j. Equivalently, we write

$$Q(T,N,V) = q_1^{N_1}q_2^{N_2}\exp(-\beta B)\sum\nolimits_{X} g(N_1,N_2,X)\exp(-\beta zwX) \tag{C.6}$$

where the sum Σ_X is now over the number of configurations having X type 1–2 pairs with the degeneracy factor $g(N_1,N_2,X)$ for each configuration labeled with (N_1,N_2,X).

In the zeroth approximation we consider that molecules are distributed randomly over the lattice, in spite of their interaction with others, which we restrict to their z nearest-neighbors. For the number of type 1–2 pairs zX we then get

$$zX = N_1 \cdot z\frac{N_2}{N} = N_1 \cdot z\frac{N-N_1}{N} = zN_1 - z\frac{N_1^2}{N} \quad \text{or equivalently} \tag{C.7}$$

$$zX = N_1 \cdot z(N_2/N) = N(N_1/N)\cdot z(N_2/N) = zNx_1x_2 \tag{C.8}$$

The first expression was used in Chapter 8, but here we will use the second expression. The next step is to calculate $g(N_1,N_2,X)$. Considering the distribution N_1 molecules of type 1 and N_2 of type 2 over N cells, one obtains for the number of configurations

$$g(N_1, N_2, X) = N!/(N_1!N_2!) \tag{C.9}$$

Therefore, the partition function becomes

$$Q(T,N,V) = q_1^{N_1} q_2^{N_2} \exp(-\beta B)\frac{N!}{N_1!N_2!}\exp(-\beta Nzwx_1x_2) \tag{C.10}$$

so that for the Helmholtz energy the result is

$$F = -kT\ln Q(T,N,V) = -kT\ln(q_1^{N_1}q_2^{N_2}) + B - kT\ln\frac{N!}{N_1!N_2!} + Nzwx_1x_2 \tag{C.11}$$

$$= -kT\ln(q_1^{N_1}q_2^{N_2}) + B + NkT(x_1\ln x_1 + x_2\ln x_2) + Nzwx_1x_2 \tag{C.12}$$

Consequently, the mixing Helmholtz energy reads

$$\Delta_{\text{mix}}F = Nzwx_1x_2 + NkT(x_1\ln x_1 + x_2\ln x_2) \tag{C.13}$$

As a check, one can calculate the internal energy of mixing $\Delta_{\text{mix}}U$ energy from $\Delta_{\text{mix}}F$ via $U = \partial(F/T)/\partial(1/T)$ which leads, as expected, to

$$\Delta_{\text{mix}}U = Nzwx_1x_2 \tag{C.14}$$

The coexistence line between a phase-separated system and a homogeneous system can be calculated from $\partial\Delta_{\text{mix}}G/\partial x_2 = \partial\Delta_{\text{mix}}F/\partial x_2 = 0$. This leads to

$$\ln(x_2/x_1) = \beta zw(2x_2 - 1) \tag{C.15}$$

Using the transformation $s = 2x_2 - 1$, the final expression is

$$\beta w = s^{-1}\ln[(1+s)/(1-s)] = 2s^{-1}\tanh^{-1}s \qquad \text{or} \qquad s = \tanh\left(\frac{1}{2}\beta zws\right) \tag{C.16}$$

The *critical temperature* T_{cri} is the temperature where the molecules start to order, and we determine T_{cri} by considering that $\tanh(\frac{1}{2}\beta zws)$ is a continuously increasing function of s passing through $s = 0$. Hence, there is always the solution $s = 0$ but for $\text{d}[\tanh(\frac{1}{2}\beta zws)]/\text{d}s|_{s=0} > 1$, two other solutions occur (Figure C.1). Because this derivative equals $\frac{1}{2}\beta zw$, we find $T_{\text{cri}} = zw/2k$. The composition as a function of T can be found by solving Eq. (C.16), either numerically or graphically, and the result is shown in Figure C.2a.

The associated heat capacity C_V can be found from $C_V = \partial U/\partial T$, with $U = -Q_N^{-1}\partial Q_N/\partial\beta$ resulting in (after some calculation and plotted in Figure C.2b)

$$C_V = \frac{2Nks^2(T_{\text{cri}}/T)^2}{\cosh^2(sT_{\text{cri}}/T) - (T_{\text{cri}}/T)} \tag{C.17}$$

Note that C_V shows a finite jump at $T = T_{\text{cri}}$ with a maximum of $3Nk$.

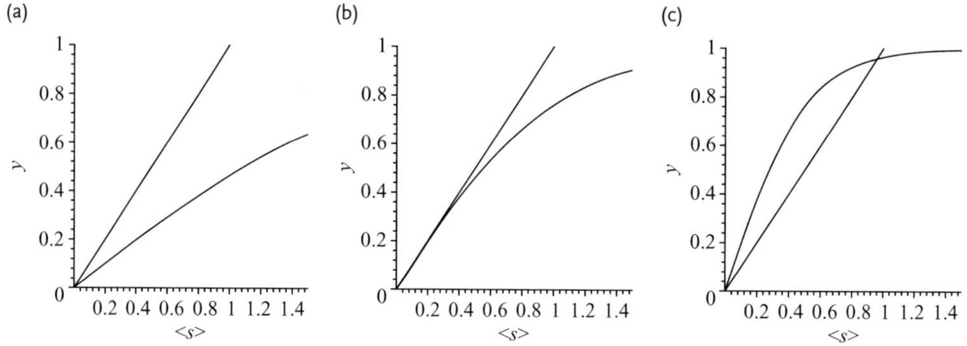

Figure C.1 The curves $y = s$ and $y = \tanh(as)$ with (a) $a = 0.5$, (b) $a = 1.0$, and (c) $a = 2.0$, showing for the zeroth order solution, respectively, the homogeneous state ($T > T_{cri}$), the critical state ($T = T_{cri}$), and the phase-separated state ($T < T_{cri}$). All curves are plotted for $s \geq 0$ only.

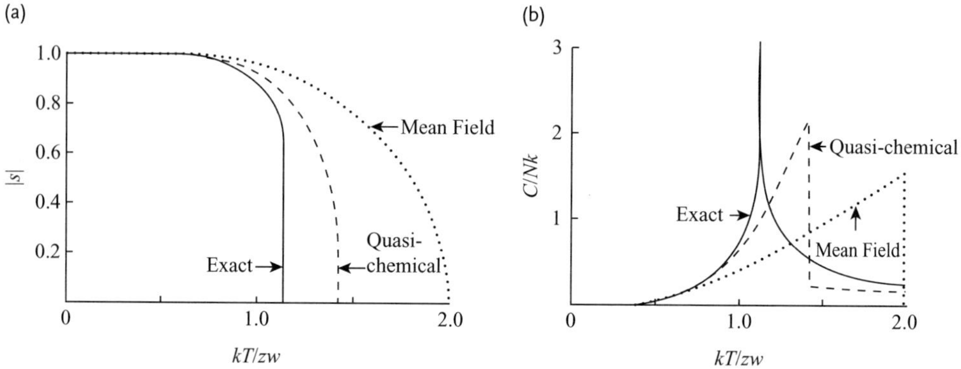

Figure C.2 The composition $|s| = |2x_2 - 1|$ where phase separation occurs (a) and the associated heat capacity C/Nk (b) as a function of kT/zw for the 2D square lattice, showing the zeroth order (mean field), the first order (quasi-chemical), and the exact solution.

C.3 The First Approximation or Quasi-Chemical Solution

In the zeroth approximation, single molecules are distributed independently and randomly over the lattice. Improvement is possible by distributing randomly pairs of molecules instead of single molecules. This solution is known as the *first approximation* or *quasi-chemical solution*[3].

3) Also denoted as the Bethe–Peierls approximation. For further details, see Refs [9] and [10].

C.3.1
Pair Distributions

Let us consider an independent distribution of pairs 1–1, 1–2, 2–1 and 2–2 over the lattice. Any such a distribution is characterized by the numbers N_1, N_2 and X, and this leads in the usual way to the number of possible configurations

$$h(N_1, N_2, X) = \frac{\left(\frac{1}{2}zN\right)!}{\left[\frac{1}{2}z(N_1 - X)\right]!\left(\frac{1}{2}zX\right)!\left(\frac{1}{2}zX\right)!\left[\frac{1}{2}z(N_2 - X)\right]!} \tag{C.18}$$

where we have used a separate entry for the 1–2 and 2–1 pairs in order to avoid the use of a symmetry number. This result cannot be correct, as the sum over all configurations zX should yield $N!/N_1!N_2!$. However, we can – at least approximately – remedy this defect by introducing a normalization factor $c(N_1,N_2)$, and write for

$$g(N_1, N_2, X) = c(N_1, N_2)h(N_1, N_2, X) \quad \text{so that} \tag{C.19}$$

$$G(N_1, N_2) = \sum\nolimits_X g(N_1, N_2, X) = c(N_1, N_2)\sum\nolimits_X h(N_1, N_2, X) \tag{C.20}$$

We evaluate the sum G by taking its largest term (see Justification 5.2), and this term, labeled by X^*, is obtained by setting $\partial G/\partial X = 0$. The result is, after some straightforward manipulation,

$$X^{*2} = (N_1 - X^*)(N_2 - X^*) \qquad \text{or} \qquad X^* = N_1N_2\,/\,(N_1 + N_2) \tag{C.21}$$

This implies that the sum G for X^*, labeled $G^*(N_1,N_2)$, becomes

$$G^*(N_1, N_2) = \frac{c(N_1, N_2)\left(\frac{1}{2}zN\right)!}{\left[\frac{1}{2}z(N_1 - X^*)\right]!\left(\frac{1}{2}zX^*\right)!\left(\frac{1}{2}zX^*\right)!\left[\frac{1}{2}z(N_2 - X^*)\right]!} \tag{C.22}$$

If we sum $g(N_1,N_2,X)$ over all values of X, we must regain the total number of ways of placing N_1 molecules of type 1 and N_2 of type 2 over $N = N_1 + N_2$ sites. In other words, we also have $G = N!/N_1!N_2!$ and we obtain for $c(N_1,N_2)$

$$c(N_1, N_2) = N!/N_1!N_2!h(n_1, n_2, X^*) \tag{C.23}$$

Therefore, $g(N_1,N_2,X) = N!h(N_1,N_2,X)/N_1!N_2!h(n_1,n_2,X^*)$ or, in full

$$g(N_1, N_2, X) = \frac{N!}{N_1!N_2!}\,\frac{\left[\frac{1}{2}z(N_1 - X^*)\right]!\left(\frac{1}{2}zX^*\right)!\left(\frac{1}{2}zX^*\right)!\left[\frac{1}{2}z(N_2 - X^*)\right]!}{\left[\frac{1}{2}z(N_1 - X)\right]!\left(\frac{1}{2}zX\right)!\left(\frac{1}{2}zX\right)!\left[\frac{1}{2}z(N_2 - X)\right]!} \tag{C.24}$$

The configurational partition sum Q becomes

$$Q = \sum_X g(N_1, N_2, X)\exp[-\beta(\tfrac{1}{2}z\varepsilon_{11}N_1 + \tfrac{1}{2}z\varepsilon_{22}N_2 + zwX)] \tag{C.25}$$

with, as usual, $\beta = 1/kT$. The maximum term trick is now played again and hence we seek the maximum term of Q, labeled $\bar{X}$, by setting $\partial Q/\partial X = 0$, or equivalently, by setting $\partial \ln Q/\partial X = 0$. After a straightforward calculation we obtain

$$\bar{X}^2 = (N_1 - \bar{X})(N_2 - \bar{X})\exp(-2\beta w) \tag{C.26}$$

For an explicit solution we define $\eta^2 \equiv \exp(2\beta w)$ and the parameter α by

$$\bar{X} = \frac{N_1 N_2}{N_1 + N_2}\frac{2}{\alpha + 1} \tag{C.27}$$

so that we can transform Eq. (C.26) to $\alpha^2 - (1 - 4x_1x_2) = 4\eta^2x_1x_2$ (or $\alpha^2 - (1 - 2x_2)^2 = 4\eta_2x_1x_2$), which has the solution $\alpha = [1 + 4x_1x_2(\eta^2 - 1)]^{1/2}$ with $x_j = N_j/(N_1 + N_2)$. The energy of mixing becomes $\Delta_{\text{mix}}U = 2x_1x_2zwN/(\alpha + 1)$. Note that for $\alpha = 1$ we regain the zeroth approximation.

C.3.2
The Helmholtz Energy

The configurational Helmholtz energy F is given by $F = -kT\ln Q$ and becomes

$$\begin{aligned} F &= -kT\ln g(N_1, N_2, \bar{X}) + \left(\frac{1}{2}N_1z\varepsilon_{11} + \frac{1}{2}N_2z\varepsilon_{22} + \bar{X}zw\right) \\ &= -kT\ln\left[\frac{N!h(N_1, N_2, \bar{X})}{N_1!N_2!h(N_1, N_2, X^*)}\right] + \left(\frac{1}{2}N_1z\varepsilon_{11} + \frac{1}{2}N_2z\varepsilon_{22} + \bar{X}zw\right) \end{aligned} \tag{C.28}$$

The evaluation of F is complex, and it is convenient to calculate first the chemical potential $\mu_j = \partial F/\partial N_j$. If this calculation is handled without thinking, it is a straightforward but tedious task. However, if we realize that $\bar{X}$ is obtained from $\partial F/\partial \bar{X} = 0$, the evaluation becomes straightforward. This implies that all terms resulting from $\partial F/\partial \bar{X}$ taken together cancel. Moreover, X^* is obtained from $\partial \ln g/\partial X = 0$, implying that all terms from $\partial \ln g/\partial X^*$ taken together also cancel. This renders the differentiation of F a relatively simple task, leading to

$$\beta\mu_j = \beta\partial F/\partial N_j = \ln\left(\frac{N_j}{N_1 + N_2}\right) + \frac{1}{2}z\ln\left(\frac{N_j - \bar{X}}{N_\alpha - X^*}\right) + \frac{1}{2}z\varepsilon_{jj} \tag{C.29}$$

To obtain $\mu_j - \mu_j^\circ$, we realize that μ_j° is obtained from Eq. (C.28), again via $\mu_j = \partial F/\partial N_j$, but now with $N_2 = 0$, leading to $\mu_j^\circ = \tfrac{1}{2}z\beta\varepsilon_{jj}$. Writing $F = N_1\mu_1 + N_2\mu_2 = N(x_1\mu_1 + x_2\mu_2)$, we obtain for the molar Helmholtz energy of mixing

$$\frac{\Delta_{\text{mix}}F_{\text{m}}}{RT} = x_1\ln x_1 + x_2\ln x_2 + \frac{1}{2}z\left[x_1\ln\left(\frac{N_1 - \bar{X}}{N_1 - X^*}\right) + x_2\ln\left(\frac{N_2 - \bar{X}}{N_2 - X^*}\right)\right] \tag{C.30}$$

Using the (absolute) activity $\lambda_j = \exp(\beta\mu_j)$ and, as before, assuming that the activity can be replaced by the partial pressure, the activity coefficient γ becomes

$$\lambda_j/(\lambda_j^\circ x_j) = P_j/(P_j^\circ x_j) = \gamma_j = [(N_j - \bar{X})/(N_j - X^*)]^{\frac{1}{2}z} \tag{C.31}$$

In all these expressions X^* and $\bar{X}$ are given by Eqs (C.21) and (C.27), respectively.

C.3.3
Critical Mixing

The above expression for $\Delta_{mix}F_m$ can be made more explicit by writing the various contributions in terms of $\alpha = [1 + 4x_1x_2(\eta^2 - 1)]^{1/2}$ with $\eta^2 \equiv \exp(2\beta w)$, $x_1 = N_1/(N_1 + N_2)$ and $x_2 = N_2/(N_1 + N_2)$. We find

$$\bar{X}/N = 2x_1x_2/(\alpha+1) \qquad X^*/N = x_1x_2 \tag{C.32}$$

$$(N_1 - \bar{X})/N = x_1(\alpha+1-2x_2)/(\alpha+1) \qquad (N_1 - X^*)/N = x_1^2 \tag{C.33}$$

$$(N_2 - \bar{X})/N = x_2(\alpha-1+2x_2)/(\alpha+1) \qquad (N_2 - X^*)/N = x_2^2 \tag{C.34}$$

so that $\Delta_{mix}F_m$ becomes

$$\frac{\Delta_{mix}F_m}{RT} = x_1\ln x_1 + x_2\ln x_2 + \frac{1}{2}z\left[x_1\ln\left(\frac{\alpha+1-2x_2}{x_1(\alpha+1)}\right) + x_2\ln\left(\frac{\alpha-1+2x_2}{x_2(\alpha+1)}\right)\right] \tag{C.35}$$

A similar substitution can be made for $\lambda_j = (l_j^\circ/P_j^\circ)P_j$ and P_j, leading to

$$\frac{P_1}{P_1^\circ x_1} = \gamma_1 = \left[\frac{\alpha+1-2x_2}{x_1(\alpha+1)}\right]^{\frac{1}{2}z} \quad \text{and} \quad \frac{P_2}{P_2^\circ x_2} = \gamma_2 = \left[\frac{\alpha-1+2x_2)}{x_2(\alpha+1)}\right]^{\frac{1}{2}z} \tag{C.36}$$

The critical point, due to symmetry, still occurs at $x_1 = x_2 = ½$. To obtain the expression for T_{cri}, we first derive the expression for the coexistence line, determined by $P_1/P_1^\circ = P_2/P_2^\circ$. Upon substitution of the expression for P_j/P_j° we obtain

$$(\alpha-1+2x_2)/(\alpha+1-2x_2) = (x_2/x_1)^{(z-2)/z} \equiv r^{(z-2)/z} \equiv \gamma \tag{C.37}$$

where the molecular ratio $r \equiv x_2/x_1$ and the abbreviation γ are introduced. Solving α yields $\alpha/(1 - 2x_2) = (1 + \gamma)/(1 - \gamma)$, and thus leads to

$$\alpha-1+2x_2 = 2\gamma(1-2x_2)/(1-\gamma) \quad \text{and} \quad \alpha+1-2x_2 = 2(1-2x_2)/(1-\gamma) \tag{C.38}$$

Multiplying these two expressions leads to

$$\alpha^2 - (1-2x_2)^2 = 4\gamma(1-2x_2)^2/(1-\gamma)^2 \tag{C.39}$$

Comparing with the transform of Eq. (C.26), $\alpha^2 - (1 - 2x_2)^2 = 4\eta^2x_1x_2$, we obtain

$$\eta = (\gamma/x_1x_2)^{1/2}(1-2x_2)/(1-\gamma) \quad \text{or} \quad \eta = e^{\beta w} = (1-r)/(r^{1/z} - r^{(z-1)/z}) \tag{C.40}$$

Note that if $r = r_1$ is a solution of Eq. (C.36), the other one is $r_2 = 1/r_1$.

The critical temperature T_{cri} is now obtained by putting $r = 1 + \delta$ in Eq. (C.40), expanding in powers of δ and taking the limit $\delta \to 0$. We thus obtain

$$\eta_{cri} = \exp(w / kT_{cri}) = z / (z-2) \quad \text{or} \quad zw / kT_{cri} = z \ln[z / (z-2)] \tag{C.41}$$

For the SC lattice ($z = 6$) we have $zw/kT_{cri} = 2.433$, while for the BCC lattice ($z = 8$) and FCC lattice ($z = 12$) we obtain $zw/kT_{cri} = 2.301$ and $zw/kT_{cri} = 2.188$, respectively. Note that if we take $z = \infty$, we regain the zeroth approximation $zw/kT_{cri} = 2$.

Calculating, as before, the internal energy via $U = \partial(F/T)/\partial(1/T)$ is considerably more complex as for zeroth approximation and leads to

$$\Delta_{mix}U = 2Nzwx_1x_2 / (\alpha + 1) \tag{C.42}$$

So far w was considered to be temperature independent but considering w as a temperature-dependent parameter presents no specific problems, and by using the definition $u \equiv w - T(\partial w/\partial T)$, one can show that in this case one obtains $\Delta_{mix}U = 2Nzux_1x_2/(\alpha + 1)$.

C.4 Final Remarks

As stated in Section C.2, only a (partial) exact solution exists for 2D lattices. For a square lattice this solution reads $zw/kT_{cri} = 3.523$, to be compared with $zw/kT_{cri} = 2.773$ for the first and $zw/kT_{cri} = 2.00$ for the zeroth approximation. The results from the exact solution for s and C_V are shown, as are the results for the zeroth and first approximation, in Figure C.2. From these plots it becomes clear that considerable improvement is obtained by using the first approximation, but also that the exact solution is still far from being approached. Finally, the higher the number of dimensions, the better the mean field solution (see Chapter 16), which implies that the situation for 3D is somewhat better than for 2D.

References

1 Yang, C.N. and Lee, T.D. (1952) *Phys. Rev.*, **87**, 404 and 410.

2 Huang, K. (1987) *Statistical Mechanics*, 2nd edn, John Wiley & Sons, Inc., New York.

3 Ising, E. (1925) *Z. Phys.*, **31**, 253.

4 Brush, S.G. (1967) *Rev. Mod. Phys.*, **39**, 883.

5 Frenkel, J. (1946) *Kinetic Theory of Liquids*, Oxford University Press, Oxford. See also Dover Publishers reprint (1953).

6 (a) Eyring, H. (1936) *J. Chem. Phys.*, **4**, 283; (b) Cernuschi, F. and Eyring, H. (1939) *J. Chem. Phys.*, **7**, 547.

7 Onsager, L. (1944) *Phys. Rev.*, **65**, 117.

8 Stanley, H.E. (1971) *Introduction to Phase Transitions and Critical Phenomena*, Oxford University Press, New York.

9 Fowler, R.H. and Guggenheim, E.A. (1939) *Statistical thermodynamics*, Cambridge University Press.

10 Guggenheim, E.A. (1952) *Mixtures*, Clarendon, Oxford.

Appendix D
Elements of Electrostatics

In this appendix we review briefly some aspects of electrostatics that are relevant to our discussion on polar liquids. Free use has been made of the literature indicated in the bibliography.

D.1
Coulomb, Gauss, Poisson, and Laplace

In vacuum a charge q_1 at a distance $r = |\boldsymbol{r}|$ of another charge q_2 experiences a force f given by *Coulomb's law*, using $\mathcal{C} \equiv (4\pi\varepsilon_0)^{-1}$ with ε_0 the permittivity of vacuum,

$$f = \mathcal{C}q_1q_2\boldsymbol{r}/r^3 \tag{D.1}$$

The *field* E is the force a test charge of unit charge experiences from a charge q at position r so that we have

$$\boldsymbol{E} = \mathcal{C}q\boldsymbol{r}/r^3 \tag{D.2}$$

The energy increment $\mathrm{d}w_{12}(r)$ of a charge q_1 in the field of a charge q_2 is given by $dw_{12}(r) = -q_1\boldsymbol{E}_2 \cdot \mathrm{d}\boldsymbol{r}$ and accordingly the (potential or) *Coulomb energy* reads

$$w_{12}(r) = -q_1\int_\infty^r \boldsymbol{E}_2 \cdot \mathrm{d}\boldsymbol{r} = -\mathcal{C}q_1q_2\int_\infty^r r^{-3}\boldsymbol{r} \cdot \mathrm{d}\boldsymbol{r} = \mathcal{C}q_1q_2r^{-1} \equiv q_1\phi_2 \tag{D.3}$$

where the *potential* $\phi = \mathcal{C}q/r$, as defined by $\boldsymbol{E} \equiv -\nabla\phi$, is introduced. As conventional, the force is thus $\boldsymbol{f} = -\nabla w$.

A pair of opposite and equal charges q_1 and $q_2 = -q_1$, separated by a (small) vector $\boldsymbol{d}$ pointing from q_2 to q_1, is called a *dipole*, characterized by dipole moment $\boldsymbol{\mu} = q_1\boldsymbol{d}$. The potential of such a dipole fixed in space at large distance r from the center of the dipole is (Figure D.1)

$$\phi = \phi_1 + \phi_2 = \mathcal{C}q_1\Big/\left|\boldsymbol{r} - \frac{1}{2}\boldsymbol{d}\right| + \mathcal{C}q_2\Big/\left|\boldsymbol{r} + \frac{1}{2}\boldsymbol{d}\right| \cong \mathcal{C}\,\boldsymbol{\mu} \cdot \boldsymbol{r}/r^3 \tag{D.4}$$

The field E is then given by

$$\boldsymbol{E} = -\nabla\phi = -\mathcal{C}\,\boldsymbol{\mu} \cdot \nabla\frac{\boldsymbol{r}}{r^3} = -\frac{\mathcal{C}\,\boldsymbol{\mu}}{r^3}\nabla\boldsymbol{r} - \boldsymbol{\mu} \cdot \boldsymbol{r}\nabla\frac{1}{r^3} = \frac{\mathcal{C}}{r^5}[3(\boldsymbol{\mu} \cdot \boldsymbol{r})\boldsymbol{r} - (\boldsymbol{r} \cdot \boldsymbol{r})\boldsymbol{\mu}] \tag{D.5}$$

Liquid-State Physical Chemistry, First Edition. Gijsbertus de With.

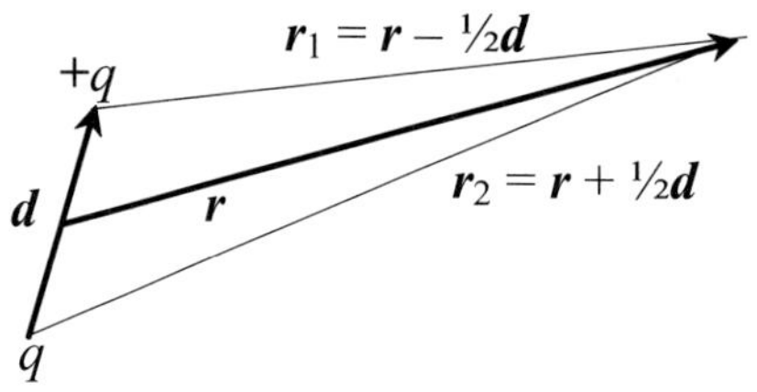

Figure D.1 The potential of a dipole $\mu = qd$ at distance r.

The Coulomb energy of a dipole in a field with potential ϕ can be calculated to be $w = \mu \cdot \nabla\phi = -\mu \cdot E$ (see Chapter 3).

Consider now an arbitrary closed surface S with surface elements $\mathrm{d}\boldsymbol{S} = \boldsymbol{n}\mathrm{d}S$, located at position r from the origin, where $\boldsymbol{n}$ is the positive outward normal vector of the surface element and $\mathrm{d}S$ its area. Because the scalar product $r \cdot \mathrm{d}S/r$ is the projection of $\mathrm{d}S$ on a sphere of radius r, the product $r \cdot \mathrm{d}S/r^3$ is the projection on a unit sphere centered at the origin. Hence, if S encloses the origin, the closed surface integral

$$\int_S \boldsymbol{E} \cdot \mathrm{d}\boldsymbol{S} = -\int_S \nabla\phi \cdot \mathrm{d}\boldsymbol{S} = -\int_S \mathcal{C}q\nabla\frac{1}{r} \cdot \mathrm{d}\boldsymbol{S} = \mathcal{C}q\int_S \frac{1}{r^3}\boldsymbol{r} \cdot \mathrm{d}\boldsymbol{S}$$

$$= \mathcal{C}q\int_S \frac{1}{r^3}\boldsymbol{r} \cdot \boldsymbol{n}\mathrm{d}S = \mathcal{C}q\int_0^1 \frac{r}{r^3}4\pi r^2 \mathrm{d}r = 4\pi\mathcal{C}q = \frac{q}{\varepsilon_0} \tag{D.6}$$

If S does not enclose the origin, $\boldsymbol{r}$ cuts S at two points, the contributions from these two intersections cancel and the integral is zero. It follows that the integral $\int_S \boldsymbol{E} \cdot \mathrm{d}\boldsymbol{S}$ is q/ε_0 if S does contain q and zero otherwise. For a general charge distribution ρ within S we obtain *Gauss' law*

$$\int_S \boldsymbol{E} \cdot \mathrm{d}\boldsymbol{S} = \varepsilon_0^{-1}\int \rho \mathrm{d}V \tag{D.7}$$

In order to go from this global description to a local one, we use the Gauss divergence theorem and find *Poisson's equation*

$$\int \boldsymbol{E} \cdot \mathrm{d}\boldsymbol{S} = \int \nabla \cdot \boldsymbol{E} \mathrm{d}V = \varepsilon_0^{-1}\int \rho \mathrm{d}V \quad \text{or} \quad \nabla \cdot \boldsymbol{E} = \varepsilon_0^{-1}\rho \tag{D.8}$$

Using $E = -\nabla\phi$ we obtain

$$\nabla^2\phi = -\varepsilon_0^{-1}\rho \tag{D.9}$$

reducing to *Laplace's equation* $\nabla^2\phi = 0$ if $\rho = 0$, which is the usual case for a dielectric material. For a distribution of volume charge ρ and surface charge σ, the potential becomes

$$\phi(\boldsymbol{r}) = \mathcal{C}\int_\xi \frac{\rho(\xi)}{|\boldsymbol{r}-\xi|}\mathrm{d}V + \mathcal{C}\int_\xi \frac{\sigma(\xi)}{|\boldsymbol{r}-\xi|}\mathrm{d}S \tag{D.10}$$

Molecules are polarizable and therefore a field induces in a molecule a dipole moment $\mu_{ind} = \alpha E$, where the proportionality factor α is a material-dependent parameter called the *polarizability*. For macroscopic matter this occurs also, and we have the *polarization* $P = \varepsilon_0 \chi E$ with the *susceptibility* χ. Since P represents a volumetric distribution of induced dipoles we have for the associated potential

$$\begin{aligned}\phi(r) &= -\int_\xi \frac{P(\xi)\cdot x}{x^3}\mathrm{d}V = \int_\xi P(\xi)\cdot\nabla x^{-1}\mathrm{d}V = \int_\xi \nabla\cdot[P(\xi)r^{-1}]\mathrm{d}V - \int_\xi r^{-1}\nabla\cdot P\mathrm{d}V \\ &= \int_\xi x^{-1}P(\xi)\cdot\mathrm{d}S - \int_\xi x^{-1}\nabla\cdot P(\xi)\mathrm{d}V \end{aligned} \tag{D.11}$$

where $x = r - \xi$ and $x = |x|$ is used. The second line states that the potential is analogous to one caused by a surface charge $n\cdot P$ and a volume charge $-\nabla\cdot P$. Using the same reasoning as before, and going again from a global to a local description via the Gauss divergence theorem using the volume charge density $\rho - \nabla\cdot P$, we have within the dielectric, since the surface charges do not enter Poisson's equation,

$$\nabla\cdot E = \varepsilon_0^{-1}(\rho - \nabla\cdot P) \quad \text{or} \quad \nabla\cdot(\varepsilon_0 E + P) = \rho \quad \text{or} \quad \nabla\cdot D = \rho \tag{D.12}$$

with the *dielectric displacement* $D \equiv \varepsilon_0 E + P = (1+\chi)\varepsilon_0 E \equiv \varepsilon_r\varepsilon_0 E \equiv \varepsilon E$, thus introducing the *relative permittivity* (or *dielectric constant*) $\varepsilon_r = 1 + \chi$, and *permittivity* $\varepsilon \equiv \varepsilon_r\varepsilon_0$. Outside the dielectric, $P = 0$ and $\nabla\cdot D = \rho$ reduces to $\varepsilon_0\nabla\cdot E = \rho$.

So, the dielectric displacement satisfies the condition $\nabla\cdot D = \rho$, and $\varepsilon_0\nabla\cdot\varepsilon_r\nabla\phi = -\rho$ is the governing equation for the resulting potential ϕ. From this equation we can derive the requirement on the continuity of D across the boundary between media by writing the integral equivalent of $\nabla\cdot D = \rho$ to obtain

$$\int_V \nabla\cdot D\mathrm{d}V = \int_S D\cdot\mathrm{d}S = \int_V \rho\mathrm{d}V \tag{D.13}$$

Applying this relation to an infinitesimally small pill-box-shaped volume with a bottom and top surface with area A parallel to the boundary on each side of the boundary (Figure D.2), we have

$$A[(D_n)_2 - (D_n)_1] = \int_V \rho\mathrm{d}V \tag{D.14}$$

where D_n is the component of D normal to the boundary (with the direction from medium 1 to 2 as positive). If the charge density at the boundary remains

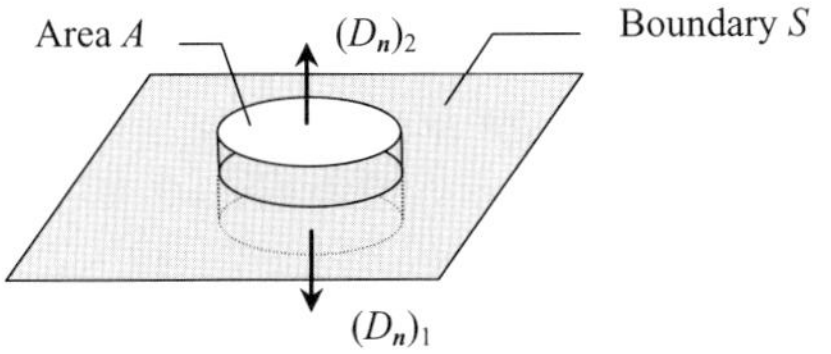

Figure D.2 The continuity of D at the boundary S.

finite, the integral tends to zero when volume of the pill-box V goes to zero. Hence,

$$(D_n)_2 - (D_n)_1 = 0 \tag{D.15}$$

If a surface charge is present we obtain

$$(D_n)_2 - (D_n)_1 = \sigma \tag{D.16}$$

In conclusion, the normal component D_n of $\boldsymbol{D}$ across a boundary between two dielectrics is continuous in the absence of surface charge, but if a surface charge is σ is present this component jumps by σ.

D.2
A Dielectric Sphere in a Dielectric Matrix

To illustrate the use of the above theory, we now calculate the potential field and some associated relations outside (domain 1) and inside (domain 2) a sphere with radius a and relative permittivity ε_2 embedded in a matrix with relative permittivity ε_1 to which a field $\boldsymbol{E}$ along the z-axis is applied by a set of external charges that would be uniform if $\varepsilon_1 = \varepsilon_2$. Generally, it is most convenient to use a coordinate system associated with the boundary conditions and for spherical entities these are the spherical coordinates r, θ, φ defined in terms of the Cartesian coordinates x, y, z by

$$x = r\sin\theta\cos\varphi, \quad y = r\sin\theta\sin\varphi, \quad z = r\cos\theta \tag{D.17}$$

In these coordinates the Laplace operator operating on a scalar reads

$$\nabla^2(\cdot) = \frac{1}{r^2}\frac{\partial}{\partial r}\left(r^2\frac{\partial(\cdot)}{\partial r}\right) + \frac{1}{r\sin\theta}\frac{\partial}{\partial\theta}\left(\sin\theta\frac{\partial(\cdot)}{\partial\theta}\right) + \frac{1}{r^2\sin^2\theta}\frac{\partial^2(\cdot)}{\partial\varphi^2} \tag{D.18}$$

Since we interested in an axially symmetric system, we have $\phi = \phi(r,\theta)$ and use a trial solution in the form $\phi(r,\theta) = R(r)\Theta(\theta)$ from which we obtain

$$\frac{1}{R}\frac{\partial}{\partial r}\left(r^2\frac{\partial R}{\partial r}\right) + \frac{1}{\Theta\sin\theta}\frac{\partial}{\partial\theta}\left(\sin\theta\frac{\partial\Theta}{\partial\theta}\right) = 0 \tag{D.19}$$

As the two parts involve different spatial variables, this expression can be only satisfied if each part equals a constant, say λ, and therefore

$$\frac{\partial}{\partial r}\left(r^2\frac{\partial R}{\partial r}\right) - \lambda R = 0 \quad \text{and} \quad \frac{1}{\sin\theta}\frac{\partial}{\partial\theta}\left(\sin\theta\frac{\partial\Theta}{\partial\theta}\right) + \lambda\Theta = 0 \tag{D.20}$$

By substituting $x = \cos\theta$ and $\Theta(\theta) = y(x)$, the equation for Θ transforms to *Legendre's equation*, of which the generic solutions are $y(x) = A_\nu P_\nu(x) + B_\nu Q_\nu(x)$, where P_ν and Q_ν are the Legendre functions of the first and second type. The label ν is determined by $\lambda = \nu(\nu + 1)$, and A_ν and B_ν are arbitrary constants. The full generic solution must be used when the z-axis ($x = \cos\theta = \pm 1$) is excluded from the region where the potential has to be calculated. In our case, however, the axis has to be included. It appears that the $P_\nu(x)$ for fractional ν and the $Q_\nu(x)$ for all values of ν have a singularity at $x = \pm 1$, that is, at $\theta = 0$ and $\theta = \pi$. The consequence is that all B_ν coefficients must be zero because all $Q_\nu(x)$ functions lead to singularities. Moreover, ν can only take integer values n, so $\lambda = n(n+1)$ and $\Theta(\theta) = \Sigma_n A_n P_n(\cos\theta)$. The first few Legendre polynomials $P_n(\cos\theta)$ are

$$P_0 = 1, \quad P_1 = \cos\theta, \quad P_2 = (3\cos^2\theta - 1)/2, \quad P_3 = (5\cos^3\theta - 3\cos\theta)/2 \tag{D.21}$$

With $\lambda = n(n+1)$, the solution for the equation for R becomes

$$R = A_n r^n + B_n / r^{n+1} \tag{D.22}$$

Hence, the generic solution for the Laplace operator for axial symmetry, recalling that outside the sphere is domain (1) and inside the sphere is domain (2), becomes

$$\phi_1 = \sum_{n=0}^{n=\infty}\left(A_n r^n + \frac{B_n}{r^{n+1}}\right)P_n(\cos\theta) \quad \text{and} \quad \phi_2 = \sum_{n=0}^{n=\infty}\left(C_n r^n + \frac{D_n}{r^{n+1}}\right)P_n(\cos\theta) \tag{D.23}$$

To find the specific solution we must add the boundary conditions reading:

I) $(\phi_1)_{r\to\infty} = -Er\cos\theta$ because the field gradient far removed from the origin should approach the applied field $\boldsymbol{E}$.

II) $(\phi_1)_{r=a} = (\phi_2)_{r=a}$ as ϕ should be continuous across the boundary.

III) $\varepsilon_1(\partial\phi_1/\partial r)_{r=a} = \varepsilon_2(\partial\phi_2/\partial r)_{r=a}$ because at the boundary the normal dielectric displacement component D_n should be continuous.

IV) ϕ_2 is finite at $r = 0$ since the potential may not have a singularity at $r = 0$.

Satisfying boundary condition IV can only be achieved if $D_n = 0$ and condition I requires $A_1 = -E = -|\boldsymbol{E}|$, while all other $A_n = 0$. Conditions II and III say that, except for $n = 1$,

$$B_n / a^{n+1} = C_n a^n \quad \text{and} \quad -\varepsilon_1(n+1)B_n / a^{n+2} = \varepsilon_2 n C_n / a^{n-1} \tag{D.24}$$

From these equations it follows that $B_n = C_n = 0$ for all $n \neq 1$. For $n = 1$ it is found that

$$B_1 / a^2 - Ea = C_1 a \qquad \text{and} \qquad \varepsilon_1(2B_1 / a^3 + E) = -\varepsilon_2 C_1 \tag{D.25}$$

Therefore

$$B_1 = \frac{\varepsilon_2 - \varepsilon_1}{\varepsilon_2 + 2\varepsilon_1}a^3 E \qquad \text{and} \qquad C_1 = -\frac{3\varepsilon_1}{\varepsilon_2 + 2\varepsilon_1}E \tag{D.26}$$

and ϕ_1 and ϕ_2 become

$$\phi_1 = \left(\frac{\varepsilon_2 - \varepsilon_1}{\varepsilon_2 + 2\varepsilon_1}\frac{a^3}{r^3} - 1\right)Ez \qquad \text{and} \qquad \phi_2 = -\frac{3\varepsilon_1}{\varepsilon_2 + 2\varepsilon_1}Ez \tag{D.27}$$

Because the potential due to the external charges is $\phi = -Ez$, the contributions $\phi_1' = \phi_1 - \phi$ and $\phi_2' = \phi_2 - \phi$ due to the polarization become

$$\phi_1' = \frac{\varepsilon_2 - \varepsilon_1}{\varepsilon_2 + 2\varepsilon_1} \frac{a^3}{r^3} Ez \qquad \text{and} \qquad \phi_2' = \frac{\varepsilon_2 - \varepsilon_1}{\varepsilon_2 + 2\varepsilon_1} Ez \tag{D.28}$$

The expression for ϕ_1' is identical to that of an equivalent ideal dipole $\boldsymbol{\mu}_{\text{equ}}$ at the center of the sphere, surrounded by vacuum with the dipole vector directed along the external field $\boldsymbol{E}$ (the z-axis), and given by

$$\boldsymbol{\mu}_{\text{equ}} = \frac{\varepsilon_2 - \varepsilon_1}{\varepsilon_2 + 2\varepsilon_1} \frac{a^3}{\mathcal{C}} \boldsymbol{E} \tag{D.29}$$

The field $\boldsymbol{E}_2'$ inside the sphere due to the polarization and the total field $\boldsymbol{E}_2$ inside the sphere are described by, respectively,

$$\boldsymbol{E}_2' = -\frac{\varepsilon_2 - \varepsilon_1}{\varepsilon_2 + 2\varepsilon_1} \boldsymbol{E} \qquad \text{and} \qquad \boldsymbol{E}_2 = \frac{3\varepsilon_1}{\varepsilon_2 + 2\varepsilon_1} \boldsymbol{E} \tag{D.30}$$

In case the sphere is a cavity so that $\varepsilon_1 = \varepsilon$ and $\varepsilon_2 = 1$, the total field $\boldsymbol{E}_2$ becomes the *cavity field* $\boldsymbol{E}_{\text{cav}} = 3\varepsilon\boldsymbol{E}/(2\varepsilon + 1)$.

D.3
A Dipole in a Spherical Cavity

If in a spherical cavity in an infinite matrix a dipole $\boldsymbol{\mu}$ is placed at the center, this dipole will polarize the surrounding matter and the resulting polarization will give rise to a field at the dipole, called the reaction field $\boldsymbol{R}$. The reaction field will be proportional to $\boldsymbol{\mu}$, having the same direction as $\boldsymbol{\mu}$ so that $\boldsymbol{R} = f\boldsymbol{\mu}$. We choose the direction of the z-axis along the dipole vector so that we can employ again the axi-symmetric solutions [Eq. (D.23)] to the Laplace equation. In this case, the boundary conditions are, using $\varepsilon_1 = \varepsilon$ and $\varepsilon_2 = 1$:

$$\text{I: } (\phi_1)_{r\to\infty} = 0, \quad \text{II: } (\phi_1)_{r=a} = (\phi_2)_{r=a}, \quad \text{III: } \varepsilon(\partial\phi_1/\partial r)_{r=a} = (\partial\phi_2/\partial r)_{r=a} \tag{D.31}$$

The D-terms in ϕ_2 are due to the charges inside the cavity. Because the potential of an ideal dipole along the z-axis is $\phi = \mathcal{C}\mu r^{-2}\cos\theta$, the contribution $D_n = 0$ for all n except $n = 1$, which reads $D_1 = \mathcal{C}\mu$. For ϕ_2, we thus have

$$\phi_2 = \sum_{n=0}^{n=\infty} C_n r^n P_n(\cos\theta) + \mathcal{C}\mu r^{-2}\cos\theta \tag{D.32}$$

From condition I, we conclude that $A_n = 0$ for all n, so that ϕ_1 becomes

$$\phi_1 = \sum_{n=0}^{n=\infty} \frac{B_n}{r^{n+1}} P_n(\cos\theta) \tag{D.33}$$

From conditions II and III we obtain for all $n \neq 1$,

$$B_n / a^{n+1} = C_n a^n \qquad \text{and} \qquad -\varepsilon(n+1)B_n / a^{n+2} = nC_n / a^{n-1} \tag{D.34}$$

which can be satisfied only if $B_n = C_n = 0$ for $n \neq 1$, while for $n = 1$ the result is

$$B_1 / a^2 = C_1 a + \mathcal{C}\mu / a^2 \qquad \text{and} \qquad -2\varepsilon B_1 / a^3 = C_1 - 2\mathcal{C}\mu / a^3 \tag{D.35}$$

Solving for B_1 and C_1 leads to

$$B_1 = \frac{3}{2\varepsilon+1}\mathcal{C}\mu \qquad \text{and} \qquad C_1 = -\frac{2(\varepsilon-1)}{2\varepsilon+1}\frac{\mathcal{C}\mu}{a^3} \tag{D.36}$$

Substitution in Eqs (D.32) and (D.33) results in

$$\phi_1 = \frac{3}{2\varepsilon+1}\frac{\mathcal{C}\mu}{r^2}\cos\theta \qquad \text{and} \qquad \phi_2 = \frac{\mathcal{C}\mu}{r^2}\cos\theta - \frac{2(\varepsilon-1)}{2\varepsilon+1}\frac{\mathcal{C}\mu}{a^3}r\cos\theta \tag{D.37}$$

Because the potential due to the dipole is given by $\phi = \mathcal{C}\mu r^{-2}\cos\theta$, we see from the first part of Eq. (D.37) that formally the potential in the dielectric matrix can be described as the field of an effective (or virtual) dipole at the center of the cavity given by

$$\mu_{\text{eff}} = \frac{3\varepsilon}{2\varepsilon+1}\mu \tag{D.38}$$

From the second part of Eq. (D.37) we see that the field in the cavity is the field of the dipole in vacuum plus the uniform reaction field $\boldsymbol{R}$ given by

$$\boldsymbol{R} = f\boldsymbol{\mu} \equiv \frac{2(\varepsilon-1)}{2\varepsilon+1}\frac{\boldsymbol{\mu}}{a^3} \tag{D.39}$$

Using again the potential due to the dipole given by $\phi = \mathcal{C}\mu r^{-2}\cos\theta$, the contributions $\phi_1' = \phi_1 - \phi$ and $\phi_2' = \phi_2 - \phi$ due to the surface charges become

$$\phi_1' = -\frac{2(\varepsilon-1)}{2\varepsilon+1}\frac{\mathcal{C}\mu}{r^2}\cos\theta \quad \text{and} \quad \phi_2' = -\frac{2(\varepsilon-1)}{2\varepsilon+1}\frac{\mathcal{C}\mu}{a^3}r\cos\theta \tag{D.40}$$

These potentials can be described by a dipole $\boldsymbol{m}$ given by

$$\boldsymbol{m} = -\frac{2(\varepsilon-1)}{2\varepsilon+1}\boldsymbol{\mu} \tag{D.41}$$

The results in Eqs (D.30) and (D.39) are used in Chapter 10.

Further Reading

Böttcher, C.J.F. (1973) *Theory of Electric Polarization*, vol. 1, 2nd edn, Elsevier, Amsterdam.

Jackson, J.D. (1999) *Classical Electrodynamics*, 3rd edn, John Wiley & Sons, Inc., New York.

Robinson, F.N.H. (1973) *Electromagnetism, Oxford Physics Series*, Clarendon Press, Oxford.

Stratton, J.A. (1941) *Electromagnetic Theory*, McGraw-Hill.

Appendix E
Data

Table E.1 Lennard-Jones parameters ε and σ for several molecules from simulations.[a)]

Compound	(ε/k) (K)	σ (10^{-12} m)	Compound	(ε/k) (K)	σ (10^{-12} m)
Ar	111.84	362.3	CH_4	140.42	401.5
Kr	154.87	389.5	CD_4	138.16	402.3
Xe	213.96	426.0	C_2H_6	216.12	478.2
O_2	113.27	365.4	nC_3H_8	255.18	547.1
N_2	91.85	391.9	nC_4H_{10}	287.20	608.1
F_2	104.29	357.1	nC_5H_{12}	309.75	670.9
Cl_2	296.27	448.5	nC_6H_{14}	327.47	731.9
CF_4	155.31	491.1	nC_7H_{16}	340.97	790.2
CCl_4	378.86	624.1	cC_3H_6	277.46	509.2
SiH_4	193.65	453.9	cC_5H_{10}	346.11	610.0
SiF_4	140.14	669.2	C_6H_6	377.46	617.4
$SiCl_4$	338.01	668.2	C_2H_2	209.11	463.5
$TiCl_4$	417.49	672.5	C_2H_4	200.78	458.9
SF_6	206.85	578.3	CO_2	201.71	444.4
UF_6	323.61	637.9	CS_2	388.81	498.3

a) Data from Ref. [1], obtained from fitting MD simulation results of the pressure at low density and the Soave equation of state.

Table E.2 Lennard-Jones parameters ε and σ for several molecules from experimental data.[a)]

Compound	(ε/k) (K)	σ (10^{-12} m)	Compound	(ε/k) (K)	σ (10^{-12} m)
Ne	35.60	274.9	CH_4	148.2 (137*)	381.7 (382.2*)
Ar	119.8	340.5	C_2H_6	230*	441.8*
Kr	171	360	nC_3H_8	254*	506.1*
Xe	221	410	nC_4H_{10}	410*	499.7*
O_2	117.5	358	nC_5H_{12}	345*	576.9*
N_2	95.05	369.8	nC_6H_{14}	413*	590.9*
F_2	112*	365.3	CH_3OH	507*	358.5*
Cl_2	357*	411.5*	C_2H_5OH	391*	445.5*
Br_2	520*	426.8*	C_6H_6	440*	527*
I_2	550*	498.2*	C_2H_2	185*	422.1*
CH_3Cl	855	337.5	C_2H_4	205*	423.2*
CH_2Cl_2	406	475.9	CO_2	189	448.6
$CHCl_3$	327	543.0	CS_2	488*	443.8*
CCl_4	327*	588.1*	SiF_4	149	559
CF_4	152.5	470	SF_6	200.9	551

a) Data from Ref. [2], obtained from the viscosity (*) or from the second virial coefficient.

Table E.3 Thermodynamic vapor pressure constants a, b, and c for several compounds.[a)]

Molecule	a	b	c	L_0 (cal·mol^{-1})
Ar	14.059	2.814	498	2278
CS_2	15.9206	2.90	1844	8444
H_2O	20.9586 (17.443)	4.0843 (3.868)	2825.3 (2795)	12926 (13611)
H_2S	11.778	1.510	1145	5248
NH_3	17.718	3.406	1612.5	7377
CH_4	14.9640	3.283	598.4	2739
CH_3Cl	22.0114	5.133	1701	7786
CH_2Cl_2	20.2795	4.2857	2098.1	9601
$CHCl_3$	12.2999	3.9158	2179.1	9974
CCl_4	24.3085	5.669	2452.6	11220
C_2H_6	16.1506	3.332	1055	4831
nC_3H_8	15.772	3.003	1325	6066
nC_4H_{10}	19.877	4.376	1729	7912
nC_5H_{12}	22.752	5.253	2099	9611
nC_6H_{14}	25.281	6.013	2465	11284
nC_8H_{18}	30.040	7.396	3162	14473
CH_3OH	22.43	4.634	2661	12180
nC_4H_9OH	40.21	10.35	4100	18750
$nC_5H_{11}OH$	46.49	12.42	4580	20940
$nC_6H_{13}OH$	51.00	13.80	5068	23170
$nC_7H_{15}OH$	56.20	15.41	5580	25510
$nC_8H_{17}OH$	65.21	18.40	6190	28300
C_6H_6	20.818 (26.075)	4.7793 (6.203)	2388 (2610)	10927 (11940)
$(CH_3)_3N$	30.094	7.978	2142	10030
C_6H_5COOH	30.172	6.720	4714	21567
$(CH_3)_3CO$	19.557	3.966	2203	10078

$\log P_{sat}$ (mmHg) $= a - b\log T - c/T$), where T [K] and $L_0 = 2.303Rc$ with R the gas constant.
a) Data from Ref. [3].

Table E.4 Trouton constant *C* for several molecules.[a)]

Molecule	*C*	Molecule	*C*	Molecule	*C*
Ar	9.0	CH_4	8.8	H_2S	10.6
Kr	9.1	C_2H_6	9.6	SO_2	11.4
Xe	9.2	n-C_4H_{10}	9.9	CH_3Cl	10.4
Rn	9.4	i-C_4H_{10}	9.8	$CHCl_3$	10.6
N_2	8.7	n-C_7H_{16}	10.3	$(CH_3)_2O$	10.4
O_2	9.1	c-C_6H_{12}	10.2	NH_3	11.7
Cl_2	10.3	C_6H_6	10.5	H_2O	13.1
CO	8.9	HCl	10.3	CH_3OH	12.6
N_2O	10.8	HBr	10.3	C_2H_5OH	13.2

a) Data from Ref. [4].

Table E.5 Vogel constants *A*, *B*, and *C* for the viscosity η of some polar liquids.[a)]

	−*A* (Pa s)	*B* (°C)	−*C* (°C)	% error	*T*-range (°C)
Ammonia	1.7520	218.76	50.701	0.76	−66/+40
Methylamine	1.3634	126.389	102.886	0.32	−70/−10
Water	1.5668	230.298	146.797	0.51	−10/+160
Methanol	1.6807	354.876	48.585	2.05	−98/+50
Ethanol	2.4401	774.414	−15.249	2.66	−98/+70
n-Propanol	2.4907	725.903	37.474	1.10	0/+70
n-Butanol	3.0037	1033.306	−4.3828	0.80	−51/+100
Glycerine	2.8834	997.586	128.481	4.50	−42/+30
Ethylene glycol	1.5923	438.064	141.617	0.18	+20/+100

$\ln\eta = A + B/(t/°C + C)$, where t is temperature in °C.
a) Data from Ref. [5].

Table E.6 Dipole moment μ and polarizability volume $\alpha' = \alpha/4\pi\varepsilon_0$ for some compounds.[a)]

Compound	μ (D)	α' (10^{-30} m^{-3})	Compound	μ (D)	α' (10^{-30} m^{-3})
He	0	0.201	$(CH_3)_2CO$	2.84	6.33
Ar	0	1.63	CH_3OCH_3	1.29	5.16
C_6H_6	0	10.3	CH_3OH	1.71	3.23
$C_6H_5CH_3$	0.36	–	C_2H_5OH	1.69	–
C_6H_5Cl	1.70	12.25	HF	1.91	0.80
CH_4	0	2.60	HCl	1.03	2.63
CH_3Cl	1.86	4.56	HBr	0.79	3.61
CH_2Cl_2	1.58	6.48	HI	0.38	5.45
$CHCl_3$	1.02	8.23	N_2	0	1.76
CCl_4	0	10.5	O_2	0	1.60
H_2O	1.84	1.49	Cl_2	0	4.61
NH_3	1.47	2.26	CO	0.10	1.95
$N(CH_3)_3$	0.60	8.29	CO_2	0	2.65

a) Dipole moment data from Ref. [6].

Table E.7 Compressibility κ and expansivity α for some solvents at 20 °C.[a)]

Solvent	ρ (g cm^{-3})	κ_S (10^{11} Pa^{-1})	α_P (10^3 K^{-1})	Solvent	ρ (g cm^{-3})	κ_S (10^{11} Pa^{-1})	α_P (10^3 K^{-1})
n-Hexane	0.654	130	1.36	Water	0.997	45.5	0.257
n-Heptane	0.684	108	–	Methanol	0.792	100	–
n-Octane	0.703	99.3	1.15	Ethanol	0.798	91.1	1.12
Carbon tetrachloride	1.595	71.4	1.22	*n*-Butyl alcohol	0.810	76.9	–
Benzene	0.878	64.9	1.23	Acetone	0.792	89	–
Toluene	0.866	65.5	1.06	Diethyl ether	0.714	138	1.43
Formamide	1.1292	–	59.1				

a) Data for ρ and κ_S from Ref. [7] (ultrasonic methods determine κ_S, but κ_S can be converted to κ_T using Eq. (2.24)). Data for α_P from Ref. [8].

Table E.8 Critical data for several organic compounds.[a)]

Name	Molar mass	ω	T_{cri} (K)	P_{cri} (bar)	V_{cri} (cm^3 mol^{-1})	n_D	T_n (K)
Acetone	58.080	0.307	508.2	47.01	209.	1.3590	329.4
MEK	72.11	0.329	535.6	41.5	267.	1.3788	267
Dimethylether	46.07	0.192	400.0	53.7	178	–	249
Diethylether	74.123	0.281	466.7	36.40	280.	1.353	307.6
Carbon tetrafluoride	88.00	0.191	227.6	37.4	140.	1.0005	145
Carbon tetrachloride	153.822	0.193	556.4	45.60	276.	1.4601	349.8
Chloroform	119.377	0.222	536.4	54.72	239.	1.4459	334.3
Water	18.015	0.345	647.1	220.55	55.9	1.3330	373.2
Methanol	32.042	0.564	512.6	80.97	118.	1.33	337.9
Ethanol	46.069	0.645	513.9	61.48	167.	1.36	351.4
1-Propanol	60.096	0.622	536.8	51.75	219.	1.387	370.4
i-Propanol	60.096	–	508.3	47.6	220	1.3776	356
1-Butanol	74.123	0.594	563.1	44.23	275.	1.399	390.8
Benzene	78.114	0.210	562.2	48.98	259.	–	353.2
Toluene	92.141	0.262	591.8	41.06	316.	1.399	383.8
Ethylbenzene	106.167	0.303	617.2	36.06	374.	–	409.4
Aniline	93.13	0.384	699	53.1	274.	–	457
Pyridine	79.1	0.24	620.0	56.3	254.	1.5093	254
o-Xylene	106.167	0.310	630.3	37.34	369.	1.50545	417.6
m-Xylene	106.167	0.326	617.1	35.36	376.	1.49722	412.3
p-Xylene	106.167	0.322	616.2	35.11	379.	1.49582	411.5
Phenol	94.113	0.444	694.3	61.30	229.	–	455.0
Methane	16.04	0.011	190.6	45.19	98.7	–	111.6
Ethane	30.07	0.100	305.4	48.83	148.	–	184.6
Propane	44.1	0.153	369.85	42.48	200.0	–	231.1
Cyclopropane	42.08	0.264	397.8	54.9	210	–	240
Butane	58.123	0.200	425.1	37.96	255.	–	272.7
Cyclobutane	56.107	0.209	459.9	49.8	210	–	286
Pentane	72.150	0.252	469.7	33.70	313.	1.358	309.2
Cyclopentane	70.134	0.196	511.8	45.02	258.	–	322.4
Neopentane	72.15	0.197	433.8	31.96	311.1	–	282.6
Hexane	86.177	0.301	507.6	30.25	371.	1.375	341.9
Cyclohexane	84.161	0.210	553.6	40.73	308.	1.4262	353.9
Heptane	100.204	0.350	540.2	27.40	428.	1.387	371.6
Octane	114.231	0.400	568.7	24.90	486.	1.398	398.8
Nonane	128.258	0.444	594.6	22.90	544.	1.405	424.0
Acetaldehyde	44.05	0.303	461	56	154	–	293
Acetonitrile	41.05	0.309	508.1	47.0	209	1.344	354
Chlorobenzene	112.56	0.249	632.4	45.2	308	–	404
Trimethylamine	59.11	0.195	433.2	40.7	254	–	270

a) Data from Refs [9] and [10].
MEK, methylethyl ketone.

Table E.9 Physical data for some solvents at 20 °C.[a)]

Solvent	ε	μ (D)	R_M (cm^3)	γ (mN m^{-2})	ρ (g cm^{-3})	t_n (°C)
Acetic acid	6.19	1.74	13	26.9	1.0497	117.9
Acetic anhydride	21	2.8	22.4	31.9	1.075	140
Acetone	20.7	2.88	16.2	22.9	0.792	56.3
Acetonitrile	3.44	3.44	0	28.45	0.7768	81.6
Aniline	6.8	1.55	–	42.8	1.022	184.4
Benzene	2.28	0	26.2	28.2	0.878	80.1
Benzyl alcohol	13.1	1.71	32.5	39	1.042	205.2
Bromobenzene	5.4	1.7	34	35.7	1.495	156
Bromoform	4.3	1.3	0	31.7	2.8776	149.6
Butanol	17.5	1.66	0	24.3	1.0269	117.7
i-Butanol	17.7	1.64	22.4	22.5	0.798	107.7
t-Butanol	10.9	1.66	22.2	20	0.7812	82.4
Carbon disulfide	2.6	0	–	31.5	1.2566	46.26
Carbon tetrachloride	2.23	0	26.7	26.2	1.5842	76.8
Chlorobenzene	5.62	1.69	31.1	32.7	1.11	131
Chloroform	4.81	1.04	21.5	26.5	1.4799	61.2
Cyclohexane	2.02	0	27.8	24.4	0.7739	80.7
Dichloromethane	8.93	1.60	16.3	27.3	1.3168	39.8
Diethyl ether	4.34	1.15	22.5	16.5	0.7078	34.6
N,N-Dimethylacetamide	37.8	3.81	24.4	33.3	0.937	165
N,N-Dimethylformamide	36.7	3.86	20	35.2	0.9445	153
Dimethyl sulfoxide	49	3.96	20.1	42.8	1.0958	189
1,4-Dioxane	2.21	0	21.7	32.9	1.0269	101.3
Ethanol	24.3	1.69	14.9	21.8	0.798	78.3
Ethyl acetate	6.02	1.78	22.1	23.2	0.8945	77.1
Ethylene glycol	37.7	2.28	14.5	48.1	1.1097	197.3
Formamide	110	3.7	10.7	58.5	1.1292	210.5
Formic acid	58	1.41	8.6	37.1	1.2131	100.7
Glycerol	42.5	2.56	20.5	62.5	1.2582	290
n-Heptane	1.92	0	34.6	19.8	0.684	98.5
n-Hexane	1.89	0.08	29.9	17.9	0.654	68.7
Methanol	32.6	1.7	8.2	22.4	0.792	64.7
Methyl acetate	6.7	1.72	17.5	24.1	0.932	56.9
n-Methylpyrrolidone	33	4.1	–	–	0.703	202
Nitrobenzene	35	4.22	32.9	42.8	1.1987	210.8
Nitromethane	38.6	3.46	12.5	50.7	1.1312	101.2
n-Pentane	1.84	0	25.3	15.5	0.626	36.2
Propanol	20.1	1.68	17.5	23.4	0.804	97.5
i-Propanol	18.3	1.66	17.6	20.8	0.781	82.3
Pyridine	12.3	2.19	24.1	36.3	0.982	115.3
Tetrahydrofuran	7.32	1.63	19.9	26.9	0.8844	66
Toluene	2.38	0.36	31.1	27.9	0.866	110.6
Water	78.5	1.84	3.7	71.8	0.997	100

a) Data for ε, μ, R_m and γ from Ref. [11].
t_n represents the normal boiling point.

Table E.10 Van der Waals constants *a* and *b* for several compounds.[a]

Compound	*a* (atm l² mol⁻²)	*b* (10⁻² l mol⁻¹)	Compound	*a* (atm l² mol⁻²)	*b* (10⁻² l mol⁻¹)
He	0.03457	2.370	C_6H_{14}	24.71	17.35
Ne	0.2135	1.709	H_2O	5.536	3.049
Ar	1.363	3.219	H_2S	4.490	4.287
Kr	2.349	3.978	CH_3OH	9.649	6.702
Xe	4.250	5.105	C_2H_5OH	12.18	8.407
C_6H_6	18.24	11.54	CH_3Cl	7.570	6.483
$C_6H_5CH_3$	24.38	14.63	CCl_4	19.75	12.81
C_6H_5F	20.19	12.86	$SiCl_4$	4.377	5.786
C_6H_5Cl	25.77	14.53	$(CH_3)_2C{=}O$	14.09	9.94
C_6H_5Br	28.94	15.39	CO	1.505	3.985
C_6H_5I	33.52	16.56	CO_2	3.640	4.267
CH_4	2.283	4.278	H_2	0.2476	2.661
C_2H_6	5.562	6.380	O_2	1.378	3.183
C_3H_8	8.779	8.445	N_2	1.408	3.913
C_4H_{10}	14.66	12.26	Cl_2	6.579	5.622
C_5H_{12}	19.26	14.6	NH_3	4.225	3.707

a) Data from Ref. [12].

Table E.11 Critical data for several inorganic compounds.[a]

Name	Molar mass	ω	T_{cri} (K)	P_{cri} (bar)	V_{cri} (cm³ mol⁻¹)	Z_{cri}	T_n (K)
Ne	20.180	−0.016	44.44	27.605	41.70	0.312	27.1
Ar	39.948	−0.002	150.9	48.98	74.57	0.291	87.3
Kr	83.800	−0.002	209.4	55.00	91.20	0.288	119.7
Xe	131.29	0.002	289.7	58.40	118.0	0.286	165.0
N_2	28.014	0.037	126.2	33.98	91.10	0.289	77.4
O_2	31.999	0.022	154.6	50.43	73.4	0.288	90.2
F_2	37.997	0.051	144.3	52.2	66.2	0.288	85.0
Cl_2	70.905	0.069	417.2	77.00	124.	0.275	239.1
Br_2	159.81	0.119	584.1	103.0	135	0.269	331.9
I_2	253.81	0.229	819	117.0	155.	0.266	457.6
CO	28.01	0.045	132.9	34.94	93.10	0.292	81.7
CO_2	40.01	0.225	304.1	73.74	94.07	0.247	–
CS_2	76.14	0.115	552	79.0	170	0.293	319
HF	20.01	0.329	461.0	65.0	69.0	0.117	292.7
HCl	36.46	0.133	324.7	83.1	81.0	0.249	188.2
HBr	80.91	0.069	363.2	85.1	–	–	206.4
HI	127.91	0.038	423.9	90.00	132.7	0.303	237.6
NH_3	17.031	0.255	405.4	113.53	72.47	0.255	239.8
H_2S	34.08	0.090	373.4	89.63	98.00	0.283	212.8

a) Data mainly from Ref. [13].

Table E.12 Antoine constants *A*, *B*, and *C* for the vapor pressure P_{sat} of several compounds.[a)]

Name	Formula	*A*	*B*	*C*	Range (°C)	$\Delta_{vap}H_n$ (kJ mol^{-1})	t_n (°C)
Acetone	C_3H_6O	14.3145	2756.22	228.060	−27 to 77	29.10	56.2
MEK	C_4H_8O	14.1343	2838.4	218.690	−8 to 103	31.30	79.6
Butyl ether	$C_4H_{10}O$	14.0735	2511.29	231.200	−43 to 55	26.52	34.4
Carbon tetrachloride	CCl_4	14.0572	2914.23	232.148	−14 to 101	29.82	76.6
Chloroform	$CHCl_3$	13.7324	2548.74	218.552	−23 to 84	29.94	61.1
Dichloromethane	CH_2Cl_2	13.9891	2463.912	223.240	−38 to 60	28.06	39.7
Water	H_2O	16.3872	3885.70	230.170	0 to 200	40.66	100.0
Methanol	CH_4O	16.5785	3638.27	239.500	−11 to 83	35.21	64.7
Ethanol	C_2H_6O	16.8958	3795.17	230.918	3 to 96	38.56	78.2
1-Propanol	C_3H_8O	16.1154	3483.67	205.807	20 to 116	41.44	97.2
2-Propanol	C_3H_8O	16.6796	3640.20	219.610	8 to 100	39.85	82.2
1-Butanol	$C_4H_{10}O$	15.3144	3212.43	182.739	37 to 138	43.29	117.6
2-Butanol	$C_4H_{10}O$	15.1989	3026.03	186.500	25 to 120	40.75	99.5
Iso-butanol	$C_4H_{10}O$	14.6047	2740.95	166.670	30 to 128	41.82	107.8
Tert-butanol	$C_4H_{10}O$	14.8445	2658.29	177.650	10 to 101	39.07	82.3
Benzene	C_6H_6	13.7819	2726.81	217.572	6 to 104	30.73	80.0
Toluene	C_7H_8	13.9320	3056.96	217.625	13 to 136	33.18	110.6
Ethylbenzene	C_8H_{10}	13.9726	3259.93	212.300	33 to 163	35.57	136.2
Phenol	C_6H_6O	14.4387	3507.80	175.400	20 to 208	46.18	181.8
o-Xylene	C_8H_{10}	14.0415	3358.79	212.041	40 to 172	36.24	144.4
m-Xylene	C_8H_{10}	14.1387	3381.81	216.120	35 to 166	35.66	139.1
p-Xylene	C_8H_{10}	14.0579	3331.45	214.627	35 to 166	35.67	138.3
Phenol	C_6H_6O	14.4387	3507.80	175.400	80 to 208	46.18	181.8
Butane	C_4H_{10}	13.6608	2154.70	238.789	−73 to 19	22.44	−0.5
Iso-butane	C_4H_{10}	13.8254	2181.79	248.870	−83 to 7	21.30	−11.9
Pentane	C_5H_{12}	13.7667	2451.88	232.014	−45 to 58	25.79	36
Cyclo-pentane	C_5H_{10}	13.9727	2653.90	234.510	−35 to 71	27.30	49.2
Hexane	C_6H_{14}	13.8193	2696.04	224.317	−19 to 92	28.85	68.7
Cyclo-hexane	C_6H_{12}	13.6568	2723.44	220.618	9 to 105	29.97	80.7
Heptane	C_7H_{16}	13.8622	2910.26	216.432	4 to 123	31.77	98.4
Octane	C_8H_{18}	13.9346	3123.13	209.635	26 to 152	43.41	125.6
Iso-octane	C_8H_{18}	13.6703	2896.31	220.767	2 to 125	30.79	99.2
Nonane	C_9H_{20}	13.9854	3311.19	202.694	46 to 178	36.91	150.8
Decane	$C_{10}H_{22}$	13.9748	3442.76	193.858	65 to 203	38.75	174.1
Acetonitrile	C_2H_3H	14.8950	3413.10	250.253	−27 to 81	30.19	81.6
Acetic acid	$C_2H_4O_2$	15.0717	3580.80	224.650	24 to 142	23.70	117.9

$\ln P_{sat}/\text{kPa} = A - B/(t/°\text{C} +)$, where t is the temperature in °C.
$\Delta_{vap}H_n$ is the enthalpy of evaporation, and t_n the normal boiling temperature.
a) Data from Ref. [9].
MEK, methylethyl ketone.

Table E.13 Solubility products K_{SP} of various salts.[a)]

Compound	K_{SP} at 286 K	Compound	K_{SP} at 286 K	Compound	K_{SP} at 286 K
BaF_2	1.7×10^{-6}	$BaSO_4$	1×10^{-10}	ZnS[b)]	1×10^{-23}
CaF_2	3.4×10^{-11}	$SrSO_4$	2.8×10^{-7}	CdS	8×10^{-27}
MgF_2	6.5×10^{-9}	$CaSO_4$	2×10^{-4}	HgS	2×10^{-52}
PbF_2	3.6×10^{-8}	$PbSO_4$	2×10^{-8}	CuS	8.5×10^{-36}
AgCl	1×10^{-10}	$AlOH_3$	2×10^{-33}	NiS	2×10^{-21}
CuCl	1.0×10^{-6}	$MgOH_2$	1.2×10^{-11}	CoS	8×10^{-22}
$HgCl_2$	2.0×10^{-18}	$FeOH_3$	1.1×10^{-36}	FeS	3.7×10^{-18}
$PbCl_2$	1.7×10^{-5}	$FeOH_2$	1.6×10^{-14}	MnS	2.5×10^{-10}
AgI	1.5×10^{-16}	$MnOH_2$	4×10^{-14}	Ag_2S	1.6×10^{-49}
CuI	5×10^{-12}	$NiOH_2$	4×10^{-14}	$BaCO_3$	5×10^{-9}
HgI_2	1.2×10^{-28}	$CuOH_2$	1×10^{-19}	$CaCO_3$	1×10^{-8}
PbI_2	1.4×10^{-8}	$ZnOH_2$	1.8×10^{-14}	$MgCO_3$	2.6×10^{-5}
$BaCrO_4$	2.4×10^{-10}	AgOH	1×10^{-8}	$PbCO_3$	3.3×10^{-14}
$PbCrO_4$	1.8×10^{-14}				
Ag_2CrO_4	1.2×10^{-12}				

a) Data from Ref. [14].
b) Sulfides often crystallize in more than one form, with different solubility products. Values used may differ by as much as 10^9.

Table E.14 Ionic radii according to Pauling r_P and Shannon-Prewitt r_{SP} for IV and VI coordination.[a)]

Ion	r_P (pm)	$^{IV}r_{SP}$ (pm)	$^{VI}r_{SP}$ (pm)	Ion	r_P (pm)	$^{IV}r_{SP}$ (pm)	$^{VI}r_{SP}$ (pm)
Li^+	60	73	90 (88)	B^{3+}	20	25 (26)	41
Na^+	95	113	116	Al^{3+}	50	53	67.5 (67.0)
K^+	133	151	152	Sc^{3+}	81	–	88.5 (87.0)
Rb^+	148	–	166 (163)	Ga^{3+}	62	61	76
Cs^+	169	–	181 (184)	Y^{3+}	93	–	104 (103.2)
Be^{2+}	31	41	59 (45)	In^{3+}	81	76	94 (3)
Mg^{2+}	65	71 (63)	86.0	Fe^{3+}		63 HS	69 LS
Ca^{2+}	99	–	114			–	78.5 HS
Zn^{2+}	74	74	88.0 (88.5)	F^-	136	117	119
Sr^{2+}	113	–	132 (130)	Cl^-	181	–	167
Ba^{2+}	135	–	149 (150)	Br^-	195	–	182
Fe^{2+}		77 HS	75 LS	I^-	216	–	206
		–	92 HS (91)	O^{2-}	140	124	126
Mn^{2+}		80 HS	81 LS	S^{2-}	184	–	170
			97 HS (96)	Se^{2-}	198	–	184
Ni^{2+}		69	84 (83)	Te^{2-}	221	–	207

a) Data from Refs [15] and [16].
For VI coordination, no brackets indicate the same values reported in 1969 as in 1976; otherwise data in brackets refer to the 1969 report.
LS, low-spin; HS, high-spin.

References

1 Cuadros, F., Cachadina, I., and Ahamuda, W. (1996) *Mol. Eng.*, **6**, 319.
2 Hirschfelder, J.O., Curtiss, C.F., and Bird, R.B. (1954) *Molecular Theory of Gases and Liquids*, John Wiley & Sons, Inc., New York.
3 Moelwyn-Hughes, E.A. (1961) *Physical Chemistry*, 2nd edn, Pergamon, Oxford.
4 Rowlinson, J.S. and Swinton, F.L. (1982) *Liquids and Liquid Mixtures*, 3rd edn, Butterworth, London.
5 Reid, R.C., Prausnitz, J.M., and Sherwood, T.K. (1977) *The Properties of Gases and Liquids*, 3rd edn, McGraw-Hill, New York.
6 *Landolt-Börnstein Tabellen* (1951) vol. I, part 3, Springer, Berlin.
7 Bergmann, L. (1954) *Der Ultraschall*, 6th edn, Hirzel Verlag, Stuttgart.
8 Marcus, Y. (1977) *Introduction to Liquid State Chemistry*, John Wiley & Sons, Ltd, London.
9 Smith, J.M., Van Ness, H.C. and Abbott, M.M. (2005) *Introduction to Chemical Engineering Thermodynamics*, 7th edn, McGraw-Hill, New York.
10 Pitzer, K.S. (1995) *Thermodynamics*, 3rd edn, McGraw-Hill, New York.
11 Connors, K.A. (1990) *Chemical Kinetics*, Wiley-VCH Verlag GmbH, Weinheim, Germany.
12 Weast, R.C. (ed.) (1972) *Handbook of Chemistry and Physics*, 53rd edn, Chemical Rubber Company, Cleveland.
13 Poling, B.E., Prausnitz, J.M., and O'Connell, J.P. (2001) *The Properties of Gases and Liquids*, 4th and 5th edn, McGraw-Hill.
14 Gray, H.B. and Haight, G.P. (1967) *Basic Principles of Chemistry*, Benjamin, New York.
15 Pauling, L. (1960) *The Nature of the Chemical Bond*, 3rd edn, Cornell University Press, Ithaca.
16 (a) Shannon, R.D. and Prewitt, C.T. (1969) *Acta Crystallogr.*, **B25**, 925; (b) Shannon, R.D. (1976) *Acta Crystallogr.*, **A32**, 751.

Appendix F
Numerical Answers to Selected Problems

3.11 Ar: 129 K, HBr: 467 K.

3.16 Taking $\alpha_1/\alpha_2 = 2$, we obtain the following results rendering the Berthelot approximation doubtful.

σ_{11}/σ_{22}	$\varepsilon_{11}/\varepsilon_{22}$	Berthelot	Eq. (3.32)
1	1	1.00	0.80
1	2	1.41	1.33
2	1	1.00	0.33
2	2	1.41	0.34

3.17 $U_{coh} = 14.3\,\text{kJ/mol}$ and $H_{coh} = U_{coh} + RT = 16.8\,\text{kJ/mol}$ and $U_{coh} = 28.9\,\text{kJ/mol}$ and $H_{coh} = U_{coh} + RT = 31.4\,\text{kJ/mol}$.

3.18 The relative contribution is $(11/32)\alpha/\sigma^3 = 0.011$.

5.3 $Z = 1 + \exp(-\varepsilon/kT)$, $p_1 = 0.73$, $p_2 = 0.27$, $U = \varepsilon/[\exp(\varepsilon/kT) + 1]$, $C_V = [\exp(\varepsilon/kT) + 1]^{-2}(\varepsilon^2/kT^2)\exp(\varepsilon/kT)$.

5.11 $\Lambda(\text{He}) \cong 4.3\,\text{Å}$ with $\rho^{-1/3} \cong 3.7\,\text{Å} \rightarrow$ classical approximation not valid; $\Lambda(\text{Ar}) \cong 0.30\,\text{Å}$ with $\rho^{-1/3} \cong 4.0\,\text{Å} \rightarrow$ classical approximation valid.

5.21 Assuming deviations less than 5% are allowed, one has $1 + \theta/3T < 1.05$ or $T > 7\theta$.

6.1 $\rho_{rel} = 0.984 \cdot \rho_{glass}/\rho_{FCC} = 0.984 \cdot 0.637/0.741 = 0.85$. Fig. 6.3 provides $0.9 < V^* < 1.1$, hence $0.77 < \rho_{rel} < 0.95$.

6.3

	SC		BCC		FCC		HCP	
Shell	$(r/a)^2$	N_r	$(r/a)^2$	N_r	$(r/a)^2$	N_r	$(r/a)^2$	N_r
1	1	6	1	8	1	12	1	12
2	2	12	1⅓	6	2	6	2	6
3	3	8	2⅔	12	3	24	2⅔	2
4	4	6	3⅔	24	4	12	3	18
5	5	24	4	8	5	24	3⅔	12
V/N	a^3		$(4/3\sqrt{3})a^3$		$(1/\sqrt{2})a^3$		$(1/\sqrt{2})a^3$	

6.4 $N^{(1)} = 9.2$, $N^{(2)} = 44.4$.

6.6 a) Nearest neighbour Li-Cl ≅ 2.0 Å, next nearest neighbour Li-Cl ≅ 5.3 Å
b) Nearest neighbour Li-Li ≅ Cl-Cl ≅ 3.7 Å.

7.5 $PV/KT = 1 + \Sigma_{n=1}(n^2 + 3n)\eta^n$.

7.6 $\nu = 1.70$; $\beta P_{vdW} = \rho(1 + 4v_0\rho + 16v_0^2\rho^2 + \ldots)$; $\beta P_{CS} = \rho(1 + 4v_0\rho + 10v_0^2\rho^2 + \ldots)$;

Liquid-State Physical Chemistry, First Edition. Gijsbertus de With.

8.2 a) $v_L/v_G = a/(a-d)$
b) $V_f = (4\pi/3)(a-d)^3 = (4\pi a^3/3)(v_G/v_L)^3 = 0.008\ (4\pi a^3/3)$

10.3 H_2O: $\varepsilon_r = 78.5$, CH_3OH: $\varepsilon_r = 31.6$, C_2H_5OH: $\varepsilon_r = 24.3$.

10.5 $\alpha' = 10.3\,\text{Å}^3$, $\mu = 1.57\,\text{D}$.

10.7 $\alpha' = 6.4\,\text{Å}^3$, $\mu = 2.8\,\text{D}$.

10.10 $\alpha' = 13.8 \times 10^{-30}\,\text{m}^3/\text{mol}$, $\mu = 0.34\,\text{D}$ using the Debye equation. The data point at −110 °C is also at the line for the liquid, indicating that the molecule still rotates in the solid phase. However, α' is large and μ is small as compared to independent data ($\alpha' = 3.23 \times 10^{-30}\,\text{m}^3/\text{mol}$, $\mu = 1.7\,\text{D}$) illustrating the effect of hydrogen bonding.

10.11 $\mu^*/\mu = 1.26$.

11.5 $V_1 = A + B + C[x_1(2 - x_1)]$, $V_2 = A - Cx_1^2$.

11.6 $V_1 = bx_1^2$.

11.10 a) $\Delta H = zwx_1(1 - x_1) = 800\,\text{J/mol}$ at $x_1 = 0.5$, hence $zw = 3200$ and $w = 320\,\text{J/mol}$.
b) $\ln \gamma_2 = (wz/kT)x_1^2 = 0.0801$ or $\gamma_2 = 1.083$.

11.12 Since the expressions are symmetric in x_1 and x_2, yes.

12.2 $\Delta_{sol}H = 4\,\text{kJ mol}^{-1} > 0$, hence $\Delta T > 0$. $T\Delta S > 4\,\text{kJ mol}^{-1}$ or $\Delta S > 13.3\,\text{J K}^{-1}\,\text{mol}^{-1}$.

12.3 $\Delta S = -(z^2e^2N_A/8\pi\varepsilon_0 r)\varepsilon_r^{-2}\partial\varepsilon/\partial T$.

12.9 Assuming the same structure as in Fig. 12.4, the angle is 40°.

12.11 $\Delta H = A - DT^2$, $\Delta S = C - 2DT$, $\Delta C_P = -2DT$.

12.13 $\kappa^{-1} \cong 1.8\,\text{nm}$.

12.17 $t_H = 3 \cdot 10^{-7}\,\text{s}$.

12.18 a) $\Delta T = 0.31\,°\text{C}$; b) $\Lambda_m = 0.0273\,\Omega^{-1}\,\text{mol}^{-1}\,\text{m}^2$; c) $I = 14\,\text{mA}$; d) $I = 20\,\text{mA}$ using $\eta(H_2O) = 1 \cdot 10^{-3}\,\text{Pa s}$, $a(Ca^{2+}) = 1.0\,\text{Å}$ and $a(I^-) = 2.15\,\text{Å}$; e) $t(Ca^{2+}) = 0.44$; f) 2.4 K/hour.

13.1 A: $L = 10\,\text{nm}$, $X = 1.0\,\text{nm}$ B: $L = 100\,\text{nm}$, $X = 3.16\,\text{nm}$
C: $L = 100\,\text{nm}$, $X = 5.47\,\text{nm}$ D: $L = 100\,\text{nm}$, $X = 13.2\,\text{nm}$

13.9 $\delta = 18.2\,\text{MPa}^{1/2}$.

13.10 $\delta = 18.3\,\text{MPa}^{1/2}$.

13.11 $\phi_1 = 0.10$, $\phi_2 = 0.41$, $\phi_3 = 0.49$.

15.1 $u^{(\sigma)} = 26.1\,\text{mJ m}^{-2}$.

15.8 $\gamma = 0.0264\,\text{N/m}$. The reasonable good agreement for this calculation of cyclohexane with the experimental value 0.0247 N/m is probably fortuitous.

Index

Note: Page references in *italics* refer to Figures; those in **bold** refer to Tables